A. Crookes

A. Crookes

INTERMEDIATE
ALGEBRA

INTERMEDIATE

ALGEBRA

SECOND
EDITION

Jack Barker   James Rogers   James Van Dyke

Portland Community College
Portland, Oregon

SAUNDERS COLLEGE PUBLISHING
A Harcourt Brace Jovanovich College Publisher

FORT WORTH   PHILADELPHIA   SAN DIEGO
NEW YORK   ORLANDO   AUSTIN   SAN ANTONIO
TORONTO   MONTREAL   LONDON   SYDNEY   TOKYO

Text Typeface • *Times Roman*
Compositor • *York Graphic Services, Inc.*
Acquisitions Editor • *Robert B. Stern*
Developmental Editor • *Ellen Newman*
Managing Editor • *Carol Field*
Copy Editor • *York Production Services*
Manager of Art and Design • *Carol Bleistine*
Text Designer • *York Production Services*
Cover Designer • *Lawrence R. Didona*
Art Creation • *York Production Services*
Director of EDP • *Tim Frelick*
Production Manager • *Bob Butler*
Marketing Manager • *Monica Wilson*
Cover Credit • *Rectangles overlapping circle, © 1991 Bishop/PHOTOTAKE, NYC*

Printed in the United States of America

INTERMEDIATE ALGEBRA, second edition

ISBN 0-03-072856-8

Library of Congress Catalog Card Number: 91-38452

234   039   98765432

To our families:

Mary, Ken, Norm, and Linda Barker
Elinore Rogers, John, Heather, and Eddie Fincher, Paul, Patty, and Pamela Hurst, Terry, Becky, and Perry Washington, Jim, Michelle, and Tyler Raible
Carol, Dan, Claudia, Avalon, Tom, Larry, Karla, Greg, and Ann Van Dyke

PREFACE

Intermediate Algebra is a text that serves as a review of elementary algebra together with a study of the traditional topics of intermediate algebra. It is designed for students at any level so that they can gain the skills necessary to fulfill competency requirements, achieve adequate scores on placement exams for entrance to certain professional schools, prepare for technical mathematics courses, prepare for problem solving in applied courses, and complete prerequisites for college algebra or business courses.

Changes in the Second Edition The second edition contains the following new features:

- Each chapter begins with a Preview to introduce the student to the material discussed in the chapter.
- The text of each section has been rewritten so that real-life applications and examples with solutions are integrated into the discussion.
- The step-by-step strategies printed in blue to the right of the example have been expanded and clarified to give students more help in solving problems.
- A four-color system is used to highlight the various pedagogical features of the text, such as definitions, cautions, procedures, properties, and formulas. Color is also used to distinguish between two lines that are graphed simultaneously. The complete color system is described in more detail on p. xxiii.
- Exercises have been completely revised, and all problems have been checked by two accuracy reviewers. Every section of the text now contains State Your Understanding and Challenge Exercises. Maintain Your Skills Exercises are now keyed to sections of the text.
- Each chapter concludes with a Chapter Summary, Chapter Review Exercises (keyed to section number and objective), a True–False Concept Review, and a Chapter Test. In addition, there are Cumulative Reviews following Chapters 3, 6, and 9.
- Particular attention has been paid to the testing requirements for various states (e.g., ELM, TASP, CLAST, etc.). Please see pages xv–xvii for additional information on how the text meets the requirements for these states.

A complete pedagogical system that is designed to motivate the student and make mathematics more accessible includes the following features:

Chapter Preview Each chapter begins with a Preview to introduce the student to the material and show how it is integrated with the study of algebra.

Objectives Objectives are identified at the beginning of each section to help students focus on skills to be learned.

Examples with Solutions Examples are used to illustrate the concepts explained within the section. As each example is worked out, there is a step-by-step explanation that expands on the procedures and the thinking necessary to work the problem. The examples also illustrate shortcuts. Where applicable, caution comments about common errors are included, highlighted in red for the student.

Calculator Examples Some examples contain problems solved by using a calculator. These examples, set off with a calculator symbol, demonstrate how to use a calculator and signal to the student that these problems are suited for calculator practice. However, the use of the calculator is left to the discretion of the instructor and/or the student. Nowhere is the use of the calculator required, and all sections of the text can be studied without a calculator.

Pedagogical Use of Color This text uses color to highlight and distinguish definitions, caution comments, procedures, properties, and formulas. Color is also used to distinguish between two lines that are graphed simultaneously. The complete color system is described in more detail on page xxiii.

EXERCISES

The second edition of *Intermediate Algebra* continues to organize the exercises in terms of difficulty. As before, "A" indicates that the problems are relatively easy; the "A" problems have been increased in number and can be used as class or "oral" exercises. The "B" problems may require paper and pencil. The "C" problems are more difficult and in some instances offer a challenge for the advanced students. At the discretion of the instructor, the "C" problems may provide calculator drill for the students.

Applications Whenever appropriate, the "D" problems (word problems) have been included so that the student constantly practices translating word phrases and statements into mathematical equations. The word problems are realistic and are taken from geometry, science, business, and economics. The structure of problem solving is emphasized throughout the text, beginning with the simple conversion of word phrases into mathematical expressions.

Maintain Your Skills The purpose of this portion of the exercises is to review material previously covered. The exercises in each Maintain Your Skills section have been referenced to the section or sections reviewed.

State Your Understanding Each set of exercises at the end of the section contains exercises that require the students to explain their answer in words, thus addressing the trend of writing across the curriculum.

Challenge Exercises These exercises appear at the end of each section to provide problems for the more capable students to solve.

In all exercise sets, an effort has been made to pair the problems so that the set of odd-numbered problems is equivalent in kind and difficulty to the set of even-numbered problems. Answers to the odd-numbered exercises are provided in the back of the text.

Problem Solving One of the most important features of the new edition is the emphasis on problem solving. We feel very strongly about this aspect of the text because students need to learn how to read a word problem and then think about how to obtain the solution. Because of this we have included several learning aids to help the student with the problem-solving process. In each example section a strategy solution column has been included that contains a step-by-step explanation to provide the student with a good model to use when trying to work the many different exercises in the text. Another pedagogical device that is very helpful to the student in problem solving is the procedure box, which provides instructions on how to solve a specific kind of problem, for example, how to divide using synthetic division. In addition, we have also included exercises at the end of each section entitled ''State Your Understanding'' where the student is required to explain how the problem is solved rather than give a numerical answer. This type of problem helps the students to explain their understanding of the mathematics and the applications presented. Many educators now agree that if students can write and think about a problem clearly, they most likely understand the concepts involved.

Within the text itself we have devoted an entire section to solving word problems. This material is found in Chapter 1 on p. 68. A flow chart shows how to translate words into mathematical equations, along with a complete explanation of how the process works. The strategy solution column following each example is broken down into more manageable pieces for solution.

Chapter Summary Each chapter concludes with a comprehensive review, which includes definitions and strategies learned in the chapter. Each item is keyed to the page number where it is presented in the text for easy reference.

Chapter Review Exercises These exercises are included at the end of the chapter and review the objectives to ensure that the student has attained a level of proficiency and is comfortable proceeding to the next chapter. All exercises are keyed to sections and objectives so the student can refer to the text for assistance. Answers to the odd-numbered exercises can be found in the back of the book.

True–False Concept Review Serves as a check on the students' understanding of the new concepts presented in the chapter. Answers to all of the questions appear in the back of the book.

Chapter Test Each chapter concludes with a Chapter Test as a final review of student comprehension. The test contains representative problems from the entire chapter; answers to all of the test items appear in the back of the book.

Cumulative Review Cumulative review exercises are included after Chapters 3, 6, and 9 to help reinforce concepts and improve skills. Answers to all cumulative review exercises appear in the back of the book.

Timetable The text is intended to be used in a one-quarter or one-semester course. In a Math Lab or Learning Center the time lines are variable. Other variations are possible, but preparation for more difficult algebra courses is the intent.

State Requirements Particular attention has been paid to the testing requirements for various states (e.g., TASP, ELM, CLAST, etc.). Please see pages xv–xvii for additional information on how this text meets requirements for your state.

CONTENT OVERVIEW

In **Chapter 1** real numbers and their properties are reviewed. Fundamental operations and order of operations are presented. Solution of linear equations, literal equations, formulas for a specific variable, and inequalities are reviewed. The chapter ends with a discussion on the applications of algebra to real-life situations and a logical and systematic approach to their solutions.

In **Chapter 2** the topics begin with a review of the laws of exponents applied to integer exponents. These are used to develop the procedures for multiplying and dividing monomials. Section 2.3 presents scientific notation as an extensively used and practical application of exponents. Polynomials are defined and classified in Section 2.4. The fundamental operations on polynomials are shown in Sections 2.5, 2.6, and 2.7. These include the special products whose recognition is vital for future work in factoring. The remainder of the chapter is devoted to review of factoring techniques for all polynomials with special emphasis on trinomials in Section 2.9 and special cases in Section 2.10. Finally, Section 2.11 summarizes factoring by offering a variety of polynomials.

In **Chapter 3** rational expressions are presented. Polynomial operations are relied upon to lead into the operations on rational expressions. Care is taken to note the restrictions necessary on the domain of the variable(s) in order to assure that the rational expressions named are equivalent. Complex fractions are simplified using either the Basic Principle of Fractions or by performing the indicated division. Finally division of polynomials is covered with special note as to how it relates to factoring a polynomial. Synthetic division prepares the student for a future course in theory of equations.

In **Chapter 4** the relationships between roots, radicals, and rational exponents are shown. After establishing the identity of $x^{1/2}$ and $\sqrt{x}$ the presentation continues with operations on radicals. Radicals are simplified, added, subtracted, multiplied, and divided (rationalize the denominator). When simplifying, the necessity for restricting the domain of the variable is pointed out. In the absence of restrictions we are careful to point out the need for absolute value notation. After solving radical equations the number system is extended to complex numbers. After defining complex numbers the basic operations on complex numbers are performed.

In **Chapter 5** solutions of quadratic equations, absolute value equations, equations that are quadratic in form, and absolute value, quadratic, and rational inequalities are presented. Quadratic equations are solved by factoring, square roots, completing the square, and the quadratic formula. Equations with complex roots are included. Care is taken to restrict the domain of the variables to exclude extraneous roots. Identification of critical numbers on the number line forms the basis for the solution for quadratic and rational inequalities.

In **Chapter 6** linear equations in two variables and their graphs are covered. There follows a discussion on the properties of the graphs including slope, intercepts, distance formula, and useful forms of the linear equation. Identification of parallel and perpendicular lines using the slope is studied. Graphing linear inequalities follows from the prior work on equations. The chapter ends with the application of variation, presenting direct, inverse, and joint variations.

In **Chapter 7** linear systems of equations in two or three variables are solved. Solution by substitution, linear combinations, Cramer's Rule, and matrices are illustrated. Determinants are introduced to support Cramer's Rule.

In **Chapter 8** the conic sections are covered. The parabola is presented in detail with a complete discussion of vertices and maximum or minimum values. Other conic sections are treated in Sections 8.2 and 8.3. The chapter concludes with the solutions of systems of two equations involving quadratics. In 8.4 the systems contain one linear and one quadratic equation. In 8.5, the systems contain two quadratic equations. The graphs of the systems are used to illustrate the common solutions.

In **Chapter 9** relations and functions are defined and then symbolized using set notation with related ranges and domains. Functional notation is defined and is used throughout. Operations with functions, that is, addition, subtraction, multiplication, division, and composition of functions are discussed at great length. The inverse of a relation and one-to-one functions are defined. The inverse of one-to-one functions is applied in the study of exponential functions and logarithmic functions in Chapter 10.

In **Chapter 10** exponential and logarithmic functions are presented. Properties of logarithms are given. Logarithms are found using a calculator. Base e and base ten logs are found and students are given a method of finding a logarithm of a number with any positive base. Equations involving variable exponents and/or logarithms are solved.

In **Chapter 11** functions of counting numbers are presented. Emphasis is placed on progressions, both arithmetic and geometric. The general binomial expansion and finding a specified term in the expansion concludes the coverage.

A new **Appendix** on **Sets** is included for those instructors who wish to cover sets because of interest or state requirements. In that appendix, you will find complete coverage of sets, including the complement, union, and intersection of the set, along with examples for the student to study and problems for the student to work.

ANCILLARY ITEMS

The following supplements are available to accompany this text:

Instructor's Resource Manual This manual contains complete, worked-out solutions to all of the problems in the text, to complement this Annotated Instructor's Edition, which contains answers to all exercises in red next to the problem and teaching tips in the margin.

Instructor's Testing Manual This manual contains six written tests for each chapter, a final examination, and a Diagnostic Test. Half of the tests have multiple-choice questions and half use open-ended questions to provide flexibility in testing. Also included in this manual is a Printed Test Bank containing multiple-choice tests generated from the Computerized Test Bank. These questions are keyed to the section number and objective of the text, and are graded in difficulty.

ExaMaster™ Computerized Test Bank Available for Apple II, IBM, and Macintosh computers, this test bank contains over 2,000 questions, representing every section of the text. Keyed to section number, objective, and level of difficulty, these questions can be combined to create customized tests. The instructor can add and edit questions, and separate grading keys are provided.

MathCue Interactive Software This program disk contains practice problems from each section of the textbook. Using an interactive approach, the software provides students with an alternate way to learn the material and, at the same time, provides the student with more individualized attention. The program will automatically advance to the next level of difficulty once the student has successfully solved a few problems; the student may also ask to see the solution to check his or her understanding of the process used. The software is keyed to the textbook and will refer the student to the appropriate section of the text if an incorrect answer is input. A useful tool to check skills and to identify and correct any difficulties in finding solutions, this software is available for the Apple II, Macintosh and IBM PC microcomputers.

MathCue Solution Finder Software Available for IBM and Macintosh, this software allows students to input their own questions through the use of an expert system, a branch of artificial intelligence. Students may check their answers or receive help as if they were working with a tutor. The software will refer the student to the appropriate section of the text and will record the number of problems entered and number of correct answers given. Featuring a function grapher, the students can zoom in and out, evaluate a function at a point, graph up to four functions simultaneously, and save and retrieve function setups via disk files.

Videotapes A complete set of videotapes (23 hours) is available to give added assistance or to serve as a quick review of the book. The tapes review problem-solving methods and guide students through practice problems; students can stop the tape to work the problems and begin it again to check their solutions. Keyed to the text and providing coverage of every section of the book, these tapes offer another approach to mastery of the given topic.

Student Solutions Manual This manual contains worked-out solutions to one quarter of the problems in the exercise sets (every other odd-numbered problem) to help the student learn and practice the techniques used in solving problems.

ACKNOWLEDGMENTS

The authors appreciate the unfailing patience and continuing support of their wives, Mary Barker, Elinore Rogers, and Carol Van Dyke, who made the completion of this work possible. Thanks go to Richard Davis of Community College of Allegheny County for his help in revising the exercises. We also thank our colleagues for their help and suggestions for the improved second edition.

We are grateful to Bob Stern and Ellen Newman of Saunders College Publishing for their suggestions during the preparation of the text and to Kirsten Kauffman of York Production Services. We would also like to express our gratitude to the following reviewers for their many excellent contributions to the development of the text:

Robert Billups, Citrus College

Charles J. Clare, Diablo Valley College

Linda Holden, Indiana University

Herbert Kasube, Bradley University

Kenneth F. Klopfenstein, Colorado State University

Theodore Lai, Hudson County Community College

Harvey Lambert, University of Nevada, Reno

John C. Neuenfeldt, University of Wisconsin, Stout

Joanne Peeples, El Paso Community College

Arthur Schwartz, Mercer County Community College

Barbara Jane Sparks, Camden County College

Special thanks to Jean Moran of Avila College and to Linda Farish of North Lake College for their accuracy reviews of all the problems and exercises in the text.

We would also like to thank the following people for their excellent work on the various ancillary items that accompany Intermediate Algebra:

Linda Farish, North Lake College (Instructor's Resource Manual)

Mark Serebransky, Camden County College and John Garlow, Tarrant County Junior College (Instructor's Testing Manual)

George W. Bergeman, Northern Virginia Community College (MathCue Interactive Software and MathCue Solution Finder Software)

Bob Finnell and Hollis Adams, Portland Community College (Videotapes)

Grace Malaney, Donnelly College (Student Solutions Manual)

John Garlow, Tarrant County Junior College (Computerized Test Bank)

Jack Barker
Jim Rogers
Jim Van Dyke

ELM MATHEMATICAL SKILLS

The following table lists the California ELM MATHEMATICAL SKILLS and where coverage of these skills can be found in the text. Skills not covered in this text can be found in Fundamentals of Mathematics, 5th Edition, or Elementary Algebra, 2nd Edition. Location of skills are indicated by text section or chapter.

Skill	Location in Text
Exponentiation and square roots	2.1, 2.2, 4.2
Applications (averages, percents, word problems)	In applications throughout the text
Simplifying polynomials by grouping (1 and 2 variables)	1.4
Evaluating polynomials (1 and 2 variables)	1.6
Addition and subtraction of polynomials	2.5
Multiplication of a polynomial by a monomial	2.6
Multiplying two binomials	2.6
Squaring a binomial	2.7
Divide a polynomial by a monomial (no remainder)	3.6
Divide a polynomial by a linear binomial	3.6
Factor out the GCF from a polynomial	2.8
Factor a trinomial	2.9, 2.11
Factor the difference of squares	2.10
Simplify a rational expression by cancelling common factors	3.1
Addition and subtraction of rational expressions	3.3
Multiplication and division of rational expressions	3.2
Simplification of complex fractions	3.4
Positive exponents	2.1, 4.1
Laws of exponents (positive)	2.1
Simplifying an expression (+ exponents)	2.1
Integer exponents	2.2
Laws of exponents (integers)	2.2
Simplifying an expression (integer exponents)	2.2
Scientific notation	2.3
Radical sign (square roots)	4.2
Simplify products under a radical	4.3
Addition and subtraction of radicals	4.4
Multiplication of radicals	4.5
Solving radical equations	4.7
Solving linear equations, one variable, numerical coefficients	1.7, Ch. 2, Ch. 3
Solving linear equations, one variable, literal coefficients	1.8
Ratio, proportion and variance	6.5
Solving equations reducible to linear	Ch. 3, 4.7, 5.4
Solving linear inequalities: 1 variable numerical coefficients	1.9
Solving 2 equations, 2 unknowns, numerical coefficients by substitution	7.1
Solving 2 equations, 2 unknowns, numerical coefficients by elimination	7.2
Solving quadratic equations from factors	2.8, 2.9, 2.10, 2.11
Solving quadratic equations by factoring	2.8, 2.9, 2.10, 2.11
Graphing points on number lines	1.1, 1.9, 5.6, 5.7
Graphing linear inequalities (one unknown)	1.9
Graphing points in the coordinate plane	6.1
Graphing linear equations: $y = mx$, $y = b$, $x = b$	6.2, 6.3

TASP MATHEMATICS SKILLS

The following table lists the Texas TASP MATHEMATICS SKILLS and where coverage of these skills can be found in the text. Skills not covered in this text can be found in Fundamentals of Mathematics, 5th Edition, or Elementary Algebra, 2nd Edition. Location of skills are indicated by text section or chapter.

Skill	Location in Text
Use number concepts and computation skills.	Ch. 1
Solve word problems involving integers, fractions, or decimals (including percents, ratios, and proportions).	1.10, 3.5, throughout the text
Interpret information from a graph, table or chart	Ch. 6
Graph numbers or number relationships.	Ch. 6
Solve one- and two-variable equations.	Ch. 7, 8.4, 8.5, throughout the text
Solve word problems involving one and two variables.	In applications throughout the text
Understand operations with algebraic expressions.	Ch. 2, Ch. 3
Solve problems involving quadratic equations.	Ch. 5, Ch. 8
Solve problems involving geometric figures.	In applications throughout the text

CLAST MATHEMATICAL SKILLS

The following table lists the Florida CLAST MATHEMATICAL SKILLS and where coverage of these skills can be found in the text. Skills not covered in this text can be found in Fundamentals of Mathematics, 5th Edition, or Elementary Algebra, 2nd Edition. Location of skills are indicated by text section or chapter.

Skill	Location in Text
2A1 —Recognizes the meaning of exponents	2.1, 2.2, 2.3
2A4 —Determines the order-relation between magnitudes	1.9
4B1 —Solves real-world problems involving perimeters, areas, and volumes of geometric figures	In applications throughout the text
4B2 —Solves real-world problems involving the Pythagorean property	5.1
1C1a—Adds and subtracts real numbers	1.3, 1.4
1C1b—Multiplies and divides real numbers	1.5
1C2 —Applies the order-of-operations agreement to computations involving numbers and variables	1.6
1C3 —Uses scientific notation in calculations involving very large or very small numbers	2.3
1C4 —Solves linear equations and inequalities	1.7, 1.8, 1.9
1C5 —Uses given formulas to compute results when geometric measurements are not involved	In applications throughout the text
1C6 —Finds particular values of a function	9.1
1C7 —Factors a quadratic expression	2.8, 2.9, 2.10, 2.11
1C8 —Finds the roots of a quadratic equation	5.1, 5.2, 5.3
2C1 —Recognizes and uses properties of operations	1.2
2C2 —Determines whether a particular number is among the solutions of a given equation or inequality	5.6, 5.7
2C3 —Recognizes statements and conditions of proportionality and variation	6.5
2C4 —Recognizes regions of the coordinate plane which correspond to specific conditions	6.1, 6.2, 6.3, 6.4, 8.1, 8.2, 8.3
3C2 —Selects applicable properties for solving equations and inequalities	Throughout the text
4C1 —Solves real-world problems inviting the use of variables, aside from commonly used geometric formulas	In applications throughout the text

CONTENTS

CHAPTER   QUADRATIC AND OTHER
   5   EQUATIONS AND INEQUALITIES
           323

CHAPTER   LINEAR EQUATIONS AND
   6   GRAPHING 404

CHAPTER   SYSTEMS OF LINEAR
   7   EQUATIONS AND INEQUALITIES
           478

CHAPTER   SECOND-DEGREE EQUATIONS
   8   IN TWO VARIABLES 545

CHAPTER   FUNCTIONS 614
   9

PEDAGOGICAL USE OF COLOR

The various colors in the text figures are used to improve clarity and understanding. Any figures with three-dimensional representations are shown in various colors to make them as realistic as possible. Color is used in those graphs where different lines are being plotted simulanteously and need to be distinguished.

In addition to the use of color in the figures, the pedagogical system in the text has been enhanced with color as well. We have used the following colors to distinguish the various pedagogical features:

■ **PROPERTY**

■ **DEFINITION**

■ **PROCEDURE**

■ **CAUTION**

■ RULE

■ FORMULA

Table Title

Table Head

INTRODUCING THE BOOK

Previews introduce the student to the material and show its relevance to the study of algebra.

PREVIEW

In chapter 1 real numbers and their properties are reviewed. Fundamental operations and order of operations are presented. Solutions of linear equations, literal equations, formulas for a specific variable, and inequalities are reviewed. (These are all topics of elementary algebra.) The chapter ends with a discussion on the applications of algebra to real-life situations and a logical systematic approach to their solution.

Understanding the material in this chapter is essential if one is to be prepared to study intermediate algebra. Each of the topics will be used extensively throughout the text.

1.1

**REAL NUMBERS:
A REVIEW**

OBJECTIVES

1. Identify subsets.
2. Find the union and intersection of two sets.
3. Identify subsets of real numbers.
4. Classify equations.
5. Order real numbers.

Objectives are identified at the beginning of each section to help students focus on skills to be learned.

In this section, we review the basic terms and definitions that form the foundation of algebra.

■ DEFINITION

Set

A *set* is a collection or group of objects. The objects contained in the set are called members or elements of the set.

Definitions are highlighted with blue-green.

A set can be described in several ways. Two of these are listing the elements and giving a verbal description. In algebra, sets usually have numbers as members. Braces are used to denote sets, so $A = \{1, 3, 5, 7, 9\}$ is the set A that contains the elements 1, 3, 5, 7, and 9.

■ DEFINITION

Empty Set
∅

The *empty set* or *null set* is a set that contains no elements. The symbol used to designate the empty or null set is ∅.

2

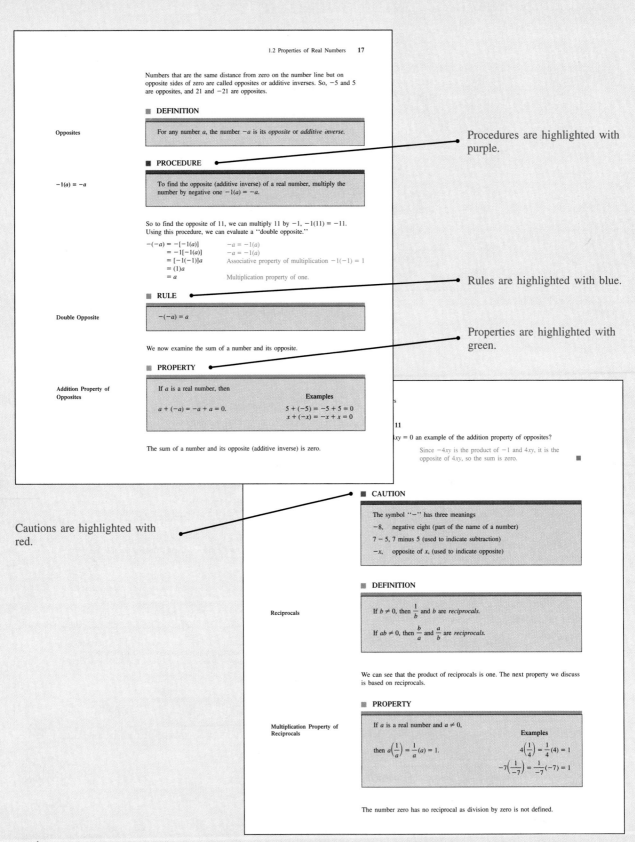

Numbers that are the same distance from zero on the number line but on opposite sides of zero are called opposites or additive inverses. So, -5 and 5 are opposites, and 21 and -21 are opposites.

■ **DEFINITION**

Opposites

> For any number a, the number $-a$ is its *opposite* or *additive inverse*.

■ **PROCEDURE**

$-1(a) = -a$

> To find the opposite (additive inverse) of a real number, multiply the number by negative one $-1(a) = -a$.

So to find the opposite of 11, we can multiply 11 by -1, $-1(11) = -11$. Using this procedure, we can evaluate a "double opposite."

$$\begin{aligned}
-(-a) &= -[-1(a)] & & -a = -1(a) \\
&= -1[-1(a)] & & -a = -1(a) \\
&= [-1(-1)]a & & \text{Associative property of multiplication } -1(-1) = 1 \\
&= (1)a \\
&= a & & \text{Multiplication property of one.}
\end{aligned}$$

■ **RULE**

Double Opposite

> $-(-a) = a$

We now examine the sum of a number and its opposite.

■ **PROPERTY**

Addition Property of Opposites

> If a is a real number, then
>
> $a + (-a) = -a + a = 0.$
>
> **Examples**
> $5 + (-5) = -5 + 5 = 0$
> $x + (-x) = -x + x = 0$

The sum of a number and its opposite (additive inverse) is zero.

$xy = 0$ an example of the addition property of opposites?

Since $-4xy$ is the product of -1 and $4xy$, it is the opposite of $4xy$, so the sum is zero.   ■

■ **CAUTION**

> The symbol "$-$" has three meanings
>
> -8, negative eight (part of the name of a number)
>
> $7 - 5$, 7 minus 5 (used to indicate subtraction)
>
> $-x$, opposite of x, (used to indicate opposite)

■ **DEFINITION**

Reciprocals

> If $b \neq 0$, then $\dfrac{1}{b}$ and b are *reciprocals*.
>
> If $ab \neq 0$, then $\dfrac{b}{a}$ and $\dfrac{a}{b}$ are *reciprocals*.

We can see that the product of reciprocals is one. The next property we discuss is based on reciprocals.

■ **PROPERTY**

Multiplication Property of Reciprocals

> If a is a real number and $a \neq 0$,
>
> then $a\left(\dfrac{1}{a}\right) = \dfrac{1}{a}(a) = 1.$
>
> **Examples**
> $4\left(\dfrac{1}{4}\right) = \dfrac{1}{4}(4) = 1$
> $-7\left(\dfrac{1}{-7}\right) = \dfrac{1}{-7}(-7) = 1$

The number zero has no reciprocal as division by zero is not defined.

Procedures are highlighted with purple.

Rules are highlighted with blue.

Properties are highlighted with green.

Cautions are highlighted with red.

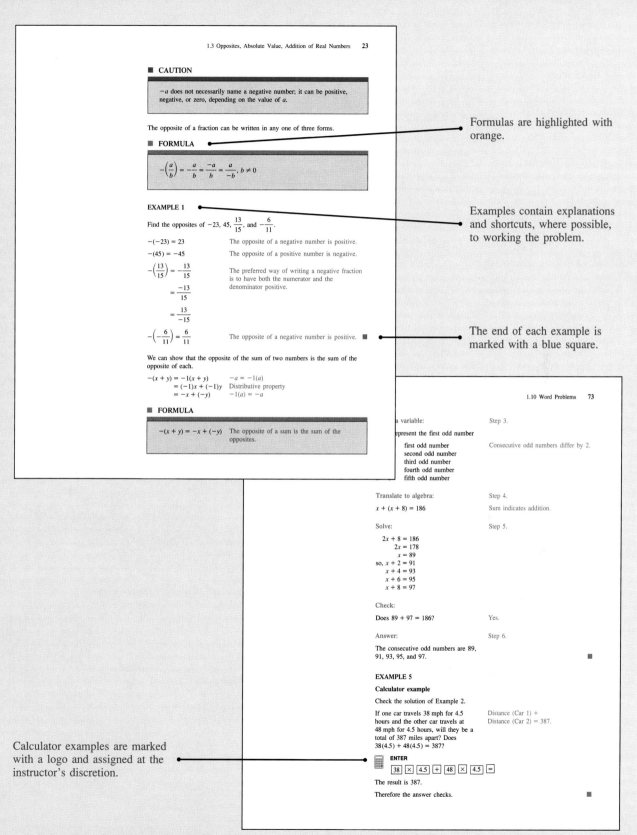

■ CAUTION

$-a$ does not necessarily name a negative number; it can be positive, negative, or zero, depending on the value of a.

The opposite of a fraction can be written in any one of three forms.

■ FORMULA

$$-\left(\frac{a}{b}\right) = -\frac{a}{b} = \frac{-a}{b} = \frac{a}{-b}, \, b \neq 0$$

Formulas are highlighted with orange.

EXAMPLE 1

Find the opposites of -23, 45, $\frac{13}{15}$, and $-\frac{6}{11}$.

$-(-23) = 23$ The opposite of a negative number is positive.

$-(45) = -45$ The opposite of a positive number is negative.

$-\left(\frac{13}{15}\right) = -\frac{13}{15}$ The preferred way of writing a negative fraction is to have both the numerator and the denominator positive.

$\phantom{-\left(\frac{13}{15}\right)} = \frac{-13}{15}$

$\phantom{-\left(\frac{13}{15}\right)} = \frac{13}{-15}$

$-\left(-\frac{6}{11}\right) = \frac{6}{11}$ The opposite of a negative number is positive. ■

Examples contain explanations and shortcuts, where possible, to working the problem.

We can show that the opposite of the sum of two numbers is the sum of the opposite of each.

$-(x + y) = -1(x + y)$ $-a = -1(a)$
$ = (-1)x + (-1)y$ Distributive property
$ = -x + (-y)$ $-1(a) = -a$

■ FORMULA

$-(x + y) = -x + (-y)$ The opposite of a sum is the sum of the opposites.

The end of each example is marked with a blue square.

a variable:

epresent the first odd number

     first odd number
     second odd number
     third odd number
     fourth odd number
     fifth odd number

Step 3.

Consecutive odd numbers differ by 2.

Translate to algebra:

$x + (x + 8) = 186$

Step 4.

Sum indicates addition.

Solve:

$2x + 8 = 186$
$2x = 178$
$x = 89$
so, $x + 2 = 91$
$x + 4 = 93$
$x + 6 = 95$
$x + 8 = 97$

Step 5.

Check:

Does $89 + 97 = 186$?

Yes.

Answer:

Step 6.

The consecutive odd numbers are 89, 91, 93, 95, and 97. ■

EXAMPLE 5

Calculator example

Check the solution of Example 2.

If one car travels 38 mph for 4.5 hours and the other car travels at 48 mph for 4.5 hours, will they be a total of 387 miles apart? Does $38(4.5) + 48(4.5) = 387$?

Distance (Car 1) + Distance (Car 2) = 387.

▦ ENTER

[38] [×] [4.5] [+] [48] [×] [4.5] [=]

The result is 387.

Therefore the answer checks. ■

Calculator examples are marked with a logo and assigned at the instructor's discretion.

EXAMPLE 13

True or false? $18 \neq 11 + 7$

False   The symbol "$\neq$" indicates that the two expressions do not name the same number. However, $18 = 11 + 7$. ∎

EXAMPLE 14

Use $=$, $<$, or $>$ in the box to make $0.5 \ \square \ -8.3$ a true statement.

$0.5 \ \boxed{>} \ -8.3$ Any positive number is greater than a negative number. ∎

Exercise 1.1

A

True or false:

1. $\{3, 5, 9\}$ is a subset of $\{0, 1, 2, 3, 4, 5, 6, 7, 8, 9\}$.

2. $\{-4, 0, 4\}$ is a subset of $\{-5, -3, -1, 0, 1, 3, 5\}$.

3. If $A = \{a, b, c, d, e\}$ and $B = \{a, d, e\}$, then B is a subset of A.

4. If $T = \left\{\frac{1}{2}, \frac{1}{4}, \frac{1}{8}, \frac{1}{16}\right\}$ and $S = \left\{\frac{1}{2}, \frac{1}{8}, \frac{1}{16}\right\}$, then T is a subset of S.

5. $\{0, 3, 5\}$ is the intersection of the sets $\{0, 1, 2, 3, 4, 5, 6\}$ and $\{-1, 0, 2, 3, 4, 5, 10\}$.

6. $\{5, 10, 15\}$ is the intersection of the sets $\{-5, 0, 5, 10, 15, 20, 25\}$ and $\{5, 6, 7, 8, 9, 10, 11, 12, 13, 14, 15\}$.

7. $\{3, 9, 7, 8\}$ is the union of $\{3, 9\}$ and $\{7, 8\}$.

8. $\{-1, 0, 1, 3, 5\}$ is the union of $\{-1, 0, 1, 3, 5\}$ and $\{-1, 0, 1, 3, 4, 5\}$.

9. $\{-12, 7, 0, -34, 681\}$ is a subset of the integers.

10. $\left\{0.25, -8.6, -\frac{3}{4}, 0, \frac{5}{9}\right\}$ is a subset of the rational numbers.

11. $\left\{-9, -3, 1, 4, \frac{4}{5}, 7.8 \ \sqrt{2}\right\}$ is a subset of the rational numbers.

12. $\{0, 6, 16, 25, 101\}$ is a subset of the natural numbers.

13. $3x = 14$ is a conditional equation.

14. $3x + 2 = 2 + 3x$ is an identity.

15. $x^2 = x$ is a contradiction.

16. $x = x - 2$ is a contradiction.

17. $-72 < -5$ 18. $0 < -51$ 19. $\frac{1}{2} < -\frac{2}{3}$ 20. $5.2 > -2.1$

13. $x - 12$ 14. $x + 9$

$-4(-7 + 8) - 9$ 17. $-3 \cdot 4 - 2(9 - 12)$

$-2)^2(4) - 4(5)^2$ 20. $(4)^2(5) - 14(-2)^2$

B

Simplify:

21. $5x + 9 - (3 - 2x)$ 22. $7a - 5 - (5a + 7)$

23. $4(3x - 1) - 2(5x + 2)$ 24. $2(3a + 4) + 6(-3a - 7)$

25. $4a - 3b - (3b + 4a)2$ 26. $9y + 4z + (-5z - 7y)$

27. $4 \cdot 3a - 2(4a + 4) - 2(-2a)$ 28. $3a(-4) - 5(2a - 5) - 3(4a)$

29. $3(-6x - 7) - 4(-2x + 6)$ 30. $-2(5c - 7) + 6(12 - 3c)$

Evaluate the following if $a = -2$ and $b = 2$:

31. $-5a - 3(2a - 3b)$ 32. $-4b + 4(3a - b)$ 33. $2a + 3b - 5(a + 1)$

34. $-4a - 2b + 2(a - 2)$ 35. $-3a - 4b - 2(a)^2$ 36. $6b - 9a + 4(b)^2$

37. $4a - 3b - 2(a + b)$ 38. $-3a + 4b + 5(a - b)$ 39. $a^2 + b^2 - 10$

40. $(a + b)^2 - 10$

C

Perform the indicated operations:

41. $-2(8 - 12) - 3[2(8 - 4)] - 3$ 42. $5(13 + 4) - 8[3(12 - 9)] + 4$

43. $3 - 2(-3) - [(-8 + 2 \cdot 5) - 4(8 - 5)] - 2[12 + (-4)]$

44. $-8 + 4(-6) - [(8 - 4 \cdot 9) - 9(12 - 6)] - 2[-8 - 12]$

45. $-5 + 3(-2)^2 - [(4 - 5)^2 - 8(-2 - 4)]$ 46. $7 - 5(3)^2 - [2(2 - 5)^2 - 3(-4 + 7)]$

Simplify:

47. $[8 - (4x - 3)] - [x - (5x - 10)]$ 48. $[4 + (3b - 7)] - [3b - (2b + 6)]$

49. $9(x - 2) - 2[3(2x - 1) - 3(-2x + 3)]$ 50. $-3(2b + 1) - 4[-2(b + 2) - 4(3b + 2)]$

51. $-5(3x + 2) - 3[-2(x + 1) - 3(2x + 5)]$ 52. $-3(2y - 1) + 3[-4(2y + 5) + 8(y - 2)]$

Evaluate each of the following if $x = -5$ and $y = 6$:

53. $9(x - 2y) - [3(2x - y) - 6(-3x + 2y)]$ 54. $-5(x + y) - [-4(5x + 2y) - 3(2x - 5y)]$

55. $(x - y)(x + y)(x + y)(x - y)$ 56. $(x + 2y)(x - 2y)(x + y)(x + y)$

Exercises are graded in difficulty; A Exercises are relatively easy and may be done mentally.

The "B" exercises may require paper and pencil.

These problems are more challenging. At the discretion of the instructor, "C" problems may sometimes be solved with a calculator.

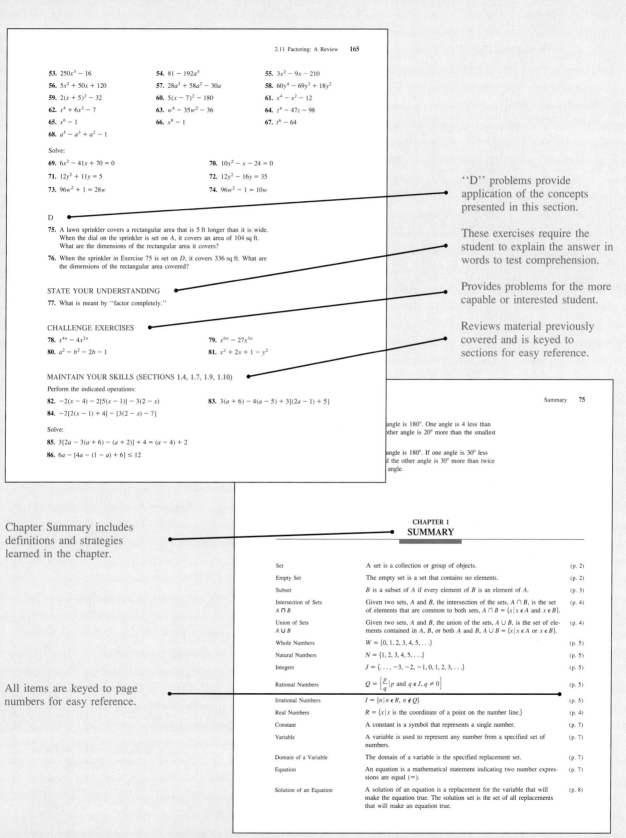

53. $250x^3 - 16$

54. $81 - 192a^3$

55. $3x^2 - 9x - 210$

56. $5x^2 + 50x + 120$

57. $28a^3 + 58a^2 - 30a$

58. $60y^4 - 69y^3 + 18y^2$

59. $2(x + 5)^2 - 32$

60. $5(x - 7)^2 - 180$

61. $x^4 - x^2 - 12$

62. $x^4 + 6x^2 - 7$

63. $w^4 - 35w^2 - 36$

64. $z^4 - 47z - 98$

65. $x^6 - 1$

66. $x^8 - 1$

67. $t^6 - 64$

68. $a^5 - a^3 + a^2 - 1$

Solve:

69. $6x^2 - 41x + 70 = 0$

70. $10x^2 - x - 24 = 0$

71. $12y^2 + 11y = 5$

72. $12y^2 - 16y = 35$

73. $96w^2 + 1 = 28w$

74. $96w^2 - 1 = 10w$

D

75. A lawn sprinkler covers a rectangular area that is 5 ft longer than it is wide. When the dial on the sprinkler is set on A, it covers an area of 104 sq ft. What are the dimensions of the rectangular area it covers?

76. When the sprinkler in Exercise 75 is set on D, it covers 336 sq ft. What are the dimensions of the rectangular area covered?

STATE YOUR UNDERSTANDING

77. What is meant by "factor completely."

CHALLENGE EXERCISES

78. $x^{4n} - 4x^{2n}$

79. $x^{6n} - 27x^{3n}$

80. $a^2 - b^2 - 2b - 1$

81. $x^2 + 2x + 1 - y^2$

MAINTAIN YOUR SKILLS (SECTIONS 1.4, 1.7, 1.9, 1.10)

Perform the indicated operations:

82. $-2(x - 4) - 2[5(x - 1)] - 3(2 - x)$

83. $3(a + 6) - 4(a - 5) + 3[(2a - 1) + 5]$

84. $-2[2(x - 1) + 4] - [3(2 - x) - 7]$

Solve:

85. $3[2a - 3(a + 6) - (a + 2)] + 4 = (a - 4) + 2$

86. $6a - [4a - (1 - a) + 6] \leq 12$

"D" problems provide application of the concepts presented in this section.

These exercises require the student to explain the answer in words to test comprehension.

Provides problems for the more capable or interested student.

Reviews material previously covered and is keyed to sections for easy reference.

---

angle is 180°. One angle is 4 less than
other angle is 20° more than the smallest

angle is 180°. If one angle is 30° less
d the other angle is 30° more than twice
angle.

Chapter Summary includes definitions and strategies learned in the chapter.

CHAPTER 1
SUMMARY

Set	A set is a collection or group of objects.	(p. 2)
Empty Set	The empty set is a set that contains no elements.	(p. 2)
Subset	B is a subset of A if every element of B is an element of A.	(p. 3)
Intersection of Sets $A \cap B$	Given two sets, A and B, the intersection of the sets, $A \cap B$, is the set of elements that are common to both sets, $A \cap B = \{x \,|\, x \in A$ and $x \in B\}$.	(p. 4)
Union of Sets $A \cup B$	Given two sets, A and B, the union of the sets, $A \cup B$, is the set of elements contained in A, B, or both A and B, $A \cup B = \{x \,|\, x \in A$ or $x \in B\}$.	(p. 4)
Whole Numbers	$W = \{0, 1, 2, 3, 4, 5, \ldots\}$	(p. 5)
Natural Numbers	$N = \{1, 2, 3, 4, 5, \ldots\}$	(p. 5)
Integers	$J = \{\ldots, -3, -2, -1, 0, 1, 2, 3, \ldots\}$	(p. 5)
Rational Numbers	$Q = \left\{\dfrac{p}{q} \,\middle|\, p$ and $q \in J, q \neq 0\right\}$	(p. 5)
Irrational Numbers	$I = \{n \,|\, n \in R, n \notin Q\}$	(p. 5)
Real Numbers	$R = \{x \,|\, x$ is the coordinate of a point on the number line.$\}$	(p. 4)
Constant	A constant is a symbol that represents a single number.	(p. 7)
Variable	A variable is used to represent any number from a specified set of numbers.	(p. 7)
Domain of a Variable	The domain of a variable is the specified replacement set.	(p. 7)
Equation	An equation is a mathematical statement indicating two number expressions are equal (=).	(p. 7)
Solution of an Equation	A solution of an equation is a replacement for the variable that will make the equation true. The solution set is the set of all replacements that will make an equation true.	(p. 8)

All items are keyed to page numbers for easy reference.

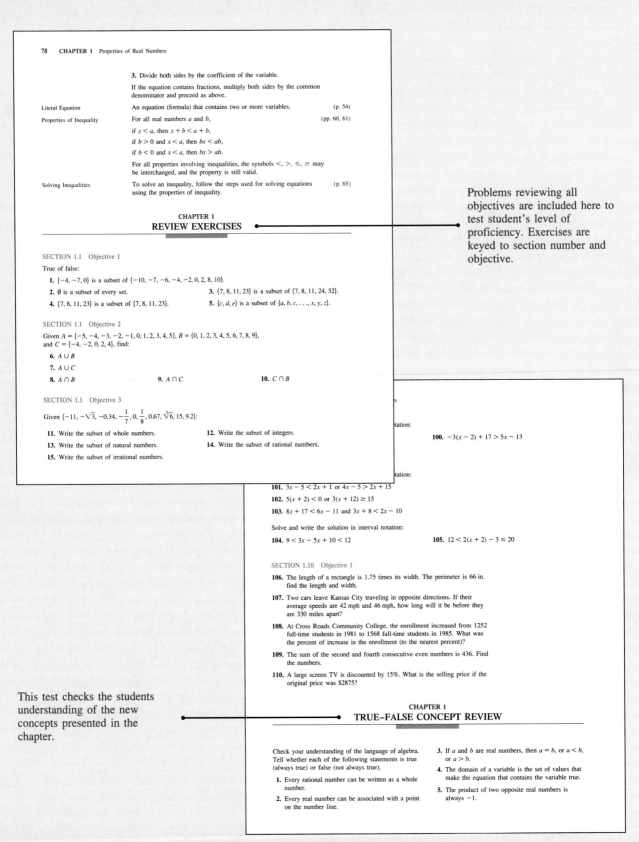

3. Divide both sides by the coefficient of the variable.

If the equation contains fractions, multiply both sides by the common denominator and proceed as above.

Literal Equation	An equation (formula) that contains two or more variables.	(p. 54)
Properties of Inequality	For all real numbers a and b,	(pp. 60, 61)

if $x < a$, then $x + b < a + b$,

if $b > 0$ and $x < a$, then $bx < ab$,

if $b < 0$ and $x < a$, then $bx > ab$.

For all properties involving inequalities, the symbols $<$, $>$, $\leq$, $\geq$ may be interchanged, and the property is still valid.

Solving Inequalities	To solve an inequality, follow the steps used for solving equations using the properties of inequality.	(p. 65)

CHAPTER 1
REVIEW EXERCISES

Problems reviewing all objectives are included here to test student's level of proficiency. Exercises are keyed to section number and objective.

SECTION 1.1   Objective 1

True of false:

1. $\{-4, -7, 0\}$ is a subset of $\{-10, -7, -6, -4, -2, 0, 2, 8, 10\}$.

2. $\emptyset$ is a subset of every set.

3. $\{7, 8, 11, 23\}$ is a subset of $\{7, 8, 11, 24, 32\}$.

4. $\{7, 8, 11, 23\}$ is a subset of $\{7, 8, 11, 23\}$.

5. $\{c, d, e\}$ is a subset of $\{a, b, c, \ldots, x, y, z\}$.

SECTION 1.1   Objective 2

Given $A = \{-5, -4, -3, -2, -1, 0, 1, 2, 3, 4, 5\}$, $B = \{0, 1, 2, 3, 4, 5, 6, 7, 8, 9\}$, and $C = \{-4, -2, 0, 2, 4\}$, find:

6. $A \cup B$

7. $A \cup C$

8. $A \cap B$ **9.** $A \cap C$ **10.** $C \cap B$

SECTION 1.1   Objective 3

Given $\{-11, -\sqrt{3}, -0.34, -\frac{1}{7}, 0, \frac{1}{8}, 0.67, \sqrt[3]{6}, 15, 9.2\}$:

11. Write the subset of whole numbers.

12. Write the subset of integers.

13. Write the subset of natural numbers.

14. Write the subset of rational numbers.

15. Write the subset of irrational numbers.

100. $-3(x - 2) + 17 > 5x - 13$

101. $3x - 5 < 2x + 1$ or $4x - 5 > 2x + 15$

102. $5(x + 2) < 0$ or $3(x + 12) \geq 15$

103. $8x + 17 < 6x - 11$ and $3x + 8 < 2x - 10$

Solve and write the solution in interval notation:

104. $9 < 3x - 5x + 10 < 12$ **105.** $12 < 2(x + 2) - 3 \leq 20$

SECTION 1.10   Objective 1

106. The length of a rectangle is 1.75 times its width. The perimeter is 66 in. find the length and width.

107. Two cars leave Kansas City traveling in opposite directions. If their average speeds are 42 mph and 46 mph, how long will it be before they are 330 miles apart?

108. At Cross Roads Community College, the enrollment increased from 1252 full-time students in 1981 to 1568 full-time students in 1985. What was the percent of increase in the enrollment (to the nearest percent)?

109. The sum of the second and fourth consecutive even numbers is 436. Find the numbers.

110. A large screen TV is discounted by 15%. What is the selling price if the original price was $2875?

This test checks the students understanding of the new concepts presented in the chapter.

CHAPTER 1
TRUE–FALSE CONCEPT REVIEW

Check your understanding of the language of algebra. Tell whether each of the following statements is true (always true) or false (not always true).

1. Every rational number can be written as a whole number.

2. Every real number can be associated with a point on the number line.

3. If a and b are real numbers, then $a = b$, or $a < b$, or $a > b$.

4. The domain of a variable is the set of values that make the equation that contains the variable true.

5. The product of two opposite real numbers is always -1.

6. The commutative property of multiplication allows us to interchange the position of any two consecutive factors.

7. Division by zero is not defined.

8. The reciprocal of a whole number cannot be a whole number.

9. The absolute value of a real number is never negative.

10. Like terms are terms involving the same variables.

11. The sum of a number and its additive inverse is 1.

12. To subtract two real numbers, add their opposites.

13. The product of two real numbers is always positive or negative.

14. If a and b are both negative real numbers, then $ab = |a| \cdot |b|$.

15. The graph of $x < 7$ is a half-open interval.

16. When multiplying each member of an inequality by a negative number, the sense of the inequality is reversed.

17. On the number line, $|x|$ is the distance between x and $-x$.

18. The sum of a positive number and a negative number is negative.

19. The product of a positive number and a negative number is negative.

20. Some equations are false and have no solution.

21. If $x < a$, then $a > x$.

22. The inequality $x \le x$ is an identity (always true).

23. If $a < x < b$, then $a > x > b$.

24. The intersection of two sets is never the empty set.

25. If $d \in B$ and B is a subset of A, then $d \in A$.

CHAPTER 1
TEST

This test is a final review of student comprehension.

1. Add: $-\dfrac{7}{40} + \dfrac{3}{50}$

2. Name the property of real numbers illustrated by the following: $40.3(1) = 1(40.3) = 40.3$

3. Simplify: $3(3a + b) - 4(6a - 3b)$

4. Simplify: $-4(7 + 3)^2 - 5(8 - 2^2)$

5. True or false?
   a. $-18.65 < -23.7$
   b. $0.0034 > 0$
   c. $-5.02 < 0.003$

6. Solve: $3x + 2(2x - 6) \ge 9$

7. Given $A = \{a, b, c, d, e\}$ and $B = \{c, d, e, f, g\}$, find $A \cup B$.

8. Simplify: $9 \cdot 5^2 + 6(8 - 12)^2$

9. Combine terms: $3b - 12b + 6b - b$

10. Name the property of real numbers illustrated by the following: $7x + (-7x) = 0$

11. Given $\left\{-7, -3.6, -\dfrac{3}{5}, 1, 9, 12.7, \sqrt{19}\right\}$, list the subset of
    a. integers
    b. irrational numbers
    c. real numbers

12. Solve

13. Simpl

14. Comb

15. Name the f

Students can take a Cumulative Test after Chapters 3, 6, and 9 to reinforce concepts and determine their progress.

CHAPTERS 1–3
CUMULATIVE REVIEW

CHAPTER 1

Simplify:

1. $4(-2) - 5[3(8 - 12)]$

2. $(-12)(-3) - (-2)(9 - 4^2)$

Evaluate each of the following if $x = -5$ and $y = -3$:

3. $5(3x - y) - x[x + y(x - y)]$

4. $3x^2 - y^2 + 4(2x - 3y)$

Solve:

5. $4x - 12 - 8x - 9 = 2x + 3$

6. $12a + 15 - 9a - 12 = 5a - 8$

7. $3x - [5x - (19 - x) - x] = x - (3x - 1)$

8. $y - \{3y - [5y - (12 - 2y) + 4] - 8\} = 0$

Solve for the indicated variable:

9. $3a + 6b = 12$, a

10. $8x - 32y = 40$; x

11. $x = \dfrac{2a - 6b}{2}$; a

12. $A = \dfrac{1}{2}h(b_1 + b_2)$; b_2

Solve and write the solution in interval notation:

13. $-8 < 3x + 4 < 10$

14. $-15 \le 5x + 10 \le 20$

15. $\dfrac{x}{5} - \dfrac{x - 2}{3} < 1$

16. $(4x - 10) - (3x + 4) > 8 + 2x$

Solve the following:

17. The product of 3 and 12 more than a number is 4 more than 7 times the number. What is the number?

18. The product of 4 and the difference of a number and 8 is 16 more than three times the number. What is the number?

19. The length of a rectangle is 1.75 times its width. The perimeter is 66 in. Find the length and width.

20. Two cars leave Kansas City traveling in opposite directions. If their average speeds are 42 mph and 46 mph, how long will it be before they are 330 miles apart?

CHAPTER 2

Perform the indicated operations:

21. $(5a - 7b + 2) + (3a - 2b + 8)$

22. $(4x^2 - 3x - 12) - (-2x^2 - 5x - 4)$

23. $(7s^2 - 7t + 12) - (13s^2 + 2t - 4)$

24. $(-3a + b - c) + (2a - 3b + 4c) + (8a - 2b - 3c)$

25. $14 - [(-2w + 13) - (5w - 12) + 6] - (9w + 1)$

Multiply:

26. $-5x(2x + y + 3)$

27. $-8a(-2a - 3b - 4c)$

28. $(3y + 7)(4y - 11)$

29. $(2x + y - 1)(x + 2y + 1)$

30. $(3t^2 - 4)(t^2 - 11t + 5)$

31. $(x - 3)(x + 4) + (2x - 5)(x + 1)$

32. $(13t + 14m)(13t - 14m)$

33. $(2x - 3)^2 - (x + 1)^2$

1

PROPERTIES OF REAL NUMBERS

In many locations there are adjacent mountains and lowlands where the altitude—the distance above or below sea level—changes very rapidly. Deserts with nearby mountains are prime examples of this change. Height, or altitude above sea level, can be expressed as a positive integer and altitude below sea level as a negative integer. In Exercise 66, Section 1.4, the highest point is 1500 ft above sea level ($+1500$) and the lowest point is 75 ft below sea level (-75). The change in altitude between the two points is found by finding the difference between the lowest point and the highest point, $1500 - (-75)$. *(Edward S. Ross/Phototake)*

PREVIEW

In chapter 1 real numbers and their properties are reviewed. Fundamental operations and order of operations are presented. Solutions of linear equations, literal equations, formulas for a specific variable, and inequalities are reviewed. (These are all topics of elementary algebra.) The chapter ends with a discussion on the applications of algebra to real-life situations and a logical systematic approach to their solution.

Understanding the material in this chapter is essential if one is to be prepared to study intermediate algebra. Each of the topics will be used extensively throughout the text.

1.1

REAL NUMBERS: A REVIEW

OBJECTIVES

1. Identify subsets.

2. Find the union and intersection of two sets.

3. Identify subsets of real numbers.

4. Classify equations.

5. Order real numbers.

In this section, we review the basic terms and definitions that form the foundation of algebra.

■ DEFINITION

Set

A *set* is a collection or group of objects. The objects contained in the set are called members or elements of the set.

A set can be described in several ways. Two of these are listing the elements and giving a verbal description. In algebra, sets usually have numbers as members. Braces are used to denote sets, so $A = \{1, 3, 5, 7, 9\}$ is the set A that contains the elements 1, 3, 5, 7, and 9.

■ DEFINITION

Empty Set

$\emptyset$

The *empty set* or *null set* is a set that contains no elements. The symbol used to designate the empty or null set is $\emptyset$.

New sets can be formed by using elements from a given set. These new sets are called subsets.

■ DEFINITION

Subset

> B is a *subset* of A if every element of B is an element of A.

EXAMPLE 1

Given $B = \{2, 4, 6, 8\}$, is B a subset of $A = \{1, 2, 3, 4, 5, 6, 7, 8, 9, 0\}$?

Yes

Since every element of B is an element of A, B is a subset of A.   ■

EXAMPLE 2

Write the subset of even numbers in the set $\{3, 13, 25, 71, 101\}$.

$\emptyset$

There are no even numbers in the set, so the subset is empty ($\emptyset$ is a subset of every set).   ■

In Example 2, although the set $\emptyset$ contains no elements, it satisfies the condition for a subset. Every element of the set $\emptyset$ is an element of the given set.

■ CAUTION

> Do not use the symbol $\{\emptyset\}$ to represent the empty set.

When variables are used to hold the place of numbers, sets can also be designated using "set builder notation." For example, the equation

$$A = \underbrace{\quad}_{A \text{ is the set}} \underbrace{\{x}_{\text{of all } x} \underbrace{\mid}_{\text{such that}} \underbrace{x > 10\}}_{x \text{ is a number larger than 10}} \text{ is read:}$$

The symbol " ϵ " is used to indicate that an element is a member of a set. The expression $x \in A$ is read "x is an element of A," "x is a member of A," or "x is in A." The symbol " $\notin$ " indicates that an element is not a member of the set. So given the set,

$D = \{a, b, c, d, e, f, g\}$

$c \in D$ and $s \notin D$

are both true statements.

■ DEFINITION

Intersection of Sets
$A \cap B$

Given two sets, A and B, the *intersection* of the sets, $A \cap B$, is the set of elements that are common to both sets,

$A \cap B = \{x \mid x \in A \text{ and } x \in B\}.$

EXAMPLE 3

Given $A = \{1, 3, 5, 7, 9, 11\}$ and $B = \{3, 4, 5\}$, find the intersection of the sets.

$A \cap B = \{3, 5\}$ The elements 3 and 5 are the only elements common to both A and B. ■

■ DEFINITION

Union of Sets
$A \cup B$

Given two sets, A and B, the *union* of the sets, $A \cup B$, is the set of elements contained in A, B, or both A and B,

$A \cup B = \{x \mid x \in A \text{ or } x \in B\}.$

EXAMPLE 4

Given $A = \{1, 3, 5, 7, 9, 11\}$ and $B = \{3, 4, 5\}$, find the union of the sets.

$A \cup B = \{1, 3, 4, 5, 7, 9, 11\}$ The union of the sets contains all the elements that are in A or B. ■

The set of numbers we will use as we begin the study of intermediate algebra is the set of real numbers. Real numbers are those that can be associated with the points on the number line. The number that is associated with a given point is called the **coordinate** of the point.

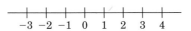

Each number shown below the line is the coordinate of the point on the line directly above.

■ DEFINITION

Real Numbers $R = \{x \mid x \text{ is the coordinate of a point}$
                              on the number line$\}$

You should be familiar with several major subsets of the real numbers. They are redefined here for your convenience.

■ DEFINITION

Natural Numbers	$N = \{1, 2, 3, 4, 5, \ldots\}$
Whole Numbers	$W = \{0, 1, 2, 3, 4, 5, \ldots\}$
Integers	$J = \{\ldots, -3, -2, -1, 0, 1, 2, 3, \ldots\}$
Rational Numbers	$Q = \{\frac{p}{q} \mid p \text{ and } q \in J, q \neq 0\}$
Irrational Numbers	$I = \{n \mid n \in R, n \notin Q\}$

Subsets that represent all of the positive or negative integers, rational numbers, irrational numbers, or real numbers can be represented by one of the raised symbols $+$ or $-$ to the right of the letter designating the set. So,

R^+ represents the set of positive real numbers, and

I^- represents the set of negative irrational numbers.

Rational numbers can be written in both fraction form and decimal form where the decimal terminates or repeats.

EXAMPLE 5

Write the $\frac{p}{q}$ form of the rational number 1.35.

$$1.35 = \frac{135}{100} = \frac{27}{20}$$ Expressed as a fraction, $1.35 = 1\frac{35}{100} = \frac{135}{100}$.

A repeating decimal is a decimal in which a block of digits repeats indefinitely. These are the decimal representations of fractions that have prime factors other than 2 and 5 in the denominator. For instance,

$$\frac{1}{3} = 0.3333\ldots = 0.\overline{3}$$ The digit 3 repeats. A bar over the digit(s) indicates that it repeats.

$$\frac{3}{7} = 0.428571428571\ldots$$ The sequence 428571 repeats.
$$= 0.\overline{428571}$$

Irrational numbers have decimal representations that neither repeat nor terminate. Some irrational numbers are,

$$\pi = 3.141592654\ldots$$ The decimal does not end, and there is no sequence of digits that repeats.

$\sqrt{2} = 1.4142135\ldots$

0.010010001... Although the digits form a pattern, they are not repeating.

$\sqrt{n}$ An irrational number if n is positive and not a perfect square. ∎

The relationship between the various subsets of real numbers is shown in the following chart.

Real Numbers

Rational Numbers (Ratios of Integers)	Irrational Numbers (Nonrepeating, non-terminating decimals)
$\dfrac{3}{7}, -\dfrac{2}{9}, \dfrac{17}{1}, 0.78, -3.42, \dfrac{22}{7}$	
Integers (Whole numbers and their opposites) $3, -3, 29, -82$	Represented by special symbols $\sqrt{2}, \pi, e, \sqrt[3]{7}$
Whole Numbers $0, 1, 2, 3, \ldots$	
Natural Numbers $1, 2, 3, \ldots$	

Given $\left\{0, -3, 17, -\dfrac{1}{2}, 0.63, 5, -32, -\dfrac{11}{7}, \sqrt{2}\right\}$

EXAMPLE 6

List the subset of rational numbers.

$\left\{0, -3, 17, -\dfrac{1}{2}, 0.63, 5, -32, -\dfrac{11}{7}\right\}$ Any number that can be written as a fraction in which both the numerator and denominator are integers and the denominator is not zero is a rational number. ∎

EXAMPLE 7

List the subset of irrational numbers.

$\{\sqrt{2}\}$ An irrational number is a real number that is not rational.   ■

EXAMPLE 8

List the subset of integers.

$\{0, -3, 17, 5, -32\}$ Integers are either whole numbers or their opposites.   ■

EXAMPLE 9

List the subset of natural numbers.

$\{17, 5\}$ Natural numbers are the nonzero whole numbers.   ■

Vocabulary that you will need to recall as we begin our study of intermediate algebra are:

■ **DEFINITION**

Constant
Variable
Domain

> A *constant* is a symbol that represents a single number.
>
> A *variable* represents any number in a specified set of numbers.
>
> The *domain* of a variable is the specified replacement set.

If x represents a whole number less than 10, the domain of x is the set $\{0, 1, 2, 3, 4, 5, 6, 7, 8, 9\}$.

■ **DEFINITION**

Equation

> An *equation* is a mathematical statement indicating that two number expressions are equal ($=$).

So $3x = 7y$ indicates that $3x$ and $7y$ name the same number. An equation can be true or false.

$4 = 2(2)$ is true.

$6 = 11 - 4$ is false.

$4x = 12$ is true when x is replaced by 3 and false when x is replaced by 5.

■ **DEFINITION**

Solution of an Equation

> A *solution* of an equation is a replacement for the variable that will make the equation true. The *solution set* is the set of all replacements that will make an equation true.

Equations can be classified as one of three types:

1. **Identity:** $3 = 3$ or $x = x$, the statement is always true.
2. **Conditional:** $2x = 6$, true for some but not all replacements of x.
3. **Contradiction:** $x + 1 = x$, the statement is never true.

EXAMPLE 10

Classify each of the following equations: $3(4) = 0$, $6x = 3x$, $8 - 5 = 3$, $2x = 2x$, and $6x = 24$.

$3(4) = 0$ Contradiction $3(4) = 12$, not 0.

$6x = 3x$ Conditional   True if $x = 0$, false for all other values.

$8 - 5 = 3$ Identity $8 - 5$ is 3, so it is always true.

$2x = 2x$ Identity   Regardless of the replacement for x, the equation is true.

$6x = 24$ Conditional   True if $x = 4$, false for all other values. ■

Algebra can be compared to a game in that there are rules or properties to be observed or played by. The properties to be discussed are properties of equality.

■ **PROPERTY**

Properties of Equality

> If a, b, and c are real numbers,
>
> $a = a$ Reflexive Property
>
> If $a = b$, then $b = a$.   Symmetric Property
>
> If $a = b$, and $b = c$, then $a = c$.   Transitive Property
>
> If $a = b$, then b can replace a in any equation and its validity remains unchanged.   Substitution Property

The reflexive property tells us that a number or expression is equal to itself. The symmetric property indicates that members (sides) of an equation may be interchanged and the validity remains unchanged. The transitive property tells us that if a first number is equal to a second number and the second number is in turn equal to a third, then the first and third numbers are also equal.

The real numbers are said to be **ordered.** This order can be shown on a **number line.** On a number line, those numbers to the right of zero are **positive,** and those to the left are **negative.** Zero is neither positive nor negative.

The graph of the set

$$\left\{-5, -4, -3.5, -3, -\sqrt{5}, -2, -1, -\frac{1}{2}, 0, 1, \frac{1}{2}, \sqrt{2}, 2, \frac{5}{2}, 3, \frac{13}{4}, 4, 4.5, 5\right\}$$

is shown on the number line. Each member of the set is the coordinate of the point on the graph.

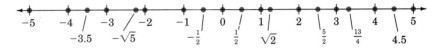

Inequality symbols show the order of the real numbers. The symbols $<$, less than; $\leq$, less than or equal to; $>$, greater than; $\geq$, greater than or equal to; and $\neq$, not equal to, are symbols of inequality.

■ DEFINITION

Less Than $<$

$a < b$ if and only if there is a positive p such that $a + p = b$.

The definition implies that a is to the left of b on the number line. This definition leads to the following method of determining whether one number is less than another one. Graph the two numbers on the number line. The number whose graph is on the left is the smaller. From the number line, we conclude that:

1. All negative numbers are less than zero.

2. All positive numbers are greater than zero.

3. Any negative number is less than any positive number.

4. Any positive number is greater than any negative number.

EXAMPLE 11

True or false? $-5 < 3$

True   A negative number is less than a positive number.   ■

EXAMPLE 12

True or false? $-4 < -8$

False -8 is to the left of -4 on the number line.   ■

EXAMPLE 13

True or false? $18 \neq 11 + 7$

False   The symbol "$\neq$" indicates that the two expressions do
                             not name the same number. However, $18 = 11 + 7$.   ■

EXAMPLE 14

Use $=$, $<$, or $>$ in the box to make $0.5 \boxed{} -8.3$ a true statement.

$0.5 \boxed{>} -8.3$ Any positive number is greater than a negative number.   ■

EXERCISE   1.1

A

True or false:

1. $\{3, 5, 9\}$ is a subset of $\{0, 1, 2, 3, 4, 5, 6, 7, 8, 9\}$.

2. $\{-4, 0, 4\}$ is a subset of $\{-5, -3, -1, 0, 1, 3, 5\}$.

3. If $A = \{a, b, c, d, e\}$ and $B = \{a, d, e\}$, then B is a subset of A.

4. If $T = \left\{\dfrac{1}{2}, \dfrac{1}{4}, \dfrac{1}{8}, \dfrac{1}{16}\right\}$ and $S = \left\{\dfrac{1}{2}, \dfrac{1}{8}, \dfrac{1}{16}\right\}$, then T is a subset of S.

5. $\{0, 3, 5\}$ is the intersection of the sets $\{0, 1, 2, 3, 4, 5, 6\}$ and $\{-1, 0, 2, 3, 4, 5, 10\}$.

6. $\{5, 10, 15\}$ is the intersection of the sets $\{-5, 0, 5, 10, 15, 20, 25\}$ and $\{5, 6, 7, 8, 9, 10, 11, 12, 13, 14, 15\}$.

7. $\{3, 9, 7, 8\}$ is the union of $\{3, 9\}$ and $\{7, 8\}$.

8. $\{-1, 0, 1, 3, 5\}$ is the union of $\{-1, 0, 1, 3, 5\}$ and $\{-1, 0, 1, 3, 4, 5\}$.

9. $\{-12, 7, 0, -34, 681\}$ is a subset of the integers.

10. $\left\{0.25, -8.6, -\dfrac{3}{4}, 0, \dfrac{5}{9}\right\}$ is a subset of the rational numbers.

11. $\left\{-9, -3, 1, 4, \dfrac{4}{5}, 7.8 \sqrt{2}\right\}$ is a subset of the rational numbers.

12. $\{0, 6, 16, 25, 101\}$ is a subset of the natural numbers.

13. $3x = 14$ is a conditional equation.

14. $3x + 2 = 2 + 3x$ is an identity.

15. $x^2 = x$ is a contradiction.

16. $x = x - 2$ is a contradiction.

17. $-72 < -5$ **18.** $0 < -51$ **19.** $\dfrac{1}{2} < -\dfrac{2}{3}$ **20.** $5.2 > -2.1$

B

For exercises 21–26, given $A = \{6, 12, 15, 91, 106, 210\}$ and $B = \{-5, 0, 6, 10, 91, 300\}$:

21. Is B a subset of A?

22. Is $\{6, 91, 210\}$ a subset of B.

23. Find $A \cap B$.

24. Find $A \cup B$.

25. Is $6 \in A$?

26. Is $106 \in B$?

For exercises 27–32, given $A = \left\{-8, -\sqrt{3}, -\dfrac{2}{3}, -0.16, 0, 5, \dfrac{16}{3}, \sqrt{39}, 42, 55.8\right\}$:

27. List the subset of integers.

28. List the subset of rational numbers.

29. List the subset of irrational numbers.

30. List the subset of natural numbers.

31. List the subset of whole numbers.

32. List the subset of positive rational numbers.

For exercises 33–36, given $B = \left\{-1.01, -\dfrac{21}{4}, 2.3, \sqrt[3]{9}, -0.21\overline{21}, \dfrac{4}{5}, 0.78, 62, -29\right\}$:

33. List the subset of integers.

34. List the subset of rational numbers.

35. List the subset of irrational numbers.

36. List the subset of natural numbers.

For exercises 37–40, use $=$, $<$, or $>$ in the box to make the statement true.

37. $5 \,\square\, 12$

38. $-\dfrac{7}{8} \,\square\, 0$

39. $0 \,\square\, -\dfrac{8}{9}$

40. $-12 \,\square\, -16$

C

For exercises 41–50, given $A = \{-3, -2, -1, 0, 1, 2, 3\}$, $B = \{0, 1, 2, 3\}$, $C = \{-8, -4, -2\}$, and $D = \{-3, -1, 0, 1, 3, 5, 7\}$, find:

41. $A \cup B$

42. $B \cup D$

43. $C \cap A$

44. $B \cap C$

45. $A \cup D$

46. $B \cup C$

47. $C \cap D$ **48.** $A \cap D$

49. Is C a subset of the integers?

50. Is B a subset of the natural numbers?

True or false:

51. $0.36\overline{36}$ is an irrational number.   **52.** $0.36036003600036. . .$ is an irrational number.

53. $5x = 7x + 3$ is a conditional equation.   **54.** $-2x - 4 = -4 - 2x$ is an identity.

55. $-28.6 < -27.99$ **56.** $-121 \neq -121$

57. $\{2, 10, 15, 20, 25\}$ is a subset of the rational numbers.

58. $\left\{-\dfrac{1}{3}, -\dfrac{3}{4}, \dfrac{2}{3}, \dfrac{15}{16}\right\}$ is a subset of the rational numbers.

59. $\emptyset$ is a subset of every set.   **60.** The intersection of $\{a, b, c\}$ and $\{d, e\}$ is $\emptyset$.

STATE YOUR UNDERSTANDING

61. The symmetric property of equality states if $a = b$ then $b = a$. State this in
your own words.

CHALLENGE EXERCISES

Given $A = \{0, 1, 2, 3\}$ and $B = \{9, 10, 12\}$

62. $A \cup \emptyset = ?$ **63.** $A \cap \emptyset = ?$ **64.** $B \cap \emptyset = ?$ **65.** $B \cup \emptyset = ?$

1.2

PROPERTIES OF REAL NUMBERS

OBJECTIVE

1. Identify properties of real numbers.

The four basic operations of arithmetic and algebra are reviewed in the
following table:

Operation	Symbol	Answer
Addition	$+$	Sum
Subtraction	$-$	Difference
Multiplication	$\times$ or $\cdot$	Product
Division	$\div$	Quotient

These operations are called binary operations.

■ DEFINITION

Binary Operation

A *binary operation* associates two numbers of a set with another.

If the associated number is a member of the same set, the set is said to be *closed* with respect to the operation.

EXAMPLE 1

Is the set of integers closed with respect to the operation of addition?

Yes, it is closed.   The sum of any two integers is an integer, for example, $3 + (-5) = -2$ and $-5 + (-1) = -6$. ■

EXAMPLE 2

Is the set of whole numbers closed with respect to subtraction?

No   The difference of two whole numbers is not always a whole number, for example, $4 - 10$ is not a whole number. ■

In algebra, when expressing multiplication between a number and a variable or between variables, we do not use a symbol for multiplication. For example, $3x$ is understood to mean 3 times x. Similarly, the expression ab means a times b.

We often use grouping symbols when writing expressions that involve more than one operation. These symbols are parentheses ( ), brackets [ ], the fraction bar, —, and braces { }. The grouping symbols indicate that the operations inside them are performed before other operations are done.

The operations of addition and multiplication applied to real numbers obey certain properties. We will now explore these properties.

■ PROPERTY

Closure Properties of Addition and Multiplication

If a and b are real numbers, then	Examples
$a + b$ is a real number	$11 + 9 = 20$
	$-2 + (-5) = -7$
and	
ab is a real number.	$9 \cdot 5 = 45$
	$\left(-\dfrac{1}{2}\right)(-4) = 2$

The closure properties state that the sum and product of two real numbers is itself a real number. The set of real numbers is said to be "closed" with respect to addition and multiplication.

■ **PROPERTY**

Commutative Properties of Addition and Multiplication

If a and b are real numbers, then

Examples

$a + b = b + a$

$$-4 + 7 = 7 + (-4)$$
$$x + y = y + x$$

and

$ab = ba.$

$$-1 \cdot 2 = 2(-1)$$

$$xy = yx$$

The commutative properties state that the order in which two real numbers are added or multiplied does not affect the sum or product. Addition and multiplication are said to be "commutative."

EXAMPLE 3

Is $3x + 2y = 2y + 3x$ an example of the commutative property of addition?

Yes   Only the order in which the terms are added has been changed; the commutative property of addition allows this.   ■

EXAMPLE 4

Is $3x + 2y = 3x + y(2)$ an example of the commutative property of multiplication?

Yes   The order in the multiplication $2y$ has been changed to $y(2)$; the commutative property of multiplication allows this.   ■

■ **PROPERTY**

Associative Properties of Addition and Multiplication

If a, b, and c are real numbers, then

Examples

$(a + b) + c = a + (b + c)$

$$(-2 + 4) + 1 = -2 + (4 + 1)$$
$$x + (y + z) = (x + y) + z$$

and

$(ab)c = a(bc)$

$$(-2 \cdot 4)5 = -2(4 \cdot 5)$$

The associative properties state that when adding or multiplying three or more real numbers, we can start by adding or multiplying any two consecutive addends or factors. When only one operation is involved, no parentheses are needed, so $2 + (3 + 4) = 2 + 3 + 4$ and $w(xz) = wxz$. Both addition and multiplication are ''associative.''

EXAMPLE 5

Is $3x + (2y + 7) = (2y + 7) + 3x$ an example of the associative property of addition?

No   This is an example of the commutative property of addition. The order of the terms has been changed. ■

EXAMPLE 6

Is $(5x)y^2 = 5(xy^2)$ an example of the associative property of multiplication?

Yes   The left side indicates the multiplication of the first two factors is done first. The right side indicates that we multiply the last two factors first. This is the associative property of multiplication. ■

■ **PROPERTY**

Distributive Property

If a, b, and c are real numbers, then

$$a(b + c) = ab + ac$$

Examples

$2(3 + 4) = 2(3) + 2(4)$
$-3(2 + 5) = -3(2) + -3(5)$

and

$$(b + c)a = ba + ca$$

$(x + 3)y = xy + 3y$

The distributive property enables us to change a product such as $a(b + c)$ to a sum $ab + ac$ and vica versa. Technically, the name for this property is the Distributive Property of Multiplication over Addition.

EXAMPLE 7

Is $2x(3y - 6) = (3y - 6)(2x)$ an example of the distributive property?

No   The order of multiplication has been changed. This is an example of the commutative property of multiplication. ■

EXAMPLE 8

Is $(2x + 5y)z = 2x(z) + 5y(z)$ an example of the distributive property?

Yes   Multiplication has been "distributed" over the addition. This is an example of the distributive property.   ■

The Distributive Property can be expanded to include more values. It is valid for a finite number of addends inside the set of parentheses. For example,

$$a(p + q + r + s + t) = ap + aq + ar + as + at$$

■ **PROPERTY**

Properties of Zero and One

If a is a real number, then	Examples
$a + 0 = a$ and $0 + a = a$ (Addition Property of Zero)	$2 + 0 = 0 + 2 = 2$ $x + 0 = 0 + x = x$
and	
$a \cdot 1 = a$ and $1 \cdot a = a$ (Multiplication Property of One)	$-8(1) = 1(-8) = -8$ $b \cdot 1 = 1 \cdot b = b$
and	
$a \cdot 0 = 0$ and $0 \cdot a = 0$ (Multiplication Property of Zero)	$7(0) = 0(7) = 0$ $m \cdot 0 = 0 \cdot m = 0$

Zero is called the "additive identity" because when zero is added to any real number the sum is the real number. One is called the "multiplicative identity" because when any real number is multiplied by one the product is the real number.

EXAMPLE 9

Is $4(3)(0) = 0$ an example of the multiplication property of zero.

Yes   The property states that zero times any real number is zero.   ■

EXAMPLE 10

Is $5x(1) = 1(5x)$ an example of the multiplication property of one?

No   The factors have been interchanged, so it is an example of the commutative property of multiplication. The multiplication property of one says $5x(1) = 5x$.   ■

Numbers that are the same distance from zero on the number line but on opposite sides of zero are called opposites or additive inverses. So, -5 and 5 are opposites, and 21 and -21 are opposites.

■ **DEFINITION**

Opposites

> For any number a, the number $-a$ is its *opposite* or *additive inverse*.

■ **PROCEDURE**

$-1(a) = -a$

> To find the opposite (additive inverse) of a real number, multiply the number by negative one $-1(a) = -a$.

So to find the opposite of 11, we can multiply 11 by -1, $-1(11) = -11$. Using this procedure, we can evaluate a "double opposite."

$$
\begin{aligned}
-(-a) &= -[-1(a)] & -a = -1(a) \\
&= -1[-1(a)] & -a = -1(a) \\
&= [-1(-1)]a & \text{Associative property of multiplication } -1(-1) = 1 \\
&= (1)a \\
&= a & \text{Multiplication property of one.}
\end{aligned}
$$

■ **RULE**

Double Opposite

> $-(-a) = a$

We now examine the sum of a number and its opposite.

■ **PROPERTY**

Addition Property of Opposites

> If a is a real number, then
>
> **Examples**
>
> $a + (-a) = -a + a = 0.$
>
> $5 + (-5) = -5 + 5 = 0$
> $x + (-x) = -x + x = 0$

The sum of a number and its opposite (additive inverse) is zero.

EXAMPLE 11

Is $-4xy + 4xy = 0$ an example of the addition property of opposites?

Yes

Since $-4xy$ is the product of -1 and $4xy$, it is the opposite of $4xy$, so the sum is zero. ■

■ **CAUTION**

The symbol "$-$" has three meanings

-8,   negative eight (part of the name of a number)

$7 - 5$, 7 minus 5 (used to indicate subtraction)

$-x$,   opposite of x, (used to indicate opposite)

■ **DEFINITION**

Reciprocals

If $b \neq 0$, then $\dfrac{1}{b}$ and b are *reciprocals*.

If $ab \neq 0$, then $\dfrac{b}{a}$ and $\dfrac{a}{b}$ are *reciprocals*.

We can see that the product of reciprocals is one. The next property we discuss is based on reciprocals.

■ **PROPERTY**

Multiplication Property of Reciprocals

If a is a real number and $a \neq 0$,

Examples

then $a\left(\dfrac{1}{a}\right) = \dfrac{1}{a}(a) = 1.$

$$4\left(\dfrac{1}{4}\right) = \dfrac{1}{4}(4) = 1$$

$$-7\left(\dfrac{1}{-7}\right) = \dfrac{1}{-7}(-7) = 1$$

The number zero has no reciprocal as division by zero is not defined.

EXAMPLE 12

Is $-\dfrac{1}{3}$ the reciprocal of 3?

No   The product, $\left(-\dfrac{1}{3}\right)(3) = -1$, is not 1, so $\dfrac{1}{3}$ is the

reciprocal, not $-\dfrac{1}{3}$.   ∎

■ PROPERTY

Trichotomy Property

For real numbers a and b, exactly one of the following is true	
	Examples
$a < b$, $a = b$, or $a > b$.	$6 = 9$ False
	$6 < 9$ True
	$6 > 9$ False

The trichotomy property states that given a and b on the number line, that a and b are coordinates of the same point or a is to the right or to the left of b.

■ PROPERTY

Transitive Property of Less Than

For real numbers a, b, and c,	
	Examples
if $a < b$, and $b < c$, then $a < c$.	$-5 < 8$ and $8 < 10.5$
	so $-5 < 10.5$

The transitive property can easily be seen from the number line:

$$\underset{a}{\mid} \overset{a<b}{} \underset{b}{\mid} \overset{b<c}{} \underset{c}{\mid}$$

$a < c$ (a is to the left of c)

EXAMPLE 13

Is the statement "if $-3 < -1$ and $-4 < -1$ then $-3 < -4$" an example of the transitive property?

No   The statement indicates that both -3 and -4 are less than -1 but does not imply an ordering of -3 and -4.   ∎

These properties play a major role in the study of algebra. You should memorize them and become very familiar with their meanings.

EXERCISE 1.2

A

The following are examples of what properties of real numbers?

1. $-8 + 4 = 4 + (-8)$

2. $5(-3) = -3 \cdot 5$

3. $-5(2 \cdot 3) = (-5(2)) \cdot 3$

4. $(-12 + 8) + 9 = -12 + (8 + 9)$

5. $-8 \cdot 1 = -8$

6. $-8 \cdot 0 = 0$

7. $2 + 3 = 5$

8. $4 \cdot 5 = 20$

9. $3 + x = x + 3$

10. $2(xy) = (2x)y$

11. $xy = yx$

12. $1(2z) = 2z$

13. $0 + (-32) = -32$

14. $5 + (-5) = 0$

15. $(-8 + 3) \cdot 2 = (-8)(2) + 3(2)$

16. $6(3 + 2) = 6(3) + 6(2)$

True or false:

17. 4 is the reciprocal of $\dfrac{1}{4}$.

18. -12 and $\dfrac{1}{12}$ are reciprocals of each other.

19. Zero is its own reciprocal.

20. If $a + b = 0$, then a and b are opposites.

B

True or false:

21. $2 \cdot 3 + 4 = 2 \cdot 3 + 2 \cdot 4$ is an example of the distributive property.

22. $3(2 \cdot 4) = (3 \cdot 2)4$ is an example of the distributive property.

23. 5 is the reciprocal of $\dfrac{1}{5}$.

24. -5 is the reciprocal of $-\dfrac{1}{5}$.

25. $7 + (-7) = 0$ is an example of the addition property of opposites.

26. $-(5 + 7) = -1(5 + 7)$ is an example of finding an opposite.

27. $2a + (-2a) = 0$ is an example of the addition property of opposites.

28. $(4x + 3) \cdot 0 = 0$ is an example of the multiplication property of zero.

29. $3(x + y) = (3x) + y$ is an example of the associative property of addition.

30. $3(x + y) = (x + y)3$ is an example of the distributive property.

31. $9(x + y) = 9x + 9y$ is an example of the distributive property.

32. $9 + (x + y) = (9 + x) + (9 + y)$ is an example of the distributive property.

33. $1(x - 2) = (x - 2)(1)$ is an example of the multiplication property of one.

34. If $\left(\dfrac{1}{3}\right)(3) = 1$, then $\dfrac{1}{3}$ and 3 are reciprocals.

The following are examples of what properties of real numbers?

35. $5 \cdot 1 = 1 \cdot 5 = 5$

36. $5 \cdot 0 = 0 \cdot 5 = 0$

37. $(7 + 9)4 = 7 \cdot 4 + 9 \cdot 4$

38. $(12 + 4)5 = 12 \cdot 5 + 4 \cdot 5$

39. $(-12)(0) = (0)(-12) = 0$

40. $0 + 14 = 14 + 0 = 14$

41. $\left(\dfrac{2}{x}\right)\left(\dfrac{x}{2}\right) = 1, \; x \neq 0$

42. $5a + (-5a) = 0$

43. $(5x + 2y) + 3z = 5x + (2y + 3z)$

44. $(7x + 3y) + 2 = 2 + (7x + 3y)$

C

For exercises 45–47, justify each step in the simplification of the expression:

45. $3 + 4(-8 + 8) = 3 + 4(0)$

46. $ = 3 + 0$

47. $ = 3$

For exercises 48–52, justify each step in the simplification of the expression.

48. $3 + 1(-3 + 8) = 3 + 1(-3) + 1(8)$

49. $ = 3 + (-3) + 8$

50. $ = [3 + (-3)] + 8$

51. $ = 0 + 8$

52. $ = 8$

For exercises 53–55, justify each step in the simplification of the expression.

53. $-12 + 4(3 + 2) = -12 + 12 + 8$

54. $ = 0 + 8$

55. $ = 8$

For exercises 56–57, justify each step in the simplification of the expression.

56. $-(2x + 2y) = -1(2x + 2y)$

57. $ = -2x + (-2y)$

For exercises 58–60, justify each step in the simplification of the expression.

58. $7[3 + (-3)] + 12 = 7 \cdot 0 + 12$

59. $ = 0 + 12$

60. $ = 12$

For exercises 61–65, justify each step in the simplification of the expression.

61. $5 + 2(3 + 2(1 + -1)) = 5 + 2(3 + 2(0))$.

62. $= 5 + 2(3 + 0)$

63. $= 5 + 2(3)$

64. $= 5 + 6$

65. $= 11$

STATE YOUR UNDERSTANDING

66. What does it mean for a set to be closed with respect to addition?

CHALLENGE EXERCISES

67. Is the set $\{0, 1\}$ closed with respect to addition?

68. Is the set $\{0, 1\}$ closed with respect to multiplication?

69. Is the set $\{1, 2\}$ closed with respect to addition?

70. Is the set $\{1, 2\}$ closed with respect to multiplication?

71. Is the set of even numbers closed with respect to addition?

72. Is the set of even numbers closed with respect to multiplication?

1.3

OPPOSITES, ABSOLUTE VALUE, ADDITION OF REAL NUMBERS

OBJECTIVES

1. Write the opposite of a real number.

2. Find the absolute value of a real number.

3. Add real numbers.

The opposite of a positive number is a negative number, and the opposite of a negative number is a positive number. The symbol -9 can be read "negative nine," "the opposite of nine," or "the additive inverse of nine." However, the symbol "$-a$" should not be read "negative a." The expression "$-a$" should be read "the opposite of a" or "the additive inverse of a" since a can be replaced by a positive number, a negative number, or zero.

■ **CAUTION**

$-a$ does not necessarily name a negative number; it can be positive, negative, or zero, depending on the value of a.

The opposite of a fraction can be written in any one of three forms.

■ **FORMULA**

$$-\left(\frac{a}{b}\right) = -\frac{a}{b} = \frac{-a}{b} = \frac{a}{-b}, \ b \neq 0$$

EXAMPLE 1

Find the opposites of -23, 45, $\dfrac{13}{15}$, and $-\dfrac{6}{11}$.

$-(^-23) = 23$ The opposite of a negative number is positive.

$-(45) = -45$ The opposite of a positive number is negative.

$-\left(\dfrac{13}{15}\right) = -\dfrac{13}{15}$ The preferred way of writing a negative fraction is to have both the numerator and the denominator positive.

$\qquad = \dfrac{-13}{15}$

$\qquad = \dfrac{13}{-15}$

$-\left(-\dfrac{6}{11}\right) = \dfrac{6}{11}$ The opposite of a negative number is positive. ■

We can show that the opposite of the sum of two numbers is the sum of the opposite of each.

$-(x + y) = -1(x + y)$ $-a = -1(a)$

$\qquad\qquad = (-1)x + (-1)y$ Distributive property

$\qquad\qquad = -x + (-y)$ $-1(a) = -a$

■ **FORMULA**

$-(x + y) = -x + (-y)$ The opposite of a sum is the sum of the opposites.

EXAMPLE 2

Find the opposites of $4x + 32$ and $7b + (-19)$.

$-(4x + 32) = -4x + (-32)$ To find the opposite of an expression, find the opposite of each term or multiply each term by -1.

$-[7b + (-19)] = -7b + [-(-19)]$ The sum of the opposite of each addend.
$\qquad\qquad\quad = -7b + 19$ ■

There are times in algebra when we are interested only in the distance a number is from zero on the number line. In this case, it does not matter whether the number is positive or negative.

■ **DEFINITION**

Absolute Value

> The *absolute value* of a real number is the distance between that number and zero on the number line. The symbol $|a|$ is read, "the absolute value of a."

This definition leads us to the following:

If a is a positive number, $|a|$ is a.

If a is a negative number, $|a|$ is the opposite of a.

If a is zero, $|a|$ is zero.

■ **FORMULA**

> $|a| = a$ if $a \geq 0$
>
> $|a| = -a$ if $a < 0$
>
> Note that $|a|$ is nonnegative.

EXAMPLE 3

Find the value of $|5|$, $|-2|$, $|0|$, and $|-96|$.

$|5| = 5$ The absolute value of a positive number is itself.

$|-2| = -(-2) = 2$ The absolute value of a negative number is its opposite.

$|0| = 0$ The absolute value of zero is zero.

$|-96| = 96$ Take the opposite of -96 mentally. ■

Here are the rules for adding real numbers:

■ RULE

Addition of Real Numbers

1. To find the sum of two numbers of like sign, add their absolute values and use the common sign.
2. To find the sum of a positive and a negative, find the difference of their absolute values and use the sign of the number with the larger absolute value.

The sum, $-11 + 5$, is found by subtracting the absolute values and using the sign of the number with the larger absolute value.

$$|-11| = 11 \qquad |5| = 5 \qquad 11 - 5 = 6$$

Since $|-11| > |5|$, the answer is negative (11 has larger absolute value). So,

$$-11 + 5 = -(|-11| - |5|) = -(6) = -6$$

EXAMPLE 4

Add: $-18 + -27$

$$-18 + (-27) = -45$$

The sum of two negative numbers is negative.   ■

EXAMPLE 5

Add: $148 + (-56)$

$$|148| - |-56| = 92$$

To find the sum of a positive and a negative number, subtract their absolute values and use the sign of the number with larger absolute value.

$$148 + (-56) = 92$$

Since $|148| > |-56|$, the sum is positive.   ■

EXAMPLE 6

Add: $\dfrac{5}{16} + \left(-\dfrac{3}{4}\right) + \dfrac{1}{3}$

$$\frac{5}{16} = \frac{15}{48}, \quad -\frac{3}{4} = -\frac{36}{48}, \quad \frac{1}{3} = \frac{16}{48}$$

To add fractions, first rewrite using a common denominator.

$$\frac{5}{16} + \left(-\frac{3}{4}\right) + \frac{1}{3} = \frac{15}{48} + \left(-\frac{36}{48}\right) + \frac{16}{48}$$

$$= \frac{15}{48} + \frac{-36}{48} + \frac{16}{48} \quad \text{Since} \quad -\frac{36}{48} = \frac{-36}{48}$$

$$= \frac{15 + (-36) + 16}{48} \qquad \text{Add the numerators and retain the common denominator.}$$

$$= -\frac{5}{48} \qquad \text{Simplify.} \qquad \blacksquare$$

EXAMPLE 7

Add: $-49 + 35 + (-18) + (-33) + 28$

$-49 + 35 + (-18) + (-33) + 28$ Using the associative and
$= [(-49) + (-18) + (-33)] + (35 + 28)$ commutative properties of addition,
$= -100 + 63$ we can add the negative and the
$= -37$ positive numbers. The sum is
                     negative as $|-100| > |63|$. $\blacksquare$

You should memorize the rules for addition and be able to perform addition quickly.

EXERCISE   1.3

A

Perform the indicated operations:

1. $|-6|$
2. $|-8|$
3. $-(-6)$
4. $-(-12)$
5. $-|-4|$
6. $-|16|$
7. $9 + (-3)$
8. $12 + (-6)$
9. $-8 + (-2)$
10. $-4 + (-7)$
11. $-8 + 8$
12. $-12 + 4$
13. $-(4 + 1)$
14. $-(8 + 2)$
15. $24 + (-12)$
16. $38 + (-26)$
17. $|8 + 7|$
18. $|4 + 9|$
19. $|-8 + 2|$
20. $|-7 + 12|$

B

21. $-(5 + 12)$
22. $-(8 + 3)$
23. $-(-7 + 5)$
24. $-(-8 + 12)$
25. $-8 + (-3) + 5$
26. $-7 + (-2) + 12$
27. $(-5 + 4) + 1$
28. $12 + [-8 + (-4)]$
29. $-6 + (-8) + (-12)$
30. $-9 + (-7) + (-14)$
31. $-(x + 2)$
32. $-(y + 4)$
33. $-[x + (-2)]$
34. $-[y + (-4)]$
35. $-((-a) + (-b))$
36. $-((-5) + (-y))$
37. $-[2x + 0(3y - 4)]$
38. $|-6 + 0(5x - 7y)|$
39. $-|a + b|,\ a > 0,\ b > 0$
40. $-|a + b|,\ a < 0,\ b < 0$

C

41. $-(15 + 12) + (-36)$

42. $-[-45 + (-7)] + (-9)$

43. $-15 + 9 + (-7) + (-12)$

44. $-75 + (-13) + (-64)$

45. $-(3.2) + (-7.5) + (1.9)$

46. $(-7.62) + (-0.4) + (-1.7)$

47. $-(-8.3 + 5.2) + [-(-0.7) + (-0.21)]$

48. $(-7.3 + 3.2) + [-(-0.8) + 1.9]$

49. $-[-|-4| + (-|3|)] + [3 + (-2)]$

50. $-[-7 + |-4|] + [-4 + (-3)]$

51. $-(-8 + 12) + (-|-4| + 9) + [12 + (-6)]$

52. $-(-9 + 11) + (-|-2| + 7) + [8 + (-7)]$

53. $-|-4 + 7| + |-16 + 12| + |-12 + 2|$

54. $-|-7 + 8| + |-24 + 16| + |14 + 2|$

55. $|-5 + 10| + (-|-7 + (-8)|) + |-10 + 5|$

56. $-|-4| + |6 + (-10)| + (-|8 + (-4)|)$

57. $|5 + (-8)| + |-16 + (-4)| + (-|-5 + (-9)|)$

58. $-|-11 + (-12)| + |10 + 4| + |-8 + (-2)|$

59. $|-6.2 + 3.1| + (-|-2.5 + 5.5|) + |-5.2 + (-4.5)|$

60. $-|2.1 + (-4.6)| + |-5.9 + (-4.7)| + |-2.5 + 7.9|$

D

61. The checking account of the Square Hole Donut Company showed credits of $894.72 and debits of $674.72 (−$674.72) for one day. What increase or decrease will the balance of the account show?

62. The checking account of the See-More Video Company showed debits of $1394.72 (−$1394.72) and credits of $627.48 for one day. What increase or decrease will the balance of the account show?

63. What is the net worth of a business that listed assets of $35,000 and liabilities of $17,821 (−$17,821)? (*Hint:* The net worth is the sum of the assets and the liabilities.)

64. What is the net worth of a business whose assets are $47,541 and liabilities are −$38,714?

65. Jan Long has a checking account balance of $641.32. She wrote checks for $49.50 and $36.47, then made a deposit of $50.00. What is her current balance?

66. The Clancy family has a net income of $2156 a month. If they spend $650 on rent, $425 on food, $150 on bills, and $125 on gas, how much do they have left?

67. A plane is flying at an elevation of 8525 feet. It then drops 1210 feet and rises 325 feet. What is the resultant elevation?

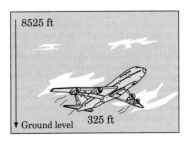

Figure for Exercise 67

68. Marie's checking account has a balance of $569.72. She makes deposits of $150.00 and $75.00, then writes checks for $14.20, $31.75, and $59.25. What is her present balance?

69. Harry and Tom formed a partnership. They agree to split all earnings and expenses equally between each other. If Harry earned $172.50 and Tom earned $156.25, what is the net worth of the partnership if Harry's expenses were $33.15 and Tom's expenses were $28.35?

STATE YOUR UNDERSTANDING

70. Why is the absolute value of a real number never negative?

CHALLENGE EXERCISES

71. Is $|a + b| = |a| + |b|$ always true? If the answer is no, try to give an example when it is true.

72. Is $|a + 0| = |a|$ always true. If the answer is no, give an example when it is false.

73. Is $|a + b| \leq |a| + |b|$ always true. If the answer is no, give an example when it is false.

74. Is $|a + (-a)| = |a| + |-a|$ always true? If the answer is no, give an example when it is true.

75. Is $|a| = |-a|$ always true? If the answer is no, give an example when it is false.

76. Is $-(a + b) = -a + b$ ever true? If the answer is yes, state when.

1.4

◼

SUBTRACTION OF REAL NUMBERS AND COMBINING TERMS

OBJECTIVES

1. Subtract two real numbers.

2. Combine like terms.

Recall from elementary algebra that subtraction will undo addition. For example, $18 + 7 - 7 = 18$. We say that addition and subtraction are inverse operations.

■ DEFINITION

Inverse operations are operations that have opposite effects.

So, $x + y - y = x$. Subtracting y has the opposite effect of ("undoes") adding y. Using this idea, we convert the subtraction of two numbers to addition.

■ DEFINITION

Subtraction of Real Numbers

If a and b are real numbers, then $a - b = a + (-b)$.

So to subtract real numbers, add the opposite or additive inverse of the numbers being subtracted.

EXAMPLE 1

Subtract: $9 - 12$

$$9 - 12 = 9 + (-12)$$
$$= -3$$

Convert the subtraction to addition. Add the opposite of 12.   ■

EXAMPLE 2

Subtract: $-12 - (-3)$

$$-12 - (-3) = -12 + [-(-3)]$$
$$= -12 + 3$$
$$= -9$$

Convert the subtraction to addition. $-(-3) = 3$.   ■

EXAMPLE 3

Subtract: $\dfrac{3}{14} - \dfrac{5}{7}$

$$\frac{3}{14} - \frac{5}{7} = \frac{3}{14} - \frac{10}{14}$$

Find a common denominator.

$$= \frac{3}{14} + \left(-\frac{10}{14}\right)$$

Convert subtraction to addition.

$$= \frac{3}{14} + \frac{-10}{14}$$

$$-\frac{a}{b} = \frac{-a}{b}$$

$$= \frac{-7}{14} = -\frac{7}{14}$$

■

■ **CAUTION**

> Subtraction is neither commutative nor associative.

It is frequently useful or necessary to simplify an algebraic expression. One way of simplifying is to combine like terms. From elementary algebra, we recall the following ideas about terms.

Term	-102	A number
	t	A variable
	$-3xy$	The product of numbers and variables
	$\dfrac{3}{11}x^2y^3$	The quotient of numbers and variables
Like terms	$3x$ and $6x$	Terms in which the variable factors
	$2xy$ and xy	are the same
	$-3abc$ and $\dfrac{2}{3}abc$	
Unlike terms	$3x$ and $5y$	Variable factors are not the same
	$-3x^2$ and $-3x$	
Numerical coefficient	4 in $4rst$	The numerical factor
	1 in xy	
	-3 in $-3xy^2$	

Since subtraction is defined in terms of addition, we can extend the distributive property to include subtraction.

■ **PROPERTY**

> $a(b - c) = ab - ac$
>
> and
>
> $(b - c)a = ba - ca$

The distributive property is used to combine like terms.

■ **DEFINITION**

Combine Like Terms

> To *combine like terms* means to add or subtract the terms.

For instance,

$43x + 29x = (43 + 29)x = 72x$

■ PROCEDURE

> To combine like terms:
>
> 1. Use the distributive property to factor, then find the sum or difference of the numerical coefficients.
> 2. Write the sum or difference and the common variable factors as an indicated product.

EXAMPLE 4

Combine like terms: $7x + 9x$

$7x + 9x = (7 + 9)x$ Factor using the distributive property.

$= 16x$ Add the coefficients. We usually do the steps mentally and write $7x + 9x = 16x$. ■

EXAMPLE 5

Combine like terms: $5ab - 7ab$

$5ab - 7ab = (5 - 7)ab$ Factor and subtract the coefficients.
$= -2ab$ ■

EXAMPLE 6

Combine like terms: $16a + 13b - 22a - 41b$

$16a + 13b - 22a - 41b = (16a - 22a) + (13b - 41b)$ Group the like terms using the commutative and associative properties.

$= -6a + (-28b)$ Combine like terms.

$= -6a - 28b$ Simplify. ■

Unlike terms cannot be combined. In Example 6, $-6a$ and $-28b$ are unlike terms and thus cannot be combined.

EXERCISE 1.4

A

Perform the indicated operations (subtract or combine like terms):

1. $12 - 6$

2. $15 - 8$

3. $-8 - 9$

4. $-4 - 12$

5. $-8 - (-4)$

6. $-12 - (-6)$

7. $-21 - (-23)$

8. $-16 - (-25)$

9. $7a + 4a$

10. $12a + 5a$

11. $-3b + 5b$

12. $-6xy + 9xy$

13. $-8abc - 9abc$

14. $-15xyz - 13xyz$

15. $14xy - (-21xy)$

16. $23ab - (-32ab)$

17. $115 - 291$

18. $274 - 365$

19. $-82 - 99$

20. $-64 - 88$

B

21. $\dfrac{5}{8} - \dfrac{1}{4}$

22. $-\dfrac{5}{6} - \dfrac{1}{3}$

23. $-0.3 - 0.7$

24. $-0.5 - 0.8$

25. $-0.5 - (-0.4)$

26. $2.7 - (-3.2)$

27. $4xy + 5xy + 7xy$

28. $3y + 2y + 9y$

29. $6x + 5x - 7x$

30. $3b + 9b - 15b$

31. $4a - 3a + 6a$

32. $8y - 9y - 6y$

33. $6x + 8 + 12x$

34. $5a - 6 + 4a$

35. $-6x - 5x - 8x$

36. $-4y + 10y - 8y$

37. $-8.7 - 14.6$

38. $-9.4 - 15.1$

39. $-19.4 - (-38.62)$

40. $-49.6 - (-72.61)$

C

41. $22.5 - (-16.3)$

42. $-48.9 - (-16.2)$

43. $4x + 3y + 2x + 5y$

44. $6ab + 9bc + 7ab + 6bc$

45. $4x + 7y - 8x - 4y$

46. $14x + 9y - 7x - 12y$

47. $\dfrac{1}{2}x - \dfrac{1}{8}y + \dfrac{3}{4}x - \dfrac{1}{2}y$

48. $-\dfrac{1}{3}x + \dfrac{2}{3}y + \dfrac{5}{6}x - \dfrac{7}{3}y$

49. $-\dfrac{1}{3}a - \dfrac{1}{4}b - \dfrac{1}{12}a - \dfrac{1}{2}b$

50. $-\dfrac{2}{3}a - \dfrac{3}{4}b - \dfrac{1}{9}a - \dfrac{1}{6}b$

51. $-0.8x - 0.7y - 0.12x - 0.3y$

52. $-0.21a - 0.32b - 0.11a - 0.24b$

53. $3.52y - 12 + 6.3z + 14$

54. $0.7a - 15 + 7.2b - 12$

55. $2.4x + 8 - 5.2x - 11$

56. $-3.2y + 16 - 8.4y - 24$

57. $-8.3x + 5.9x - 3.7x - 10.3x + 15.2x$

58. $17.4y - 3.6y - 5.7y + 1.9y + 3.8y$

59. $-(3x - 7y + 2) + 4x - 2y + 7$

60. $-(-5x - 3y - 7) + 2x + 5y - 9$

D

61. What was the change in temperature (difference) from a low of $-8°$F to a high of $40°$F?

62. What is the change in temperature (difference) from a low of $-20°$ to a high of $35°$?

63. What is the change in temperature (difference) from a high of $45°$ to a low of $-15°$?

64. What is the difference in altitude of 100 ft below sea level $(-100$ ft) and 5000 above sea level?

65. What is the change in altitude from 250 ft below sea level (-250) to 600 ft above sea level?

66. What is the change in altitude from 1500 ft above sea level to 75 feet below sea level (-75)?

67. If a checking account has a balance of $325.54 and a check for $379.50 is written, what is the balance?

68. If a checking account has a balance of $95.36 and a check for $99.72 is written, what is the balance?

69. If a check is written for the amount of $85.42 on an account with a balance of $96.57, what is the new balance?

70. If a check is written for $45.21 on an account with a balance of $75.45, what is the new balance?

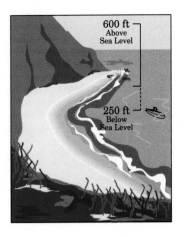

Figure for Exercise 65

STATE YOUR UNDERSTANDING

71. What are the three uses for the minus sign?

CHALLENGE EXERCISES

72. Is $a - b = b - a$ ever true? If the answer is yes, state when it is true.

73. Is $(a - b) - c = a - (b - c)$ ever true? If the answer is yes, state when it is true.

1.5

RECIPROCALS, MULTIPLICATION, AND DIVISION OF REAL NUMBERS

OBJECTIVES

1. Multiply real numbers.

2. Find the reciprocal of a real number.

3. Divide real numbers.

4. Multiply or divide terms by real numbers.

Here are the rules for multiplying real numbers.

■ **RULE**

$(+)(+) = +$
$(+)(-) = -$
$(-)(+) = -$
$(-)(-) = +$

1. The product of two positive numbers is positive.

2. The product of a positive number and a negative number in either order is negative.

3. The product of two negative numbers is positive.

For shortcuts, remember that multiplication of real numbers is commutative and associative.

EXAMPLE 1

Multiply: $-4(5)$

$$-4(5) = -20$$

The product of a negative number and a positive number is negative. ■

EXAMPLE 2

Multiply: $\left(-\dfrac{3}{7}\right)\left(-\dfrac{28}{33}\right)$

$$\left(-\dfrac{3}{7}\right)\left(-\dfrac{28}{33}\right) = \left(-\dfrac{\overset{1}{\cancel{3}}}{\underset{1}{7}}\right)\left(-\dfrac{\overset{4}{\cancel{28}}}{\underset{11}{\cancel{33}}}\right)$$

Reduce (divide out the common factors).

$$= \dfrac{4}{11}$$

Multiply the numerators and multiply the denominators. The product of two negative numbers is positive. ■

EXAMPLE 3

Multiply: $(-14)^2$

$(-14)^2 = 196$ The square of a negative number is positive.   ■

■ **CAUTION**

> Do not confuse -2^2 and $(-2)^2$. They do not mean the same thing. The expression $(-2)^2 = (-2)(-2) = 4$. However, $-2^2 = -(2)(2) = -4$. In general, $(-b)^2$ means that $-b$ is squared, $(-b)^2 = b^2$, while $-b^2$ means the opposite of "b squared," $-b^2 = -(b)^2$. So $(-b)^2 \neq -b^2$.

Recall from elementary algebra that division will undo multiplication. For example, $12(4) \div 4 = 12$. We say that multiplication and division are inverse operations.

So, $x(y) \div y = x$. Dividing by y has the opposite effect of ("undoes") multiplying by y. Using this idea, we can convert the division of two numbers to multiplication.

■ **DEFINITION**

Division of Real Numbers

> If a and b are real numbers, $b \neq 0$, then $\dfrac{a}{b} = a \div b = a\left(\dfrac{1}{b}\right)$.

So to divide two numbers, multiply by the reciprocal (multiplicative inverse) of the divisor. Recall that the reciprocal of $\dfrac{a}{b}$ is $\dfrac{b}{a}$ and the product $\left(\dfrac{a}{b}\right)\left(\dfrac{b}{a}\right) = 1$.

EXAMPLE 4

Find the reciprocals of 3, -2, $\dfrac{11}{7}$, and $-\dfrac{5}{13}$.

The reciprocal of 3 is $\dfrac{1}{3}$.   Think of 3 as $\dfrac{3}{1}$. The reciprocal of $\dfrac{b}{1}$ is $\dfrac{1}{b}$.

The reciprocal of -2 is $-\dfrac{1}{2}$, $\dfrac{1}{-2}$, or $\dfrac{-1}{2}$.

The reciprocal of $\dfrac{11}{7}$ is $\dfrac{7}{11}$.   The reciprocal of $\dfrac{a}{b}$ is $\dfrac{b}{a}$.

The reciprocal of $-\dfrac{5}{13}$ is $-\dfrac{13}{5}$, $\dfrac{-13}{5}$, or $\dfrac{13}{-5}$.   ■

EXAMPLE 5

Divide: $-17 \div 3$

$$-17 \div 3 = -17\left(\frac{1}{3}\right)$$ Convert division to multiplication.

$$= -\frac{17}{3}$$ Simplify.   ■

EXAMPLE 6

Divide: $-128 \div (-32)$

$$-128 \div (-32) = \frac{-128}{-32}$$ Write the division in fraction form.

$$= 4$$ Simplify. As in multiplication, the quotient of two negative real numbers is positive.   ■

Recall that zero has no reciprocal, so $\frac{1}{0}$ has no meaning. Since division is defined in terms of the reciprocal, it follows that DIVISION BY ZERO IS NOT DEFINED.

Three points to remember about division are:

1. Division is not commutative, $a \div b \neq b \div a$.

2. Division is not associative, $(a \div b) \div c \neq a \div (b \div c)$.

3. DIVISION BY ZERO IS NOT DEFINED.

■ **CAUTION**

$a \div b = \dfrac{a}{b}$ is not defined when $b = 0$.

The following procedures are used to multiply or divide a term by a real number.

■ **PROCEDURE**

1. To multiply a term by a number, multiply the number times the coefficient of the term using the associative and commutative properties of multiplication.

2. To divide a term by a real number, divide the coefficient of the term by the real number.

EXAMPLE 7

Multiply: $24(32x)$

$24(32x) = (24 \cdot 32)x$ Use the associative property to group the numbers.

$\qquad = 768x$ Multiply.   ■

EXAMPLE 8

Multiply: $(-28y)(-18)$

$(-28y)(-18) = [(-28)(-18)]y$ Group the numbers.

$\qquad = 504y$ Simplify.   ■

EXAMPLE 9

Divide: $-480c \div 12$

$-480c \div 12 = \dfrac{-480c}{12}$ Write the division as a fraction.

$\qquad = -40c$ Divide the coefficient by 12.   ■

Using the distributive property, we can multiply sums of terms by a number.

■ **PROCEDURE**

To multiply a real number times the sum or difference of two or more terms:

1. Multiply using the distributive property.
2. Simplify.

EXAMPLE 10

Multiply: $2(x + 3)$

$2(x + 3) = 2(x) + 2(3)$ Distributive property.

$\qquad = 2x + 6$ Simplify.   ■

EXAMPLE 11

Multiply: $-4(-3x + 2)$

$-4(-3x + 2) = -4(-3x) + (-4)(2)$ Distributive property.

$\qquad = 12x + (-8)$ Simplify.

$\qquad = 12x - 8$ Write as subtraction.   ■

EXAMPLE 12

Multiply: $-8(5a - 3b - c)$

$$-8(5a - 3b - c) = -8(5a) - (-8)(3b) - (-8)c \qquad \text{Distributive property.}$$
$$= -40a - (-24b) - (-8c) \qquad \text{Multiply.}$$
$$= -40a + 24b + 8c \qquad \text{Simplify.} \qquad ■$$

EXERCISE 1.5

A

Find the reciprocal of each of the following:

1. 6 **2.** -7 **3.** $\dfrac{1}{3}$ **4.** $-\dfrac{3}{4}$

Perform the indicated operations:

5. $(-7)(-5)$ **6.** $(-8)(-7)$ **7.** $(-14)(5)$

8. $(-12)(4)$ **9.** $-12(0)$ **10.** $0(-15)$

11. $-22(1)$ **12.** $1(-14)$ **13.** $-64 \div 8$

14. $-72 \div 9$ **15.** $(-24) \div (-6)$ **16.** $(-56) \div (-8)$

17. $(-10x) \div (-2)$ **18.** $(-14y) \div (-7)$ **19.** -20^2

20. $(-20)^2$

B

Find the reciprocal of each of the following:

21. $-\dfrac{11}{12}$ **22.** $\dfrac{25}{3}$ **23.** $1\dfrac{1}{2}$ **24.** $-\left(1\dfrac{3}{5}\right)$

Perform the indicated operations:

25. $(-2)(-3)4$ **26.** $(-3)(-4)5$ **27.** $(-2)(-3)(-4)$

28. $(-4)(-5)(-2)$ **29.** $\dfrac{-50}{-5}$ **30.** $\dfrac{-76}{-4}$

31. $-147 \div 21$ **32.** $-150 \div 25$ **33.** $(-3a)(4)(-2)$

34. $(-5x)(2)(-5)$ **35.** $(-16)(-3b)(2)$ **36.** $(-15)(4y)(3)$

37. $3x \div 5$ **38.** $-5a \div 6$ **39.** $(-2^2)(-3)(-5)$

40. $(-2)^2(-3)(-5)$ **41.** $-(3)^2(-2)(5)$ **42.** $(16)(-3)^2(-2)$

43. $(-2)(-5)(-3)(-1)$ **44.** $(-3)(2)(-4)(6)$

C

Find the reciprocal of each of the following:

45. 0.25 **46.** 0.3 **47.** -0.55 **48.** -2.4

Perform the indicated operations:

49. $[(6x)(-5)] \div 2$ **50.** $[(-8x)(9)] \div 3$ **51.** $[-18y \div 3](-6)$

52. $\dfrac{(-3x)(-2)}{5}$ **53.** $\dfrac{(-5z)(2)}{7}$ **54.** $-2(3b - 4)$

55. $-4(-2a - 1)$ **56.** $5(-3a + 5)$ **57.** $6(-2x + 4)$

58. $-5(3x + 7)$ **59.** $-8(2x + 5)$ **60.** $-3(-3x - 4y + 2)$

61. $-5(-5a + 2b - 3)$ **62.** $-2(4x - 8y - 3z)$ **63.** $-6(-3w - 4y - 9z)$

64. $-4(2x - 5y + 1)$ **65.** $-2a(2x + 4y - 3)$ **66.** $-3x(-2 + 4a - 5b)$

D

67. Use the formula $C = \dfrac{5}{9}(F - 32)$ to change 50°F to a Celsius measure.

68. Use the formula in Exercise 67 to change -13°F to a Celsius measure.

69. Use the formula in Exercise 67 to change 14°F to a Celsius measure.

70. A stock lost 2 points (-2) for each of five consecutive days. What was the total loss for that time?

71. A stock lost 1.25 points (-1.25) for each of three consecutive days. What was the total loss for that time?

72. In a business, a cash flow can be either positive or negative. If the Fillum-Up service station had an average cash flow of $-\$20$ a day for 10 days, what was the cash flow for that period?

73. If the average cash flow of the Super Duper Cookie Company is $-\$125.00$ a day for five days, what was the cash flow for that period?

74. The temperature in Fairbanks, Alaska, was dropping 1.5° every hour (-1.5). If the temperature was 10° at 4:00 P.M., what was the temperature at 8 P.M.?

75. A submarine is diving at a rate of 20 ft every minute (-20). If the submarine was originally at 100 ft below sea level (-100), what is its elevation after ten minutes?

76. A plane is descending at a rate of 150 ft every minute (-150). If it was originally 1020 ft above sea level, what is its elevation after five minutes?

STATE YOUR UNDERSTANDING

77. What does it mean for two numbers to be reciprocals?

CHALLENGE EXERCISES

78. Is $x = x^2$ ever true? If the answer is yes, give an example.

79. Is $x < x^2$ ever true? If the answer is true, give an example.

80. Is $\dfrac{a}{b} = \dfrac{b}{a}$ ever true? If the answer is yes, give an example.

81. Is it possible for a number to be its reciprocal? If the answer is yes, give an example.

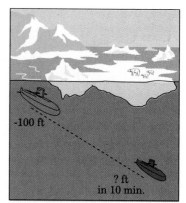

Figure for Exercise 75

1.6 ■ **ORDER OF OPERATIONS**	**OBJECTIVES** ▬ 1. Simplify expressions using the rules for order of operations. 2. Evaluate expressions.

When an algebraic expression contains more than one operation, you must take care to avoid the possibility of ambiguity. In this section, we state and use the agreement for simplifying multiple operations problems. Here the word *simplify* means to perform the indicated operations where possible.

■ PROCEDURE

> In an algebraic expression with two or more operations, perform the operations in the following order:
>
> 1. GROUPING SYMBOLS—Perform operations included by grouping symbols first (parentheses, fraction bar, and so on) according to Steps 2, 3, and 4.
>
> 2. EXPONENTS—Perform operations indicated by exponents.

3. MULTIPLY and DIVIDE—Perform multiplication and division as they appear from left to right.

4. ADD and SUBTRACT—Perform addition and subtraction as they appear from left to right.

A phrase that may be used to help you remember the order of operations is "Please Excuse My Dear Aunt Sally": Parentheses, Exponents, Multiply, Divide, Add, and Subtract.

EXAMPLE 1

Simplify: $14(15 + 12)$

$14(15 + 12) = 14(27)$ Do the addition inside the parentheses first.

$\quad = 378$ ■

EXAMPLE 2

Simplify: $4(15) + 32$

$4(15) + 32 = 60 + 32$ Do multiplication before addition.

$\quad = 92$ ■

EXAMPLE 3

Simplify: $(4)^2(-3) + (3 \cdot 4)^2$

$(4)^2(-3) + (3 \cdot 4)^2 = (4)^2(-3) + (12)^2$ Multiply inside parentheses first.

$\quad = 16(-3) + 144$ Find the squares next.

$\quad = -48 + 144$ Multiply.

$\quad = 96$ Add.   ■

If expressions contain both like and unlike terms, use the order of operations and combine the like terms.

EXAMPLE 4

Simplify: $3x + 2(x + 3)$

$3x + 2(x + 3) = 3x + 2x + 6$ Since we cannot add x and 3, use the distributive property to clear the parentheses.

$\quad = 5x + 6$ Combine like terms.   ■

EXAMPLE 5

Simplify: $2(x + 5y) + 3(x - 2y)$

$$2(x + 5y) + 3(x - 2y) = 2x + 10y + 3x - 6y \qquad \text{Distributive property.}$$
$$= 5x + 4y \qquad \text{Combine like terms.} \quad \blacksquare$$

When more than one set of grouping symbols is used in a problem, you must take special care when simplifying. If a set of grouping symbols is inside another set, start with the innermost and work out.

EXAMPLE 6

Simplify: $5 + 2[3x - (x + 1) + 4]$

$$5 + 2[3x - (x + 1) + 4] = 5 + 2[3x - x - 1 + 4] \qquad -(x + 1) = -x - 1$$
$$= 5 + 2[2x + 3] \qquad \text{Combine terms inside the brackets.}$$
$$= 5 + 4x + 6 \qquad \text{Distributive property.}$$
$$= 4x + 11 \qquad \text{Combine terms.} \quad \blacksquare$$

The order of operations plays an important role in evaluating expressions. To evaluate an expression, we use the substitution property of equality. Recall that if a and b are real numbers and $a = b$, then b can replace a in any expression, and the value of the expression will be unchanged.

■ PROCEDURE

In general, to evaluate an expression:

1. Replace the variables with the designated real numbers.
2. Simplify, following the agreements for order of operations and grouping symbols.

EXAMPLE 7

Evaluate: $3x + 2[2(3x - 1) - 5]$ when $x = -2$

$$3x + 2[2(3x - 1) - 5] = 3(-2) + 2[2(3(-2) - 1) - 5] \qquad \text{Substitute } -2 \text{ for } x$$
$$= -6 + 2[2(-6 - 1) - 5] \qquad \text{and perform the indicated operations.}$$
$$= -6 + 2[2(-7) - 5]$$
$$= -6 + 2[-14 - 5]$$

$$= -6 + 2[-19]$$

$$= -6 + (-38)$$

$$= -44 \qquad \blacksquare$$

EXAMPLE 8

Evaluate: $(3x - 2)^2 - (4x + 1)^2$ when $x = 2$

$(3x - 2)^2 - (4x + 1)^2 = (3 \cdot 2 - 2)^2 - (4 \cdot 2 + 1)^2$ Replace the variable with 2 and perform the indicated operations.

$$= (6 - 2)^2 - (8 + 1)^2$$

$$= (4)^2 - (9)^2$$

$$= 16 - 81$$

$$= -65 \qquad \blacksquare$$

EXAMPLE 9

To convert degrees Celsius to degrees Fahrenheit, we use the formula $F = \dfrac{9}{5}C + 32$. Convert 80°C to degrees Fahrenheit.

$F = \dfrac{9}{5}C + 32$ Formula.

$= \dfrac{9}{5}(80) + 32$ Replace C with 80 and perform the indicated operations.

$$= 144 + 32$$

$$= 176$$

Therefore, 80°C = 176°F. $\qquad \blacksquare$

EXERCISE 1.6

A

Perform the indicated operations:

1. $-5(-4 + 3)$ **2.** $-8(-7 + 6)$ **3.** $-5(9) + 9$

4. $-6(8) + 4$ **5.** $(2 + 3)^2$ **6.** $(4 + 1)^2$

Simplify:

7. $4(3x - 1) + 3x$ **8.** $3(3a + 2) - 2a$

9. $4 \cdot 3x - 2 \cdot 5x$ **10.** $8 \cdot 5b - 4 \cdot 3b$

Evaluate the following if $x = -5$:

11. $-4x + 2$ **12.** $-5x + 6$ **13.** $x - 12$ **14.** $x + 9$

Perform the indicated operations:

15. $12 - 3(-8 + 4) - 7$ **16.** $8 - 4(-7 + 8) - 9$ **17.** $-3 \cdot 4 - 2(9 - 12)$

18. $(-8)(-2) - 4(-6 + 12)$ **19.** $(-2)^2(4) - 4(5)^2$ **20.** $(4)^2(5) - 14(-2)^2$

B

Simplify:

21. $5x + 9 - (3 - 2x)$ **22.** $7a - 5 - (5a + 7)$

23. $4(3x - 1) - 2(5x + 2)$ **24.** $2(3a + 4) + 6(-3a - 7)$

25. $4a - 3b - (3b + 4a)2$ **26.** $9y + 4z + (-5z - 7y)$

27. $4 \cdot 3a - 2(4a + 4) - 2(-2a)$ **28.** $3a(-4) - 5(2a - 5) - 3(4a)$

29. $3(-6x - 7) - 4(-2x + 6)$ **30.** $-2(5c - 7) + 6(12 - 3c)$

Evaluate the following if $a = -2$ and $b = 2$:

31. $-5a - 3(2a - 3b)$ **32.** $-4b + 4(3a - b)$ **33.** $2a + 3b - 5(a + 1)$

34. $-4a - 2b + 2(a - 2)$ **35.** $-3a - 4b - 2(a)^2$ **36.** $6b - 9a + 4(b)^2$

37. $4a - 3b - 2(a + b)$ **38.** $-3a + 4b + 5(a - b)$ **39.** $a^2 + b^2 - 10$

40. $(a + b)^2 - 10$

C

Perform the indicated operations:

41. $-2(8 - 12) - 3[2(8 - 4)] - 3$ **42.** $5(13 + 4) - 8[3(12 - 9)] + 4$

43. $3 - 2(-3) - [(-8 + 2 \cdot 5) - 4(8 - 5)] - 2[12 + (-4)]$

44. $-8 + 4(-6) - [(8 - 4 \cdot 9) - 9(12 - 6)] - 2[-8 - 12]$

45. $-5 + 3(-2)^2 - [(4 - 5)^2 - 8(-2 - 4)]$ **46.** $7 - 5(3)^2 - [2(2 - 5)^2 - 3(-4 + 7)]$

Simplify:

47. $[8 - (4x - 3)] - [x - (5x - 10)]$ **48.** $[4 + (3b - 7)] - [3b - (2b + 6)]$

49. $9(x - 2) - 2[3(2x - 1) - 3(-2x + 3)]$ **50.** $-3(2b + 1) - 4[-2(b + 2) - 4(3b + 2)]$

51. $-5(3x + 2) - 3[-2(x + 1) - 3(2x + 5)]$ **52.** $-3(2y - 1) + 3[-4(2y + 5) + 8(y - 2)]$

Evaluate each of the following if $x = -5$ and $y = 6$:

53. $9(x - 2y) - [3(2x - y) - 6(-3x + 2y)]$ **54.** $-5(x + y) - [-4(5x + 2y) - 3(2x - 5y)]$

55. $(x - y)(x + y)(x + y)(x - y)$ **56.** $(x + 2y)(x - 2y)(x + y)(x + y)$

57. $(2x - y)(x + y)(-3x - 4y)$

58. $2(5x - 2y) - 3(2x + 4y) + (x + 2y)(2x - y)$

59. $(x - y)(2x + y)(-2x + y)(x + y)$

60. $-3(x + y)(2x + y) - (x + y)(x - y)$

D

61. Use the formula $F = \dfrac{9}{5}C + 32$ to convert 50°C to degrees Fahrenheit.

62. Convert -20°C to degrees Fahrenheit. (Use the formula in Exercise 61.)

63. Convert -35°C to degrees Fahrenheit. (Use the formula in Exercise 61.)

64. The total price of an article purchased on the monthly deferred-payment plan is $T = D + pm$, where T is the total price, D is the down payment, p is the payment per month, and m is the number of months to pay. Find the total price of an automobile if a down payment of \$500 was made and there are 36 monthly payments of \$78 each.

65. Find the total price of a TV if the down payment is \$50 and there are 24 monthly payments of \$29.18. (Use the formula in Exercise 64.)

66. The formula for the perimeter of a rectangle is $P = 2L + 2W$, where L is the length and W is the width. Find the perimeter of a rectangle with a length of 13 m and a width of 6 m.

67. Find the perimeter of a rectangle with a length of 8 yd and a width of 5 yd. (Use the formula in Exercise 66.)

68. The formula for the distance that an object falls is $s = 16t^2$, where s represents the distance in feet and t represents time. How far has an object fallen in two seconds?

69. How far does an object fall after three seconds? (Use the formula in Exercise 68.)

70. The formula for the area of a triangle is $A = \dfrac{1}{2}bh$, where A represents the area, b the base, and h the height. Find the area of a triangle with a base of 8 in. and a height of 10 in.

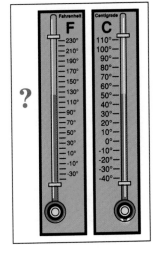

Figure for Exercise 61

STATE YOUR UNDERSTANDING

71. What is the agreement for performing the operations in an algebraic expression which contains two or more operations?

CHALLENGE EXERCISES

72. Is $4 + 8 \cdot x = (4 + 8)x$ ever true? If the answer is yes, give an example.

73. Is $4 - 8 \cdot x = (4 - 8)x$ ever true? If the answer is yes, give an example.

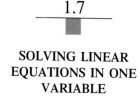

1.7

SOLVING LINEAR EQUATIONS IN ONE VARIABLE

1. Solve linear equations in one variable.

In this section we study linear equations in one variable.

■ DEFINITION

A *linear equation in one variable* is an equation that can be written in the form $ax + b = c$.

To solve an equation means to find the replacement set for the variable that will make the equation true. Recall that the domain of a variable is the set of replacements for the variable.

■ DEFINITION

A *solution* or *root* of an equation is an element from the domain of the variable that makes the equation true. The set of all solutions or roots of an equation is called the solution set of the equation.

To solve an equation, we use the properties of equality to write a series of equivalent equations until we reach one of the form

$$x = r \quad \text{or} \quad r = x.$$

The solution set for the equation is $\{r\}$.

■ DEFINITION

Two or more equations are *equivalent* if and only if they have the same solution set.

Two properties of equations are:

■ PROPERTY

Addition Property of Equality

For all real numbers a and b, if $x = a$, then $x + b = a + b$.

So if $x = 28$, then $x + 10 = 28 + 10$ is an equivalent equation. Since subtraction is defined as adding the opposite, this property includes subtraction. So $x = a$ and $x - b = a - b$ are equivalent.

■ PROPERTY

Multiplication Property of Equality

> For all real numbers a and b, $b \neq 0$; if $x = a$, then $bx = ba$.

So if $x = -14$, then $4x = 4(-14)$ is an equivalent equation. Since division is defined as multiplying by the reciprocal, this property includes division. So

$$x = a \text{ and } \frac{x}{b} = \frac{a}{b} \text{ are equivalent when } b \neq 0.$$

The next two examples illustrate these properties.

EXAMPLE 1

Solve: $x + 42 = 78$

$x + 42 = 78$	Original equation.
$x + 42 - 42 = 78 - 42$	Subtract 42 from both sides. (Subtraction property of equality)
$x = 36$	Simplify.

Check:

$x + 42 = 78$
$36 + 42 = 78$

The solution set is $\{36\}$.

Check by substituting the value in the original equation.

EXAMPLE 2

Solve: $-6x = 108$

$-6x = 108$	Original equation.
$\dfrac{-6x}{-6} = \dfrac{108}{-6}$	Divide both sides by -6. (Division property of equality)
$x = -18$	Simplify.

Check:

$$-6x = 108$$ Original equation.

$$(-6)(-18) = 108$$ Substitute.

$$108 = 108$$ True.

The solution set is $\{-18\}$.   ■

In general to solve an equation:

■ PROCEDURE

To solve an equation with no fractions:

1. Simplify each side of the equation by performing the indicated operations.
2. Isolate the terms containing the variable on one side of the equation by using the addition property of equality.
3. Isolate the constant terms on the other side using the addition property of equality. Combine like terms.
4. Divide both sides of the equation by the coefficient of the variable.
5. The solution can now be determined from the resulting equation. (Some of the above steps may not be needed.)

If an equation contains fractions:

1. Multiply both sides by the common denominator.
2. Follow the steps above.

EXAMPLE 3

Solve: $9x - 15 = 5x - 40$

$$9x - 5x - 15 = 5x - 5x - 40$$ Subtract $5x$ from both sides.

$$4x - 15 = -40$$ Combine terms.

$$4x - 15 + 15 = -40 + 15$$ Add 15 to both sides.

$$4x = -25$$ Combine terms.

$$\frac{4x}{4} = \frac{-25}{4}$$ Divide both sides by 4.

$$x = -\frac{25}{4}$$ Simplify.

Check:

 We use a calculator to check.

$9x - 15 = 5x - 40$	Original equation.

$$9x - 15 = 9\left(\frac{-25}{4}\right) - 15$$

Evaluate the left side first. Replace x with $-\dfrac{25}{4}$.

9	×	25	+/−	÷	4	−	15	=

Press the keys in the order indicated.

The result is -71.25.

$$5x - 40 = 5\left(\frac{-25}{4}\right) - 40$$

Evaluate the right side next. If -71.25 is the result, the statement is true.

5	×	25	+/−	÷	4	−	40	=

Press the keys in the order indicated.

The result is -71.25.

The solution set is $\left\{-\dfrac{25}{4}\right\}$. ∎

When equations contain grouping symbols, it is often quicker to simplify the equation to the form $ax + b = cx + d$ before using the addition or multiplication properties of equality.

EXAMPLE 4

Solve: $5 - [x - (6 - 3x) + 1] = x - 10$

$5 - [x - 6 + 3x + 1] = x - 10$	Remove parentheses by subtraction.
$5 - [4x - 5] = x - 10$	Combine terms.
$5 - 4x + 5 = x - 10$	Remove brackets by subtraction.
$-4x + 10 = x - 10$	Combine terms.
$-4x - x + 10 = x - x - 10$	Subtract x from both sides.
$-5x + 10 = -10$	Combine terms.
$-5x + 10 - 10 = -10 - 10$	Subtract 10 from both sides.
$-5x = -20$	Combine terms.
$\dfrac{-5x}{-5} = \dfrac{-20}{-5}$	Divide both sides by -5.
$x = 4$	Simplify.

Check:

$$5 - [4 - (6 - 3 \cdot 4) + 1] = 4 - 10 \qquad \text{Replace } x \text{ in the original equation with 4.}$$

$$5 - [4 - (6 - 12) + 1] = -6$$

$$5 - [4 - (-6) + 1] = -6$$

$$5 - [11] = -6$$

The solution set is $\{4\}$.   ■

Recall that equations can be classified by the number of solutions. If the domain of the variable is specified to be R, the following table shows the type of equation and the type of solution.

Equation	Example	Solution Set
Identity	$3(x + 4) = 3x + 12$	R (the domain)
Contradiction	$x = x + 2$	$\emptyset$ (No solution)
Conditional Equation	$3x - 4 = 14$	A non empty subset of R

EXAMPLE 5

Solve: $3x + 1 = 3x - 4$

$$3x - 3x + 1 = 3x - 3x - 4 \qquad \text{Subtract } 3x \text{ from both sides.}$$

$$1 = -4 \qquad \text{The equation is a contradiction.}$$

The solution set is $\emptyset$.   ■

Solutions of equations have many applications in business and industry. Many of these will be found throughout the text.

EXAMPLE 6

The cost per minute of advertising on the local television station is $150 more than five times the cost of a minute on a local radio station. If an advertiser buys one minute in each medium and spends a total of $990, what is the cost per minute in each?

We use an organized approach to solving applications. See Section 1.10.

Simpler word form:

$$\left(\begin{array}{c} \text{cost per minute} \\ \text{on radio} \end{array} \right) + \left(\begin{array}{c} \text{cost per minute} \\ \text{on TV} \end{array} \right) = 990$$

Select variable:

Let x = cost of one minute on radio.
Then $5x + 150$ = cost of one minute
on TV.

Use a variable and an expression
containing the variable to represent the
two costs.

Translate to algebra:

$$x + (5x + 150) = 990$$
$$6x + 150 = 990$$
$$6x = 840$$
$$x = 140$$

If $x = 140$, then
$5x + 150 = 5(140) + 150 = 850$

We need to find the cost on TV.

Check:

Radio cost: $140
  TV cost: 850
Total cost: $990

Answer:

It costs $140 per minute on radio and $850 per minute on TV.   ■

EXERCISE 1.7

A

Solve:

1. $x + 2 = 5$

2. $x + 3 = 7$

3. $x - 8 = 4$

4. $x - 5 = 3$

5. $-2x = -4$

6. $-5x = 10$

7. $-x = 5$

8. $-x = -5$

9. $-x + 1 = 3$

10. $-x - 1 = 3$

11. $x + 7 = 11$

12. $x - 19 = 21$

13. $-x + 15 = 36$

14. $-x - 27 = 33$

15. $x = x$

16. $-x = -x$

17. $x = x + 1$

18. $x = x - 1$

19. $3x = 2x$

20. $-4x = 2x$

B

Solve:

21. $2x + 3 = 3 + 2x$

22. $4x + 10 = 10 + 4x$

23. $5x - 4 = 11$

24. $3x - 7 = 14$

25. $2x + 7 = 5x + 4 - 3x$

26. $x + 48 + x = 2x$

27. $6t + (3t - 7) = 21$

28. $6t - (3t - 7) = 21$

29. $5b + 3(b - 6) = 35$

30. $5b - 3(b - 6) = 35$

31. $4x - 7 = 8x - 12$

32. $9y + 36 = -6y + 11$

33. $4x - 43 = 9x - 40$

34. $9a - 5 = 7a - 10$

35. $(x + 6) - (2x + 7) = 3x - 9$

36. $(4a - 3) - (2a + 7) = 6 - a$

37. $5x + 6 - (10 - 2x) = (7x + 6) - (6x - 2)$

38. $5y - 8 - (y + 2) = (8y + 12) + (-6y - 4)$

39. $-10a + 6 + (6a - 4) = (4a + 2) - (12a + 8)$

40. $-16b - 15 + (10b - 5) = (6b + 2) - (2b - 8)$

C

Solve:

41. $2x - [3x - (8x + 2) + 3] + 6 = 0$

42. $5y - [2y - (y + 4) - 7] + 3 = 0$

43. $2x - [7x - (8 - 2x) + x] = 2 - (x + 3)$

44. $3y - [5y - (9 - y) - y] = y - (2y + 1)$

45. $x - \{3x - [2x - (8 - x) + 2] + 3\} - 6 = 0$

46. $5y - \{2y - [3y - (y + 1) - 3] - 4\} - 7 = 0$

47. $2[5x - 7(x + 2) - 4(x + 1)] - 3(2x + 3) = 9$

48. $5[-2x + 6(-x - 2) + (7x - 5)] + 9(2 - x) = 17$

49. $6x - 8(2x + 4) = 8x - 12(6 - x)$

50. $2x - 6(5x - 12) = 14x - 9(8 - 2x)$

51. $-8a - [-6a - (-2a + 4) - 15] = 5[2a - 3(a + 1)]$

52. $9y - [-4y - (6y - 8) + 12] = 4[-2(y - 3)]$

53. $-[3(x - 2) + 6] - [-2(x - 1) + 5] = -2[(x + 4) - 1]$

54. $-5(-2x + 3) - 3(3x - 2) = 3(-4x - 2) + 5(3x - 5)$

55. $2x + 5 = 3(-2x + 1) - 4[2(x + 4) - 3]$

56. $4(2x + 3) - 2(3x + 1) = -5[2(x + 1) - 3]$

57. $2(x - 3) - 3(2x - 7) = 4[3(x + 1) - 2]$

58. $-3(x + 2) + 6(x - 5) = -[2(x - 3) + 1]$

59. $3[2(x - 5) - 4(x + 3)] = -(2x - 5)$

60. $7 - 5[-(x + 8) + 3(x - 1)] = -5(x + 6)$

D

61. The cost of advertising for a minute on television is $250 more than 3 times the cost for a minute on the radio. If an advertiser spends $1000 for one minute of television time and two minutes of radio time, what is the cost per minute of each?

62. The cost of an automobile is $500 more than 3 times the cost of a motor bike. If both can be purchased for a total of $2500, what does each cost?

63. If one number is 8 more than 5 times another and their sum is -10, what are the two numbers?

64. If one number is 7 less than 3 times another and their sum is 25, what are the two numbers?

65. The cost of a television set is $50 more than twice the cost of a microwave oven. If the total cost for both is $650, what is the cost of the microwave oven?

66. The total cost of a washer and dryer is $749. If the washer costs $49 more than the dryer, what is the cost of the dryer?

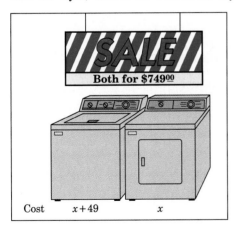

Figure for Exercise 66

67. Happy Car Auto Repair charges $32.50 per hour for labor. If Rick's total repair bill came to $116.75 with $35.50 for parts, how many hours did they work on Rick's car?

68. The temperature in Phoenix, Arizona, was one degree less than three times the temperature in Fairbanks, Alaska. If the temperature in Phoenix is 89°F, what is the temperature in Fairbanks?

69. One number is two more than three times another. If the sum of the two numbers is 22, find the two numbers.

70. If one number is three less than three times another number and their sum is -23, find the two numbers.

STATE YOUR UNDERSTANDING

71. What is meant by the domain of the variable?

CHALLENGE EXERCISES

72. Given the equation $4x + 3b = -1$, find a value of b so that when solving for x, the solution set is $\{2\}$.

73. Given the equation $2x - 3b = 0$, find a value for b so that when solving for x, the solution set is $\{-9\}$.

74. The temperature in Kelvin units, T, is given by the formula

$T = t + 273.15$

where t is the temperature measured in Celsius units. If liquified nitrogen boils at 77K, what is the temperature in °C?

75. Oxygen freezes at 55.15K. What is this temperature measured in Celsius units? (Use the formula in exercise 74.)

1.8

LITERAL EQUATIONS AND FORMULAS

OBJECTIVE

1. Solve literal equations and formulas.

The application of equations to business and science often involves equations that contain more than one variable.

■ DEFINITION

Literal Equation

A *literal equation* is an equation that contains two or more variables.

A formula is a literal equation that expresses a rule.

The properties of equality also apply to literal equations. Since literal equations contain more than one variable, it is necessary to designate the variable for which we are to solve.

EXAMPLE 1

Solve for x: $2x - 3y = 11y$

$2x - 3y + 3y = 11y + 3y$	To isolate the term containing the variable x, add $3y$ to both sides.
$2x = 14y$	Combine terms.
$\dfrac{2x}{2} = \dfrac{14y}{2}$	Divide both sides by 2.
$x = 7y$	■

In this text, we will leave the solution of literal equations and formulas in equation form.

■ **CAUTION**

> The equation is not solved if the specified variable appears on both sides. $t = at + 4$ is not solved for t.

The following example gives the formula for the circumference of a circle in terms of the diameter, and it is solved for the diameter in terms of the circumference.

EXAMPLE 2

Solve for d: $C = \pi d$

$C = \pi d$	Original formula.
$\dfrac{C}{\pi} = \dfrac{\pi d}{\pi}$	Divide both sides by π.
$\dfrac{C}{\pi} = d$	Simplify.
or $d = \dfrac{C}{\pi}$	Symmetric property of equality. ■

The next example asks for the width of a rectangle in terms of the perimeter and its length.

EXAMPLE 3

Solve for w: $P = 2w + 2l$

$$2w + 2l = P \qquad\qquad \text{Symmetric property of equality.}$$

$$2w + 2l - 2l = P - 2l \qquad\qquad \text{Subtract } 2l \text{ from both sides.}$$

$$2w = P - 2l$$

$$w = \frac{P - 2l}{2} \qquad\qquad \text{Divide both sides by 2.}$$

or

$$w = \frac{P}{2} - l \qquad\qquad \text{Divide each term of the numerator by 2.} \quad\blacksquare$$

More complicated literal equations take additional steps to solve.

EXAMPLE 4

Solve for C: $A = \dfrac{B}{C + d}$

$$(C + d)A = (C + d) \cdot \frac{B}{C + d} \qquad\qquad \text{To eliminate the fraction, we multiply both}$$
$$\text{sides by } (C + d).$$

$$AC + Ad = B$$

$$AC = B - Ad \qquad\qquad \text{Subtract } Ad \text{ from both sides.}$$

$$C = \frac{B - Ad}{A} \qquad\qquad \text{Divide both sides by } A. \qquad\qquad \blacksquare$$

EXAMPLE 5

Jamie knows that the formula to change temperature from degrees Fahrenheit to degrees Celsius is $C = \dfrac{5}{9}(F - 32)$. A series of problems in her science class require her to change Celsius temperatures to Fahrenheit, hence she needs to solve the given formula for F. Solve the formula for F.

$$C = \frac{5}{9}(F - 32) \qquad\qquad \text{Original formula.}$$

$$9C = 5(F - 32) \qquad\qquad \text{Multiply both sides by 9.}$$

$$9C = 5F - 160$$

$$9C + 160 = 5F \qquad\qquad \text{Add 160 to both sides.}$$

$$\frac{9C + 160}{5} = F. \qquad \text{Divide both sides by 5.}$$

$$\frac{9}{5}C + 32 = F$$

The formula is usually written, $F = \dfrac{9}{5}C + 32$. ■

EXERCISE 1.8

A

Solve for the indicated variable:

1. $d = rt$; t

2. $V = \ell wh$; h

3. $i = prt$; r

4. $C = 2\pi r$; π

5. $V = \pi r^2 h$; h

6. $2S = gt^2$; g

7. $A = \dfrac{1}{2}bh$; b

8. $V = \dfrac{1}{3}\pi r^2 h$; h

9. $F = \dfrac{GMm}{r^2}$; m

10. $F = \dfrac{mv^2}{r}$; m

11. $x + y = 12$; y

12. $x - y = 7$; x

13. $y = mx - 3$; x

14. $y = mx + 4$; m

15. $4a = 2b + 6$; b

16. $6y = 3x - 9$; x

17. $2s - 3t = 8$; t

18. $9w - 3z = 11$; z

19. $7D + 8C = 2C - 5$; C

20. $5C - 3D - 5 = 4D$; D

B

21. $ax - 5y = 2$; x

22. $5x + by = 12$; y

23. $A = P + Prt$; t

24. $A = P + Prt$; r

25. $PV = nRT$; n

26. $PV = nRT$; R

27. $S = 2\pi r + 2\pi r^2 h$; h

28. $S = 2wh + 2w\ell + 2\ell h$; h

29. $A = \dfrac{B - C}{n}$; B

30. $Z = \dfrac{X - Y}{a}$; X

31. $A = \dfrac{B - C}{n}$; C **32.** $Z = \dfrac{X - Y}{a}$; Y

33. $L = f + (n - 1)d$; d **34.** $A = P(1 + rt)$; r

35. $A = \pi r^2 + 2\pi rh$; h **36.** $A = \pi rs + \pi r$; s

37. $A(P - 3t) = d$; P **38.** $A(P - 3t) = d$; t

39. $7(2C - d) = d + 3$; C **40.** $7(2C - d) = d + 3$; d

C

41. $P = 2L + 2W$; L **42.** $v = gt + s$; t

43. $F = ma + kx$; x **44.** $A = \pi r^2 + xy$; y

45. $s = \dfrac{1}{2}gt^2 + vt + x$; g **46.** $E = \dfrac{1}{2}mv^2 + P$; m

47. $W = (ma + kx)d$; x **48.** $V = \pi r^2 x + xyz$; z

49. $y = \dfrac{4}{5}x - 5$; x **50.** $y = \dfrac{2}{7}x + 4$; x

51. $I = \dfrac{E}{R + r}$; R **52.** $I = \dfrac{E}{R + r}$; r

53. $2A = h(b + B)$; B **54.** $2A = h(b + B)$; b

55. $S = \dfrac{1}{2}(a + b + c)$; c **56.** $S = \dfrac{1}{2}(a + b + c)$; a

57. $S = \dfrac{1}{2}(a + b) - \dfrac{1}{2}(b - c)$; a **58.** $S = \dfrac{1}{2}(a + b) - \dfrac{1}{2}(2b - c)$; b

59. $3(S - 2L) + 6 = 5L + 7$; S **60.** $3(S - 2L) + 6 = 5L + 7$; L

STATE YOUR UNDERSTANDING

61. What is a literal equation?

CHALLENGE EXERCISES

62. The estimated demand (D) for a TV set depends upon the price (P), in dollars, is given by the formula

$D = -20P + 10,000$

At what price should a TV set sell to expect a demand of 5200 sets?

63. Using the formula in Exercise 62, find the price for which a TV set should sell to expect a demand of 4800 sets.

64. To change Fahrenheit units (F) to Celsius units (C) the formula is

$$C = \frac{5}{9}(F - 32).$$

To convert Celsius units to Kelvin Units (K) the formula is

K = C + 273.15

What Fahrenheit temperature corresponds to 273.15K?

65. Using the formulas in Exercise 64, what Kelvin temperature corresponds to −40°F?

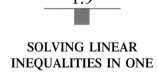

1.9

SOLVING LINEAR INEQUALITIES IN ONE VARIABLE

OBJECTIVES

1. Solve linear inequalities in one variable.

2. Solve compound inequalities.

In this section we study linear inequalities in one variable.

■ DEFINITION

> A linear inequality in one variable is an inequality that can be written in one of the following forms: $ax + b < c$ or $ax + b > c$.

To solve an inequality means to find all the replacements for the variable that will make the inequality true. We solve inequalities in much the same manner as we do equations. The basic idea is to use the properties of inequality to write a series of equivalent inequalities (inequalities that have the same solution set) until we reach one of the form

$$x < a, \quad x > a, \quad x \le a, \quad \text{or} \quad x \ge a.$$

■ PROPERTY

Addition Property of Inequality

For all real numbers a and b, if $x < a$, then $x + b < a + b$.

This property states that the same real number may be added to both sides of an inequality and the result is an equivalent inequality. Since subtraction is defined as adding the opposite, $x < a$ and $x - b < a - b$ are also equivalent.

EXAMPLE 1

Solve: $x - 32 < -25$

$x < -25 + 32$	Add 32 to both sides.
$x < 7$	Simplify. The inequality, $x < 7$, is equivalent to the original; thus, we conclude that the solution set contains all real numbers less than 7.
The solution set is $\{x \mid x < 7\}$.	Set builder notation. ■

The solution set of the inequality can be graphed on a number line by shading all points to the left of 7. An unshaded dot is used at 7 to indicate that 7 is not included in the solution set.

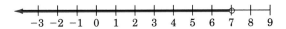

The arrowhead on the left indicates that all numbers to the left are included. If 7 is included in the solution set, as in $x \leq 7$, a shaded dot is used at 7.

A third way of showing the solution set of an inequality is to use interval notation. This notation is especially useful because it is brief and parallels the interval (line segment) on the number line. An **open interval** does not include its endpoints. A **closed interval** includes its endpoints. A **half-open interval** includes only one of its endpoints.

Here are some examples of the ways to write (or show) the solution of an inequality. Become familiar with each notation. You will be asked to write solutions in all three forms.

Interval Notation	Set Notation	Number Line Graph
Closed Interval		
$[3, 7]$	$\{x \mid 3 \leq x \leq 7\}$	

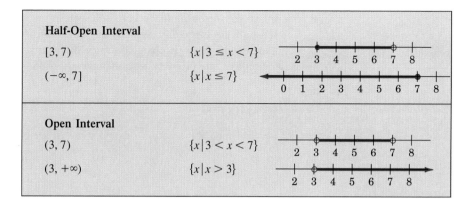

Half-Open Interval		
$[3, 7)$	$\{x \mid 3 \le x < 7\}$	
$(-\infty, 7]$	$\{x \mid x \le 7\}$	
Open Interval		
$(3, 7)$	$\{x \mid 3 < x < 7\}$	
$(3, +\infty)$	$\{x \mid x > 3\}$	

In interval notation, if the endpoint of the interval is included, a bracket is used. If the endpoint of the interval is not included, a parenthesis is used. The symbols "$-\infty$" and "$+\infty$" do not name numbers. They are used to indicate that the interval has no boundary in that direction. The entire number line is the interval $(-\infty, +\infty)$.

■ PROPERTY

For all properties involving inequalities, the symbols $<$, $>$, $\le$, $\ge$ may be interchanged, and the property is still valid.

EXAMPLE 2

Solve. State the solution in interval notation: $x + 3 > 8$

$x > 8 - 3$ Subtract 3 from both sides.

$x > 5$

The solution set is the open interval, $(5, +\infty)$.   ■

The multiplication property of inequality does not parallel the multiplication property of equality. For instance,

If $2 < 4$, then $2(0) < 4(0)$ or $0 < 0$ is false, and

if $2 < 4$, then $2(-1) < 4(-1)$ or $-2 < -4$ is false.   ■

■ PROPERTY

Multiplication Property of Inequality

For all real numbers a, b, and x with $b > 0$,
if $x < a$, then $bx < ab$.

For all real numbers a, b, and x with $b < 0$,
if $x < a$, then $bx > ab$.

In words, if you multiply both sides of an inequality by a positive number, the sense of the inequality remains the same, but if you multiply by a negative number, the sense is changed. Since division is defined as multiplying by the reciprocal, this property also includes division.

EXAMPLE 3

Solve. Graph the solution set and write in set builder notation: $2x < 6$

$2x < 6$	Original inequality.
$x < 3$	Divide both sides by 2.

$$\xleftarrow{\quad\quad\quad\quad} \underset{-2\ -1\ \ 0\ \ 1\ \ 2\ \ 3\ \ 4\ \ 5\ \ 6}{\overset{\circ}{\mid\quad\mid\quad\mid\quad\mid\quad\mid\quad\mid\quad\mid\quad\mid\quad\mid}}$$ Graph.

$\{x\,|\,x < 3\}$	Set builder notation. ▪

EXAMPLE 4

Solve. Graph the solution set and write in interval notation: $-\dfrac{x}{2} < \dfrac{3}{2}$

$$-2\left(-\frac{x}{2}\right) > -2\left(\frac{3}{2}\right)$$ Multiply both sides by -2. Note that the inequality sign is reversed.

$$x > -3$$

$$\underset{-5\ -4\ -3\ -2\ -1\ \ 0\ \ 1\ \ 2\ \ 3\ \ 4}{\overset{\circ}{\mid\quad\mid\quad\mid\quad\mid\quad\mid\quad\mid\quad\mid\quad\mid\quad\mid\quad\mid}}\xrightarrow{\quad\quad}$$ Graph. Note the unshaded dot at -3 indicating that -3 is not included.

$$(-3, -\infty)$$ Open interval notation. ▪

More difficult inequalities are solved by using combinations of the properties of inequalities.

EXAMPLE 5

Solve. Graph the solution and write in set builder notation: $3x + 7 \geq -2x - 8$

$3x \geq -2x - 15$	Subtract 7 from both sides.
$5x \geq -15$	Add $2x$ to both sides.
$x \geq -3$	Divide both sides by 5.

$$\underset{-4\ -3\ -2\ -1\ \ 0\ \ 1\ \ 2\ \ 3\ \ 4\ \ 5\ \ 6\ \ 7}{\overset{\bullet}{\mid\quad\mid\quad\mid\quad\mid\quad\mid\quad\mid\quad\mid\quad\mid\quad\mid\quad\mid\quad\mid\quad\mid}}\xrightarrow{\quad\quad}$$ Note the shaded dot at -3 indicates -3 is included.

$$[-3, +\infty)$$ Half-open interval. The bracket indicates that -3 is included. ▪

EXAMPLE 6

Solve. Write the solution set in set builder notation: $3x - 16 > 8x + 4$

$3x > 8x + 20$	Add 16 to both sides.
$-5x > 20$	Subtract $8x$ from both sides.
$x < -4$	Divide both sides by -5. Note that the sign of the inequality is reversed.

The solution set is $\{x \mid x < -4\}$.   ■

■ **CAUTION**

> When multiplying or dividing both sides by a negative number, the direction of the inequality changes. Thus, $-x < 4$ is equivalent to $x > -4$ (but not $x < -4$).

There are many situations in life where one is faced with boundaries that can be translated to inequalities.

EXAMPLE 7

Jean has a budget of $550 to spend on her parent's anniversary reception. The caterer charges a base fee of $100 plus $8.50 per person at the reception. What is the maximum number of people Jean can have at the reception?

Simpler word form:

Cost of reception $\leq$ $550

Select variable:

Let x be the number of people. The cost is then $100 + 8.50x$	The cost is $100 plus $8.50 per person.

Translate to algebra:

$100 + 8.50x \leq 550$

Solve:

$8.5x \leq 450$	Subtract 100 from both sides.
$x \leq 52.94$	Divide both sides by 8.5 and round to the nearest hundredth.

Answer:

Jean can have at most 52 people at the reception.   ■

Compound inequalities are formed by joining two or more inequalities with the word *and* or the word *or*. For example, $a < x < b$ is read, "*a* is less than *x* and *x* is less than *b*." The solution set is the set of all numbers in the interval on the number line between *a* and *b*. So, $-5 < x < 3$ describes the open interval $(-5, 3)$ and has the graph,

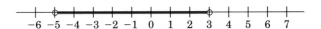

The compound inequality using the word "and" can be represented by the intersection of the solution sets of $x > a$ and $x < b$.

$$\{x \mid x > a\} \cap \{x \mid x < b\}$$

If the variable is not isolated in a compound inequality, it can be isolated by using the properties of inequality.

EXAMPLE 8

Solve. Graph the solution and write it in interval notation: $-9 \le 2x + 3 < 9$

$-12 \le 2x < 6$	Subtract 3 from each member.
$-6 \le x < 3$	Divide each member by 2.

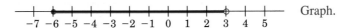   Graph.

The solution is the interval $[-6, 3)$.   ∎

If the compound inequality is formed using the word "or," it can be represented as the union of the solution sets of the two inequalities, $\{x \mid x < a\} \cup \{x \mid x > b\}$.

EXAMPLE 9

Solve. Graph the solution and write it in interval notation: $3x - 4 < 5$ or $2x + 1 \ge 9$

$3x - 4 < 5$	or	$2x + 1 \ge 9$ Solve each inequality.
$3x < 9$		$2x \ge 8$
$x < 3$	or	$x \ge 4$

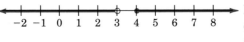   Draw the graph of each inequality on the same number line. All shaded points are in the solution.

$(-\infty, 3) \cup [4, +\infty)$ Interval notation.   ∎

The procedure summarizes the methods given for solving inequalities.

■ PROCEDURE

If the inequality does not contain fractions:

1. Simplify each side of the inequality by performing the indicated operations.

2. Isolate the terms containing the variable on one side using the addition (subtraction) property of inequality.

3. Isolate the constant terms on the other side using the addition (subtraction) property of inequality. Combine terms.

4. Divide both sides of the inequality by the coefficient of the variable using the division property of inequality. If the coefficient is negative, the sense of the inequality must be reversed.

5. The solution can now be determined from the resulting inequality. (Not all of the steps are needed for every inequality.)

If the inequality contains fractions:

1. Multiply both sides by the common denominator. If the common denominator is negative, the sense of the inequality must be reversed.

2. Follow the above steps.

An inequality can be classified by the number of solutions it has. If the domain of the variable is specified to be R, the following table shows the type of inequality and the type of solution.

Solution Set	Classification	Example
R (the domain)	Identity	$x^2 \geq 0$
A non empty subset of R	Conditional	$x \geq 3$
$\emptyset$ (no solution)	Contradiction	$x > x + 1$

EXERCISE   1.9

A

Solve and graph the solution set:

1. $x < -5$ **2.** $a > -2$ **3.** $b + 3 > 4$

4. $c - 2 < 3$

5. $-y \leq 2$

6. $-x \geq 1$

7. $-2x \leq 8$

8. $-5x \leq -10$

9. $-x + 4 > 0$

10. $-x - 3 < 0$

Solve and write the solution set in interval notation:

11. $x - 8 > 4$

12. $x - 7 < 2$

13. $2x < 8$

14. $3x > 6$

15. $-4x < 16$

16. $-5x > 15$

Solve and write the solution in set builder notation:

17. $2x + 3 < 13$

18. $4x - 5 > 15$

19. $-2x + 6 \geq 10$

20. $-3x - 5 \leq 13$

B

21. $-3x + 6 \geq 12$

22. $-6x - 5 \leq 25$

Solve and graph the solution set:

23. $5a - 9 > 21$

24. $7a - 12 < 9$

25. $4x + 2 \leq 3x + 2$

26. $6y - 7 \geq 5y - 7$

27. $-x + 7 > x + 5$

28. $-3x - 4 < 3x + 8$

Solve and write the solution set in set builder notation:

29. $-5(2x + 3) \geq 4(x - 2)$

30. $3(-6x - 4) \leq -2(3x + 3) - 3(4x + 2)$

31. $(5x + 2) - (3x + 8) \geq 2(x + 5)$

32. $-2(x + 5) - 6(2x + 1) \leq 7(2x - 3)$

33. $-3x + 6 < 5x - 9$

34. $4x - 7 < 7x + 4$

Solve and write the solution in interval notation:

35. $2(x + 6) - 5 < 8$

36. $3(x - 1) + 4 < -4$

37. $5 - (2x + 7) \geq -1$

38. $6 - (3x - 5) \leq 7$

39. $2(x + 1) > 3(x - 5)$

40. $-3(x + 6) < 2(2x + 1)$

C

Solve and graph the solution set:

41. $5x - (3x + 2) \geq x - 9$

42. $3x - (x - 6) \leq x + 8$

43. $(6x - 11) - (5x + 6) \geq 3x + 1$

44. $(11x - 9) - (7x + 2) \leq 7x - 14$

45. $4x - [5 - (2x - 3)] - 6(x + 5) \geq 10$

46. $-10x - [-3 - 5(2x + 2)] \leq -8$

Solve and write the solution set in interval notation:

47. $9 < 2x + 5 < 17$

48. $7 < 3x - 8 < 19$

49. $-12 \leq 5x - 3 < 22$

50. $-16 < 6x + 2 \leq 2$

51. $-7 \leq 4x + 9 \leq 13$

52. $-40 \leq 6x + 2 \leq -34$

53. $25 \leq 7x + 3 \leq 66$

54. $48 \leq 4x - 8 \leq 72$

55. $-6 \leq -3x - 3 \leq 0$

56. $-4 \leq -4x + 8 \leq 6$

Solve and write the solution in set builder notation.

57. $3x - 9 < 6$ or $2x + 1 > 13$

58. $-2x + 7 < 7$ or $-3x - 5 > 10$

59. $2(x + 1) \leq 5$ and $-2(x + 1) \leq 5$

60. $3(x - 2) + 6 \geq 0$ or $2(x + 3) - 7 \geq 0$

61. If a bus company charges $20 plus 12 cents per mile for a ticket, could a trip of 800 miles be taken if the person had at most $120 to spend?

62. Referring to Exercise 61, could a trip of 1000 miles be taken if the person had $115 to spend?

63. An elevator can lift at most 3000 lb (can safely lift 3000 lb or less). How many 98-lb sacks of cement can be lifted safely if the elevator operator weighs 158 lb?

64. Referring to Exercise 63, how many 98-lb sacks can be lifted safely if the elevator can lift at most 4000 lb and the elevator operator weighs 178 lb?

65. Jack can spend, at most, $55 on paint. If a gallon can of paint costs $7.50 per can, what is the maximum number of cans of paint Jack can purchase?

66. Button Pushing Electronics manufactures calculators. It costs $10.50 to make each calculator and $6000 a month for advertising expenses. If the budget for manufacture and advertising cannot exceed $50,000, what is the maximum number of calculators that Button Pushing Electronics can produce?

67. Jill's scores on her math exams are 70%, 75%, and 84%. She has one more exam to take. To receive a grade of B, Jill's average must be at least 80%. What is the lowest score that Jill can receive and still get her B?

68. Jack scored 81%, 88%, and 78% on his first three history exams. He needs an average score of at least 85% on his four exams to receive a grade of B. What is the lowest score he can achieve on the next exam to receive a grade of B?

69. It costs $3.50 for the first pound and $1.75 per pound after the first pound for a messenger to deliver a package. Will Charlie be able to get his 5-lb package delivered if he has only $10.00?

70. Greg wants to send roses to his wife, Carol. The flowers cost $1.50 per rose with a delivery charge of $2.00. If Greg can spend a maximum of $25.00, what is the greatest number of roses he can send to his wife?

STATE YOUR UNDERSTANDING

71. How does solving a linear inequality in one variable differ from solving a linear equation in one variable?

CHALLENGE EXERCISES

72. If $-8 < x < 12$, what statement is true regarding $-x$?

73. If $9 < x < 25$, what statement is true regarding $-x$?

74. If one-fourth of a number is between -12 and 24, what are the possible numbers that will satisfy this condition?

75. If one-third of a number is between -8 and -2, what are the possible numbers that will satisfy this condition?

1.10

WORD PROBLEMS

OBJECTIVE

1. Solve word problems. These are called "application," "verbal problems," "applied problems," "story problems," or "stated problems."

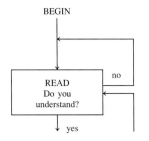

Translating words to algebra requires a lot of practice. The following is a list of suggested steps that can help you make the translation. On the left is a flow chart to follow. The path for the answer to each step is shown. On the right is a step-by-step description of the flow chart.

1. Read the problem. First read the problem through quickly once or twice to get the general idea. Then read the problem through slowly several times to sort out the information. Read especially to identify the question. (What problem are we asked to solve?)

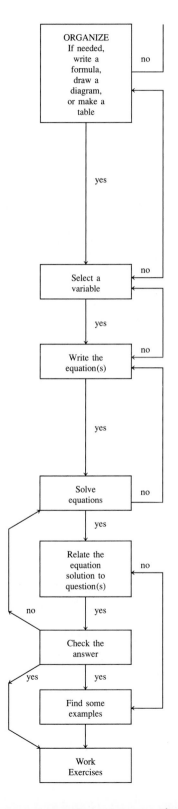

2. Organize the information. The information in a problem can be arranged in different ways (from the original) to make it more useful. For instance,
   a. Make a drawing and label it with the problem information (see Example 1). Look for relationships between values (for instance, sides of triangles, speeds of planes, and rates of work).
   b. Make a table and label the columns and rows so the information can be written in the spaces (see Example 2.)
   c. Write a shorter, simpler word statement that expresses the relationships stated (often suggested) by the problem. This is done by comparing or relating the data given. (See Example 2.)
   d. Use a formula. Some formulas are contained within the problem. Other formulas are commonly found in geometry, business, science, social science, and other fields. (See Example 1.)
   Your organization may use any or all of these.

3. Select a variable (such as x or n) to represent one of the values that is asked for (unknown). Write an explicit definition of this basic variable. Use this variable to write algebraic expressions to represent other unknowns (if there are others). In some problems two variables might be used (see Chapter 7).

4. Translate the information into algebra.
   a. Use a drawing or table to write an equation. The left and right sides are two (different) expressions that represent the same value. (See Examples 1 and 2.)
   b. Write an equation from a simpler word statement. (See Example 3.)
   c. Substitute known values into an appropriate formula to get an equation. (See Example 1.)

5. Use the properties of equations to solve the equation or system of equations.

6. Translate the solution of the equation(s) to the answer. Write in words the answer to the question(s) asked in the problem. (See all the examples.)

7. Use the original problem (not the equation) to check. If the equation is incorrect, this kind of check will help you make this discovery. (See Example 1.)

8. Study the examples in the text, then go to the set of Exercises and work similar problems. In this way you will become better skilled at problem-solving.

The order of steps 2, 3, and 4 may vary, depending on the nature of the problem and your basic understanding of the concept involved in the solution.

EXAMPLE 1

The length of a rectangle is 2.5 times its width. The perimeter is 77 m. Find the length and width.

The problem describes the relation between the length and width of the rectangle. The perimeter is given. The question asks for the length and width.

Step 1.

Make a drawing.

Step 2.

Select variable:

Step 3.

Let w represent the width. The length is $2.5w$.

Represent the width and length with a variable or an expression that contains the variable. Since the length is 2.5 times the width, the length is $2.5w$.

Formula:

Step 4(c).

$P = 2\ell + 2w$
$P = 77, \ell = 2.5w$

Formula for the perimeter of a rectangle.

$77 = 2(2.5w) + 2w$

Substitute into the formula.

Solve:

Step 5.

$77 = 5w + 2w$
$77 = 7w$
$11 = w$

Answer:

Step 6.

The width is 11 m and the length is 27.5 m since $\ell = 2.5(11)$.

Check:

Step 7.

$11 + 27.5 + 11 + 27.5 = 77$
$77 = 77$

Is the perimeter of the rectangle 77 m?

The width is 11 m and the length is 27.5 m   ■

EXAMPLE 2

Two cars leave Sioux City traveling in opposite directions. If their average speeds are 38 mph and 48 mph, how long will it be until they are 387 miles apart?

The problem states the speeds of two cars and the distance between them. The question asks for the *number of hours* they travel.

Step 1.

Make a table.

Step 2.

	RATE (speed)	TIME	DISTANCE
Car 1	38		
Car 2	48		

Neither car goes 387 miles alone, so this number does not show in the table.

Simpler word form:

Distance (Car 1) +
Distance (Car 2) = Total Distance

Select variable:

Step 3.

Let t represent the number of hours until the cars are 387 miles apart.

t will represent the number of hours each car travels.

RATE	TIME	DISTANCE
38	t	$38t$
48	t	$48t$

The numerical distance that each car travels does not show in the table. We know only the total distance. The formula $d = rt$ is used to represent the distance each car travels.

Translate to algebra:

Step 4(b).

$38t + 48t = 387$

Car 1 travels $38t$ miles and car 2 travels $48t$ miles. Total distance is 387 miles.

Solve:

Step 5.

$$38t + 48t = 387$$
$$86t = 387$$
$$t = 4.5$$

Answer:

Step 6.

In 4.5 hours the cars will be 387 miles apart.

Check:

Step 7.

See Example 5.

■

EXAMPLE 3

The enrollment at Johnson High School dropped from 847 in 1980 to 522 in 1985. What was the percent of decrease in enrollment (to the nearest percent)?

The problem gives two enrollment figures. The question asks for a comparison (in percent) between the decrease and the first enrollment.

Step 1.
Note that the decrease in enrollment is 325.

Simpler word statement:

Step 2.

What percent of 847 is 325?

The first enrollment was 847.

Select a variable:

Step 3.

Let x represent the (rate of) percent.

Translate to algebra:

$x(847) = 325$

Step 4.
"of" signifies multiply and "is" means equal.

Solve:

Step 5.

$$x = \frac{325}{847}$$

Divide and round to the nearest hundredth.

$x \approx 0.38$

The symbol "$\approx$" means approximately equal to.

Answer:

Step 6.

The enrollment declined approximately 38%.

Check:

Step 7.

$(38\%)(847) \approx 322 \quad 847 - 322 = 525$
$(39\%)(847) \approx 330 \quad 847 - 330 = 517$

We check 38% and 39%.
We determine that 38% is closer.

The percent of enrollment decline is approximately 38%.   ■

EXAMPLE 4

The sum of the first and fifth of five consecutive odd numbers is 186. Find the numbers.

Consecutive odd numbers differ by 2. We are given the sum of the first and fifth of these numbers.

Step 1.

Simpler word statement:

Step 2.

The sum of the first and fifth consecutive odd numbers is 186.

Select a variable:	Step 3.
Let x represent the first odd number	

x	first odd number	Consecutive odd numbers differ by 2.
$x + 2$	second odd number	
$x + 4$	third odd number	
$x + 6$	fourth odd number	
$x + 8$	fifth odd number	

Translate to algebra:	Step 4.
$x + (x + 8) = 186$	Sum indicates addition.

Solve:	Step 5.

$$2x + 8 = 186$$
$$2x = 178$$
$$x = 89$$
$$\text{so, } x + 2 = 91$$
$$x + 4 = 93$$
$$x + 6 = 95$$
$$x + 8 = 97$$

Check:	
Does $89 + 97 = 186$?	Yes.

Answer:	Step 6.
The consecutive odd numbers,are 89, 91, 93, 95, and 97.	■

EXAMPLE 5

Calculator example

Check the solution of Example 2.

If one car travels 38 mph for 4.5 hours and the other car travels at 48 mph for 4.5 hours, will they be a total of 387 miles apart? Does $38(4.5) + 48(4.5) = 387$?

Distance (Car 1) +
Distance (Car 2) = 387.

ENTER

| 38 | × | 4.5 | + | 48 | × | 4.5 | = |

The result is 387.

Therefore the answer checks.   ■

EXERCISE   1.10

1. The length of a rectangle is 4 times its width. The perimeter is 78 feet. Find the length and width.

2. The length of a rectangle is 5 times its width. The perimeter is 78 yards. Find the length and width.

3. The width of a rectangle is half of its length. The perimeter is 47 cm more than the length. Find the area of the rectangle. (*Hint:* First find the length and width.)

4. The width of a rectangle is one fourth of its length. The perimeter is 27 inches more than the length. Find the area of the rectangle.

5. Two cars leave Austin traveling in opposite directions. If their average speeds are 42 mph and 54 mph, how long will it be until they are 312 miles apart?

6. Two planes leave Little Rock traveling in opposite directions. If their average speeds are 248 mph and 282 mph, how long will it be until they are 1484 miles apart?

7. The enrollment at Four Star High School dropped from 622 in 1980 to 493 in 1983. What was the percent decrease in enrollment (to the nearest percent)?

8. The registration at Mid State University decreased from 2037 in 1981 to 1954 in 1983. What was the percent decrease in registration (to the nearest percent)?

9. Myrna's baby girl weighed 5.75 pounds at birth. At age 7 months she weighed 16.25 pounds. What was the percent of increase in the 7 months (to the nearest percent)?

10. A store sells its merchandise for cost plus 10%. What is the selling price of a cassette/radio that costs the store $89.50?

11. Future Stereo Stores is discounting a CD player by 15%. What is the selling price if the original price was $375?

12. Two cars leave Portland, Oregon, at the same time, headed in the same direction. One travels at 55 miles per hr while the other travels at 65 miles per hr. How far apart will they be after 4 hours?

13. The first side of a triangle is 15 inches. The third side is 3 less than 2 times the second side. If the perimeter is 51 inches, find the lengths of the second and third sides.

14. The second side of a triangle is one inch more than twice the length of the shortest side, and the remaining side is three times the shortest side. If the perimeter is 37 inches, find the lengths of the three sides.

Figure for Exercise 11

15. The sum of the three angles of a triangle is 180°. One angle is 4 less than twice the smallest angle, while the other angle is 20° more than the smallest angle. Find the smallest angle.

16. The sum of the three angles of a triangle is 180°. If one angle is 30° less than six times the smallest angle and the other angle is 30° more than twice the smallest angle, find the smallest angle.

CHAPTER 1
SUMMARY

Set	A set is a collection or group of objects.	(p. 2)
Empty Set	The empty set is a set that contains no elements.	(p. 2)
Subset	B is a subset of A if every element of B is an element of A.	(p. 3)
Intersection of Sets $A \cap B$	Given two sets, A and B, the intersection of the sets, $A \cap B$, is the set of elements that are common to both sets, $A \cap B = \{x \mid x \in A \text{ and } x \in B\}$.	(p. 4)
Union of Sets $A \cup B$	Given two sets, A and B, the union of the sets, $A \cup B$, is the set of elements contained in A, B, or both A and B, $A \cup B = \{x \mid x \in A \text{ or } x \in B\}$.	(p. 4)
Whole Numbers	$W = \{0, 1, 2, 3, 4, 5, \ldots\}$	(p. 5)
Natural Numbers	$N = \{1, 2, 3, 4, 5, \ldots\}$	(p. 5)
Integers	$J = \{\ldots, -3, -2, -1, 0, 1, 2, 3, \ldots\}$	(p. 5)
Rational Numbers	$Q = \left\{ \dfrac{p}{q} \mid p \text{ and } q \in J, q \neq 0 \right\}$	(p. 5)
Irrational Numbers	$I = \{n \mid n \in R, n \notin Q\}$	(p. 5)
Real Numbers	$R = \{x \mid x$ is the coordinate of a point on the number line.$\}$	(p. 4)
Constant	A constant is a symbol that represents a single number.	(p. 7)
Variable	A variable is used to represent any number from a specified set of numbers.	(p. 7)
Domain of a Variable	The domain of a variable is the specified replacement set.	(p. 7)
Equation	An equation is a mathematical statement indicating two number expressions are equal ($=$).	(p. 7)
Solution of an Equation	A solution of an equation is a replacement for the variable that will make the equation true. The solution set is the set of all replacements that will make an equation true.	(p. 8)

Types of Equations	Identity—The statement is true for all replacements of the variable from the domain.	(p. 8)
	Conditional—The statement is not true for some replacement of the variable.	
	Contradiction—The statement is never true.	
Properties of Equality	Reflexive property— $a = a$.	(p. 8)
	Symmetric property—If $a = b$, then $b = a$.	
	Transitive property—If $a = b$ and $b = c$, then $a = c$.	
	Substitution property—If $a = b$, then b can replace a in any equation, and its validity remains unchanged.	
Less Than <	$a < b$ if and only if there is a positive number p such that $a + p = b$.	(p. 9)
Binary Operation	A binary operation associates two numbers of a set with another.	(p. 13)
Closure Properties	If a and b are real numbers, then $a + b$ is a real number and ab is a real number.	(p. 13)
Commutative Properties	If a and b are real numbers, then $a + b = b + a$ and $ab = ba$.	(p. 14)
Associative Properties	If a, b, and c are real numbers, then $(a + b) + c = a + (b + c)$ and $(ab)c = a(bc)$.	(p. 14)
Distributive Property	If a, b, and c are real numbers, then $a(b + c) = ab + ac$ and $(b + c)a = ba + ca$.	(p. 15)
Properties of 0 and 1	If a is a real number, then $a + 0 = 0 + a = a$, and $a \cdot 1 = 1 \cdot a = a$, and $a \cdot 0 = 0 \cdot a = 0$.	(p. 16)
Opposites	For any real number a, the number $-a$ is its opposite or additive inverse.	(p. 17)
Finding Opposites	To find the opposite of a, multiply a by negative one. $-1(a) = -a$ and $-(-a) = a$.	(p. 17)
Addition Property of Opposites	If a is a real number, then $a + (-a) = -a + a = 0$.	(p. 17)
Reciprocals	If $b \neq 0$, then $\dfrac{1}{b}$ and b are reciprocals.	(p. 18)
	If $ab \neq 0$, then $\dfrac{b}{a}$ and $\dfrac{a}{b}$ are reciprocals.	
Multiplying Reciprocals	If a is a real number and $a \neq 0$, then $a\left(\dfrac{1}{a}\right) = \left(\dfrac{1}{a}\right)a = 1$.	(p. 18)
Trichotomy Property	For real numbers a and b, exactly one of the following is true: $a < b$, $a = b$, or $a > b$.	(p. 19)
Transitive Property of Inequality	For real numbers a, b, and c, if $a < b$ and $b < c$, $a < c$.	(p. 19)

Opposite of a Fraction	$-\left(\dfrac{a}{b}\right) = -\dfrac{a}{b} = \dfrac{-a}{b} = \dfrac{a}{-b},\ b \neq 0$	(p. 23)
Opposite of a Sum	$-(x + y) = (-x) + (-y)$	(p. 23)
Absolute Value	$\|a\| = a$ if $a \geq 0$	(p. 24)
	$\|a\| = -a$ if $a < 0$	
Addition of Real Numbers	To find the sum of two numbers of like sign, add their absolute values, and use the common sign. To find the sum of a positive and a negative, find the difference of their absolute values, and use the sign of the number with the larger absolute value.	(p. 25)
Subtraction of Real Numbers	If a and b are real numbers, then $a - b = a + (-b)$.	(p. 29)
Combine Like Terms	Add or subtract the numerical coefficients, $3x + 5x = 8x$ and $6x - 9x = -3x$.	(p. 30)
Multiplication of Real Numbers	$(+)(+) = +$	(p. 34)
	$(+)(-) = -$	
	$(-)(+) = -$	
	$(-)(-) = +$	
Division of Real Numbers	If a and b are real numbers, $b \neq 0$, then $\dfrac{a}{b} = a \div b = a\left(\dfrac{1}{b}\right)$.	(p. 35)
Multiplying or Dividing a Term by a Real Number	To multiply or divide a term by a real number, multiply or divide the coefficient by the real number.	(p. 36)
Order of Operations	In an algebraic expression with two or more operations, perform the operations in the following order:	(p. 40)

1. Grouping symbols—Perform operations included by grouping symbols first.

2. Exponents—Perform operations indicated by exponents.

3. Multiply and divide—Perform multiplication and division as they appear from left to right.

4. Add and subtract—Perform addition and subtraction as they appear from left to right.

Evaluate an Expression	To evaluate an expression, replace the variables by the designated real numbers and simplify.	(p. 42)
Properties of Equality	For all real numbers a and b, if $x = a$, then $x + b = a + b$, and if $b \neq 0$, then $bx = ab$.	(pp. 46, 47)
	To solve an equation with no fractions:	(p. 48)

1. Simplify each side of the equation by performing the indicated operations.

2. Isolate the terms containing the variable on one side and the constant terms on the other side of the equation by using the addition property of equality. Combine terms.

3. Divide both sides by the coefficient of the variable.

If the equation contains fractions, multiply both sides by the common denominator and proceed as above.

Literal Equation	An equation (formula) that contains two or more variables.	(p. 54)
Properties of Inequality	For all real numbers a and b,	(pp. 60, 61)

if $x < a$, then $x + b < a + b$,

if $b > 0$ and $x < a$, then $bx < ab$,

if $b < 0$ and $x < a$, then $bx > ab$.

For all properties involving inequalities, the symbols $<$, $>$, $\leq$, $\geq$ may be interchanged, and the property is still valid.

Solving Inequalities   To solve an inequality, follow the steps used for solving equations   (p. 65)
using the properties of inequality.

CHAPTER 1
REVIEW EXERCISES

SECTION 1.1   Objective 1

True of false:

1. $\{-4, -7, 0\}$ is a subset of $\{-10, -7, -6, -4, -2, 0, 2, 8, 10\}$.

2. $\emptyset$ is a subset of every set.

3. $\{7, 8, 11, 23\}$ is a subset of $\{7, 8, 11, 24, 32\}$.

4. $\{7, 8, 11, 23\}$ is a subset of $\{7, 8, 11, 23\}$.

5. $\{c, d, e\}$ is a subset of $\{a, b, c, \ldots, x, y, z\}$.

SECTION 1.1   Objective 2

Given $A = \{-5, -4, -3, -2, -1, 0, 1, 2, 3, 4, 5\}$, $B = \{0, 1, 2, 3, 4, 5, 6, 7, 8, 9\}$, and $C = \{-4, -2, 0, 2, 4\}$, find:

6. $A \cup B$

7. $A \cup C$

8. $A \cap B$ **9.** $A \cap C$ **10.** $C \cap B$

SECTION 1.1   Objective 3

Given $\{-11, -\sqrt{3}, -0.34, -\frac{1}{7}, 0, \frac{1}{8}, 0.67, \sqrt[3]{6}, 15, 9.2\}$:

11. Write the subset of whole numbers.

12. Write the subset of integers.

13. Write the subset of natural numbers.

14. Write the subset of rational numbers.

15. Write the subset of irrational numbers.

SECTION 1.1   Objective 4

Classify each of the following equations as an identity, a conditional, or a contradiction:

16. $3x = 3x$

17. $2x + 3 = 5$

18. $2x - 5 = 2x$

19. $7x + 8 = 8 + 7x$

20. $3x + 1 = 5x + 1$

SECTION 1.1   Objective 5

True or false:

21. $-6.3 < -5.9$

22. $\dfrac{16}{5} > \dfrac{13}{7}$

23. $9.99 < 1.05$

24. $0 > -14.8$

25. $-99.5 > -98.6$

SECTION 1.2   Objective 1

The following are examples of what property of real numbers:

26. $3x + 7 = 7 + 3x$

27. $-6(x + 9) = -6(x) + (-6)(9)$

28. $4y + (3z - 9) = (3z - 9) + 4y$

29. $0(-0.23) = 0$

30. $4(3y) = (4 \cdot 3)y$

SECTION 1.3   Objective 1

Write the opposite of each of the following:

31. -7

32. $\dfrac{12}{7}$

33. -0.67

34. 1.67

35. $2x + 3$

SECTION 1.3   Objective 2

Find the value of each of the following:

36. $|-3|$

37. $|0.75|$

38. $\left|-\dfrac{7}{3}\right|$

39. $|3x|, \ x > 0$

40. $|3x|, \ x < 0$

SECTION 1.3   Objective 3

Add:

41. $3 + (-7) + 10$

42. $8 + (-19) + (-45) + 14 + (-21)$

43. $7 + |5 + (-10)|$

44. $-3 + |-17 + 12| + (-56) + 45$

45. $-(-56) + 23 + (-|13 - 26|) + 12$

SECTION 1.4   Objective 1

Subtract:

46. $65 - 98$

47. $-18 - 43$

48. $45 - (15 - 17)$

49. $-45 - |34 - 65| - 15$

50. $15 - [23 - 18 - (-45)] - 12$

SECTION 1.4   Objective 2

Combine:

51. $3x - 17x + 13x$

52. $25y - 67y - 12y + 72y$

53. $17ab - 23ab + 51ab - 32ab$

54. $12b - 13b - (2b - 5b)$

55. $\dfrac{3}{4}x - \dfrac{3}{5}x + \dfrac{3}{20}x - \dfrac{7}{10}x$

SECTION 1.5   Objective 1

Multiply:

56. $(-5)(-12)$

57. $(-17)(4)$

58. $(-3)(-4)(5)(-1)(-1)$

59. $(-0.25)(0.3)$

60. $(-1)(-1)(-1)(-1)(-1)(-1)(-2)(-2)(-3)$

SECTION 1.5   Objective 2

Write the reciprocal of each of the following:

61. -5

62. 7

63. $\dfrac{4}{5}$

64. $-\dfrac{13}{8}$

65. 0.35

SECTION 1.5   Objective 3

Divide:

66. $(-25) \div (-5)$

67. $33 \div (-11)$

68. $(-65) \div 13$

69. $\dfrac{-75}{15}$

70. $\dfrac{-110}{-22}$

SECTION 1.5   Objective 4

Perform the indicated operations. Assume the domain of each variable is R^+:

71. $-12(3x)$

72. $0.3(-0.4y)$

73. $-124ab \div (-4)$

74. $|45x(4)| \div (-6)$

75. $|-6a(-3)| \div 6$

SECTION 1.6   Objective 1

76. $-3(5) + 7(-2)$

77. $9(-14) + (-6) - 15$

78. $23 - 5(-15 + 12) - 8$

79. $(3)^2(6) - 15(-2)^3$

80. $-3(5) - 3(15 - 21) - 18 - (-23)$

SECTION 1.6   Objective 2

Evaluate the following if $a = -3$, $b = 4$, and $c = -1$:

81. abc

82. $ab - bc$

83. $a + b - ac$

84. $ac - bc - ab$

85. $a^2 - b^2 - c^2$

SECTION 1.7   Objective 1

Solve:

86. $5x - 18 = 3x + 44$

87. $3x - 5x + 23 = -4x + 33$

88. $4x - (7x + 9) = 5$

89. $5x - (-3x + 15) = 25$

90. $\dfrac{1}{2}x - \dfrac{2}{3} = \dfrac{1}{3}x + \dfrac{3}{4}$

SECTION 1.8   Objective 1

Solve for the indicated variable:

91. $i = prt$; t

92. $4a - 3b + 2 = 14$; b

93. $F = \dfrac{GMm}{r^2}$; G

94. $S = 2wh + 2w\ell + 2\ell h$; w

95. $W = (ma + kx)d$; a

SECTION 1.9   Objective 1

Solve and write the solution in interval notation:

96. $3x - 15 < 18$

97. $5x - 27 \geq -3x - 51$

98. $5x - (14x + 27) < 8$

Solve and write the solution in set builder notation:

99. $(3x - 7) - (7x + 9) \leq 2x - 1$

100. $-3(x - 2) + 17 > 5x - 13$

SECTION 1.9   Objective 2

Solve and write the solution in set builder notation:

101. $3x - 5 < 2x + 1$ or $4x - 5 > 2x + 15$

102. $5(x + 2) < 0$ or $3(x + 12) \geq 15$

103. $8x + 17 < 6x - 11$ and $3x + 8 < 2x - 10$

Solve and write the solution in interval notation:

104. $9 < 3x - 5x + 10 < 12$

105. $12 < 2(x + 2) - 3 \leq 20$

SECTION 1.10   Objective 1

106. The length of a rectangle is 1.75 times its width. The perimeter is 66 in. find the length and width.

107. Two cars leave Kansas City traveling in opposite directions. If their average speeds are 42 mph and 46 mph, how long will it be before they are 330 miles apart?

108. At Cross Roads Community College, the enrollment increased from 1252 full-time students in 1981 to 1568 full-time students in 1985. What was the percent of increase in the enrollment (to the nearest percent)?

109. The sum of the second and fourth consecutive even numbers is 436. Find the numbers.

110. A large screen TV is discounted by 15%. What is the selling price if the original price was $2875?

CHAPTER 1
TRUE–FALSE CONCEPT REVIEW

Check your understanding of the language of algebra. Tell whether each of the following statements is true (always true) or false (not always true).

1. Every rational number can be written as a whole number.

2. Every real number can be associated with a point on the number line.

3. If a and b are real numbers, then $a = b$, or $a < b$, or $a > b$.

4. The domain of a variable is the set of values that make the equation that contains the variable true.

5. The product of two opposite real numbers is always -1.

6. The commutative property of multiplication allows us to interchange the position of any two consecutive factors.

7. Division by zero is not defined.

8. The reciprocal of a whole number cannot be a whole number.

9. The absolute value of a real number is never negative.

10. Like terms are terms involving the same variables.

11. The sum of a number and its additive inverse is 1.

12. To subtract two real numbers, add their opposites.

13. The product of two real numbers is always positive or negative.

14. If a and b are both negative real numbers, then $ab = |a| \cdot |b|$.

15. The graph of $x < 7$ is a half-open interval.

16. When multiplying each member of an inequality by a negative number, the sense of the inequality is reversed.

17. On the number line, $|x|$ is the distance between x and $-x$.

18. The sum of a positive number and a negative number is negative.

19. The product of a positive number and a negative number is negative.

20. Some equations are false and have no solution.

21. If $x < a$, then $a > x$.

22. The inequality $x \le x$ is an identity (always true).

23. If $a < x < b$, then $a > x > b$.

24. The intersection of two sets is never the empty set.

25. If $d \in B$ and B is a subset of A, then $d \in A$.

CHAPTER 1
TEST

1. Add: $-\dfrac{7}{40} + \dfrac{3}{50}$

2. Name the property of real numbers illustrated by the following: $40.3(1) = 1(40.3) = 40.3$

3. Simplify: $3(3a + b) - 4(6a - 3b)$

4. Simplify: $-4(7 + 3)^2 - 5(8 - 2^2)$

5. True or false?
   a. $-18.65 < -23.7$
   b. $0.0034 > 0$
   c. $-5.02 < 0.003$

6. Solve: $3x + 2(2x - 6) \ge 9$

7. Given $A = \{a, b, c, d, e\}$ and $B = \{c, d, e, f, g\}$, find $A \cup B$.

8. Simplify: $9 \cdot 5^2 + 6(8 - 12)^2$

9. Combine terms: $3b - 12b + 6b - b$

10. Name the property of real numbers illustrated by the following: $7x + (-7x) = 0$

11. Given $\{-7, -3.6, -\dfrac{3}{5}, 1, 9, 12.7, \sqrt{19}\}$, list the subset of
    a. integers
    b. irrational numbers
    c. real numbers

12. Solve: $9x + 3 = 2x - 4$

13. Simplify: $3x + 4y - 2(2x - 3y)$

14. Combine terms: $13y - 23z + 14z - 18y$

15. Name the property of real numbers illustrated by the following: $25.78 + 110 = 110 + 25.78$

16. Subtract: $-\dfrac{5}{3} - \left(-\dfrac{7}{8}\right)$

17. Divide: $\dfrac{-29.6}{8}$

18. Name the property of real numbers illustrated by the following: $(10x - 17) \cdot 0 = 0$

19. Find the absolute value: $|-34.7 + 18.6|$

20. Solve: $3(3y - 1) + 10y - 9 = 2\left(2y + \dfrac{5}{2}\right)$

21. Solve for n: $L = f + (n - 1)d$

22. Name the property of real numbers illustrated by the following: $-5(6 \cdot 9.6) = (6 \cdot 9.6)(-5)$

23. Simplify: $-2(a + 3b) - 4(-3a - b)$

24. Add: $-16.7 + (-12.5) + (-4.2)$

25. Given $S = \{-3, -2, -1, 0, 1, 2, 3\}$ and $T = \{-2, 0, 2, 4, 6\}$, find $S \cap T$.

26. Combine terms: $8x - 16x - 45x + 23x$

27. Name the property of real numbers illustrated by the following: $0.3(6x + 10) = 0.3(6x) + 0.3(10)$

28. Simplify: $-8(-3 + 7) - 5(-6 - 42)$

29. Solve: $(3a - 5) - (7a + 12) = 2a + 13$

30. Subtract: $36.73 - 53.76$

31. Simplify: $6 \cdot 2 - 8(3 + 4)$

32. Multiply: $(-2)(-15)(-3.6)$

33. Combine terms: $4x - 8.3y - 3.7x - 2.1y$

34. Simplify: $-5(3.1x - 5.2y) - 3(2.1y - 3.3x)$

35. Solve: $-42 \le 5x + 3 < 8$

36. Solve for y: $4x = -2(y - 6)$

37. Solve: $2x - (5x + 7) \ge x - 9$

38. Solve: $4x + 3 = 2x + 5$

39. The perimeter of a rectangle is 150 ft. If the length is 3 more than twice the width, find the dimensions.

40. Two bicyclists leave the same starting point traveling in opposite directions. If their average speeds are 12 mph and 15 mph, how long will it take them to be 108 miles apart?

POLYNOMIALS AND RELATED EQUATIONS

Products of polynomials are often used in the solution of problems involving real-life situations. In Exercise 62, Section 2.7, the photo of the Golden Gate Bridge is surrounded by a frame 2 inches wide. We have two apparent areas, the area of the photo and the area of photo and frame combined. The frame surrounds both sides of the photo so that the length and width of the photo is increased by twice the width of the frame to get the dimensions of the combined area. Since we are given the increase in area we know that: (area of framed picture) − (area of picture) = 48. Each of these areas involves the product of polynomials. *(Fran Heyl Associates)*

PREVIEW

Chapter 2 opens with a review of the laws of exponents applied to integer exponents. These are used to develop the procedures for multiplying and dividing monomials. Section 2.3 presents scientific notation as an extensively used and practical application of exponents. Polynomials are defined and classified in section 2.4. The fundamental operations on polynomials are shown in sections 2.5, 2.6, and 2.7. These include the special products whose recognition is vital for future work in factoring. The remainder of the chapter is devoted to review of factoring techniques for all polynomials with special emphasis on trinomials in section 2.9 and special cases in section 2.10. Finally section 2.11 summarizes factoring by offering a variety of polynomials.

2.1

POSITIVE INTEGER EXPONENTS AND MULTIPLICATION OF MONOMIALS

OBJECTIVES

1. Use the properties of exponents to multiply monomials.

2. Use the properties of exponents to raise monomials to powers.

Exponents are used to write products of identical factors in an abbreviated fashion. Here is an exponential form of a number.

EXPONENT
$$\downarrow$$
$$3^4 = 3 \cdot 3 \cdot 3 \cdot 3 = 81$$
$$\uparrow \qquad\qquad\qquad \uparrow$$
POWER OR VALUE

■ DEFINITION

Exponent

> An *exponent* that is a whole number greater than one indicates how many times the base is to be used as a factor. In general,
>
> $b^1 = b$ and
> $b^n = \underbrace{b \cdot b \cdot b \cdot b \dots b}_{n \text{ FACTORS OF } b}$, where n is a positive integer, $n > 1$.

The expression 3^4 is read "three to the fourth power" and indicates that three is used as a factor four times.

When working with exponential expressions (terms containing exponents), there are several properties that provide shortcuts for multiplication and division of monomials.

■ DEFINITION

Monomial

A *monomial* is a real number or the product of a real number and variables raised to positive integer powers.

The terms 2, $3x$, $5yz$, $\dfrac{2x}{13}$ or $\dfrac{2}{13}x$, and $5x^3y^2$ are monomials.

The properties covering multiplication are used in this section. Those for division are covered later. Remember that we use the symbol J^+ to represent the set of positive integers (whole numbers) and R to represent the set of real numbers.

■ PROPERTY

First Property of Exponents

$b^m \cdot b^n = b^{m+n}$, where $b \in R$, m and $n \in J^+$. Multiplying like bases.

To multiply two exponential expressions with the same base, add the exponents and keep the common base.

At the present, we will deal only with positive exponents, but we will see later that the same property can be applied when the exponents are negative or rational numbers.

EXAMPLE 1

Multiply the monomials: $(6x^{14})(-3x^3)$

$(6x^{14})(-3x^3) = [6 \cdot (-3)][x^{14} \cdot x^3]$ Group the factors that can be simplified. This is an example of the use of the commutative and associative properties of multiplication. Then we use the properties of exponents to simplify.

$= -18x^{14+3}$ Add the exponents.

$= -18x^{17}$ ■

EXAMPLE 2

Multiply the monomials: $(3x)(2x^2)(-6x^4)$

$(3x)(2x^2)(-6x^4) = [3(2)(-6)][x \cdot x^2 \cdot x^4]$ Group the constants and group the variables.

$= -36x^{1+2+4} = -36x^7$ Add the exponents.   ■

EXAMPLE 3

The volume of a cylinder is given by the formula

$$V = \pi r^2 h,$$

where r represents the radius, and h represents the height of the cylinder. If the number of units in the height of a certain cylinder is the cube of the number of units in the radius. Express the volume in terms of the radius. Find the volume when $r = 3$ in. ($\pi \approx 3.14$).

$V = \pi r^2 h = \pi r^2 (r^3)$	To find the volume of the cylinder in terms or r, replace h by r^3 and multiply.
$V = \pi r^5$	This is the volume in terms of the radius for this certain cylinder.
$V \approx 3.14(3)^5$	To find the volume when $r = 3$, replace r by 3 and π by 3.14 and multiply.
≈ 763.02 cubic inches	

The second property of exponents provides us with a shortcut for raising a power to a power.

■ PROPERTY

Second Property of Exponents

$(b^m)^n = b^{mn}$, where $b \in R$, m and $n \in J^+$. Raise a power to a power.

To raise an exponential expression to a power, multiply the exponents.

The expression $(8^3)^2$ can be simplified using either of the two properties.

$(8^3)^2 = (8^3)(8^3) = 8^{3+3} = 8^6$ Add the exponents.

$(8^3)^2 = 8^{2(3)} = 8^6$ Multiply the exponents.

This is why we often refer to these properties as shortcuts.

EXAMPLE 4

Raise the monomial y^{12} to the fourth power: $(y^{12})^4$

$(y^{12})^4 = y^{12 \cdot 4} = y^{48}$ Multiply the exponents.   ■

The third property can be helpful when we have a monomial raised to a power.

■ PROPERTY

Third Property of Exponents

$(ab)^m = a^m b^m$, where $a \in R$, m and $n \in J^+$. Raise a product to a power.

To raise a product to a power, raise each factor to the power.

EXAMPLE 5

Simplify: $(-7x^2y^5)^3$

$(-7x^2y^5)^3 = (-7)^3(x^2)^3(y^5)^3$ Raise each factor in parentheses, independently, to the third power.

$\qquad = -343x^{2\cdot3}y^{5\cdot3}$ Multiply the exponents when raising a power to a power.

$\qquad = -343x^6y^{15}$ ∎

EXAMPLE 6

Simplify: $(-3a^2b^2)^2$

$(-3a^2b^2)^2 = (-3)^2(a^2)^2(b^2)^2$ Raise each factor to the power.
$\qquad = 9a^4b^4$ Raise a power to a power.   ∎

■ **CAUTION**

You must take care not to confuse $-b^2$ and $(-b)^2$.

$-b^2 = -(b \cdot b)$ Find b^2, then take the opposite of b^2.

$(-b)^2 = (-b)(-b) = b^2$ Find the opposite of b, then square it.

■ **DEFINITION**

In general,

$-b^n$ means the opposite of b^n.

$(-b)^n$ means take the opposite of b, and then raise it to the nth power.

EXAMPLE 7

Evaluate: -6^4, $(-6)^4$, -5^3, and $(-5)^3$

$-6^4 = -(6^4) = -1296$

$(-6)^4 = (-6)(-6)(-6)(-6) = 1296$

$-5^3 = -(5)^3 = -125$

$(-5)^3 = (-5)(-5)(-5) = -125$ ∎

The same properties apply when the exponents are variables.

EXAMPLE 8

Simplify: $(x^a)(x^{a+2})$

$(x^a)(x^{a+2}) = a^{a+(a+2)} = a^{2a+2}$

Add the exponents.   ∎

EXERCISE 2.1

A

Multiply:

1. $r^7 \cdot r^4 \cdot r$

2. $s \cdot s^2 \cdot s^5$

3. $(5a^2)(6a^3)(2a)$

4. $(-3b^4)(-2b^3)(-2b)$

5. $(a^4)^4$

6. $(b^3)^6$

7. $(3t^3)^2$

8. $(4v^4)^3$

9. $(-3x^2y^3)^2$

10. $(-5a^4b^5)^3$

11. $(-5z)(-4z^7)$

12. $(-16m)(-5m^4)$

13. $(x^3y)(xy^3)(x^2y^2)$

14. $(a^4b)(ab^2)(a^3b^3)$

15. $(2x^2)^4$

16. $(3y^3)^3$

17. $(4t^4)^3$

18. $(2y^5)^3$

19. $(11s^4)^2$

20. $(15x^{10})^2$

B

21. $(2x^{11})(-3x^{12})(5x^{15})$

22. $(-4a^6)(-5a^{10})(3a^{11})$

23. $(-5x^5y^6z^7)(-3x^6y^2z^7)$

24. $(7a^3b^2c)(-4a^6b^6c^6)$

25. $(-14p^2q^3)^2$

26. $(-15r^6s^4)^2$

27. $(-8p^6q^4r^3)^3$

28. $(-3m^4n^3p)^5$

29. $(t^2)^3(t^3)^2$

30. $(-w^3)^4(w^2)^5$

31. $(2x)^2(3x)^3$

32. $(4x^2)^2(5x)^2$

33. $(-2x)^2(-2x)^3$

34. $(-3x^2)^2(-3x^2)^3$

35. $(cy^2)^2(cy^3)^3$

36. $(a^2b)^2(2a^2b)^3$

37. $(xy)^2(x^2y^2)^2$

38. $(m^3n)^3(mn^2)^2$

39. $(ab)^4(a^2b)^4$

40. $(c^2d^4)^5(c^3d)^4$

C

41. $(-2xy)^3(4x^2y^3)$

42. $(-3x^2y)^3(-2xy)^2$

43. $(3a^2b^2c)(-2a^3b^3c^2)^2$

44. $(-5a^3b^3)^2(2a^2b)^3$

45. $(2t)^2(-3t^3)^3(-t^2)^4$

46. $(-w^3)^3(2w^4)^4(-3w^3)^2$

47. $(-4x^2y^2z)^2(-3x^3y^2)(2y^3z^3)^4$

48. $(3a^2b^2c)^3(-a^3b^3c)^3(-2a^4bc^2)^3$

49. $(x + y)^2(x + y)^4$

50. $(a - b)^5(a - b)^6$

51. $(2^t)(2^{-t+5})$

52. $(3^{w+1})(3^{-w+3})$

53. $(2t^{4n})(t)^{2n-2}$

54. $(-3)^2(x^{2n-1})(x^{3n+5})$

55. $(5y)(y^{5n+6})(y^{2n-3})$

56. $(3b)(b^{2n})(b^{4n+3})$

57. $(2x)^2(x^{m+4})(x^{2m-3})$

58. $(3y)^2(y^{t-5})(y^{3t-2})$

59. $(w^{t+1})^2(w^{t-3})$

60. $(z^{s+6})^2(z^{s-8})$

D

61. If a special cylinder has a height that is the fifth power of the radius ($h = r^5$), express the volume of the cylinder in terms of the radius. Use the formula $V = \pi r^2 h$.

62. Another cylinder has a height that is the square of the radius. Express its volume in terms of the radius. Use the formula $V = \pi r^2 h$. *Compute for* π

$6, 41$

63. The volume of a triangular prism is

$$V = \frac{1}{2}abh,$$

where $\frac{1}{2}ab$ represents the area of the triangular end and h represents the height. In a certain prism, b is the cube of a, and h is the cube of a. Express the volume in terms of a.

64. In another triangular prism, a is the square of b, and h is the sixth power of b. Express the volume in terms of b.

65. The area of a circle is $A = \pi r^2$, where r is the radius of the circle. The radius is half the diameter, $r = \frac{1}{2}d$. Find a formula for the area in terms of the diameter.

66. The volume of a box is given by the formula $V = LWH$, where L is the length, W is the width, and H is the height. Suppose that $L = 2W$ and $H = \frac{1}{2}W^2$. Find a formula for the volume of the box in terms of W.

67. The area of a triangle is $A = \frac{1}{2}bh$, where b is the base and h is the height. Suppose that the height is twice the cube of the base. Express the area of the triangle in terms of the base.

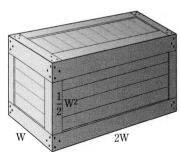

Figure for Exercise 66

68. The volume of a sphere is given by the formula $V = \dfrac{4}{3}\pi r^3$, where r is the radius. If the radius is half the diameter, find an expression for the volume of the sphere in terms of the diameter.

69. The volume of a cone is given by the formula $V = \dfrac{1}{3}\pi r^2 h$, where r is the radius and h is the height. If the radius is three times the cube of the height, find an expression for the volume of the cone in terms of the height.

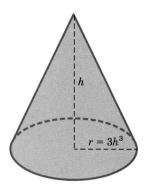

70. Using the formula in Exercise 69, find an expression for the volume of the cone if the radius is two times the fourth power of the height.

Figure for Exercise 69

STATE YOUR UNDERSTANDING

71. What is the difference in the meaning of -8^z and $(-8)^z$?

CHALLENGE EXERCISES

Simplify, assume all variables represent positive integers.

72. $5^a \cdot 5^b \cdot 5^c$ **73.** $8^x \cdot 8^y \cdot 8^z$ **74.** $(12^n)^2$ **75.** $(5^a)^{b+c}$

MAINTAIN YOUR SKILLS (SECTIONS 1.3, 1.4)

Simplify (add, subtract, or combine like terms):

76. $(79) - (48) + (21) - (73)$

77. $(-18.4) + (-21.7) - (13.84) - (-21.52)$

78. $-75 - 84 + 39 - 50 + 8$

79. $\dfrac{7}{8} + \left(-\dfrac{2}{3}\right) - \left(\dfrac{4}{5}\right) - \left(-\dfrac{9}{10}\right)$

80. $76ab - 19bc - 71ab - 67bc$

81. $-\dfrac{4}{5}x - \dfrac{3}{4}y + \dfrac{1}{2}x - \dfrac{1}{2}y$

82. What is the change in temperature from a low of $-13°C$ to a high of $21°C$?

83. What is the difference in altitude from a height of 3000 ft to a depth below sea level of 318 ft (-318 ft)?

2.2

**INTEGER EXPONENTS
AND DIVISION OF
MONOMIALS**

OBJECTIVES

1. Simplify expressions that contain negative integer exponents.

2. Divide monomials.

The definition of zero and negative exponents can be understood from a study of the following patterns.

Simplify Using the Basic Principle of Fractions	Simplify by Subtracting Exponents
$\dfrac{y^7}{y^5} = \dfrac{y}{y} \cdot \dfrac{y}{y} \cdot \dfrac{y}{y} \cdot \dfrac{y}{y} \cdot \dfrac{y}{y} \cdot y \cdot y = y^2$	$\dfrac{y^7}{y^5} = y^{7-5} = y^2$
$\dfrac{y^7}{y^6} = \dfrac{y}{y} \cdot \dfrac{y}{y} \cdot \dfrac{y}{y} \cdot \dfrac{y}{y} \cdot \dfrac{y}{y} \cdot \dfrac{y}{y} \cdot y = y$	$\dfrac{y^7}{y^6} = y^{7-6} = y^1$
$\dfrac{y^7}{y^7} = \dfrac{y}{y} \cdot \dfrac{y}{y} \cdot \dfrac{y}{y} \cdot \dfrac{y}{y} \cdot \dfrac{y}{y} \cdot \dfrac{y}{y} \cdot \dfrac{y}{y} = 1$	$\dfrac{y^7}{y^7} = y^{7-7} = y^0$
$\dfrac{y^7}{y^8} = \dfrac{y}{y} \cdot \dfrac{y}{y} \cdot \dfrac{y}{y} \cdot \dfrac{y}{y} \cdot \dfrac{y}{y} \cdot \dfrac{y}{y} \cdot \dfrac{y}{y} \cdot \dfrac{1}{y} = \dfrac{1}{y}$	$\dfrac{y^7}{y^8} = y^{7-8} = y^{-1}$
$\dfrac{y^7}{y^9} = \dfrac{y}{y} \cdot \dfrac{y}{y} \cdot \dfrac{y}{y} \cdot \dfrac{y}{y} \cdot \dfrac{y}{y} \cdot \dfrac{y}{y} \cdot \dfrac{y}{y} \cdot \dfrac{1}{y} \cdot \dfrac{1}{y} = \dfrac{1}{y^2}$	$\dfrac{y^7}{y^9} = y^{7-9} = y^{-2}$

Consistent with the pattern, we make the following definitions.

■ **DEFINITION**

Zero Exponent

Any number, except 0, to the zero power is equal to one.

$y^0 = 1, \ y \neq 0$.

■ **CAUTION**

In the expression y^0, the base y cannot be zero because division by zero is not defined. $\dfrac{0^7}{0^7}$ and 0^0 have no meaning (value).

Similarly,

■ **DEFINITION**

Negative Exponent

Any number, except 0, to the opposite of m is equal to the reciprocal of y^m.

$$y^{-m} = \frac{1}{y^m}, y^m = \frac{1}{y^{-m}}, y \neq 0, y \in R, m \in J^+$$

EXAMPLE 1

Simplify: $6^2 \cdot 7^0$

$6^2 \cdot 7^0 = 36 \cdot 1 = 36$ $6^2 = 6 \cdot 6$ and $7^0 = 1$ ■

EXAMPLE 2

Simplify: 4^{-3} and $\dfrac{1}{5^{-2}}$

$4^{-3} = \dfrac{1}{4^3}$

The exponent, -3, has two effects on the base. First, the negative sign in an exponent denotes a reciprocal.

$= \dfrac{1}{4 \cdot 4 \cdot 4} = \dfrac{1}{64}$

Second, the 3 indicates a repeated factor, $4 \cdot 4 \cdot 4$, in the denominator.

$\dfrac{1}{5^{-2}} = 5^2 = 25$

■ **CAUTION**

The exponent, -2, does *not* mean that the base is a negative number.

■

The three properties of exponents in the previous section are also valid for expressions containing negative exponents.

EXAMPLE 3

Multiply: $x^3 \cdot x^{-5}$

$x^3 \cdot x^{-5} = x^{3+(-5)}$ Add the exponents.

$= x^{-2} = \dfrac{1}{x^2}$ ■

EXAMPLE 4

Multiply: $(x^{-3})^{-2}$

$(x^{-3})^{-2} = x^{(-3)(-2)} = x^6$ Raise a power to a power.   ■

EXAMPLE 5

Multiply: $(x^3y^{-3})^2$

$(x^3y^{-3})^2 = (x^3)^2(y^{-3})^2$ Raise a product to a power.

$\qquad\qquad = x^6y^{-6}$ Raise a power to a power.

$\qquad\qquad = x^6 \cdot \dfrac{1}{y^6} = \dfrac{x^6}{y^6}$ Definition of negative exponents.   ■

The use of negative exponents gives rise to two more exponent properties.

■ **PROPERTY**

Fourth Property of Exponents

> $\dfrac{b^m}{b^n} = b^{m-n}$, where $b \in R$, m and $n \in J^+$, $b \neq 0$. Dividing like bases.
>
> To divide two exponential expressions with the same base, subtract the exponents and keep the common base.

EXAMPLE 6

Divide: $\dfrac{y^{16}}{y^8} = y^{16-8} = y^8$ Divide like bases.   ■

EXAMPLE 7

Divide: $\dfrac{y^6}{y^8} = y^{6-8} = y^{-2}$ Divide like bases.

$\qquad\qquad = \dfrac{1}{y^2}$ Definition of negative exponent.   ■

This property, also, can be viewed as a short cut. Example 7 is a short cut for $\dfrac{y^6}{y^8} = \dfrac{\cancel{yyyyyy}}{\cancel{yyyyyy}yy} = \dfrac{1}{y^2}$.

The fifth property of exponents does for division what the third property does for multiplication.

Fifth Property of Exponents

■ PROPERTY

$\left(\dfrac{a}{b}\right)^m = \dfrac{a^m}{b^m}$, where $a, b \in R$, $m \in J^+$, and $b \neq 0$. Raise a quotient to a power.

To raise a quotient to a power, raise both the numerator and the denominator to the power.

EXAMPLE 8

Multiply: $\left(\dfrac{x^3}{y^4}\right)^3$

$$\left(\dfrac{x^3}{y^4}\right)^3 = \dfrac{(x^3)^3}{(y^4)^3}$$ Raise a quotient to a power.

$$= \dfrac{x^9}{y^{12}}$$ Raise a power to a power.   ■

The fourth and fifth properties are frequently used to divide monomials. This is done by dividing the numerical coefficients and dividing variable factors with common bases.

EXAMPLE 9

Divide: $\dfrac{14x^2y^3}{7xy^4}$

$$\dfrac{14x^2y^3}{7xy^4} = \dfrac{14}{7} \cdot \dfrac{x^2}{x} \cdot \dfrac{y^3}{y^4}$$ Divide the numbers and divide the variable factors with the same base.

$$= 2x^{2-1}y^{3-4}$$ Divide the variable factors by subtracting the exponents of the variables.

$$= 2xy^{-1} = \dfrac{2x}{y}$$ Definition of negative exponents.   ■

A negative exponent does not influence the sign of a factor. For instance,
$$(-2)^{-3} = \dfrac{1}{(-2)^3} = \dfrac{1}{(-8)} = -\dfrac{1}{8} \text{ and } -x^{-2} = -\dfrac{1}{x^2}.$$

The following table summarizes the properties of integral exponents. The properties are used to multiply and divide monomials.

	Description	Algebraic Form
First Property	Multiplying powers with a common base	$b^m \cdot b^n = b^{m+n}$
Second Property	Raising a power to a power	$(b^m)^n = b^{mn}$
Third Property	Raising a product to a power	$(ab)^m = a^m b^m$
Fourth Property	Dividing powers with a common base	$\dfrac{b^m}{b^n} = b^{m-n}$
Fifth Property	Raising a quotient to a power	$\left(\dfrac{a}{b}\right)^m = \dfrac{a^m}{b^m}$
Zero exponent	Any nonzero number to the zero power is 1	$b^0 = 1$
Negative exponent	Any nonzero number raised to a negative power is the reciprocal of the number raised to the corresponding positive power	$b^{-m} = \dfrac{1}{b^m}$

EXERCISE 2.2

A

Simplify, using only positive exponents:

1. 8^{-2}

2. $\dfrac{1}{4^{-2}}$

3. $s^5 \cdot s^{-6}$

4. $t^7 \cdot t^{-9}$

5. $a^{-1}b^2$

6. $c^{-2}d^5$

7. $(r^{-1}s^2)(r^5 s^{-4})$

8. $(vw^{-3})(v^{-2}w^7)$

9. $\dfrac{a^{-3}b}{a^{-7}b^{-2}}$

10. $\dfrac{b^{-5}c^2}{bc^{-3}}$

11. $\dfrac{3^{-2}}{4}$

12. $\dfrac{3}{4^{-2}}$

13. $(a^{-1}b^2)^{-2}$

14. $(ab^{-2})^2$

15. $(x^{-1}y)^{-1}(x^{-2})$

16. $(x^{-2}y^{-2})^{-1}x^2$

17. $\dfrac{x^{-1}y^{-1}}{w^{-1}}$

18. $\dfrac{a^{-2}b^{-1}}{c^{-2}}$

19. $\dfrac{x^{-1}y}{x^{-1}y^{-1}}$

20. $\dfrac{a^2 b^{-1}}{a^{-2}b^{-1}}$

B

21. $(2x^{-3}y^2)(-6x^2y^{-4})$

22. $(3a^{-6}b^7)(4a^{-5}b^{-3})$

23. $(5x^2y^{-3})^{-2}$

24. $(7a^3b^{-5})^{-3}$

25. $(2x^2y^{-3})(-3x^{-2}y)^{-3}$

26. $(-5a^3b^{-2})^2(2a^{-3}b^{-1})$

27. $\dfrac{50x^8 y^{-10}}{10x^7 y^{-2}}$

28. $\dfrac{-30a^{-5}b^{-3}}{-5a^2 b^{-3}}$

29. $\left(\dfrac{m^{-1}}{n}\right)^{-1}$

30. $\left(\dfrac{p^{-1}q}{q^{-1}}\right)^{-1}$

31. $\left(\dfrac{x^{-1}}{y^{-1}}\right)^{-1}$

32. $\left(\dfrac{w}{z^{-1}}\right)^{-1}$

33. $\left(\dfrac{a^5}{b^2}\right)^{-4}$

34. $\left(\dfrac{x^7}{y}\right)^{-5}$

35. $\left(\dfrac{2x^3}{y^{-2}}\right)^{-3}$

36. $\left(\dfrac{5x^{-2}}{y^3}\right)^{-2}$

37. $\left(\dfrac{x^2}{3y^{-3}}\right)^{2}$

38. $\left(\dfrac{a^{-3}}{2b^2}\right)^{3}$

39. $\left(\dfrac{x^{-2}}{3y^{-3}}\right)^{-2}$

40. $\left(\dfrac{a^{-3}}{2b^{-2}}\right)^{-3}$

C

41. $(2x^2)^{-1}(2x^2)^{-2}(2x^2)^{-3}$

42. $(3a^{-2})^2(3a^{-2})^3(3a^{-2})^{-2}$

43. $\left(\dfrac{5xy^6}{3z^4}\right)^{-2}$

44. $\left(\dfrac{2a^3 b^4}{3c^5 d}\right)^{-4}$

45. $\dfrac{(5x^2 y^{-3})^2}{(10x^{-2} y^3)^{-2}}$

46. $\dfrac{(4a^4 b^{-3})^{-2}}{(2a^{-3} b^4)^3}$

47. $\dfrac{(6x^2)^{-2}(3x^3)^2}{(2x^3)^{-4}}$

48. $\dfrac{(5ab)^{-2}(2a^3 b^{-3})^2}{(5a^3 b^2)^{-1}}$

49. $\left(\dfrac{x^2 y^{-3}}{2^{-1}x^{-1}}\right)^{-1}\left(\dfrac{x^{-3}y^2}{3y^{-1}}\right)^{-2}$

50. $\left(\dfrac{a^4 b^{-2}}{2a^{-1}}\right)^{-2}\left(\dfrac{a^{-1}b^3}{3^{-1}a^{-2}}\right)^{2}$

51. $\left(\dfrac{p^{-1}}{q^{-2}}\right)^{-1}\left(\dfrac{p^2}{q^{-1}}\right)^{2}\left(\dfrac{p^{-1}q^{-1}}{p^{-2}}\right)^{-2}$

52. $\left(\dfrac{r}{s^{-1}}\right)^{-2}\left(\dfrac{r^2}{s^{-2}}\right)^{2}\left(\dfrac{r^{-1}s^{-2}}{rs}\right)^{-3}$

53. $(x + a)^{-2}(x + a)^4(x + a)$

54. $(c - d)^{-3}(c - d)^2(c - d)^4$

55. $(x + y)^m(x + y)^2$

56. $(s - t)^{-3}(s - t)^n$

57. $\dfrac{(2r - 3)^5}{(2r - 3)^{-2}}$

58. $\dfrac{(8t + 9)^{-6}}{(8t + 9)^{-2}}$

59. $\dfrac{(x + 3)^{-2}}{(x - 4)^{-1}} \cdot \dfrac{(x - 4)^2}{(x + 3)^{-3}}$

60. $\dfrac{(y + 2)^{-5}}{(2y - 1)^{-3}} \cdot \dfrac{(2y - 1)^{-2}}{(y + 2)^3}$

D

61. The pH of a solution with a hydrogen ion concentration of 4.6×10^{-1} is approximately 0.3. Write the hydrogen ion concentration in decimal form.

62. The pH of a solution with a hydrogen ion concentration of 3.7×10^{-4} is approximately 3.4. Write the hydrogen ion concentration in decimal form.

pH = 0.3

Figure for Exercise 61

63. A solution with pH of 5.3 has a hydrogen ion concentration of 4.9×10^{-6}. Write this concentration in decimal form.

64. A solution with pH of 1.0 has a hydrogen ion concentration of 9.4×10^{-2}. Write this concentration in decimal form.

65. The pH of a solution with a hydrogen ion concentration of 7.3×10^{-3} is approximately 2.1. Write the hydrogen ion concentration in decimal form.

66. A solution with a pH of 5.5 has a hydrogen ion concentration of 2.9×10^{-6}. Write this concentration in decimal form.

67. The wave length of an x-ray is 1×10^{-6} centimeters. Write this length in decimal form.

68. The radius of a red blood corpuscle is about 3.8×10^{-5} centimeters. Write this distance in decimal form.

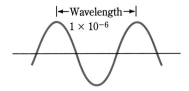

Figure for Exercise 67

STATE YOUR UNDERSTANDING

69. Explain the difference between $(xy)^2$ and $(x + y)^2$.

CHALLENGE EXERCISES

70. Show by example that $(x + y)^2 \neq x^2 + y^2$

71. Show by example that $(x + y)^{-1} \neq x^{-1} + y^{-1}$

MAINTAIN YOUR SKILLS (SECTIONS 1.4, 1.7, 1.9, 1.10)

Solve:

72. $x - \{3x - [2 - 3(5 - x)] + 6\} < 2x + 5$

73. $3(x - 2) - 4 \geq x - (x + 3)$

74. $4(x - 5) - 6(5 - 3x) = 5(4x - 6)$

Add or subtract as indicated:

75. $3x^2 - 4x + 6 - x^2 - 5x - 7 + x^2 - 4x - 11$

76. $4 - \{x - [x - (x - 4)]\}$

77. $5y - \{4y - 5[6 - 7(y - 8)] - 9\} - 10$

78. A discount store sells merchandise for 18% off the list price. What is the list price for a television set that sells for $651.90?

79. A stereo shop sells merchandise for cost plus 18%. What is the cost of a compact disk player that sells for $349.28?

<u>**2.3**</u>

SCIENTIFIC NOTATION

<u>OBJECTIVES</u>

1. Write a number in scientific notation.

2. Change scientific notation to place value notation.

3. Perform operations on numbers written in scientific notation.

Our daily newspapers and television news programs show us the use of very large numbers (the number of dollars in the national budget) and very small numbers (the size of disease-causing viruses). Scientific notation is a way of writing these numbers that requires less space. Furthermore, scientific notation gives us the ability to show greater precision in less space, as on the display of a hand-held calculator.

■ **DEFINITION**

Scientific Notation

A number written in *scientific notation* has the form

$a \times 10^m$,

where $1 \leq a < 10$ (a is a number between 1 and 10 or $a = 1$) and m is an integer.

The number 347,000,000 can be written in scientific notation in three steps:

Step 1:   3.47 is between 1 and 10   Move the decimal point eight places to the left so that we have a number between 1 and 10.

3.47000000

Step 2:   100,000,000 or 10^8 Select the power of 10 that, when multiplied times 3.47, will restore the value to 347,000,000.

Step 3: $347{,}000{,}000 = 3.47 \times 10^8$ Write the indicated product of the numbers found in Steps 1 and 2.

The number 0.0756 can be written in scientific notation in three steps:

Step 1:   7.56 is between 1 and 10   The decimal point has been moved two places right.

0.0756

Step 2: $\dfrac{1}{100} = \dfrac{1}{10^2}$ or 10^{-2} Dividing by 100 will restore the value to 0.0756.

Step 3: $0.0756 = 7.56 \times 10^{-2}$

■ PROCEDURE

To write a number in scientific notation:

Step 1: Move the decimal left or right so that the result is 1 or a number between 1 and 10.

Step 2: Find the power of 10 that will restore the original value after multiplying. The exponent of 10 is the number of positions and the direction the decimal point is moved in order to restore it to its original position: negative to the left, positive to the right.

Step 3: Write the original number as the product of the number in Step 1 and the power of 10 in Step 2.

To change a number in scientific notation to place value notation, do the indicated multiplication.

The display of a calculator or computer shows scientific notation in a distinctive manner.

Scientific Notation	Calculator Display	Computer Display
3.65×10^{11}	3.65 11	3.65 E11
6.91×10^{-12}	6.91 −12	6.91 E-12

EXAMPLE 1

Write 38,900 in scientific notation.

3.89 is between 1 and 10   Step 1. Move the decimal point left four places.

3.89 times 10,000 or 10^4 is 38,900   Step 2. The product of the number and the fourth power of 10 is the original value.

$38,900 = 3.89 \times 10^4$ Step 3.   ■

EXAMPLE 2

Write 0.000207 in scientific notation.

2.07 is between 1 and 10   Step 1. Move the decimal point right four places.

2.07 times $\dfrac{1}{10,000}$ is 0.000207

Step 2.   The product of the number and $\dfrac{1}{10}$ four times $\Big($or by the fourth power of $\dfrac{1}{10}\Big)$ is the original value.

0.000207 $= 2.07 \times 10^{-4}$

Step 3.   ∎

EXAMPLE 3

Write 357.06×10^3 is scientific notation.

357.06×10^3 is not in scientific notation

357.06 is not between 1 and 10.

3.5706 is between 1 and 10

Step 1.   Move the decimal two places to the left.

$357.06 \times 10^3 = 3.50706 \times 10^2 \times 10^3$

Step 2.   The product of 3.5706 and 10^2 is 357.06.

$357.06 \times 10^3 = 3.5706 \times 10^5$

Property 1 of exponents: $10^2 \times 10^3 = 10^{2+3}$ ∎

EXAMPLE 4

Write 9.28×10^5 in place value notation.

$9.28 \times 10^5 = 928,000$

Move the decimal five places to the right. This is equivalent to multiplying by 10 five times.   ∎

EXAMPLE 5

Write 9.28×10^{-6} in place value notation.

$9.28 \times 10^{-6} = 0.00000928$

Move the decimal six places to the left. This is equivalent to multiplying by $\dfrac{1}{10}$ six times.   ∎

Operations on large and small numbers are often done more efficiently using scientific notation with or without a calculator.

EXAMPLE 6

Perform the indicated operations using scientific notation and the properties of exponents. Write the result in both scientific notation and place value notation.

$(2.4 \times 10^7)(3.9 \times 10^{-3})$

$= (2.4 \times 3.9)(10^7 \times 10^{-3})$ Use the commutative and associative properties of multiplication to group the decimals and powers of 10.

$= 9.36 \times 10^4$ Multiply the decimals and add the exponents of 10.

$= 93{,}600$ Place value notation.   ■

EXAMPLE 7

Perform the indicated operations using scientific notation and the properties of exponents. Write the result in both scientific notation and place value notation.

$$\frac{(6.32 \times 10^6)(0.78 \times 10^{-1})}{(6.5 \times 10^4)(31.6 \times 10^{-8})}$$

$$= \frac{(6.32)(0.78)}{(6.5)(31.6)} \times \frac{10^{6+(-1)}}{10^{4+(-8)}}$$ Separate the division of the powers of 10 and the other numbers. Add the exponents.

$$= 0.024 \times \frac{10^5}{10^{-4}}$$ Perform the operations for each fraction.

$= 0.024 \times 10^9$ Subtract the exponents. This product is not in scientific notation since 0.024 is not between 1 and 10.

$= 2.4 \times 10^{-2} \times 10^9$

$= 2.4 \times 10^7$ Scientific notation.

$= 24{,}000{,}000$ Place value notation.   ■

EXAMPLE 8

Calculator Example

Perform the indicated operations using a scientific calculator. Write the result in both scientific notation (two decimal places) and place value notation:

$$\frac{(6.34 \times 10^6)(0.78 \times 10^{-1})}{(8.5 \times 10^4)(42.1 \times 10^{-8})}$$

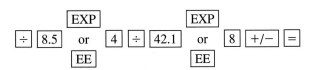

The display should read

13819197.98 or 1.3819   07   The display 1.3819   07 indicates 1.3819×10^7.

1.38×10^7 Scientific notation rounded to two places.

13,800,000   Place value notation.

■

EXERCISE   2.3

A

Write in scientific notation:

1. 50,000

2. 520,000

3. 0.00004

4. 0.0000043

5. 430,000

6. 4,300,000

7. 0.0000009

8. 0.00091

9. 825,000

10. 0.0825

Change to place value notation:

11. 9.3×10^2

12. 9.0×10^{-2}

13. 8.91×10^4

14. 8.91×10^{-4}

15. 2.32×10^{-3}

16. 2.32×10^3

17. 6.7×10^{-6}

18. 6.0×10^6

19. 3.142×10^5

20. 3.142×10^{-5}

B

Write in scientific notation:

21. 377,000

22. 0.00122

23. 0.0000701

24. 80,600,000

25. 611,000,000

26. 0.000116

Change to place value notation:

27. 3.21×10^{-1}

28. 1.23×10^1

29. 6.89×10^4

30. 9.68×10^{-3}

Perform the indicated operations using scientific notation and the laws of exponents. Write the result in scientific notations:

31. $(3.4 \times 10^6)(7 \times 10^2)$

32. $(4.7 \times 10^5)(5 \times 10^3)$

33. $(5.4 \times 10^{-3})(2 \times 10^{-4})$

34. $(4.2 \times 10^{-5})(5 \times 10^{-2})$

35. $(24,000,000)(0.0006)$

36. $(3,800,000)(0.000004)$

37. $\dfrac{8.1 \times 10^{-5}}{3.0 \times 10^2}$

38. $\dfrac{6.3 \times 10^4}{2 \times 10^{-3}}$

39. $\dfrac{240,000}{0.0008}$

40. $\dfrac{0.000048}{1200}$

C

Write in scientific notation:

41. 3784

42. 0.2894

43. 34480

44. 0.07762

45. 0.000484

46. 2,876,000

Change to place value notation:

47. 3.84×10^7

48. 4.38×10^{-6}

49. 2.36×10^{-9}

50. 6.35×10^{10}

Perform the indicated operations using scientific notation and the laws of exponents. Write the result in both scientific notation and place value notation:

51. $\dfrac{2.552 \times 10^3}{8.8 \times 10^{-2}}$

52. $\dfrac{1.65 \times 10^2}{2.5 \times 10^{-6}}$

53. $\dfrac{(7.2 \times 10^2)(3.6 \times 10^{-5})}{(4 \times 10^{-1})(5.4 \times 10^3)}$

54. $\dfrac{(8.1 \times 10^{-5})(1.24 \times 10^3)}{(5.4 \times 10^{-9})(3.1 \times 10^{-2})}$

Perform the indicated operations. Write the result in both scientific notation (two decimal places) and place value notation:

55. $\dfrac{(58 \times 10^{-3})(0.61 \times 10^5)}{(18.7 \times 10^{-5})(9 \times 10^2)}$

56. $\dfrac{(0.058 \times 10^7)(6.1 \times 10^{-3})}{(1.87 \times 10^{-5})(8.6 \times 10^3)}$

57. $\dfrac{16,000,000}{(0.00815)(80,000)}$

58. $\dfrac{76,300,000}{(20,300)(0.0177)}$

59. $\dfrac{(0.381)(44,000,000)}{2,300,000}$

60. $\dfrac{(0.0098)(88,000)}{1,700,000}$

D

Write in scientific notation:

61. In one year, light travels approximately 5,870,000,000,000 miles.

62. The king bolt bushings of an early-model automobile had a clearance of 0.0005 in.

63. A bat emits high-pitched sounds (too high for humans to hear) of approximately 50,000 cycles per second.

64. The age of a 22-year-old student is approximately 694,000,000 seconds.

65. The length of a long x-ray is approximately 0.000001 centimeter.

66. The length of a short x-ray is approximately 0.000000007 centimeter.

Write in place value notation:

67. The shortest wavelength of visible light is approximately 4×10^{-5} centimeters.

68. A certain x-ray has a length of 1.3×10^{-7} centimeters.

69. The distance from Earth to the nearest star is approximately 2.55×10^{13} miles.

70. The average number of red blood cells in the human adult is approximately 2.0×10^{12}.

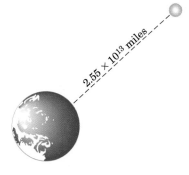

Figure for Exercise 69

Write in scientific notation:

71. The speed of light is approximately 300,000,000 meters per second.

72. The diameter of the sun is approximately 864,000 miles.

STATE YOUR UNDERSTANDING

73. Why is 18.25 not written in scientific notation?

CHALLENGE EXERCISES

74. Avogadro's number is approximately 602,200,000,000,000,000,000,000, write this in scientific notation.

75. One picometer $= 1.00 \times 10^{-12}\, m$. Write this in place value notation.

76. The speed of light in a vacuum is approximately 2.998×10^{8} meters/sec. Write this in place value notation.

77. One Joule $= 0.23901$ calorie, write the number of calories in scientific notation.

MAINTAIN YOUR SKILLS (SECTIONS 1.3, 1.4, 1.9, 1.10)

Add or Subtract:

78. $(13) + (-56) + (-102) + 43 + (-14)$

79. $(-145.87) - (25.34)$

Combine terms:

80. $0.94x - 0.73y + 0.91 - 0.25x + 1.03y + 2.11 + 0.33x - 2.05y + 1.33$

81. $3.7p - 2.6pq + 1 - 2.9p + 3.8pq + q + 7.3pq + 2.1q - 3.4$

Solve:

82. $5(2x - 6) - 3(2x + 7) \leq 5(x - 4)$

83. $3(6x - 2) - 5(4x - 1) > 7(x - 3)$

84. The population of Hillsboro went from 18,750 to 30,000 in ten years. What was the percent of increase?

85. Phil's weight dropped from 227 pounds to 183 pounds in six months. What was the percent of weight loss, to the nearest percent?

2.4

POLYNOMIALS (PROPERTIES AND CLASSIFICATIONS)

OBJECTIVES

1. Classify a polynomial by the number of its terms.

2. Determine the degree of a polynomial.

Polynomial

■ DEFINITION

A *polynomial* is a finite sum of terms whose variables have positive integral exponents and in which no variables occur in a denominator. The general form of a polynomial in x is

$$a_n x^n + a_{n-1} x^{n-1} + \cdots + a_2 x^2 + a_1 x + a_0,$$

where the coefficients are real numbers and n is a positive integer.

All the following expressions are examples of polynomials:

$$4x^2 + 7x + 4, \qquad 18x^5 + 22x^3 + 3x + 55, \qquad 15, \qquad 2w, \qquad t$$

Polynomials may contain subtraction symbols also since every subtraction can be written as an indicated sum.

$$2x^3 - 3x + 7 = 2x^3 + (-3x) + 7, \qquad s - t = s + (-t),$$
$$-3x^2 - 6x - 4 = -3x^2 + (-6x) + (5)$$

Two useful ways of classifying polynomials are (1) by the number of terms in the polynomials and (2) by the degree of the polynomial.

■ DEFINITION

Monomial

Binomial

Trinomial

A polynomial that contains

1. one term is a monomial.
2. two terms is a binomial.
3. three terms is a trinomial.

For example,

$$3, \quad 2x, \quad \sqrt{7}\,x, \quad 4x^2, \quad -8y^5, \quad \text{and} \quad \frac{x^2y}{7}$$

are monomials;

$$3x + 2y, \quad x^3 + 1, \quad \text{and} \quad t - 34$$

are binomials, and

$$x^2 + 2x + 1, \quad ab - 3bc + 4ac, \quad \text{and} \quad 7a^4 - 2a^2 + 17$$

are trinomials.

EXAMPLE 1

Classify the polynomial $-\dfrac{3}{4}y$ by number of terms.

$-\dfrac{3}{4}y$ is a monomial

This polynomial has one term that is the product of a rational number $\left(-\dfrac{3}{4}\right)$ and a variable (y). ■

EXAMPLE 2

Classify the polynomial $6x - 3y^2 - 2z^3$ by number of terms.

$6x - 3y^2 - 2z^3$ is a trinomial

This polynomial has three terms. The three terms are $6x$, $-3y^2$, and $-2z^3$. ■

To give meaning to the phrase "degree of a polynomial," we first define the degree of a monomial.

■ DEFINITION

Degree of a Monomial

> The degree of a monomial is the exponent of the variable when the monomial contains one variable. The degree of a monomial with two or more variables is the sum of the exponents of the variables. A monomial that is a nonzero constant has degree 0. The degree of the number 0 is not defined.

EXAMPLE 3

Classify the monomial $-\dfrac{3}{4}y$ by degree.

$-\dfrac{3}{4}y$ has degree 1 (linear)   The degree of y is one (y^1).   ■

EXAMPLE 4

Classify the monomial $7ab^2$ by degree.

$7ab^2$ has degree 3 (cubic)   The degree of a is 1 (a^1), and the degree of b is 2 (b^2). The sum of the exponents is the degree of the monomial, $1 + 2 = 3$.   ■

The degree of a polynomial is dependent upon the degree of its terms.

■ DEFINITION

Degree of a Polynomial

> The *degree of a polynomial* is the degree of the term with the largest degree.

■ DEFINITION

Linear Polynomial

Quadratic Polynomial

Cubic Polynomial

> A polynomial with
>
> 1. degree one is called a linear polynomial.
> 2. degree two is called a quadratic polynomial.
> 3. degree three is called a cubic polynomial.

EXAMPLE 5

Classify the polynomial $6x - 3y^2 - 2z^3$ by degree.

$6x - 3y^2 - 2z^3$ has degree 3 (cubic)   The first term, $6x$, has degree 1. The second term, $-3y^2$, has degree 2. The third term, $-2z^3$, has degree 3. The largest of these is 3. ∎

EXAMPLE 6

Classify the polynomial $x^4y - 3xy^2 + 5x - 8$ by number of terms and by degree.

$x^4y - 3xy^2 + 5x - 8$ is a four-term polynomial   There are four terms: xy^4, $-3xy^2$, $5x$, and -8. The term x^4y has degree $4 + 1 = 5$. The term $-3xy^2$ has degree $1 + 2 = 3$. The term $5x$ has degree 1. The constant term, -8, has degree 0.

$x^4y - 3xy^2 + 5x - 8$ has degree 5   The largest degree of these terms is 5. ∎

EXAMPLE 7

Classify the polynomial 11 by number of terms and by degree.

11 is a monomial   Any single constant or variable is a one-term polynomial.

11 has degree 0   A nonzero constant is defined to have degree zero. We seldom have occasion to do so, but since $x^0 = 1$ if $x \neq 0$, 11 can be written $11 = 11x^0$. ∎

The following table illustrates the common classifications.

Polynomial	Classification by Terms	Classification by Degree
$x^3 + 8$	Binomial	Degree 3 or cubic
$x^2 - 6x - 8$	Trinomial	Degree 2 or quadratic
x^5	Monomial	Degree 5
$x^6 + 3x^2 - 4x + 1$	Polynomial	Degree 6
$3x + 2$	Binomial	Degree 1 or linear

These classifications are used throughout this and subsequent chapters when dealing with polynomials and equations that contain polynomials.

EXERCISE 2.4

A–D

Classify the following polynomials by number of terms and by degree:

	Polynomial	Number of Terms	Degree
1.	$4x + 2$		
2.	$\dfrac{2}{3}x^2y$		
3.	$5x^2 + 8x - 3$		
4.	$5x^5 + 17$		
5.	16		
6.	$9x - 12$		
7.	$15x^2 + 3x - 8$		
8.	$4x^5 - 3x^4 + 2x^2 - 8x - 1$		
9.	$13x^3 - 4$		
10.	-116		
11.	$3a + 2b - 3c + 4d$		
12.	$3x^3 + 5x - 12$		
13.	$\dfrac{3}{4}x - \dfrac{5}{8}y + 9z$		
14.	$0.8x^3 - 0.7xy - 3x$		
15.	$16x + 3x^5 + 8x^2 + 1$		
16.	$27x^2 - 50xy + y^2$		
17.	$5xy + 8x - 9y$		
18.	$4w - 7x + 9t + 16$		
19.	$x^2y + 4sy^2 + 2x^2 - y^2$		
20.	$2a^3b + 3a^2b - 5ab + 7ab^2 + 2$		

STATE YOUR UNDERSTANDING

21. What is a polynomial?

22. What is a binomial?

23. What is a trinomial?

CHALLENGE EXERCISES

24. Why is the degree of $xy^2 + 3x^2 - 7$ equal to 3 instead of 2?

25. Can the degree of a polynomial be negative?

MAINTAIN YOUR SKILLS (SECTIONS 1.1, 1.2)

26. True or false? $-77 < -139$

27. True or false? $-\dfrac{22}{3} < -\dfrac{36}{5}$

28. True or false? -3.8 is a rational number.

29. True or false? -3.8 is a real number.

30. The following is an example of what property of real numbers?
$$-\frac{4}{5}\left(\frac{7}{7}\right) = -\frac{4}{5}$$

31. The following is an example of what property of real numbers?
$(x + 3)(m + n) = (m + n)(x + 3)$

32. The following is an example of what property of real numbers?
$(x + 3)(m + n) = (x + 3)m + (x + 3)n$

33. What is the reciprocal of $1\dfrac{3}{4}$?

2.5

ADDITION AND SUBTRACTION OF POLYNOMIALS

OBJECTIVES

1. Add polynomials.

2. Subtract polynomials.

3. Solve related equations.

Polynomials can be combined by addition and subtraction. If the polynomials contain like terms, the sum or difference can be simplified.

In Chapter 1, we learned that like terms have the same variable factors. To add polynomials, we first use the commutative and associative properties of addition to group the like terms.

■ PROCEDURE

To add two or more polynomials:

1. Group the like terms.

2. Add the like terms.

EXAMPLE 1

Add: $(x^2 + 2x - 1) + (3x^2 - 4x + 6) + (-2x^2 + 5x - 10)$

$= (x^2 + 3x^2 - 2x^2) + (2x - 4x + 5x) + (-1 + 6 - 10)$ Group the like terms. In this one step, several properties of algebra have been applied: the definition of subtraction and the commutative and associative properties of addition.

$= (1 + 3 - 2)x^2 + (2 - 4 + 5)x + (-1 + 6 - 10)$ Distributive property.

$= 2x^2 + 3x - 5$ Simplify. ■

The addition can be shortened by doing the second step mentally.

EXAMPLE 2

Add: $(6a + 3b - 8) + (-2a - 4b + 7) + (8a - 9b - 11)$

$= (6a - 2a + 8a) + (3b - 4b - 9b) + (-8 + 7 - 11)$ Group like terms.

$= 12a - 10b - 12$ Simplify.   ■

The polynomials can also be written vertically (based on the associative and commutative properties).

EXAMPLE 3

Add: $(-4x^3 + 3x^2 - 2x + 10) + (2x^3 - 8x - 21) + (4x^2 + 7x + 5)$

$$\begin{array}{r} -4x^3 + 3x^2 - 2x + 10 \\ 2x^3 \qquad - 8x - 21 \\ 4x^2 + 7x + \ 5 \\ \hline -2x^3 + 7x^2 - 3x - \ 6 \end{array}$$

Like terms are grouped in columns.   ■

It is recommended that you practice both forms, horizontal and vertical, for adding and subtracting polynomials.

The definition for subtracting polynomials is similar to that for subtracting real numbers.

■ **DEFINITION**

Subtracting Polynomials

> To subtract two polynomials P and Q, add the opposite of the polynomial to be subtracted.
>
> $P - Q = P + (-Q)$

The opposite of a polynomial, as well as a number, can be found by multiplying the polynomial by -1. The opposite of $2x^2 - 8x + 6$

$= -(2x^2 - 8x + 6)$

$= -1(2x^2 - 8x + 6)$

$= -2x^2 + 8x - 6$

EXAMPLE 4

Subtract: $(3x^2 - 7x - 9) - (2x^2 - 8x + 6)$

$= (3x^2 - 7x - 9) + [-(2x^2 - 8x + 6)]$ Use the definition

$= (3x^2 - 7x - 9) + (-2x^2 + 8x - 6)$ of subtraction to change to an

$= (3x^2 - 2x^2) + (-7x + 8x) + (-9 - 6)$ addition problem. Then group and

$= x^2 + x - 15$ combine like terms.   ■

EXAMPLE 5

Subtract $5x^3 - 7x + 9$ from $14x^3 + 6x^2 - 12$.

Using the vertical format:

$$
\begin{array}{l}
14x^3 + 6x^2 - 12 \\
\underline{5x^3 - 7x + 9} \\
9x^3 + 6x^2 + 7x - 21
\end{array}
$$

The polynomial we subtract *from* is written above the polynomial we subtract. Subtract by adding the opposite of the term being subtracted. ∎

As with the order of operations, when adding or subtracting polynomials where grouping symbols are used, work from the inside out.

EXAMPLE 6

Simplify: $3y - [2y - (8 - 5y) + 2] - 13$

$= 3y - [2y + (-8 + 5y) + 2] - 13$ — Subtract by adding the opposite.

$= 3y - [(2y + 5y) + (-8 + 2)] - 13$ — Group the like terms inside brackets.

$= 3y - [7y - 6] - 13$ — Simplify.

$= 3y + [-7y + 6] - 13$ — Again, subtract by adding the opposite.

$= (3y - 7y) + (6 - 13)$ — Remove brackets and group like terms.

$= -4y - 7$ — Simplify. ∎

Equations that contain the sums or difference of polynomials can be solved by first combining the like terms.

EXAMPLE 7

Solve: $(3x - 2) - (2x + 6) - (2x + 7) = 9$

$(3x - 2) + (-2x - 6) + (-2x - 7) = 9$ — Subtract by adding the opposite.

$(3x - 2x - 2x) + (-2 - 6 - 7) = 9$ — Group like terms on the left side.

$-x - 15 = 9$ — Add the terms on the left side.

$-x = 24$ — Add 15 to both sides.

$x = -24$ — Multiply both sides by -1. ∎

Check:

$$[3(-24) - 2] - [2(-24) + 6] - [2(-24) + 7] = 9 \qquad \text{Substitute } -24 \text{ for } x.$$
$$[-72 - 2] - [-48 + 6] - [-48 + 7] = 9$$
$$[-74] - [-42] - [-41] = 9$$
$$-74 + 42 + 41 = 9$$
$$-32 + 41 = 9$$

The solution set is $\{-24\}$. ■

EXAMPLE 8

Solve: $(3x^2 - 5x + 1) - (2x^2 - 4x + 6) = x^2 - 3x + 3$

$(3x^2 - 5x + 1) + (-2x^2 + 4x - 6) = x^2 - 3x + 3$	Subtract by adding the opposite.
$(3x^2 - 2x^2) + (-5x + 4x) + (1 - 6) = x^2 - 3x + 3$	Group like terms on the left.
$x^2 - x - 5 = x^2 - 3x + 3$	Add like terms on the left.
$(x^2 - x^2) + (-x + 3x) = 3 + 5$	Solve, as usual, by isolating variable terms on the left and constant terms on the right.
$2x = 8$	
$x = 4$	

The solution set is $\{4\}$. 　　　　　　　　　The check is left for the
　　　　　　　　　　　　　　　　　　　student. ■

In Example 8, the terms with x^2 drop out. Chapter 5 deals with equations in which squared terms do not drop out.

In business, cost equations can contain polynomials.

EXAMPLE 9

The cost of manufacturing q widgets at the Super Whammy Widget Factory, using people to assemble the widgets, is given by

$C = 0.5q^2 - 25q + 4000$,   where q is the number of widgets.

The cost is reduced when robots are used to assemble the widgets and is given by

$C = 0.25q^2 - 30q + 5000$.

The savings that result from using robots is the difference in the two costs and is given by

$S = (0.5q^2 - 25q + 4000) - (0.25q^2 - 30q + 5000)$.

Simplify the right side of the formula and calculate the savings on the production of 100 widgets.

$S = (0.5q^2 - 25q + 4000) + (-0.25q^2 + 30q - 5000)$ To simplify the right side of the formula, subtract by adding the opposite.

$S = (0.5q^2 - 0.25q^2) + (-25q + 30q) + (4000 - 5000)$ Group the like terms.

$S = 0.25q^2 + 5q - 1000$ Simplify by combining.   ■

To find the savings on 100 widgets, substitute $q = 100$ into the formula and evaluate.

$S = 0.25(100)^2 + 5(100) - 1000$

$= 2500 + 500 - 1000$

$= 2000$

The savings is $2000.

Exercise 2.5

A

Add or subtract as indicated:

1. $(4x + 3y) + (6x + 12y)$

2. $(8y + 16) + (7y - 5)$

3. $(x^2 + 6x + 10) + (3x^2 - 4x - 6)$

4. $(11x + 5y) - (6x + 3y)$

5. $(19x - 20) - (11x + 16)$

6. $(7a^2 + 13a + 20) - (5a^2 + 10a + 6)$

Add:

7. $8x + 6y - 9$
   $3x - 5y + 8$
   $4x - 2y + 7$

8. $3a - 2b + 6$
   $7a - 6b - 7$
   $-11a + 12b + 5$

Subtract:

9. $16x^2 + 5x + 20$
   $5x^2 + 2x + 6$

10. $8c - 5d + 8$
   $3c + 2d + 6$

Add:

11. $\left(\dfrac{1}{2}x^2 + x - \dfrac{1}{3}y\right) + \left(-\dfrac{1}{3}x^2 - \dfrac{1}{3}x - \dfrac{1}{6}y\right)$

12. $\left(-\dfrac{3}{8}a + b - c\right) + \left(\dfrac{1}{2}a - \dfrac{1}{3}b - \dfrac{1}{2}c\right)$

13. $(2.7p - 2.4q + 8.4) + (-3.8p + 3.1q - 2.7)$

14. $(-1.6x - 5.5y + z) + (1.2x - 1.2y - 1.1z)$

15. Subtract $2x + 4w$ from $7x - 3w$.

16. Subtract $8a^2 - 6$ from $5a^2 - 2$.

Add or subtract as indicated:

17. $(4r - 2s + 3) - (-3r + 2s - 4) + (8r - 5s - 1)$

18. $(4r - 2s + 3) + (-3r + 2s - 4) - (8r - 5s - 1)$

19. $(3x^2 + 4) + (2x - 5) - (7x^2 + 3x - 10)$

20. $(3x^2 + 4) - (2x - 5) + (7x^2 + 3x - 10)$

B

Add or subtract as indicated:

21. $(5x^3 + 4x^2 - 5x) + (3x^2 + 2x + 1) - (8x - 5)$

22. $(2x + 7y + 3w - 6z) + (8x - 3w - 6z) - (3y - 8z + 2x)$

23. $(17x^2y - 9xy + 4xy^2) - (13x^2y - 12xy - 15xy^2) - (6xy - 3xy^2 + 2)$

24. $(4a + 3b - 9) - (2a - 7b - 9) - (10a + 11b - 10)$

25. Add: $4a^2 - 3ab + b^2$, $12ab - 7b^2 + 2$,
$3a^2 - 6ab + 7b^2 - 9$, $5a^2 + 7b^2 + 10$

26. Add: $3x^4 + 8x - 9$, $2x^2 + 11$, $-4x^3 + 5x^2 - 10$,
$-4x^4 + 6x - 20$, $x + 11$

27. Subtract $b^2 - 4a^2$ from $2a^2 - 3b^2$.

28. Subtract $-8t^2 + 5$ from $-2t^2 + 5$.

29. Subtract $4a^4 - a^3 - a^2$ from $7a^4 - 3a^3 + a$.

30. Subtract $4p^2 + 7p - 9$ from $2p^3 - p^2 + p$.

31. Subtract $0.3x - 0.5y - 0.6$ from $0.5x + y - 0.7$.

32. Subtract $0.2m + 0.25n - 0.5$ from $0.75m - 0.5n + 0.3$.

33. Subtract $2x + 3$ and $8 - x$ from $4x - 2$.

34. Subtract $4a - b$ and $2b - 7a$ from $-3a + 11b$.

Solve:

35. $(5x + 2) - (3x - 5) = (6x - 5) + (-7x + 2)$

36. $(-3x - 5) - (5x + 2) = (7x + 2) - (3x - 1)$

37. $(5x + 2) - (4x + 1) - 3x = (4x - 5) - (-3x + 2)$

38. $(10x + 2) - (8x - 1) + x = (5x - 3) - (x - 1)$

39. $(3y - 7) - (1 - 4y) = y - (3 - y)$

40. $(11 - y) - (2y + 4) = 15 - (y - 8)$

C

Add or subtract as indicated:

41. $\left(\dfrac{1}{5}p + 2q - \dfrac{1}{2}\right) + \left(\dfrac{3}{2}p - \dfrac{3}{4}q - 1\right) - \left(-\dfrac{1}{2}p - \dfrac{1}{2}q - \dfrac{1}{2}\right)$

42. $\left(\dfrac{3}{5}y^2 + \dfrac{1}{4}y - \dfrac{3}{20}\right) - \left(\dfrac{1}{2}y^2 - \dfrac{1}{5}y + \dfrac{3}{10}\right) - \left(-\dfrac{3}{4}y^2 + \dfrac{11}{10}y - \dfrac{17}{20}\right)$

43. $(0.73r - 0.44s - 0.18t) - (0.81r + 0.23s - 0.38t) + (-0.21r + s - 0.91t)$

44. $(0.12a - 0.181b + 0.31c) - (-0.42a - 0.08b + 0.55c) - (0.75a - 0.201b - 0.38c)$

45. $x^2 - [2x^2 - (5x^2 + 6x) - x + 8] + 3$

46. $4a - [-14a - (3a - 2) + 6] + a - 3$

47. $2y - \{3y - [5y - (8 - 4y) + 6] + 2y\} - 8$

48. $16x - \{8 - [13x - (12 - 2x) - 9x] - 4x\} + 11$

49. Add $7x + 2y$ to the difference of $6x - y$ and $2x + 3y$.

50. Subtract $8a - b$ from the sum of $2a + 2b$ and $4a - 7b$.

51. From $(7r^2 - 5r - 3)$ subtract the sum of $(r^2 - 7)$ and $(r - 4r^2 + 1)$.

52. From the sum of $(2a - 7b + 1)$ and $(8b - a)$ subtract $(3 - 4b)$.

53. Subtract $(x - y)$ from the difference of $(2x + 3y)$ and $(7x - 8y)$.

54. Subtract $(2x + 3y)$ from the difference of $(7x - 8y)$ and $(x - y)$.

Solve:

55. $(5x^2 + 2x + 5) - (2x^2 + 4x - 4) = 3x^2 - 4x + 5$

56. $(-3x^2 - 2x + 1) - (-6x^2 + 3x - 2) = (2x^2 + 1) + (x^2 + x)$

57. $2x^2 - [-3x^2 + (5x^2 + 2x + 1)] = -3x + 5$

58. $10x^2 - [4x^2 + (6x^2 - 3x + 5)] = 2x - 7$

59. $7 - [x - (3 - x)] = [2x - (x + 1)] - 1$

60. $x - [7 - (x - 3)] = [2 - (2x + 1)] + 9$

D

61. The cost of manufacturing n brushes is given by $C = \dfrac{1}{5}n^2 - 14n + 20$. The cost of marketing the same items is given by $C = 5n^2 - 30n + 50$. Find the formula for the total cost of manufacturing and marketing the brushes. Find the total cost when $n = 50$.

62. The cost of manufacturing n carburetors before automation is given by $C = 15{,}000 - 1000n + n^2$. The cost after automation is given by $C = 1000 - 750n + \frac{1}{2}n^2$. Write the formula for the savings, S, due to automation. Find the savings on manufacturing 50 carburetors.

63. The cost of manufacturing x thousand calculators is $C = 2x^2 - x + 1000$. The revenue generated from selling x thousand calculators is $R = 3x^2 + x - 500$. Find a formula for the profit. (Profit $=$ Revenue $-$ Cost)

64. If the cost of manufacturing x thousand teddy bears is $C = 3x^2 - 2x + 600$, and the revenue from selling x thousand teddy bears is $R = 2x^2 + 3x + 500$, find a formula for the profit in selling x thousand teddy bears. (Profit $=$ Revenue $-$ Cost)

65. The volume of one container is given by the formula $V_1 = 2x^3 - 3x^2 + x + 40$ (where x is the length of one side), and the volume of a second container is $V_2 = x^3 + 3x^2 - 5x + 10$. Find a formula for the combined volume, V, of the two containers. Find the combined volume if $x = 2$.

66. One machine can produce $5t^2 - 4t + 10$ toys per day (where t is the number of hours). A second machine can produce $3t^2 - 2t + 50$ toys per day. Find a formula for the number of toys, T, produced per day when both machines are working.

67. The number of bolts that Nut & Bolts Co. can produce in t hours is $N = 5t^2 - 3t + 500$. The number of defective bolts is $D = t^2 + 150$. Find a formula for the number of good bolts, B, the company can produce in t hours. How many good bolts can the company produce during the first hour?

68. The number of cookies the Cookie Factory can produce in t hours is $C = 2t^2 - t + 50$. The number of broken cookies in t hours is $B = 10t + 35$. Find a formula for the number of good cookies, G, they can produce in t hours.

69. Girl scout troop #52 can sell $\frac{3}{2}t^2 - 2t + 20$ boxes of cookies in t days. Troop #63 can sell $2t^2 - 4t + 10$ boxes of cookies in t days. Find a formula for the total number of boxes of cookies, T, both troops can sell in t days. How many boxes will both troops have sold in 2 days?

70. In t days, Bright Light Bulb Co. can produce $3t^2 - t + 100$ light bulbs. If $\frac{1}{2}t^2 - 2t$ bulbs are defective, find a formula for the number of good light bulbs, B, produced in t days. Find the number of light bulbs produced in 4 days.

STATE YOUR UNDERSTANDING

71. What is the procedure for subtracting polynomials?

CHALLENGE EXERCISES

Add or subtract as indicated, assume all variables represent positive integers:

72. $(x^{3n} + 2x^{2n} + 3x^n + 5) + (3x^{3n} + x^{2n} + 2x^n - 3)$

73. $(4x^{2n} + 5x^n - 12) - (2x^{2n} - 5x^n - 4)$

74. $(x^{3n} - 2x^{2n} + x^n + 4) - (3x^{3n} + 2x^{2n} + x^n + 5)$

75. $(2x^n + 4x^2 - 12x + 1) + (4x^n + 8x - 3)$

MAINTAIN YOUR SKILLS (SECTIONS 1.3, 1.4)

Perform the indicated operations:

76. $72 + (-89) + (-23) + (-19)$

77. $(128) + (43) + (-219) + (-71)$

78. $(-68) + (-193) + (-79) + (108)$

79. $|-13 - 57|$

80. $-[(-13) + (42) + (-29)]$

81. $-[415 + (-329) + (-38) + (18)]$

82. $-|(41) + (-28)| + |(-29) + (-3)|$

83. $229 + (-301) + [-|-118 + 32|]$

2.6

MULTIPLICATION OF POLYNOMIALS

OBJECTIVES

1. Multiply polynomials.

2. Solve related equations.

The product of polynomials can be rewritten as the multiplication of monomials by using the distributive property.

$$3x(x^2 + 3x - 9) = \boxed{3x} \cdot x^2 + \boxed{3x} \cdot 3x + \boxed{3x} \cdot (-9) \qquad \text{Distributive property.}$$
$$= 3x^3 + 9x^2 - 27x \qquad \text{Multiply.}$$

EXAMPLE 1

Multiply: $12xy^2(5xy^2 + 13x - 8y)$

$$\boxed{12xy^2} \, (5xy^2 + 13x - 8y)$$
$$= \boxed{12xy^2} \cdot 5xy^2 + \boxed{12xy^2} \cdot 13x - \boxed{12xy^2} \cdot 8y \qquad \text{Distributive property.}$$
$$= 60x^2y^4 + 156x^2y^2 - 96xy^3 \qquad \text{Multiply.} \qquad \blacksquare$$

If both polynomials contain two or more terms, it may be necessary to use the distributive property more than once.

EXAMPLE 2

Multiply: $(w - 14)(w + 7)$

$(w - 14)\,(w + 7)$

$= (w - 14)\,(w) + (w - 14)\,(7)$ Distributive property. Here the factor that is "distributed" is a binomial, $(w - 14)$.

$= w\,(w) - 14\,(w) + w\,(7) - 14\,(7)$

$= w^2 - 14w + 7w - 98 = w^2 - 7w - 98$ Multiply.   ∎

EXAMPLE 3

Multiply: $(x + 1)(2x^2 + 3x - 5)$

$(x + 1)\,(2x^2 + 3x - 5)$

$= (x + 1)\,(2x^2) + (x + 1)\,(3x) + (x + 1)\,(-5)$ Distributive property.

$= x \cdot 2x^2 + 1 \cdot 2x^2 + x \cdot (3x) + 1 \cdot (3x) +$
$\quad x \cdot (-5) + 1 \cdot (-5)$ Again we use the distributive property. This time the factors that are distributed are $2x^2$, $3x$, and -5.

$= 2x^3 + 2x^2 + 3x^2 + 3x - 5x - 5$ Multiply.

$= 2x^3 + 5x^2 - 2x - 5$ Combine the like terms.   ∎

■ PROCEDURE

To multiply two polynomials:

1. Rewrite the product using the distributive property until the result is the sum of products of monomials.
2. Multiply.
3. Combine the like terms.

EXAMPLE 4

Multiply: $(x + 3)(2x - 1)$

$= (x + 3) \cdot 2x + (x + 3) \cdot (-1)$ Distributive property.

$= x \cdot 2x + 3 \cdot 2x + x \cdot (-1) + 3 \cdot (-1)$ Distributive property, again.

$$= 2x^2 + 6x - x - 3 \qquad \text{Multiply.}$$

$$= 2x^2 + 5x - 3 \qquad \text{Simplify.} \qquad \blacksquare$$

■ CAUTION

A common error is to confuse addition and multiplication of monomials. $(-2x^2) + (14x^2) = 12x^2$ and $(-2x^2)(14x^2) = -28x^4$. The terms $3x^2$ and $7x$ can be multiplied $(21x^3)$, but they cannot be combined since they are not like terms $(3x^2 + 7x \neq 10x^3)$.

Equations that contain the products of polynomials can be solved by first multiplying the polynomials.

EXAMPLE 5

Solve: $x(x - 3) - 2[2x - x(2x + 1)] = 5x(x - 7) + 15$

$$x^2 - 3x - 2[2x - 2x^2 - x] = 5x^2 - 35x + 15 \qquad \text{Multiply on both sides to remove the parentheses.}$$

$$x^2 - 3x - 2[x - 2x^2] = 5x^2 - 35x + 15 \qquad \text{Combine like terms in brackets.}$$

$$x^2 - 3x - 2x + 4x^2 = 5x^2 - 35x + 15 \qquad \text{Multiply to remove brackets.}$$

$$5x^2 - 5x = 5x^2 - 35x + 15 \qquad \text{Combine the like terms.}$$

$$30x = 15 \qquad \text{Solve for } x.$$

$$x = \frac{1}{2} \text{ or } 0.5$$

The solution set is $\{0.5\}$. $\qquad \blacksquare$

It is not uncommon for the solution of problems to require the multiplication of polynomials.

EXAMPLE 6

Solutions of 30% pesticide spray and 60% pesticide spray are to be mixed to form a solution that is 42% pesticide spray. How many gallons of each spray must be used to make 20 gallons of the 42% pesticide spray?

Simpler word form:

$$\left(\begin{matrix}\text{Gallons of pesticide} \\ \text{in 30\% solution}\end{matrix}\right) + \left(\begin{matrix}\text{Gallons of pesticide} \\ \text{in 60\% solution}\end{matrix}\right)$$

The amount of pesticide in the 42% solution must equal the sum of the amounts in the other two solutions.

$$= \left(\begin{matrix}\text{Gallons of pesticide} \\ \text{in 42\% solution}\end{matrix}\right)$$

Select a variable:

Let x represent the number of gallons of the 30% solution, then $20 - x$ will represent the number of gallons of the 60% solution.

The total amount of the mixture is 20 gallons.

Solution	30%	60%	42%
Gallons of solution	x	$20 - x$	20
Gallons of pesticide	$0.3x$	$0.6(20 - x)$	$0.42(20)$

A chart will help organize the information. The formula for amount is $A = RB$, where R is the rate of percent and B is the base. The gallons (amount) of pesticide in each case are found by multiplying the percent of solution times the gallons (base) of solution.

Translate to algebra:

$0.3x + 0.6(20 - x) = 0.42(20)$

The number of gallons of pesticide in the two solutions equals the amount in the mixture.

$30x + 60(20 - x) = 42(20)$

Multiply both sides by 100 to clear the decimals.

$$-30x = -360$$
$$x = 12$$

Number of gallons of 30% solution.

$20 - x = 8$

Number of gallons of 60% solution.

Answer:

So, 12 gallons of 30% pesticide solution mixed with 8 gallons of 60% pesticide solution will make 20 gallons of 42% pesticide solution. ■

EXAMPLE 7

A bank thief heads for Canada after a robbery, a trip that will take seven hours. The resident agent of the FBI, Juan, leaves one hour after the thief, following the same route. If the thief averages 48 mph and Juan averages 57 mph, will the thief be caught before he reaches Canada?

Simpler word form:

Distance of thief = Distance of Juan

If Juan is to catch the thief, he must travel the same distance within 7 hours.

Select a variable:

Let t represent Juan's time.
Then $t + 1$ represents the thief's time.

	Rate	Time	Distance
Thief	48	$t + 1$	$48(t + 1)$
Juan	57	t	$57t$

A table is often useful to organize the information in a problem.

Translate to algebra:

$48(t + 1) = 57t$

The distances ($D = rt$) are equal.

Solve:

$48t + 48 = 57t$
$48 = 9t$
$5\dfrac{1}{3} = t$

Juan's time.

$t + 1 = 5\dfrac{1}{3} + 1 = 6\dfrac{1}{3}$

Thief's time.

Answer:

Juan will catch the thief 5 hours and 20 minutes after Juan leaves, or 6 hours and 20 minutes after the thief leaves. Since the thief's time is less than 7 hours, he will be caught before he gets to Canada. ∎

EXERCISE 2.6

A

Multiply:

1. $2(x + 3)$

2. $3(y + 1)$

3. $x(x - 1)$

4. $y(y - 3)$

5. $2(3x + 7)$

6. $10(9y - 8)$

7. $-6(2a - 3b)$

8. $-11(2c - 5d)$

9. $y(y^2 + 3y - 8)$

10. $c(c^2 - 2c + 6)$

11. $2a(a^2 - 3a + 6)$

12. $5b(3b^2 - 7b + 4)$

13. $3xy(x^2 - 6xy + y^2)$

14. $-4cd(c^2 - 3cd - d^2)$

15. $-3abc(2a - 3b + c - 1)$

16. $-6xyz(3x - y + 2z - 5)$

17. $-a^3(-2a^4 + 5a^3 - 3a - 1)$

18. $-2x^2(4x^4 - 8x^2 - 5x + 1)$

19. $13x^2y(12x^2y - 11xy^2 + 8xy)$

20. $16ab^2(15ab^2 - 9a^2b - 16ab)$

B

21. $(x - 3)(x + 4)$

22. $(a + 6)(a + 13)$

23. $(2x - y)(2x - 3y)$

24. $(x^2 + 3x)(x^2 + 5x)$

25. $(3m - 7t)(t - 4m)$

26. $(2a + 6b)(3b - a)$

27. $(a + b)(a + b + 3)$

28. $(x - y)(x + y - 6)$

29. $(x^2 + 1)(2x^2 - 7x + 5)$

30. $(x^2 - 3)(3x^2 - 5x + 2)$

31. $(y^2 + 3)(4y^3 + y^2 - 3)$

32. $(a^2 + b)(2a^2 + a - 2b)$

Solve:

33. $3(x + 2) - 2(2x + 9) = 4(x + 2)$

34. $2a - 3(a + 6) = 24 - 4(3 + a)$

35. $a(a + 2) - 3[2a - a(a - 7)] = 4a(a - 1) + 21$

36. $y - 3y[2 - (4 + y)] = 3y(y - 6) + 50$

37. $2x^2 + 18 = (2x + 3)(x + 5)$

38. $5x^2 + 1 = (5x - 1)(x + 2)$

39. $(2x + 5)(x - 4) = (8 - x)(3 - 2x)$

40. $(3 + x)(2 + 3x) = (3x - 4)(x - 1)$

C

Multiply:

41. $(a + b - x)(a - b + x)$

42. $(2y - z + x)(2y + z - x)$

43. $(c + d + 4)(c + d + 5)$

44. $(2x + y - 3)(2x + y + 4)$

45. $(2x + 3y)(x^2 - 3xy + y^2 + 6)$

46. $(a - 2b)(a^2 + 3ab - b^2 - 3)$

47. $(x + y + z)^2$

48. $(a - b + c)^2$

49. $(x + 1)(x - 2)(x + 3)$

50. $(2y + 1)(3y - 1)(4y + 1)$

51. $(x^2 + 2)(5x^3 + 3x^2 - 2x - 1)$

52. $(3y^2 - 1)(3y^3 - 2y^2 + y - 2)$

Solve:

53. $x(x + 3) + (x + 2)(x - 7) = 2x(x + 3) - 12$

54. $(a - 3)(a + 3) + (4 - a)(4 + a) = 3 - 2(a + 6)$

55. $(x - 1)(x + 1) - 2(3 + x)(x - 1) = -(x + 4)(x + 1)$

56. $5 - (a + 4)(a - 5) = (3 - a)(2 + a) - 19a$

57. $3x(4x - 3) - (2x - 1)(3x + 2) = 2x(3x - 4)$

58. $-2x(3x - 4) + (2x + 1)(x - 1) = -4x(x - 2)$

59. $4x(x - 1) + (2x + 1)(2x - 1) = -8x(1 - x)$

60. $-x(2 - 4x) + (2x + 3)(2x - 3) = 2x(4x - 5)$

D

61. Mr. Ewing averaged 50 mph on his recent trip to Houston. On the return trip, he took a scenic route part of the way and took 2 hours longer and averaged only 42 mph. If the return trip was 30 miles longer, how long did it take Mr. Ewing to drive each direction?

(handwritten) $D_g = r_g t_g = 50 t_g$
$30 + D_g = 42(t_g + 2)$
$42(t_g + 2) - 30 = 50 t_g$
$t_g = 6\,3/4$

62. A marathon runner runs 4.6 miles in 45 minutes. If she averaged 6 mph on the first part of the course and 6.5 mph on the rest, how long did she run at each average speed?

(handwritten) $D = n_1 t_1 + n_2 t_2$ or
$4.6 = 6.5 t_1 + 6 t_2$ and
$t_1 + t_2 = 45\,min = \frac{3}{4}\,hr$
$t_1 = \frac{1}{10}\,hr$

63. Oil costing $1.30 per quart is to be mixed with oil costing $1.60 per quart to make 90 quarts of a mixture that would cost $1.42 per quart. How much of each grade should be used?

(handwritten) $\left(\dfrac{1.30}{x}\right) + \left(\dfrac{1.60}{90-x}\right) = \left(\dfrac{1.42}{90}\right)$
$1.3x - 1.6x + 144 = 127.8$
$x = 54$

64. A radiator on Mr. Benz's new Mercedes holds 4 gallons and is now filled with 10% antifreeze. How much of the 10% solution must be drained and replaced with 100% antifreeze so that the radiator will contain a 25% solution?

(handwritten) $\left(\dfrac{x\,gal}{100\%}\right) + \left(\dfrac{4-x}{1.0\%}\right) = \left(\dfrac{4}{25\%}\right)$ $x = \frac{2}{3}\,gal$

65. A 15-pound candy mixture selling for $1.90 per pound is made up of caramels costing $2.00 per pound and hard candy costing $1.75 per pound. How many pounds of each kind of candy make up the mixture?

(handwritten) $\left(\dfrac{x}{2}\right) + \left(\dfrac{15-x}{1.75}\right) = \left(\dfrac{15}{1.90}\right)$ $x = 9$

66. How many ml of water must be added to 10 ml of a 45% solution of HCL to obtain a 15% solution of HCL?

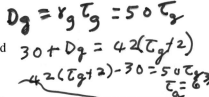

(handwritten) $\left(\dfrac{x}{0\%\,water}\right) + \left(\dfrac{10\,ml}{55\%\,water}\right) = \left(\dfrac{x+10}{85\%\,water}\right)$ or $\left(\dfrac{10}{45}\right)\% = \dfrac{x+10}{15\%}$
$4.5 = .15x + 1.5$ $x = 20$

67. Linda jogs to a lake at a rate of 4 miles/hr, then walks back at a rate of 1.5 miles/hr. If it takes her one hour longer to walk back, how long did it take her to jog to the lake?

(handwritten) $D = r \cdot t = 4 \cdot t_1$
$D = 1.5(t_1 + 1)$ or $4t_1 = 1.5(t_1 + 1)$
$t_1 = 3/5\,hr$

68. How many liters of a 20% solution of alcohol must be added to 4 liters of a 50% solution to obtain a 30% solution of alcohol?

(handwritten) $\left(\dfrac{x}{20\%}\right) + \left(\dfrac{4}{50\%}\right) = \left(\dfrac{x+4}{30\%}\right)$ $x = 8$

69. Mari rode her bicycle to work at a rate of 6 miles/hr and walked home at a rate of 2 miles/hr. If it took her 2 hours longer to walk home, how long did it take her to ride her bicycle to work?

(handwritten) $D = n \cdot t = 6t_1$ and $D = 2(t_1 + 2)$ or $6t_1 = 2(t_1 + 2)$
$t_1 = 1$

70. The Nutty Nut Store sells a mixture of peanuts and cashews for $3.50 per pound. Mr. Nutty packages the mixture into 5-pound bags. If peanuts cost $2.00 per pound and cashews cost $4.50 per pound, find the number of pounds of peanuts and the number of pounds of cashews in each 5-pound bag.

(handwritten) $\left(\dfrac{x}{2.00}\right) + \left(\dfrac{5-x}{4.50}\right) = \left(\dfrac{5}{3.50}\right)$

R UNDERSTANDING

. what is the procedure for multiplying a trinomial by a binomial?

CHALLENGE EXERCISES

Multiply, assume that all variables name positive integers:

72. $x^n(x^3 + 2x^2 + 3x + 1)$

73. $(x^n + y^n)^2$

74. $(2x^n + b^n)^2$

75. $x^{-n}(x^{2n} - x^n + 2)$

MAINTAIN YOUR SKILLS (SECTION 1.5)

Multiply or divide as indicated:

76. $(-13)(-5)(-2)$

77. $(-256) \div (-50)$

78. $(-8)(-3x)(4)$

79. $(-15)(-4)(-7y)$

80. $112t \div (-35)$

81. $-17(-2x + 8w)$

82. Change $-31°F$ to Celsius measure.

83. A new restaurant had a total cash flow of $-\$372$ for the first 12 days after it opened. What was the average cash flow per day?

2.7

SPECIAL PRODUCTS

OBJECTIVES

1. Find the product of two binomials: $(ax + b)(cx + d)$.

2. Find the product of conjugate binomials: $(a - b)(a + b)$.

3. Square a binomial: $(a + b)^2$.

4. Solve related equations.

Although the product of any two binomials can be found by the methods of the last section, there are shortcuts for multiplying. The shortcut for the distributive property used to find the product of any two binomials is called the FOIL method. The binomial product is labeled as shown.

FOIL Method

F   L F   L
$(3x + 4)(2x + 5)$
O   I I   O

F refers to the first terms.
O refers to the outside terms.
I refers to the inside terms.
L refers to the last terms.

The sum of the products of F's, O's, I's, and L's is the product.

$$\overset{\text{F}\quad\text{L}\;\;\text{F}\quad\text{L}\qquad\text{F}\qquad\text{O}\qquad\text{I}\qquad\text{L}}{(3x+4)(2x+5) = 3x \cdot 2x + 3x \cdot 5 + 4 \cdot 2x + 4 \cdot 5}$$
$$\overset{\text{O}\quad\;\text{I}\;\;\text{I}\quad\;\;\text{O}}{} = 6x^2 + 15x + 8x + 20$$
$$= 6x^2 + 23x + 20$$

EXAMPLE 1

Multiply: $(x + 3)(x - 6)$

$$\overset{\qquad\qquad\qquad\text{F}\quad\;\;\text{O}\quad\;\;\text{I}\quad\;\text{L}}{(x + 3)(x - 6) = x^2 - 6x + 3x - 18}$$
$$= x^2 - 3x - 18 \qquad\qquad \text{Simplify.} \qquad\blacksquare$$

The special products of this section appear often in algebra, so it is worth the time now to practice them in order to save time later.

EXAMPLE 2

Multiply: $(5y - 9)(2y + 7)$

$$\overset{\qquad\qquad\qquad\text{F}\quad\;\;\;\text{O}\qquad\text{I}\quad\;\;\text{L}}{(5y - 9)(2y + 7) = 10y^2 + 35y - 18y - 63} \qquad \text{The product of two binomials}$$
$$= 10y^2 + 17y - 63 \qquad\qquad \begin{array}{l}\text{contains four terms. Like terms}\\ \text{can be combined.} \qquad\blacksquare\end{array}$$

The shortcut for conjugate binomials is even quicker.

■ DEFINITION

Conjugate Binomials

> Two binomials are called *conjugate binomials* if one is the sum and the other is the difference of the same two terms.

Each of these illustrates a pair of conjugate binomials:

$$\begin{array}{ll} w + 3 & w - 3 \\ 2x - 3y & 2x + 3y \\ m + \dfrac{2}{3} & m - \dfrac{2}{3} \\ x^2 - 7y^3 & x^2 + 7y^3 \end{array}$$

The product of conjugate binomials, $(a + b)(a - b)$, is special because in the FOIL product the "O" and the "I" terms are opposites and so have sum zero.

$$(2x - 3)(2x + 3) = 2x \cdot 2x + 2x \cdot 3 - 3 \cdot 2x - 3 \cdot 3$$
$$= 4x^2 + 6x - 6x - 9$$
$$= 4x^2 - 9$$

The pattern is:

■ FORMULA

Product of Conjugate Binomials

$$(a + b)(a - b) = a^2 - b^2$$

Formula for the product of conjugate binomials.

EXAMPLE 3

Multiply: $(3y + 5)(3y - 5)$

$(3y + 5)(3y - 5) = 9y^2 - 25$

Since we recognize these to be conjugate binomials, we multiply only the first and last terms, aware that the middle terms will cancel. ■

EXAMPLE 4

Multiply: $(2ab - 3y^3)(2ab + 3y^3)$

$= 4a^2b^2 - 9y^6$

This is the product of conjugates. Notice that $y^3 \cdot y^3$ is y^6, *not* y^9. ■

Since the "outer" and "inner" terms of the product of conjugates have sum 0, the product always has two terms. These terms are the squares of the first and second terms of each of the binomials. Such a product is called "the difference of two squares."

■ DEFINITION

Difference of Two Squares

A binomial is called the *difference of two square* if it contains two squared terms separated by a minus sign.

We can see from using the FOIL shortcut that the square of a binomial follows these patterns.

■ FORMULA

Square of a Binomial

$$(a + b)(a + b) = a^2 + 2ab + b^2$$

Formulas for the square of binomials.

$$(a - b)(a - b) = a^2 - 2ab + b^2$$

Other squares can be found by substituting into the patterns.

EXAMPLE 5

Multiply: $(2y + 7)^2$

$= (2y)^2 + 2(2y)(7) + (7)^2$

$= 4y^2 + 28y + 49$

In words, to square a binomial, square the first term, double the product of the first and last terms, and square the last term. Note that the first and last terms are always positive. ■

EXAMPLE 6

Multiply: $(12a - 11b)^2$

$= (12a)^2 - 2(12a)(11b) + (11b)^2$

$= 144a^2 - 264ab + 121b^2$ ■

EXAMPLE 7

Multiply: $(a - b + 4)(a - b - 4)$

$= [(a - b) + 4][(a - b) - 4]$

$= (a - b)^2 - 16$

If $(a - b)$ is grouped as a single term, the product can be found by multiplying the conjugate binomials with terms $(a - b)$ and 4.

$= a^2 - 2ab + b^2 - 16$

Square $(a - b)$. ■

EXAMPLE 8

Solve: $(x - 8)(x + 8) = (x - 13)^2 + 14$

$x^2 - 64 = x^2 - 26x + 169 + 14$

$x^2 - 64 = x^2 - 26x + 183$

$26x = 247$

$x = 9.5 \text{ or } \dfrac{19}{2}$

Multiply on both sides to remove the parentheses. Simplify.

■

The solution set is $\{9.5\}$ or $\left\{\dfrac{19}{2}\right\}$.

Example 9 is an application that requires the skill of multiplying polynomials.

EXAMPLE 9

The city of Toronto increased both the length and width of a city flower garden by 12 meters. Because of this increase in the dimensions, the area was increased by 468 square meters. If the length was originally 3 meters longer than the width, what are the new dimensions of the flower garden?

Simpler word form:

Original garden area + 468 square meters = New garden area

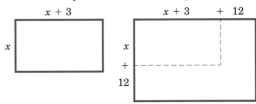

Select a variable:

Let x represent the width of the original garden. Then $x + 3$ represents the
length of the original garden.

The formula for the area of a
rectangle is $A = \ell w$.

Translate to algebra:

$$x(x + 3) + 468 = (x + 12)(x + 15)$$
$$x^2 + 3x + 468 = x^2 + 27x + 180$$
$$-24x = -288$$
$$x = 12$$
$$x + 3 = 15$$

Answer:

The dimensions of the original garden were 12 meters by 15 meters. The dimensions of the new garden are 24 meters by 27 meters.

EXERCISE 2.7

A

Multiply:

1. $(x + 3)(x + 4)$

2. $(a + 1)(a + 2)$

3. $(y + 5)(y - 7)$

4. $(w - 11)(w + 8)$

5. $(b - 1)(b - 4)$

6. $(c + 6)(c + 8)$

7. $(y - 8)(y - 3)$

8. $(x - 2)(x - 10)$

9. $(x + 6)(x - 7)$

10. $(y + 5)(y - 9)$

11. $(a - 10)(a + 10)$

12. $(b - 12)(b + 12)$

13. $(x - 9)^2$

14. $(a + 10)^2$

15. $(w - 3)^2$

16. $(y + 7)^2$

17. $(2x + 11)(2x - 11)$

18. $(3x + 12)(3x - 12)$

19. $(p + 13)^2$

20. $(q - 16)^2$

B

21. $(3x - 5)(x - 6)$

22. $(4y + 1)(y + 2)$

23. $(2a + 5)(a - 3)$

24. $(b + 6)(4b - 3)$

25. $(3x + 7)(5x - 6)$

26. $(2y + 9)(3y - 1)$

27. $(2a - 5)(2a + 5)$

28. $(3y + 7)(3y - 7)$

29. $(2a + 5)^2$

30. $(3y - 7)^2$

31. $(3x + 5y)^2$

32. $(4a - 3b)^2$

33. $(8w - 13)^2$

34. $(11t + 16)^2$

Solve:

35. $(x + 3)(x + 5) = (x + 7)(x - 5) - 4$

36. $(y - 5)(y + 5) = (y + 2)^2 + 1$

37. $(3t - 4)(3t + 4) = (3t + 2)^2 + 4$

38. $(2t - 3)^2 = (4t + 1)(t - 3)$

39. $(4 + 5x)(4 - 5x) = 22 - (5x + 3)^2$

40. $(6x - 1)^2 - (6x + 1)(6x - 1) = 6$

C

Multiply:

41. $(5x - 11)(3x + 9)$

42. $(8y + 7)(6y - 10)$

43. $(0.6x - 0.5)(0.3x - 0.2)$

44. $(0.7y + 0.8)(0.3y + 0.5)$

45. $\left(\dfrac{1}{2}x + \dfrac{1}{3}\right)\left(\dfrac{1}{3}x + \dfrac{1}{2}\right)$

46. $\left(\dfrac{1}{5}y - \dfrac{1}{2}\right)\left(\dfrac{1}{4}y - \dfrac{1}{5}\right)$

47. $[(a + b) - 8][(a + b) + 8]$

48. $[(c + 3) - d][(c + 3) + d]$

49. $[(2x + y) - 5]^2$

50. $[(3x - 2y) + 4]^2$

51. $[(3x - y) + 4][(3x - y) + 3]$

52. $[(2a + 3b) - 5][(2a + 3b) + 2]$

53. $[(a + b) + (c + d)][(a + b) - (c + d)]$

54. $[(w + 4)(w - 4)]^2$

Solve:

55. $(x - 7)^2 = (x + 6)^2 - 13$

56. $(x + 10)^2 - 108 = (x - 8)^2$

57. $(2x - 3)(3x + 4) + (3x - 1)(x + 1) = (3x + 5)^2 + 20$

58. $(2y + 5)^2 - (2y - 6)(2y + 6) = 3y - 7$

59. $(3x - 2)(3x + 2) - (3x - 2)^2 = 0$

60. $(3x - 2)^2 - (3x + 2)(3x - 2) = 0$

$x + 525 = $ ⬚ $x+5$ $x = 50$

building lot had to be increased by 5 feet on each side to meet
If this increase added 525 square feet to the area, what was the
length of a side of the original lot?

62. A photo of the Golden Gate Bridge is 2 inches longer than it is wide.
When a frame with 2-inch-wide sides is used, the area of the photo and
frame is 48 square inches larger than the area of the photo alone. Find the
dimensions of the photo. $x(x+2)+48 = (x+4)(x+6)$ $x = 3$

63. John is 5 years older than Mary is now. In three years, the product of their
ages will be 150 more than the product of their ages now. How old is each
now? $x = $ Mary's age now $(x+3)(x+8) = 150 + x(x+5)$ $x = 21$
$x+5 = $ John's age now

64. The product of Maude's age 5 years ago and 5 years from now is equal to
the square of her age 1 year ago. How old is Maude now? $x = $ age now
$(x-5)(x+5) = (x-1)^2$

65. The length of a vegetable garden plot is to be increased by 2 ft while the
width is increased by 4 ft. These changes will increase the area by 100 sq ft.
If the length was originally 5 ft more than the width, find the new
dimensions of the garden plot. $xy+100 = (x+2)(y+4)$
$x = y+5$ $y = 12$

66. The length of a frame must be increased by 3 inches and the width must be
decreased by 2 inches for the frame to fit a painting. These changes will
increase the area enclosed by the frame by 10 sq in. If the length of the
original frame was 5 in. more than the width, find the new dimensions of
the frame.

67. Jim is ten years older than Pam. In two years, the product of their ages will
be 48 more than the product of their ages now. How old are Jim and Pam
now? $(x+2)(x+12) = 48 + x(x+10)$

68. Alice is four years younger than her sister, Ann. Three years ago, the
product of their ages was seventy-five years less than the product of their
present ages. How old are the two girls now? $(x-3)(x+1) = x(x+4)-75$

69. Johnny is six years older than his brother, Jason. Two years ago, the $(x+4)(x-2) = x(x+6)-20$
product of their ages was 20 less than the product of their present ages.
How old are the boys now?

70. Mr. Jackson decided to add ten yards to the length of his pasture fence and
five yards to the width. This will result in an increase of 225 sq yds in
fenced area. If the original length was 15 yd more than the width, find the
new dimensions of the fence.
$LW + 225 = (L+10)(W+5)$
$L = 15 + W$

Figure for Exercise 62

STATE YOUR UNDERSTANDING

71. What are conjugate binomials?

CHALLENGE EXERCISES

72. Is the statement $(x + y)^2 = x^2 + y^2$ ever true? If yes, give an example.

73. Give an example to show that $(x - y)^2 \neq x^2 - y^2$.

74. Multiply: $(x^n + 1)(x^n - 1)$. Assume n is a positive integer.

75. Multiply: $(2x^a - 3)(2x^a + 3)$

76. Multiply: $(a + b)(a^2 - ab + b^2)$

77. Multiply: $(a - b)(a^2 + ab + b^2)$

MAINTAIN YOUR SKILLS (SECTION 1.6)

Perform the indicated operations:

78. $(-8)^2(14) - 6(15)^2$

79. $14(23x - 10) - 12(5x - 11)$

80. $24p - 13q - (23q + 24p)5$

81. Evaluate $-5a - 3(2a - 3b)$ if $a = -12$ and $b = 15$.

Perform the indicated operations:

82. $15(3 + 14) - 7[11(18 - 23)] + 16$

83. $-18 + 14(-6) - [(18 - 14 \cdot 6) - 19(18 - 15)] - 12(-18 - 7)$

84. The total price of an article purchased on the monthly deferred-payment plan is $T = D + pm$, where T is the total price, D is the down payment, p is the payment per month, and m is the number of months to pay. Find the total price of a stereo system if a down payment of \$155.90 was made and there are eighteen monthly payments of \$29.95.

85. Find the total price of a set of collector's plates if the down payment is \$12.75 and there are fifteen monthly payments of \$18.75.

<table>
<tr><td>

2.8

▪

COMMON FACTOR AND FACTORING BY GROUPING

</td><td>

OBJECTIVES

1. Factor the GCF from a polynomial.

2. Factor a polynomial by grouping.

3. Solve related equations.

</td></tr>
</table>

To factor a polynomial means to write it as a product of two or more polynomials. To factor a polynomial is the inverse of multiplication. In this sense, factoring is similar to division, but unlike division, when factoring we write *both* the divisor and the quotient (factors).

$$x^2 - 5x + 6 = \underbrace{(x - 3)(x - 2)}_{\text{FACTORS}}$$

The factors are $(x - 3)$ and $(x - 2)$. The product of these factors is equal to the original polynomial.   ■

■ **PROCEDURE**

Common Monomial Factor

To factor a polynomial when each term contains a common factor, write the product of

1. the common factor and

2. the polynomial that remains when the common factor is "factored out," or "taken out," using the distributive property.

The following table shows the two-step process: (1) identify the common factor, and (2) rewrite using the distributive property.

Polynomial	Step 1	Factors (Step 2)
$ax + ay =$	$a\,x + a\,y =$	$a\,(x + y)$
$5x^2 + 10x =$	$5x \cdot x + 5x \cdot 2 =$	$5x\,(x + 2)$
$8a^2b + 6ab - 10b =$	$2b \cdot 4a^2 + 2b \cdot 3a + 2b \cdot (-5) =$	$2b\,(4a^2 + 3a - 5)$

EXAMPLE 1

Factor: $14ac - 7ad$

$14ac - 7ad = 7a(2c) - 7a(d)$

$= 7a(2c - d)$

The number 7 is a common factor, and the variable a is a common factor. We use $7a$ as the common monomial factor. The monomial, $7a$, is called the GCF (greatest common factor) or HCF (highest common factor). It would not be incorrect to use $-7a$ as the common factor and write $-7a(-2c + d)$. It is common practice, however, to use the positive factor.   ■

■ **CAUTION**

Example 1 is not completely factored if we write $7(2ac - ad)$ or $a(14c - 7d)$.

To determine when a polynomial is completely factored, we need two definitions.

■ DEFINITION

Polynomial over Integers

A *polynomial over the integers* is a polynomial that has only integers for coefficients of its terms.

■ DEFINITION

Prime Polynomial

A *prime polynomial* is a polynomial that has exactly two factors that are polynomials over the integers, the number 1 and the polynomial itself.

The polynomials $7x + 10$ and $3x - 4y$ are both prime polynomials.

Thus $2bc - bd$ is not a prime polynomial since the variable b is a factor, and the polynomial $14c - 7d$ is not prime since the number 7 is a factor.

EXAMPLE 2

Factor: $16m^2n + 36mn + 12mn^5$

$16m^2n + 36mn + 12mn^5$

$= 4mn(4m) + 4mn(9) + 4mn(3n^4)$

$= 4mn(4m + 9 + 3n^4)$

The greatest common factor is $4mn$. The trinomial in parentheses cannot be factored further. It is a prime polynomial. ■

A polynomial is completely factored when it is written as the product of a real number and polynomials that are prime polynomials.

The idea of a common monomial factor can be extended to common polynomial factors. Some polynomials with four or six terms can be factored by grouping (or associating) terms with a common factor.

EXAMPLE 3

Factor: $ax + bx + 5a + 5b$

$ax + bx + 5a + 5b = (ax + bx) + (5a + 5b)$

The first two terms have common factor x, and the last two terms have common factor 5. We group the terms by writing them in parentheses.

$= x(a + b) + 5(a + b)$

Factor each group.

$$= (x + 5)(a + b)$$

The terms $x(a + b)$ and $5(a + b)$ have common factor $(a + b)$ in the same way that xP and $5P$ have the common factor P, and it is factored in the same way. ∎

■ CAUTION

The polynomial $x(a + b) + 5(a + b)$ is *not* in factored form since the expression is the sum of polynomials, not the product of polynomials.

Example 4 suggests an alternative procedure for factoring by grouping.

EXAMPLE 4

Factor: $x^2 + 6x + xy + 6y - 2x - 12$

$$x^2 + 6x + xy + 6y - 2x - 12$$

$= (x^2 + 6x) + (xy + 6y) + (-2x - 12)$ Group the terms in pairs.

$= x(x + 6) + y(x + 6) - 2(x + 6)$ Factor each group.

$= xP + yP - 2P$ Substitute $P = (x + 6)$.

$= (x + y - 2)P$ Factor, using the distributive property.

$= (x + y - 2)(x + 6)$ Substitute $P = (x + 6)$.

or

$(x + 6)(x + y - 2)$ Either of these two forms is correct. ∎

■ PROCEDURE

Grouping

To factor a polynomial with an even number of terms by grouping:

1. Group the terms of the polynomial.

2. Factor out the greatest common factor (GCF) from each group using the distributive property.

3. Factor out the common polynomial factor from each term using the distributive property.

Polynomial	Step 1	Step 2	Factors (Step 3)
$ax + ay + bx + by$	$(ax + ay) + (bx + by)$	$a(x + y) + b(x + y)$	$(a + b)(x + y)$
$6x^2 + 4x + 15x + 10$	$(6x^2 + 4x) + (15x + 10)$	$2x(3x + 2) + 5(3x + 2)$	$(2x + 5)(3x + 2)$
$5x^2 + 15x - x - 3$	$(5x^2 + 15x) + (-x - 3)$	$5x(x + 3) - 1(x + 3)$	$(5x - 1)(x + 3)$
$18y^2 - 24y - 21y + 28$	$(18y^2 - 24y) + (-21y + 28)$	$6y(3y - 4) - 7(3y - 4)$	$(6y - 7)(3y - 4)$

EXAMPLE 5

Factor: $4ax + 3by + 4ay + 3bx$

$4ax + 3by + 4ay + 3bx$ — The four terms have no common factor other than 1. There is no common factor in the first pair or the second pair of terms. There is a common factor of $4a$ in the first and third terms.

$= (4ax + 4ay) + (3bx + 3by)$ — We use the associative and commutative properties of addition to group the terms.

$= 4a(x + y) + 3b(x + y)$ — Factor each group.

$= (4a + 3b)(x + y)$ — Now $(x + y)$ is a common factor.   ∎

EXAMPLE 6

Factor: $x^2 + 6b - 7y + 11$

Prime polynomial — No two terms have a common factor.   ∎

A major application of factoring is in the solution of equations containing polynomials. The zero-product property is used after factoring.

■ PROPERTY

Zero-product Property

If $A \cdot B = 0$ then $A = 0$ or $B = 0$.

EXAMPLE 7

Solve: $(x + 3)(x - 2) = 0$

$x + 3 = 0$ or $x - 2 = 0$ — Zero-product property.

$x = -3$ or $x = 2$

The solution set is $\{-3, 2\}$. — The check is left for the student.   ∎

The zero-product property states that either factor may be zero, $A = 0$ *or* $B = 0$. The word, *or,* is used the same way in mathematics and on computers. It signifies that either statement may be true or both may be true. The solution set of the compound statement $A = 0$ or $B = 0$ is the union of the solutions sets of each of the two parts.

EXAMPLE 8

Solve: $11x^2 - 44x = 0$

$11x(x - 4) = 0$	Factor the left side.
$11x = 0$ or $x - 4 = 0$	Set each factor equal to 0 using the zero-product property.
$x = 0$ or $x = 4$	Solve for x.
The solution set is $\{0, 4\}$.	The check is left for the student. ■

■ CAUTION

Zero is one of the solutions. Be sure to include *all* solutions in the solution set.

EXAMPLE 9

Solve: $12x^2 - 9x + 6 = 8x$

$12x^2 - 9x - 8x + 6 = 0$	Since the original equation does not have a zero on the right side, we cannot use the zero-product property until we subtract $8x$ from both sides.
$(12x^2 - 9x) + (-8x + 6) = 0$	Group the terms on the left. If we do not combine the like terms, we can factor by grouping. In Section 2.9, we factor *after* the like terms are combined.
$3x(4x - 3) - 2(4x - 3) = 0$	For the pair $(-8x + 6)$, we choose $-2(4x - 3)$ rather than $2(-4x + 3)$ so that the binomial factors will be the same.
$(4x - 3)(3x - 2) = 0$	Factor the left side.
$4x - 3 = 0$ or $3x - 2 = 0$	Zero-product property.
$x = \dfrac{3}{4}$ or $x = \dfrac{2}{3}$	Solve for x.
The solution set is $\left\{\dfrac{2}{3}, \dfrac{3}{4}\right\}$.	The check is left for the student. ■

In Example 10 a formula from an application in business can be solved by factoring.

EXAMPLE 10

The average cost per unit of manufacturing q units of furniture is given by $C = \frac{1}{10}q^2 - 3q + 150$. How many units must be manufactured to have an average cost of $150?

Formula:

$$C = \frac{1}{10}q^2 - 3q + 150$$

Substitute:

$$150 = \frac{1}{10}q^2 - 3q + 150 \qquad \text{Substitute } C = 150 \text{ into the formula.}$$

Solve:

$$0 = \frac{1}{10}q^2 - 3q \qquad \text{We need zero on one side or the other so that we can use the zero-product property.}$$

$$0 = q\left(\frac{1}{10}q - 3\right) \qquad \text{Factor the right side.}$$

$$q = 0 \quad \text{or} \quad \frac{1}{10}q - 3 = 0$$

$$q = 0 \quad \text{or} \qquad q = 30$$

Answer:

The number of units manufactured is 30.

For this problem, the solution $q = 0$ is rejected. We cannot discuss average cost per unit if no units are manufactured. ■

EXERCISE 2.8

A

Factor:

1. $12m - 12n$

2. $17y - 17z$

3. $16a - 32$

4. $14x - 42$

5. $6ab - 12ac + 18ad$

6. $7y^4 - 21y^2z + 14y^3z^2$

7. $cy + dy + 4c + 4d$

8. $mn - 2n + 4m - 8$

9. $3xy + 6y + xz + 2z$

10. $8ab + 4b + 6a + 3$

11. $6xy + 8y + 9xz + 12z$

12. $8ax - 20a + 10bx - 25b$

13. $ax + bx + cx + ay + by + cy$

14. $tx + ty + tz - 7x - 7y - 7z$

Solve:

15. $(x - 5)(x - 9) = 0$

16. $(x - 2)(x - 8) = 0$

17. $(2x - 1)(4x - 1) = 0$

18. $(3x + 1)(3x + 2) = 0$

19. $(3x - 4)(4x + 3) = 0$

20. $(5x - 2)(x + 5) = 0$

B

Factor:

21. $18x^2y^2 - 30xy^2$

22. $12a^2b^2 - 14a^2b$

23. $2a^2 + 2ab - 14a$

24. $3bc - 3c^2 + 6c$

25. $20x^2 - 15x + 8xy - 6y$

26. $18xy - 14x + 27y - 21$

27. $5x^2 + 10x + 6x + 12$

28. $6a^2 - 10a + 21a - 35$

29. $2x^2y - 6x^2 + y - 3$

30. $4wx^2 + 5w - 4x^2 - 5$

31. $3x + 4y + 2w + 7$

32. $8a - 4y - xy + 4$

Solve:

33. $3x^2 + 18x = 0$

34. $9x^2 + 45x = 0$

35. $20x^2 + 25x + 12x + 15 = 0$

36. $6x^2 - 3x - 10x + 5 = 0$

37. $14x^2 - 35x + 6x - 15 = 0$

38. $15x^2 - 25x - 12x + 20 = 0$

39. $4x^2 + 28x - 7x - 49 = 0$

40. $5x^2 + 45x - 9x - 81 = 0$

C

Factor:

41. $6x^3y^2z - 9x^2y^3z - 24x^2y^2z^2$

42. $15a^2b^2c^3 - 10ab^3c^3 + 30ab^2c^3$

43. $36x^7y^6 - 84x^5y^9 + 72x^5y^6$

44. $75a^6b^6 + 90a^5b^7 - 15a^4b^8$

45. $42x^2 + 35xy - 18xy - 15y^2$

46. $6a^2 + 15ab - 14ab - 35b^2$

47. $5x^3 + 25x^2 + 10x + 8yx^2 + 40yx + 16y$

48. $6xy^2 + 15xy + 3x - 4y^3 - 10y^2 - 2y$

49. $2a^2x + 2bx + 2cx - 3a^2 - 3b - 3c$

50. $16wy^2 - 8wy + 16w + 14y^2 - 7y + 14$

Solve:

51. $x^2 + 3x - 5x - 15 = 0$

52. $x^2 + 7x - 9x - 63 = 0$

53. $x^2 - 5x - 11x + 55 = 0$

54. $x^2 - x - 13x + 13 = 0$

55. $x^2 - ax = 0$

56. $x^2 + cx = 0$

57. $x^2 - 2ax + 3ax - 6a^2 = 0$

58. $x^2 - 3cx + 5ax - 15ac = 0$

59. $x^2 + 5x - 2bx - 10b = 0$

60. $x^2 + 9cx - 11x - 99c = 0$

D

61. If the average profit (p) per set of dishes sold is given by
$p = \dfrac{1}{4}q^2 - 36q + 42$, where q is the number of sets sold, how many sets
must be sold to have an average profit of \$42?

62. If the average profit (p) per car sold is given by $p = \dfrac{1}{5}c^2 - 40c + 410$,
where c is the number of cars sold, how many cars must be sold to have an
average profit of \$410?

63. The equation $d = 16t^2$ represents the distance, from the top of a cliff, that a
dropped object has fallen. How far does the object fall in 4 seconds? If the
cliff is 400 feet high, how long does it take the object to fall to the bottom
of the cliff?

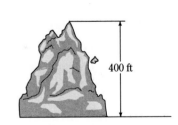

Figure for Exercise 63

64. For Exercise 63, how far does the object fall in 2 seconds? If the cliff is
784 feet high, how long does it take the object to fall to the bottom?

65. Given the equation $s = 16t - 4t^2$, where s is the height of a falling object
and t is the time the object falls, how long does it take the object to hit the
ground? ($s = 0$)

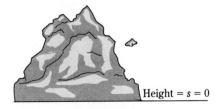

Figure for Exercise 65

66. Given the equation $s = 36t - 16t^2$, where s is the height of a falling object
and t is the time it falls, how long does it take the object to hit the ground?
($s = 0$)

67. Given the equation $s = 48t - 16t^2$, where s is the height of a falling object
and t is the time it falls, how long does it take the object to hit the ground?
($s = 0$)

68. The revenue gained from selling x pairs of shoes is given by $R = 100x^2 - 500x + 600$. Find the number of pairs of shoes the company needs to sell to bring in $600.

69. The cost of manufacturing x toy soldiers is $C = 2x^2 - 1600x + 5000$. How many soldiers can be produced at a cost of $5000?

70. The profit obtained from selling x books is given by $P = 3x^2 - 60x + 150$. How many books must be sold to reach a profit of $150?

STATE YOUR UNDERSTANDING

71. What is the procedure for factoring by grouping?

CHALLENGE EXERCISES

Factor, assume all variables represent positive integers:

72. $x^{3n} + x^{2n} + x^n$

73. $x^{3n} - x^{2n}$

74. $x^{n+2} + x^n$

75. $x^{2n+3} + x^{n+3}$

76. Use factoring by grouping to show that
$$xw - xz + yz - yw = (x - y)(w - z) \text{ or } (y - x)(z - w)$$

MAINTAIN YOUR SKILLS (SECTIONS 1.6, 1.7, 1.10)

Perform the indicated operations:

77. $6 \cdot 3 + 4(8 \cdot 6 - 2) + 3 \cdot 2$

78. $-6 + 4 \cdot 3^2 - [6(-5) + 3(8 - 5)]$

Solve:

79. $(x - 3) - (4x - 5) - (3 - 2x) = 7$

80. $(5a - 3) + (6 - 2a) = 8a + 10$

81. $x - 2(5x + 3) = 4 - 7x$

82. $a - 2(3x - a) = 3(2x - 3a)$

83. Arnold is twice as old as Betty, and Cindy is five years younger than Betty. In three years, Arnold will be three times as old as Cindy. How old is Betty now?

84. The cost of a round-trip airfare from Portland, Oregon, to New York City is $10 less than twice the cost of a round-trip airfare from Portland, Oregon, to Honolulu, Hawaii. If both fares can be purchased for a total of $770, what does each fare cost?

2.9

FACTORING TRINOMIALS

1. Factor trinomials of the form $x^2 + bx + c$.

2. Factor trinomials of the form $ax^2 + bx + c$, $a \neq 0$.

3. Solve related equations.

To factor the trinomial $x^2 + 12x + 27$ means to find two binomials whose product is $x^2 + 12x + 27$. If we believe the factors to be $(x + 3)(x + 9)$, we can check by using FOIL to multiply.

$$(x + 3)(x + 9) = \overset{\text{F}}{x^2} + \overset{\text{O}}{9x} + \overset{\text{I}}{3x} + \overset{\text{L}}{27}$$

$$= x^2 + (9 + 3)x + 27$$

$$= x^2 + 12x + 27$$

The example above shows that to factor $x^2 + 12x + 27$, we need to reverse the FOIL process by finding two integers m and n whose sum is 12 and whose product is 27. Since $3 + 9 = 12$ and $(3)(9) = 27$, the factors are $(x + 3)(x + 9)$.

EXAMPLE 1

Factor: $x^2 + 13x + 42$

$(x + m)(x + n) = x^2 + 13x + 42$ We look for two integers, m and n, such that $mn = 42$ and $m + n = 13$.

$mn = 42$	$m + n = 13$
$1 \cdot 42$	43
$2 \cdot 21$	23
$3 \cdot 14$	17
$6 \cdot 7$	13 ←

The pairs of factors that give a positive product and a positive sum are listed.

The numbers we want are 6 and 7.

$x^2 + 13x + 42 = (x + 6)(x + 7)$ ■

The list shown in Example 1 shows all possible combinations of m and n that have a product of 42. When there are a large number of such combinations, it may be quicker to try some before filling out the entire list.

EXAMPLE 2

Factor: $x^2 + 26x + 144$

$(x + 12)(x + 12)$ Without writing down all factors starting with $1 \cdot 144$, we begin by trying $m = 12$ and $n = 12$. Note that $12 = \sqrt{144}$.

$$(x + 12)(x + 12) = x^2 + 24x + 144$$

These factors do not have the product $x^2 + 26x + 144$, but they are close. This is an example of the "trial and error" method. We need to look for other factors, but we need not look blindly. It seems reasonable that since $24x$ is close to the desired middle term, $26x$, the factors should be close to 12. We quickly find that 10 and 11 are not factors of 144, so we try 9 and 16.

$$(x + 9)(x + 16) = x^2 + 25x + 144$$

These factors are closer to $24x$.

$$(x + 8)(x + 18) = x^2 + 26x + 144$$

These are the correct factors.   ■

Students must decide for themselves whether to begin by listing all the factors or by trying the first combinations that come to mind. It is not uncommon to use both methods, deciding for each problem which method is quicker.

■ **PROCEDURE**

Factoring
$x^2 + bx + c$

To factor a polynomial of the form $x^2 + bx + c$:

1. Find integers m and n such that $mn = c$ and $m + n = b$.

2. The factors are $(x + m)(x + n)$. If no such integers exist, we say that the polynomial is prime.

EXAMPLE 3

Factor: $x^2 - 5x - 100$

Prime polynomial

There are no integers whose product is -100 and whose sum is -5.   ■

Trinomials where the coefficient of x^2 (the lead coefficient) is not 1 can be factored in a similar manner. In this case, in $ax^2 + bx + c$, the product of m and n is ac. We can see this by looking at some products.

$$(2x - 1)(3x - 7) = 6x^2 - 3x - 14x + 7,$$

where $a = 6$, $c = 7$, $m = -3$, and $n = -14$ so that $mn = (-3)(-14) = 42$ and $ac = 6(7) = 42$.

$$(3x - 4)(5x + 3) = 15x^2 - 20x + 9x - 12,$$

where $mn = -20(9) = -180$ and $ac = 15(-12) = -180$.

So, to factor $6x^2 - 5x - 25$, find m and n such that $mn = 6(-25)$ and $m + n = -5$. Then rewrite the trinomial in the form $6x^2 + mx + nx - 25$ and factor by grouping.

EXAMPLE 4

Factor: $6x^2 - 5x - 25$

$mn = -150$	$m + n = -5$
$1 \cdot (-150)$	-149
$2 \cdot (-75)$	-73
$3 \cdot (-50)$	-47
$5 \cdot (-30)$	-25
$6 \cdot (-25)$	-19
$10 \cdot (-15)$	$-5 \leftarrow$

Since the sum is negative, choose m and n so that the one with the larger absolute value is negative. Hence, we need not write down values such as $-2(75)$ or $-6(25)$.

$$6x^2 - 5x - 25 = 6x^2 + 10x - 15x - 25$$
$$= (6x^2 + 10x) + (-15x - 25)$$
$$= 2x(3x + 5) - 5(3x + 5)$$
$$= (2x - 5)(3x + 5)$$

Rewrite the middle term as a sum and factor by grouping.

■ PROCEDURE

Factoring $ax^2 + bx + c$

To factor a polynomial of the form $ax^2 + bx + c$, $a \neq 0$:

1. Find integers m and n such that $mn = ac$ and $m + n = b$.
2. Write the polynomial in the form $ax^2 + mx + nx + c$.
3. Factor by grouping.

EXAMPLE 5

Factor: $20x^2 + 31x - 7$

$mn = 20(-7) = -140$	$m + n = 31$
$-1 \cdot 140$	139
$-2 \cdot 70$	68
$-4 \cdot 35$	$31 \leftarrow$

Since the sum is negative, choose m and n so that the one with the larger absolute value is positive.

$$20x^2 - 4x + 35x - 7 = 4x(5x - 1) + 7(5x - 1)$$
$$= (4x + 7)(5x - 1)$$

Factor by grouping.

Trial and error can also be employed to find the factors of a polynomial whose leading coefficient is not 1. To factor $2x^2 + x - 21$, we make these observations:

1. Concentrate on the first and last terms.

2. $2x^2$ is $2x \cdot x$

3. $21 = 3 \cdot 7$ or $1 \cdot 21$

4. The last term, -21, is negative, so try -7 and 3.

$(2x - 7)(x + 3) = 2x^2 - x - 21$

The middle term is positive not negative, so try reversing the signs.

$(2x + 7)(x - 3) = 2x^2 + x - 21$

These are the correct factors.

EXAMPLE 6

Factor: $18y^2 - 67y + 14$

$mn = 18(14) = 252$	$m + n = -67$	Some possible combinations
$-1(-252)$	-253	are listed. With such large
$-2(-126)$	-128	numbers, trial and error is
$-3(-84)$	-87	time consuming.
$-4(-63)$	-67	Use this pair.

$18y^2 - 67y + 14 = (18y^2 - 4y) + (-63y + 14)$ Rewrite the middle term
$67y = 4y + 63y$, and factor

$\qquad\qquad\qquad = 2y(9y - 2) - 7(9y - 2)$ by grouping.

$\qquad\qquad\qquad = (2y - 7)(9y - 2)$ The check by multiplication
is left for the student. ∎

With one side of an equation factored and the other side zero, we can use the zero-product property to solve the equation.

EXAMPLE 7

Solve: $x^2 + 9x = 22$

$x^2 + 9x - 22 = 0$

$(x - 2)(x + 11) = 0$ Factor the left side.

$x - 2 = 0 \quad \text{or} \quad x + 11 = 0$ Zero-product property.

$\qquad x = 2 \quad \text{or} \qquad x = -11$ The check is left for the
student.

The solution set is $\{-11, 2\}$. ∎

Many applications in the real world can be solved using the factoring of polynomials and the zero-product property.

EXAMPLE 8

A strip of metal 39 inches wide is to be bent into a trough with a rectangular cross section with open top. The area of the cross section is to be 175 square inches. Find the width and depth of the trough.

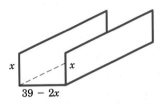

Simpler word form:

Area of cross section = Width of trough times depth of trough	The cross section is a rectangle. $A = \ell w$

Select a variable:

Let x represent the depth of the trough, then $39 - 2x$ represents the width of the trough.	The 39-inch width must be bent into a three-sided trough with two vertical sides and a bottom.

Translate to algebra:

$175 = x(39 - 2x)$	Substitute into the formula, $A = \ell w$.

Solve:

$175 = 39x - 2x^2$	
$2x^2 - 39x + 175 = 0$	Write the equation in the form $ax^2 + bx + c = 0$.
$(2x - 25)(x - 7) = 0$	Factor the left side.
$2x - 25 = 0 \quad$ or $\quad x - 7 = 0$ $\qquad x = 12.5 \quad$ or $\qquad x = 7$	The value of x gives us the depth of the trough. There are two possible solutions.

Now find the width.

$39 - 2x = 14 \quad$ or $\quad 39 - 2x = 25$	Width of trough.

Check:

First solution. Does $12.5(14) = 175$?	Yes.
Second solution. Does $7(25) = 175$?	Yes.

Answer:

The metal can be bent in two ways, both of which have a cross section of 175 square inches. Either we have a depth of 7 inches and a width of 25 inches or a depth of 12.5 inches and a width of 14 inches.   ■

EXERCISE   2.9

A

Factor:

1. $x^2 + 12x + 35$

2. $x^2 + 11x + 30$

3. $x^2 - 10x + 21$

4. $y^2 - 12y + 32$

5. $a^2 - 7a - 18$

6. $b^2 - 7b - 44$

7. $a^2 + 4a - 21$

8. $c^2 + 6c - 55$

9. $a^2 - 4a - 10$

10. $c^2 + 4c - 18$

11. $5y^2 + 16y + 3$

12. $4a^2 + 3a - 1$

Solve:

13. $x^2 - 3x + 2 = 0$

14. $a^2 - 6a + 8 = 0$

15. $x^2 - 2x - 8 = 0$

16. $x^2 - 11x + 30 = 0$

17. $x^2 + 16x + 63 = 0$

18. $x^2 - 16x + 63 = 0$

19. $x^2 + 2x - 63 = 0$

20. $x^2 - 2x - 63 = 0$

B

Factor:

21. $x^2 + x - 110$

22. $x^2 - 9x - 90$

23. $a^2 - 20a + 96$

24. $b^2 + 20b + 84$

25. $2a^2 + 9a - 18$

26. $3b^2 + 8b - 35$

27. $5x^2 - 13x - 6$

28. $7x^2 - 32x - 15$

29. $2t^2 - 9t - 35$

30. $2t^2 - 3t - 35$

Solve:

31. $x^2 - 6x - 72 = 0$

32. $y^2 - 8y - 105 = 0$

33. $9x^2 - 5x - 4 = 0$

34. $7y^2 - 3y - 10 = 0$

35. $3b^2 - b - 2 = 0$

36. $2x^2 - 5x + 3 = 0$

37. $12x^2 + 7x - 12 = 0$

38. $24x^2 + 14x - 3 = 0$

39. $2x^2 - 23x + 11 = 0$

40. $2x^2 - 27x - 14 = 0$

C

Factor:

41. $x^2 - 6x - 216$

42. $y^2 - 6y - 280$

43. $x^2 - 28x + 195$

44. $a^2 + 24a + 128$

45. $6y^2 + 38y + 55$

46. $8a^2 - 35a + 35$

47. $15b^2 - 17b - 18$

48. $18c^2 - 3c - 55$

49. $(x + 2y)^2 - (x + 2y) - 42$

50. $(p + q)^2 + 10(p + q) + 21$

51. $(2a + 1)^2 + 3(2a + 1) - 18$

52. $(w + 4)^2 + 20(w + 4) + 64$

53. $(6t - 1)^2 - 17(6t - 1) - 38$

54. $(7w + 5)^2 + 31(7w + 5) - 66$

Solve:

55. $x^2 - 23x + 112 = 0$

56. $y^2 + 37y + 330 = 0$

57. $10x^2 - 31x - 63 = 0$

58. $15y^2 - 32y - 60 = 0$

59. $20a^2 + 3a - 56 = 0$

60. $20b^2 + 32b - 45 = 0$

D

61. The perimeter of a framed mirror is 36 inches. If the area of the framed mirror is 72 square inches, find the dimensions of the frame.

62. The perimeter of a rectangular building lot is 250 ft. If the area of the lot is 3850 sq ft, find the dimensions of the lot.

63. What are the dimensions of a building lot that contains 5200 sq ft and took 290 ft of fence to enclose it?

64. A strip of metal 26 inches wide is to be bent into a trough of rectangular cross section (the top is open). The area of the cross section is 84 sq in. Find the width and depth of the trough. $(A = \ell w)$

A = 72 in²

P = 36 in

Figure for Exercise 61

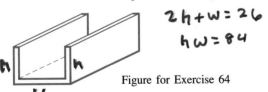

Figure for Exercise 64

65. A strip of metal 30 inches wide is to be bent into a rectangular conduit that is enclosed on all four sides. If the area of the cross section is 50 sq in., what are the dimensions of the cross section?

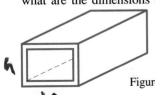

Figure for Exercise 65

STATE YOUR UNDERSTANDING

66. Explain the procedure for factoring a trinomial of the form $x^2 + bx + c$.

CHALLENGE EXERCISES

Factor the following, assume all variables represent positive integers:

67. $x^{2n} - 3x^n + 2$

68. $x^{2n} + 4x^n + 3$

69. $x^4 + 8x^2 + 12$

70. $x^4 - 5x^2 - 6$

MAINTAIN YOUR SKILLS (SECTIONS 1.7, 1.8)

Solve for x:

71. $7x - 6x = 2x + 4$

72. $7x - 6x = 2x + 4x$

73. $6x - 7x = 2x - 3x$

74. $7x - 6x = 3x - 2x + 4$

75. $15x - [12x - (x + 14) - 27] + 43 = 0$

76. $2x - [7x - (18 - 2x) + x] = 12 - (x + 33)$

77. $t = \dfrac{3}{5}x - 9$

78. $3G = t(x + T)$

2.10

FACTORING SPECIAL CASES

OBJECTIVES

1. Factor the difference of two squares.

2. Factor a perfect square trinomial.

3. Factor the sum or difference of two cubes.

4. Solve related equations.

There are three special cases of factoring that are useful to know. Each of these three can be handled by the methods you have already learned, that is, by using the distributive property or by trial and error. However, these patterns occur so often in future work that it is more expedient to treat them separately.

The first is the difference of two squares. We learned in Section 2.7 that the product of conjugate binomials is itself a binomial, which is the difference of two squares. $(x + y)(x - y) = x^2 - y^2$, so using the symmetric property of equality, we have:

■ **FORMULA**

Difference of Two Squares

$$x^2 - y^2 = (x + y)(x - y)$$

EXAMPLE 1

Factor: $x^2 - 16$

$x^2 - 16 = x^2 - 4^2$

The binomial $x^2 - 16$ is the difference of the squares of x and 4.

$\quad\quad = (x + 4)(x - 4)$

The factors are conjugate binomials. ■

EXAMPLE 2

Factor: $9a^2b^2 - 169$

$9a^2b^2 - 169 = (3ab)^2 - (13)^2$

$\quad\quad = (3ab + 13)(3ab - 13)$

This binomial is the difference of the squares of $3ab$ and 13. ■

The second special case is the perfect square trinomial. A perfect square trinomial is formed by squaring a binomial.

$(a + b)^2 = a^2 + 2ab + b^2$ and $(a - b)^2 = a^2 - 2ab + b^2$, so again, with the symmetric property of equality, we have:

■ **FORMULA**

Perfect Square Trinomial

$$a^2 + 2ab + b^2 = (a + b)^2$$
$$a^2 - 2ab + b^2 = (a - b)^2$$

EXAMPLE 3

Factor: $x^2 + 14x + 49$

$x^2 + 14x + 49 = x^2 + 2(7x) + 7^2$

$\quad\quad = (x + 7)^2$

When the first term and last term of a trinomial are squares, in this case x^2 and 49, we look at the middle term to determine whether it is twice the product of the squared expressions, in this case $2 \cdot x \cdot 7$. If so, it is a perfect square trinomial. ■

EXAMPLE 4

Factor: $16w^2 - 40w + 25$

$$16w^2 - 40w + 25 = (4w)^2 + 2(4w)(-5) + (-5)^2$$

$$= (4w - 5)^2$$

Since the middle term, $-40w$, is twice the product of the squared terms $4w$ and -5, the trinomial is a perfect square trinomial. ■

EXAMPLE 5

Factor: $(a^2 - 6a + 9) - b^2$

$$a^2 - 6a + 9 = (a - 3)^2$$

The expression in parentheses is a perfect square trinomial.

$$(a - 3)^2 - b^2$$

This expression is the difference of the squares of $a - 3$ and b.

$$(a - 3 + b)(a - 3 - b)$$

If you do not easily see the pattern for the difference of two squares, it may help to make a substitution. Let $(a - 3) = P$. Then write $P^2 - b^2 = (P + b)(P - b)$, and replace P by $a - 3$. ■

The third special case is the sum or difference of two cubes. The patterns for these are:

■ **FORMULA**

Sum and Difference of Two Cubes

$$a^3 + b^3 = (a + b)(a^2 - ab + b^2)$$
$$a^3 - b^3 = (a - b)(a^2 + ab + b^2)$$

These formulas can be checked by multiplying the polynomials on the right (see Exercises 71 and 72). The following observations may help you remember these formulas:

1. The sum or difference of two cubes is the product of a binomial and a trinomial.

2. The binomial is the sum or difference of the values that are cubed.

3. The trinomial is the square of the first term of the binomial, plus negative one times the product of the two terms of the binomial, plus the square of the second term of the binomial.

EXAMPLE 6

Factor: $a^3 + 64$

$= a^3 + 4^3$	The sum of the cubes of a and 4.
$= (a + 4)(a^2 - 4a + 16)$	The factors are a binomial, $a + 4$, and a trinomial, which is the square of a plus $-1(a)(4)$ plus the square of 4. ■

EXAMPLE 7

Factor: $8c^3 - 125$

$= (2c)^3 - 5^3 = (2c)^3 + (-5)^3$	The difference of the cubes of $2c$ and 5.
$= (2c - 5)(4c^2 + 10c + 25)$	The factors are a binomial, $2c - 5$, and a trinomial, which is the square of $2c$ plus $-1(2c)(-5)$ plus the square of 5. ■

■ **CAUTION**

Factoring
$x^2 + y^2$

> The sum of two squares that have no common factor is a prime polynomial.

The sum of two squares is not one of the special cases of this section. The polynomial $x^2 + 100$ is the sum of two squares. In attempting to factor $x^2 + 100$ by trial and error, we have

$(x + 10)(x + 10) = x^2 + 20x + 100$

$(x - 10)(x - 10) = x^2 - 20x + 100$

$(x + 10)(x - 10) = x^2 - 100$ The polynomial $x^2 + 100$ is prime.

We see that none of the reasonable factors will produce the sum of two squares. Using only real numbers, no factors can be found. At a later time, you will learn how to factor this polynomial.

With one side of an equation factored and the other side zero, we can use the zero-product property to solve the equation.

EXAMPLE 8

Solve: $x^2 - 144 = 0$

$(x - 12)(x + 12) = 0$	Factor the left side.
$x - 12 = 0$ or $x + 12 = 0$	Zero-product property.
$x = 12$ or $x = -12$	The check is left for the student.

The solution set is $\{-12, 12\}$.   ■

EXAMPLE 9

A variable electrical current is given by $i = t^2 - 10t + 28$. If t is the number of seconds, in how many seconds will the current (i) equal 3 amperes?

Formula

$i = t^2 - 10t + 28$

Substitute:

$3 = t^2 - 10t + 28$	Substitute $i = 3$.

Solve:

$t^2 - 10t + 25 = 0$	Rewrite the equation with 0 on one side.
$(t - 5)^2 = 0$	The left side is a perfect square trinomial.
$t - 5 = 0$ or $t - 5 = 0$	Zero-product property.
$t = 5$ or $t = 5$	This is sometimes called a *double root*.

Answer:

The current will be 3 amperes in 5 seconds.   ■

EXERCISE 2.10

A

Factor if possible:

1. $a^2 - 1$	**2.** $b^2 - 144$	**3.** $x^2 - 9$
4. $m^2 - n^2$	**5.** $y^2 + 16$	**6.** $n^2 + 25$

7. $x^2 + 8x + 16$

8. $x^2 + 12x + 36$

9. $c^2 + 6c + 9$

10. $a^2 + 26a + 169$

11. $y^2 - 28y + 196$

12. $w^2 - 34w + 289$

13. $x^3 - 1$

14. $c^3 + 1$

15. $y^3 - 125$

16. $b^3 + 27$

Solve:

17. $x^2 - 49 = 0$

18. $y^2 - 100 = 0$

19. $w^2 - 256 = 0$

20. $t^2 - 400 = 0$

B

Factor if possible:

21. $16a^2 + 49$

22. $25c^2 + 36$

23. $9a^2b^2 - 64c^2$

24. $81x^2y^2 - 16z^2$

25. $16y^2 + 24y + 9$

26. $9y^2 - 30y + 25$

27. $4a^2 - 28a + 49$

28. $25c^2 + 20c + 4$

29. $a^3b^3 - c^3$

30. $x^3y^3 + z^3$

31. $8w^3 + 1$

32. $64t^3 - 1$

Solve:

33. $4x^2 - 25 = 0$

34. $25y^2 - 81 = 0$

35. $x^2 + 6x + 9 = 0$

36. $w^2 - 8w + 16 = 0$

37. $4x^2 - 36x = -81$

38. $9y^2 + 48y = -64$

39. $w^2 + 32w = -256$

40. $t^2 + 400 = 40t$

C

Factor if possible:

41. $81x^4 - 16y^4$

42. $625a^4 - b^4$

43. $(a + b)^2 - c^2$

44. $a^2 - (b + c)^2$

45. $36a^2b^2 - 132abc + 121c^2$

46. $81x^2y^2 + 234xyz + 169z^2$

47. $100y^2 - 260y + 169$

48. $225a^2 + 330a + 121$

49. $64a^3 - 27b^3$

50. $125x^3 + 8y^3$

51. $1000c^3d^3 - 343$

52. $216m^3n^3 + 1$

53. $(3x + y)^2 - (2x + 5)^2$

54. $(2w - 3)^2 - (x + 7y)^2$

55. $(a - b)^2 - (x - y)^2$

56. $(a - 7)^2 - (x - 8)^2$

57. $(x^2 - 10x + 25) - y^2$

58. $(w^2 - 12w + 36) - 9x^2$

59. $144 - (4t^2 + 4st + s^2)$

60. $100 - (x^2 + 18xy + 81y^2)$

Solve:

61. $4x^2 - 169 = 0$

62. $25x^2 - 144 = 0$

63. $4x^2 - 28x + 49 = 0$

64. $25x^2 - 20x + 4 = 0$

D

65. A variable electrical current is given by $i = t^2 - 16t + 75$. If t is in seconds, in how many seconds will the current (i) equal 11 amperes?

66. A variable electrical current is given by $i = t^2 - 14t + 75$. If t is in seconds, in how many seconds will the current (i) equal 26 amperes?

67. The square of Elsa's age in 12 years minus 24 times her age in 24 years is nine. How old is Elsa today?

68. If 20 times Chang's age in 25 years is subtracted from the square of his age ten years from now, the result is zero. How old is Chang today?

69. During a 12-hour shift at the Clear Picture television assembly plant, the number of sets assembled in a given hour is shown by $N = 140 + 18t - t^2$, $t > 1$, where N is the number of sets and t is the hour of the shift. During what hour are 221 sets assembled?

70. The number of loaves of bread the Good-For-The-Tummy Bakery bakes in a given hour of a 24-hour day is given by $N = 500 + 24t - t^2$, $t > 1$, where N is the number of loaves and t is the hour of the day. During which hour are 644 loaves baked?

71. Check the formula $a^3 + b^3 = (a + b)(a^2 - ab + b^2)$ by multiplying the polynomials on the right.

72. Check the formula $a^3 - b^3 = (a - b)(a^2 + ab + b^2)$ by multiplying the polynomials on the right.

STATE YOUR UNDERSTANDING

73. How do we recognize when a polynomial is a perfect square trinomial?

CHALLENGE EXERCISES

Factor, assume that all variables represent positive integers:

74. $x^{2n} - 1$

75. $x^{3n} + 1$

76. $x^{4n} - 1$

77. $x^{3n} - 1$

78. Use factoring by grouping to show that

$$a^2 - b^2 + 2bc - c^2 = (a - b + c)(a + b - c)$$

MAINTAIN YOUR SKILLS (SECTIONS 1.4, 1.7, 1.9, 1.10)

Perform the indicated operations:

79. $6 - (5a + 2) - (3a - 6) + 8$ **80.** $6 - [5a + 2(3a - 6) + 8]$

Solve:

81. $2y - [y - (5 - y) + 6] = -8$ **82.** $3y - 18 < 7y + 14$

83. $18 - 3y < 14 + 7y$ **84.** $3x - 2 - (2x + 3) \geq 11$

85. If a new pickup is advertised as having a load capacity of 3000 pounds, what is the maximum number of 160-lb crates it can carry?

86. A flatbed trailer has a load capacity of 3 tons. What is the maximum number of 225-lb crates it can carry if it is already loaded with seventeen 175-lb crates?

2.11

■

FACTORING: A REVIEW

OBJECTIVES

1. Review all methods of factoring a polynomial.

2. Solve related equations.

In this section, we use all the factoring techniques that we have studied. Recall that a polynomial is said to be completely factored when it is written as the product of a real number and prime polynomials. The polynomial $ab^4 - 16a$ has four factors when completely factored.

EXAMPLE 1

Factor: $ab^4 - 16a$

$= a(b^4 - 16)$ Monomial factor.

$= a(b^2 + 4)(b^2 - 4)$ Difference of two squares.

$= a(b^2 + 4)(b + 2)(b - 2)$ Difference of two squares.   ■

■ PROCEDURE

To factor a polynomial completely:

1. Factor out the greatest common factor (GCF), if any.

2. If the polynomial or any factor is a binomial, check to see if it is the difference of two squares or if it is the sum or difference of two cubes.

3. If the polynomial or any factor is a trinomial, factor by the appropriate method, if possible.

4. If the polynomial contains four or more terms, try factoring by grouping.

If the polynomial cannot be factored by the preceding methods, it is a prime polynomial. Other methods of factoring are studied in more advanced mathematics courses.

EXAMPLE 2

Factor: $4x^3y + 6x^2y + 8xy$

■ CAUTION

Always look for the greatest common factor (GCF) first.

$= 2xy(2x^2 + 3x + 4)$

$= 2xy(2x^2 + 3x + 4)$

The GCF is $2xy$. We next attempt to factor the trinomial in parenthesis by trial and error. Since $(2x + 4)(x + 1)$, $(2x + 1)(x + 4)$, and $(2x + 2)(x + 2)$ are not factors, we conclude that the trinomial is prime.   ■

EXAMPLE 3

Factor: $x^2 + 12x - 45$

$= (x + 15)(x - 3)$

There is no common factor (GCF $= 1$). There are only three pairs of factors for 45: $1 \cdot 45$, $3 \cdot 15$, and $5 \cdot 9$. The product, -45, is negative, so one integer is positive and one is negative. By inspection, we can quickly determine the factors.   ■

EXAMPLE 4

Factor: $3ax + 3ay - 5bx - 5by$

$= (3ax + 3ay) + (-5bx - 5by)$ Since there are four terms, try grouping.

$= 3a(x + y) - 5b(x + y)$ The common factor is $(x + y)$.

$= (3a - 5b)(x + y)$ ∎

EXAMPLE 5

Factor: $(x + 2)^2 - 16$

$= (x + 2 + 4)(x + 2 - 4)$ Treat $(x + 2)$ as if it were a single variable, say P. We can think:

$$(x + 2)^2 - 16 = P^2 - 16$$
$$= (P + 4)(P - 4)$$
$$= (x + 2 + 4)(x + 2 - 4)$$

$= (x + 6)(x - 2)$ Simplify.   ∎

EXAMPLE 6

Factor: $56x^2 - 59x + 15$

$mn = 840$	$m + n = -59$
$-1(-840)$	-841
$-14(-60)$	-74
$-15(-56)$	-71
$-20(-42)$	-62
$-21(-40)$	-61
$-24(-35)$	-59

This trinomial takes more time to factor because the numbers are larger and have more factors. Use the mn chart. We show only negative factors of 840 since the sum is negative and the product is positive. The sum -841 is a long way from -59, so we skip to $(-14)(-60)$ instead of listing *all* the possibilities.

$= 56x^2 - 24x - 35x + 15$ Rewrite $-59x = -24x - 35x$.

$= (56x^2 - 24x) + (-35x + 15)$ Factor by grouping.

$= 8x(7x - 3) - 5(7x - 3)$

$= (8x - 5)(7x - 3)$ ∎

EXAMPLE 7

Factor: $100x^2 - 20x + 1$

$= (10x)^2 - 20x + 1^2$ Since the first and last terms are squares, check to see if the middle term is twice the product of the terms that are squared. It is.

$= (10x)^2 - 2(10x)(1) + 1^2$

$= (10x - 1)^2$ This is a perfect square trinomial.   ∎

EXAMPLE 8

Factor: $5x^2 - 60x + 135$

$= 5(x^2 - 12x + 27)$	The GCF is 5.
$= 5(x - 3)(x - 9)$	The trinomial can be factored by trial and error. ∎

EXAMPLE 9

Factor: $128a^3 - 250$

$= 2(64a^3 - 125)$	The numbers 128 and 250 are neither squares
$= 2[(4a)^3 - (5)^3]$	nor cubes, so it may seem hopeless at first. *But* they have a GCF of 2. The numbers 64 and 125 are cubes. The polynomial in brackets is the difference of two cubes, so we can factor further.
$= 2(4a - 5)(16a^2 + 20a + 25)$	Remember that the second factor of the difference of two cubes is the square of the first term, $4a$, plus negative one times the product of the two terms, $4a$ and -5, plus the square of the second term, -5.

EXAMPLE 10

Factor: $x^4 - 16$

$= (x^2 + 4)(x^2 - 4)$	Factor the difference of two squares. We are not finished yet because the second factor is *also* the difference of two squares. Remember that the sum of two squares $x^2 + 4$ is prime. ∎
$= (x^2 + 4)(x + 2)(x - 2)$	

Equations with factorable polynomials can be solved using the zero-product property.

EXAMPLE 11

Solve: $3x^2 + 16x + 3 = -2$

$3x^2 + 16x + 5 = 0$	Rewrite the equation with zero on one side.
$(3x + 1)(x + 5) = 0$	Factor the left side.
$3x + 1 = 0$ or $x + 5 = 0$	Zero-product property.
$x = -\dfrac{1}{3}$ or $x = -5$	Solve.

The solution set is $\left\{ -\dfrac{1}{3}, -5 \right\}$.   The check is left for the student. ∎

EXAMPLE 12

A lawn sprinkler waters a rectangular area of lawn that is 3 feet longer than it is wide. If the manufacturer says that by setting the dial on C it will cover 270 square feet, what are the dimensions of the watered area?

Formula:

$\ell w = A$

Since we know the area of the lawn that is sprinkled, we can use the formula for the area of a rectangle.

Select a variable:

Let x represent the width of the rectangle, then $x + 3$ represents the length.

Make a drawing.

$$
\begin{array}{c}
x + 3 \\
\boxed{ A = 270 } \\
\end{array}
$$

x $A = 270$

Substitute:

$(x + 3)x = 270$

Substitute x for w, $x + 3$ for ℓ, and 270 for a in the formula.

Solve:

$$x^2 + 3x = 270$$

$x^2 + 3x - 270 = 0$ Rewrite so that one side is zero.

$(x - 15)(x + 18) = 0$ Factor.

$x - 15 = 0$ or $x + 18 = 0$

$\phantom{x - 15 = 0 \text{ or } } x = 15$ or $x = -18$

The negative answer is rejected since rectangles do not have negative measure.

$w = x = 15$ and $\ell = x + 3 = 18$

Check:

$(18)(15) = 270$

Answer:

The dimension of the watered area is 15 ft by 18 ft.   ■

Exercise 2.11

A

Factor if possible:

1. $9c + 6b - 15$

2. $14abc + 7bc + 21c$

3. $x^2 - 8x$

4. $12y^2 - 4y$

5. $x^2 - 8x + 12$

6. $x^2 - 15x + 50$

7. $x^2 - 100$

8. $a^2 - 144$

9. $a^2 + 81$

10. $y^2 + 49$

11. $a^2 + 7a + 12$

12. $b^2 + 9b + 20$

Solve:

13. $4x^2 - 8x = 0$

14. $11a^2 - 33a = 0$

15. $x^2 - 16 = 0$

16. $y^2 - 25 = 0$

17. $x^2 - 8x + 15 = 0$

18. $x^2 - 16x + 15 = 0$

B

Factor if possible:

19. $16x^2y - 8xy^2 + 12xy$

20. $14ab^3 - 21ab^2 - 28a^2b^2$

21. $x^2 - 16x + 55$

22. $a^2 + 13a + 42$

23. $c^2 - 4c - 45$

24. $y^2 + 3y - 54$

25. $3x^2 + 9x - 30$

26. $5x^2 - 5x - 100$

27. $49y^2 - 36$

28. $4a^2b^2 - 9c^2$

29. $25x^2 + 1$

30. $16x^2y^2 + z^2$

31. $x^3 + 64$

32. $c^3 - 64$

33. $1000c^3 - d^3$

34. $64y^3 - 27$

35. $4a^3 - a^2b + 4a - b$

36. $6abc + 3c - 8abd - 4d$

37. $2x^2 + 13x + 21$

38. $5x^2 + 13x - 6$

39. $6x^2 + 11x - 10$

40. $9x^2 - 27x + 20$

Solve:

41. $x^2 - 11x + 24 = 0$

42. $x^2 + 9x + 18 = 0$

43. $y^2 - 81 = 0$

44. $4a^2 - 25 = 0$

45. $6x^2 + 6x - 12 = 0$

46. $7x^2 + 7x - 42 = 0$

C

Factor completely:

47. $12c^2 + 11cd - 15d^2$

48. $21x^2 - 41x + 18$

49. $36w^2 + 60w + 25$

50. $16q^2 - 72q + 81$

51. $5x^2 - 180$

52. $7y^3 - 63y$

53. $250x^3 - 16$

54. $81 - 192a^3$

55. $3x^2 - 9x - 210$

56. $5x^2 + 50x + 120$

57. $28a^3 + 58a^2 - 30a$

58. $60y^4 - 69y^3 + 18y^2$

59. $2(x + 5)^2 - 32$

60. $5(x - 7)^2 - 180$

61. $x^4 - x^2 - 12$

62. $x^4 + 6x^2 - 7$

63. $w^4 - 35w^2 - 36$

64. $z^4 - 47z - 98$

65. $x^6 - 1$

66. $x^8 - 1$

67. $t^6 - 64$

68. $a^5 - a^3 + a^2 - 1$

Solve:

69. $6x^2 - 41x + 70 = 0$

70. $10x^2 - x - 24 = 0$

71. $12y^2 + 11y = 5$

72. $12y^2 - 16y = 35$

73. $96w^2 + 1 = 28w$

74. $96w^2 - 1 = 10w$

D

75. A lawn sprinkler covers a rectangular area that is 5 ft longer than it is wide. When the dial on the sprinkler is set on *A*, it covers an area of 104 sq ft. What are the dimensions of the rectangular area it covers?

76. When the sprinkler in Exercise 75 is set on *D*, it covers 336 sq ft. What are the dimensions of the rectangular area covered?

W

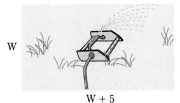

W + 5

Figure for Exercise 75

STATE YOUR UNDERSTANDING

77. What is meant by "factor completely."

CHALLENGE EXERCISES

78. $x^{4n} - 4x^{2n}$

79. $x^{6n} - 27x^{3n}$

80. $a^2 - b^2 - 2b - 1$

81. $x^2 + 2x + 1 - y^2$

MAINTAIN YOUR SKILLS (SECTIONS 1.4, 1.7, 1.9, 1.10)

Perform the indicated operations:

82. $-2(x - 4) - 2[5(x - 1)] - 3(2 - x)$

83. $3(a + 6) - 4(a - 5) + 3[(2a - 1) + 5]$

84. $-2[2(x - 1) + 4] - [3(2 - x) - 7]$

Solve:

85. $3[2a - 3(a + 6) - (a + 2)] + 4 = (a - 4) + 2$

86. $6a - [4a - (1 - a) + 6] \leq 12$

87. $4a - [6a - (a - 1) + 7] \geq -12$

88. Two planes leave Tempe traveling in opposite directions. If their average speeds are 420 mph and 525 mph, how long will it be until they are 1260 miles apart?

89. Two planes leave Fairbanks traveling in the same direction. If their average speeds are 420 mph and 525 mph, how long will it be until they are 1260 miles apart?

CHAPTER 2
SUMMARY

The properties and definitions of exponents are used to simplify algebraic expressions involving exponents. These properties and definitions frequently are used as shortcuts for steps that would be more time consuming without them.

Multiplying Like Bases	$b^m \cdot b^n = b^{m+n}$	$b \in R, m, n \in J$	(p. 87)
Raising a Power to a Power	$(b^m)^n = b^{mn}$	$b \in R, m, n \in J$	(p. 88)
Raising a Product to a Power	$(ab)^m = a^m b^m$	a and $b \in R, m \in J$	(p. 88)
Dividing Like Bases	$\dfrac{b^m}{b^n} = b^{m-n}, b \neq 0$	a and $b \in R, m, n \in J$	(p. 95)
Raising a Quotient to a Power	$\left(\dfrac{a}{b}\right)^m = \dfrac{a^m}{b^m}, b \neq 0$	a and $b \in R, m \in J$	(p. 96)
Zero Exponent	$b^0 = 1, b \neq 0$	$b \in R$	(p. 93)
Negative Exponent	$b^{-m} = \dfrac{1}{b^m}$ and	$b \in R, m \in J$	(p. 94)
	$\dfrac{1}{b^{-m}} = b^m, b \neq 0$		

Scientific notation is used to save space in print and on calculator and computer displays to symbolize very large and very small numbers.

Scientific Notation	$a \times 10^m$	$1 \leq a < 10$ (a is number between 1 and 10 or $a = 1$) and m is an integer.	(p. 100)
Monomial	A polynomial that contains one term		(p. 108)
Binomial	A polynomial that contains two terms		(p. 108)
Trinomial	A polynomial that contains three terms		(p. 108)
Linear Polynomial	A polynomial of degree one		(p. 109)

Quadratic Polynomial	A polynomial of degree two	(p. 109)
Cubic Polynomial	A polynomial of degree three	(p. 109)
Adding Polynomials	Group and add the like terms	(p. 113)
Subtracting Polynomials	Add the opposite of the polynomial being subtracted. $P - Q = P + (-Q)$	(p. 114)
Multiplying Polynomials	Use the distributive property until the result is the sum of products of monomials. Then multiply and combine the like terms. FOIL can be used as a shortcut if the factors are binomials.	(pp. 122, 128)
Special Products		
Conjugate Binomials	$(a + b)(a - b) = a^2 - b^2$	(p. 129)
Squares of Binomials	$(a + b)(a + b) = a^2 + 2ab + b^2$ and $(a - b)(a - b) = a^2 - 2ab + b^2$	(p. 130)
Solving Quadratic Equations	Factor and use the zero-product property. If $A \cdot B = 0$, then $A = 0$ or $B = 0$.	(p. 139)
To Factor a Polynomial Completely	**1.** Factor out the greatest common factor (GCF), if any.	(p. 136)
	2. If any factor is a binomial, check to see if it is the difference of two squares or if it is the sum or difference of two cubes.	(pp. 153, 154)
	3. If any factor is a trinomial, factor by the appropriate method.	(pp. 146, 147)
	4. If the polynomial contains four or more terms, factor by grouping.	(p. 138)
Special Patterns		
Difference of Two Squares	$x^2 - y^2 = (x + y)(x - y)$	(p. 153)
Perfect Squares	$a^2 + 2ab + b^2 = (a + b)^2$ $a^2 - 2ab + b^2 = (a - b)^2$	(p. 153)
Sum of Two Cubes	$a^3 + b^3 = (a + b)(a^2 - ab + b^2)$	(p. 154)
Difference of Two Cubes	$a^3 - b^3 = (a - b)(a^2 + ab + b^2)$	(p. 154)

CHAPTER 2
REVIEW EXERCISES

SECTION 2.1   Objective 1

Multiply.

1. $t^5 \cdot t^9 \cdot t$

2. $(6m^2)(2m^6)(m)$

3. $(-6)^{12}(-6)^6(-6)^9$

4. $(2)^{10}(2^{10})(2)^5$

5. $(11r^4s^3t)(-9rs^2t^3)$

6. $(15mn)(7m^2n^2)(-m^3n)$

SECTION 2.1   Objective 2

Multiply.

7. $(14x^5y^3)^2$

8. $(-7t^6v^7)^3$

9. $(7^w)(7^w)$

10. $(8^{p+1})(8^{p+4})$

11. $(4t^2)^3(2t^3)^4$

12. $(a^{4t})^2(a^{2t})^2$

SECTION 2.2   Objective 1

Simplify using only positive exponents.

13. 13^{-2}

14. $x^{-1}y^{-2}z$

15. $\dfrac{1}{14^{-2}}$

16. $\dfrac{1}{a^{-1}b^{-2}c^3}$

17. $(c^{-1}d^2)^{-2}$

18. $(r^2t^{-3})^{-3}$

SECTION 2.2   Objective 2

Simplify using only positive exponents.

19. $\left(\dfrac{x^{-1}}{y}\right)^{-1}$

20. $\left(\dfrac{m}{n^{-1}}\right)^{-1}$

21. $\dfrac{5^{-1}x^{-3}y^{-2}}{(3xy)^{-4}}$

22. $\dfrac{(a^2b^{-3})^{-1}}{a^3b^{-2}}$

23. $\dfrac{(x+7)^{-1}(x+7)^3}{(x+7)^{-a}}$

24. $\dfrac{(y-5)^{-3}}{(y-5)^{-2}} \cdot (y-5)^b$

SECTION 2.3   Objective 1

Write in scientific notation.

25. 36,000,000

26. 3,600,000,000

27. 0.00036

28. 0.036

29. 73.22×10^5

30. 732.2×10^{-4}

SECTION 2.3   Objective 2

Write in place value notation.

31. 8.34×10^7

32. 8.34×10^3

33. 8.34×10^{-4}

34. 8.34×10^{-7}

35. 834×10^{-6}

36. 0.00000834×10^5

SECTION 2.3   Objective 3

Perform the indicated operations. Write the result in scientific notation.

37. $(0.00028)(950,000)$

38. $\dfrac{17,480}{0.0023}$

39. $\dfrac{0.001748}{23,000}$

SECTION 2.4   Objectives 1 and 2

Classify each of the following polynomials by number of terms and by degree.

40. $0.32x + 9.7$

41. $17 - t^2$

42. $1 + 2x + 3y^3$

43. $a^6 - b^2 - x$

44. $-14ab^2xy^3$

45. 6

SECTION 2.5   Objectives 1 and 2

Simplify by adding or subtracting as indicated.

46. $(9y - 3) + (2y + 5)$

47. $(15a^2 - a + 19) + (9a^2 - 4a - 3)$

48. $(9y - 3) - (2y + 5)$

49. $(15a^2 - a + 19) - (9a^2 - 4a - 3)$

50. $4y + [2y - (8y + 7)]$

51. $[3 - (4w - 1)] + [w - (7w - 2)]$

52. $4y - [2y - (8y + 7)]$

53. $[3 - (4w - 1)] - [w - (7w - 2)]$

SECTION 2.5   Objective 3

Solve.

54. $x - (14 - x) = 4$

55. $y - (2y - 3) = 4 - 2y$

56. $(6x - 21) - 3 = 2 - (1 - x)$

57. $5 - (x - 3) = -4 - (9 - 2x)$

58. $(5x - 7) - (35x + 60) + 6 = -1$

59. $(2x^2 - 13x - 7) - (2x^2 - 6) - 24 = 1$

SECTION 2.6   Objective 1

Multiply.

60. $4x^2y(xy^2 - 8xy + 12x^2y)$

61. $-13rs^2t^3(3r - 7s + 9t - 1)$

62. $(t + 15)(t + 6)$

63. $(v + 14)(v + 9)$

64. $(y^2 - 3a)(y^2 + 7a)$

65. $(w^3 + 9b)(w^3 - 12b)$

66. $(x^2 - 2y)(x + 2xy - 3y)$

67. $(x^2 - 2y)(x - 2y + 3)$

SECTION 2.6   Objective 2

Solve.

68. $7(x - 3) = 4(2x - 3)$

69. $10x - 3 = 3(5x + 2) - x$

70. $2(x - 2) + 36 = 3(3x - 1)$

71. $6(5x + 8) = 46(5x - 2)$

72. $10x - 2(5 - x) = 14$

73. $(x + 3)(x - 2) = (x - 1)(x + 5) - 13$

SECTION 2.7   Objective 1

Multiply.

74. $(x + 9)(x + 11)$ **75.** $(y + 12)(y + 6)$ **76.** $(2w - 9)(w + 12)$

77. $(4w + 7)(w - 8)$ **78.** $(6t + 1)(2t - 5)$ **79.** $(7t - 2)(4t - 7)$

SECTION 2.7   Objective 2

Multiply.

80. $(x + 11)(x - 11)$ **81.** $(y + 14)(y - 14)$ **82.** $(w - 18)(w + 18)$

83. $(t - 21)(t + 21)$ **84.** $(7a + 12)(7a - 12)$ **85.** $(4b - 15)(4b + 15)$

SECTION 2.7   Objective 3

Multiply.

86. $(x + 15)^2$ **87.** $(y + 11)^2$ **88.** $(2w - 9)^2$

89. $(7t - 8)^2$ **90.** $(12a + 1)^2$ **91.** $(13a - 2)^2$

SECTION 2.7   Objective 4

Solve.

92. $(x + 1)(x + 8) = (x + 1)(x - 8) + 8$ **93.** $(x - 3)(x - 11) = (x - 3)(x + 11) + 11$

94. $(x + 7)(x - 7) = (x + 7)^2 - 21$ **95.** $(x + 7)(x - 7) = (x - 7)^2 + 7$

96. $(4t - 1)^2 = (8t + 3)(2t - 1)$ **97.** $(4t + 1)^2 = (8t - 5)(2t + 1)$

SECTION 2.8   Objective 1

Factor.

98. $13xy^3 + 78x^3y$ **99.** $17w^2x^2 + 51wx^2$ **100.** $8a^2b^3 - 12a^4b + 4a^2b$

101. $30m^3n^3 + 72m^3n^2 - 6m^2n^2$ **102.** $t(3t - 7) + 6(3t - 7)$ **103.** $x(2x - 5y) - 3y(2x - 5y)$

SECTION 2.8   Objective 2

Factor.

104. $pq + pr - nq - nr$ **105.** $st - 3t + sm - 3m$ **106.** $w^2 + wt - sw - st$

107. $b^2 - bp - bq + pq$ **108.** $35 + 49w + 25x + 35wx$ **109.** $1 - 8s + 9t - 72st$

SECTION 2.8   Objective 3

Solve.

110. $13x^2 - 65x = 0$ **111.** $13y^2 + 91y = 0$ **112.** $(7x - 91)(6x - 54) = 0$

113. $(8x + 88)(9x + 108) = 0$ **114.** $x^2 - 11x - 8x + 88 = 0$ **115.** $x^2 + 5x - 13x - 65 = 0$

SECTION 2.9   Objective 1

Factor.

116. $y^2 + y - 2$

117. $z^2 - 2z - 24$

118. $w^2 - w - 56$

119. $t^2 + t - 110$

120. $a^2 - 10a - 75$

121. $b^2 - 28b + 75$

SECTION 2.9   Objective 2

Factor.

122. $2x^2 - 9x + 4$

123. $5y^2 - 26y + 5$

124. $6t^2 - 17t + 12$

125. $8w^2 + 54w + 81$

126. $6x^2 + 15x - 9$

127. $21a^2 - 67a + 42$

SECTION 2.9   Objective 3

Solve.

128. $x^2 - 30x - 1000 = 0$

129. $y^2 + 15y - 1000 = 0$

130. $6y^2 + y - 12 = 0$

131. $6w^2 + 71w - 12 = 0$

132. $3x^2 - 43x + 14 = 0$

133. $3x^2 + 49x + 16 = 0$

SECTION 2.10   Objective 1

Factor.

134. $16m^2n^2 - 169$

135. $25p^2q^2 - 196$

136. $1 - 100r^2$

137. $4 - 121t^2$

138. $16 - (2x + 3y)^2$

139. $169 - (2a - 7b)^2$

SECTION 2.10   Objective 2

Factor.

140. $x^2 + 28x + 196$

141. $y^2 - 26y + 169$

142. $4t^2 - 36t + 81$

143. $25p^2 + 110p + 121$

144. $1 + 20b + 100b^2$

145. $36 + 84w + 49w^2$

SECTION 2.10   Objective 3

Factor.

146. $x^3 + 216$

147. $y^3 + 512$

148. $1 - 125t^3$

149. $1 - 343m^3$

150. $8 - 125w^3$

151. $27 + 1000t^3$

SECTION 2.10   Objective 4

Solve.

152. $16x^2 - 25 = 0$

153. $25y^2 - 16 = 0$

154. $4a^2 - 44a + 121 = 0$

155. $169p^2 + 78p + 9 = 0$

SECTION 2.11   Objective 1

Factor.

156. $3m^2 - 300$

157. $5w^4 - 125y^2$

158. $3a^3 - 3000$

159. $40p^3 + 135$

160. $7ab + 42b + 7ac + 42c$

161. $4x^2st - 24x^2t + 4x^2s - 24x^2$

162. $a^2 - 7ax - 18x^2$

163. $36y^3 + 106y^2 - 70y$

SECTION 2.11   Objective 2

Solve.

164. $9x^2 - 108x = 0$

165. $9x^2 - 100 = 0$

166. $x^2 - 2x - 143 = 0$

167. $t^2 + 2t - 168 = 0$

168. $13a^2 + 26a - 624 = 0$

169. $17b^2 - 34b - 1071 = 0$

170. $19y^2 - 266y + 931 = 0$

CHAPTER 2
TRUE–FALSE CONCEPT REVIEW

Check your understanding of the language of algebra. Tell whether each of the following statements is true (always true) or false (not always true).

1. All monomials contain variable factors.

2. Sixty-four is the sixth power of two.

3. If a and b are whole numbers, then $(x^a)^b = x^{a+b}$.

4. Any negative number raised to the zero power is equal to one.

5. a^{-1} is the multiplicative inverse of a.

6. A number is written in scientific notation, $a \times 10^m$, only when a is a real number between one and ten, $1 < a < 10$.

7. Polynomials contain only terms that are monomials.

8. The sum of two or more trinomials is a trinomial.

9. A cubic polynomial is a polynomial that contains three terms.

10. The opposite of a polynomial is found by writing the sum of the opposites of each term of the polynomial.

11. The distributive property enables the product of two polynomials to be found by multiplying each term of one polynomial by the other polynomial.

12. The product of conjugate binomials is the difference of two squares.

13. If $AB = 0$, then A and B are both equal to zero.

14. Every polynomial is an algebraic expression.

15. Every algebraic expression is a polynomial.

16. The sum or difference of two polynomials is always a polynomial.

17. The quotient of two monomials is always a monomial.

18. When two terms are multiplied, the exponents of the variables are added.

19. The FOIL shortcut for multiplying binomials is based on the distributive property.

20. The square of a binomial is a trinomial.

21. The product of conjugate binomials is a trinomial.

22. If x is negative, x^{-1} is positive.

23. Scientific notation for large and small numbers is used only by scientists.

24. In scientific notation, $a \times 10^m$, the exponent, m, is positive.

25. Since factoring reverses the effect of multiplication, it is related to division.

CHAPTER 2
TEST

1. Multiply: $(-3x^3y^4)^4$

2. Multiply: $(3a - 4)(2a + 9)$

3. Solve: $2x^2 + 3x + 28 = 18x$

4. Factor: $16c^2 - 24c + 9$

5. Write in scientific notation: 47,800,000

6. Add: $(15a^2 - 16ab + b^2) + (-4a^2 + 12ab - 9b^2) + (-a^2 - 8ab + 12b^2)$

7. Simplify using only positive exponents: $(-2a^4b^{-3}c^2)^{-3}$

8. Subtract: $(3q^2 - 8qr + 3r^2) - (-q^2 - 6qr + 5r^2)$

9. True or false: $x^3y - y + 1$ is a trinomial.

10. Multiply: $(12x^4y^2z^3)(-3x^4yz^5)$

11. Factor: $a^2 - 7a - 78$

12. Factor: $10c^2d - 15c^2d^2 + 5cd^2$

13. Solve:
$4x - 2x[2 - 5(x + 6)] = 10x(x + 12) + 30$

14. Subtract:
$(-13s^2 - 14st + 15t^2) - (7s^2 + 16st - 13t^2)$

15. Simplify using only positive exponents:
$(-3x^{-2}y^2z^{-3})^{-2}$

16. Solve: $3x^2 + 14x + 10 = 15$

17. Factor: $27c^3 + 125$

18. Factor: $7xy + 21y - 5x - 15$

19. Write in place value notation: 6.2×10^{-3}

20. Factor: $15b^2 + 29b - 14$

21. Solve: $3x^2 - 14x + 15 = 0$

22. Solve: $4p^2 - 81 = 0$

23. Simplify using only positive exponents:
$$\frac{51a^{-3}y^{-2}}{-3a^{-4}y^3}$$

24. Multiply: $(y - 5)(y^2 - 3y - 4)$

25. What is the degree of the polynomial $13xy^4 + 6x^3y^3 - 8y^5$?

26. Multiply: $(5a + 4b)(5a - 4b)$

27. Factor: $121x^2y^2 - 81$

28. Multiply: $(7m + 4)^2$

29. A photograph that is two inches longer than it is wide is framed with a 2-inch-wide frame. If the total area of the photograph and frame is 120 square inches, what are the dimensions of the photograph?

30. In order to accommodate a concert at the Civic Auditorium, five feet of additional stage was added to the depth of the existing rectangular stage. If the old stage was 20 feet wider than it was deep and the new stage contains 1350 square feet, what are the dimensions of the new stage?

CHAPTER

3

RATIONAL EXPRESSIONS AND RELATED EQUATIONS

Distance, rate, and time problems frequently can be solved using equations involving rational expressions, since time $= \dfrac{\text{distance}}{\text{rate}}$ and rate $= \dfrac{\text{distance}}{\text{time}}$. In Exercise 63, Section 3.5, Linda and Carol are jogging at different rates and therefore cover different distances in the same time. If x represents Linda's rate then $x - 2$ represents Carol's rate, since she is jogging 2 mph slower than Linda. Using the distance each covers, the time of each can be expressed in terms of the distance and rate. Since they jogged the same length of time, these rational expressions are equal and the resulting equation can be solved.
(© David R. Frazier/Photolibrary)

175

PREVIEW

In this chapter rational expressions are presented. Polynomial operations are relied upon to lead into the operations on rational expressions. Care is taken to note the restrictions necessary on the domain of the variable(s) in order to assure that the rational expressions named are equivalent. Complex fractions are simplified using either the Basic Principle of Fractions or by performing the indicated division. Finally division of polynomials is covered with special note as to how it relates to factoring a polynomial. Synthetic division prepares the student for a future course in theory of equations.

3.1

PROPERTIES OF RATIONAL EXPRESSIONS

OBJECTIVES

1. Determine the values for which a rational expression is not defined.

2. Build rational expressions.

3. Reduce rational expressions to lowest terms.

Rational numbers are defined as the set $Q = \left\{ \dfrac{p}{q} \mid p \text{ and } q \in J,\ q \neq 0 \right\}$. A rational number is the indicated quotient of two integers, p and q with $q \neq 0$. We define a rational expression in a similar way.

■ DEFINITION

Rational Expressions

> A *rational expression* is a ratio or indicated quotient of two polynomials: $\dfrac{P}{Q}$ is a rational expression when P and Q are polynomials and $Q \neq 0$.

Some examples of rational expressions are

$$\frac{5x}{10} \qquad \frac{x+3}{y},\ y \neq 0 \qquad \frac{m-3}{m^2-9},\ m \neq \pm 3 \qquad x^2+1 \text{ or } \frac{x^2+1}{1}$$

Rational expressions are sometimes called fractions of algebra.

The *domain of the variables* in a rational expression is the set of real numbers for which the expression is defined: Division by zero is excluded. In the rational expression $\dfrac{m-3}{m^2-9}$, m cannot equal 3 or -3.

EXAMPLE 1

For what value of x is $\dfrac{2x}{x+2}$ undefined?

If $x + 2 = 0$, then $x = -2$.   Set $x + 2 = 0$ and solve.

The expression is not defined when $x = -2$.   ■

EXAMPLE 2

For what values of x is $\dfrac{x+3}{x-7}$ undefined?

If $x - 7 = 0$, then $x = 7$.   Set $x - 7 = 0$ and solve.

The expression is not defined when $x = 7$.   ■

EXAMPLE 3

For what values of x is $\dfrac{x^2 - x + 7}{x^2 - 5x + 6}$ undefined?

$$x^2 - 5x + 6 = 0$$

$$(x - 2)(x - 3) = 0$$

Set $x^2 - 5x + 6 = 0$ and solve. Factor and use the zero-product property.

$$x = 2 \qquad x = 3$$
So $x \neq 2$ or $x \neq 3$.   ■

In operations involving rational expressions, it is often necessary to build or reduce rational expressions.

■ PROPERTY

Basic Principle of Fractions

$$\frac{A}{B} = \frac{AC}{BC}, \quad BC \neq 0$$

To build a rational expression, use the basic principle of fractions to multiply both the numerator and the denominator by the same nonzero factor.

EXAMPLE 4

Build the fraction $\dfrac{x+2}{x+5}$ to have a denominator of $x^2 + 2x - 15$.

$$x^2 + 2x - 15 = (x - 3)(x + 5) \qquad \text{Factor the given denominator.}$$

$$\frac{x+2}{x+5} = \frac{(x+2)(x-3)}{(x+5)(x-3)}$$

The needed factor is $(x-3)$. Multiply both numerator and denominator by it.

$$= \frac{x^2-x-6}{x^2+2x-15}$$

■

■ PROCEDURE

To build an expression to an equivalent expression with a given denominator:

1. Factor the given denominator to find the needed factor.
2. Multiply both the numerator and the denominator by the required factor.
3. Multiply.

EXAMPLE 5

Find the missing numerator. State the restrictions on the variable:

$$\frac{2}{x+3} = \frac{?}{x^2+8x+15}$$

$$x \neq -3, -5$$

$x+3 \neq 0$ and $x^2+8x+15 = (x+3)(x+5) \neq 0$, so $x \neq -3$ and $x \neq -5$.

$$\frac{2}{x+3} = \frac{2(x+5)}{(x+3)(x+5)}$$

Since $(x+5)$ is introduced as a factor in the denominator, use the basic principle of fractions to introduce it in the numerator.

$$= \frac{2x+10}{x^2+8x+15}, \, x \neq -3, -5$$

Multiply. ■

We reduce rational expressions in the same manner that rational numbers are reduced.

■ PROCEDURE

To reduce a rational expression:

1. Factor the expressions in the numerator and denominator if necessary.

> 2. Divide out the common factors.
>
> $$\frac{AC}{BC} = \frac{A}{B}, \ BC \neq 0$$

A rational expression is said to be *reduced to lowest terms* when all common integral factors and common polynomial factors have been eliminated from the numerator and the denominator.

EXAMPLE 6

Reduce $\dfrac{x^2 + 8x + 12}{x^2 + x - 30}$ to lowest terms.

$$\frac{x^2 + 8x + 12}{x^2 + x - 30} = \frac{(x + 2)\overset{1}{\cancel{(x + 6)}}}{\underset{1}{\cancel{(x + 6)}}(x - 5)} \qquad$$ Factor both the numerator and denominator.

$$= \frac{x + 2}{x - 5} \qquad$$ Reduce by dividing both the numerator and denominator by $x + 6$. Since the numerator and denominator have no other factors in common, the rational expression is reduced to lowest terms. ∎

Although the step showing the common factors eliminated or crossed out is sometimes referred to as canceling, it is recommended that you avoid using that word to prevent errors. A better idea of the process is described by the word *reducing* or by the phrase *replace the common factors by 1* or by the phrase *divide both the numerator and denominator by their common factor.*

In Example 6, the rational expression $\dfrac{x + 2}{x - 5}$ cannot be reduced because x is not a factor of the numerator and denominator.

$$\frac{x + 2}{x - 5} \neq -\frac{2}{5}.$$

■ CAUTION

> Only factors can be eliminated by reducing.

A rational expression can also be reduced if it has a pair of factors in the numerator and denominator that are the opposite of each other. This is accomplished by factoring -1 from one of the factors.

EXAMPLE 7

Reduce $\dfrac{xy - 3y}{3t - tx}$ to lowest terms. State the restrictions on the variables.

$\dfrac{xy - 3y}{3t - tx} = \dfrac{y(x - 3)}{t(3 - x)}$

Factor the numerator and denominator. The factors $x - 3$ and $3 - x$ are opposites.

$\dfrac{-1y(3 - x)}{t(3 - x)}$

We factor -1 from $x - 3$.
$x - 3 = -1(3 - x)$.
Note that the denominator is zero if $x = 3$ or $t = 0$.

$= \dfrac{\overset{1}{-1y(3 - x)}}{\underset{1}{t(3 - x)}}$

Reduce.

$= -\dfrac{y}{t}, \quad x \neq 3, \quad t \neq 0$ ∎

EXAMPLE 8

Reduce $\dfrac{x^2 - 9}{3x^2 + 9x}$ to lowest terms. State the restrictions on the variable.

$x \neq 0, -3$ $3x^2 + 9x = 3x(x + 3) \neq 0.$

$\dfrac{x^2 - 9}{3x^2 + 9x} = \dfrac{(x - 3)\overset{1}{(x + 3)}}{3x\underset{1}{(x + 3)}}$

$= \dfrac{x - 3}{3x} \quad x \neq 0, -3$

■ **CAUTION**

Neither x nor 3 can be divided out since neither is a factor of the numerator.

■

EXAMPLE 9

Reduce $\dfrac{x^2 + 10x + 16}{2x^2 + 19x + 24}$ to lowest terms. State the restrictions on the variable.

$$\frac{x^2 + 10x + 16}{2x^2 + 19x + 24} = \frac{(x + 2)\cancel{(x + 8)}^{1}}{(2x + 3)\cancel{(x + 8)}_{1}}$$

Factor the numerator and denominator. State the restrictions on the variable and reduce.

$$= \frac{x + 2}{2x + 3} \quad x \neq -8, \; -\frac{3}{2}$$ ■

EXERCISE 3.1

A

For what value of the variable is each of the following undefined?

1. $\dfrac{5x}{x - 3}$

2. $\dfrac{8x}{x + 4}$

3. $\dfrac{2x - 7}{x + 6}$

4. $\dfrac{3x + 2}{x - 9}$

5. $\dfrac{3x + 1}{(x - 2)(x + 5)}$

6. $\dfrac{4x + 3}{(x - 9)(x - 7)}$

7. $\dfrac{x^2 + 8x - 9}{(x - 1)(x + 9)}$

8. $\dfrac{x^2 - 5x + 6}{(x - 2)(x - 3)}$

9. $\dfrac{x + 8}{x^2 - 8x - 20}$

10. $\dfrac{x + 5}{x^2 + 6x - 16}$

11. $\dfrac{x^2 + 7x + 9}{3x^2 - x - 2}$

12. $\dfrac{x^2 + 12x - 12}{4x^2 - 11x - 3}$

Build these rational expressions by finding the missing numerator. You do not need to state the variable restrictions:

13. $\dfrac{m}{x} = \dfrac{?}{x^2}$

14. $\dfrac{3x}{4y} = \dfrac{?}{8yz}$

15. $\dfrac{3}{x - 4} = \dfrac{?}{3x - 12}$

16. $\dfrac{5}{y + 7} = \dfrac{?}{4y + 28}$

17. $\dfrac{10}{2w + 7} = \dfrac{?}{2w^2 + 7w}$

18. $\dfrac{7}{3x - 2y} = \dfrac{?}{3x^2 - 2xy}$

19. $\dfrac{3x + 2}{5x - 1} = \dfrac{?}{15x - 3}$

20. $\dfrac{5m + 4}{2a + 5} = \dfrac{?}{10a^2 + 25a}$

B

Reduce. You do not need to state the variable restrictions:

21. $\dfrac{3abc}{18abx}$

22. $\dfrac{11xyz}{44wxy}$

23. $\dfrac{-8x^2y^2}{-10x^3y}$

24. $\dfrac{-10r^2st}{-25rst^2}$

25. $\dfrac{5(y - 2)}{6(y - 2)}$

26. $\dfrac{5y(z^2 - 1)}{3z(z^2 - 1)}$

27. $\dfrac{4a - 16}{3b(a - 4)}$

28. $\dfrac{5x - 10}{3x(x - 2)}$

Build these rational expressions by finding the missing numerator. You do not
need to state the variable restrictions:

29. $\dfrac{-3}{6mn} = \dfrac{?}{-12m^2n}$

30. $\dfrac{4a}{-3rs} = \dfrac{?}{15r^2s}$

31. $\dfrac{3}{x} = \dfrac{?}{4x^2y}$

32. $\dfrac{9}{5v} = \dfrac{?}{35vw^2}$

33. $\dfrac{3}{x-1} = \dfrac{?}{x^2-1}$

34. $\dfrac{5}{x-7} = \dfrac{?}{x^2-49}$

35. $\dfrac{2x}{x-4} = \dfrac{?}{x^2-5x+4}$

36. $\dfrac{6y}{x-8} = \dfrac{?}{x^2-9x+8}$

Reduce. State the restrictions on the variables so that the denominator is not zero:

37. $\dfrac{54a^2b^2c^3}{72a^4bc^2}$

38. $\dfrac{-60x^4y^6z^3}{44x^3y^3z^3}$

39. $\dfrac{3x-12}{x^2-6x+8}$

40. $\dfrac{4w+20}{w^2+8w+15}$

C

41. $\dfrac{x^2-1}{4x+4}$

42. $\dfrac{x^2-4}{5x-10}$

43. $\dfrac{2x^2-18}{2x+6}$

44. $\dfrac{10x^2-250}{5x-25}$

Build these rational expressions. State the restrictions on the variables so that
the denominator is not zero:

45. $\dfrac{m+2}{2m+7} = \dfrac{?}{4m^2-49}$

46. $\dfrac{2m-5}{3m+1} = \dfrac{?}{9m^2-1}$

47. $\dfrac{-7}{y+2} = \dfrac{?}{y^2+4y+4}$

48. $\dfrac{-8}{c+9} = \dfrac{?}{c^2+18c+81}$

49. $\dfrac{7x+2}{2x-1} = \dfrac{?}{6x^2-5x+1}$

50. $\dfrac{3y-4}{3y-2} = \dfrac{?}{3y^2+4y-4}$

51. $\dfrac{2y+1}{y+1} = \dfrac{?}{y^3+1}$

52. $\dfrac{-7}{x-2} = \dfrac{?}{x^3-8}$

Reduce. You do not need to state the variable restrictions:

53. $\dfrac{x^2-16}{x^2-x-12}$

54. $\dfrac{a^2-36}{a^2+13a+42}$

55. $\dfrac{3x^2+7x+2}{x^2-3x-10}$

56. $\dfrac{5x^2+5x-30}{x^2-x-12}$

57. $\dfrac{2x^2 - x - 15}{2x^2 + 7x + 5}$

58. $\dfrac{3x^2 - x - 14}{5x^2 + 7x - 6}$

59. $\dfrac{a^2 - 9}{a^3 + 27}$

60. $\dfrac{b^2 - 4b + 4}{b^3 - 8}$

61. $\dfrac{x^2 - 2x - 3}{3x^3 + 3}$

62. $\dfrac{y^2 - 3y - 10}{2y^3 - 250}$

STATE YOUR UNDERSTANDING

63. Explain the difference between building a fraction and reducing a fraction.

64. Why is it necessary to restrict variables in the denominators of rational expressions?

65. Explain why the following rational expression will not reduce in the manner indicated.

$$\frac{x - 3}{x + 1} = -3$$

CHALLENGE EXERCISES

66. The average acceleration $(\bar{a})$ during a time interval is defined as the change in velocity (ΔV) divided by the time interval (t) during which the change occurs. It is given by the formula

$$(\bar{a}) = \frac{\Delta V}{t}$$

Write a rational expression that will represent the average acceleration if the change in velocity is $x - 5$ m/sec and the time interval is $x + 4$ seconds.

67. Use the formula of Exercise 66 and write a rational expression to represent the average acceleration if the change in velocity is $2x + 3$ and the time interval is $x + 3$.

68. Tensil stress (S) on a bar is defined as the ratio of the magnitude of the force (F) applied along the bar (perpendicular to the cross section) to the cross sectional area. It is given by the formula

$$S = \frac{F}{A}$$

Write a rational expression that will represent the tensil stress on a bar if a force of $x^2 + 3x + 5$ Newtons is applied and the cross-sectional area is $x^2 + x + 1$ cm^2

69. Using the formula of Exercise 68, write a rational expression that will represent the tensil stress on a bar if a force of $x^2 - 2x - 8$ Newtons is applied and the cross-sectional area is $x^2 + 1$.

MAINTAIN YOUR SKILLS (SECTION 1.5)

Perform the indicated operations:

70. $\dfrac{8}{9} \cdot \dfrac{7}{11}$

71. $\dfrac{15}{19} \cdot \left(-\dfrac{3}{11}\right)$

72. $\left(-\dfrac{7}{9}\right) \cdot \left(-\dfrac{3}{11}\right)$

73. $\left(-\dfrac{39}{45}\right) \cdot \left(\dfrac{9}{26}\right)$

74. $\left(-\dfrac{9}{10}\right) \div \left(-\dfrac{7}{11}\right)$

75. $\dfrac{14}{15} \div \left(-\dfrac{11}{8}\right)$

76. $\dfrac{35}{48} \div \dfrac{7}{12}$.

77. $-\dfrac{46}{39} \div \dfrac{23}{13}$

3.2

MULTIPLICATION AND DIVISION OF RATIONAL EXPRESSIONS

OBJECTIVE

1. Multiply and divide rational expressions.

Multiplication and division of rational expressions are similar to multiplication and division of fractions.

■ RULE

> To multiply two rational expressions, multiply the numerators and multiply the denominators.
>
> $$\frac{A}{B} \cdot \frac{C}{D} = \frac{AC}{BD}, \quad BD \neq 0$$

As with fractions, the multiplication is easier if we reduce first.

■ PROCEDURE

> To multiply rational expressions:
>
> 1. If possible, factor the numerators and denominators.
> 2. Reduce if possible.
> 3. Multiply the numerators and denominators.

EXAMPLE 1

Multiply and reduce. You do not need to state variable restrictions.

$$\frac{4x^2}{5y^2} \cdot \frac{10y}{12x}$$

$$= \frac{4 \cdot x \cdot x}{5 \cdot y \cdot y} \cdot \frac{5 \cdot 2 \cdot y}{4 \cdot 3 \cdot x}$$

Step 1. Factor each numerator and each denominator.

$$= \frac{\overset{1}{\cancel{4}} \cdot \overset{1}{\cancel{x}} \cdot x}{\underset{1}{\cancel{5}} \cdot \underset{1}{\cancel{y}} \cdot y} \cdot \frac{\overset{1}{\cancel{5}} \cdot 2 \cdot \overset{1}{\cancel{y}}}{\underset{1}{\cancel{4}} \cdot 3 \cdot \underset{1}{\cancel{x}}}$$

Step 2. Reduce.

$$= \frac{2x}{3y}$$

Step 3. Multiply the numerators and denominators. ∎

EXAMPLE 2

Multiply and reduce. State restrictions on the variable.

$$\frac{x + 3}{x - 6} \cdot \frac{2}{x + 3}$$

$x - 6 \neq 0$ and $x + 3 \neq 0$. So $x \neq 6$ and $x \neq -3$.

$$\frac{\overset{1}{\cancel{x + 3}}}{x - 6} \cdot \frac{2}{\underset{1}{\cancel{x + 3}}} = \frac{2}{x - 6}, \; x \neq -3, 6$$ ∎

■ **CAUTION**

> We can reduce *only* within the same fraction or across the multiplication sign when the numerators and denominators are in factored form.

EXAMPLE 3

Multiply and reduce. State the restrictions on the variable.

$$\frac{3x + 6}{x + 5} \cdot \frac{x^2 + 2x - 15}{x^2 - 4}$$

$$x \neq -5, -2, 2$$

$x + 5 \neq 0$ and $x^2 - 4 = (x - 2)(x + 2) \neq 0$

$$= \frac{3(x + 2)}{x + 5} \cdot \frac{(x + 5)(x - 3)}{(x - 2)(x + 2)}$$

Step 1. Factor the numerators and denominators and state the restrictions on the variable.

$$= \frac{3\cancel{(x+2)}^{1}}{\cancel{x+5}_{1}} \cdot \frac{\cancel{(x+5)}^{1}(x-3)}{(x-2)\cancel{(x+2)}_{1}}$$

Step 2. Reduce.

$$= \frac{3(x-3)}{x-2}$$

Step 3. Multiply the numerators and denominators.

$$= \frac{3x-9}{x-2}, \; x \neq -5, -2, 2$$ ∎

To do example 4, factor by grouping.

EXAMPLE 4

Multiply and reduce. State the restrictions on the variable.

$$\frac{cx + 4x + 5c + 20}{ac + ad + bc + bd} \cdot \frac{ax + bx + ay + by}{x^2 + xy + 5x + 5y}$$

$$= \frac{\cancel{(x+5)}^{1}(c+4)}{\cancel{(a+b)}_{1}(c+d)} \cdot \frac{\cancel{(x+y)}^{1}\cancel{(a+b)}^{1}}{\cancel{(x+5)}_{1}\cancel{(x+y)}_{1}}$$

$$= \frac{c+4}{c+d}, \; c \neq -d, \; a \neq -b, \; x \neq -5, -y$$ ∎

Even though a polynomial is not written in fraction form, we can write it with denominator of 1.

EXAMPLE 5

Multiply and reduce. State the restrictions on the variable.

$$(x^2 + 5x + 6) \cdot \frac{x^2 - 1}{x^2 + 4x + 3}$$

$$= \frac{x^2 + 5x + 6}{1} \cdot \frac{x^2 - 1}{x^2 + 4x + 3}$$

Write as the polynomial as a fraction with denominator one.

$$= \frac{(x+2)\cancel{(x+3)}^{1}}{1} \cdot \frac{(x-1)\cancel{(x+1)}^{1}}{\cancel{(x+3)}_{1}\cancel{(x+1)}_{1}}, \; x \neq -3, 1$$

Factor, state restrictions, and reduce.

$$= \frac{(x+2)(x-1)}{1}, \; x \neq -3, -1$$

Multiply.

$$= (x+2)(x-1), \; x \neq -3, -1$$
$$\text{or}$$
$$= x^2 + x - 2, \; x \neq -3, -1$$ ∎

Recall that two numbers are called *reciprocals* if their product is one. This is also true for rational expressions. The reciprocal of a fraction can be found by "inverting" the fraction. So the reciprocal of $\dfrac{3}{x}$ is $\dfrac{x}{3}$, provided $x \neq 0$. Zero does not have a reciprocal.

■ RULE

To divide two rational expressions, multiply the first by the reciprocal of the divisor.

$$\frac{A}{B} \div \frac{C}{D} = \frac{A}{B} \cdot \frac{D}{C} = \frac{AD}{BC}, \; BCD \neq 0$$

■ PROCEDURE

To divide two rational expressions:

1. Change the division to multiplication by multiplying by the reciprocal of the divisor (the second fraction).
2. Multiply.

EXAMPLE 6

Divide and reduce. State the restrictions on the variables.

$$\frac{8x^2y}{15xy^3} \div \frac{16x^2}{25y}$$

$15xy^3 \neq 0, \; 25y \neq 0, \; \text{and} \; 16x^2 \neq 0$ — Since each of these expressions is a divisor, none may be zero.

$x \neq 0 \quad \text{and} \quad y \neq 0$

$$\frac{8x^2y}{15xy^3} \div \frac{16x^2}{25y} = \frac{8x^2y}{15xy^3} \cdot \frac{25y}{16x^2}$$

Step 1. Change to multiplication. That is, multiply by the reciprocal of the divisor.

$$= \frac{\overset{1}{\cancel{8}} \cdot \overset{1}{\cancel{x^2}} \cdot \overset{1}{\cancel{y}}}{\underset{1}{\cancel{3}} \cdot 3 \cdot x \cdot y \cdot \underset{1}{\cancel{y}} \cdot \underset{1}{\cancel{y}}} \cdot \frac{\overset{1}{\cancel{3}} \cdot 5 \cdot \overset{1}{\cancel{y}}}{\underset{1}{\cancel{8}} \cdot 2 \cdot \underset{1}{\cancel{x^2}}}$$

Follow the procedure for multiplication.

$$= \frac{5}{6xy}, \; x \neq 0, \; y \neq 0 \qquad \blacksquare$$

EXAMPLE 7

Divide and reduce. State the restrictions on the variable.

$$\frac{2x^2 + 3x - 5}{3x^2 - 4x + 1} \div \frac{2x^2 + 9x + 10}{3x^2 + 11x - 4}$$

$$= \frac{2x^2 + 3x - 5}{3x^2 - 4x + 1} \cdot \frac{3x^2 + 11x - 4}{2x^2 + 9x + 10}$$

$$= \frac{(2x + 5)(x - 1)}{(3x - 1)(x - 1)} \cdot \frac{(3x - 1)(x + 4)}{(2x + 5)(x + 2)}$$

Rewrite the division as multiplication. Factor and restrict the variable.

$$x \neq -4, -2, -\frac{5}{2}, \frac{1}{3}, 1$$

Recall in $\frac{A}{B} \div \frac{C}{D}$, $BCD \neq 0$, so

$3x - 1 \neq 0$, $x + 4 \neq 0$, $x - 1 \neq 0$, $2x + 5 \neq 0$, and $x + 2 \neq 0$.

$$= \frac{\overset{1}{\cancel{(2x+5)}}\overset{1}{\cancel{(x-1)}}}{\underset{1}{\cancel{(3x-1)}}\underset{1}{\cancel{(x-1)}}} \cdot \frac{\overset{1}{\cancel{(3x-1)}}(x + 4)}{\underset{1}{\cancel{(2x+5)}}(x + 2)}$$

$$= \frac{x + 4}{x + 2}, x \neq -4, -2, -\frac{5}{2}, \frac{1}{3}, 1$$

Reduce and multiply.   ■

When only multiplication and division of real numbers appear in an exercise, the operations are performed in their order of appearance. The same is true for rational expressions.

EXAMPLE 8

Perform the indicated operations and reduce. State the restrictions on the variable.

$$\frac{x}{2 - x} \cdot \frac{-x - 3}{x^3} \div \frac{x + 3}{x - 2}$$

$$= \frac{x}{2 - x} \cdot \frac{-x - 3}{x^3} \cdot \frac{x - 2}{x + 3}$$

Rewrite division as multiplication. Note that $-x - 3$ and $x + 3$ are opposites. $2 - x \neq 0$, $x^3 \neq 0$, $x - 2 \neq 0$, and $x + 3 \neq 0$

The restrictions on the variable are $x \neq -3, 0, 2$.

$$= \frac{\overset{1}{\cancel{x}}}{-1\underset{1}{\cancel{(x-2)}}} \cdot \frac{-1\overset{1}{\cancel{(x+3)}}}{\underset{1}{\cancel{x} \cdot x \cdot x}} \cdot \frac{\overset{1}{\cancel{(x-2)}}}{\underset{1}{\cancel{(x+3)}}}$$

Reduce.

$$= \frac{-1}{-x^2}$$

Multiply.

$$= \frac{1}{x^2}, x \neq -3, 0, 2$$

Simplify.   ■

EXERCISE 3.2

A

Multiply and simplify. You do not need to state the variable restrictions:

1. $\dfrac{y}{8} \cdot \dfrac{16}{w}$

2. $\dfrac{c^2}{9} \cdot \dfrac{54}{d}$

3. $\dfrac{3}{7} \cdot \dfrac{x-1}{x+2}$

4. $\dfrac{-4}{5} \cdot \dfrac{-2x+3}{x-8}$

5. $\dfrac{a}{12} \cdot \dfrac{-6}{a-3}$

6. $\dfrac{-5}{y} \cdot \dfrac{y+5}{25}$

7. $\dfrac{p+2}{4} \cdot \dfrac{3}{p+2}$

8. $\dfrac{13}{x-y} \cdot \dfrac{x-y}{15}$

9. $\dfrac{2a-b}{4} \cdot \dfrac{12}{2a-b}$

10. $\dfrac{x+3y}{18} \cdot \dfrac{24}{x+3y}$

Divide and simplify. State the restrictions on the variables:

11. $6 \div \dfrac{x}{5}$

12. $\dfrac{4}{9} \div a$

13. $\dfrac{7y}{3} \div \dfrac{3z}{4}$

14. $\dfrac{5}{6b} \div \dfrac{3b}{5}$

15. $(x-4) \div \dfrac{x-3}{2}$

16. $(w+7) \div \dfrac{w+1}{3}$

17. $\dfrac{t-4}{6} \div 5$

18. $\dfrac{4}{3x-10} \div \dfrac{1}{4}$

19. $\dfrac{5}{2x+3y} \div \dfrac{10}{2x+3y}$

20. $\dfrac{4a-3b}{8} \div \dfrac{4a-3b}{16}$

B

Multiply or divide and simplify. You do not need to state the variable restrictions:

21. $\dfrac{6ax}{5by} \cdot \dfrac{10y}{18x}$

22. $\dfrac{12mn}{5n^2} \cdot \dfrac{30m^2}{4mn}$

23. $\dfrac{5xy}{10xz} \div \dfrac{15y}{3z}$

24. $\dfrac{18ab}{5c} \div \dfrac{6b}{15d}$

25. $\dfrac{7}{3(x+2)} \cdot \dfrac{(x+2)}{14}$

26. $\dfrac{2(y+3)}{11} \cdot \dfrac{-33}{(y+3)}$

27. $\dfrac{3}{x-2} \div \dfrac{6}{2-x}$

28. $\dfrac{1-y}{18} \div \dfrac{y-1}{6}$

29. $(x^2 - 8x - 9)\left[\dfrac{12}{3x-27}\right]$

30. $(25a^2 - 4b^2)\left[\dfrac{16}{20a+8b}\right]$

Multiply or divide and simplify. State the restrictions on the variables:

31. $\dfrac{3x+6}{2x-4} \cdot \dfrac{5x-10}{6x+12}$

32. $\dfrac{8x+24}{3x-3} \cdot \dfrac{x^2-x}{6x+18}$

33. $\dfrac{5x-3}{y^2-y} \cdot \dfrac{8y-8}{3-5x}$

34. $\dfrac{9y-27}{3z-33} \cdot \dfrac{z-11}{6-2y}$

35. $\dfrac{2z-26}{z^2-5z} \div \dfrac{3z-39}{z^2+3z}$

36. $\dfrac{16a-8}{a^2+5a} \div \dfrac{6a-3}{a^2-5a}$

37. $\dfrac{2x+6}{x-5} \div \dfrac{x^2-9}{x^2-10x+25}$

38. $\dfrac{x^2+3x+2}{x^2-3x+2} \div \dfrac{x+2}{x-1}$

39. $(a^2-a-12) \div \dfrac{a-4}{3}$

40. $(x^2-3xy-4y^2) \div \dfrac{x+y}{a}$

C

Multiply or divide and simplify. You do not need to state the variable restrictions:

41. $\dfrac{5y+10}{y^2-4} \cdot \dfrac{y^2-7y+10}{8y-40}$

42. $\dfrac{-3y-6}{y^2-64} \cdot \dfrac{y^2-7y-8}{4y+8}$

43. $\dfrac{x^2-49}{9-x^2} \cdot \dfrac{x^2-10x+21}{5x-35}$

44. $\dfrac{-x-5}{3x^2+3x-60} \cdot \dfrac{16-x^2}{x^2-25}$

45. $\dfrac{x^2+9x+18}{x^2+2x-15} \div \dfrac{2x+12}{3x+15}$

46. $\dfrac{x^2+3x-10}{7x+14} \div \dfrac{14x+70}{x^2-4}$

47. $\dfrac{10x^2+21x-10}{2x^2-7x-30} \cdot \dfrac{x^2+3x-54}{15x^2-11x+2}$

48. $\dfrac{11x^2-32x-3}{x^2-8x+16} \cdot \dfrac{x^2-2x-3}{11x^2+12x+1}$

49. $\dfrac{6x^2+39x+63}{10x^2-17x+3} \div \dfrac{6x^2+33x+45}{10x^2+33x-7}$

50. $\dfrac{2x^2-3x-27}{16x^2-70x-9} \div \dfrac{8x^2+23x-3}{8x^2+9x+1}$

51. $\dfrac{x^2-9}{x^2+6x+9} \cdot \dfrac{x^4-16}{x^3-27} \cdot \dfrac{x^2-2x-15}{x^2-4}$

52. $\dfrac{x^3+8}{2x^2+5x-7} \cdot \dfrac{2x^2+3x-14}{x^2-2x+4} \div \dfrac{x^2-4}{x^2-x-2}$

Multiply or divide and simplify. State the restrictions on the variables:

53. $\dfrac{3y^2+6y+12}{9y-y^3} \cdot \dfrac{y^3+9y}{y^3-8}$

54. $\dfrac{3y^3+81}{y^3+4y} \cdot \dfrac{y^2+4y}{6y^2-18y+54}$

55. $\dfrac{x^4+2x^2-24}{x^3+9x^2+14x} \cdot \dfrac{x^3+5x^2}{x^4+6x^2} \cdot \dfrac{x^3+10x^2+21x}{x^2+3x-10}$

56. $\dfrac{x^3+8}{2x^2-9x-56} \cdot \dfrac{x^3-8x^2}{x^4-16} \div \dfrac{x^3+4x^2+4x}{x^4+7x^2+12}$

Multiply or divide and simplify.

57. $\dfrac{ac + ad + bc + bd}{ac - ad + bc - bd} \cdot \dfrac{c^2 - cd + bc - bd}{ac + ad - bc - bd}$

58. $\dfrac{x^2 + xz + xy + yz}{ax + az + bx + bz} \cdot \dfrac{ax - ay + bx - by}{x^2 - xz + xy - yz}$

59. $\dfrac{ac - ad - bc + bd}{2ac + 2ad + 3bc + 3bd} \div \dfrac{ac - ad - bc + bd}{4ac + 6ad + 6bc + 9bd}$

60. $\dfrac{3bd - 3be - 2cd + 2ce}{4bd + d^2 - 4be - de} \div \dfrac{3ab - 6b^2 - 2ac + 4bc}{4ab - 8b^2 + ad - 2bd}$

61. $\dfrac{6x^2 + 11x - 10}{4x^2 - 7x - 2} \div \dfrac{x^2 - 3x - 40}{x^2 + 3x - 10} \div \dfrac{2x^2 - 3x - 20}{4x^2 - 31x - 8}$

62. $\dfrac{6a^2 + ab - 2b^2}{4a^2 - 5ab + b^2} \div \dfrac{3a^2 + 14ab + 8b^2}{a^2 + 3ab - 4b^2} \cdot \dfrac{8a^2 + 2ab - b^2}{2a^2 + ab - b^2}$

63. $\dfrac{2x^3 - 8x}{8x^2 - 28x - 16} \cdot \dfrac{32x^2 - 8}{10x^3 + 35x^2 - 20x} \cdot (10x^2 - 160)$

64. $(a^3 - 9a) \cdot \dfrac{a - 2}{a^2 + 2a - 15} \div \dfrac{a^3 + 8a^2 + 15a}{a^2 + 4a - 5}$

STATE YOUR UNDERSTANDING

65. Explain the difference in the processes for multiplication and division of rational expressions.

CHALLENGE EXERCISES

Perform the indicated operations, reduce to lowest terms. You do not need to state the restrictions on the variables. Assume that all exponents are positive integers.

66. $\dfrac{x^{2n} + 5x^n + 4}{x^{2n} - 1} \cdot \dfrac{2x^{2n} + x^n - 3}{x^{2n} + 3x^n - 4}$

67. $\dfrac{x^{4n} - 1}{x^{2n} - 3x^n - 10} \cdot \dfrac{x^{2n} + 4x^n + 4}{x^{2n} - 1}$

68. $\dfrac{x^{2n} - 2x^n - 15}{x^{2n} - 3x^n - 10} \div \dfrac{x^{2n} + 7x^n + 12}{x^{2n} + 6x^n + 8}$

69. $\dfrac{x^{3n} - 8}{x^{2n} - 16} \div \dfrac{x^{2n} + 2x^n + 4}{x^n + 4}$

MAINTAIN YOUR SKILLS (SECTIONS 1.3, 1.4)

Add or subtract as indicated:

70. $\dfrac{4}{5} + \dfrac{7}{8}$

71. $\dfrac{8}{9} + \dfrac{5}{12}$

72. $\dfrac{6}{7} - \dfrac{2}{3}$

73. $\dfrac{11}{9} - \dfrac{8}{7}$

74. $-\dfrac{1}{2} + \dfrac{7}{8}$

75. $-\dfrac{11}{5} + \left(-\dfrac{15}{2}\right)$

76. $\dfrac{11}{12} - \left(-\dfrac{14}{15}\right)$

77. $\dfrac{5}{7} - \dfrac{32}{21}$

3.3

ADDITION AND SUBTRACTION OF RATIONAL EXPRESSIONS

OBJECTIVES

1. Find the LCD of two or more rational expressions.

2. Add and subtract rational expressions.

Addition and subtraction of rational expressions is similar to addition and subtraction of fractions; that is, if they have a common denominator you add (or subtract) the numerators and retain the common denominator.

■ DEFINITION

The *least common denominator* (LCD) of two or more rational expressions is the polynomial with the least number of factors that is a multiple of each of the denominators.

■ PROCEDURE

To find the LCD of two or more rational expressions:

1. Factor each denominator completely.

2. Write the product of the highest power of each different factor.

EXAMPLE 1

Find the LCD of $\dfrac{1}{3xy}$, $\dfrac{a}{6x^2}$, and $\dfrac{3xz}{2y^2}$.

$3xy = 3 \cdot x \cdot y$

$6x^2 = 2 \cdot 3 \cdot x^2$ Factor each expression.

$2y^2 = 2 \cdot y^2$

The different factors are:

2 with highest exponent (power) of 1

3 with highest exponent (power) of 1

x with highest exponent (power) of 2

y with highest exponent (power) of 2

The LCD is $2 \cdot 3 \cdot x^2 \cdot y^2 = 6x^2y^2$.

The LCD is the product of the highest power of each of the four factors.   ■

EXAMPLE 2

Find the LCD of $\dfrac{x + 8}{x^2 + 2x + 1}$, $\dfrac{x - 7}{x^2 - 2x + 1}$, and $\dfrac{3x}{x^2 - 1}$.

$x^2 + 2x + 1 = (x + 1)^2$ Factor each denominator.

$x^2 - 2x + 1 = (x - 1)^2$

$x^2 - 1 \quad\;\; = (x - 1)(x + 1)$

The different factors are:

$x + 1$ with a largest exponent (power) of 2

$x - 1$ with a largest exponent (power) of 2

The LCD $= (x + 1)^2(x - 1)^2$.   The LCD is the product of the highest power of each of the two factors. ■

EXAMPLE 3

Find the LCD of $\dfrac{2x - 7}{x^2 + 5x + 6}$ and $\dfrac{x - 4}{x^2 - 4}$.

$x^2 + 5x + 6 = (x + 3)(x + 2)$ Factor each of the denominators.
$x^2 - 4 = (x + 2)(x - 2)$

The different factors are:

$x + 3$ with highest exponent (power) of 1

$x + 2$ with highest exponent (power) of 1

$x - 2$ with highest exponent (power) of 1

The LCD is $(x + 3)(x + 2)(x - 2)$.   The LCD is the product of the highest power of each of the three factors. ■

Addition of rational expressions is defined in the same way as addition of rational numbers. We discuss addition of rational expressions in three stages.

■ **PROCEDURE**

1. Common denominators.

$$\frac{A}{C} + \frac{B}{C} = \frac{A + B}{C}, \; C \neq 0$$

In words, to add two or more rational expressions with common denominators, add the numerators and write that sum over the common denominator.

EXAMPLE 4

Add $\dfrac{6}{2x + 1} + \dfrac{5}{2x + 1}$. State the restrictions on the variable.

$$= \frac{11}{2x + 1}, \ x \neq -\frac{1}{2}$$ Add the numerators and retain the common denominator. The restriction is the value of x that makes the denominator zero. ∎

EXAMPLE 5

Add $\dfrac{x}{x^2 - 1} + \dfrac{1}{x^2 - 1}$. State the restrictions on the variable.

$x \neq \pm 1$ These are the values of x that make the denominators zero.

$$= \frac{x + 1}{x^2 - 1}$$ Since the denominators are common, add the numerators.

$$= \frac{\overset{1}{\cancel{x + 1}}}{(x - 1)\underset{1}{\cancel{(x + 1)}}}$$ Factor the denominator and reduce.

$$= \frac{1}{x + 1}, \ x \neq \pm 1$$ ∎

■ **PROCEDURE**

> 2. Opposite denominators.
>
> $$\frac{A}{C} + \frac{B}{-C} = \frac{A}{C} + \frac{-B}{C} = \frac{A + (-B)}{C} = \frac{A - B}{C}, \ C \neq 0$$

In words, if the denominators are opposites, we can use either as a common denominator. We first multiply either fraction by $\dfrac{-1}{-1}$. Then add as before.

EXAMPLE 6

Add $\dfrac{7x}{x - 2y} + \dfrac{3x}{2y - x}$. State the restrictions on the variable.

$x \neq 2y$ If $x = 2y$, the denominators will be zero.

$$= \frac{7x}{x - 2y} + \left(\frac{-1}{-1}\right)\left(\frac{3x}{2y - x}\right)$$ Note $x - 2y$ and $2y - x$ are opposites. We choose to use $x - 2y$ for the LCD.

$$= \frac{7x}{x - 2y} + \frac{-3x}{x - 2y}$$

The fractions now have a common denominator.

$$= \frac{7x + (-3x)}{x - 2y}$$

$$= \frac{4x}{x - 2y}, \ x \neq 2y$$

If we had chosen $2y - x$ for the LCD, the (equivalent) answer is $\frac{-4x}{2y - x}, \ x \neq 2y$. ■

EXAMPLE 7

Add $\frac{3}{a - b} + \frac{x + 1}{a - b} + \frac{x - 1}{b - a}$. State the restrictions on the variable.

$a \neq b$

Set the denominators equal to zero to find the restrictions.

$$= \frac{3}{a - b} + \frac{x + 1}{a - b} + \left(\frac{-1}{-1}\right)\left(\frac{x - 1}{b - a}\right)$$

$a - b$ and $b - a$ are opposites. We use $a - b$ for the LCD.

$$= \frac{3}{a - b} + \frac{x + 1}{a - b} + \frac{-x + 1}{a - b}$$

Multiply.

$$= \frac{3 + x + 1 - x + 1}{a - b}$$

Add.

$$= \frac{5}{a - b}, \ a \neq b$$

■

■ PROCEDURE

3. Unlike denominators.
   If two expressions have unlike denominators, we build both fractions to have a common denominator and add as before.

$$\frac{A}{B} + \frac{C}{D} = \frac{AD}{BD} + \frac{BC}{BD} = \frac{AD + BC}{BD}, \ BD \neq 0$$

To add two rational expressions with different denominators, the first step is write equivalent fractions with a common denominator.

EXAMPLE 8

Add $\dfrac{3x}{y} + \dfrac{2}{x}$. State the restrictions on the variables.

$$= \dfrac{3x\,x}{y\,x} + \dfrac{2\,y}{x\,y}$$

Build each fraction to have a common denominator. The factors used to do this are shaded.

$$= \dfrac{3x^2}{xy} + \dfrac{2y}{xy}$$

$$= \dfrac{3x^2 + 2y}{xy}, \; xy \neq 0$$

Neither x nor y can be zero. ■

EXAMPLE 9

Add $\dfrac{x - 1}{x + 2} + \dfrac{3}{x + 3}$.

$$= \dfrac{(x - 1)\,(x + 3)}{(x + 2)\,(x + 3)} + \dfrac{3\,(x + 2)}{(x + 2)\,(x + 3)}$$

By observation the LCD is $(x + 2)(x + 3)$. Build each fraction to have a common denominator.

$$= \dfrac{x^2 + 2x - 3}{(x + 2)(x + 3)} + \dfrac{3x + 6}{(x + 2)(x + 3)}$$

Multiply.

$$= \dfrac{x^2 + 2x - 3 + 3x + 6}{(x + 2)(x + 3)}$$

Add.

$$= \dfrac{x^2 + 5x + 3}{(x + 2)(x + 3)}$$

Simplify.

$$= \dfrac{x^2 + 5x + 3}{x^2 + 5x + 6}$$

Alternative form of the answer. However, we usually leave the denominator in factored form. ■

Subtraction of rational expressions is defined in the same way as subtraction of numbers and polynomials.

■ **PROCEDURE**

Subtraction of Rational Expressions

$$\dfrac{A}{B} - \dfrac{C}{D} = \dfrac{A}{B} + \dfrac{-C}{D} = \dfrac{AD - BC}{BD}, \; BD \neq 0$$

In words, to subtract rational expressions, *add the opposite of the rational expression being subtracted*. The opposite of a rational expression is found by writing the opposite of the polynomial in either the numerator or the denominator (see Section 2.5).

$$-\frac{1}{x-1} = \frac{-1}{x-1}, x \neq 1 \quad \text{or}$$

$$-\frac{1}{x-1} = \frac{1}{-(x-1)} = \frac{1}{1-x}, x \neq 1$$

We use this to change subtraction exercises to addition.

EXAMPLE 10

Subtract $\dfrac{x}{x+3} - \dfrac{5}{x+3}$. State the restrictions on the variable.

$$x \neq -3 \qquad\qquad \text{The denominators are zero when } x = -3.$$

$$= \frac{x}{x+3} + \frac{-5}{x+3} \qquad \text{Change subtraction to addition.}$$

$$= \frac{x-5}{x+3}, x \neq -3 \qquad \text{Subtract the numerators and retain the common denominator} \quad\blacksquare$$

EXAMPLE 11

Subtract $\dfrac{2}{x-3} - \dfrac{1}{x-1}$. State the restrictions on the variable.

$$x \neq 1, 3 \qquad\qquad \text{The denominators will be zero when } x = 1 \text{ or } x = 3.$$

$$= \frac{2}{x-3} + \frac{-1}{x-1} \qquad \text{Change subtraction to addition.}$$

By observation the LCD is $(x-3)(x-1)$.

$$= \frac{2(x-1)}{(x-3)(x-1)} + \frac{-1(x-3)}{(x-3)(x-1)} \qquad \text{Build each to have a common denominator.}$$

$$= \frac{2x - 2 - x + 3}{(x-3)(x-1)} \qquad \text{Multiply and add the numerators.}$$

$$= \frac{x+1}{(x-3)(x-1)}, x \neq 1, 3 \qquad \text{Simplify.} \quad\blacksquare$$

When only addition and subtraction of real numbers appear in the same problem, the operations are performed in the order of their appearance. The same procedure is used when adding or subtracting rational expressions.

EXAMPLE 12

Perform the indicated operations. State the restrictions on the variables.

$$\frac{2}{x} + \frac{1}{y} - \frac{a}{z}$$

$xyz \neq 0$	The denominators are zero if any of x, y, or z are zero.
$= \dfrac{2}{x} + \dfrac{1}{y} + \dfrac{-a}{z}$	Change subtraction to addition. The LCD is xyz.
$= \dfrac{2yz}{xyz} + \dfrac{1xz}{xyz} + \dfrac{-axy}{xyz}$	Build each fraction to have a common denominator.
$= \dfrac{2yz + xz - axy}{xyz}, \; xyz \neq 0$	Combine numerators and retain the denominator. ∎

EXAMPLE 13

Perform the indicated operations. State the restrictions on the variable.

$$\frac{3}{x^2 - 4} - \frac{4}{x^2 - 3x - 10} + \frac{1}{x^2 - 7x + 10}$$

To determine the LCD, we must factor each denominator.

$$x^2 - 4 = (x - 2)(x + 2)$$
$$x^2 - 3x - 10 = (x - 5)(x + 2)$$
$$x^2 - 7x + 10 = (x - 5)(x - 2)$$

The LCD of the denominators is $(x - 2)(x + 2)(x - 5)$.

$= \dfrac{3}{x^2 - 4} + \dfrac{-4}{x^2 - 3x - 10} + \dfrac{1}{x^2 - 7x + 10}$	Change subtraction to addition.

$$= \frac{3(x - 5)}{(x - 2)(x + 2)(x - 5)} + \frac{-4(x - 2)}{(x - 2)(x + 2)(x - 5)} + \frac{1(x + 2)}{(x - 2)(x + 2)(x - 5)}$$

$$= \frac{3x - 15}{(x - 2)(x + 2)(x - 5)} + \frac{-4x + 8}{(x - 2)(x + 2)(x - 5)} + \frac{x + 2}{(x - 2)(x + 2)(x - 5)}$$

$$= \frac{3x - 15 - 4x + 8 + x + 2}{(x - 2)(x + 2)(x - 5)}$$

$$= \frac{-5}{(x - 2)(x + 2)(x - 5)}, \; x \neq \pm 2, 5$$

or

$$= \frac{5}{(x-2)(x+2)(x-5)}, \ x \neq \pm 2, \ 5$$

Either form of the answer is acceptable.   ■

Negative exponents lead to rational expressions, as shown in Example 14.

EXAMPLE 14

Perform the indicated operations. State the restrictions on the variable.

$$x^{-2} - 2x^{-1} + 1$$

$$\frac{1}{x^2} - \frac{2}{x} + \frac{1}{1}$$

Write in fraction form with positive exponents.

$$x \neq 0$$

The restriction is made by observation.

$$\frac{1}{x^2} - \frac{2}{x} + \frac{1}{1} = \frac{1}{x^2} + \frac{-2}{x} + \frac{1}{1}$$

Change subtraction to addition. The LCD is x^2.

$$= \frac{1}{x^2} + \frac{-2x}{x^2} + \frac{x^2}{x^2}$$

Build each fraction to a common denominator of x^2.

$$= \frac{1 - 2x + x^2}{x^2}$$

Add the numerators and retain the common denominator.   ■

EXAMPLE 15

Perform the indicated operations. State the restrictions on the variable.

$$(a+5)^{-1} + (a-2)^{-1} - (a-1)^{-1}$$

$$= \frac{1}{a+5} + \frac{1}{a-2} - \frac{1}{a-1}, \ a \neq -5, \ 1, \ 2$$

Rewrite with positive exponents. The LCD is $(a+5)(a-2)(a-1)$.

$$= \frac{1(a-2)(a-1)}{(a+5)(a-2)(a-1)} + \frac{1(a+5)(a-1)}{(a+5)(a-2)(a-1)} + \frac{-1(a+5)(a-2)}{(a+5)(a-2)(a-1)}$$

Build each fraction to have a common denominator. Change subtraction to addition.

$$= \frac{a^2 - 3a + 2}{(a+5)(a-2)(a-1)} + \frac{a^2 + 4a - 5}{(a+5)(a-2)(a-1)} + \frac{-1(a^2 + 3a - 10)}{(a+5)(a-2)(a-1)}$$

Multiply in each numerator.

$$= \frac{a^2 - 3a + 2 + a^2 + 4a - 5 - a^2 - 3a + 10}{(a + 5)(a - 2)(a - 1)}$$

Add numerators and retain the common denominator.

$$= \frac{a^2 - 2a + 7}{(a + 5)(a - 2)(a - 1)}, \ a \neq -5, 1, 2$$ ∎

EXERCISE 3.3

A

Find the LCD of the fractions:

1. $\dfrac{1}{5ab} + \dfrac{1}{5ac}$

2. $\dfrac{3}{4xy} + \dfrac{5}{6yz}$

3. $\dfrac{3y}{6y - 12} - \dfrac{2y}{y - 2}$

4. $\dfrac{7}{3x + 12} - \dfrac{5}{x + 4}$

Add or subtract. Reduce if possible. You do not need to state the variable restrictions:

5. $\dfrac{3}{x} + \dfrac{6}{x} - \dfrac{5}{x}$

6. $\dfrac{2y}{3x} + \dfrac{5y}{3x} - \dfrac{6y}{3x}$

7. $\dfrac{w}{7} + \dfrac{w}{14}$

8. $\dfrac{t}{7} + \dfrac{t}{21}$

9. $\dfrac{a + 1}{3} - \dfrac{a + 2}{2}$

10. $\dfrac{b - 1}{5} - \dfrac{b + 3}{2}$

11. $\dfrac{1}{2(a + 1)} - \dfrac{1}{3(a + 1)}$

12. $\dfrac{1}{4(c - 7)} - \dfrac{1}{3(c - 7)}$

13. $\dfrac{1}{5ab} + \dfrac{1}{5ac}$

14. $\dfrac{3}{4xy} + \dfrac{5}{6yz}$

15. $\dfrac{3y}{6y - 12} - \dfrac{2y}{y - 2}$

16. $\dfrac{7}{3x + 12} - \dfrac{5}{x + 4}$

17. $\dfrac{5}{2y - 2} + \dfrac{4}{3y - 3}$

18. $\dfrac{6}{10y + 5} - \dfrac{5}{8y + 4}$

19. $-2 + \dfrac{5}{3x - 2}$

20. $7 - \dfrac{4}{2x - 3}$

B

Find the LCD of the following fractions:

21. $\dfrac{5x}{x-2} + \dfrac{6}{2-x}$

22. $\dfrac{4}{2y-1} + \dfrac{5}{1-2y}$

23. $\dfrac{1}{3} + \dfrac{1}{x} + \dfrac{3}{x-1}$

24. $\dfrac{3}{5} + \dfrac{1}{y} + \dfrac{y}{y+5}$

Add or subtract. Simplify if possible. You do not need to state the variable restrictions:

25. $\dfrac{7}{w-4} + \dfrac{1}{4-w}$

26. $\dfrac{2}{x-y} + \dfrac{5}{y-x}$

27. $\dfrac{2}{7x} + \dfrac{11}{14x} - \dfrac{2}{21x}$

28. $\dfrac{7}{10y} - \dfrac{7}{15y} - \dfrac{11}{6y}$

29. $\dfrac{3t+4}{5} + t - 1$

30. $\dfrac{2s-1}{4} - s + 2$

31. $\dfrac{5x}{x-2} + \dfrac{6}{2-x}$

32. $\dfrac{4}{2y-1} + \dfrac{5}{1-2y}$

33. $\dfrac{3}{x+6} + \dfrac{5}{2x-3}$

34. $\dfrac{7}{m-8} - \dfrac{5}{m+1}$

Add or subtract. Reduce if possible. State the restrictions on the variables:

35. $\dfrac{1}{3} + \dfrac{1}{x} + \dfrac{3}{x-1}$

36. $\dfrac{3}{5} + \dfrac{1}{y} + \dfrac{y}{y+5}$

37. $\dfrac{1}{2x-3} - \dfrac{x+6}{x+3}$

38. $\dfrac{2x-1}{x+7} - \dfrac{x+3}{x-2}$

39. $\dfrac{10}{y^2-2y} - \dfrac{5}{y^2-4}$

40. $\dfrac{5}{x^2-1} + \dfrac{3}{x^2+x}$

C

Find the LCD of the following fractions:

41. $\dfrac{x+1}{x^2+2x-15} + \dfrac{2x+1}{x^2-4x+3}$

42. $\dfrac{2x-3}{x^2+x-6} - \dfrac{x+3}{x^2-x-2}$

43. $\dfrac{6}{x^2+x-6} + \dfrac{5}{x^2+2x-8} + \dfrac{4}{x^2+7x+12}$

44. $\dfrac{y+6}{y^2-4y+3} - \dfrac{y-3}{y^2+5y-6} + \dfrac{y-1}{y^2+3y-18}$

Add or subtract. Reduce if possible. State the restrictions on the variables:

45. $\dfrac{2}{x+3} - \dfrac{3}{x-5} + \dfrac{x}{x^2 - 2x - 15}$

46. $\dfrac{5}{x-7} + \dfrac{4}{x+2} - \dfrac{3x}{x^2 - 5x - 14}$

47. $\dfrac{2x}{x^2 - 3x - 10} + \dfrac{4}{x^2 - 4}$

48. $\dfrac{3y}{y^2 - 9} - \dfrac{3}{y^2 - 5y - 24}$

49. $\dfrac{6}{x^2 + x - 6} - \dfrac{5}{x^2 + 2x - 8} + \dfrac{4}{x^2 + 7x + 12}$

50. $\dfrac{3}{a^2 + a - 2} - \dfrac{5}{a^2 - 4a + 3} - \dfrac{2}{a^2 - a - 6}$

51. $\dfrac{x+1}{2x^2 - 9x - 5} + \dfrac{1-x}{x^2 - 4x - 5} - \dfrac{x+2}{2x^2 + 3x + 1}$

52. $\dfrac{y+6}{y^2 - 4y + 3} - \dfrac{y-3}{y^2 + 5y - 6} + \dfrac{y-1}{y^2 + 3y - 18}$

53. $\dfrac{1}{x} - \dfrac{2}{x+2} - \dfrac{2}{x^2 + 3x + 2}$

54. $\dfrac{1}{y} - \dfrac{3}{y+3} - \dfrac{3}{y^2 + 5y + 6}$

55. $\dfrac{3y}{1-y} - \dfrac{2}{y^2 - 1}$

56. $\dfrac{2}{y^2 - y - 6} - \dfrac{1}{3 + 2y - y^2}$

57. $\dfrac{p}{q^{-1}} - \dfrac{p^{-1}}{q}$

58. $\dfrac{p^{-1}}{q} - pq^{-1}$

59. $\dfrac{(x+2)^{-1}}{x} + x$

60. $\dfrac{(y+1)}{(y-3)^{-1}} - y^{-1}$

Perform the indicated operations. You do not need to state the variable restrictions:

61. $x^{-1} + (x+1)^{-1} - x^{-2}$

62. $x^{-1} + (x-1)^{-1} - (x^2 - 1)^{-1}$

63. $(x+2)^{-1} + (x-2)^{-1}$

64. $(x^2 - 9)^{-1} + (x+3)^{-1} - (x+3)^{-1}$

65. $(x^2 + x - 6)^{-1} - 2(x^2 - 4)^{-1} + 3(x^2 + 5x + 6)^{-1}$

66. $2(x^2 - 7x + 12)^{-1} + (x^2 - 2x - 8)^{-1} + 3(x^2 - x - 6)^{-1}$

D

67. The formula for the total resistance (R), in ohms, of two resistors connected in parallel is given by

$$\frac{1}{R} = \frac{1}{r_1} + \frac{1}{r_2},$$

where r_1 is the resistance of the first resistor and r_2 is the resistance of the second resistor. Combine the terms on the right side of the formula.

68. The formula for the total resistance (R), in ohms, of three resistors connected in parallel is given by

$$\frac{1}{R} = \frac{1}{r_1} + \frac{1}{r_2} + \frac{1}{r_3},$$

where r_1 is the resistance of the first resistor, r_2 is the resistance of the second resistor, and r_3 is the resistance of the third resistor. Combine the terms on the right side of the formula.

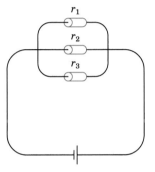

Figure for Exercise 68

STATE YOUR UNDERSTANDING

69. Define least common denominator (LCD).

CHALLENGE EXERCISES

Perform the indicated operations. You need not state the restrictions on the variables.

70. $\dfrac{1}{x^2} + (x - 1)^{-1} - \dfrac{3}{x^2 - 1}$

71. $\left(\dfrac{x + 2}{x + 1}\right)^{-1} + \dfrac{3}{x - 2} - (x^2 - 4)^{-1}$

MAINTAIN YOUR SKILLS (SECTION 1.7)

Solve the following equations:

72. $3x - 5 = 2x + 7$

73. $6x + 7 = 7x - 1$

74. $3x - 9 + 8x = x + 9$

75. $4x - 12 - 7x = 2x - 5$

76. $(x - 2)(x - 3) = (x + 5)(x + 3)$

77. $(2x + 4)(x + 4) = (x - 5)(2x + 5)$

78. $(3x + 5)(2x + 3) = (x - 5)(6x + 7)$

79. $(x - 5)^2 = (x + 2)(x + 3)$

3.4

COMPLEX FRACTIONS

1. Simplify complex fractions.

Complex Fraction

A *complex fraction* is a fraction that contains a fraction in the numerator, the denominator, or both. A complex fraction is a quotient of two fractions or rational expressions.

Some examples of complex fractions are

$$\frac{3}{\dfrac{1}{x} + 5}, \qquad \frac{\dfrac{2}{x} + 3}{7}, \quad \text{and} \quad \frac{1 + \dfrac{1}{1 + \dfrac{1}{x}}}{x}.$$

There are two procedures for simplifying complex fractions.

■ PROCEDURE I

Simplifying Complex Fractions

$$\frac{\dfrac{A}{B}}{\dfrac{C}{D}} = \frac{A}{B} \div \frac{C}{D} = \frac{A}{B} \cdot \frac{D}{C}, \quad BCD \neq 0$$

Write the complex fraction as a division problem, do the division, and simplify.

EXAMPLE 1

Simplify. State the restrictions on the variable.

$$\frac{\dfrac{1}{x}}{\dfrac{3}{y}}$$

$$xy \neq 0$$

The restrictions are a result of $x \neq 0$, $y \neq 0$, and $\dfrac{3}{y} \neq 0$. No denominator can be zero.

$$\frac{\dfrac{1}{x}}{\dfrac{3}{y}} = \frac{1}{x} \div \frac{3}{y}$$

Write the fraction as a division exercise.

$$= \frac{1}{x} \cdot \frac{y}{3}$$

Multiply by the reciprocal.

$$= \frac{y}{3x}, \; xy \neq 0$$

Multiply.   ∎

EXAMPLE 2

Simplify. State the restrictions on the variable.

$$\frac{1 + \dfrac{1}{x}}{1 - \dfrac{1}{y}}$$

$$xy \neq 0, \; y \neq 1$$

Since no denominator can be zero, $x \neq 0$, $y \neq 0$. Also $1 - \dfrac{1}{y} \neq 0$, implying $y \neq 1$.

$$\frac{1 + \dfrac{1}{x}}{1 - \dfrac{1}{y}} = \frac{\dfrac{x+1}{x}}{\dfrac{y-1}{y}}$$

Perform the operations in the numerator (add) and denominator (subtract). This follows the order of operations because a fraction bar groups terms in the same way as parentheses.

$$= \frac{x+1}{x} \div \frac{y-1}{y}$$

Rewrite as a division exercise.

$$= \frac{x+1}{x} \cdot \frac{y}{y-1}$$

Multiply by the reciprocal.

$$= \frac{xy + y}{xy - x}, \; xy \neq 0, \; y \neq 1$$ ∎

Example 3 is more complicated.

EXAMPLE 3

Simplify. State the restrictions on the variable.

$$\frac{\dfrac{x}{x+1} + \dfrac{1}{x-1}}{\dfrac{1}{x-1} + \dfrac{1}{x+1}}$$

$$x \neq -1, 0, 1$$

Restrict the variable to avoid division by zero. The restriction $x \neq 0$ is easier to see in the final simplified form.

We work this example in a different order to show different ways of doing the work.

$$= \left[\frac{x}{x+1} + \frac{1}{x-1}\right] \div \left[\frac{1}{x-1} + \frac{1}{x+1}\right]$$

First, rewrite as a division exercise.

$$= \frac{x^2 + 1}{(x+1)(x-1)} \div \frac{2x}{(x+1)(x-1)}$$

Next, add inside the brackets.

$$= \frac{x^2 + 1}{(x+1)(x-1)} \cdot \frac{(x+1)(x-1)}{2x}$$

Then multiply by the reciprocal.

$$= \frac{x^2 + 1}{2x}, \; x \neq -1, 0, 1.$$ ∎

The second method of simplifying complex fractions uses the basic principle of fractions.

■ **PROCEDURE II**

Simplifying Complex Fractions

$$\frac{\dfrac{A}{B}}{\dfrac{C}{D}} = \frac{BD\left[\dfrac{A}{B}\right]}{BD\left[\dfrac{C}{D}\right]} = \frac{AD}{BC}$$

Multiply both the numerator and denominator of the complex fraction by the LCD of the fractions within the expression.

EXAMPLE 4

Simplify. State the restrictions on the variable.

$$\frac{\dfrac{1}{6}}{\dfrac{3}{y}}$$

$$y \neq 0$$

$$\frac{\dfrac{1}{6}}{\dfrac{3}{y}} = \frac{6y \cdot \dfrac{1}{6}}{6y \cdot \dfrac{3}{y}}$$

Multiply the numerator and denominator by the LCD ($6y$) of both denominators.

$$= \frac{y}{18}, \; y \neq 0$$ ∎

EXAMPLE 5

Simplify. State the restrictions on the variable.

$$\frac{\dfrac{x}{3} - \dfrac{1}{2}}{\dfrac{x}{4} + \dfrac{1}{6}}$$

$$x \neq -\frac{2}{3}$$

Since the denominator of the complex fraction cannot be zero, we have $\dfrac{x}{4} + \dfrac{1}{6} \neq 0$, which results in the restriction $x \neq -\dfrac{2}{3}$.

$$\frac{\dfrac{x}{3} - \dfrac{1}{2}}{\dfrac{x}{4} + \dfrac{1}{6}} = \frac{12\left[\dfrac{x}{3} - \dfrac{1}{2}\right]}{12\left[\dfrac{x}{4} + \dfrac{1}{6}\right]}$$

Multiply both numerator and denominator by the LCD (12) and simplify.

$$= \frac{4x - 6}{3x + 2}, \; x \neq -\frac{2}{3} \qquad \blacksquare$$

We now rework Example 3 using Procedure II.

EXAMPLE 6

Simplify. State the restrictions on the variable.

$$\frac{\dfrac{x}{x + 1} + \dfrac{1}{x - 1}}{\dfrac{1}{x - 1} + \dfrac{1}{x + 1}}$$

Same restrictions as Example 3.

$$x \neq -1, 0, 1$$

$$= \frac{(x + 1)(x - 1)\left[\dfrac{x}{x + 1} + \dfrac{1}{x - 1}\right]}{(x + 1)(x - 1)\left[\dfrac{1}{x - 1} + \dfrac{1}{x + 1}\right]}$$

Multiply the numerator and denominator by the LCD of the denominators, $(x + 1)(x - 1)$.

$$= \frac{\dfrac{x(x + 1)(x - 1)}{x + 1} + \dfrac{(x + 1)(x - 1)}{x - 1}}{\dfrac{(x + 1)(x - 1)}{x - 1} + \dfrac{(x + 1)(x - 1)}{x + 1}}$$

$$= \frac{x^2 - x + x + 1}{x + 1 + x - 1}$$ Multiply.

$$= \frac{x^2 + 1}{2x}, \; x \neq -1, 0, 1$$ Simplify.

Check Using A Calculator

We can check that the fractions are equivalent by substituting a value for x in the original fraction and the simplified fraction to see if they are equal. Let $x = 5$.

$$\frac{\dfrac{x}{x + 1} + \dfrac{1}{x - 1}}{\dfrac{1}{x - 1} + \dfrac{1}{x + 1}}$$ becomes

$$\frac{\dfrac{5}{5 + 1} + \dfrac{1}{5 - 1}}{\dfrac{1}{5 - 1} + \dfrac{1}{5 + 1}} = \frac{\dfrac{5}{6} + \dfrac{1}{4}}{\dfrac{1}{4} + \dfrac{1}{6}}$$ Substitute 5 for x in the original fraction and simplify.

Enter	Display	
5 ÷ 6 + 1 ÷ 4 =	1.0833333	The value of the numerator.
÷ (1 ÷ 4 + 1 ÷ 6) =	2.6	The value of the fraction.

$$\frac{x^2 + 1}{2x} = \frac{(5)^2 + 1}{2(5)}$$ Substitute 5 for x in the simplified fraction.

Enter	Display	
5 x^2 + 1 =	26.	The value of the numerator.
÷ (2 × 5) =	2.6	The value of the fraction.

So the fractions are equivalent. ■

EXAMPLE 7

Simplify. State the restrictions on the variable.

$$\frac{3}{2 - \dfrac{1}{1 - \dfrac{2}{a}}}$$

$$\frac{3}{2 - \dfrac{1}{1 - \dfrac{2}{a}}} = \frac{3}{2 - \left[\dfrac{a}{a}\right]\left[\dfrac{1}{1 - \dfrac{2}{a}}\right]}$$

Multiply $\dfrac{1}{1 - \dfrac{2}{a}}$ by $\dfrac{a}{a}$ to eliminate the complex fraction in the denominator.

$$= \frac{3}{2 - \dfrac{a}{a - 2}}$$

Multiply.

$$= \frac{a - 2}{a - 2} \cdot \left[\frac{3}{2 - \dfrac{a}{a - 2}}\right]$$

Multiply the numerator and denominator by $a - 2$.

$$= \frac{3(a - 2)}{2(a - 2) - a}$$

Simplify.

$$= \frac{3a - 6}{2a - 4 - a}$$

Simplify.

$$= \frac{3a - 6}{a - 4}, \ a \neq 0, 2, 4$$

The restrictions are listed here because it is easier to simplify first and then check back for restrictions. ∎

EXERCISE 3.4

A

Simplify. You do not need to state the variable restrictions:

1. $\dfrac{\dfrac{3}{x}}{\dfrac{6}{y}}$

2. $\dfrac{\dfrac{5}{m}}{\dfrac{m}{3}}$

3. $\dfrac{3 + \dfrac{4}{5}}{2 + \dfrac{1}{5}}$

4. $\dfrac{6 - \dfrac{3}{4}}{8 - \dfrac{1}{4}}$

5. $\dfrac{\dfrac{1}{x + 1}}{\dfrac{1}{x - 1}}$

6. $\dfrac{\dfrac{2}{x + 6}}{\dfrac{3}{2x - 1}}$

7. $\dfrac{\dfrac{m}{3} + 2}{n}$

8. $\dfrac{5 - \dfrac{a}{4}}{b}$

9. $\dfrac{q - \dfrac{1}{3}}{p + \dfrac{1}{3}}$

10. $\dfrac{t - \dfrac{1}{4}}{s + \dfrac{1}{8}}$

11. $\dfrac{\dfrac{w + 1}{2} + 1}{\dfrac{w + 2}{3} + 1}$

12. $\dfrac{\dfrac{x + 4}{3} - 2}{\dfrac{x - 3}{6} - 4}$

Simplify. State the restrictions on the variables:

13. $\dfrac{2 - \dfrac{1}{a}}{3 + \dfrac{1}{a}}$

14. $\dfrac{5 + \dfrac{2}{b}}{6 - \dfrac{2}{b}}$

15. $\dfrac{x^{-1} + 2}{x^{-1} - 3}$

16. $\dfrac{3a^{-1} + 2}{2a^{-1} - 3}$

B

Simplify. You do not need to state the variable restrictions:

17. $\dfrac{\dfrac{x}{3} + 2}{\dfrac{x + 1}{3}}$

18. $\dfrac{4 + \dfrac{6}{y + 1}}{\dfrac{3}{y + 1}}$

19. $\dfrac{\dfrac{x}{y + 2}}{\dfrac{x}{y + 2} + 5}$

20. $\dfrac{\dfrac{m}{n - 5}}{2 - \dfrac{m}{n - 5}}$

21. $\dfrac{t - \dfrac{t^2}{w}}{t + \dfrac{t}{w^2}}$

22. $\dfrac{\dfrac{p}{(q + 1)} - 1}{\dfrac{p^2}{(q + 1)^2} + 1}$

23. $\dfrac{\dfrac{(2x + 1)}{5} + \dfrac{x + 1}{3}}{\dfrac{x - 1}{15}}$

24. $\dfrac{\dfrac{(x + 5)}{4} - \dfrac{x - 1}{2}}{\dfrac{x + 3}{8}}$

25. $\dfrac{\dfrac{5y - 8}{3}}{\dfrac{y + 1}{4} - \dfrac{y - 1}{6}}$

26. $\dfrac{\dfrac{3x - 7}{8}}{\dfrac{x - 2}{4} + \dfrac{x - 1}{12}}$

27. $\dfrac{\dfrac{3a}{7} + \dfrac{b}{14}}{\dfrac{3a}{28} - \dfrac{b}{4}}$

28. $\dfrac{\dfrac{3}{b} - \dfrac{4}{3a}}{\dfrac{6}{a} + \dfrac{5}{2b}}$

Simplify. State the restrictions on the variables:

29. $\dfrac{2a^{-1} - 3b^{-1}}{b - a}$

30. $\dfrac{y + x}{5x^{-1} - 4y^{-1}}$

31. $\dfrac{(x + 2)^{-1} + 3}{(x + 2)^{-1} + 4}$

32. $\dfrac{4 - (x - 1)^{-1}}{5 + (x - 1)^{-1}}$

C

Simplify. You do not need to state the variable restrictions:

33. $\dfrac{\dfrac{1}{a+b} - \dfrac{1}{a-b}}{\dfrac{1}{a+b} + \dfrac{1}{a-b}}$

34. $\dfrac{\dfrac{2}{x+3} + \dfrac{3}{x-3}}{\dfrac{4}{x+3} - \dfrac{1}{x-3}}$

35. $\dfrac{y+6+\dfrac{1}{y-2}}{y-3+\dfrac{1}{y-2}}$

36. $\dfrac{b-5-\dfrac{1}{b+6}}{b-2+\dfrac{1}{b+6}}$

37. $\dfrac{\dfrac{2}{a-1} + \dfrac{1}{a+1}}{\dfrac{6}{a^2-1}}$

38. $\dfrac{\dfrac{5}{x^2-1}}{\dfrac{2}{x-1} - \dfrac{3}{x+1}}$

39. $\dfrac{2}{1+\dfrac{3}{1+\dfrac{1}{y}}}$

40. $\dfrac{x}{2-\dfrac{5}{1-\dfrac{3}{x}}}$

41. $[(x+2)^{-1} - (x-2)^{-1}]^{-1}$

42. $(x(x+1)^{-1} + x^{-1}(x+1))^{-1}$

Simplify:

43. $\dfrac{1-\dfrac{1}{1-\dfrac{1}{x}}}{1+\dfrac{1}{1+\dfrac{1}{x}}}$

44. $\dfrac{\dfrac{1}{y} - \dfrac{1}{1-\dfrac{2}{y}}}{\dfrac{1}{y} - \dfrac{1}{1-\dfrac{3}{y}}}$

45. $\dfrac{2a+\dfrac{1}{6-\dfrac{1}{a}}}{3a+\dfrac{2}{6+\dfrac{1}{a}}}$

46. $\dfrac{5x-\dfrac{2}{3-\dfrac{1}{x}}}{2x+\dfrac{1}{3+\dfrac{1}{x}}}$

47. $\dfrac{1-\dfrac{1}{x^2-1}}{x-1+\dfrac{1}{1-\dfrac{1}{x-1}}}$

48. $\dfrac{2-\dfrac{1}{x^2-4}}{x+1+\dfrac{1}{1-\dfrac{1}{x-2}}}$

49. $\dfrac{1+\dfrac{1}{1+1}}{1+\dfrac{1}{1+\dfrac{1}{1+1}}}$

50. $\dfrac{1+\dfrac{2}{1+2}}{1+\dfrac{1}{1+\dfrac{1}{2+3}}}$

D

51. Show that the formula $\dfrac{1}{R} = \dfrac{1}{r_1} + \dfrac{1}{r_2}$ is equivalent to the formula

$$R = \dfrac{1}{\dfrac{1}{r_1} + \dfrac{1}{r_2}}.$$

52. Show that the formula $\dfrac{1}{R} = \dfrac{1}{r_1} + \dfrac{1}{r_2} + \dfrac{1}{r_3}$ is equivalent to the formula

$$R = \dfrac{1}{\dfrac{1}{r_1} + \dfrac{1}{r_2} + \dfrac{1}{r_3}}.$$

53. Show that $\dfrac{x^{-1} + 3^{-1}}{x^{-1} - 3^{-1}} \neq \dfrac{(x+3)^{-1}}{(x-3)^{-1}}$.

STATE YOUR UNDERSTANDING

54. What is a complex fraction?

CHALLENGE EXERCISES

Simplify. You need not state the restrictions on the variables. Assume that all exponents are positive integers.

55. $(x^n - 1)^{-1} - (x^n + 1)^{-1} - (x^{2n} - 1)^{-1}$

56. $(x^{2n} + 4)^{-1} + (x^{2n} - 4)^{-1} - 8(x^{4n} - 16)^{-1}$

MAINTAIN YOUR SKILLS (SECTION 1.7)

57. $\dfrac{3}{4}x + \dfrac{3}{4} = \dfrac{1}{2}x - 4$

58. $\dfrac{2}{3}a - \dfrac{1}{4} = \dfrac{3}{4}a + \dfrac{2}{3}$

59. $\dfrac{4}{5}x - \dfrac{7}{8} = \dfrac{1}{2}x - \dfrac{11}{12}$

60. $\dfrac{5}{6}b - \dfrac{9}{10} - \dfrac{2}{5} = \dfrac{1}{15}b + 1$

61. $\dfrac{7}{9}y - 1 + \dfrac{2}{3}y = 12$

62. $\dfrac{15}{16}z + \dfrac{7}{8} - \dfrac{1}{2}z = \dfrac{9}{16} - \dfrac{1}{4}z$

63. $\dfrac{9}{10}a + \dfrac{3}{4} - 3a = 2a + \dfrac{9}{10}$

64. $\dfrac{25}{32}x - \dfrac{13}{16} + \dfrac{1}{4}x = \dfrac{5}{64}$

3.5

**EQUATIONS
CONTAINING
RATIONAL
EXPRESSIONS**

OBJECTIVE

1. Solve equations containing rational expressions.

To solve an equation that contains one or more rational expressions, we first *clear the fractions*. This means write an equation that is equivalent to the given equation that contains no rational expressions.

■ PROCEDURE

To solve an equation containing rational expressions:

1. Multiply both sides of the equation by the LCD of the fractions. This will clear the fractions.

2. Solve the resulting equation by the usual methods.

■ CAUTION

Multiplying each side of a conditional equation by zero does not always yield an equivalent equation.

$x = 5$ True only for $x = 5$.

$0 \cdot x = 0 \cdot 5$ Not equivalent.

$0 \cdot x = 0$ True for all x.

To determine whether the solution set of the resulting equation is the same as the solution set of the original equation, we must:

1. Check all solutions in the original equation, or

2. Compare the solutions to the list of restricted values.

Any solution for which a denominator has value zero must be discarded.

EXAMPLE 1

Solve: $\dfrac{1}{x} + \dfrac{1}{3} = \dfrac{1}{2}$

$x \neq 0$ List the restrictions on x.

$$6x\left[\frac{1}{x} + \frac{1}{3}\right] = 6x\left[\frac{1}{2}\right]$$

Multiply both sides by $6x$, the LCD of the denominators.

$$6x\left[\frac{1}{x}\right] + 6x\left[\frac{1}{3}\right] = 6x\left[\frac{1}{2}\right]$$

Multiply each term by $6x$. Use the distributive property.

$$6 + 2x = 3x$$

Multiply.

$$6 = x$$

Check:

$$\frac{1}{6} + \frac{1}{3} = \frac{1}{2}$$

Substitute 6 for x in the original equation.

$$\frac{1}{6} + \frac{2}{6} = \frac{1}{2}$$

Write each fraction with a common denominator of 6.

$$\frac{3}{6} = \frac{1}{2}$$

Add.

$$\frac{1}{2} = \frac{1}{2}$$

The solution set is $\{6\}$.   ■

EXAMPLE 2

Solve: $\dfrac{1}{x - 2} + \dfrac{1}{x + 2} = \dfrac{4}{x^2 - 4}$

$$x \neq -2, 2$$

List the restrictions on x.

$$(x - 2)(x + 2)\left[\frac{1}{x - 2} + \frac{1}{x + 2}\right]$$

$$= (x - 2)(x + 2)\left[\frac{4}{x^2 - 4}\right]$$

Multiply both sides by the LCD of the denominators. The LCD is $(x - 2)(x + 2)$.

$$x + 2 + x - 2 = 4$$

$$2x = 4$$

$$x = 2$$

Since 2 is one of the restricted values for the variable, the solution set is $\emptyset$.   ■

EXAMPLE 3

Solve: $\dfrac{1}{x + 4} + \dfrac{1}{x - 1} = \dfrac{x + 5}{x^2 + 3x - 4}$

$x + 4 = 0$ or $x - 1 = 0$ To find the restrictions on x set each
$\quad\ x = -4$ $\ x = 1$ denominator equal to 0.

or

$x^2 + 3x - 4 = 0$
$(x + 4)(x - 1) = 0$

$x + 4 = 0$ or $x - 1 = 0$
$\quad\ x = -4$ $\ x = 1$

Therefore, $x \neq -4, 1$.   List the restrictions on x.

$(x + 4)(x - 1)\left[\dfrac{1}{x + 4} + \dfrac{1}{x - 1}\right]$ Multiply each side by the LCD of the
fractions.

$= (x + 4)(x - 1)\left[\dfrac{x + 5}{x^2 + 3x - 4}\right]$

$x - 1 + x + 4 = x + 5$
$\qquad\qquad\quad x = 2$

Since 2 does not appear in the list of restrictions, 2 is the solution if we have made no errors.

Check using a calculator.

$\dfrac{1}{x + 4} + \dfrac{1}{x - 1}$

when $x = 2$, $\dfrac{1}{2 + 4} + \dfrac{1}{2 - 1}$

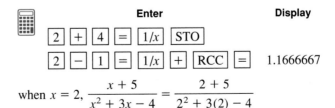

	Enter	**Display**
	2 + 4 = 1/x STO	
	2 − 1 = 1/x + RCC =	1.1666667

when $x = 2$, $\dfrac{x + 5}{x^2 + 3x - 4} = \dfrac{2 + 5}{2^2 + 3(2) - 4}$

	Enter	**Display**
	2 x^2 + 3 × 2 − 4 = STO	
	2 + 5 = ÷ RCC =	1.1666667

So, $x = 2$ checks.

The solution set is $\{2\}$.

■ CAUTION

> The steps for adding and subtracting fractions are *different* from those used for solving equations even though the LCD is used for both.

EXAMPLE 4

Combine: $\dfrac{x-2}{x} - \dfrac{x-3}{x-6} - \dfrac{1}{x}$

$x \neq 0, 6$　　List the restrictions on the variable.

$= \dfrac{x-6}{x-6} \cdot \dfrac{x-2}{x} - \dfrac{x}{x} \cdot \dfrac{x-3}{x-6} - \dfrac{x-6}{x-6} \cdot \dfrac{1}{x}$　　Build each fraction to a common denominator. The LCD is $x(x-6)$.

$= \dfrac{x^2 - 8x + 12}{x(x-6)} - \dfrac{x^2 - 3x}{x(x-6)} - \dfrac{x-6}{x(x-6)}$　　Add the numerators and retain the common denominator.

$= \dfrac{x^2 - 8x + 12 - (x^2 - 3x) - (x-6)}{x(x-6)}$

$= \dfrac{-6x + 18}{x(x-6)}, \ x \neq 0, 6$

■ CAUTION

> The answer is a fraction. The denominator is not eliminated.

Solve: $\dfrac{x-2}{x} - \dfrac{x-3}{x-6} - \dfrac{1}{x} = 0$

$x \neq 0, 6$　　Restrict the variable.

$x(x-6)\left[\dfrac{x-2}{x} - \dfrac{x-3}{x-6} - \dfrac{1}{x}\right] = x(x-6)(0)$　　Multiply each side by the LCD.

$(x-6)(x-2) - x(x-3) - 1(x-6) = 0$　　Multiply.

$x^2 - 8x + 12 - (x^2 - 3x) - (x-6) = 0$

$$-6x + 18 = 0$$

$$x = 3$$ The check is left for the student.

The solution set is $\{3\}$. ∎

In the solution of fractional equations, we eliminate the fractions in the second step. This is not possible when combining fractions.

When working with formulas and equations that contain more than one variable, it is convenient to solve for one of the variables.

EXAMPLE 5

Solve: $\dfrac{1}{t} + \dfrac{1}{s} = \dfrac{1}{2}$ for t.

$$2st\left[\frac{1}{t} + \frac{1}{s}\right] = 2st\left[\frac{1}{2}\right]$$ Multiply each side by the LCD of t, s, and 2.

$$2s + 2t = st$$ Multiply.

$$2s = st - 2t$$

$$2s = t(s - 2)$$ Factor the right side to isolate t.

$$\frac{2s}{s - 2} = t, \; st \neq 0, \; s \neq 2$$ ∎

Example 6 is an application involving rational expressions.

EXAMPLE 6

Martha and Mary can build a fence in 20 hours. Martha can build the fence by herself in 9 hours less time than it takes Mary. How many hours would it take each of them alone?

Simpler word form:

$$\begin{pmatrix} \text{Fraction of} \\ \text{fencing done} \\ \text{by Martha in} \\ \text{1 hour} \end{pmatrix} + \begin{pmatrix} \text{Fraction of} \\ \text{fencing done} \\ \text{by Mary in} \\ \text{1 hour} \end{pmatrix} = \begin{pmatrix} \text{Fraction of} \\ \text{fencing done} \\ \text{by both in} \\ \text{1 hour} \end{pmatrix}$$

The part of the fence that Martha built plus the part that Mary built equals the total built in one hour.

Select variable:

Let t represent the time it takes Mary to build the fence alone.

	Time	Fraction Done in 1 Hour
Martha	$t - 9$	$\dfrac{1}{t - 9}$
Mary	t	$\dfrac{1}{t}$
Both	20	$\dfrac{1}{20}$

Use a chart to help organize the information in a usable form.

Martha can do the job in 9 hours less time.

Time it takes working together.

Translate to algebra:

$$\frac{1}{t - 9} + \frac{1}{t} = \frac{1}{20}$$

Replace the parts of the simpler word form with the appropriate fraction from the chart.

Solve:

$$20t(t - 9)\left(\frac{1}{t - 9} + \frac{1}{t}\right) = 20t(t - 9)\left(\frac{1}{20}\right) \quad t \neq 0, 9$$

Multiply each side by the LCD.

$$20t + 20t - 180 = t^2 - 9t$$

Multiply.

$$t^2 - 49t + 180 = 0$$

$$(t - 4)(t - 45) = 0$$

Factor.

$$t = 4 \quad \text{or} \quad t = 45$$

Solve using the zero-product law.

Martha takes 9 hours less time, therefore, $45 - 9 = 36$.

The solution $x = 4$ is rejected because $4 - 9 = -5$ is negative. The variable t representing time is a positive number in this application.

So, Mary takes 45 hours alone and Martha takes 36 hours alone to build the fence. ■

EXERCISE 3.5

A

Solve:

1. $\dfrac{x}{4} + \dfrac{1}{3} = \dfrac{x}{12}$

2. $\dfrac{w}{15} - \dfrac{1}{6} = \dfrac{w}{3}$

3. $\dfrac{5}{x} + \dfrac{3}{2x} = \dfrac{-1}{4}$

4. $\dfrac{2}{3a} + \dfrac{1}{5} = \dfrac{3}{a}$

5. $\dfrac{5}{t-2} = 4$

6. $\dfrac{-2}{3t+2} = 4$

7. $\dfrac{2}{x-5} = \dfrac{3}{2x+1}$

8. $\dfrac{-4}{x+2} = \dfrac{2}{2x-5}$

9. $\dfrac{4}{x+3} = \dfrac{5}{2x+7}$

10. $\dfrac{-3}{y-5} = \dfrac{2}{y+3}$

11. $\dfrac{7}{3y+7} = \dfrac{-2}{2y-9}$

12. $\dfrac{5}{4x+3} = \dfrac{2}{4x-3}$

13. $\dfrac{4}{x-3} + \dfrac{3}{x-3} = 5$

14. $\dfrac{6}{y-5} - \dfrac{4}{y-5} = 8$

15. $\dfrac{5}{x-1} - \dfrac{3}{2} = 1$

16. $\dfrac{3}{x+1} + \dfrac{2}{5} = 1$

17. $\dfrac{2}{x+4} + \dfrac{3}{x+4} = 10$

18. $\dfrac{7}{x-8} - \dfrac{3}{x-8} = 2$

19. $\dfrac{3}{x+4} + \dfrac{5}{x+4} = \dfrac{1}{2}$

20. $\dfrac{7}{x-8} + \dfrac{3}{x-8} = \dfrac{2}{3}$

B

21. $\dfrac{1}{x+2} - \dfrac{3}{4} = \dfrac{7}{x+2}$

22. $\dfrac{12}{x+2} + \dfrac{3}{x} = 2$

23. $\dfrac{1}{x-6} + \dfrac{1}{x+4} + \dfrac{10}{x^2-2x-24} = 0$

24. $\dfrac{1}{z+5} - \dfrac{2}{z-2} + \dfrac{14}{z^2+3z-10} = 0$

25. $\dfrac{3}{y+3} - \dfrac{9}{y^2-9} = \dfrac{2}{y-3}$

26. $\dfrac{7w+1}{w^2-9} - \dfrac{5}{w-3} = \dfrac{10}{w+3}$

27. $\dfrac{3}{2-x} - \dfrac{7}{x-2} = 4$

28. $\dfrac{y}{y-3} + \dfrac{1}{3-y} = 3$

29. $\dfrac{4}{x-3} - \dfrac{8}{x^2-9} = \dfrac{3}{3-x}$

30. $\dfrac{7}{y-2} + \dfrac{5}{y^2-4} = \dfrac{2}{2-y}$

31. Solve for c: $\dfrac{1}{a} + \dfrac{1}{b} = \dfrac{1}{c}$

32. Solve for c: $\dfrac{2}{a-c} + \dfrac{2}{a} = 1$

33. Solve for a: $\dfrac{3}{b+a} - \dfrac{2}{b} = 5$

34. Solve for z: $\dfrac{1}{xy} + \dfrac{1}{xz} - \dfrac{1}{yz} = 3$

35. Solve for b: $\dfrac{3}{x-b} + \dfrac{2}{x+b} = \dfrac{c}{x^2 - b^2}$

36. Solve for a: $\dfrac{5}{a+b} + \dfrac{3}{a-b} = \dfrac{a}{a^2 - b^2}$

Solve:

37. $\dfrac{7}{x+2} - \dfrac{3}{x+4} = \dfrac{2}{x^2 + 6x + 8}$

38. $\dfrac{5}{x-3} + \dfrac{2}{x-5} = \dfrac{4}{x^2 - 8x + 15}$

39. $\dfrac{9}{x+1} + \dfrac{3}{2x+1} = \dfrac{-30}{2x^2 + 3x + 1}$

40. $\dfrac{12}{2x+3} + \dfrac{4}{3x-2} = \dfrac{16}{6x^2 + 5x - 6}$

C

41. $\dfrac{1}{x^2 + 4x + 3} - \dfrac{1}{x^2 - 5x - 6} = \dfrac{1}{x^2 - 3x - 18}$

42. $\dfrac{3}{x^2 - 25} + \dfrac{1}{x^2 - x - 30} = \dfrac{2}{x^2 - 11x + 30}$

43. $\dfrac{3}{x^2 + 4x - 21} - \dfrac{5}{x^2 - 8x + 15} = \dfrac{8}{x^2 + 2x - 35}$

44. $\dfrac{6}{2x^2 + 13x - 24} + \dfrac{1}{3x^2 + 23x - 8} = \dfrac{3}{6x^2 - 11x + 3}$

45. $\dfrac{x+3}{x^2 - x - 2} + \dfrac{x+6}{x^2 + 3x + 2} = \dfrac{2x-1}{x^2 - 4}$

46. $\dfrac{y+1}{y^2 - 2y - 24} = \dfrac{2y+5}{y^2 - 9y + 18} - \dfrac{y+2}{y^2 + y - 12}$

47. $\dfrac{3x+1}{2x^2 + 7x - 4} - \dfrac{x-1}{2x^2 - 7x + 3} = \dfrac{x+5}{x^2 + x - 12}$

48. $\dfrac{2x-1}{3x^2 + 13x - 10} + \dfrac{4x}{3x^2 + 16x - 12} = \dfrac{2x+7}{x^2 + 11x + 30}$

49. $\dfrac{4+x}{x^2 - 2x - 8} - \dfrac{3+2x}{2x^2 + 5x + 2} = \dfrac{7}{2x^2 - 7x - 4}$

50. $\dfrac{4}{4x^2 + 4x - 3} + \dfrac{x+2}{2x^2 + 5x - 3} = \dfrac{x+1}{2x^2 + 9x + 9}$

51. $\dfrac{2}{a^2 - a - 12} + \dfrac{1}{a^2 - 9} = \dfrac{2}{a^2 - 7a + 12}$

52. $\dfrac{1}{b^2 + 5b + 4} + \dfrac{1}{b^2 + 3b - 4} = \dfrac{-3}{b^2 - 1}$

Hint: In Exercises 53–56, you need to recall the definition of a negative exponent.

53. $x^{-1} + 5^{-1} = 3x^{-1} + 1$

54. $3(x + 2)^{-1} + 5(x - 8)^{-1} = \dfrac{2}{x^2 - 6x - 16}$

55. $3(x - 3)^{-1} + 4(x + 3)^{-1} = 4(x^2 - 9)^{-1}$

56. $8(x + 5)^{-1} + 3(x - 2)^{-1} = 15(x^2 + 3x - 10)^{-1}$

Hint: In Exercises 57–60, simplify the complex fraction before attempting to solve the equation.

57. $\dfrac{\dfrac{5}{x} + \dfrac{2}{3}}{\dfrac{3}{x}} = 1$

58. $\dfrac{\dfrac{3}{4} + \dfrac{1}{x}}{\dfrac{2}{x}} = -1$

59. $\dfrac{\dfrac{1}{2} - \dfrac{3}{x}}{\dfrac{5}{x}} = 1$

60. $\dfrac{\dfrac{7}{3} + \dfrac{2}{x}}{\dfrac{9}{x}} = 1$

D

61. Frank and Freda can pick a quart of blueberries in 12 minutes. Frank alone takes 18 minutes longer than Freda. How long does it take each of them to pick a quart?

x = time for Freda
x + 18 = time for Frank.
Part of job done in one minute
$\frac{1}{x} + \frac{1}{x+18} = \frac{1}{12}$

62. Harry can paint a room in 4 hours. It takes Tom 6 hours to paint the same room. How would it take to paint the room if Harry and Tom paint the room together? $\frac{1}{4} + \frac{1}{6} = \frac{1}{x}$ *x = # hours for both*

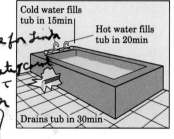

Cold water fills tub in 15min

Hot water fills tub in 20min

Drains tub in 30min

Figure for Exercise 64

63. Linda and Carol start their jogging together at the park. Carol jogs 2 miles/hr slower than Linda. If Linda can jog 6 miles in the same time that Carol jogs 4 miles, find how fast Linda jogs.

A: Note for Linda
$n_L - 2 = $ *rate Carol*
$D = n \cdot t$ $6 = n_L \cdot t$ *and* $4 = (n_L - 2) \cdot t$ *or* $\frac{6}{n_L} = t$ *and* $\frac{4}{n_L - 2} = t$

64. If the hot water faucet is left on alone, it takes 20 minutes to fill the bath tub. It takes only 15 minutes to fill the tub if the cold water faucet is left on alone. It it takes 30 minutes to drain a full bath tub, how long will it take to fill the tub if both faucets are left on and the drain is open?

In one minute, amt filled $= \frac{1}{20} + \frac{1}{15} - \frac{1}{30} = \frac{3}{60} + \frac{4}{60} - \frac{2}{60} = \frac{5}{60} = \frac{1}{12}$

$\frac{6}{n_L} = \frac{4}{n_L - 2}$ *or* $4n_L = 6n_L - 12$

$2n_L = 12$

$n_L = 6$

so it takes 12 minutes to fill

65. Charlie and Jim decide to drive to Seattle. They both leave from the same place at the same time, but Charlie drives 5 miles/hr faster than Jim. If Charlie has traveled 65 miles in the same time that Jim has traveled 60 miles, find how fast each was driving.

66. One number is three times another. The sum of their reciprocals is $\dfrac{2}{9}$. Find the numbers. $x, 3x$ $\frac{1}{x} + \frac{1}{3x} = \frac{2}{9}$

67. One number is five times another. If the sum of their reciprocals is $\dfrac{3}{5}$, find the numbers. $x, 5x$ $\frac{1}{x} + \frac{1}{5x} = \frac{3}{5}$

[handwritten: $q = x$ $q - 7$ in one minute $\frac{1}{x} + \frac{1}{x-7} = \frac{1}{12}$]

68. One pipe can empty a tank in 7 minutes less time than a smaller pipe. If both pipes can empty the tank in 12 minutes, how long would each pipe take alone?

[handwritten: $m \cdot c = ^{\$}900$ $n = days$ $c = cost/day$ $(m+3)(c-15) = 900$]

69. Ariel saved $900 for vacation. She figures that by spending $15 a day less, she can stay away 3 days longer than originally planned. How many days did she originally plan to be gone? $\left(Hint: \dfrac{\text{total cost}}{\text{number of days}} = \text{cost per day} \right)$

[handwritten: $m = \#$ $c = cost$ orig $m \cdot c = 180$ $(m-5)(c+3) = 180$ $c = 9$]

70. The Tinee Miniature Club arranges with a restaurant to pay $180 for a banquet. They were to split the cost equally, but five people did not attend, so each of those who did attend had to pay $3 extra. How many club members attended?

71. Solve the formula $\dfrac{1}{R} = \dfrac{1}{r_1} + \dfrac{1}{r_2}$ for R.

72. Solve the formula $\dfrac{1}{R} = \dfrac{1}{r_1} + \dfrac{1}{r_2} + \dfrac{1}{r_3}$ for R.

STATE YOUR UNDERSTANDING

73. Explain the procedure for solving an equation containing rational expressions.

CHALLENGE EXERCISES

Solve:

74. $2(x - 3)^{-1} + 4(x + 3)^{-1} = 36(x^2 - 9)^{-1}$

75. $5(a - 2)^{-1} + 3(a + 3)^{-1} = 81(a^2 + a - 6)^{-1}$

76. $(x + 2)^{-1} - 3(x - 3)^{-1} + 2(x - 1)^{-1} = -8(x^2 - 4x + 3)^{-1}$

77. $(x + 5)^{-1} - (x + 2)^{-1} + (x - 2)^{-1} = x(x^2 - 4)^{-1}$

MAINTAIN YOUR SKILLS (SECTIONS 1.8, 1.9, 2.1, 2.6)

Solve for w:

78. $ax - 4 = bw + 3$

79. $aw - 4 = bw + x$

80. $8w - 1 \le w + 16$

81. $1 - 8w < w + 16$

Multiply:

82. $(-b^2c^3)^2(-a^3b^3c^5)^3$

83. $(4a^4b^5)^4(-3a^2b)^2(a^5b)^4$

84. $-8xyz(6x^2 - 3y + 2yz - 2z^2)$

85. $(b - 5)(2b^2 - 4b + 5)$

3.6

DIVISION OF POLYNOMIALS

OBJECTIVES

1. Divide a polynomial by a monomial.

2. Divide a polynomial by a polynomial of two or more terms.

In mathematics, we have several ways of writing a division problem. Two of the more common are fractions and the symbolism of long division. If P and Q are polynomials, we can write:

$$\frac{P}{Q} \quad \text{or} \quad P/Q \quad \text{or} \quad Q\overline{)P}, \quad \text{or} \quad P \div Q, Q \neq 0$$

Recall that Q is the *divisor* and P is the *dividend*. The answer is the *quotient*.

In all cases, we assume that the domains of the variables in the divisor are restricted so that the divisor is not zero.

■ PROCEDURE

To divide a polynomial by a monomial:

1. Write the division in fraction form.

2. Use the distributive property and distribute the division over each term. (Write as a sum of fractions.)

3. If possible, reduce each resulting term (fraction).

$$\frac{A + B}{C} = \frac{A}{C} + \frac{B}{C}, C \neq 0$$

EXAMPLE 1

Divide: $(24x^4 - 36x^3 + 18x^2 - 12x) \div 6x$

$$= \frac{24x^4 - 36x^3 + 18x^2 - 12x}{6x}$$
Step 1. Write as a fraction.

$$= \frac{24x^4}{6x} - \frac{36x^3}{6x} + \frac{18x^2}{6x} - \frac{12x}{6x}$$
Step 2. Use the distributive property to write as the sum of fractions.

$$= 4x^3 - 6x^2 + 3x - 2$$
Step 3. Reduce.   ■

EXAMPLE 2

Divide: $(14x^2y^2 - 18xy^3 + 3x^3y) \div 2xy$

$= \dfrac{(14x^2y^2 - 18xy^3 + 3x^3y)}{2xy}$ Step 1.   Write as a fraction.

$= \dfrac{14x^2y^2}{2xy} - \dfrac{18xy^3}{2xy} + \dfrac{3x^3y}{2xy}$ Step 2.   Write as the sum of fractions.

$= 7xy - 9y^2 + \dfrac{3}{2}x^2$ Step 3.   Reduce.   ■

EXAMPLE 3

Divide: $\dfrac{18a^2b^4 + 15ab^2 - 24a^3b}{3a^2b^2}$.   Step 1.   Write the result with positive exponents.

$= \dfrac{18a^2b^4}{3a^2b^2} + \dfrac{15ab^2}{3a^2b^2} - \dfrac{24a^3b}{3a^2b^2}$ Step 2.   Write as the sum of fractions.

$= 6b^2 + \dfrac{5}{a} - \dfrac{8a}{b}$ Step 3.   Reduce.   ■

In Example 3, the result is not a polynomial since there is a variable in the denominators of the last two terms.

The division of a polynomial by a binomial follows the same pattern as long division of whole numbers.

EXAMPLE 4

Divide: $(x^2 + 8x + 15) \div (x + 3)$

$x + 3 \overline{)x^2 + 8x + 15}$ Think: What times x will give a product of x^2?

$\begin{array}{r} x \\ x + 3 \overline{)x^2 + 8x + 15} \\ \underline{x^2 + 3x} \downarrow \\ 5x + 15 \end{array}$ Place x over the x column and multiply $x(x + 3)$. Then subtract $x^2 + 3x$. The result is $5x$. Bring down the next term, 15.

Don't forget to add the opposite when subtracting.

Think: What times x will give a product of $5x$?

$$
\begin{array}{r}
x + 5 \\
x + 3 \overline{)\, x^2 + 8x + 15\,} \\
\underline{x^2 + 3x} \\
5x + 15 \\
\underline{5x + 15} \\
0
\end{array}
$$

Place the 5 over the last column and multiply $5(x + 3)$. Then subtract $5x + 15$.

Check:

$$(x + 3)(x + 5) = x^2 + 8x + 15.$$

Check by multiplication.

The quotient is $x + 5$. ■

As with division of whole numbers, division of polynomials does not always have a remainder of zero. If the remainder is not zero, it is shown as $\dfrac{\text{remainder}}{\text{divisor}}$. A remainder is identified when it has a degree *less than* the degree of the divisor.

EXAMPLE 5

Divide: $x - 5 \overline{)\, x^2 - 2x + 9\,}$

$$
\begin{array}{r}
x + 3 \\
x - 5 \overline{)\, x^2 - 2x + 9\,} \\
\underline{x^2 - 5x} \downarrow \\
3x + 9 \\
\underline{3x - 15} \\
24
\end{array}
$$

$x \cdot x = x^2 \cdot x(x - 5) = x^2 - 5x.$
Subtract $x^2 - 5x$ from $x^2 - 2x$.
Don't forget to add the opposite.
Bring down 9.
$3 \cdot x = 3x$. Multiply and subtract.
Since the degree of 24 is zero, and the degree of the divisor is one, we have a remainder of 24.

Check:

$$
\begin{aligned}
(x - 5)(x + 3) + 24 &= x^2 - 2x - 15 + 24 \\
&= x^2 - 2x + 9
\end{aligned}
$$

To check, we multiply $(x - 5)(x + 3)$ and add the remainder, 24.

So, the quotient is $x + 3 + \dfrac{24}{x - 5}$. ■

In Example 6 the lead coefficients are not one.

EXAMPLE 6

Divide: $2x - 5\overline{)6x^2 - 7x - 24}$

$$
\begin{array}{r}
3x + 4 \\
2x - 5\overline{)6x^2 - 7x - 24} \\
\underline{6x^2 - 15x} \downarrow \\
8x - 24 \\
\underline{8x - 20} \\
-4
\end{array}
$$

$3x(2x - 5) = 6x^2 - 15x$
Subtract.
Bring down the next term, -24.
$4(2x - 5) = 8x - 20$
Subtract. The degree of -4 is zero, and the degree of the divisor is one. We have a remainder of -4.

The quotient is $3x + 4 - \dfrac{4}{2x - 5}$.   The remainder is written using subtraction.

When dividing a polynomial by a polynomial of two or more terms, be sure that each polynomial is written in descending order.

EXAMPLE 7

Divide: $x - 3\overline{)x^3 + 15 - 11x - x^2}$

$$
\begin{array}{r}
x^2 + 2x - 5 \\
x - 3\overline{)x^3 - x^2 - 11x + 15} \\
\underline{x^3 - 3x^2} \downarrow \\
2x^2 - 11x \\
\underline{2x^2 - 6x} \downarrow \\
-5x + 15 \\
\underline{-5x + 15} \\
0
\end{array}
$$

Rewrite the dividend in descending order.

$x^2(x - 3) = x^3 - 3x^2$

Bring down $-11x$.
$2x(x - 3) = 2x^2 - 6x$
Bring down 15.
$-5(x - 3) = -5x + 15$

The quotient is $x^2 + 2x - 5$.

If some of the powers are missing in the dividend, it is useful to insert zeros to replace them.

EXAMPLE 8

Divide: $x - 3\overline{)x^4 - 81}$

$x - 3\overline{)x^4 - 81}$ The terms containing x^3, x^2, and x are missing.

$x - 3\overline{)x^4 + 0x^3 + 0x^2 + 0x - 81}$ We add $0x^3$, $0x^2$, and $0x$ for the missing terms.

$$\begin{array}{r} x^3 + 3x^2 + 9x + 27 \\ x - 3{\overline{\smash{\big)}\,x^4 + 0x^3 + 0x^2 + 0x - 81}} \\ \underline{x^4 - 3x^3} \\ 3x^3 + 0x^2 \\ \underline{3x^3 - 9x^2} \\ 9x^2 + 0x \\ \underline{9x^2 - 27x} \\ 27x - 81 \\ \underline{27x - 81} \\ 0 \end{array}$$

$x^3(x - 3) = x^4 - 3x^3$
Subtract. Bring down $0x^2$.
$3x^2(x - 3) = 3x^3 - 9x^2$
Subtract. Bring down $0x$.
$9x(x - 3) = 9x^2 - 27x$
Subtract. Bring down -81.

So, the quotient is $x^3 + 3x^2 + 9x + 27$. ■

The remainder in Example 8 is zero. This tells us that $x - 3$ is a factor of $x^4 - 81$ because $x^4 - 81 = (x - 3)(x^3 + 3x^2 + 9x + 27)$. A divisor is a factor of the dividend when the remainder is zero. We use this fact to answer the question in Example 9.

EXAMPLE 9

Is $2x - 3$ a factor of $6x^2 - 5x - 6$?

$$\begin{array}{r} 3x + 2 \\ 2x - 3{\overline{\smash{\big)}\,6x^2 - 5x - 6}} \\ \underline{6x^2 - 9x} \\ 4x - 6 \\ \underline{4x - 6} \\ 0 \end{array}$$

If $2x - 3$ is a factor, then when $6x^2 - 5x - 6$ is divided by $2x - 3$ the remainder must be zero. If the remainder is not zero, then it is not a factor.

The remainder is zero, therefore $2x - 3$ is a factor of $6x^2 - 5x - 6$. ■

EXAMPLE 10

Factor $2x^3 + 15x^2 + 31x + 12$ completely if it is known that $x + 4$ is one of its factors.

$$\begin{array}{r} 2x^2 + 7x + 3 \\ x + 4{\overline{\smash{\big)}\,2x^3 + 15x^2 + 31x + 12}} \\ \underline{2x^3 + 8x^2} \\ 7x^2 + 31x \\ \underline{7x^2 + 28x} \\ 3x + 12 \\ \underline{3x + 12} \\ 0 \end{array}$$

Since $x + 4$ is a factor of the polynomial, the division has zero remainder.

So,

$2x^3 + 15x^2 + 31x + 12 = (x + 4)(2x^2 + 7x + 3)$

$= (x + 4)(x + 3)(2x + 1)$.

We now factor the quadratic polynomial. ■

EXERCISE 3.6

A

1. $\dfrac{5x^3 + 15x^2 + 10x}{5x}$

2. $\dfrac{9x^3 + 27x^2 - 18x}{3x}$

3. $\dfrac{18a^5 - 27a^3 + 9a^2}{9a^2}$

4. $\dfrac{30b^6 - 24b^5 - 18b^3}{6b^3}$

5. $\dfrac{14x^3 - 28x^2 - 6x}{7x}$

6. $\dfrac{12y^5 - 15y^3 - 6y^2}{6y^2}$

7. $\dfrac{20x^3y^2z^2 - 18x^2y^3z^2 + 14x^2y^2z^3}{-2x^2yz^2}$

8. $\dfrac{4a^3b^3c^3 + 16a^2b^2c^2 - 8a^3b^2c^2}{-4a^2b^2c^2}$

9. $\dfrac{10x^2y^2 - 15xy^3 - 20y}{5xy}$

10. $\dfrac{36a^3b^3 + 12ab^2 - 24a^2b}{12a^2b^2}$

11. $\dfrac{-14x^4y^5z - 16x^3yz^4 - 20xyz}{-2x^2y^3z^2}$

12. $\dfrac{-45a^3b^3c^3 + 50a^2b^2c^2 - 15abc}{15a^2bc}$

Divide:

13. $a - 3 \overline{)a^2 - 3a}$

14. $b - 5 \overline{)b^2 - 5b}$

15. $c - d \overline{)ac - ad}$

16. $x + y \overline{)3x^2 + 3xy}$

17. $6x + 11 \overline{)18x^2 + 33x}$

18. $7y + 2 \overline{)49y^2 + 14y}$

19. $4y + 5 \overline{)12y^3 + 15y^2}$

20. $2y - 9 \overline{)10y^3 - 45y^2}$

B

21. $x + 2 \overline{)x^2 + 5x + 6}$

22. $x + 4 \overline{)x^2 + 7x + 12}$

23. $(x^2 + 3x + 2) \div (x + 1)$

24. $(x^2 - 4x + 4) \div (x - 2)$

25. $(x^2 + 7x - 18) \div (x + 9)$

26. $(x^2 + 15x + 54) \div (x + 6)$

27. $(x^2 - 23x - 50) \div (x + 2)$

28. $(x^2 + 14x - 51) \div (x + 17)$

29. $x - 3 \overline{)2x^2 + x - 21}$

30. $x + 4 \overline{)5x^2 + 14x - 24}$

31. $x + 2 \overline{)x^2 - 8x - 3}$

32. $x - 5 \overline{)2x^2 + 3x + 1}$

33. $2x + 3 \overline{)6x^2 + 23x + 21}$

34. $2y - 5 \overline{)12y^2 - 16y - 35}$

35. $4x - 9 \overline{)4x^2 + 15x - 50}$

36. $3x + 7 \overline{)15x^2 + 8x - 60}$

37. $(2x^2 - 5x + 16) \div (2x + 1)$

38. $(5x^2 - 4x + 12) \div (x + 7)$

39. $(6x^2 - 3x - 5) \div (2x + 3)$

40. $(24x^2 - 15x + 1) \div (8x + 3)$

C

41. $(x^3 + 3x^2 + 3x + 1) \div (x + 1)$

42. $(x^3 - 3x^2 + 3x - 1) \div (x - 1)$

43. $(x^3 + 6x^2 + 12x + 8) \div (x + 2)$

44. $(x^3 - 9x^2 + 27x - 27) \div (x - 3)$

45. $x - 2 \overline{)x^3 - 2x^2 + 3x - 6}$

46. $x + 7 \overline{)x^3 + 7x^2 - x - 7}$

47. $x + 3 \overline{)x^5 + 3x^4 + x^2 + 18}$

48. $2x - 1 \overline{)2x^4 - x^3 + 16x^2 - 4}$

49. $3x + 1 \overline{)6x^3 + 5x^2 - 2x + 1}$

50. $x + 5 \overline{)x^4 - 2x^3 - 35x^2 - x + 1}$

51. $x + 1 \overline{)x^5 + 1}$

52. $x - 1 \overline{)x^7 - 1}$

53. $(x^6 + 1) \div (x + 1)$

54. $(x^7 + x^5 - x^4 + 2) \div (x - 1)$

D

55. Is $x + 2$ a factor of $7x^2 + 23x + 18$? If so, write the factors.

56. Is $2x - 5$ a factor of $12x^2 + 16x - 33$? If so, write the factors.

57. Is $x + 5$ a factor of $x^3 + 10x^2 + 36x + 55$? If so, write the factors.

58. Is $x - 1$ a factor of $x^5 - 1$? If so, write the factors.

59. Is $x - 1$ a factor of $x^6 + 1$? If so, write the factors.

60. Is $x + 2$ a factor of $x^5 + 32$? If so, write the factors.

61. Factor $x^3 + 9x^2 + 26x + 24$ completely if $x + 4$ is one of its factors.

62. Factor $x^3 - 3x^2 - 10x + 24$ completely if $x - 4$ is one of its factors.

63. Factor $6x^3 - x^2 - 19x - 6$ completely if $x - 2$ is one of its factors.

64. Factor $24x^3 - 2x^2 - 31x - 12$ completely if $2x + 1$ is one of its factors.

65. Factor $x^5 + 3x^4 - 16x - 48$ completely if $x + 3$ is one of its factors.

66. Factor $x^7 - x^6 - x + 1$ completely if $x - 1$ is one of its factors.

STATE YOUR UNDERSTANDING

67. How does one know when the division of polynomials is completed?

CHALLENGE EXERCISES

Divide, assume all exponents are positive integers:

68. $(x^{2n} - x^n - 6) \div (x^n + 2)$

69. $(x^{2n} + 4x^n + 3) \div (x^n + 1)$

70. $(x^{2n} - x^n + 12) \div (x^n - 3)$

71. $(2x^{2n} + 6x^n + 9) \div (x^n + 3)$

72. $(x^{3n} - 3x^{2n} + x^n - 12) \div (x^n + 2)$

73. Find a value for k so that if $x^3 + 6x^2 - x - k$ is divided by $x - 2$, the remainder is 1.

MAINTAIN YOUR SKILLS (SECTIONS 2.6, 2.7)

Multiply:

74. $(x + 3)(x^2 - 3x + 9)$

75. $(6x - 1)(36x^2 + 6x + 1)$

76. $(x + 2)(x - 2)(x^2 + 4)$

77. $(2y - 1)(2y + 1)(4y^2 + 1)$

78. $(x^2 + 2x - 1)(x^2 - 2x - 1)$

79. $(p^3 + 2p - 1)(p^3 - 2p + 1)$

80. $(x - 3)^3$

81. $(3y + 2)^3$

3.7

SYNTHETIC DIVISION

OBJECTIVE

1. Divide a polynomial by a binomial of the form $x - a$ using synthetic division.

Just as there is a shortcut for division of whole numbers by some prime numbers, there is a shortcut for division of polynomials by divisors of the form $x - a$, where a is a constant. This shortcut is called *synthetic division*.

Consider the following division:

$$
\begin{array}{r}
x^2 + 5x + 5 \\
x - 3{\overline{\smash{\big)}\,x^3 + 2x^2 - 10x + 9}} \\
\underline{x^3 - 3x^2} \\
5x^2 - 10x \\
\underline{5x^2 - 15x} \\
5x + 9 \\
\underline{5x - 15} \\
24
\end{array}
$$

The following versions of the above division show the development for the format for synthetic division.

$$
\begin{array}{r}
1 + 5 + 5 \\
1 - 3{\overline{\smash{\big)}\,1 + 2 - 10 + 9}} \\
\underline{1 - 3} \\
5 - 10 \\
\underline{5 - 15} \\
5 + 9 \\
\underline{5 - 15} \\
24
\end{array}
$$

If we always make sure the polynomial is in descending order, we can omit the variables and still perform the division.

$$
\begin{array}{r}
1 + 5 + 5 \\
-3\overline{)1 + 2 - 10 + 9} \\
\underline{-3} \\
5 - 10 \\
\underline{-15} \\
5 + 9 \\
\underline{-15} \\
24
\end{array}
$$

Each term in the partial quotient is chosen so that when it is multiplied by the x in $x - a$ and subtracted from the dividend the difference is zero, so this repetition can be omitted.

$$
\begin{array}{r}
1 + 5 + 5 \\
3\overline{)1 + 2 - 10 + 9} \\
\underline{3} \\
5 - 10 \\
\underline{15} \\
5 + 9 \\
\underline{15} \\
24
\end{array}
$$

In each step, we had to subtract, but by changing the sign of the divisor, all the signs are changed, and we can add.

$$
\begin{array}{r}
1 + 5 + 5 \\
3\overline{)1 + 2 - 10 + 9} \\
\underline{3 + 15 + 15} \\
5 + 5 + 24
\end{array}
$$

By not bringing the -10 and the 9 down and by writing the products immediately below, we save additional space.

$$
\begin{array}{r}
3\overline{)1 + 2 - 10 + 9} \\
\underline{3 + 15 + 15} \\
1 + 5 + 5 + 24
\end{array}
$$

Notice that the last row contains the coefficients of the quotient except for the first term. By bringing the 1 down, the last row becomes the coefficients of the quotient.

$$
x^2 + 5x + 5 + \frac{24}{x - 3}
$$

The degree of the quotient is one less than the degree of the dividend. The last term of the last row is the remainder. ■

■ PROCEDURE

To divide using synthetic division:

1. Write the divisor in the form $x - a$.

2. Write down the coefficients of the dividend in descending order using zeros for any missing terms.

3. Instead of dividing by $x - a$, divide by a.

4. Bring down the first coefficient, multiply it by a, and add the product to the second coefficient.

5. Multiply a by the sum found in step 3 and add the product to the next coefficient. Continue this process until the division is complete.

6. The quotient has the same coefficients as the last row of the division. The first term on the left has degree one less than the highest degree of the dividend. The last term is the remainder.

■ CAUTION

This process known as synthetic division is defined *only* for a divisor of the form $x - a$. The lead coefficient is one, and the degree of the divisor is one.

EXAMPLE 1

Divide using synthetic division: $(x^2 - x - 42) \div (x + 6)$

$$-6 \underline{\begin{array}{r} 1 - 1 - 42 \\ -6 + 42 \\ \hline 1 - 7 + 0 \end{array}}$$

Remember to divide by -6 since $x + 6 = x - (-6)$.

The quotient is $x - 7$.   The remainder is zero.   ■

In long division of polynomials, when there are terms missing, zeros can be inserted for the missing terms.

EXAMPLE 2

Divide using synthetic division: $(x^5 - 3x^4 - x + 3) \div (x - 3)$

$$3 \underline{\begin{array}{r} 1 - 3 + 0 + 0 - 1 + 3 \\ +3 + 0 + 0 + 0 - 3 \\ \hline 1 + 0 + 0 + 0 - 1 + 0 \end{array}}$$

Zeros are inserted for the missing terms x^2 and x.

The quotient is $x^4 - 1$.   ■

When the remainder is not zero the last term of the quotient is a fraction.

EXAMPLE 3

Divide using synthetic division: $(x^3 - x^2 + 7x - 13) \div (x + 2)$

$$-2 \underline{\begin{array}{r} 1 - 1 + 7 - 13 \\ -2 + 6 - 26 \\ \hline 1 - 3 + 13 - 39 \end{array}}$$

Write the remainder over the divisor as the last term of the quotient.

The quotient is $x^2 - 3x + 13 - \dfrac{39}{x + 2}$.   ■

Synthetic division is used in algebra to factor and solve polynomial equations.

EXAMPLE 4

Is $x - 6$ a factor of $x^3 - 9x^2 + 23x - 30$?

$$6\,|\,1 - 9 + 23 - 30$$
$$\underline{+ 6 - 18 + 30}$$
$$1 - 3 + 5 + 0$$

If $x - 6$ is a factor, the remainder will be zero. If the remainder is not zero, it is not a factor.

$x - 6$ is a factor.

Another factor is $x^2 - 3x + 5$.   ■

When using synthetic division, if the remainder is zero, the divisor is a factor of the dividend. Synthetic division can also be used to determine whether a given number is a solution of a polynomial equation.

$$x^2 + 5x + 6 = 0$$

$$(x + 2)(x + 3) = 0 \qquad \text{Factor.}$$

So, $x = -2$ *or* $x = -3$

The solution set is $\{-3, -2\}$.

If we divide $x^2 + 5x + 6$ by $x + 2$ using synthetic division we have:

$$-2\,|\,1 \qquad 5 \qquad 6$$
$$\underline{ \quad -2 \quad -6}$$
$$1 \qquad 3 \qquad 0$$

Or if we divide $x^2 + 5x + 6$ by $x + 3$:

$$-3\,|\,1 \qquad 5 \qquad 6$$
$$\underline{ \quad -3 \quad -6}$$
$$1 \qquad 2 \qquad 0$$

In each case, the remainder is zero. In general, an equation of degree two or more has a for a root when the polynomial is divided by $x - a$ and the remainder is zero.

EXAMPLE 5

Is 2 a solution of $x^3 + 5x^2 - 2x - 24 = 0$?

$$2\,|\,1 \qquad 5 \qquad -2 \qquad -24$$
$$\underline{ \quad 2 \qquad 14 \qquad 24}$$
$$1 \qquad 7 \qquad 12 \qquad 0$$

Divide the polynomial (the left side of the equation) by 2.
The remainder is zero.

Yes, 2 is a solution.   ■

EXERCISE 3.7

A

Divide. Use synthetic division:

1. $(x^2 - 2x - 15) \div (x - 5)$

2. $(x^2 + 19x + 78) \div (x + 6)$

3. $(2x^2 + 5x - 63) \div (x + 7)$

4. $(3x^2 + 7x - 66) \div (x + 6)$

5. $(4x^2 - 31x + 55) \div (x - 5)$

6. $(7x^2 - 44x + 45) \div (x - 5)$

7. $(3x^2 - 11x - 104) \div (x - 8)$

8. $(7x^2 - 55x - 72) \div (x - 9)$

9. $(x^3 + 2x^2 - 8x + 5) \div (x - 1)$

10. $(x^3 + 8x^2 + 11x - 2) \div (x + 2)$

11. $(x^3 - 8x^2 + 16x + 72) \div (x + 2)$

12. $(x^3 + 5x^2 - x - 20) \div (x + 4)$

B

Divide. Use synthetic division:

13. $(x^2 - 3x + 7) \div (x + 2)$

14. $(x^2 + 5x - 6) \div (x + 3)$

15. $(x^2 - 8x + 9) \div (x + 4)$

16. $(x^2 + 6x - 9) \div (x - 2)$

17. $(x^3 - 5x^2 + 12x - 15) \div (x - 3)$

18. $(x^3 + x^2 + x + 7) \div (x + 2)$

19. $(x^3 + x^2 - 19x + 1) \div (x + 5)$

20. $(x^3 + x^2 + x + 7) \div (x + 2)$

21. $(2x^3 - 5x^2 + 4x + 3) \div (x - 1)$

22. $(3x^3 - 7x^2 + 8x - 14) \div (x - 2)$

23. $(2x^3 - 13x + 15) \div (x + 3)$

24. $(3x^3 - 41x + 28) \div (x + 4)$

C

Divide. Use synthetic division:

25. $(x^3 - 3x - 7) \div (x + 2)$

26. $(x^3 + 2x + 9) \div (x - 3)$

27. $(x^4 - 2x^3 - 8x^2 + 26x + 15) \div (x + 3)$

28. $(x^4 + x^3 + 3x^2 - 24x + 12) \div (x - 2)$

29. $(2x^4 - 13x^3 + 17x^2 + 18x - 24) \div (x - 4)$

30. $(3x^4 + 5x^3 - 9x^2 + 9x - 27) \div (x + 3)$

31. $(x^6 - 7x^5 + 12x^4 - 2x^3 + 6x^2 + 6x - 18) \div (x - 3)$

32. $(x^5 - 9x^4 + 20x^3 - 3x^2 + 19x - 25) \div (x - 4)$

33. $(x^4 - 16) \div (x + 2)$

34. $(x^6 - 729) \div (x - 3)$

35. $(x^7 - 2) \div (x - 1)$

36. $(x^5 + 3) \div (x + 1)$

D

37. Is $x - 5$ a factor of $x^5 - 19x^3 + 20x^2 - 19x + 5$?

38. Is $x + 7$ a factor of $x^4 + 4x^3 - 12x^2 + 56x - 49$?

39. Is $x + 6$ a factor of $x^4 + 6x^3 - 3x^2 - 15x + 12$?

40. Is $x - 8$ a factor of $x^5 - 8x^4 - 2x^3 + 16x^2 + 3x - 48$?

41. Is $x + 2$ a factor of $2x^2 - 5x - 3$? If so, write the factors.

42. Is $x - 1$ a factor of $x^2 + 2x - 3$? If so, write the factors.

43. Is $x + 1$ a factor of $x^3 + 3x^2 + 3x + 1$? If so, write the factors.

44. Is $x - 2$ a factor of $x^3 - 6x^2 + 12x - 8$? If so, write the factors.

45. Is 1 a solution of $4x^2 - x - 3 = 0$?

46. Is -2 a solution of $7x^2 + 23x + 18 = 0$?

47. Is -1 a solution of $x^5 - 1 = 0$?

48. Is -2 a solution of $x^4 - 16 = 0$?

49. Is -5 a solution of $x^3 + 10x^2 + 36x + 55 = 0$?

50. Is 3 a solution of $x^3 + 2x^2 - 10x + 9 = 0$?

STATE YOUR UNDERSTANDING

51. When is it necessary to insert zeros in the dividend when doing synthetic division?

CHALLENGE EXERCISES

Use synthetic division to divide the following. Assume all exponents are positive integers.

52. $(x^{3n} + 2x^{2n} - x^n + 4) \div (x^n + 2)$

53. $(x^{4n} - 5x^{3n} - 10x^{2n} - x^n + 15) \div (x^n - 1)$

54. $(3x^3 - 25x^2 + 44x - 12) \div (3x - 1)$ Hint: Divide by $\left(x - \dfrac{1}{3}\right)$, then divide that result by 3.

55. $(4x^3 + 37x^2 + 81x + 18) \div (4x + 1)$

MAINTAIN YOUR SKILLS (SECTIONS 2.8, 2.9, 2.10, 3.1)

Factor completely:

56. $20x^4 - 1620$

57. $x^4 - x^2 - 12$

58. $3x^6 - 192$

59. $x^3 + 5x^2 - x - 5$

60. $5x^5 + 135x^2$

61. $mx^2 + nx^2 - 9m - 9n$

Reduce:

62. $\dfrac{m^2 - 16}{4m^2 + 16m}$

63. $\dfrac{x^3 - 1}{7x^3 + 7x^2 + 7x}$

CHAPTER 3
SUMMARY

Rational Numbers	The set of all numbers that can be written as the indicated quotient of two integers.	(p. 176)

$$Q = \left\{ \frac{p}{q} \middle| p \in J, q \in J, q \neq 0 \right\}$$

Rational Expressions	The set of all algebraic expressions that can be written as the indicated quotient of two polynomials.	(p. 176)

$\dfrac{P}{Q}$ is a rational expression where P and Q are polynomials, $Q \neq 0$.

Variable Restrictions in Rational Expressions	If variables appear in the denominator of a rational expression, the variable must be restricted so that the denominator is not equal to zero.	(p. 176)
Build a Rational Expression	Use the basic principle of fractions. If A, B, and C are polynomials.	(p. 178)

$$\frac{A}{B} = \frac{AC}{BC}, BC \neq 0.$$

Reduce a Rational Expression	Use the basic principle of fractions. If A, B, and C are polynomials,	(p. 178)

$$\frac{AC}{BC} = \frac{A}{B}, BC \neq 0.$$

Multiplication of Rational Expressions	Multiply the numerators and write that product over the product of the denominators. If A, B, and C are polynomials,	(p. 184)

$$\frac{A}{B} \cdot \frac{C}{D} = \frac{AC}{BD}, BD \neq 0.$$

Division of Rational Expressions	Rewrite the division problem as a multiplication problem. That is, multiply by the reciprocal of the divisor. If A, B, and C are polynomials,	(p. 187)

$$\frac{A}{B} \div \frac{C}{D} = \frac{A}{B} \cdot \frac{D}{C}, BCD \neq 0.$$

Least Common Denominator (LCD) of Rational Expressions	The polynomial with the least number of factors that is a multiple of each denominator.	(p. 192)
Addition of Two Rational Expressions With a Common Denominator	Find the sum of the numerators and write that sum over the common denominator. If A, B, and C are polynomials,	(p. 193)

$$\frac{A}{C} + \frac{B}{C} = \frac{A + B}{C}, C \neq 0.$$

Addition of Rational Expressions With Opposite Denominators	Use one denominator as the common denominator, multiply the other fraction by $\dfrac{-1}{-1}$, then add. If A, B, and C are polynomials,	(p. 194)

$$\frac{A}{C} + \frac{B}{-C} = \frac{A}{C} + \frac{-B}{C} = \frac{A - B}{C}, C \neq 0.$$

Addition of Rational Expressions Without a Common Denominator	Build both fractions to have a common denominator and then add. If A, B, and C are polynomials,	(p. 195)

$$\frac{A}{B} + \frac{C}{D} = \frac{AD + BC}{BD}, BD \neq 0.$$

Subtraction of Two Rational Expressions	Add the opposite of the rational expression being subtracted. If A, B, C, and D are polynomials,	(p. 196)

$$\frac{A}{B} - \frac{C}{D} = \frac{A}{B} + \frac{-C}{D} = \frac{AD - BC}{BD}, BD \neq 0.$$

Complex Fraction	A fraction that contains a fraction in the numerator, the denominator, or both.	(p. 204)
A Method of Simplifying a Complex Fraction	Do the indicated division and simplify.	(p. 204)

$$\frac{\dfrac{A}{B}}{\dfrac{C}{D}} = \frac{A}{B} \div \frac{C}{D} = \frac{A}{B} \cdot \frac{D}{C}, BCD \neq 0$$

Another Method of Simplifying a Complex Fraction	Multiply the numerator and denominator of the complex fraction by the (p. 206) LCD of the fractions within the expression.

$$\frac{\dfrac{A}{B}}{\dfrac{C}{D}} = \frac{BD\left[\dfrac{A}{B}\right]}{BD\left[\dfrac{C}{D}\right]} = \frac{AD}{BC}, BCD \neq 0$$

Solve an Equation Containing Rational Expressions	**1.** Multiply both sides of the equation by the LCD of the denominators (p. 213) to clear the fractions.
	2. Solve the resulting equation.
Divide a Polynomial by a Monomial	**1.** Write the division problem as a fraction. (p. 223)
	2. Use the distributive property to distribute the division over each term.
	3. Write the division as the sum of fractions and reduce each fraction.
Divide a Polynomial by a Polynomial of Two or More Terms	**1.** Write each polynomial in descending order, inserting zeros for any (p. 224) missing terms.
	2. Following the pattern of long division of whole numbers.
Synthetic Division	Used to divide polynomials when the lead coefficient is one and the (p. 231) degree of the divisor is one. The divisor must be of the form $x - a$.

CHAPTER 3
REVIEW EXERCISES

SECTION 3.1   Objective 1

For what values of the variable are the following rational expressions undefined?

1. $\dfrac{a}{a-4}$

2. $\dfrac{b}{b+9}$

3. $\dfrac{x+3}{x}$

4. $\dfrac{y-2}{y}$

5. $\dfrac{4x+2}{5}$

6. $\dfrac{3x-5}{9}$

7. $\dfrac{2a-3}{a^2-a-6}$

8. $\dfrac{3x+5}{x^2-2x-8}$

9. $\dfrac{x^2+2x+5}{x^2+6x+5}$

10. $\dfrac{a^2+5a+12}{a^2+7a+12}$

11. $\dfrac{x^2-4}{x^2-4x+4}$

12. $\dfrac{a^2-16}{a^2+8a+16}$

SECTION 3.1   Objective 2

Build each of the following by finding the missing numerator. You do not need
to state the variable restrictions.

13. $\dfrac{12}{x} = \dfrac{?}{x^3}$

14. $\dfrac{5}{a} = \dfrac{?}{a^4}$

15. $\dfrac{3}{a-b} = \dfrac{?}{a^2-b^2}$

16. $\dfrac{9}{x-2} = \dfrac{?}{x^2-4}$

17. $\dfrac{3x}{x-5} = \dfrac{?}{x^2-5x}$

18. $\dfrac{6y}{y+4} = \dfrac{?}{y^2+4y}$

19. $\dfrac{2a+b}{3a+b} = \dfrac{?}{3a^2+4ab+b^2}$

20. $\dfrac{x+y}{2x+y} = \dfrac{?}{2x^2+3xy+y^2}$

21. $\dfrac{x+y}{x-y} = \dfrac{?}{x^3-y^3}$

22. $\dfrac{a-b}{a+b} = \dfrac{?}{a^3+b^3}$

23. $\dfrac{5x+2}{3x-5} = \dfrac{?}{6x^2-x-15}$

24. $\dfrac{4a+3b}{2a-5b} = \dfrac{?}{8a^2-14ab-15b^2}$

SECTION 3.1   Objective 3

Reduce. You do not need to state the variable restrictions.

25. $\dfrac{18a^3}{36}$

26. $\dfrac{24b^2}{48}$

27. $\dfrac{4a^2}{16a^5}$

28. $\dfrac{8x^4}{24x^2}$

29. $\dfrac{12x+18}{30x-24}$

30. $\dfrac{15a-36}{12a-9}$

31. $\dfrac{14a^2+8a}{10a^2+12ab}$

32. $\dfrac{4x^2y+10xy^2}{8x^2y+6xy^2}$

33. $\dfrac{x^2+2x+1}{x^2-3x-4}$

34. $\dfrac{x^2+6x+9}{x^2+x-6}$

35. $\dfrac{a^2-b^2}{a^3-b^3}$

36. $\dfrac{x^2-16}{x^3-64}$

37. $\dfrac{y^2-y-20}{y^2+y-30}$

38. $\dfrac{a^2+2a-15}{a^2-7a+12}$

39. $\dfrac{a-b}{b-a}$

40. $\dfrac{2a-5}{5-2a}$

41. $\dfrac{ax-bx+ay-by}{ax-bx-ay+by}$

42. $\dfrac{x^2+2x+bx+2b}{ax+ab-bx-b^2}$

43. $\dfrac{x^2+x-6}{2b-2x-bx+x^2}$

44. $\dfrac{x^2-6x+9}{9-x^2}$

SECTION 3.2 Objective 1

Multiply or divide and simplify. You do not need to state the variable restrictions.

45. $\dfrac{x}{9} \cdot \dfrac{18}{y}$

46. $\dfrac{x^2}{8} \cdot \dfrac{64}{x}$

47. $\dfrac{x+5}{3} \cdot \dfrac{5}{x+5}$

48. $\dfrac{a-b}{6} \cdot \dfrac{7}{a-b}$

49. $\dfrac{7a}{6} \div \dfrac{14}{3b}$

50. $\dfrac{9}{4b} \div \dfrac{3a}{2}$

51. $\dfrac{1-x}{12} \div \dfrac{x-1}{8}$

52. $\dfrac{a-5}{14} \div \dfrac{5-a}{7}$

53. $\dfrac{4x+8}{3x-6} \cdot \dfrac{5x-10}{8x+16}$

54. $\dfrac{9a-18}{4a-36} \cdot \dfrac{3a-27}{4a-8}$

55. $\dfrac{8a-16}{a^2-4a} \div \dfrac{a^2-4}{a^2-16}$

56. $\dfrac{3bx-b}{ab-2a} \div \dfrac{2-6x}{4b-8}$

57. $\dfrac{5a+20}{a^2-16} \cdot \dfrac{a^2+a-20}{a^2+9a+20}$

58. $\dfrac{4x+8}{x^2-4} \cdot \dfrac{x^2+5x-14}{x^2+10x+21}$

59. $\dfrac{2a^2+5a+2}{4a^2-1} \cdot \dfrac{3-5a-2a^2}{a^2+5a+6}$

60. $\dfrac{4-x^2}{2x^2+3x-2} \cdot \dfrac{2x^2+5x-3}{x^2+x-6}$

61. $\dfrac{8x^2+6x+1}{4x^2+21x+5} \div \dfrac{x^2+5x+6}{x^2+7x+10}$

62. $\dfrac{a^3-b^3}{2a^2b+3ab^2+b^3} \div \dfrac{a^2-b^2}{4a^2b+8ab^2+4b^3}$

63. $\dfrac{x^3-1}{x^2+2x+1} \cdot \dfrac{x^2+6x+5}{x^2+10x+25} \div \dfrac{x^2+x+1}{x^2-25}$

64. $\dfrac{6x^2-13x-5}{x^2+7x} \cdot \dfrac{x^3+6x^2-7x}{2x-5} \div \dfrac{3x^2-8x-3}{x^2-5x+6}$

65. $\dfrac{ab+ac+b^2+bc}{a^3-a^2b-2ab^2} \cdot \dfrac{a^3-3a^2b+2ab^2}{bc+bd+c^2+cd} \div \dfrac{a^2-b^2}{ac+ad+bc+bd}$

SECTION 3.3 Objective 1

Find the LCD of the fractions.

66. $\dfrac{2}{3xy} + \dfrac{8}{5xz}$

67. $\dfrac{4}{16ab} + \dfrac{3}{12bc}$

68. $\dfrac{6}{a-b} - \dfrac{4}{b-a}$

69. $\dfrac{3}{2y-3} - \dfrac{5}{3-2y}$

70. $\dfrac{1}{4} + \dfrac{1}{a} + \dfrac{3}{a-1}$

71. $\dfrac{2}{5} + \dfrac{1}{z} + \dfrac{3}{z+2}$

72. $\dfrac{x}{x+1} - \dfrac{y}{x+3}$

73. $\dfrac{b}{a+2} + \dfrac{c}{a-3}$

74. $\dfrac{8}{x^2-1} + \dfrac{9}{x^2+x} + \dfrac{5}{x-1}$

75. $\dfrac{6}{x^2+x-6} + \dfrac{8}{x^2+2x-8} + \dfrac{5}{x^2+7x+12}$

SECTION 3.3   Objective 2

Add or subtract. You do not need to state the variable restrictions.

76. $\dfrac{1}{2} + \dfrac{1}{x} + \dfrac{3}{x + 1}$

77. $\dfrac{2}{5} + \dfrac{1}{y} + \dfrac{y}{y - 2}$

78. $\dfrac{1}{3x + 1} - \dfrac{x + 1}{x + 3}$

79. $\dfrac{2}{2x + 3} - \dfrac{x + 3}{x - 1}$

80. $\dfrac{15}{x^2 - 2x} - \dfrac{3}{x^2 - 4}$

81. $\dfrac{7}{a^2 - b^2} - \dfrac{3}{a^2 - ab}$

82. $\dfrac{x + 1}{x^2 + 2x - 15} + \dfrac{2x + 1}{x^2 - 4x + 3}$

83. $\dfrac{2x - 3}{x^2 + x - 6} - \dfrac{x - 3}{x^2 - x - 2}$

84. $\dfrac{6}{x^2 + x - 6} + \dfrac{5}{x^2 + 2x - 8} + \dfrac{4}{x^2 + 7x + 12}$

85. $\dfrac{y + 6}{y^2 - 4y + 3} + \dfrac{y - 3}{y^2 + 5y - 6} - \dfrac{y - 1}{y^2 + 3y - 18}$

86. $\dfrac{3x}{6x + 2} - \dfrac{1}{6x^2 + 2x} - \dfrac{1}{x}$

87. $\dfrac{-a}{3a + 6} + \dfrac{4}{3a^2 + 6a} - \dfrac{3}{a}$

88. $\dfrac{y}{3y + 6} - \dfrac{4}{3y^2 + 6y} - \dfrac{1}{y}$

89. $(x + y)^{-1} + (x - y)^{-1}$

90. $2(x + 2)^{-1} - 3(x + 2)^{-1}$

SECTION 3.4   Objective 1

Simplify. You do not need to state the variable restrictions.

91. $\dfrac{\dfrac{3}{5}}{\dfrac{7}{10}}$

92. $\dfrac{\dfrac{2}{3}}{\dfrac{8}{9}}$

93. $\dfrac{\dfrac{x}{y}}{\dfrac{x^2}{y}}$

94. $\dfrac{\dfrac{a}{b}}{\dfrac{a}{b^2}}$

95. $\dfrac{\dfrac{3x^2y}{ab}}{\dfrac{9x^3y^2}{4ab}}$

96. $\dfrac{2 + \dfrac{1}{x}}{3 - \dfrac{1}{x}}$

97. $\dfrac{\dfrac{2}{x - y}}{\dfrac{4}{x^2 - 2xy + y^2}}$

98. $\dfrac{\dfrac{x}{2} + \dfrac{y}{4}}{\dfrac{x}{3} + \dfrac{y}{6}}$

99. $\dfrac{\dfrac{2}{b} - \dfrac{3}{c}}{\dfrac{4c^2 - 9b^2}{bc}}$

100. $\dfrac{1 + \dfrac{1}{x}}{1 - \dfrac{1}{1 - \dfrac{1}{x}}}$

SECTION 3.5   Objective 1

Solve the following equations.

101. $\dfrac{2a}{a-1} - \dfrac{a}{3a-3} = \dfrac{10}{2a-2}$

102. $\dfrac{1}{m-1} + \dfrac{2}{m+1} = \dfrac{5}{m^2-1}$

103. $\dfrac{4}{x+2} = \dfrac{5}{x-1}$

104. $\dfrac{8}{x-3} = \dfrac{2}{x+4}$

105. $\dfrac{x}{x^2-4x+3} + \dfrac{2}{x^2-3x} = \dfrac{1}{x}$

106. $\dfrac{1}{x-3} + \dfrac{1}{x-4} = \dfrac{5}{x^2-7x+12}$

107. $6 + \dfrac{5}{x} - \dfrac{6}{x^2} = 0$

108. $8 + \dfrac{6}{x} = \dfrac{5}{x^2}$

109. $\dfrac{1}{x+1} + \dfrac{1}{x+2} = \dfrac{5}{6}$

110. $\dfrac{1}{x+1} + \dfrac{1}{x+2} = \dfrac{7}{12}$

SECTION 3.6   Objective 1

Divide.

111. $\dfrac{25a^3b - 15ab^2 - 20ab}{5ab}$

112. $\dfrac{12m^3 + 15m^2 - 18m}{3m}$

113. $\dfrac{5x^2 - 7x - 4}{x-2}$

114. $\dfrac{3x^2 - 8x - 7}{x-1}$

115. $4x + 5\overline{)8x^2 + 6x - 5}$

116. $2x + 3\overline{)6x^2 + 5x - 6}$

117. $a - 5\overline{)a^3 - 125}$

118. $b + 6\overline{)b^3 + 216}$

119. Factor $2a^3 + 7a^2 - 7a - 30$ if one factor is $2a + 5$.

120. Factor $3x^3 + 10x^2 - 51x - 18$ if one factor is $3x + 1$.

SECTION 3.7   Objective 1

Use synthetic division to perform the division.

121. $(x^3 + 5x^2 - 8x - 12) \div (x + 2)$

122. $(x^4 - 3x^3 + 2x^2 - 3x + 5) \div (x - 1)$

123. $(x^4 - 4x^3 - 7x^2 + 34x - 24) \div (x - 4)$

124. $(a^4 + 2a^3 - 16a^2 - 2a + 15) \div (a + 5)$

125. $(x^4 - x^3 - 5x + 5) \div (x - 1)$

126. $(x^3 + 2x^2 - 3x - 4) \div (x - 1)$

127. Is 3 a solution of the equation $x^4 - x^3 + x^2 - 3x - 54 = 0$?

128. Is -2 a solution of the equation $2x^3 - 3x^2 + 4x + 40 = 0$?

CHAPTER 3
TRUE–FALSE CONCEPT REVIEW

Check your understanding of the language of algebra and arithmetic. Tell whether each of the following statements is true (always true) or false (not always true).

1. A reduced rational expression has the same value as the original.

2. A rational expression cannot have zero as one of its values.

3. Since division by zero is not defined, every rational expression with variables in the denominator must have restrictions.

4. The inverse (opposite) of reducing a rational expression is building a rational expression.

5. Two rational expressions may be multiplied or divided without finding their common denominator.

6. The LCD (least common denominator) of two fractions is used only to add or subtract them.

7. In order to add two rational expressions, they must have a common denominator.

8. The LCD of two or more denominators is the product of the smallest number of factors that is a multiple of each of the original polynomials.

9. The same procedure can be used to find the LCM of two polynomials as for two whole numbers.

10. Complex fractions can always be treated as the division of fractions.

11. An equation that contains rational expressions can always be simplified to an equivalent equation that contains no fractions.

12. To reduce a rational expression means to eliminate the common variables in the numerator and denominator.

13. The quotient of two polynomials always has degree that is less than the degree of the dividend.

14. If two rational expressions are reciprocals, their product is either 1 or -1.

15. To subtract two rational expressions, you can multiply the one being subtracted by -1 and then add that product to the other.

16. When building two or more rational expressions to equivalent expressions with a common denominator, the common denominator is always positive.

17. The divisor in a synthetic division problem must be of the form $x - a$.

18. If the numerator and denominator of a rational expression have no common integral factors and no common polynomial factors, then it cannot be reduced.

19. The remainder that results from a synthetic division is always the same as the remainder that would result if the division were done by long division.

20. If the remainder of the polynomial is 0 when the polynomial is divided by $x - a$, then $x - a$ is a factor of the polynomial.

CHAPTER 3
TEST

---

1. Reduce: $\dfrac{45a^5b^3}{54a^2b^5}$

2. Find the missing numerator:

$\dfrac{2t}{t+3} = \dfrac{?}{2t^2 + 5t - 3}$

3. Find the LCD of the denominators:

$\dfrac{5}{3a^3b} + \dfrac{7}{18ab^2} - \dfrac{2}{9a^2b^2}$

4. Simplify: $\dfrac{\dfrac{1}{c} - \dfrac{1}{3} - \dfrac{1}{2c}}{\dfrac{5}{c} - \dfrac{1}{2} - \dfrac{1}{6c}}$

5. Add: $\dfrac{5}{2x-2} + \dfrac{2x}{3-2x}$

6. Multiply and simplify: $\dfrac{y^2 - 7y + 10}{y^2 - 16} \cdot$

$\dfrac{y^2 + 9y + 20}{y^2 - 25}$

7. Solve: $\dfrac{5x}{x+1} - 5 = \dfrac{6}{x}$

8. Divide: $4x - 1 \overline{)12x^3 + 13x^2 - 1}$

9. Divide and simplify: $\dfrac{52st}{3t^3} \div \dfrac{-26}{s^3t^5}$

10. Find the missing numerator: $\dfrac{3x^2}{4y} = \dfrac{?}{56x^2y^3}$

11. Find the LCD of the denominators:

$\dfrac{x}{x^2 + 8x + 12} + \dfrac{2}{x^2 - 6x - 16} - \dfrac{2x}{x^2 - 2x - 48}$

12. Subtract: $\dfrac{y}{y^2 - 4y - 45} - \dfrac{2}{y^2 - 81}$

13. Reduce: $\dfrac{6x^2 - 40x - 14}{2x^2 - 98}$

14. Combine: $\dfrac{3x}{x-1} - \dfrac{2}{x+1} + \dfrac{2}{x^2 - 1}$

15. Multiply and simplify: $\dfrac{20a^2b^3c}{-3a^2b^2} \cdot \dfrac{-7a^3b^3}{35b^4c}$

16. Solve: $2 - \dfrac{4}{x+5} = \dfrac{2x}{x-4}$

17. Divide and simplify: $\dfrac{c^2 - 13c + 42}{2c^2 - 24c} \div$

$\dfrac{c^2 - 4c - 21}{c^2 - 9c - 36}$

18. Divide: $(x^4 + 3x^3 - 3x^2 - 8x + 5) \div (x + 3)$

19. Clara and Clint are the new trainers for the ''Hot Shots,'' a local basketball team. Clara can tape the team in 50 minutes, and Clint can do the same job in 70 minutes. How long will it take them to tape the team working together? (to the nearest minute).

20. The Salvation Army has $840 to buy canned food for Christmas baskets. They priced cases of food at two stores. One store would sell the food for two dollars a case less than the other one. This savings allowed them to buy an additional 14 cases of food. What was the cost of a case of food at each store?

CHAPTER 1

Simplify:

1. $4(-2) - 5[3(8 - 12)]$

2. $(-12)(-3) - (-2)(9 - 4^2)$

Evaluate each of the following if $x = -5$ and $y = -3$:

3. $5(3x - y) - x[x + y(x - y)]$

4. $3x^2 - y^2 + 4(2x - 3y)$

Solve:

5. $4x - 12 - 8x - 9 = 2x + 3$

6. $12a + 15 - 9a - 12 = 5a - 8$

7. $3x - [5x - (19 - x) - x] = x - (3x - 1)$

8. $y - \{3y - [5y - (12 - 2y) + 4] - 8\} = 0$

Solve for the indicated variable:

9. $3a + 6b = 12$, a

10. $8x - 32y = 40$; x

11. $x = \dfrac{2a - 6b}{2}$; a

12. $A = \dfrac{1}{2}h(b_1 + b_2)$; b_2

Solve and write the solution in interval notation:

13. $-8 < 3x + 4 < 10$

14. $-15 \le 5x + 10 \le 20$

15. $\dfrac{x}{5} - \dfrac{x - 2}{3} < 1$

16. $(4x - 10) - (3x + 4) > 8 + 2x$

Solve the following:

17. The product of 3 and 12 more than a number is 4 more than 7 times the number. What is the number?

18. The product of 4 and the difference of a number and 8 is 16 more than three times the number. What is the number?

19. The length of a rectangle is 1.75 times its width. The perimeter is 66 in. Find the length and width.

20. Two cars leave Kansas City traveling in opposite directions. If their average speeds are 42 mph and 46 mph, how long will it be before they are 330 miles apart?

CHAPTER 2

Perform the indicated operations:

21. $(5a - 7b + 2) + (3a - 2b + 8)$

22. $(4x^2 - 3x - 12) - (-2x^2 - 5x - 4)$

23. $(7s^2 - 7t + 12) - (13s^2 + 2t - 4)$

24. $(-3a + b - c) + (2a - 3b + 4c) + (8a - 2b - 3c)$

25. $14 - [(-2w + 13) - (5w - 12) + 6] - (9w + 1)$

Multiply:

26. $-5x(2x + y + 3)$

27. $-8a(-2a - 3b - 4c)$

28. $(3y + 7)(4y - 11)$

29. $(2x + y - 1)(x + 2y + 1)$

30. $(3t^2 - 4)(t^2 - 11t + 5)$

31. $(x - 3)(x + 4) + (2x - 5)(x + 1)$

32. $(13t + 14m)(13t - 14m)$

33. $(2x - 3)^2 - (x + 1)^2$

Factor completely:

34. $8x^2 - 4xy - 6x$

35. $x^2 - 5x - 14$

36. $ax^2 - 4x^2 + 2a - 8$

37. $(2w + 7)^2 - 100$

38. $56x^2 - 113x + 15$

39. $81w^3 + 24$

40. $4x^2 + 68x + 208$

41. $10x^2 + 15x + 9$

42. $25a^4 - 400$

Solve:

43. $x^2 + 21x - 22 = 0$

44. $2(x^2 - 1) = 3x$

45. $4x^2 - 72x + 260 = 0$

46. $3x^2 - 5x + 32 = 2x^2 - 17x$

CHAPTER 3

Perform the indicated operations. You do not need to state the restrictions on the variables:

47. $\dfrac{x^2 - 2x - 35}{4x + 24} \cdot \dfrac{x^2 + 8x + 12}{x^2 + 7x + 10}$

48. $\dfrac{24ab^3}{14b} \div \dfrac{12b^2}{3a^2}$

49. $\dfrac{4x^2 - 11x - 3}{2x^2 + 7x + 5} \div \dfrac{x^2 - 9}{2x^2 + 11x + 15}$

50. $\dfrac{16 - x^2}{x + 3} \cdot \dfrac{x^4}{x - 4} \div \dfrac{x^3 + 4x^2}{x^2 + 2x - 3}$

51. $\dfrac{3a}{6 - 2a} + \dfrac{5b}{4a - 12}$

52. $\dfrac{3}{x + 1} + \dfrac{5}{2x + 2} + x$

53. $\dfrac{x + 5}{x^2 + 6x + 8} - \dfrac{x - 3}{x^2 - 16}$

54. $\dfrac{3}{x^2 - 4} - \dfrac{4}{x^2 - 3x - 10} + \dfrac{1}{x^2 - 7x + 10}$

55. $(5x)^{-1} + 4x^{-2}$

Simplify. You do not need to state the restrictions on the variables:

56. $\dfrac{\dfrac{3}{a - 1} + \dfrac{a}{a + 2}}{\dfrac{1}{a + 2} + \dfrac{1}{a - 1}}$

57. $\dfrac{4 - a^{-2}}{3 + a^{-2}}$

58. $\dfrac{2}{3 + \dfrac{1}{1 + \dfrac{1}{x}}}$

Solve.

59. $\dfrac{3}{x - 6} + \dfrac{5}{x + 2} = \dfrac{7}{x^2 - 4x - 12}$

60. $\dfrac{2}{x + 3} + \dfrac{8}{x^2 - 2x - 15} = \dfrac{1}{x - 5}$

61. $\dfrac{y + 3}{y + 2} - \dfrac{y - 1}{y} + \dfrac{1}{y} = 0$

62. $\dfrac{1}{ab} + \dfrac{1}{ac} = \dfrac{2}{bc}$ for b.

Divide:

63. $(15x^2y - 24xy^2 - 36xy + 48) \div (3xy)$

64. $(x^2 + 5x + 4) \div (x + 1)$

65. $(x^3 - 1) \div (x - 1)$

66. Is $x + 5$ a factor of $x^3 + 8x^2 + 5x - 50$? If so, write the two factors.

Divide using synthetic division:

67. $(3x^2 + 2x - 5) \div (x - 1)$

68. $(x^5 + x^4 - 2x^2 - x + 1) \div (x + 1)$

69. $(x^3 + 3x^2 - 5x + 1) \div (x - 2)$

70. Is $x - 3$ a factor of $x^3 - 5x^2 + 7x + 3$?

CHAPTER

4

ROOTS,
RADICALS,
AND
COMPLEX
NUMBERS

A grand prix race driver and car are tested in a variety of situations, the most trying of which is negotiating the many curves along the race course. Given certain road and tire conditions, the maximum speed at which one can take a curve without skidding can be found using the radical expression, $V = \sqrt{2.5r}$, where V is the speed and r is the radius of the curve. In Exercise 68, Section 4.2, you are asked to estimate the safe speed for negotiating a curve with a radius of 90 feet. Winning grand prix drivers are those who can push their luck beyond the safe recommendations. *(Kalish/Dimaggio/Fran Heyl Associates)*

PREVIEW

In Chapter 4 the relationships of roots, radicals, and rational exponents are shown. After establishing the identity of $x^{1/2}$ and $\sqrt{x}$ we go on to operations on radicals. Radicals are simplified, added, subtracted, multiplied, and divided (rationalize the denominator). When simplifying, the necessity for restricting the domain of the variable is pointed out. In the absence of restrictions we are careful to point out the need for absolute value notation. After solving radical equations the number system is extended to complex numbers. After defining complex numbers the basic operations on complex numbers are performed.

4.1

RATIONAL EXPONENTS

OBJECTIVES

1. Simplify expressions that contain rational exponents.

2. Multiply and divide expressions that contain rational exponents.

We know that a square root of 256 is one of its two equal factors. Thus we can say that 16 or -16 is a square root of 256 since $16(16) = 256$ and $-16(-16) = 256$. We can extend this idea to include three or more equal factors.

■ DEFINITION

*n*th Root

The number r is an nth root of b if n is a positive integer and $r^n = b$.

One of three equal factors is called a third root or cube root.

6 is a cube root of 216 because $(6)(6)(6) = 216$.

One of four equal factors is called a fourth root.

6 is a fourth root of 1296 because $(6)(6)(6)(6) = 1296$.

Other examples are shown in Table 1.

TABLE 1

$3 \cdot 3 = 3^2 = 9$	3 is a square root of 9
$2 \cdot 2 \cdot 2 = 2^3 = 8$	2 is a cube root of 8
$7 \cdot 7 \cdot 7 \cdot 7 = 7^4 = 2401$	7 is a fourth root of 2401
$-11(-11) = (-11)^2 = 121$	-11 is a square root of 121
$5 \cdot 5 \cdot 5 \cdot 5 \cdot 5 = 5^5 = 3125$	5 is a fifth root of 3125
$(-1)(-1)(-1)(-1)(-1)(-1) = (-1)^6 = 1$	-1 is a sixth root of 1

We begin our study of rational exponents with positive bases and the exponents $\frac{1}{2}$, $\frac{1}{3}$, and $\frac{1}{4}$. The properties and definitions of exponents studied in Chapter 2 are also true for rational exponents. They are repeated here for reference.

	Description	Algebraic Form
Property 1	Multiplying powers with a common base	$b^m \cdot b^n = b^{m+n}$
Property 2	Raising a power to a power	$(b^m)^n = b^{mn}$
Property 3	Raising a product to a power	$(ab)^m = a^m b^m$
Property 4	Dividing powers with a common base	$\dfrac{b^m}{b^n} = b^{m-n},\ b \neq 0$
Property 5	Raising a quotient to a power	$\left(\dfrac{a}{b}\right)^m = \dfrac{a^m}{b^m},\ b \neq 0$
Zero exponent	Any nonzero number to the zero power is 1	$b^0 = 1,\ b \neq 0$
Negative exponent	Any nonzero number raised to a negative power is the reciprocal of the number raised to the corresponding positive power	$b^{-m} = \dfrac{1}{b^m},\ m \neq 0,$ $b \neq 0$

Using the first property of exponents, we can write

$9^{1/2} \cdot 9^{1/2} = 9^{1/2+1/2} = 9^1 = 9.$

Furthermore,

TABLE 2

$9^{1/2} \cdot 9^{1/2} = 9$	$9^{1/2}$ is the positive square root of 9
$8^{1/3} \cdot 8^{1/3} \cdot 8^{1/3} = 8$	$8^{1/3}$ is a cube root of 8
$2401^{1/4} \cdot 2401^{1/4} \cdot 2401^{1/4} \cdot 2401^{1/4} = 2401$	$2401^{1/4}$ is a fourth root of 2401
$-121^{1/2} \cdot -121^{1/2} = 121$ (Recall that $-a^b$ means the opposite of a^b.)	$-121^{1/2}$ is the negative square root of 121
$3125^{1/5} \cdot 3125^{1/5} \cdot 3125^{1/5} \cdot 3125^{1/5} \cdot 3125^{1/5} = 3125$	$3125^{1/5}$ is the positive fifth root of 3125

Comparing Table 1 with Table 2, we can see that:

$9^{1/2} = 3$ Since both are symbols for the positive square root of 9.

$8^{1/3} = 2$ Since both are symbols for a cube root of 8.

$2401^{1/4} = 7$ Since both are symbols for the positive fourth root of 2401.

$-121^{1/2} = -11$ Since both are symbols for the negative square root of 121.

▣ PROPERTY I

Principal *n*th Root

> If $a \geq 0$ and if n is odd, $a^{1/n}$ is one of the n identical positive factors of a.
>
> $a^{1/n}$ is called an nth root of a.
>
> If $a \geq 0$ and if n is even, $a^{1/n}$ is called the principal nth root of a, and $-a^{1/n}$ is another nth root of a.

$16^{1/4}$ is the principal (positive) fourth root of 16 and $16^{1/4} = 2$ since $2^4 = 16$.

$-25^{1/2}$ is the opposite of the principal square root of 25 and $-25^{1/2} = -5$ since $(-5)^2 = 25$.

$64^{1/3}$ is a cube root of 64 and $64^{1/3} = 4$ since $4^3 = 64$.

$-27^{1/3}$ is the opposite of the cube root of 27 and $-27^{1/3} = -3$ since $-(3)^3 = -27$.

EXAMPLE 1

Simplify: $121^{1/2}$

$121^{1/2} = 11$ The number 121 has two sets of equal factors: $(11)(11)$ and $(-11)(-11)$. The exponent, $\dfrac{1}{2}$, denotes only the positive square root. The root keys on your calculator $\boxed{\sqrt{}}$ or $\boxed{y^{1/x}}$ can help you find the root faster than by trial and error.   ∎

EXAMPLE 2

Simplify: $1089^{1/2}$

$1089^{1/2} = 33$ Since $33 \cdot 33 = 1089$.   ∎

EXAMPLE 3

Simplify: $1728^{1/3}$

$1728^{1/3} = 12$ Use the $\boxed{\sqrt[3]{}}$ key or the $\boxed{y^{1/x}}$ key on your calculator.   ∎

■ **PROPERTY II**

Negative *n*th Root

> If $a < 0$ and if n is odd, $a^{1/n}$ is one of the n identical negative factors of a.
>
> $a^{1/n}$ is called an *n*th root of a.
>
> If $a < 0$ and if n is even, $a^{1/n}$ does not represent a real number because no real number raised to an even power yields a negative number.

It is necessary to exercise caution when working with rational exponents and negative bases. Be sure to check whether the indicated root is odd or even.

$(-27)^{1/3} = -3$ and $(-32)^{1/5} = -2$ and $(-2187)^{1/7} = -3$.

$(-9)^{1/2}$ is not a real number since $(3)(3) \neq -9$ and $(-3)(-3) \neq -9$.

$(-16)^{1/4}$ is not a real number.

EXAMPLE 4

Simplify: $-64^{1/2}$

$-64^{1/2} = -8$ 　　　　　　　　　　　　$-64^{1/2} = -(64^{1/2}) = -8$ ■

EXAMPLE 5

Simplify: $(-64)^{1/2}$

$(-64)^{1/2}$ is not a real number 　　There is no positive or negative real number whose square is -64. ■

The third property of exponents leads to another important and useful property.

■ **PROPERTY III**

> If $a \geq 0$, $a^{m/n} = (a^m)^{1/n} = (a^{1/n})^m$.

This property gives us an option. When a rational exponent does not have a numerator of 1, we can first find the power (numerator) then the root (denominator), or we can first find the root (denominator) then the power (numerator).

EXAMPLE 6

Simplify: $64^{3/2}$

$64^{3/2} = (64^{1/2})^3 = 8^3 = 512$ or

$64^{3/2} = (64^3)^{1/2} = 262144^{1/2} = 512$ ■

The next property is more complex and is listed for completeness. We shall not make use of such expressions.

■ **PROPERTY IV**

> If $a < 0$, $a^{m/n}$ is ambiguous. That is, there may be more than one result.

$(-27)^{2/3} = [(-27)^{1/3}]^2 = (-3)^2 = 9$ and

$(-27)^{2/3} = [(-27)^2]^{1/3} = 729^{1/3} = 9$

but

$(-4)^{6/2} \stackrel{?}{=} [(-4)^6]^{1/2} = (4096)^{1/2} = 64$ or

$(-4)^{6/2} \stackrel{?}{=} (-4)^3 = -64$ or

$(-4)^{6/2} \stackrel{?}{=} [(-4)^{1/2}]^6$ is not yet defined because $(-4)^{1/2}$ is not a real number.

The remaining examples illustrate the use of the five properties of exponents to simplify expressions with rational exponents.

EXAMPLE 7

Simplify: $x^{2/3} \cdot x^{2/3}$ and $y^{1/2} \cdot y^{1/3}$

$x^{2/3} \cdot x^{2/3} = x^{4/3}$ When multiplying powers with the

$y^{1/2} \cdot y^{1/3} = y^{5/6}$ same base, add the exponents. ■

EXAMPLE 8

Simplify: $\dfrac{w^{7/8}}{w^{1/4}}$ and $\dfrac{6z^{1/2}}{2z^{1/6}}$

$\dfrac{w^{7/8}}{w^{1/4}} = w^{7/8 - 1/4} = w^{5/8}$ When dividing powers with the same base, subtract the exponents.

$\dfrac{6z^{1/2}}{2z^{1/6}} = 3z^{1/2 - 1/6} = 3z^{1/3}$ ■

EXAMPLE 9

Simplify: $(b^4)^{3/2}$ and $(c^{-2/3})^{-4/5}$

$(b^4)^{3/2} \quad\;\; = b^6$

When raising a power to a power, multiply the exponents.

$(c^{-2/3})^{-4/5} = c^{8/15}$

■

EXAMPLE 10

Simplify: $(x^{1/3}y^{3/4})^2$

$= (x^{1/3})^2(y^{3/4})^2$

When raising a product to a power, raise each factor to the power.

$= x^{2/3}y^{3/2}$

Raise a power to a power.

■

EXAMPLE 11

Simplify: $\left(\dfrac{x^{2/3}}{y^{3/2}}\right)^4$

$= \dfrac{(x^{2/3})^4}{(y^{3/2})^4}$

When raising a quotient to a power, raise both the numerator and denominator to the power.

$= \dfrac{x^{8/3}}{y^6}$

Raise a power to a power.

■

EXAMPLE 12

Simplify: $(x^{1/2} + y^{1/2})^2$

$= (x^{1/2})^2 + 2x^{1/2}y^{1/2} + (y^{1/2})^2$

Use the pattern for squaring a binomial: $(a + b)^2 = a^2 + 2ab + b^2$.

$= x + 2x^{1/2}y^{1/2} + y$

Raise a power to a power.

■

EXAMPLE 13

In some types of computer programs, each command is given a number. Two lines in such a program read:

Line Number	Command	or	Line Number	Command
275	$y = 9$		275	$y = 9$
280	$x = y**1.5$		280	$x = y\wedge 1.5$

In this kind of program language, the symbols $**$ and $\wedge$ are used to tell the computer that the number following is an exponent. In algebra, we write this:

$x = y^{1.5} = 9^{1.5} = 9^{3/2},$

where the exponent is the rational number $\dfrac{3}{2}$, but not an integer. What is the value of this expression?

$x = y**1.5$ or $x = y\wedge 1.5$	Computer statement 280.
$x = 9**1.5$ or $x = 9\wedge 1.5$	Substitute $y = 9$.
$x = 9^{1.5} = 9^{3/2}$	Algebra statement.
$x = (9^{1/2})^3$	Here we choose to find the root first, then raise to the power.
$x = (3)^3 = 27$	Since $9^{1/2} = 3$.

EXERCISE 4.1

A

Simplify, using only positive exponents:

1. $144^{1/2}$

2. $169^{1/2}$

3. $4^{3/2}$

4. $125^{2/3}$

5. $5^{3/5} \cdot 5^{1/5}$

6. $8^{4/9} \cdot 8^{3/9}$

7. $c^{2/5} \cdot c^{1/2}$

8. $m^{-4/5} \cdot m^{3/2}$

9. $\dfrac{x^{5/9}}{x^{2/9}}$

10. $\dfrac{y^{5/6}}{y^{-1/6}}$

11. $\dfrac{3^{4/5}}{3^{3/10}}$

12. $\dfrac{5^{7/8}}{5^{1/4}}$

13. $(10^{1/2})^3$

14. $(12^{2/3})^2$

15. $(49w^{3/2})^{1/2}$

16. $(8z^4)^{1/3}$

17. $(x^{3/2})^4$

18. $(y^{2/3})^6$

19. $(w^{3/4})^8$

20. $(t^{6/5})^{10}$

B

Simplify, using only positive exponents:

21. $3x^{1/2} \cdot 5x^{1/4}$

22. $(10a^{2/3})(-5a^{1/6})$

23. $(12b^{7/8})(3b^{-3/4})$

24. $(-2z^{5/3})(-4z^{-3/4})$

25. $(4^{1/3}x^{3/7})(4^{2/3}x^{3/7})$

26. $(6^{7/5}y^{1/2})(6^{3/5}y^{5/2})$

27. $\dfrac{24x^{1/2}}{6x^{1/3}}$

28. $\dfrac{42y^{5/6}}{7y^{1/2}}$

29. $\dfrac{-72d^{1/3}}{9d^{-1/6}}$

30. $\dfrac{48c^{1/10}}{-6c^{-2/5}}$

31. $\dfrac{5^2 p^{3/4}}{5^{-1} p^{2/3}}$

32. $\dfrac{16^{1/6}q^{-1/6}}{16^{-1/3}q^{-2/3}}$

33. $(27x^{3/4})^{2/3}$

34. $(3a^{5/3})^2$

35. $(x^{1/2}y^{2/3})^3$

36. $(a^{3/4}b^{1/2})^3$

37. $(-4c^{1/4}d^{1/2})^3$

38. $(25d^{2/5}a^{1/2})^{3/2}$

39. $\left(\dfrac{a^{3/4}}{b^{1/4}}\right)^2$

40. $\left(\dfrac{d^{5/3}}{e^{4/5}}\right)^3$

C

Simplify, using only positive exponents:

41. $(x^{1/2}y^2)(x^{2/3}y^{3/5})$

42. $(a^{1/4}b^{2/3})(a^{5/8}b^{4/9})$

43. $(d^2e^{1/4}f^3)(d^{1/3}e^{2/3}f)$

44. $(x^{5/6}y^{3/4}z^2)(x^2yz^{1/4})$

45. $(3a^{3/5}b^{7/8})^2$

46. $(2x^{1/4}y^{9/2})^3$

47. $(4x^{-3/4})^{-1/2}$

48. $(8y^{-4/5})^{-2/3}$

49. $(x^{-1/2}y^{2/3})^{-2/3}$

50. $(a^{-4/5}b^2)^{-1/2}$

51. $\dfrac{-54x^{-5/3}}{6x^{-7/2}}$

52. $\dfrac{-18a^{-3/2}}{3a^{-5/3}}$

53. $\left(\dfrac{x^{-5/6}}{z^{-2/3}}\right)^{-1/2}$

54. $\left(\dfrac{a^{-3/4}}{b^{-1/2}}\right)^{-1/4}$

55. $(a^{1/2} - b^{1/2})(a^{1/2} + b^{1/2})$

56. $(a^{1/3} + b^{1/3})^2$

57. $(a^{2/3} + 3b^{1/3})(a^{1/3} - 2b^{2/3})$

58. $(y^{1/2} + 3)(y^{1/2} - 4)$

59. $(w^{1/2} + 8)(w^{1/2} - 8)$

60. $(q^{2/3} + 7^{1/2})(q^{2/3} - 7^{1/2})$

D

61. Two lines of a computer program read:

Line Number	Command
310	$y = 16$
320	$x = y**1.5$

What is the numerical value of x?

62. Two lines of a computer program read:

Line Number	Command
1130	$w = 4$
1140	$x = w**2.5$

What is the numerical value of x?

63. A line of a computer program reads:

960    $x = 5 + 64^0.5$

What is the numerical value of x? (Recall that the order of operations in Chapter 1 states that exponential expressions are calculated first.)

64. A line of a computer program reads:

450    $x = 18 - 81^0.25$

What is the numerical value of x?

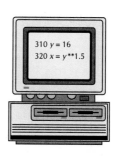

310 $y = 16$
320 $x = y**1.5$

Figure for Exercise 61

65. Two lines of a computer program read:

Line Number	Command
530	$w = 16$
540	$y = (w + 84)**1.5$

What is the numerical value of y?

66. Two lines of a computer program read:

Line Number	Command
800	$t = -200$
810	$m = (t + 325)**(1/3)$

What is the numerical value of m?

STATE YOUR UNDERSTANDING

67. Explain the meaning of a fractional exponent.

CHALLENGE EXERCISES

Simplify, assume all variable exponents represent positive integers.

68. $x^n \cdot x^{n/3}$

69. $(x^3)^{n/3} \cdot (x^{2n})^{1/2}$

70. $\dfrac{x^{5n}}{x^{n/2}}$

71. $\left[\dfrac{xy}{z^{2n}}\right]^{1/n}$

72. Show by example that $(x + y)^{1/2} \neq x^{1/2} + y^{1/2}$.

Factor as indicated:

73. $x^{3/2} + x^{1/2} = x^{1/2} (? + ?)$

74. $x - x^{3/4} = x^{1/4}(? - ?)$

MAINTAIN YOUR SKILLS (SECTIONS 2.3, 2.11)

Write in place notation:

75. 9.13×10^{-5}

76. 1.564×10^3

Calculate. Write answer in scientific notation:

77. $(8.46 \times 10^2)(5 \times 10^{-3})(0.2 \times 10^4)$

78. $\dfrac{(12.4 \times 10^4)(1.5 \times 10^{-2})}{(2.0 \times 10^5)(6.0 \times 10^{-3})}$

79. $\dfrac{(0.078 \times 10^8)(2.45 \times 10^3)}{(1.2 \times 10^6)(250 \times 10^{-5})}$

Factor completely:

80. $(x + y)^2 - 8(x + y) - 20$

81. $(x - 4)^2 - (y - 3)^2$

82. $(2x + 5)^2 - 18(2x + 5) + 81$

4.2

RADICALS

OBJECTIVES

1. Write expressions with rational exponents in equivalent radical form.

2. Write expressions with radicals in equivalent exponent form.

3. Simplify radical and exponential expressions.

In Section 4.1, we use the phrase "the nth root of a" for the symbol $a^{1/n}$. Another symbol that is more often used is called a radical. The *radical* ($\sqrt{}$) is used for the second root or square root. The meaning for $a^{1/n}$ and $\sqrt[n]{a}$ is the same. In Section 4.1, we learned that a square root of a number is one of two identical factors of that number. Thus, $a^{1/2} = \sqrt{a} = x$, $x \geq 0$ and $a \geq 0$ means $x \cdot x = a$ or, equivalently, $\sqrt{a} \cdot \sqrt{a} = a$.

■ DEFINITION

Principal Square Root

> The number $a^{1/2}$ or $\sqrt{a}$ is called the positive or principal square root of a.

The negative square root is written $-a^{1/2}$ or $-\sqrt{a}$.
The table shows the use of the radical sign to indicate square roots.

Number	Principal Square Root	Negative Square Root
361	$361^{1/2} = \sqrt{361} = 19$	$-361^{1/2} = -\sqrt{361} = -19$
49	$49^{1/2} = \sqrt{49} = 7$	$-49^{1/2} = -\sqrt{49} = -7$
576	$576^{1/2} = \sqrt{576} = 24$	$-576^{1/2} = -\sqrt{576} = -24$

Using the definition, we can check the truth of statements 1–9.

1. $\sqrt{9} = 3$ Since 9 and 3 are both positive and $3 \cdot 3 = 9$.

2. $\sqrt{225} = 15$ Since 225 and 15 are both positive and $15 \cdot 15 = 225$.

3. $\sqrt{0} = 0$ Since $0 \cdot 0 = 0$.

4. $6^{1/2} = \sqrt{6}$ Since $a^{1/2} = \sqrt{a}$.

5. $-\sqrt{64} = -8$ Since $-\sqrt{64}$ is the negative square root and $(-8)(-8) = 64$.

6. $\sqrt{50} \neq 25$ Since $(25)(25)$ is not equal to 50.

7. $\sqrt{100} \neq -10$ Since 100 and -10 are not both positive.

8. $\sqrt{-1} \neq 1$ Since $(1)(1) \neq -1$.

9. $\sqrt{-1} \neq -1$ Since $(-1)(-1) \neq -1$.

■ CAUTION

> If a is a negative number, the expression $\sqrt{a}$ does not represent a real number since no real number squared is negative.

For example, $\sqrt{-9}$ is not a real number since $(3)(3) \neq -9$ and $(-3)(-3) \neq 9$.

■ DEFINITION

Radical, Index, Radicand

> In general, the radical can be used for $a^{1/n}$ by writing
>
> $$a^{1/n} = \sqrt[n]{a},$$
>
> where $\sqrt{}$ is called the *radical*, n is called the *index* of the radical, and a is called the *radicand*.

When no index is written, it is understood to be 2. Thus,

$$\sqrt{20} = \sqrt[2]{20}.$$

EXAMPLE 1

Write in radical form: $65^{1/2}$

$65^{1/2} = \sqrt{65}$ The denominator, 2, in the exponent indicates the second or square root. The index 2 is understood to be part of the radical. Indices other than two must be written. ■

EXAMPLE 2

Write in radical form: $22x^{1/3}$

$22x^{1/3} = 22\sqrt[3]{x}$ The denominator, 3, indicates the third or cube root. The number 22 is understood to have the exponent 1, not 1/3. ■

EXAMPLE 3

Write in radical form: $(23x^2)^{1/5}$

$(23x^2)^{1/5} = \sqrt[5]{23x^2}$ The exponent 1/5 applies to both the factors in parentheses. The denominator, 5, indicates the fifth root. ■

EXAMPLE 4

Write in radical form: $7^{-1/2}$

$$7^{-1/2} = \frac{1}{7^{1/2}}$$

$$= \frac{1}{\sqrt{7}}$$

The negative exponent indicates the reciprocal of the expression $7^{1/2}$. The denominator, 2, in the exponent, indicates square root. ■

EXAMPLE 5

Write in exponent form: $\sqrt{6x}$

$$\sqrt{6x} = (6x)^{1/2} \quad \text{or} \quad 6^{1/2}x^{1/2}$$

Square roots are indicated by the exponent 1/2. ■

■ **CAUTION**

Radicand Variable Restrictions

> Unless otherwise stated, we assume throughout this text that all variables in radicands represent positive numbers.

EXAMPLE 6

Write in exponent form: $\sqrt[4]{10x^3}$

$$\sqrt[4]{10x^3} = (10x^3)^{1/4} \quad \text{or} \quad 10^{1/4}x^{3/4}$$

The radicand is positive. Either form is acceptable. ■

The statement $\sqrt{x^2} = x$ is one that requires careful thought. It is true only if x^2 and x are both positive (or both zero). So, if x is replaced by 9, we have

$$\sqrt{9^2} = 9 \quad \text{or} \quad \sqrt{81} = 9,$$

but if x is replaced by -9, we have

$$\sqrt{(-9)^2} \neq -9 \quad \text{or} \quad \sqrt{81} \neq -9$$

since both sides are *not* positive

The next table illustrates that the value of $\sqrt{x^2}$ is the same as the value of $|x|$.

Square Roots and Absolute Values

	Nonnegative Values		
x	$\sqrt{x^2}$	$|x|$	
12	$\sqrt{12^2} = \sqrt{144} = 12$	12	In each case, we see
7	$\sqrt{7^2} = \sqrt{49} = 7$	7	that $\sqrt{x^2}$ and $|x|$ and x
3.2	$\sqrt{3.2^2} = \sqrt{10.24} = 3.2$	3.2	have the same value

Square Roots and Absolute Values

	Negative Values		
x	$\sqrt{x^2}$	$|x|$	
-12	$\sqrt{(-12)^2} = \sqrt{144} = 12$	12	In each case, we see
-9	$\sqrt{(-9)^2} = \sqrt{81} = 9$	9	that $\sqrt{x^2}$ and $|x|$ and $-x$
$-\dfrac{2}{3}$	$\sqrt{\left(-\dfrac{2}{3}\right)^2} = \sqrt{\dfrac{4}{9}} = \dfrac{2}{3}$	$\dfrac{2}{3}$	have the same value

In all cases, we conclude that $\sqrt{x^2} = |x|$.

■ **FORMULA**

$\sqrt{x^2} = x$ is true if x is positive or zero.

$\sqrt{x^2} = -x$ is true if x is negative.

$\sqrt{x^2} = |x|$ is always true.

EXAMPLE 7

Simplify: $\sqrt{625x^2}$

$\sqrt{625x^2} = 25x$
In the absence of any comment on x, all values are understood to be positive and $(25x)^2 = 625x^2$. ■

EXAMPLE 8

Simplify: $(81y^8)^{1/4}$

$(81y^8)^{1/4} = \sqrt[4]{81y^8} = 3y^2$
Since $3^4 = 81$ and $(y^2)^4 = y^8$. ■

EXAMPLE 9

Simplify $\sqrt{25x^2}$ if x represents a real number.

$\sqrt{25x^2} = 5|x|$
The absolute bars around x are needed since x may be negative. The radical indicates the positive square root. ■

■ **CAUTION**

If x is any real number, to simplify $\sqrt{x^n}$ absolute value symbols are required when n is even and $\dfrac{n}{2}$ is an odd number.

For example, $\sqrt{x^6} = |x^3|$ *not* x^3, as we can see if we use -2 in place of x.

$$\sqrt{(-2)^6} = |(-2)^3| = 8 \quad but \quad \sqrt{(-2)^6} \neq (-2)^3 = -8$$

EXAMPLE 10

Simplify $\sqrt{x^{14}}$ and $\sqrt{x^{10}}$ if x represents a real number.

$$\sqrt{x^{14}} = |x^7| \quad and \quad \sqrt{x^{10}} = |x^5| \qquad \text{Absolute value bars in the answers assure a positive result.} \quad \blacksquare$$

■ **FORMULA**

> For all positive values of a,
> $$a^{m/n} = (\sqrt[n]{a})^m = \sqrt[n]{a^m}.$$

However the symbol $(\sqrt[n]{a})^m$ must be treated carefully. The meaning depends on the values of a, m, and n. The table below illustrates the meanings of $(\sqrt[n]{a})^m$. Compare these with $a^{m/n}$ in the previous section.

a	m	$n \neq 0$	$a^{m/n} = \left(\sqrt[n]{a}\right)^m$	Example
Nonnegative	Any integer	Any integer except 0	$\left(\sqrt[n]{a}\right)^m = \sqrt[n]{a^m}$	$8^{2/3} = \left(\sqrt[3]{8}\right)^2 = 2^2 = 4$ or $8^{2/3} = \sqrt[3]{8^2} = \sqrt[3]{64)} = 4$
Negative	Any integer	Odd	$\left(\sqrt[n]{a}\right)^m = \sqrt[n]{a^m}$	$(-32)^{3/5} = \left(\sqrt[5]{-32}\right)^3 = (-2)^3 = -8$ or $(-32)^{3/5} = \sqrt[5]{(-32)^3} = \sqrt[5]{-32768} = -8$
Negative	Any integer	Even	Not defined in this text. Different forms lead to different answers (ambiguous)	$(-25)^{2/4} \stackrel{?}{=} (-25)^{1/2} = \sqrt{-25}$, not real $(-25)^{2/4} \stackrel{?}{=} \left(\sqrt[4]{-25}\right)^2, \sqrt[4]{-25}$, not real $(-25)^{2/4} \stackrel{?}{=} \sqrt[4]{(-25)^2} = \sqrt[4]{625} = 5$

It is critical to the meaning of the exponent and the radical symbols whether the variables involved represent positive or negative numbers. Compare:

If y is positive or zero, $\sqrt{9y^2} = 3y$, but

if y is any real number, $\sqrt{9y^2} = |3y| = 3|y|$, and further,

$\sqrt{36a^4b^6} = 6a^2b^3$ is true only if b is positive or zero, whereas
$\sqrt{36a^4b^6} = 6a^2|b^3|$ is true for *all* real values of a and b.

EXAMPLE 11

Simplify $\sqrt{169x^2}$ if x represents a real number.

$\sqrt{169x^2} = 13|x|$ The positive square root of 169 is 13, and the positive square root of x^2 is $|x|$. The absolute value sign is necessary since x may be negative, and the radical indicates positive square roots only. ■

EXAMPLE 12

 Calculator example. Simplify: $\sqrt[4]{28{,}561}$

$\sqrt[4]{28{,}561} = 13$ A fourth root may be found by taking the square root twice. If your calculator has a root key, $\boxed{y^{1/x}}$ or $\boxed{\sqrt[x]{y}}$, it can be used directly for finding a positive root. ■

EXAMPLE 13

The formula $s = A^{1/2}$, where s represents the length of one side of a square and A represents the area of the square, can be used to find the length of the side if the area is given. If the area of a square is 225 in², what is the length?

$s = A^{1/2}$ Formula.

$s = 225^{1/2} = \sqrt{225}$ Substitute $A = 225$.

$s = 15$ Since $15^2 = 225$.

Each side of the square is 15 inches long. ■

EXERCISE   4.2

A

Write these expressions in radical form and simplify:

1. $32^{1/5}$ **2.** $16^{1/4}$ **3.** $4^{3/2}$ **4.** $8^{2/3}$

Write these expressions in radical form. Assume variables represent nonnegative real numbers:

5. $(47a)^{1/2}$ **6.** $(17y)^{1/3}$ **7.** $(2w)^{2/3}$ **8.** $(16m)^{1/4}$

Write these expressions with a rational exponent. Assume variables represent nonnegative real numbers:

9. $\sqrt{xy}$ **10.** $\sqrt{5t}$ **11.** $3\sqrt{ab}$ **12.** $6\sqrt{y}$

Write these expressions in radical form and simplify. Assume variables represent nonnegative real numbers:

13. $8s^{1/2}$ **14.** $(8s)^{1/2}$ **15.** $(16t)^{1/3}$ **16.** $16t^{1/3}$

If the variables represent real numbers, simplify. Use absolute value signs where necessary:

17. $\sqrt{w^2}$ **18.** $6\sqrt{r^2}$ **19.** $\sqrt{100m^6}$ **20.** $\sqrt{81x^2y^2}$

B

Write these expressions in radical form. Assume variables represent nonnegative real numbers:

21. $6x^{1/2}$ **22.** $-5y^{1/3}$ **23.** $2a^{-1/2}$ **24.** $5b^{-1/3}$

Write these expressions using rational exponents. Assume variables represent nonnegative real numbers:

25. $\sqrt{5xy}$ **26.** $\sqrt[3]{4ab}$ **27.** $\sqrt{x+1}$ **28.** $\sqrt[3]{a+b}$

Find the numerical value of each expression. Use a calculator with a radical key, $\boxed{\sqrt{}}$, or a square root table (see the Appendix).

29. $\sqrt{361}$ **30.** $\sqrt{441}$ **31.** $961^{1/2}$ **32.** $1225^{1/2}$

33. $-\sqrt[4]{81}$ **34.** $-\sqrt[4]{16}$ **35.** $\sqrt{-49}$ **36.** $\sqrt[4]{-64}$

If the variables represent real numbers, simplify. Use absolute value signs where necessary:

37. $\sqrt{4t^2}$ **38.** $\sqrt{144u^2}$ **39.** $\sqrt{64p^4}$ **40.** $\sqrt{81q^{10}}$

C

Simplify. Assume variables represent nonnegative real numbers:

41. $\sqrt{121x^4}$ **42.** $\sqrt{16y^{16}}$ **43.** $\sqrt[3]{27x^6}$ **44.** $\sqrt[3]{64y^{12}}$

Write these expressions in radical form. Assume variables represent nonnegative real numbers:

45. $(w^2+3)^{1/3}$ **46.** $(9-x^2)^{1/2}$ **47.** $(a^2+a)^{2/3}$

48. $(7-w)^{3/4}$ **49.** $(p-q)^{-1/2}$ **50.** $(s^2+4t)^{-1/3}$

Write these expressions using rational exponents. Assume variables represent nonnegative real numbers:

51. $\sqrt[3]{m^4n^2}$ **52.** $\sqrt[4]{x^3y^7}$ **53.** $\sqrt{5x^3y^5}$

54. $\sqrt[3]{9xy^4}$ **55.** $\sqrt{7-x}$ **56.** $(\sqrt[3]{x^2+5})^4$

If the variable represent real numbers, simplify. Use absolute value signs where necessary:

57. $\sqrt{(9 - x)^2}$

58. $\sqrt{(y + 3)^2}$

59. $\sqrt{a^2 - 2ab + b^2}$

60. $\sqrt{x^2 - 6x + 9}$

D

61. Find the length of one side of a square whose area is 196 m^2.

62. Find the length of one side of a square whose area is 12996 cm^2.

63. A square lot has an area of 1369 yd^2. Find the length of the frontage (one side of the square lot) in feet. (*Hint:* First find the length in yards.)

64. A square lot has an area of 484 km^2. Find the length of the frontage (one side of the square lot) in meters.

65. A circle has an area of 2025π square inches. Find the radius. (*Hint:* Area of a circle = πr^2.)

66. A circle has an area of 1296π square meters. Find the radius. (*Hint:* Area of a circle = πr^2.)

STATE YOUR UNDERSTANDING

67. If x names a real number, why is $\sqrt{x^2}$ not always equal to x?

CHALLENGE EXERCISES

68. A formula that can be used to find the approximate maximum speed, in miles per hour, at which a car can go through a curve without skidding is

$V = \sqrt{2.5r}$

where V represents the speed and r represents the radius of the curve in feet. If a curve has a radius of 1000 feet, how fast can a car be driven through the curve without skidding?

69. Using the formula in Exercise 68, how fast can a car be driven through a curve with radius of 160 feet without skidding?

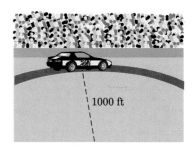

Figure for Exercise 68

70. Heron's formula for finding the area of a triangle is

$A = \sqrt{s(s - a)(s - b)(s - c)}$

where the sides of the triangle are a, b, and c and s is one-half the perimeter. That is

$s = \dfrac{1}{2}(a + b + c)$

Use Heron's formula to find the area of a triangle with sides 5 ft, 12 ft, and 13 ft.

71. Use Heron's formula to find the area of a triangle with sides 6 m, 8 m, and 10 m.

MAINTAIN YOUR SKILLS (SECTIONS 3.1, 3.2)

72. Build: $\dfrac{-8}{y - x} = \dfrac{?}{2x - 2y}$

73. Reduce: $\dfrac{98a^3b^3c^4}{40ab^5c^2}$

Multiply or divide. Reduce if possible:

74. $\dfrac{x^2 - 4}{x^2 + 4x + 4} \cdot \dfrac{x^2 + 2x + 4}{x^3 - 8}$

75. $\dfrac{y^2 + 3y + 9}{y^2 - 6y + 9} \cdot \dfrac{9 - y^2}{y^3 - 27}$

76. $\dfrac{x^3 - 8x^2 - 33x}{x^3 - 5x^2} \cdot \dfrac{x^2 - 25}{x^3 - 6x^2 - 55x}$

77. $\dfrac{6st + 4s^2}{3} \div \dfrac{6t^2 + 4st}{9}$

78. $\dfrac{k + 7}{2k + 7} \cdot \dfrac{4k^2 - 49}{k^2 - 49} \div \dfrac{k - 7}{3}$

79. $\dfrac{2p + 10}{3p - 9} \div \dfrac{5p + 5}{12p - 36} \div \dfrac{4p + 20}{pq + q}$

4.3

SIMPLIFYING AND APPROXIMATING RADICALS

OBJECTIVES

1. Simplify a radical expression.

2. Find a rational number that approximates a square root radical.

3. Simplify a radical expression by reducing its index.

A number in radical form such as $\sqrt{50}$ does not have a root that is an integer or a rational number. The number 50 is not a perfect square. Such radical expressions represent irrational numbers. These radicals are typically written in one of two ways:

1. In simplest radical form (the exact value).

2. As an approximate rational number usually in decimal form for use in a calculator or a computer or in applications.

■ DEFINITION

> An expression containing a radical is called a radical expression.

The expressions

$$\sqrt{8x^3}, \qquad \sqrt{x} + \sqrt{2y}, \qquad 6 - \sqrt[3]{7x}, \quad \text{and} \quad \sqrt{22} + 3\sqrt{3t}$$

are radical expressions.

■ RULE

Simplest Radical Form

There are two conditions that must be met before a square root radical expression is called simplest radical form.

1. The radicand has no factors that are perfect squares except the number 1.
2. No fractions are in the radicand and no radicals appear in a denominator.

The examples in the table leads us to the product and quotient properties of radicals.

Examples		
Two Radicals *Product or Quotient*	*Single Radical*	Conclusion
$\sqrt{4}\sqrt{100} = 2 \cdot 10 = 20$	$\sqrt{400} = 20$	$\sqrt{4}\sqrt{100} = \sqrt{4 \cdot 100} = \sqrt{400}$
$\sqrt{16}\sqrt{4} = 4 \cdot 2 = 8$	$\sqrt{64} = 8$	$\sqrt{16}\sqrt{4} = \sqrt{16 \cdot 4} = \sqrt{64}$
$\sqrt{a^2}\sqrt{b^2} = ab$	$\sqrt{a^2b^2} = ab$	$\sqrt{a^2}\sqrt{b^2} = \sqrt{a^2b^2}$
$\dfrac{\sqrt{9x^4}}{\sqrt{49}} = \dfrac{3x^2}{7}$	$\sqrt{\dfrac{9x^4}{49}} = \dfrac{3x^2}{7}$	$\dfrac{\sqrt{9x^4}}{\sqrt{49}} = \sqrt{\dfrac{9x^4}{49}}$
$\sqrt[3]{8} \cdot \sqrt[3]{27} = 2 \cdot 3 = 6$	$\sqrt[3]{216} = 6$	$\sqrt[3]{8} \cdot \sqrt[3]{27} = \sqrt[3]{8 \cdot 27} = \sqrt[3]{216}$

■ PROPERTY

Product Property of Radicals

If $a \geq 0$ and $b \geq 0$, then

$$\sqrt{ab} = \sqrt{a}\sqrt{b} \quad \text{and} \quad \sqrt[n]{ab} = \sqrt[n]{a}\sqrt[n]{b}.$$

■ PROPERTY

Quotient Property of Radicals

If $a \geq 0$ and $b > 0$, then

$$\sqrt{\frac{a}{b}} = \frac{\sqrt{a}}{\sqrt{b}} \quad \text{and} \quad \sqrt[n]{\frac{a}{b}} = \frac{\sqrt[n]{a}}{\sqrt[n]{b}}.$$

These basic properties of radicals are based on the laws of exponents.

$$\sqrt{ab} = (ab)^{1/2} = a^{1/2}b^{1/2} = \sqrt{a}\,\sqrt{b} \quad \text{and}$$

$$\sqrt{\frac{a}{b}} = \left(\frac{a}{b}\right)^{1/2} = \frac{a^{1/2}}{b^{1/2}} = \frac{\sqrt{a}}{\sqrt{b}}$$

In summary, we can multiply and divide radical expressions whenever the radicands are positive or when the radicals are odd roots of negative numbers. We use these basic principles to write radicals in simplest form.

EXAMPLE 1

Simplify: $\sqrt{50}$

$\sqrt{50} = \sqrt{25 \cdot 2}$ Factor the radicand into two factors so that one of them is a perfect square.

$\quad = \sqrt{25}\,\sqrt{2}$

$\quad = 5\sqrt{2}$ Simplest radical form.   ■

■ **PROCEDURE**

Simplify a Radical

To simplify a square root radical containing only multiplications, in which the variables represent positive numbers:

1. Factor the radicand so that one or more numerical factors are perfect squares and so that one or more variable factors have exponents that are multiples of 2. Continue such factoring until no square factors remain.

2. Simplify the perfect square factors.

If you cannot readily find the perfect square factor of a number, it may be found by writing the number as the product of prime factors.

EXAMPLE 2

Simplify: $\sqrt{588}$

$\sqrt{588} = \sqrt{2 \cdot 2 \cdot 3 \cdot 7 \cdot 7}$ The number is large, so it is not easy to identify the largest square factor (which is $196 = 14^2$). In this case, we use the prime factors of 588.

$\quad = \sqrt{2 \cdot 2}\,\sqrt{7 \cdot 7}\,\sqrt{3}$ Group pairs of like factors under the same radical where possible.

$\quad = 2 \cdot 7\sqrt{3}$

$\quad = 14\sqrt{3}$ Simplest radical form.   ■

When variables occur in the radicand, we must be careful that the replacements do not cause the radicand to be negative. In Example 3, x and y are assumed to be positive.

EXAMPLE 3

Simplify: $\sqrt{48x^3y^4}$

$= \sqrt{16 \cdot 3x^2 \cdot x \cdot y^4}$ Factor.

$= \sqrt{16} \sqrt{x^2} \sqrt{y^4} \sqrt{3x}$ Separate the radicals that can be simplified.

$= 4xy^2\sqrt{3x}$ Simplest radical form.   ∎

In Example 4, x and y are assumed to be real numbers.

EXAMPLE 4

Simplify $\sqrt{48x^3y^6}$ if x and y represent real numbers.

x must be nonnegative, that is, $x \geq 0$.

No restriction on y is necessary.

$\sqrt{48x^3y^6} = \sqrt{16} \sqrt{x^2} \sqrt{y^6} \sqrt{3x}$

The radicand will not be positive and the radical will not represent a real number if x is negative. By requiring that x be nonnegative, the radical expression represents a real number and can be simplified. No restriction is needed for y since y^6 is nonnegative for all real values of y.

$= 4x|y^3|\sqrt{3x},\ x \geq 0$

Absolute value bars for y^3 are required to assure that the result is nonnegative. Absolute value bars are not required for x because of the restriction.   ∎

■ **PROCEDURE**

Simplify a Radical

To simplify a square root radical containing only multiplications, in which the variables represent real numbers:

1. Restrict those variables with exponents that are odd numbers. Then follow step 1 as before.

2. After the radical has been simplified, place absolute bars around the variable factors as necessary.

EXAMPLE 5

Simplify $\sqrt{8x^6y^5}$ if x and y represent real numbers.

$y \geq 0$	The variable y has an odd number exponent, so y is restricted to nonnegative values.
$\sqrt{8x^6y^5} = \sqrt{4} \sqrt{x^6} \sqrt{y^4} \sqrt{2y},\ y \geq 0$	Factor, using perfect squares.
$\qquad = 2|x^3|y^2 \sqrt{2y},\ y \geq 0$	$\sqrt{x^6} = |x^3|$ because if x is negative, x^3 is negative, but the result must be positive. ■

Square root radicals can be simplified only when the radicand has a perfect square factor. Thus, $\sqrt{14}$, $\sqrt{30}$, and $\sqrt{85}$ cannot be simplified because none contain a perfect square factor. Furthermore, the sum of perfect squares *cannot* be simplified in the same way as factors.

$$\sqrt{16 + 9} = \sqrt{25} = 5 \quad \text{but}$$

$$\sqrt{16 + 9} \neq \sqrt{16} + \sqrt{9} = 4 + 3 = 7$$

■ **CAUTION**

> $\sqrt{a^2 + b^2}$ is *not identical* to $a + b$.

To find a rational number that approximates a radical, use a calculator with a radical key, $\boxed{\sqrt{}}$, or a square root table (see the Appendix). Round the decimal to the desired decimal place.

$\sqrt{50}\ \approx 7.07$	To the nearest hundredth.
$\sqrt{80}\ \approx 8.944$	To the nearest thousandth.
$\sqrt{165} \approx 12.85$	To the nearest hundredth.
$\sqrt{816} \approx 28.566$	To the nearest thousandth.

EXAMPLE 6

  Find the rational approximation of $\sqrt{67}$ to the nearest thousandth.

$\sqrt{67} \approx 8.185$	The value can be found using a calculator or square root table. ■

EXAMPLE 7

Find a rational fraction approximation of $\sqrt{15}$.

$\sqrt{15} \approx 3.87298$	Use a calculator or table.

$$\sqrt{15} \approx 3.9 = \frac{39}{10}$$

Each of these fractions is an approximation of $\sqrt{15}$. The fractions are listed in order of increasing accuracy.

$$\sqrt{15} \approx 3.87 = \frac{387}{100}$$

$$\sqrt{15} \approx 3.873 = \frac{3873}{1000}$$

◼

EXAMPLE 8

The Ajax Hardware Store sells a lawn sprinkler that waters a square area of lawn. If the area watered is 98 square feet, use the formula $s = \sqrt{A}$, where s is the length of a side and A is the area, to find the length of a side. Give the answer (1) in simplest radical form and (2) to the nearest tenth of a foot.

Formula:

$$s = \sqrt{A}$$

To find the length of a side of the square of lawn that is watered, substitute in the formula.

Substitute:

$$s = \sqrt{98}$$

$A = 98$

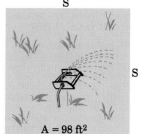

Simplify:

$$s = \sqrt{49}\sqrt{2} = 7\sqrt{2}$$

Simplest radical form.

Approximate:

$$s = \sqrt{98} \approx 9.9$$

Use a calculator or table to find the approximation.

Answer:

The length of the side is $7\sqrt{2}$ feet or approximately 9.9 feet.   ◼

The index of a radical can be reduced if powers of the radicand are factors of the index.

EXAMPLE 9

Simplify by reducing the index: $\sqrt[4]{x^2}$

$$= x^{2/4}$$

First, change from radical form to exponential form.

$$= x^{1/2}$$

Second, reduce the exponent.

$$= \sqrt{x}$$

Third, change back to radical form. This procedure depends on whether x is positive or negative. Here we assume, in the absence of a specified domain, that x is nonnegative. ∎

EXAMPLE 10

Simplify by reducing the index: $\sqrt[4]{4a^2b^2}$

$$= (4)^{1/4}(a)^{2/4}(b)^{2/4}$$ Change to exponential form.

$$= (2^2)^{1/4}(a)^{2/4}(b)^{2/4}$$ Since $4 = 2^2$.

$$= (2)^{2/4}(a)^{2/4}(b)^{2/4}$$ The property of exponents for raising a power to a power.

$$= (2)^{1/2}(a)^{1/2}(b)^{1/2}$$ Reduce.

$$= \sqrt{2ab}$$ ∎

EXERCISE 4.3

A

Simplify these radicals. Assume that all variables represent positive numbers:

1. $\sqrt{200}$

2. $\sqrt{300}$

3. $\sqrt{90}$

4. $\sqrt{99}$

5. $\sqrt{27}$

6. $\sqrt{45}$

7. $\sqrt{63}$

8. $\sqrt{18x^2}$

9. $\sqrt{40y^2}$

10. $\sqrt{50w^2x}$

11. $\sqrt{54p^3}$

12. $\sqrt{32q^3}$

13. $\sqrt{24st^2}$

14. $\sqrt{98mn^2}$

15. $\sqrt{25ab}$

16. $\sqrt{100xy}$

17. $\sqrt{49x^4}$

18. $\sqrt{121y^6}$

19. $\sqrt{18p^2q}$

20. $\sqrt{27t^4w}$

B

Use a calculator or table to find the approximate values of these radicals (to the nearest thousandth). Answers in the text were found using a calculator:

21. $\sqrt{200}$

22. $\sqrt{300}$

23. $\sqrt{90}$

24. $\sqrt{99}$

25. $\sqrt{4428}$

26. $\sqrt{15975}$

Simplify. Assume that all variables represent positive numbers:

27. $\sqrt{32x^4}$

28. $\sqrt{48y^8}$

29. $\sqrt{150m^{12}}$

30. $\sqrt{242a^{16}}$

31. $\sqrt{125a^3}$

32. $\sqrt{27x^5}$

33. $\sqrt{80y^7}$

34. $\sqrt{72x^9}$

Simplify. The variables represent real numbers. Use absolute value where necessary:

35. $\sqrt{169y^2}$ **36.** $\sqrt{196z^2}$ **37.** $\sqrt{28a^6}$ **38.** $\sqrt{108b^{10}}$

39. $\sqrt{45x^4y^6}$ **40.** $\sqrt{80p^2q^6}$

C

Simplify. Assume all variables represent positive numbers:

41. $\sqrt{175a^2b^3c^4}$ **42.** $\sqrt{171x^3y^4z^5}$ **43.** $\sqrt[3]{16y^3}$ **44.** $\sqrt[3]{54x^6}$

45. $\sqrt[3]{8x^4y^6}$ **46.** $\sqrt[3]{125x^5y^9}$ **47.** $\sqrt[4]{16a^8}$ **48.** $\sqrt[5]{64a^{10}b^8}$

49. $\sqrt{147\ell^{10}m^{11}}$ **50.** $\sqrt{162g^{12}h^{13}}$ **51.** $\sqrt[4]{16t^{16}w^9}$ **52.** $\sqrt[5]{32r^{12}s^{10}}$

Simplify. The variables represent real numbers. Use absolute value where necessary:

53. $\sqrt{48c^8d^{10}}$ **54.** $\sqrt{96m^4n^{12}}$ **55.** $\sqrt[3]{27x^3}$ **56.** $\sqrt[3]{64y^3}$

57. $\sqrt[4]{16s^4t^8}$ **58.** $\sqrt[4]{81t^4s^{16}}$

Reduce the index of the following radicals if the variables represent positive real numbers:

59. $\sqrt[4]{5^2w^2}$ **60.** $\sqrt[4]{36y^2}$ **61.** $\sqrt[4]{25m^2n^2}$ **62.** $\sqrt[4]{100r^2s^2}$

63. $\sqrt[6]{3^3c^3}$ **64.** $\sqrt[6]{8d^3}$

D

65. Find the length of one side of a square section of lawn that is watered if Ajax Hardware advertises that the sprinkler will cover an area of 192 square yards. Give the answer
   a. in simplest radical form.
   b. to the nearest tenth of a yard.

66. Find the length of one side of a square lot that has an area of 4000 square yards. Give the answer
   a. in simplest radical form.
   b. to the nearest tenth of a yard.

67. The Nice and Gooey Frosting Company advertises that one can of frosting will cover 500 square centimeters. What is the length of a side of the largest square cake that can be frosted if only the top is frosted? Give the answer
   a. in simplest radical form.
   b. to the nearest tenth of a centimeter.

68. The Nice and Gooey Frosting Company advertises that one package of frosting mix will cover 750 square centimeters. What is the length of a side

of the largest square cake that can be frosted if only the top is frosted?
Give the answer
a. in simplest radical form.
b. to the nearest tenth of a centimeter.

69. If the area of a circle is given, the radius can be approximated by the formula $r \approx \sqrt{0.318A}$. Find the radius of a circle with area 15 square meters to the nearest hundredth of a meter.

70. If the area of a circle is given, the radius can be approximated by the formula $r \approx \sqrt{0.318A}$. Find the radius of a circle with area 212 square miles to the nearest hundredth of a mile.

71. If the surface area of a sphere is given, the radius of the sphere can be approximated by the formula $r = \dfrac{1}{2}\sqrt{0.318A}$. Find the radius of a sphere with surface area 154 square inches to the nearest tenth of an inch.

72. If the surface area of a sphere is given, the radius of the sphere can be approximated by the formula $r = \dfrac{1}{2}\sqrt{0.318A}$. Find the radius of the sphere with surface area 113 square centimeters to the nearest tenth of a centimeter.

STATE YOUR UNDERSTANDING

73. In a radical expression in which a variable appears in the radicand, why is it sometimes necessary to restrict the variable?

CHALLENGE EXERCISES

Simplify, assume all variable exponents represent positive integers.

74. $\sqrt{x^{2n}}$

75. $\sqrt{x^{4n}y^{6n}}$

76. $\sqrt[3]{b^{6n}}$

77. $\sqrt[4]{y^{12n}}$

78. $\sqrt[n]{a^{6n}}$

79. $\sqrt[n]{y^{9n}}$

80. To find the approximate velocity (v) of a vehicle which is accelerating at a constant rate of acceleration (a) after a certain distance (d) is given by the formula

$$v = \sqrt{v_o + ad}$$

where v_o is the velocity just prior to the acceleration. An auto starts from rest and accelerates at the rate of 16 ft/sec^2. What is its velocity after it has traveled 100 ft?

81. If an auto is traveling at 44 ft/sec and accelerates at the rate of 10 ft/sec^2, what is its velocity after 770 ft?

MAINTAIN YOUR SKILLS (SECTIONS 3.1, 3.3)

82. Build: $\dfrac{3x + 1}{2x - 1} = \dfrac{?}{6x^2 - 5x + 1}$

83. Build: $\dfrac{2x - 5}{2x + 1} = \dfrac{?}{6x^2 + x - 1}$

84. Reduce: $\dfrac{6c^2 - 7c - 5}{9c^2 - 12c - 5}$

Add or subtract. Reduce if possible:

85. $\dfrac{10}{2x - 5} + \dfrac{6}{5 - 2x}$

86. $\dfrac{5}{3x - 8} + \dfrac{-4}{8 - 3x}$

87. $\dfrac{12}{2y + 3} + 7y$

88. $\dfrac{10}{2x - 5} - \dfrac{6}{5 - 2x}$

89. $\dfrac{5}{3x - 8} - \dfrac{-4}{8 - 3x}$

4.4

COMBINING RADICALS

OBJECTIVE

1. Combine radical expressions.

In this section, we learn how to simplify radical expressions by addition and subtraction. Recall that like terms such as $5x$ and $12x$ can be combined.

$$5x + 12x = 17x$$

■ **DEFINITION**

Like Radicals

Like radicals are radicals that have the same radicand and the same index.

The following examples illustrate like and unlike radicals.

$7\sqrt{3}, \dfrac{1}{2}\sqrt{3}, -6\sqrt{3}$ — Are like radicals.

$3\sqrt[3]{3}, -4\sqrt[3]{3}, 10\sqrt[3]{3}$ — Are like radicals.

$\sqrt{7}, \sqrt{8}, \sqrt{10}$ — Are unlike radicals since they do not have the same radicands.

$\sqrt[3]{7}, \sqrt{7}, \sqrt[4]{3}$ — Are unlike radicals since they do not have the same index.

The distributive property is used to combine like radicals in the same way it is used to combine like terms.

EXAMPLE 1

Combine: $6\sqrt{5} + 11\sqrt{5}$

$= (6 + 11)\sqrt{5}$ Factor using the distributive property.

$= 17\sqrt{5}$ Simplify. ∎

EXAMPLE 2

Combine: $\sqrt{7} + \sqrt{7}$

$= (1 + 1)\sqrt{7}$ Factor using the distributive property.

$= 2\sqrt{7}$ Simplify. ∎

EXAMPLE 3

Combine: $7\sqrt{3} + \dfrac{1}{2}\sqrt{3} - 6\sqrt{3}$

$= \left(7 + \dfrac{1}{2} - 6\right)\sqrt{3}$ Factor.

$= \dfrac{3}{2}\sqrt{3}$ Simplify. ∎

Just as unlike terms cannot be combined, unlike radicals in simplest form cannot be combined.

EXAMPLE 4

Combine: $\sqrt{7} + \sqrt{17}$

$\sqrt{7} + \sqrt{17}$ is the simplest form.   These are not like radical expressions. They do not have the same radicand. ∎

Some unlike radicals can be combined after they are simplified to like radicals.

EXAMPLE 5

Combine: $\sqrt{45} + \sqrt{80}$

$= \sqrt{9}\sqrt{5} + \sqrt{16}\sqrt{5}$ Use the product property of radicals.

$= 3\sqrt{5} + 4\sqrt{5}$ Simplify each radical.

$= (3 + 4)\sqrt{5} = 7\sqrt{5}$ Simplest form. ∎

EXAMPLE 6

Combine: $3\sqrt[3]{16} + 4\sqrt[3]{54}$

$= 3\sqrt[3]{8}\sqrt[3]{2} + 4\sqrt[3]{27}\sqrt[3]{2}$ Simplify each radical.

$= 3(2)\sqrt[3]{2} + 4(3)\sqrt[3]{2}$

$= (6 + 12)\sqrt[3]{2}$

$= 18\sqrt[3]{2}$ Simplest form. ■

EXAMPLE 7

Combine: $\sqrt{10} - \sqrt[3]{10}$

$\sqrt{10} - \sqrt[3]{10}$ is the simplest form.   These are not like radical expressions. They do not have the same index. ■

EXAMPLE 8

Combine: $\sqrt{75} - \sqrt{20}$

$= \sqrt{25}\sqrt{3} - \sqrt{4}\sqrt{5}$ Simplify both radicals.

$= 5\sqrt{3} - 2\sqrt{5}$ Simplest form.

■ **CAUTION**

The radicals $\sqrt{3}$ and $\sqrt{5}$ are unlike radical expressions and cannot be combined.

■

EXAMPLE 9

Combine: $\sqrt{300} - \sqrt{12} + \sqrt{18}$

$= 10\sqrt{3} - 2\sqrt{3} + 3\sqrt{2}$ Simplify each radical.

$= (10 - 2)\sqrt{3} + 3\sqrt{2}$

$= 8\sqrt{3} + 3\sqrt{2}$ Simplest radical form. Unlike radical expressions cannot be combined. ■

EXAMPLE 10

Combine: $9\sqrt{8x^3} + \sqrt{72x^3}$

$= 9\sqrt{4}\sqrt{x^2}\sqrt{2x} + \sqrt{36}\sqrt{x^2}\sqrt{2x}$ By convention in this text the variable x, under a radical, represents a positive number in the absence of a contrary statement.

$= 18x\sqrt{2x} + 6x\sqrt{2x}$

$= 24x\sqrt{2x}$ Combine like radicals. ■

EXAMPLE 11

In the Sunnyhill housing development, three square lots are situated so that they enclose a common triangular region. This common area is to be used by all three owners. What is the perimeter of the triangle if the lots have areas 4800 square feet, 6912 square feet, and 2352 square feet. Write the answer (1) in simplest radical form and (2) to the nearest tenth of a foot.

Formula:

$P = a + b + c$

To find the perimeter of the common ground at Sunnyhill, we must find the length of each side, then add.

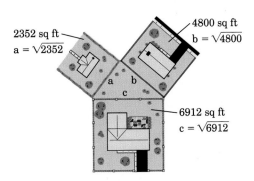

2352 sq ft
$a = \sqrt{2352}$

4800 sq ft
$b = \sqrt{4800}$

6912 sq ft
$c = \sqrt{6912}$

Substitute:

$P = \sqrt{2352} + \sqrt{4800} + \sqrt{6912}$

The formula for the side of a square is $s = \sqrt{a}$, where A is the area.

Solve:

$P = 28\sqrt{3} + 40\sqrt{3} + 48\sqrt{3}$ Simplify each radical.

$P = 116\sqrt{3}$ Simplest radical form.

$P \approx 116(1.732)$

$P \approx 200.9$ Rounded to the nearest tenth.

Answer:

The perimeter of the common triangular area is $116\sqrt{3}$ feet or, approximately, 200.9 feet ∎

EXERCISE 4.4

A

Combine: Assume that all variables represent positive numbers:

1. $\sqrt{3} + \sqrt{3}$

2. $\sqrt{11} + \sqrt{11}$

3. $2\sqrt{7} + \sqrt{7}$

4. $3\sqrt{15} + \sqrt{15}$

5. $4\sqrt{2} - 3\sqrt{2}$

6. $6\sqrt{5} - 4\sqrt{5}$

7. $11\sqrt{5} - 8\sqrt{5} + 3\sqrt{5}$

8. $14\sqrt{17} - 25\sqrt{17} + 6\sqrt{17}$

9. $-5\sqrt{35} + 10\sqrt{35} - 8\sqrt{35}$

10. $-12\sqrt{47} - 3\sqrt{47} + 6\sqrt{47}$

11. $4\sqrt{2} + \sqrt{8}$

12. $5\sqrt{3} - \sqrt{27}$

13. $\sqrt{18} - 4\sqrt{2}$

14. $-\sqrt{45} + 3\sqrt{5}$

15. $\sqrt{12} + 5\sqrt{3} - \sqrt{3}$

16. $11\sqrt{11} - \sqrt{44} + 2\sqrt{11}$

17. $\sqrt{13} + \sqrt{52} - 3\sqrt{13}$

18. $2\sqrt{14} + \sqrt{56} - \sqrt{14}$

19. $3\sqrt{20} - 7\sqrt{20} + \sqrt{80}$

20. $5\sqrt{24} - 8\sqrt{24} + \sqrt{600}$

B

Combine: Assume that all variables represent positive numbers:

21. $\sqrt{24} - 6\sqrt{6} + \sqrt{54}$

22. $\sqrt{45} - 6\sqrt{5} + \sqrt{20}$

23. $\sqrt{2x} + \sqrt{2x}$

24. $\sqrt{5x} + \sqrt{5x}$

25. $6\sqrt{3y} + 3\sqrt{3y}$

26. $8\sqrt{7t} + 2\sqrt{7t}$

27. $\sqrt{200x} - 8\sqrt{2x}$

28. $3\sqrt{3y} - \sqrt{300y}$

29. $\sqrt{25x} + \sqrt{25x}$

30. $\sqrt{49w} + \sqrt{64w}$

31. $8y\sqrt{88y} - 8\sqrt{22y^3}$

32. $4x\sqrt{6} - \sqrt{6x^2}$

33. $\sqrt{125x^3} - x\sqrt{45x} - \sqrt{20x^3}$

34. $\sqrt{8x^4} + 4\sqrt{2x^4} - x^2\sqrt{50}$

35. $\sqrt[3]{81x^3} + x\sqrt[3]{24}$

36. $\sqrt[3]{250y^3} - y\sqrt[3]{16}$

37. $\sqrt[3]{4} + \sqrt[3]{32} - 2\sqrt[3]{4}$

38. $\sqrt[3]{9} + \sqrt[3]{72} - 3\sqrt[3]{9}$

39. $4\sqrt[3]{16} + 3\sqrt[3]{16} - \sqrt[3]{54}$

40. $3\sqrt[3]{24} - \sqrt[3]{24} + 3\sqrt[3]{81}$

C

Combine: Assume that all variables represent positive numbers:

41. $\sqrt{98x^3} + 7x\sqrt{18x}$

42. $5x\sqrt{12x} - 8\sqrt{75x^3}$

43. $16\sqrt{99y^5} - 5y^2\sqrt{176y}$

44. $9\sqrt{90a^5} + 2a\sqrt{160a^3}$

45. $5y\sqrt{63y^3} + 7y\sqrt{28y^3}$

46. $7w^2\sqrt{135w} - 8w\sqrt{60w^3}$

47. $\sqrt{20a^3} + \sqrt{45a^3}$

48. $3b\sqrt{3b} + 2\sqrt{48b^3}$

49. $2\sqrt{40m^2n} - 3\sqrt{90m^2n}$

50. $\sqrt{80pq^2} - \sqrt{20pq^2}$

51. $6\sqrt{50r^2s^2} + 3\sqrt{18r^2s^2}$

52. $\sqrt[3]{24x^7} + x\sqrt[3]{81x^4}$

53. $13\sqrt{9x^2} - \sqrt{4x^2} - 3\sqrt{121x^2}$

54. $\sqrt{225ab} - \sqrt{81ab} - \sqrt{169ab}$

55. $\sqrt[4]{2a^4} + \sqrt[4]{32a^4}$

56. $\sqrt[4]{3w^8} + \sqrt[4]{243w^8}$

57. $\sqrt[4]{16x^{10}} - x\sqrt[4]{x^6}$

58. $\sqrt[4]{81y^{11}} - y\sqrt[4]{y^7}$

59. $\sqrt[3]{pq^3} - \sqrt[3]{-pr^3} - \sqrt[3]{8p}$

60. $\sqrt[3]{a^3x} + \sqrt[3]{-b^3x} + \sqrt[3]{27x}$

D

61. Find the perimeter of a triangle with sides $\sqrt{80}$ in., $\sqrt{125}$ in., and $\sqrt{45}$ in. Give the answer
   a. in simplest radical form.
   b. to the nearest tenth of an inch.

62. Find the perimeter of a pentagon with sides $\sqrt{50}$ m, $\sqrt{18}$ m, $\sqrt{50}$ m, $\sqrt{32}$ m, and $\sqrt{50}$ m. Give the answer
   a. in simplest radical form.
   b. to the nearest tenth of a meter.

63. Find the perimeter of a rectangle with length of $\sqrt{50}$ and width of $\sqrt{18}$.

64. Find the perimeter of a square with side measuring $\sqrt{98}$.

STATE YOUR UNDERSTANDING

65. Are $\sqrt{2}$ and $\sqrt[3]{2}$ like or unlike radical expressions? State the reason(s) for your answer.

CHALLENGE EXERCISES

Combine, assume all variable exponents represent positive integers.

66. $\sqrt{2} + \sqrt[4]{4} - \sqrt{8}$

67. $\sqrt[3]{2} - \sqrt[6]{4} + \sqrt[3]{16}$

68. $\sqrt{x^{2n}} + 3\sqrt[4]{x^{4n}} - \sqrt{4x^{2n}}$

69. $\sqrt[n]{x^{2n}} + \sqrt[n]{2^n x^{2n}} + \sqrt[n]{3^{2n} x^{2n}}$

MAINTAIN YOUR SKILLS (SECTIONS 3.2, 3.3)

Perform the indicated operations. Reduce if possible:

70. $2x^{-1} - 3y^{-1}$

71. $3a^{-1}b + 17ab^{-1}$

72. $\dfrac{2y^2 - 5y - 7}{3y^2 + 8y + 5} \div \dfrac{2y^2 - y - 21}{3y^2 + 14y + 15}$

73. $\dfrac{1}{4} - \dfrac{1}{x} + \dfrac{x + 2}{x - 2}$

74. $\dfrac{2}{3} - \dfrac{3}{w} + \dfrac{7 - w}{7 + w}$

75. $\dfrac{6x^3 + 6}{x^2 - 1} \cdot \dfrac{x + 1}{x^2 - x + 1} \cdot \dfrac{3x - 3}{x^2 + 2x + 1}$

76. $\dfrac{a - z}{a^3 + a^2z + az^2} \cdot \dfrac{a^2 + az}{a + z} \cdot \dfrac{a^2 + az + z^2}{a^2 - az}$

77. $\dfrac{x}{x - 5} - \dfrac{3}{x + 4}$

4.5

MULTIPLYING RADICALS

OBJECTIVE

1. Multiply radical expressions.

The product property of radicals is used to multiply radicals as well as to simplify them. Recall that the product property reads $\sqrt[n]{ab} = \sqrt[n]{a}\sqrt[n]{b}$ or, by the symmetric property of equality, $\sqrt[n]{a}\sqrt[n]{b} = \sqrt[n]{ab}$ with the appropriate restrictions on a and b depending on whether n is odd or even (see Section 4.3). Therefore,

$$\sqrt{3}\sqrt{5} = \sqrt{15}, \qquad \sqrt{7}\sqrt{6} = \sqrt{42}, \qquad (3\sqrt{5x})(3\sqrt{6y}) = 9\sqrt{30xy},$$
$$\text{and} \quad \sqrt[3]{4}\sqrt[3]{5} = \sqrt[3]{20}.$$

The product should be simplified if possible.

EXAMPLE 1

Multiply and simplify: $\sqrt{3}\sqrt{6}$

$$\sqrt{3}\sqrt{6} = \sqrt{18} \qquad\qquad \text{Multiply.}$$
$$= \sqrt{9}\sqrt{2} \qquad\qquad \text{Simplify.}$$
$$= 3\sqrt{2} \qquad\qquad \text{Simplest form.}$$ ■

It is often easier to simplify radicals before multiplying.

EXAMPLE 2

Multiply and simplify: $\sqrt{45}\sqrt{8}$

Simplify first	**Multiply first**
$\sqrt{45}\sqrt{8} = (3\sqrt{5})(2\sqrt{2})$	$\sqrt{45}\sqrt{8} = \sqrt{360}$
$= 6\sqrt{10}$	$= \sqrt{36}\sqrt{10}$
	$= 6\sqrt{10}$

In some cases, prime factoring the radicands before multiplying is easier.

EXAMPLE 3

Multiply and simplify: $\sqrt{22}\sqrt{55}$

$$\sqrt{22}\sqrt{55} = \sqrt{2 \cdot 11}\sqrt{5 \cdot 11} \qquad \text{Write the radicands in prime factored form.}$$

$$= \sqrt{2}\sqrt{11}\sqrt{5}\sqrt{11}$$

Use the product property of radicals to separate the radicals.

$$= \sqrt{11}\sqrt{11}\sqrt{2}\sqrt{5}$$
$$= 11\sqrt{10}$$

Change the order of the factors using the associative and commutative properties of multiplication, then multiply. ■

EXAMPLE 4

Multiply and simplify: $\sqrt{84}\sqrt{96}$

$$= \sqrt{2 \cdot 2 \cdot 3 \cdot 7}\sqrt{2 \cdot 2 \cdot 2 \cdot 2 \cdot 2 \cdot 3}$$

Write the radicands in prime factored form.

$$= (\sqrt{2}\sqrt{2})(\sqrt{2}\sqrt{2})(\sqrt{2}\sqrt{2})(\sqrt{3}\sqrt{3})(\sqrt{2}\sqrt{7})$$

Pair like factors since $\sqrt{a}\sqrt{a} = a$.

$$= 2 \cdot 2 \cdot 2 \cdot 3\sqrt{2}\sqrt{7}$$
$$= 24\sqrt{14}$$

Simplest form. ■

The distributive property and the FOIL shortcut for multiplying binomials can be used to multiply radical expressions containing sums and differences.

EXAMPLE 5

Multiply and simplify: $\sqrt{x}(\sqrt{x} + \sqrt{y} + 3)$

$$= \sqrt{x}\sqrt{x} + \sqrt{x}\sqrt{y} + \sqrt{x}3$$

Distributive property.

$$= x + \sqrt{xy} + 3\sqrt{x}$$

Simplify. ■

EXAMPLE 6

Multiply and simplify: $(\sqrt{y} + 3)(\sqrt{y} - 8)$

$$= (\sqrt{y})^2 - 8\sqrt{y} + 3\sqrt{y} - 24$$

The FOIL shortcut.

$$= y - 5\sqrt{y} - 24$$ ■

■ DEFINITION

Conjugate Radical Expressions

The sum and difference of the same two terms involving radicals are called conjugate radical expressions.

The radical expressions $\sqrt{5} + 2\sqrt{3}$ and $\sqrt{5} - 2\sqrt{3}$ are conjugates.

EXAMPLE 7

Multiply and simplify: $(\sqrt{a} + \sqrt{b})(\sqrt{a} - \sqrt{b})$

$= (\sqrt{a})^2 - (\sqrt{b})^2$ The pattern for multiplying conjugates.

$= a - b$ ■

EXAMPLE 8

Multiply and simplify: $(\sqrt{x} - \sqrt{y})^2$

$= (\sqrt{x})^2 - 2\sqrt{x}\sqrt{y} + (\sqrt{y})^2$ Recall the pattern for squaring a binomial.

$= x - 2\sqrt{xy} + y$

■ **CAUTION**

$$(a - b)^2 \neq a^2 + b^2 \text{ or } a^2 - b^2$$

■

Radical expressions with different index numbers can be multiplied if they are first rewritten with a common index number.

EXAMPLE 9

Multiply and simplify: $\sqrt{3} \, \sqrt[3]{4}$

$\sqrt{3} \, \sqrt[3]{4} = 3^{1/2} \cdot 4^{1/3}$ Change from radical form to exponential form.

$= 3^{3/6} \cdot 4^{2/6}$ Write the fractional exponents with a common denominator.

$= \sqrt[6]{3^3} \cdot \sqrt[6]{4^2}$ Change back to radical form. The radical expressions have the same index and can be multiplied.

$= \sqrt[6]{27} \cdot \sqrt[6]{16}$

$= \sqrt[6]{432}$ ■

EXAMPLE 10

Multiply and simplify: $\sqrt{2} \, \sqrt[4]{2}$

$\sqrt{2} \, \sqrt[4]{2} = 2^{1/2} \cdot 2^{1/4}$

$= 2^{2/4} \cdot 2^{1/4}$ The common index, 4, is the LCD of 2 and 4.

$= 2^{3/4}$ Add the exponents.

$= \sqrt[4]{2^3} = \sqrt[4]{8}$ Simplest radical form.   ■

EXAMPLE 11

The Green Thumb Nursery uses 8 ivy plants per square meter as ground cover. How many plants are needed to plant a triangular plot of ground that has a base of $\sqrt{300}$ meters and a height of $\sqrt{450}$ meters?

Select a variable:

Let p represent the number of plants.

Simpler word form:

$$\begin{pmatrix} \text{Number of} \\ \text{plants } p \end{pmatrix} = \begin{pmatrix} 8 \text{ plants} \\ \text{per square} \\ \text{meter} \end{pmatrix} \begin{pmatrix} \text{Area of plot} \\ \text{in square meters} \end{pmatrix}$$

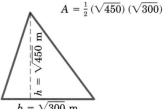

Area of triangle
$A = \frac{1}{2}bh$
$A = \frac{1}{2}(\sqrt{450})(\sqrt{300})$

$h = \sqrt{450}$ m

$b = \sqrt{300}$ m

Translate to algebra:

$$p = 8\left[\frac{1}{2}(\sqrt{450})(\sqrt{300})\right]$$

Solve:

$p = 4\sqrt{450}\sqrt{300}$

$p = 4\sqrt{5 \cdot 5 \cdot 3 \cdot 3 \cdot 2}\sqrt{2 \cdot 2 \cdot 5 \cdot 5 \cdot 3}$

$p = 4 \cdot 5 \cdot 3 \cdot 2 \cdot 5\sqrt{2 \cdot 3}$

$p = 600\sqrt{6}$

$p \approx 600(2.449)$

$p \approx 1469.4$

Answer:

The nursery will need 1470 plants to cover the area.

EXERCISE 4.5

A

Multiply and simplify. All variables represent positive numbers:

1. $\sqrt{2}\sqrt{8}$ **2.** $\sqrt{3}\sqrt{12}$ **3.** $\sqrt{6}\sqrt{2}$ **4.** $\sqrt{5}\sqrt{8}$

5. $2\sqrt{6}\sqrt{10}$ **6.** $(-3\sqrt{10})(\sqrt{5})$ **7.** $(-2\sqrt{6})(3\sqrt{2})$ **8.** $(-5\sqrt{3})(-3\sqrt{5})$

9. $\sqrt[3]{2x}\sqrt[3]{5x}$ **10.** $(5\sqrt{3})(5\sqrt{3})$ **11.** $(2\sqrt{5})^2$ **12.** $\sqrt[3]{10a}\sqrt[3]{3a}$

13. $(5\sqrt{x})(\sqrt{5x})$ **14.** $(\sqrt{7y})(7\sqrt{y})$ **15.** $(\sqrt{8a})(\sqrt{2a})(\sqrt{a})$ **16.** $(\sqrt{20b})(\sqrt{2b})(\sqrt{b})$

17. $3\sqrt{5}(5\sqrt{3} - 3\sqrt{5})$ **18.** $2\sqrt{7}(7\sqrt{2} - 2\sqrt{7})$ **19.** $(\sqrt{33})(\sqrt{22})$ **20.** $(\sqrt{21})(\sqrt{35})$

B

Multiply and simplify. All variables represent positive numbers:

21. $(3\sqrt{y})(2\sqrt{10})$ **22.** $(-4\sqrt{12})(2\sqrt{6})$ **23.** $(8\sqrt{15})(5\sqrt{20})$

24. $(-2\sqrt{18})(-3\sqrt{30})$ **25.** $(6s\sqrt{5t})(s\sqrt{10})$ **26.** $(3r\sqrt{6t})(2r\sqrt{3})$

27. $(-3\sqrt[3]{4})(2\sqrt[3]{12})$ **28.** $(4\sqrt[3]{9})(-5\sqrt[3]{12})$ **29.** $(5\sqrt{10t^2})(\sqrt{6t^3})$

30. $(8\sqrt{12x^3})(3\sqrt{6x^2})$ **31.** $\sqrt{2}(\sqrt{6} + \sqrt{2})$ **32.** $\sqrt{2}(\sqrt{10} + \sqrt{12})$

33. $\sqrt{3}(\sqrt{27} - \sqrt{12})$ **34.** $\sqrt{3}(\sqrt{6} + \sqrt{75})$ **35.** $2\sqrt{5}(\sqrt{5} - \sqrt{15})$

36. $-3\sqrt{6}(\sqrt{12} + \sqrt{18})$ **37.** $3\sqrt{7}(\sqrt{7} - \sqrt{14})$ **38.** $2\sqrt{11}(\sqrt{11} - \sqrt{33})$

39. $\sqrt{6}(2\sqrt{3} + 3\sqrt{2} - 2\sqrt{15})$ **40.** $\sqrt{10}(5\sqrt{2} - 2\sqrt{5} + 2\sqrt{15})$

C

Multiply and simplify. All variables represent positive numbers:

41. $(\sqrt{10} + \sqrt{2})(\sqrt{10} + \sqrt{2})$ **42.** $(\sqrt{6} - \sqrt{3})^2$

43. $(\sqrt{10} + \sqrt{6})(\sqrt{10} - \sqrt{6})$ **44.** $(\sqrt{11} - \sqrt{13})(\sqrt{11} + \sqrt{13})$

45. $(2\sqrt{x} + 1)(\sqrt{x} - 4)$ **46.** $(3\sqrt{y} - 1)(2\sqrt{y} + 4)$

47. $(5\sqrt{a} - 6)^2$ **48.** $(\sqrt{2y} - \sqrt{z})^2$

49. $(2\sqrt{b} - \sqrt{c})(2\sqrt{b} + \sqrt{c})$ **50.** $(3\sqrt{x} - \sqrt{2y})(3\sqrt{x} + \sqrt{2y})$

51. $\sqrt{5} \cdot \sqrt[3]{10}$ **52.** $\sqrt{6} \cdot \sqrt[3]{3}$

53. $2\sqrt{3} \cdot 5\sqrt[4]{2}$ **54.** $-3\sqrt{2} \cdot \sqrt[4]{3}$

55. $6\sqrt[3]{x} \cdot 3\sqrt[4]{x}$ **56.** $-2\sqrt[3]{w} \cdot 5\sqrt[5]{w}$

57. $\sqrt{2x} \cdot \sqrt[3]{3x} \cdot \sqrt[4]{4x}$ **58.** $\sqrt{3y} \cdot \sqrt[3]{3y} \cdot \sqrt[4]{3y}$

59. $\sqrt{2} \cdot \sqrt[3]{4} \cdot \sqrt[3]{2}$ **60.** $\sqrt{3} \cdot \sqrt[3]{9} \cdot \sqrt[3]{3}$

D

61. Find the area of a triangle with base $\sqrt{20}$ ft and height $\sqrt{15}$ ft. Give the answer
     a. in simplified radical form.
     b. to the nearest tenth of a square foot.

62. Find the area of a triangle with base $\sqrt{60}$ cm and height $\sqrt{18}$ cm. Give the answer
   a. in simplified radical form.
   b. to the nearest tenth of a square centimeter.

63. Find the area of the end of a rectangle baking pan with width $\sqrt{54}$ cm and depth $\sqrt{18}$ cm. Give the answer
   a. in simplified radical form.
   b. to the nearest tenth of a square centimeter.

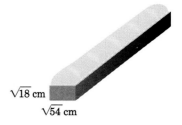

$\sqrt{18}$ cm

$\sqrt{54}$ cm

Figure for Exercise 63

64. Find the area of a rectangle with length $\sqrt{21}$ in. and width $\sqrt{14}$ in. Give the answer
   a. in simplified radical form.
   b. to the nearest tenth of a square inch.

65. Find the volume of a pyramid with a square base if the base is $\sqrt{30}$ m on a side and the height is $\sqrt{50}$ m. Express the answer in simplified radical form. The formula for the volume of a pyramid is $V = \dfrac{1}{3}b^2h$.

66. Find the volume of a pyramid with a square base if the base is $\sqrt{60}$ ft on a side and the height is $\sqrt{80}$ ft. Express the answer in simplified radical form.

67. Find the area of a trapezoid with bases of $\sqrt{45}$ dm and $\sqrt{80}$ dm and height $\sqrt{15}$ dm. Give the answer
   a. in simplified radical form.
   b. to the nearest tenth of a square decimeter.

   The formula for the area of a trapezoid is $A = \dfrac{1}{2}h(b_1 + b_2)$.

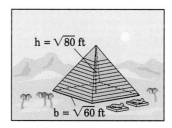

$h = \sqrt{80}$ ft

$b = \sqrt{60}$ ft

Figure for Exercise 66

68. Find the area of a trapezoid with bases of $\sqrt{75}$ dm and $\sqrt{108}$ dm and height $\sqrt{15}$ dm. Give the answer
   a. in simplified radical form.
   b. to the nearest tenth of a square decimeter.

STATE YOUR UNDERSTANDING

69. What conditions must be satisfied before two radical expressions can be multiplied?

CHALLENGE EXERCISES

Multiply, assume all variables represent positive integers.

70. $\sqrt{x} \cdot \sqrt{x^{n-1}}$

71. $\sqrt[n]{x^2} \cdot \sqrt[n]{x^{n-2}}$

72. $\sqrt[n]{4x^n} \cdot \sqrt[n]{4^{n-1}x^n}$

73. $\sqrt[n]{2x^{n+2}} \cdot \sqrt[n]{2^{n-1}x^{n-2}}$

74. $(\sqrt[4]{9} - \sqrt{2})(\sqrt[4]{9} + \sqrt{2})$

75. $(\sqrt[3]{4} - \sqrt[3]{2})(\sqrt[3]{16} + 2 + \sqrt[3]{4})$

76. $(\sqrt[3]{a} - \sqrt[3]{b})(\sqrt[3]{a^2} + \sqrt[3]{ab} + \sqrt[3]{b^2})$

77. $(\sqrt[3]{a} + \sqrt[3]{b})(\sqrt[3]{a^2} - \sqrt[3]{ab} + \sqrt[3]{b^2})$

MAINTAIN YOUR SKILLS (SECTIONS 3.3, 3.4, 3.5)

Add or subtract. Write all results with positive exponents and reduce if possible:

78. $4x^{-2}y + 5x^{-3}y^{-2} + 2xy^{-2}$

79. $\dfrac{2}{a^2 + 10a - 24} - \dfrac{1}{a^2 + 17a + 60}$

80. $3(x^2 - 8x - 33)^{-1} + 4(x^2 - 15x + 44)^{-1}$

Simplify:

81. $\dfrac{3 - \dfrac{1}{y}}{\dfrac{1}{y} + 4}$

82. $\dfrac{a - \dfrac{1}{a}}{2a + \dfrac{1}{a}}$

83. $\dfrac{(y + 3)^{-1} + 2y}{(y + 3)^{-1} - 3}$

Solve:

84. $\dfrac{7}{x - 3} - \dfrac{4}{x - 3} = 5$

85. $5(y + 4)^{-1} - 3(y + 4)^{-1} = 15$

4.6

DIVIDING RADICALS

OBJECTIVE

1. Rationalize the denominator of a radical expression.

A radical expression written in simplest form cannot contain a fraction in the radicand. We restate the conditions for simplest radical form.

1. The radicand has no factors that are perfect squares except the number 1.

2. No fractions are in the radicand, and no radicals appear in a denominator.

To divide radicals means to eliminate the radicals in the divisor. Division, here, is referred to as "rationalizing the denominator." The new fraction will contain a denominator that is an integer or a variable without radicals that is equivalent to the original fraction. To rationalize the denominator, that is, to eliminate the fraction in the radical, we use the quotient property of radicals.

EXAMPLE 1

Rationalize the denominator: $\sqrt{\dfrac{2}{3}}$

$$\sqrt{\frac{2}{3}} = \frac{\sqrt{2}}{\sqrt{3}} \qquad \text{Quotient property of radicals.}$$

$$= \frac{\sqrt{2}}{\sqrt{3}} \cdot \frac{\sqrt{3}}{\sqrt{3}} \qquad \text{Basic principle of fractions. The fraction } \frac{\sqrt{3}}{\sqrt{3}} \text{ was}$$

$$\qquad\qquad\qquad \text{chosen to make the denominator a perfect square.}$$

$$= \frac{\sqrt{6}}{3}$$

Example 2 illustrates a variation on the method above, using the same radical expression.

EXAMPLE 2

Rationalize the denominator: $\sqrt{\dfrac{2}{3}}$

$$\sqrt{\frac{2}{3}} = \sqrt{\frac{2}{3} \cdot \frac{3}{3}} \qquad \text{Basic principle of fractions. The fraction } \frac{3}{3} \text{ was}$$

$$\qquad\qquad\qquad \text{chosen to make the denominator a perfect square.}$$

$$= \sqrt{\frac{6}{9}}$$

$$= \frac{\sqrt{6}}{\sqrt{9}} \qquad \text{Quotient property of radicals.}$$

$$= \frac{\sqrt{6}}{3}$$

In Examples 1 and 2, we multiplied the original fraction by $\dfrac{\sqrt{3}}{\sqrt{3}}$ or the radicand by $\dfrac{3}{3}$, both of which have value 1. The denominator, 3, in the simplified form is a rational number. In Example 3, we use the fraction $\dfrac{\sqrt{5x}}{\sqrt{5x}}$ (with value 1) to make the radicand in the denominator a perfect square.

EXAMPLE 3

Rationalize the denominator: $\dfrac{\sqrt{3y}}{\sqrt{5x}}$

$$= \frac{\sqrt{3y}}{\sqrt{5x}} \cdot \frac{\sqrt{5x}}{\sqrt{5x}} \qquad \text{Basic principle of fractions.}$$

$$= \frac{\sqrt{15xy}}{5x} \qquad \text{Simplest radical form.}$$

EXAMPLE 4

Rationalize the denominator: $\dfrac{2\sqrt{8}}{\sqrt{10}}$

$$= \frac{2\sqrt{8}}{\sqrt{10}} \cdot \frac{\sqrt{10}}{\sqrt{10}}$$

$$= \frac{2\sqrt{80}}{10}$$

$$= \frac{2\sqrt{16}\sqrt{5}}{10} = \frac{8\sqrt{5}}{10}$$

$$= \frac{4\sqrt{5}}{5}$$

As before, we multiply the numerator and denominator by $\sqrt{10}$ to simplify. The numerator still can be simplified.

Finally, we reduce.

Alternatively, we can reduce first:

$$\frac{2\sqrt{8}}{\sqrt{10}} = \frac{2\sqrt{4}}{\sqrt{5}}$$

$$= \frac{2(2)}{\sqrt{5}} = \frac{4}{\sqrt{5}}$$

$$= \frac{4\sqrt{5}}{5}$$

■ **CAUTION**

When reducing, common factors can be eliminated only if both are inside the radicals or both are outside the radicals. So,

$$\frac{2\sqrt{8}}{\sqrt{10}} \neq \frac{\sqrt{8}}{\sqrt{5}}.$$

■

EXAMPLE 5

Rationalize the denominator: $\dfrac{3\sqrt{5}}{4\sqrt{12}}$

$$= \frac{3\sqrt{5}}{4\sqrt{12}} \cdot \frac{\sqrt{3}}{\sqrt{3}}$$

$$= \frac{3\sqrt{15}}{4\sqrt{36}}$$

$$= \frac{3\sqrt{15}}{24}$$

$$= \frac{\sqrt{15}}{8}$$

In each of the examples above, we used the denominator itself to multiply by. Here, too, we could multiply both the numerator and denominator by $4\sqrt{12}$, but such a large factor is not necessary. In the first place, the number 4 is already an integer, and secondly, $12 \cdot 3 = 36$, which is a perfect square.

Alternatively, we can simplify first:

$$\frac{3\sqrt{5}}{4\sqrt{12}} = \frac{3\sqrt{5}}{4 \cdot 2\sqrt{3}}$$

> Simplify $\sqrt{12} = 2\sqrt{3}$ before rationalizing the denominator.

$$= \frac{3\sqrt{5}}{8\sqrt{3}} \cdot \frac{\sqrt{3}}{\sqrt{3}} = \frac{3\sqrt{15}}{24}$$

$$= \frac{\sqrt{15}}{8}$$

■

EXAMPLE 6

Rationalize the denominator: $\sqrt[3]{\dfrac{2}{9}}$

$$\sqrt[3]{\frac{2}{9}} = \frac{\sqrt[3]{2}}{\sqrt[3]{9}} \cdot \frac{\sqrt[3]{3}}{\sqrt[3]{3}}$$

> Here we multiply by the smallest radical factor that will give us a perfect cube. In this case, $9(3) = 27$ is a perfect cube.

$$= \frac{\sqrt[3]{6}}{\sqrt[3]{27}} = \frac{\sqrt[3]{6}}{3}$$

■

A fraction whose denominator is an indicated sum or difference involving radicals cannot be simplified by multiplying the numerator and denominator by the expression in the denominator. For example,

$$\frac{6}{\sqrt{3} + 2} \cdot \frac{\sqrt{3} + 2}{\sqrt{3} + 2} = \frac{6\sqrt{3} + 12}{3 + 4\sqrt{3} + 4}.$$

We have not eliminated the radical. Such fractions can be simplified by multiplying by the conjugate of the denominator.

EXAMPLE 7

Rationalize the denominator: $\dfrac{6}{\sqrt{3} + 2}$

$$= \frac{6}{\sqrt{3} + 2} \cdot \frac{\sqrt{3} - 2}{\sqrt{3} - 2}$$

> Multiply numerator and denominator by conjugate of the denominator, then simplify. Remember that the product of two conjugates is the difference of two squares.

$$= \frac{6\sqrt{3} - 12}{(\sqrt{3})^2 - 2^2}$$

$$= \frac{6\sqrt{3} - 12}{3 - 4} = \frac{6\sqrt{3} - 12}{-1}$$

$$= 12 - 6\sqrt{3}$$

■

EXAMPLE 8

Rationalize the denominator: $\dfrac{\sqrt{a} + \sqrt{5}}{\sqrt{b} + \sqrt{6}}$

$= \dfrac{\sqrt{a} + \sqrt{5}}{\sqrt{b} + \sqrt{6}} \cdot \dfrac{\sqrt{b} - \sqrt{6}}{\sqrt{b} - \sqrt{6}}$ The conjugate of $\sqrt{b} + \sqrt{6}$ is $\sqrt{b} - \sqrt{6}$.

$= \dfrac{\sqrt{ab} - \sqrt{6a} + \sqrt{5b} - \sqrt{30}}{b - 6}$ ■

In the calculus, it is often desirable to rationalize the numerator of a fraction.

EXAMPLE 9

Rationalize the numerator: $\dfrac{\sqrt{2}}{3 + \sqrt{2}}$

$= \dfrac{\sqrt{2}}{3 + \sqrt{2}} \cdot \dfrac{\sqrt{2}}{\sqrt{2}}$ To eliminate the radical in the numerator
multiply by $\dfrac{\sqrt{2}}{\sqrt{2}}$.

$= \dfrac{2}{3\sqrt{2} + 2}$ The numerator is an integer as required.   ■

EXAMPLE 10

An old pendulum clock has a pendulum that is 6 feet long. The time it takes for one cycle of the pendulum is given by the formula

$$T = 2\pi \dfrac{\sqrt{L}}{\sqrt{32}},$$

where L represents the length of the pendulum in feet and T represents the time in seconds. Find the time T for one cycle. Write the answer in (1) simplest radical form and (2) to the nearest hundredth of a second.

Formula:

$$T = 2\pi \dfrac{\sqrt{L}}{\sqrt{32}}$$

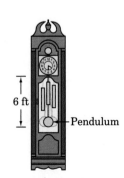

6 ft

Pendulum

Substitute:

$$T = 2\pi \dfrac{\sqrt{6}}{\sqrt{32}}$$ Substitute $L = 6$.

Simplify:

$$T = 2\pi\frac{\sqrt{6}}{\sqrt{32}} \cdot \frac{\sqrt{2}}{\sqrt{2}} = \frac{2\pi\sqrt{12}}{\sqrt{64}}$$ Rationalize the denominator.

$$T = \frac{4\pi\sqrt{3}}{8}$$

$$T = \frac{\pi\sqrt{3}}{2}$$ Simplest radical form.

$$T \approx \frac{(3.14)(1.73)}{2}$$ Substitute rational approximations for π and $\sqrt{3}$, $\pi \approx 3.14$ and $\sqrt{3} \approx 1.73$.

$$T \approx 2.72$$

Answer:

The pendulum takes $T = \dfrac{\pi\sqrt{3}}{2}$ seconds, or approximately 2.72 seconds, to complete one cycle.   ■

Exercise 4.6

A

Rationalize the denominator:

1. $\dfrac{2}{\sqrt{5}}$ **2.** $\sqrt{\dfrac{2}{5}}$ **3.** $\sqrt{\dfrac{1}{7}}$ **4.** $\dfrac{\sqrt{2}}{\sqrt{7}}$ **5.** $\sqrt{\dfrac{1}{8}}$

6. $\sqrt{\dfrac{1}{12}}$ **7.** $\dfrac{3}{\sqrt{2}}$ **8.** $\sqrt{\dfrac{3}{2}}$ **9.** $\dfrac{-3}{\sqrt{6}}$ **10.** $-\sqrt{\dfrac{5}{3}}$

11. $\dfrac{7}{\sqrt{7}}$ **12.** $\dfrac{2}{\sqrt{2}}$ **13.** $\sqrt{\dfrac{3}{8}}$ **14.** $\sqrt{\dfrac{5}{18}}$ **15.** $\sqrt{\dfrac{x}{y}}$

16. $\dfrac{x}{\sqrt{y}}$ **17.** $\dfrac{7}{\sqrt{14}}$ **18.** $\dfrac{11}{\sqrt{33}}$ **19.** $\dfrac{5}{\sqrt{50}}$ **20.** $\dfrac{6}{\sqrt{60}}$

B

Rationalize the denominator:

21. $\sqrt{\dfrac{a}{2}}$ **22.** $\dfrac{\sqrt{b}}{\sqrt{5}}$ **23.** $\dfrac{\sqrt{2}}{\sqrt{6}}$ **24.** $\dfrac{\sqrt{11}}{\sqrt{6}}$

25. $\dfrac{2\sqrt{5}}{5\sqrt{2}}$

26. $\dfrac{3\sqrt{7}}{7\sqrt{3}}$

27. $\dfrac{x}{\sqrt{x}}$

28. $\dfrac{y^2\sqrt{x}}{\sqrt{y}}$

29. $\dfrac{-2\sqrt{8}}{\sqrt{6}}$

30. $\dfrac{-3\sqrt{18}}{\sqrt{12}}$

31. $\dfrac{a}{\sqrt{ab}}$

32. $\dfrac{\sqrt{a}}{a\sqrt{b}}$

33. $\sqrt[3]{\dfrac{3}{4}}$

34. $\sqrt[3]{\dfrac{4}{3}}$

35. $\sqrt[4]{\dfrac{3}{4}}$

36. $\sqrt[4]{\dfrac{4}{3}}$

37. $\sqrt{\dfrac{3}{4}} + \sqrt{\dfrac{4}{3}} - \sqrt{\dfrac{1}{12}}$

38. $\dfrac{1}{2}\sqrt{\dfrac{4}{3}} + \dfrac{1}{3}\sqrt{\dfrac{3}{4}} + \sqrt{\dfrac{1}{12}}$

39. $\dfrac{3}{\sqrt{2}} + \sqrt{\dfrac{9}{8}} - \sqrt{\dfrac{1}{18}}$

40. $\dfrac{3}{4}\sqrt{\dfrac{9}{2}} + \dfrac{1}{2}\sqrt{\dfrac{1}{2}} - \sqrt{\dfrac{1}{8}}$

C

Rationalize the numerator:

41. $\sqrt{\dfrac{3}{2}}$

42. $\sqrt{\dfrac{7}{6}}$

43. $\dfrac{\sqrt{ab}}{\sqrt{3}}$

44. $\dfrac{3\sqrt{x}}{8}$

Rationalize the denominator:

45. $\dfrac{2 + \sqrt{3}}{\sqrt{3}}$

46. $\dfrac{2 + \sqrt{3}}{1 + \sqrt{3}}$

47. $\dfrac{\sqrt{5} - \sqrt{2}}{\sqrt{5} + \sqrt{2}}$

48. $\dfrac{\sqrt{3} + \sqrt{7}}{\sqrt{3} - \sqrt{7}}$

49. $\dfrac{\sqrt{5} + \sqrt{2}}{\sqrt{3} - \sqrt{5}}$

50. $\dfrac{\sqrt{7} - \sqrt{11}}{\sqrt{2} + \sqrt{7}}$

51. $\dfrac{2\sqrt{x}}{\sqrt{x} - \sqrt{y}}$

52. $\dfrac{-5\sqrt{c}}{\sqrt{c} + \sqrt{d}}$

53. $\dfrac{\sqrt{a}}{\sqrt{2a} + \sqrt{b}}$

54. $\dfrac{\sqrt{a} + \sqrt{b}}{\sqrt{a} - \sqrt{b}}$

55. $\dfrac{2\sqrt{5} - 3}{3\sqrt{5} + 3}$

56. $\dfrac{6 + \sqrt{11}}{4 - \sqrt{11}}$

57. $\dfrac{\sqrt{xy} - \sqrt{x}}{\sqrt{x} + \sqrt{xy}}$

58. $\dfrac{3\sqrt{ab} + \sqrt{b}}{\sqrt{ab} - 3\sqrt{b}}$

59. $\dfrac{3\sqrt{x} - \sqrt{y}}{5\sqrt{x} + 2\sqrt{y}}$

60. $\dfrac{5\sqrt{a} - 3\sqrt{b}}{2\sqrt{a} - 4\sqrt{b}}$

Rationalize the numerator:

61. $\dfrac{2 + \sqrt{3}}{\sqrt{3}}$

62. $\dfrac{4 - \sqrt{5}}{5}$

63. $\dfrac{2 + \sqrt{3}}{1 + \sqrt{3}}$

64. $\dfrac{4 - \sqrt{5}}{3 - \sqrt{5}}$

D

65. A pendulum is 1 foot long. Find the time it takes to complete one cycle. Give the answer.
   a. in simplified radical form.
   b. to the nearest tenth of a second.

66. A pendulum is 12 feet long. Find the time it takes to complete one cycle. Give the answer.
   a. in simplified radical form.
   b. to the nearest tenth of a second.

67. A pendulum is 4 inches $\left(\dfrac{1}{3} \text{ foot}\right)$ long. Find the time it takes to complete one cycle. Give the answer.
   a. in simplified radical form.
   b. to the nearest tenth of a second.

68. A pendulum is 3 inches long. Find the time it takes to complete one cycle. Give the answer.
   a. in simplified radical form.
   b. to the nearest tenth of a second.

STATE YOUR UNDERSTANDING

69. Explain what is meant by "rationalize the denominator."

CHALLENGE EXERCISES

Use the results of Exercises 76 and 77 of Section 4.5 to rationalize the denominator in each of the following.

70. $\dfrac{1}{\sqrt[3]{2} - 1}$

71. $\dfrac{2}{\sqrt[3]{4} - \sqrt[3]{2}}$

72. $\dfrac{3}{\sqrt[3]{5} + \sqrt[3]{2}}$

73. $\dfrac{1}{\sqrt[3]{8} - \sqrt[3]{4}}$

MAINTAIN YOUR SKILLS (SECTIONS 3.3, 3.4, 3.5)

Add or subtract. Write results with positive exponents and reduce if possible:

74. $2(3x - 1)^{-1} + 3(3x - 1)^{-1}$

75. $\dfrac{6b}{1 - b^2} - \dfrac{3}{b - 1}$

76. $\dfrac{6b}{b^2 - 1} + \dfrac{3}{1 - b}$

Simplify:

77. $\dfrac{\dfrac{3}{a} - \dfrac{2}{b}}{b + a}$

78. $\dfrac{2 + \dfrac{3}{xy}}{\dfrac{4}{xy} - 1}$

79. $\dfrac{2}{2 + \dfrac{1}{1 + \dfrac{2}{x}}}$

Solve:

80. $\dfrac{11}{3a^2 - 7a + 4} - \dfrac{3}{3a - 4} = \dfrac{-2}{a - 1}$

81. $\dfrac{6}{y - 5} + \dfrac{5}{2y + 1} = \dfrac{-1}{2y^2 - 9y - 5}$

4.7

EQUATIONS CONTAINING RADICALS

OBJECTIVE

1. Solve equations that contain radicals.

Equations that contain radicals can be solved by using the power property of equality.

■ **PROPERTY**

Power Property of Equality

> If $a = b$, then $a^2 = b^2$.

Using the power property, we observe that

$4 = 4$	A true statement.
$(4)^2 = (4)^2$	Square both sides.
$16 = 16$	A true statement.

We can get the same result starting with a false statement.

$-4 = 4$	A false statement.
$(-4)^2 = (4)^2$	Square both sides.
$16 = 16$	A true statement.

In either case, when each side of the equation is squared, a true statement results. However, the second case shows that squaring does not always result in equivalent equations.

■ **CAUTION**

Squaring both sides of an equation will not always yield an equivalent equation. It is possible to introduce an apparent solution that will not check.

■ **DEFINITION**

Extraneous Root

An apparent solution that is introduced when squaring both sides of an equation is an example of an *extraneous root* or *extraneous solution*.

For instance:

Given $x = 4$

$$(x)^2 = (4)^2$$ Power property.

$$x^2 = 16$$ Solve the resulting equation.

$$x = \pm 4$$

Check:

$x = 4$ or $x = -4$

$4 = 4$ $-4 = 4$ Check in the original equation.

The solution set is $\{4\}$.   The apparent solution -4 is an extraneous root.

EXAMPLE 1

Solve: $\sqrt{x + 1} = 5$

$$(\sqrt{x + 1})^2 = (5)^2$$ Square both sides.

$$x + 1 = 25$$

$$x = 24$$

Check:

$$\sqrt{24 + 1} = 5$$ Substitute 24 for x in the original equation.

$$\sqrt{25} = 5$$

The solution set is $\{24\}$.   ■

EXAMPLE 2

Solve: $\sqrt{5 - x} = -3$

$\sqrt{5 - x} = -3$ By the definition of a radical, $\sqrt{5 - x}$ must be nonnegative.

The solution set is $\emptyset$. ∎

Square root radicals cannot have negative values. Cube root radicals can have negative values; see Example 3.

EXAMPLE 3

Solve: $\sqrt[3]{x} = -2$

$(\sqrt[3]{x})^3 = (-2)^3$ Since the index is 3, cube both sides. The cube of a negative number is negative.

$\quad x = -8$

Check:

$\sqrt[3]{-8} = -2$

$\quad -2 = -2$

The solution set is $\{-2\}$. ∎

If the radical is not isolated, then do so before using the power property of equality. Otherwise, the radical will not be eliminated.

EXAMPLE 4

Solve: $\sqrt{x - 3} + 2 = 5$

$\quad \sqrt{x - 3} = 3$ Isolate the radical on the left.

$(\sqrt{x - 3})^2 = (3)^2$ Square both sides.

$\quad\quad x - 3 = 9$

$\quad\quad\quad x = 12$

Check:

$\sqrt{12 - 3} + 2 = 5$ Substitute 12 for x in the original equation.

$\quad \sqrt{9} + 2 = 5$

$\quad\quad 3 + 2 = 5$

The solution set is $\{12\}$. ∎

■ PROCEDURE

To solve a radical equation that contains a single radical:

1. Isolate the radical on one side of the equation.
2. Use the power property of equality.
3. Solve for x.
4. Check for "extraneous roots" (necessary since the equations may not be equivalent).

Sometimes the use of the power property results in an equation of higher degree.

EXAMPLE 5

Solve: $\sqrt{x + 4} = 2x - 7$

$\sqrt{x + 4} = 2x - 7$	Original equation.
$(\sqrt{x + 4})^2 = (2x - 7)^2$	Square both sides.
$x + 4 = 4x^2 - 28x + 49$	
$0 = 4x^2 - 29x + 45$	We have a quadratic equation.
$4x^2 - 29x + 45 = 0$	Symmetric property.
$(4x - 9)(x - 5) = 0$	Factor.
$4x - 9 = 0 \quad$ or $\quad x - 5 = 0$	Solve.
$x = \dfrac{9}{4} \qquad\qquad x = 5$	

Check:

$$\sqrt{\frac{9}{4} + 4} = 2 \cdot \frac{9}{4} - 7 \qquad\qquad \text{Substitute } x = \frac{9}{4}.$$

$$\sqrt{\frac{25}{4}} = \frac{9}{2} - 7$$

$$\frac{5}{2} = -\frac{5}{2} \qquad\qquad \text{Does not check; } \frac{9}{4} \text{ is an extraneous root.}$$

$$\sqrt{5 + 4} = 2 \cdot 5 - 7 \qquad\qquad \text{Substitute } x = 5.$$

$$\sqrt{9} = 10 - 7$$

$$3 = 3$$

The solution set is $\{5\}$.   ■

If two or more square root radicals occur in an equation, then continue to square both sides until all radicals have been eliminated.

■ PROCEDURE

If two or more radicals occur in an equation:

1. Isolate one of the radicals.
2. Square both sides.
3. Repeat until all radicals have been eliminated.
4. Check for "extraneous roots."

EXAMPLE 6

Solve: $\sqrt{x + 3} - \sqrt{x - 5} = 2$

$\sqrt{x + 3} - \sqrt{x - 5} = 2$	Original equation.
$\sqrt{x + 3} = 2 + \sqrt{x - 5}$	The equation contains two radicals, so isolate $\sqrt{x + 3}$ on the left side.
$(\sqrt{x + 3})^2 = (2 + \sqrt{x - 5})^2$	Square both sides to eliminate the radical on the left.
$x + 3 = 4 + 4\sqrt{x - 5} + x - 5$	
$4 = 4\sqrt{x - 5}$	Simplify.
$1 = \sqrt{x - 5}$	Divide both sides by 4.
$(1)^2 = (\sqrt{x - 5})^2$	Square both sides to eliminate the second radical.
$1 = x - 5$	
$6 = x$	

Check:

$\sqrt{6 + 3} - \sqrt{6 - 5} = 2$	Substitute $x = 6$ in the original equation.
$\sqrt{9} - \sqrt{1} = 2$	
$3 - 1 = 2$	

The solution set is $\{6\}$.   ■

EXAMPLE 7

The sum of the perimeters of two squares is 48 meters. One square has an area that is 9 times the area of the other one. What is the length of a side of each square?

Select a variable:

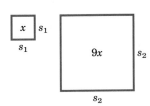

Let x represent the area of the smaller square.

$$s_1 = \sqrt{x} \qquad s_2 = \sqrt{9x}$$

Simpler word form:

$$\left(\begin{array}{c}\text{Perimeter}\\\text{of smaller}\\\text{square}\end{array}\right) + \left(\begin{array}{c}\text{Perimeter}\\\text{of larger}\\\text{square}\end{array}\right) = 48$$

Translate to algebra:

$$4\sqrt{x} + 4\sqrt{9x} = 48$$

Solve:

$$\sqrt{x} + \sqrt{9x} = 12$$
$$\sqrt{x} + 3\sqrt{x} = 12$$
$$4\sqrt{x} = 12$$
$$\sqrt{x} = 3$$
$$x = 9$$

Check:

Perimeter of smaller square = 12

Perimeter of larger square = 36

Sum of the perimeters is 48.

Answer:

The lengths of the sides of the squares are 3 m and 9 m.

Draw a diagram to show the given information.

The length of a side is the square root of the area.

$P = 4s$

Divide both sides by 4.

Simplify.

Combine radicals.

Divide both sides by 4.

Square both sides.

If 9 is the area of the smaller square, its side is $\sqrt{9} = 3$ and its perimeter is $4(3) = 12$.

Area of the larger square is $9(9) = 81$, so its side is $\sqrt{81} = 9$ and its perimeter is $4(9) = 36$.

■

EXAMPLE 8

The pressure of the stream from a fire hydrant can be calculated from the formula

$$G = 26.8d^2\sqrt{p},$$

where G represents the discharge in gallons per minute (gpm), d represents the diameter of the outlet in inches, and p represents the pressure in pounds per square inch (psi). Find the pressure (psi) of water from an outlet that is 3.5 inches in diameter and that discharges at 680 gallons per minute (round to the nearest tenth).

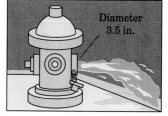

Diameter
3.5 in.

Formula:

$$G = 26.8d^2\sqrt{p}$$

Substitute:

$$680 = 26.8(3.5)^2\sqrt{p}$$

To find the pressure, substitute $G = 680$ and $d = 3.5$ into the formula.

Solve:

$$680 = 328.3\sqrt{p}$$

$$\frac{680}{1328.3} = \sqrt{p}$$ Isolate the radical.

$$2.071 \approx \sqrt{p}$$ Simplify.

$$4.289 \approx p$$ Square both sides.

Answer:

The pressure (to the nearest tenth) is 4.3 psi.   ■

EXERCISE 4.7

A

Solve:

1. $\sqrt{a} = 11$ **2.** $\sqrt{b} = 10$ **3.** $\sqrt{x} = -7$ **4.** $\sqrt{y} = -12$

5. $\sqrt{x} + 4 = 5$ **6.** $\sqrt{x} + 2 = 6$ **7.** $\sqrt{a} - 2 = 6$ **8.** $\sqrt{b} + 3 = 9$

9. $\sqrt{2x+1} = 3$ **10.** $\sqrt{2x-3} = 7$ **11.** $\sqrt[3]{2x} - 1 = 2$ **12.** $\sqrt[3]{3x} + 2 = 5$

13. $5 - \sqrt{2x} = -11$ **14.** $7 - \sqrt{5x} = -3$ **15.** $\sqrt{3x} + 13 = 10$ **16.** $\sqrt{3x} + 14 = -5$

17. $\sqrt[3]{x} = 5$ **18.** $\sqrt[3]{x} = -5$ **19.** $\sqrt[3]{x+2} = -3$ **20.** $\sqrt[4]{x+2} = 2$

B

21. $\sqrt{3-4x} = 2x$ **22.** $\sqrt{2-3x} = 3x$ **23.** $\sqrt{3a+28} = a$

24. $\sqrt{4b+5} = b$ **25.** $\sqrt{x+2} = x+2$ **26.** $\sqrt{x+2} = x-4$

27. $\sqrt{3x+3} = x-5$ **28.** $\sqrt{8x+1} = 2x-1$ **29.** $\sqrt{12x+13} = 3x-2$

30. $\sqrt{3x+1} = 2x-6$ **31.** $5\sqrt{2x+1} = 3x+3$ **32.** $3\sqrt{3x+1} = 2x-1$

33. $\sqrt{2x+5} + 5 = x$ **34.** $\sqrt{3x+39} - 7 = x$ **35.** $\sqrt{8x+24} - 3 = 2x$

36. $\sqrt{60x+51} - 4 = 6x$ **37.** $\sqrt[4]{2x-7} = \sqrt[4]{x+2}$ **38.** $\sqrt[3]{x-5} = \sqrt[3]{-3x+3}$

39. $\sqrt{x+10} = x-2$ **40.** $\sqrt{8x+1} = 2x-5$

C

41. $\sqrt{x} - \sqrt{x+9} = -1$ **42.** $\sqrt{x} - \sqrt{x+24} = -2$

43. $\sqrt{y+2} + \sqrt{y-3} = 5$ **44.** $\sqrt{y-2} + \sqrt{y+5} = 7$

45. $\sqrt{x+1} + \sqrt{x} = -7$ **46.** $\sqrt{x-2} + \sqrt{x+4} = -1$

47. $\sqrt{x} + \sqrt{x+2} = 3$ **48.** $\sqrt{y-1} + \sqrt{y+1} = 4$

49. $\sqrt{2y} + 3 = \sqrt{y-7}$ **50.** $\sqrt{y-5} = 2 + \sqrt{y+3}$

51. $\sqrt{a-5} - \sqrt{a-29} = 4$ **52.** $\sqrt{b-36} + 5 = \sqrt{b-1}$

53. $\sqrt{5w+9} = 3 + \sqrt{w}$ **54.** $\sqrt{4x+16} = 4 + \sqrt{x}$

55. $\sqrt{x+3} - \sqrt{x+2} = 1$ **56.** $\sqrt{2x+1} = \sqrt{x+1}$

57. $\sqrt{2x-1} - 2 = \sqrt{x-4}$ **58.** $\sqrt{x+6} = \sqrt{x-2} + 2$

59. $\sqrt{x^2-5x} = 6$ **60.** $\sqrt{y^2-6y} = 4$

D

61. The pressure of the stream from a fire hydrant can be calculated from the formula

$$G = 26.8d^2\sqrt{p},$$

where G represents the discharge in gallons per minute (gpm), d represents the diameter of the outlet in inches, and p represents the pressure in pounds per square inch (psi). Find the pressure (psi) of water from an outlet that is 4.5 inches in diameter and that discharges at 900 gallons per minute (round to the nearest tenth).

62. Use the formula in Exercise 61 to find the pressure of water from a fire hydrant outlet that is 3.75 inches in diameter and that discharges at 1100 gallons per minute (round to the nearest tenth).

63. The sum of the perimeters of two squares is 128 feet. One square has an area that is nine times the area of the other square. What is the area of each square?

$$4s_1 + 4s_2 = 128$$
$$9s_1^2 = s_2^2$$

64. The sum of the circumferences of two rings in a circus is 150 meters. One ring has an area that is four times the area of the other ring. What is the area of each ring?

Figure for Exercise 64

65. The interest rate (r) (compounded annually) needed to have P dollars grow to A dollars at the end of two years is given by

$$r = \sqrt{\frac{A}{P}} - 1.$$

Find the value of A if $P = 2000$ and $r = 0.1$.

66. In Exercise 65, find the value of P if $A = 6050$ and $r = 0.1$.

STATE YOUR UNDERSTANDING

67. What is an extraneous root?

CHALLENGE EXERCISES

68. The formula for finding the period (T), in seconds, of a pendulum of length (L), in feet, and the acceleration due to gravity (g) is given by the formula

$$T = 2\pi\sqrt{\frac{L}{g}}, \ g = 32 \text{ ft/sec}^2$$

What is the length (to the nearest foot) of a pendulum that has a period of 12 seconds? Let $\pi \approx 3.14$.

69. Using the formula from Exercise 68, what is the length (to the nearest foot) of a pendulum with a period of 3 seconds?

70. Solve the equation $x = y + z\sqrt{u + v}$ for v.

71. Solve the equation $x = \sqrt{y^2 - z^2}$ for y.

MAINTAIN YOUR SKILLS (SECTION 1.9)

Solve, write solutions in interval notation.

72. $13 < 2x + 5 < 27$

73. $-15 < 4x + 1 < 29$

74. $-7 < -2x + 1 < 5$

75. $9 < 2x + 5 < 17$

76. $12 \le 5x - 3 < 22$

77. $-7 \le 4x + 9 \le 13$

78. $24 \le 7x + 3 \le 66$

79. $-4 \le -5x + 1 \le 6$

4.8

COMPLEX NUMBERS

OBJECTIVES

1. Write complex numbers in standard form.

2. Add and subtract complex numbers.

3. Multiply complex numbers.

4. Divide complex numbers.

We saw in Section 4.1 that $\sqrt{-1}$ is not a real number, and as a result, the equation $x^2 = -1$ has no real solution. There is no real number whose square is -1. In order to solve such an equation, we require a new set of numbers. We use the letter i (from the word *imaginary*) as a symbol in this new set of numbers.

■ **DEFINITION**

$i^2 = -1$

> The letter i is used as a the symbol for a number whose square is negative one.
>
> $i^2 = -1$ and $i = \sqrt{-1}$

Numbers whose symbols contain $\sqrt{-1}$ are called complex numbers.

Complex Numbers

■ DEFINITION

Complex numbers are numbers that can be written in the form $a + bi$, where $a \in R$ and $b \in R$ and $i^2 = -1$.
In $a + bi$, a is called the real part and b is called the imaginary part of the complex number.

The form $a + bi$ is called the standard (or rectangular) form of a complex number.

Two complex numbers are said to be equal when their real parts are equal and their imaginary parts are equal. That is, $a + bi = c + di$ if and only if $a = c$ and $b = d$. Thus $6 + 8i = 6 + 8i$ but $2 - 3i \neq 3 - 2i$.

The table below shows examples of complex numbers.

Complex Number	Standard Form	Real Part	Imaginary Part
$3 + 4i$	$3 + 4i$	3	4
$-6 - 2i$	$-6 + (-2i)$	-6	-2
$-\sqrt{3}i$	$0 + (-\sqrt{3}i)$	0	$-\sqrt{3}$
8	$8 + 0i$	8	0

The next set of statements are results of the definition of complex numbers.

1. $i^2 = \sqrt{-1} \cdot \sqrt{-1} = -1$

2. $(-i)^2 = [-1(i)]^2 = 1(i^2) = -1$

3. If $a > 0$, then $\sqrt{-a} = \sqrt{-1}\sqrt{a} = i\sqrt{a}$

4. If $a > 0$, then $-\sqrt{-a} = -\sqrt{-1}\sqrt{a} = -i\sqrt{a}$

5. If $a > 0$, then $(\sqrt{-a})^2 = (i\sqrt{a})^2 = i^2a = -1 \cdot a = -a$

6. If $a > 0$, then $(-\sqrt{-a})^2 = (-i\sqrt{a})^2 = i^2a = -1 \cdot a = -a$

It is now possible for us to write radicals such as $\sqrt{-4}$ and $-\sqrt{-144}$ in a different form.

$$\sqrt{-4} = \sqrt{-1 \cdot 4} = \sqrt{-1}\sqrt{4} = i \cdot 2 = 2i \quad \text{and}$$

$$-\sqrt{-144} = -\sqrt{-1 \cdot 144} = -(\sqrt{-1})(\sqrt{144}) = -(i)(12) = -12i$$

EXAMPLE 1

Write in standard form: $\sqrt{-225}$

$$= \sqrt{225 \cdot -1} = \sqrt{225}\sqrt{-1} \qquad \text{Since } \sqrt{225} = 15 \text{ and } \sqrt{-1} = i.$$

$$= 15i$$

$$= 0 + 15i$$

Standard form requires a real part and an imaginary part. Here the real part is 0. ■

EXAMPLE 2

Write in standard form: $\sqrt{8} + \sqrt{-8}$

$$= \sqrt{4}\,\sqrt{2} + \sqrt{4}\,\sqrt{2}\,\sqrt{-1}$$

Simplify.

$$= 2\sqrt{2} + 2\sqrt{2}i$$

Since $\sqrt{4} = 2$ and $\sqrt{-1} = i$.
Standard form. ■

■ **DEFINITION**

Imaginary Numbers

Complex numbers with real part 0, that is, of the form $0 + bi$, are called imaginary numbers.

The numbers $2i$, $-12i$, $\sqrt{3}i$, $-\sqrt{7}i$, and $\sqrt{-b}$ with $b > 0$ are imaginary numbers. The algebra of imaginary numbers is the same as the algebra of monomials, with the exception that i^n (n an integer) can be simplified to 1, -1, i, or $-i$ as the table shows.

Power of i	Steps	Simplified Form
i^1	i^1	i
i^2	i^2	-1
i^3	$i^2 \cdot i = -1 \cdot i$	$-i$
i^4	$i^2 \cdot i^2 = (-1)(-1)$	1
i^5	$i^4 \cdot i = 1 \cdot i$	i

Higher powers of i, such as i^{29}, can be simplified by representing them as products of i, i^2, i^3, or i^4.

EXAMPLE 3

Simplify: i^{29}, i^{130}, i^{387}, i^{400}

$$i^{29} = (i^4)^7 \cdot (i) \quad = (1)^7(i) \quad = i$$

$$i^{130} = (i^4)^{32} \cdot (i^2) = (1)^{32}(-1) = -1$$

In each case, divide the power of i by 4 since $i^4 = 1$. The other factor is a power of i less than 4.

$$i^{387} = (i^4)^{96} \cdot (i^3) = (1)^{96}(-i) = -i$$

$$i^{400} = (i^4)^{100} \quad = (1)^{100} \quad = 1$$

■

EXAMPLE 4

Simplify: $(3i^3)(-6i^5)(8i^2)$

$= (3)(-6)(8)i^{10}$

$= -144(i^4)^2(i^2)$

$= -144(1)(-1)$

$= 144$

Associative and commutative laws of multiplication and property 1 of exponents.

All real numbers can be written in the form $a + 0i$ and thus can be considered a subset of the complex numbers. Numbers of the form $0 + bi$ are pure imaginary numbers. The complex numbers include all the classifications of numbers you have studied.

COMPLEX NUMBERS ($a + bi$)		
Imaginary Numbers ($0 + bi$) $-3i, i\sqrt{7}, \sqrt{-6}, -5i\sqrt{5}$ $0 + 0i$	Real Numbers ($a + 0i$)	
	Rational Numbers	Irrational Numbers
	(Ratios of Integers) $\dfrac{3}{7}, -\dfrac{2}{9}, \dfrac{17}{1}, 0.78, -4.56$ **Integers** (Whole numbers and their opposites) $3, -7, 29, -116$ **Whole Numbers** $0, 1, 2, 77, 438$ **Zero** 0	(Nonrepeating, nonterminating decimals) Represented by special symbols $\sqrt{2}, \pi, e, -\sqrt[3]{159}$

The algebraic rules for operating with complex numbers are very similar to those for polynomials and radical expressions.

■ PROCEDURE

Add and Subtract Complex Numbers

To add or subtract complex numbers, add or subtract the real parts and add or subtract the imaginary parts.

$(a + bi) + (c + di) = (a + c) + (b + d)i$

$(a + bi) - (c + di) = (a - c) + (b - d)i$

EXAMPLE 5

Add and write the result in standard form: $(7 + 12i) + (18 + 23i)$

$= (7 + 18) + (12 + 23)i$ Add.   ■

$= 25 + 35i$

EXAMPLE 6

Add: $(1 - \sqrt{-3}) + (4 + 2\sqrt{-3})$

$= (1 - \sqrt{3}i) + (4 + 2\sqrt{3}i)$ Write each complex number in standard form.

$= (1 + 4) + (-\sqrt{3} + 2\sqrt{3})i$ Add.

$= 5 + \sqrt{3}i$ Simplify.   ■

EXAMPLE 7

Subtract and write the result in standard form: $(13 + 24i) - (15 - 12i)$

$= (13 - 15) + [24 - (-12)]i$

$= [13 + (-15)] + [24 + 12]i$ Subtract.

$= -2 + 36i$ Simplify.   ■

■ PROCEDURE

Multiply Complex Numbers

To multiply complex numbers, use the same steps that are used to multiply binomials and substitute $i^2 = -1$.

$a(b + ci) = ab + aci$

$bi(c + di) = bci + bdi^2 = bci - bd = -bd + bci$

$(a + bi)(c + di) = ac + adi + bci + bdi^2$

$= ac + adi + bci - bd$

$= (ac - bd) + (ad + bc)i$

EXAMPLE 8

Multiply and write the result in standard form: $2i(3 - 2i)$

$= 6i - 4i^2$	Multiply using the distributive property in the same way as multiplying a monomial times a binomial, then simplify.
$= 6i - 4(-1)$	
$= 4 + 6i$	∎

EXAMPLE 9

Multiply and write the result in standard form: $(4 - 2i)(3 + i)$

$= 12 + 4i - 6i - 2i^2$	In standard form, complex numbers can be multiplied using FOIL, with the exception that we substitute -1 for i^2.
$= 12 - 2i - 2(-1)$	
$= 14 - 2i$	∎

■ **DEFINITION**

Conjugates

> The complex numbers $a + bi$ and $a - bi$ are called conjugate complex numbers.

The product of conjugate complex numbers is a real number since the imaginary part of the product is 0. Example 10 illustrates this.

EXAMPLE 10

Multiply and write the result in standard form: $(3 + 11i)(3 - 11i)$

$= (3)^2 - (11i)^2$	Conjugate complex numbers can be multiplied using the same pattern as conjugate binomials.
$= 130$	
$= 130 + 0i$	Standard form. ∎

To divide two complex numbers, we use a procedure similar to that for rationalizing the denominator of a fraction.

Rationalizing the Denominator	**Dividing Complex Numbers**
$\dfrac{3}{\sqrt{2}} = \dfrac{3}{\sqrt{2}} \cdot \dfrac{\sqrt{2}}{\sqrt{2}} = \dfrac{3\sqrt{2}}{2}$	$\dfrac{3}{2i} = \dfrac{3}{2i} \cdot \dfrac{i}{i} = \dfrac{3i}{2i^2} = \dfrac{3i}{2(-1)} = -\dfrac{3}{2}i$

and

$\dfrac{3}{2 + \sqrt{3}} = \dfrac{3}{2 + \sqrt{3}} \cdot \dfrac{2 - \sqrt{3}}{2 - \sqrt{3}}$	$\dfrac{3}{2 + 3i} = \dfrac{3}{2 + 3i} \cdot \dfrac{2 - 3i}{2 - 3i}$

$$= \frac{6 - 3\sqrt{3}}{4 - 3}$$

$$= 6 - 3\sqrt{3}$$

$$= \frac{6 - 9i}{4 - 9i^2}$$

$$= \frac{6 - 9i}{4 + 9}$$

$$= \frac{6}{13} - \frac{9}{13}i$$

■ PROCEDURE

Divide Complex Numbers

To divide complex numbers, write in fraction form, and multiply the numerator and denominator by the conjugate of the denominator using the basic principle of fractions.

$$\frac{a + bi}{c + di} = \frac{(a + bi)}{(c + di)} \cdot \frac{(c - di)}{(c - di)} = \frac{ac + bd}{c^2 + d^2} + \frac{-ad + bc}{c^2 + d^2}i$$

EXAMPLE 11

Divide and write the result in standard form: $(3 + i) \div (2 - i)$

$$\frac{3 + i}{2 - i} = \frac{3 + i}{2 - i} \cdot \frac{2 + i}{2 + i}$$ 　　Multiply the numerator and denominator by $2 - i$.

$$= \frac{6 + 3i + 2i + i^2}{4 - i^2}$$

$$= \frac{5 + 5i}{5}$$ 　　Simplify and write in standard form.

$$= 1 + i$$ 　　　　　■

EXAMPLE 12

Divide and write the result in standard form: $\dfrac{3 + i}{5 - 2i}$

$$= \frac{3 + i}{5 - 2i} \cdot \frac{5 + 2i}{5 + 2i}$$ 　　Multiply the numerator and denominator by the conjugate of the denominator.

$$= \frac{15 + 6i + 5i + 2i^2}{25 + 4}$$

$$= \frac{13 + 11i}{29}$$ 　　Fraction form.

$$= \frac{13}{29} + \frac{11}{29}i$$ 　　Standard form.　　■

EXAMPLE 13

Divide and write the result in standard form: $\dfrac{6 - i}{9}$

$= \dfrac{6}{9} - \dfrac{1}{9}i$ Only one step is needed to change to standard form.

$= \dfrac{2}{3} - \dfrac{1}{9}i$ ∎

EXAMPLE 14

Divide and write the result in standard form: $\dfrac{2}{6 + \sqrt{-5}}$

$= \dfrac{2}{6 + \sqrt{5}i}$ $\sqrt{-5} = \sqrt{5}i$

$= \dfrac{2}{6 + \sqrt{5}i} \cdot \dfrac{6 - \sqrt{5}i}{6 - \sqrt{5}i}$ Multiply both the numerator and denominator by the conjugate of $6 + \sqrt{5}i$.

$= \dfrac{12 - 2\sqrt{5}i}{36 + 5} = \dfrac{12 - 2\sqrt{5}i}{41}$

$= \dfrac{12}{41} - \dfrac{2\sqrt{5}}{41}i$ Standard form.   ∎

■ **CAUTION**

The product property of radicals, $\sqrt{a}\,\sqrt{b} = \sqrt{ab}$, requires a and b to be nonnegative. The property does *not* apply if a and b are negative.

Now that you are familiar with complex numbers, you can see that:

$$\sqrt{-4}\,\sqrt{-16} = 2i \cdot 4i = 8i^2 = -8 \quad \text{but} \quad \sqrt{-4}\,\sqrt{-16} \neq \sqrt{64} = 8$$

and

$$\sqrt{-6}\,\sqrt{-12} = i\sqrt{6} \cdot 2i\sqrt{3} \qquad \text{but} \quad \sqrt{-6}\,\sqrt{-12} \neq \sqrt{72} = 6\sqrt{2}$$
$$= 2i^2\sqrt{18} = -6\sqrt{2}$$

One of the first practical applications of complex numbers can be attributed to Charles Steinmetz (1865–1923). He used the complex number plane to model the behavior of electric circuits. The need arises owing to the phase relationship between voltage and current.

In the field of electronics, the letter i is traditionally used to represent the measure of current in amperes. For this reason, the letter j is used, in the formulas of electronics, to represent $\sqrt{-1}$.

EXAMPLE 15

A formula to find the total series impedance (in ohms) of resistance and capacitance in a circuit is

$$Z_T = Z_1 + Z_2,$$

where Z_T represents the total series impedance; $Z_1 = R_1 + jX_{c1}$ and $Z_2 = R_2 + jX_{c2}$, R_1 and R_2, represent the measures of resistance (in ohms); and X_{c1} and X_{c2} represent the measures of capacitive reactance (in ohms). Find the total impedance of $R_1 = 100\ \Omega$, $R_2 = 50\ \Omega$, $X_{c1} = 100\ \Omega$, and $X_{c2} = 200\ \Omega$. (The Greek letter omega, Ω, is the abbreviation for the word *ohms*.)

Formulas:

$$Z_T = Z_1 + Z_2$$
$$Z_1 = R_1 + jX_{c1}$$
$$Z_2 = R_2 + jX_{c2}$$

Substitute:

$$Z_T = (100 + j100) + (50 + j200) \qquad \text{Substitute the values } R_1 = 100,\ R_2 = 50,$$
$$X_{c1} = 100, \text{ and } X_{c2} = 200.$$

Solve:

$$Z_T = (100 + 50) + j(100 + 200)$$
$$= 150 + j300$$

Answer:

The total impedance is $150 + j300$ ohms. ∎

EXERCISE 4.8

A

Write in standard form:

1. $\sqrt{-16}$ **2.** $\sqrt{-25}$ **3.** $\sqrt{-8}$

4. $\sqrt{-27}$ **5.** $\sqrt{-18}$ **6.** $\sqrt{-48}$

7. $\sqrt{8} + \sqrt{-12}$ **8.** $\sqrt{24} - \sqrt{-32}$ **9.** $\sqrt{-75} - \sqrt{-3}$

10. $\sqrt{-72} + \sqrt{-50}$

Simplify:

11. i^{13} **12.** i^{27} **13.** i^{54}

14. i^{57} **15.** i^{101} **16.** i^{82}

Add or subtract:

17. $(4 + 2i) + (3 + 4i)$ **18.** $(2 + 5i) + (4 - 4i)$

19. $(5 + 2i) - (3 + 5i)$ **20.** $(6 - 3i) - (4 - 6i)$

21. $(4 + \sqrt{-1}) + (2 + \sqrt{-9})$ **22.** $(5 + \sqrt{-4}) + (3 + \sqrt{-16})$

B

Perform the indicated operations:

23. $(5 - 2i) - 6$ **24.** $(10 - 7i) - 15$

25. $7 - (-8 - i)$ **26.** $12 - (-6 + 2i)$

27. $(\sqrt{12} + \sqrt{-50}) + (\sqrt{75} - \sqrt{-18})$ **28.** $(-\sqrt{32} - \sqrt{-12}) + (\sqrt{8} - \sqrt{-48})$

29. $(8 + 3i) - (-3 + i)$ **30.** $(4 + 6i) - (-12 - 4i)$

31. $\sqrt{-4} \cdot \sqrt{-196}$ **32.** $\sqrt{-5} \cdot \sqrt{-45}$ **33.** $3i(2 + 5i)$

34. $-2i(3 - 2i)$ **35.** $i^{25} \cdot i^{16}$ **36.** $i^8 \cdot i^{25}$

37. $\dfrac{i^{16}}{i^{21}}$ **38.** $\dfrac{i^{41}}{i^{50}}$ **39.** i^{-12}

40. i^{-22} **41.** $\dfrac{i^{21}}{i^{-10}}$ **42.** $\dfrac{i^{-43}}{i^{14}}$

C

Perform the indicated operations:

43. $(4 + 2i)(3 - 2i)$ **44.** $(5 - 3i)(2 + 3i)$

45. $(7 + \sqrt{-16})(-2 - \sqrt{-25})$ **46.** $(-4 + \sqrt{-9})(6 - \sqrt{-81})$

47. $(2 + 3i)^2$ **48.** $(6 - 5i)^2$

49. $\dfrac{1}{2i}$ **50.** $\dfrac{-2}{3i}$

51. $\dfrac{3}{-5i}$ **52.** $\dfrac{-4}{-7i}$

53. $(\sqrt{8} + i\sqrt{12})(\sqrt{2} - i\sqrt{3})$ **54.** $(\sqrt{18} - i\sqrt{27})(\sqrt{2} + i\sqrt{3})$

55. $\dfrac{1 + i}{3 - i}$

56. $\dfrac{2 - i}{4 + i}$

57. $\dfrac{-5 - 4i}{-2 + 3i}$

58. $\dfrac{-6 - 3i}{-1 - 3i}$

59. $\dfrac{(5 + 3i)^2}{2}$

60. $\dfrac{(3 - 7i)^2}{2}$

61. $\dfrac{3}{(2 - i)^2}$

62. $\dfrac{10}{(1 - 5i)^2}$

63. $\dfrac{54 - 18i}{12i}$

64. $\dfrac{-98 + 56i}{-21i}$

65. $\dfrac{-6i}{3 + \sqrt{-3}}$

66. $\dfrac{-18i}{5 - \sqrt{-2}}$

D

Using the formula in Example 15 find the total series impedance of resistance and capacitance if:

67. $R_1 = 96.5 \ \Omega$, $X_{c1} = 105.7 \ \Omega$, $R_2 = 86.2 \ \Omega$, and $X_{c2} = 87.5 \ \Omega$

68. $R_1 = 5100 \ \Omega$, $X_{c1} = 4500 \ \Omega$, $R_2 = 3600 \ \Omega$, and $X_{c2} = 4000 \ \Omega$

69. $R_1 = 15700 \ \Omega$, $X_{c1} = 18900 \ \Omega$, $R_2 = 23800 \ \Omega$, and $X_{c2} = 30500 \ \Omega$

70. $R_1 = 819 \ \Omega$, $X_{c1} = 748 \ \Omega$, $R_2 = 588 \ \Omega$, and $X_{c2} = 980 \ \Omega$

STATE YOUR UNDERSTANDING

71. Is it possible for the set of real numbers to be considered a subset of the complex numbers? (Those numbers of the form a + bi.) If the answer is yes, how is this possible?

CHALLENGE EXERCISES

72. Evaluate $4x^2 + 4$ when $x = -i$.

73. Evaluate $4x^2 + 4$ when $x = i$.

74. Evaluate $x^2 - 3x - 1$ when $x = 1 - i$.

75. Evaluate $x^2 + 4x + 4$ when $x = 1 + i$.

76. Factor $x^2 + 16$. Hint: $16 = -(-16)$.

77. Factor $x^2 + 25$.

78. Is i a solution of the equation $4x^2 + 4 = 0$?

79. Is $-i$ a solution of the equation $4x^2 + 4 = 0$?

MAINTAIN YOUR SKILLS (SECTIONS 3.3, 3.4, 3.5)

80. $\dfrac{2a}{2a^2 + a - 3} + \dfrac{1 - a}{3a^2 - 4a + 1} - \dfrac{a - 2}{6a^2 + 7a - 3}$

Simplify:

81. $1 - \dfrac{1}{1 - \dfrac{1}{1 - x}}$

Solve:

82. $\dfrac{x}{x^2 + x - 20} = \dfrac{2x - 5}{x^2 - x - 12} - \dfrac{x + 6}{x^2 + 8x + 15}$

83. $\dfrac{y + 2}{y^2 + y - 20} = \dfrac{2y - 5}{y^2 - y - 12} - \dfrac{y + 6}{y^2 + 8y + 15}$

84. Divide: $2x - 5\overline{)2x^3 - 9x^2 + 25}$

85. Divide: $(4x^3 + 2x^2 + 2x + 2) \div (2x + 3)$

86. Jay drives the 720 miles to Springville and returns the next day. If he averages 5 mph faster on the return trip and takes 2 hours less time, what was his average speed each way?

87. The estate of Joe Riches was to have been divided equally among his grandchildren. Since three of the grandchildren could not be located, each of the remaining grandchildren received an additional \$240,000. If the estate was valued at \$10,400,000, how many grandchildren did Joe Riches have?

CHAPTER 4
SUMMARY

The properties and definitions of radicals and fractional exponents are used to simplify algebraic expressions involving radicals. These properties and definitions help us recognize different symbols for numbers whose exact value cannot be represented by a fraction or a decimal (terminating decimal).

The nth Root of a

1. If $a \geq 0$ and if n is odd, $a^{1/n}$ is one of the n identical positive factors of a.   (pp. 248, 250)

2. If $a \geq 0$ and if n is even, $a^{1/n}$ is called the principal nth root of a, and $-a^{1/n}$ is another nth root of a.

3. If $a < 0$ and if n is odd, $a^{1/n}$ is one of the n identical negative factors of a.

4. If $a < 0$ and if n is even, $a^{1/n}$ does not represent a real number because no real number raised to an even power yields a negative number.

Property of Rational Exponents	If $a \geq 0$, $a^{m/n} = (a^m)^{1/n} = (a^{1/n})^m$.	(p. 251)

If $a < 0$, the value of $a^{m/n}$ depends on m and n.

The five properties of exponents that you learned in Chapter 2 also apply to rational exponents. Radicals are alternative ways of writing expressions that contain rational exponents.

Radical Form	The number $a^{1/2}$ or $\sqrt{a}$ is called the positive or principal square root of a.	(p. 257)

In general, the radical can be used for $a^{1/n}$ by writing $a^{1/n} = \sqrt[n]{a}$, where $\sqrt{}$ is called the *radical*, n is called the *index* of the radical, and a is called the *radicand*.

Square Root Radicals	$\sqrt{x^2} = x$ is true if x is positive or zero.	(p. 260)

$\sqrt{x^2} = -x$ is true if x is negative.

$\sqrt{x^2} = |x|$ is always true.

Simplest Radical Form	**1.** The radicand has no factors that are perfect squares except the number 1.	(p. 266)

2. No fractions are in the radicand, and no radicals appear in a denominator.

Product Property of Radicals	If $a \geq 0$ and $b \geq 0$, then $\sqrt{ab} = \sqrt{a}\,\sqrt{b}$ and $\sqrt[n]{ab} = \sqrt[n]{a}\,\sqrt[n]{b}$.	(p. 266)
Quotient Property of Radicals	If $a \geq 0$ and $b > 0$, then $\sqrt{\dfrac{a}{b}} = \dfrac{\sqrt{a}}{\sqrt{b}}$ and $\sqrt[n]{\dfrac{a}{b}} = \dfrac{\sqrt[n]{a}}{\sqrt[n]{b}}$.	(p. 266)

Simplifying Radicals if Variables Represent Positive Numbers	**1.** Factor the radicand so that one or more numerical factors are perfect squares and so that one or more variable factors have exponents that are multiples of 2. Continue such factoring until no square factors remain.	(p. 267)

2. Simplify the perfect square factors.

If Variables Represent Real Numbers	**1.** Restrict those variables with exponents that are odd numbers. Then follow step 1 as before.	(p. 268)

2. After the radical has been simplified, place absolute bars around the variable factors as necessary.

Like Radicals	Like radicals are radicals that have the same radicand and the same index.	(p. 274)
Adding and Subtracting Radicals	Only like radicals can be combined by adding and subtracting.	(p. 275)
Multiplying Radicals	The product property of radicals is used to multiply radicals. The distributive property (including the FOIL shortcut for multiplying binomials) is used to multiply radical expressions containing sums and differences. Radical expressions with different index numbers can be multiplied if they are first rewritten with a common index number.	(pp. 280, 281, 282)

Dividing Radicals or Rationalizing the Denominator	To divide radicals means to eliminate the radicals in the divisor. The new fraction will contain a denominator that is an integer or a variable without radicals that is equivalent to the original fraction. To rationalize the denominator, that is, to eliminate the fraction in the radical, we use the quotient property of radicals. (p. 286)
Conjugate Radicals	The sum and difference of the same two terms involving radicals are called conjugate radical expressions. Conjugate radicals are helpful in rationalizing. (p. 281)
Extraneous Root (Extraneous Solution)	An apparent solution that is introduced when squaring both sides of an equation. (p. 295)

Containing one radical: (pp. 297, 298)

1. Isolate the radical on one side of the equation.

2. Square both sides.

3. Solve for x.

4. Check for "extraneous roots."

Containing two radicals:

1. Isolate one of the radicals.

2. Square both sides.

3. Repeat until all radicals have been eliminated.

4. Check for "extraneous roots."

Complex Numbers	Numbers that can be written in the form $a + bi$, where $a \in R$ and $b \in R$ and $i^2 = -1$. In $a + bi$, a is called the real part and b is called the imaginary part of the complex number (pp. 303, 304)
	Complex numbers provide solutions to some equations that have no real numbers and are used in some practical applications, especially in the field of electronics.
Adding and Subtracting Complex Numbers	To add or subtract complex numbers, add or subtract the real parts and add or subtract the imaginary parts. (p. 307)

$$(a + bi) + (c + di) = (a + c) + (b + d)i$$
$$(a + bi) - (c + di) = (a - c) + (b - d)i$$

Multiplying Complex Numbers	To multiply complex numbers, use the same steps that are used to multiply binomials and substitute $i^2 = -1$. (p. 307)

$$a(b + ci) = ab + aci$$
$$bi(c + di) = bci + bdi^2 = bci - bd = -bd + bci$$
$$(a + bi)(c + di) = (ac - bd) + (ad + bc)i$$

Dividing Complex Numbers	To divide complex numbers, write in fraction form, and multiply the numerator and denominator by the conjugate of the denominator using the numbers basic principle of fractions. (p. 309)

$$\frac{a + bi}{c + di} = \frac{(a + bi)}{(c + di)} \cdot \frac{(c - di)}{(c - di)} = \frac{ac + bd}{c^2 + d^2} + \frac{-ad + bc}{c^2 + d^2}i$$

CHAPTER 4
REVIEW EXERCISES

SECTION 4.1   Objective 1

Simplify.

1. $25^{1/2}$

2. $400^{1/2}$

3. $(-1)^{2/3}$

4. $(-1)^{5/3}$

5. $(0.81)^{1/2}$

6. $(0.0081)^{1/4}$

SECTION 4.1   Objective 2

Simplify using only positive exponents.

7. $(m^{4/5})(m)$

8. $(n)(n^{1/3})$

9. $(x^{4/5})(x^{1/2})$

10. $(y^{3/5})(y^{3/2})$

11. $\dfrac{p^{2/3}}{p^{3/2}}$

12. $\dfrac{q^{5/2}}{q^{1/3}}$

13. $\dfrac{144t^{7/8}}{9t^{3/4}}$

14. $\dfrac{144m^{2/3}}{16m^{1/6}}$

15. $\dfrac{-144y^{3/4}}{8y^{-3/8}}$

16. $\dfrac{144a^{-1/6}}{-48a^{-2/3}}$

17. $(4x^{1/2}y^{5/2})^2$

18. $(4p^{2/3}q^{1/2})^3$

19. $(36a^3b^{1/2})^{1/2}$

20. $(27m^2n^{6/5})^{1/3}$

21. $(2x^{1/2} + x^{1/3})(2x^{1/2} - x^{1/3})$

SECTION 4.2   Objective 1

Write these expressions in radical form and simplify. Assume variables represent positive real numbers.

22. $18^{3/2}$

23. $18^{2/3}$

24. $8y^{1/3}$

25. $(8y)^{1/3}$

26. $-9y^{1/3}$

27. $(9y)^{-1/2}$

28. $(64a)^{-1/3}$

29. $(8x^2)^{2/3}$

30. $(9y^3)^{3/2}$

SECTION 4.2   Objective 2

Write these expressions with a rational exponent. Assume variables represent positive real numbers.

31. $\sqrt{16w}$

32. $\sqrt{4a}$

33. $6\sqrt{6s}$

34. $3\sqrt{3t}$

35. $\sqrt[3]{25x^2y^4}$

36. $\sqrt[4]{10w^2z^5}$

SECTION 4.2   Objective 3

Simplify.

37. $\sqrt{196a^4}$

38. $\sqrt{400m^6}$

39. $\sqrt{144x^2y^{16}}$

40. $\sqrt{121w^6y^8}$

41. $\sqrt[3]{216a^6}$

42. $\sqrt[3]{125b^3c^{15}}$

Assume the variables represent real numbers and simplify. Use absolute value signs where necessary.

43. $\sqrt{81m^2}$

44. $\sqrt{49n^4}$

45. $\sqrt{121s^4t^6}$

46. $\sqrt{225w^4z^2}$

47. $\sqrt{x^2 - 6x + 9}$

48. $\sqrt{16y^2 + 8y + 1}$

SECTION 4.3   Objective 1

Simplify.

49. $\sqrt{500}$

50. $\sqrt{800}$

51. $\sqrt{120}$

52. $\sqrt{160}$

53. $\sqrt{175}$

54. $\sqrt{245}$

Simplify. Assume all variables represent real numbers. Use absolute value where necessary.

55. $\sqrt{200a^2}$

56. $\sqrt{242b^2}$

57. $\sqrt{300x^2y^2}$

58. $\sqrt{363w^4z^2}$

59. $\sqrt[4]{32m^4n^2}$

60. $\sqrt[4]{48s^4t^8}$

SECTION 4.3   Objective 2

Find the approximate value of each radical to four decimal places.

61. $\sqrt{500}$

62. $\sqrt{800}$

63. $\sqrt{120}$

64. $\sqrt{160}$

65. $\sqrt{175}$

66. $\sqrt{245}$

SECTION 4.3   Objective 3

Reduce the index of the following radicals if the variables represent positive real numbers.

67. $\sqrt[8]{100w^2}$

68. $\sqrt[8]{121z^2}$

69. $\sqrt[10]{32a^5}$

SECTION 4.4   Objective 1

Combine. Assume variables represent positive real numbers.

70. $3\sqrt{8} + 2\sqrt{50}$

71. $7\sqrt{98} - 3\sqrt{72}$

72. $27\sqrt{27} - 12\sqrt{12}$

73. $80\sqrt{80} - 45\sqrt{45}$

74. $2\sqrt{75} - 2\sqrt{12}$

75. $5\sqrt{45} + 2\sqrt{20}$

76. $3\sqrt{28x} + \sqrt{63x}$

77. $2\sqrt{180y} + \sqrt{125y}$

78. $3\sqrt[3]{3a} + 2\sqrt[3]{24a}$

79. $5\sqrt[3]{5mn} - \sqrt[3]{40mn}$

80. $\sqrt{8} + \sqrt{16} + \sqrt{24}$

81. $\sqrt{9} + \sqrt{18} + \sqrt{27}$

82. $2\sqrt{\dfrac{1}{2}} + 3\sqrt{\dfrac{2}{3}} + 6\sqrt{\dfrac{1}{6}}$

83. $3\sqrt{\dfrac{4}{3}} - 2\sqrt{\dfrac{3}{4}} + 3\sqrt{\dfrac{1}{12}}$

84. $5\sqrt{\dfrac{1}{10}} + 2\sqrt{10} - \sqrt{\dfrac{4}{10}}$

85. $\sqrt[3]{1029} - \sqrt[3]{375}$

86. $\sqrt[3]{\dfrac{4}{27}} + \sqrt[3]{\dfrac{1}{2}}$

87. $\dfrac{3}{\sqrt[3]{4}} + \dfrac{12}{\sqrt{3}} - 2\sqrt[4]{144}$

SECTION 4.5   Objective 1

Multiply and simplify.

88. $\sqrt{3}\,\sqrt{75}$

89. $\sqrt{5}\,\sqrt{45}$

90. $(-\sqrt{2})(-\sqrt{98})$

91. $(-\sqrt{7})(-\sqrt{28})$

92. $\sqrt[3]{4}\,\sqrt[3]{10x}$

93. $\sqrt[3]{5}\,\sqrt[3]{50y}$

94. $2\sqrt{6}(\sqrt{10} + 2\sqrt{21} - \sqrt{8})$

95. $2\sqrt{2}(2\sqrt{2} + 8\sqrt{2} - 3\sqrt{3})$

96. $\sqrt{a}(\sqrt{ab} - a\sqrt{b} + b\sqrt{a})$

97. $w\sqrt{y}(\sqrt{wy} + w\sqrt{y} - y\sqrt{w})$

98. $(\sqrt{6} + \sqrt{2})^2$

99. $(\sqrt{10} - \sqrt{2})^2$

100. $(2\sqrt{3} + 3\sqrt{2})^2$

101. $(3\sqrt{3} - 2\sqrt{2})^2$

102. $(\sqrt{2x} + \sqrt{2})(\sqrt{x} - 1)$

103. $(\sqrt{3y} + 2\sqrt{3})(\sqrt{y} - 2)$

104. $3\sqrt[3]{3}(\sqrt[3]{9} + \sqrt[3]{72})$

105. $5\sqrt[3]{2}(2\sqrt[3]{4} - \sqrt[3]{32})$

SECTION 4.6   Objective 1

Simplify by rationalizing the denominator.

106. $\dfrac{1}{\sqrt{6}}$

107. $\dfrac{6}{\sqrt{6}}$

108. $\dfrac{1}{\sqrt{98}}$

109. $\dfrac{98}{\sqrt{98}}$

110. $\dfrac{\sqrt{72}}{\sqrt{4}}$

111. $\dfrac{\sqrt{72}}{\sqrt{9}}$

112. $\dfrac{\sqrt{72}}{\sqrt{2}}$

113. $\dfrac{\sqrt{72}}{\sqrt{3}}$

114. $\sqrt{\dfrac{7}{12}}$

115. $\sqrt{\dfrac{11}{12}}$

116. $\dfrac{12}{\sqrt{8}}$

117. $\dfrac{8}{\sqrt{12}}$

118. $\dfrac{2\sqrt{18}}{3\sqrt{5}}$

119. $\dfrac{4\sqrt{8}}{5\sqrt{27}}$

120. $\dfrac{3x\sqrt{16xy}}{2\sqrt{8y}}$

121. $\dfrac{6a\sqrt{2ab^3}}{5b\sqrt{8a^3b}}$

122. $\dfrac{2}{\sqrt[3]{25}}$

123. $-\dfrac{3}{\sqrt[4]{8}}$

124. $\dfrac{3}{\sqrt{2}-1}$

125. $\dfrac{5}{4-\sqrt{5}}$

126. $\dfrac{\sqrt{2}}{\sqrt{3}+\sqrt{2}}$

127. $\dfrac{\sqrt{3}}{\sqrt{3}-\sqrt{2}}$

128. $\dfrac{1+\sqrt{5}}{\sqrt{3}+\sqrt{2}}$

129. $\dfrac{1-\sqrt{6}}{\sqrt{5}+2}$

130. $\dfrac{4+\sqrt{3}}{4-\sqrt{3}}$

131. $\dfrac{\sqrt{7}-2}{\sqrt{7}+2}$

132. $\dfrac{3\sqrt{2}}{3\sqrt{2}-\sqrt{3}}$

133. $\dfrac{1}{\sqrt{x}+\sqrt{y}}$

134. $\dfrac{3}{\sqrt{a}-\sqrt{3}}$

135. $\dfrac{x\sqrt{x}+\sqrt{y}}{x\sqrt{x}-\sqrt{y}}$

SECTION 4.7   Objective 1

Solve the following equations containing radicals.

136. $\sqrt{x}=7$

137. $\sqrt{a}=12$

138. $\sqrt{x+2}=-4$

139. $\sqrt{x-1}=-6$

140. $\sqrt{x-1}=x-3$

141. $\sqrt{x+5}=x+5$

142. $\sqrt{2x+1}=\sqrt{x+9}$

143. $\sqrt{3x+5}=\sqrt{x-3}$

144. $\sqrt{x+2}+\sqrt{x-1}=3$

145. $\sqrt{4x+1}+\sqrt{x+5}=14$

SECTION 4.8   Objective 1

Write in standard form.

146. $\sqrt{-100}$

147. $2\sqrt{-121}$

148. $\sqrt{45}+\sqrt{-54}$

Simplify.

149. i^{61}

150. i^{62}

151. i^{63}

SECTION 4.8   Objective 2

Add or subtract and write in standard form.

152. $(23+12i)+(6-3i)$

153. $(23+12i)-(6-3i)$

154. $(3-\sqrt{-9})+(4+\sqrt{-4})$

155. $(3-\sqrt{-9})-(4+\sqrt{-4})$

156. $(8-\sqrt{-18})+(-3-\sqrt{-8})$

157. $(8-\sqrt{-18})-(-3-\sqrt{-8})$

SECTION 4.8   Objective 3

Multiply and write in standard form.

158. $(3+2i)(5-4i)$

159. $(6-4i)(-1+7i)$

160. $(5 + \sqrt{-16})(3 + \sqrt{-25})$

161. $(\sqrt{9} + \sqrt{-9})(\sqrt{81} - \sqrt{-81})$

162. $(5 - 8i)^2$

163. $(6 + \sqrt{-100})^2$

SECTION 4.8   Objective 4

Divide and write in standard form.

164. $\dfrac{1}{i^5}$

165. $\dfrac{3}{i^{13}}$

166. $\dfrac{4}{1 - i}$

167. $\dfrac{3}{2 + i}$

168. $\dfrac{3 - i}{2 + 2i}$

169. $\dfrac{4 + i}{2 - 2i}$

170. $(6 + i)(2 - i)^{-1}$

171. $(3 - i)^{-1}(4 + 5i)$

CHAPTER 4
TRUE–FALSE CONCEPT REVIEW

Check your understanding of the language of algebra. Tell whether each of the following statements is true (always true) or false (not always true).

1. The eighth root of 240 is 30.

2. The nth root of a negative number is negative.

3. If $a \cdot a \cdot a \cdot a \cdot a \cdot a \cdot a = b$, then a is the seventh root of b.

4. If m is an even number, then $\sqrt[n]{a^m} \geq 0$.

5. $\sqrt{a} + \sqrt{b} = \sqrt{a + b}$ if $a \geq 0$ and $b \geq 0$.

6. A square root radical is in simplest radical form when it contains no perfect squares as factors in the radicand.

7. $\sqrt[n]{a} \cdot \sqrt[n]{b} = \sqrt[n]{ab}$

8. The product of radicals with positive radicands and different index numbers can be simplified if each radical is rewritten using a common index.

9. Rationalizing the denominator means to eliminate the radical(s) in the denominator.

10. The conjugate of $\sqrt{x} + \sqrt{y}$ is found by replacing y with its opposite.

11. $\sqrt{18x^5y^6}$ represents a real number for all x and y.

12. $\dfrac{\sqrt{a}}{\sqrt{b}} = \dfrac{\sqrt{ab}}{b}$ for $a \geq 0$ and $b \geq 0$.

13. The standard form of a complex number is $a + bi$, where $a \in R$ and $b \in R$.

14. If n is an integer, then i^n can be simplified to 1, -1, i, or $-i$.

15. Every rational number can be expressed as a complex number.

16. Products and quotients of complex numbers are complex numbers.

17. The imaginary part of the complex number $a + bi$ is b.

18. Every expression with rational exponents can be written in radical form.

19. Every radical can be written using rational exponents.

20. A radical sign always means "square root."

21. The expression under a radical sign is called the index of the radical.

22. If a is positive, then $\sqrt{a}$ always represents a real number.

23. To simplify a radical means to find a decimal value and round off.

24. The symbol $\sqrt[3]{x}$ always represents a real number.

25. A check is essential when solving radical equations because squaring both sides of an equation does not always yield an equivalent equation.

CHAPTER 4
TEST

1. Rationalize the denominator: $\dfrac{6 - \sqrt{3}}{\sqrt{18}}$

2. Simplify, a and b are real numbers: $\sqrt{108a^{10}b^8}$

3. Multiply, give answer in standard form:
$(13 + 7i)(-2 - 4i)$

4. Simplify, using only positive exponents:
$\left(\dfrac{x^{-4/7}}{z^{-1/2}}\right)^{-7/8}$

5. Simplify: $(\sqrt[3]{27x^6})^2$

6. Divide, give answer in standard form:
$\dfrac{6 - 9i}{3 - 7i}$

7. Write in radical form: $2y^{1/3}z^{2/3}$

8. Find the value to the nearest tenth: $\sqrt{240}$

9. Simplify, using only positive exponents:

$(a^{-2/3}b^{3/4})^{-1/2}$

10. Combine: $\sqrt{54} + \sqrt{216} - \sqrt{600}$

11. Solve: $\sqrt{10 - 3x} = 14$

12. Multiply and simplify, assuming b is nonnegative:
$(2\sqrt{b} - 3b)(5\sqrt{b} + 7b)$

13. Combine, assuming y is nonnegative:
$\sqrt{144y} - \sqrt{4y} + \sqrt{36y}$

14. Write using rational exponents: $\sqrt[5]{11x^2y^4}$

15. Simplify, assuming a is nonnegative: $\sqrt{242a^5}$

16. Divide, give answer in standard form:
$\dfrac{-6}{11 - 3i}$

17. Subtract: $(46 - 13i) - (-31 - 15i)$

18. Simplify: $\sqrt{648}$

19. Multiply and simplify: $\sqrt{12}(\sqrt{24} + \sqrt{15})$

20. Solve: $\sqrt{29 - x} = 7 - \sqrt{x}$

21. Rationalize the denominator, assuming c is nonnegative and $c \neq 25$: $\dfrac{2\sqrt{c} - 3}{5 - \sqrt{c}}$

22. Add: $(7 - 8i) + (5 + 2i) + (-15 - 7i)$

23. Combine, assuming y is nonnegative:
$\sqrt{45y^2} - \sqrt{180y^2} + \sqrt{20y^2}$

24. Multiply, give answer in standard form: $(7 - 8i)^2$

25. Simplify: $i^{11} \cdot i^{50}$

26. Multiply and simplify: $(\sqrt{55} - \sqrt{33})^2$

27. Find the total number of plants needed to cover a triangular area if the nursery uses 5 plants per square foot. The triangular plot is $10\sqrt{15}$ ft in height and has a base of $9\sqrt{5}$ ft. (*Hint:* First find the area of the plot to the nearest square foot.)

28. A square building lot contains 2809 sq ft. The lot is to be fenced with a cedar fence costing $1.32 per running foot. What will be the cost of the fence?

QUADRATIC AND OTHER EQUATIONS AND INEQUALITIES

The maximum height reached by a projectile shot from the ground can be expressed as a quadratic polynomial in terms of the distance traveled and the initial velocity. In Exercise 72, Section 5.2, an archer is shooting an arrow at 45° to the horizontal. Use the formula, $y = \dfrac{-32x^2}{v^2} + x$, to find the maximum height reached by the arrow. (© *The Stock Market/Brownie Harris*)

PREVIEW

In this chapter we solve quadratic equations, absolute value equations, equations that are quadratic in form, and absolute value, quadratic, and rational inequalities. Quadratic equations are solved by factoring, square roots, completing the square, and the Quadratic Formula. Equations with complex roots are included. Care is taken to restrict the domain of the variables to exclude extraneous roots. Identification of critical numbers on the number line forms the basis for the solution for quadratic and rational inequalities.

5.1

QUADRATIC EQUATIONS SOLVED BY COMPLETING THE SQUARE

OBJECTIVES

1. Solve quadratic equations that can be written in the form $x^2 = k$.

2. Solve quadratic equations by completing the square.

▪ DEFINITION

> A quadratic equation in one variable is one that can be written in the form
>
> $ax^2 + bx + c = 0,$
>
> where a, b, and c are integers and $a > 0$.

In Chapter 2, quadratic equations are solved by factoring and using the zero-product property. The *number of roots* or solutions of an equation is the same as the degree of the equation, provided that repetitions are counted as distinct roots. So an equation such as

$(x - 2)(x - 2) = 0$

is said to have two roots even though it is standard procedure to list the repeated root only once. The root is said to have *multiplicity two*. As a result of this understanding, all quadratic equations have two roots.

If $b = 0$ or $c = 0$, the equation is called an *incomplete quadratic equation.*

The incomplete quadratic equation in which $c = 0$ was solved in Chapter 2 using the zero-product property.

If $b = 0$, we get a second form of an incomplete quadratic.

$$ax^2 + bx + c = 0 \qquad \text{Standard form.}$$
$$ax^2 + c = 0 \qquad \text{If } b = 0.$$
$$ax^2 = -c$$

$$x^2 = -\frac{c}{a} \qquad \text{Solve for } x^2.$$

$$\text{or} \quad x^2 = k \qquad \text{Letting } k = -\frac{c}{a}.$$

Such incomplete quadratics can be solved by setting x equal to each of the square roots of k.

■ PROCEDURE

Solution by Square Root

To solve an incomplete quadratic in which $b = 0$:

1. Write in the form $x^2 = k$.
2. Write the equivalent equation $x = \pm\sqrt{k}$.
3. Write the solution set $\{\pm\sqrt{k}\}$.

EXAMPLE 1

Solve: $x^2 = 64$

$$x^2 = 64$$

$$x = \pm\sqrt{64} \qquad \text{If } x^2 = k, \, x = \pm\sqrt{k}.$$

$$x = \pm 8$$

Check:

$$(8)^2 = 64 \quad \text{or} \quad (-8)^2 = 64 \qquad \text{Substitute } x = 8 \text{ and } x = -8 \text{ in the original}$$
$$64 = 64 \qquad\qquad 64 = 64 \qquad \text{equation.}$$

The solution set is $\{\pm 8\}$.   ■

With this method we can also solve quadratic equations that do not have rational roots.

EXAMPLE 2

Solve: $x^2 = 12$

$$x^2 = 12$$

$$x = \pm\sqrt{12} \qquad \text{If } x^2 = k, \, x = \pm\sqrt{k}.$$

$$x = \pm 2\sqrt{3} \qquad \text{Simplify.}$$

Check:

$(2\sqrt{3})^2 = 12$ or $(-2\sqrt{3})^2 = 12$

$\qquad 12 = 12 \qquad\qquad\qquad 12 = 12$

Substitute $x = 2\sqrt{3}$ and $x = -2\sqrt{3}$ in the original equation.

The solution set is $\{\pm 2\sqrt{3}\}$.

We can apply this method to a more complicated quadratic equation. Equations of the form $(ax + b)^2 = c$ can be solved using square roots.

EXAMPLE 3

Solve: $(x - 1)^2 = 4$

$(x - 1)^2 = 4$

$x - 1 = \pm\sqrt{4}$

If $x^2 = k$, then $x = \pm\sqrt{k}$.

$x - 1 = \pm 2$

$x - 1 = 2$ or $x - 1 = -2$

$\qquad x = 3 \qquad\qquad\quad x = -1$

Rewrite as two equations by setting the left side equal to 2 or -2 and solve.

Check:

$(3 - 1)^2 = 4$ or $(-1 - 1)^2 = 4$

Substitute 3 and -1 in the original equation.

$\quad (2)^2 = 4 \qquad\qquad (-2)^2 = 4$

The solution set is $\{-1, 3\}$.

In Example 4, factor the left side and then use the square root procedure.

EXAMPLE 4

Solve: $x^2 + 6x + 9 = 18$

$(x + 3)^2 = 18$

Factor the left side.

$x + 3 = \pm\sqrt{18}$

If $x^2 = k$ then $x = \pm\sqrt{k}$.

$x + 3 = \pm 3\sqrt{2}$

Simplify the radical.

$x + 3 = 3\sqrt{2}$ or $x + 3 = -3\sqrt{2}$

Write as two equations by setting $x + 3$ equal to $3\sqrt{2}$ and $-3\sqrt{2}$.

$x = -3 + 3\sqrt{2} \qquad\qquad x = -3 - 3\sqrt{2}$

Check:

$(-3 + 3\sqrt{2})^2 + 6(-3 + 3\sqrt{2}) + 9 = 18$ Substitute $x = -3 + 3\sqrt{2}$ in the original equation.

$9 - 18\sqrt{2} + 18 - 18 + 18\sqrt{2} + 9 = 18$ Multiply.

The solution set is $\{-3 \pm 3\sqrt{2}\}$.   The check for $x = -3 - 3\sqrt{2}$ is left for the student.   ■

We can also solve equations that do not have real number solutions using the same method.

EXAMPLE 5

$(3x + 2)^2 = -5$

$\quad 3x + 2 = \pm\sqrt{-5}$ Apply the square root method.

$\quad 3x + 2 = \pm\sqrt{5}i$ $\sqrt{-5} = \sqrt{5}i$

$\quad\quad 3x = -2 \pm \sqrt{5}i$ Add -2 to both sides.

$\quad\quad\quad x = -\dfrac{2}{3} \pm \dfrac{\sqrt{5}}{3}i$

The solution set is $\left\{-\dfrac{2}{3} \pm \dfrac{\sqrt{5}}{3}i\right\}$.   The check is left for the student.   ■

The *Pythagorean theorem* describes the relationship among the sides of a right triangle. If a and b are the legs (form the right angle) and c is the hypotenuse (the side opposite the right angle), then the following relationships are true:

$a^2 + b^2 = c^2$

or

$b^2 = c^2 - a^2$

or

$a^2 = c^2 - b^2$

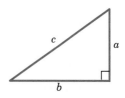

EXAMPLE 6

Find the length of side b in the following right triangle:

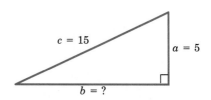

$b^2 = c^2 - a^2$ Pythagorean theorem.

$b^2 = (15)^2 - (5)^2$ Let $c = 15$ and $a = 5$.

$b^2 = 225 - 25$

$b^2 = 200$

$b = \pm\sqrt{200}$ If $x^2 = k$, then $x = \pm\sqrt{k}$.

$ = \pm 10\sqrt{2}$ $\sqrt{200} = \sqrt{100}\,\sqrt{2} = 10\sqrt{2}$

Check:

$(10\sqrt{2})^2 = (15)^2 - (5)^2$ Disregard the negative root; a length
cannot be negative.
$\phantom{(10\sqrt{2})^2} 100(2) = 225 - 25$

So the length of side b is $10\sqrt{2}$.   ■

The formula for the illumination from a light source leads to the solution of a quadratic equation by square roots.

EXAMPLE 7

The illumination (in foot candles) of a light source varies inversely as the square of the distance from the source. The formula is $Id^2 = k$ (I is the illumination and d is the distance in feet). If the constant of variation (k) is 16,000 and the illumination is 80 foot candles, what is the distance from the light source?

Formula:

$Id^2 = k$

Substitute:

$80d^2 = 16000$ To find the distance necessary to give a
light of 80 foot candles, we substitute
$k = 16{,}000$ and $I = 80$ in the formula.

Solve:

$d^2 = 200$

$d = \pm\sqrt{200}$

$d = \pm 10\sqrt{2}$

Answer:

The distance is $10\sqrt{2}$, or approximately 14.14 feet.

Since distance is not measured with negative numbers, we reject $-10\sqrt{2}$. Approximate the root using a calculator. ■

It is possible to transform any quadratic equation into the form of $(dx - e)^2 = k$. This transformation is called *completing the square*.

■ DEFINITION

> To complete the square of a quadratic polynomial of the form $x^2 + bx$ means to add a constant term so that the resulting trinomial is a perfect square trinomial.

To complete the square of $x^2 - 6x$, add 9. The trinomial $x^2 - 6x + 9$ is a perfect square trinomial.

$$x^2 - 6x + 9 = (x - 3)^2$$

The quadratic equation

$$x^2 + 6x + 6 = 0$$

cannot be solved by factoring using integers. In this form, it cannot be solved using the square root method. We change the form of the equation by completing the square.

■ PROCEDURE

Completing the Square

> To complete the square of $x^2 \pm bx$, $b \in R$:
>
> 1. Multiply the coefficient of x by $\dfrac{1}{2}$: $\left[\dfrac{1}{2} \cdot b = \dfrac{b}{2} \right]$.
>
> 2. Square the product: $\left[\dfrac{b}{2} \right]^2 = \dfrac{b^2}{4}$.
>
> 3. Add this square, $\dfrac{b^2}{4}$, to $x^2 \pm bx$: $x^2 \pm bx + \dfrac{b^2}{4}$.

The resulting polynomial is a perfect square trinomial.

$$x^2 \pm bx + \frac{b^2}{4} = \left(x \pm \frac{b}{2} \right)^2$$

The table below shows some examples.

$x^2 \pm bx$	Half of Coefficient of x	Square	Perfect Square Binomial	Factored
$x^2 + 6x$	3	9	$x^2 + 6x + 9$	$(x + 3)^2$
$x^2 - 12x$	-6	36	$x^2 - 12x + 36$	$(x - 6)^2$
$a^2 - 5a$	$-\dfrac{5}{2}$	$\dfrac{25}{4}$	$a^2 - 5a + \dfrac{25}{4}$	$\left(x - \dfrac{5}{2}\right)^2$

The next three examples are solved by completing the square.

EXAMPLE 8

Solve: $x^2 + 6x + 6 = 0$

$x^2 + 6x = -6$	Subtract 6 from both sides.
$x^2 + 6x + 9 = -6 + 9$	Add $\left(\dfrac{1}{2} \cdot 6\right)^2 = 9$ to both sides.
$(x + 3)^2 = 3$	Factor the left side.
$x + 3 = \pm\sqrt{3}$	Set $x + 3$ equal to each square root of 3.
$x + 3 = \sqrt{3}$ or $x + 3 = -\sqrt{3}$	Solve each equation.
$x = -3 + \sqrt{3}$ $x = -3 - \sqrt{3}$	

Check:

$x = -3 + \sqrt{3}$

$(-3 + \sqrt{3})^2 + 6(-3 + \sqrt{3}) + 6 = 0$

$(9 - 6\sqrt{3} + 3) + (-18 + 6\sqrt{3}) + 6 = 0$

$(9 + 3 - 18 + 6) + (-6\sqrt{3} + 6\sqrt{3}) = 0$

The check for $x = -3 - \sqrt{3}$ is left for the student.

The solution set is $\{-3 \pm \sqrt{3}\}$. ∎

EXAMPLE 9

Solve by completing the square: $x^2 + 8x + 18 = 0$

$x^2 + 8x = -18$	Subtract 18 from both sides.
$x^2 + 8x + 16 = -18 + 16$	Add $\left[\dfrac{1}{2}(8)\right]^2$ to both sides.

$$(x + 4)^2 = -2$$ — Factor the left side.

$$x + 4 = \pm\sqrt{-2}$$

$$x + 4 = \pm\sqrt{2}i$$ — $\sqrt{-2} = \sqrt{2}i$

$$x + 4 = \sqrt{2}i \quad \text{or} \quad x + 4 = -\sqrt{2}i$$ — The check is left for the student.

$$x = -4 + \sqrt{2}i \qquad x = -4 - \sqrt{2}i$$

The solution set is $\{-4 \pm \sqrt{2}i\}$. ∎

If the coefficient of x^2 is not 1, we change the procedure slightly. Simply divide each side of the equation by the lead coefficient, a, and proceed as before.

EXAMPLE 10

Solve by completing the square: $2x^2 + 5x + 2 = 0$

$$x^2 + \frac{5}{2}x + 1 = 0$$ — Divide both sides by 2.

$$x^2 + \frac{5}{2}x = -1$$ — Subtract 1 from both sides.

$$x^2 + \frac{5}{2}x + \frac{25}{16} = -1 + \frac{25}{16}$$ — Complete the square by adding $\left[\frac{1}{2} \cdot \frac{5}{2}\right]^2 = \frac{25}{16}$ to both sides.

$$\left(x + \frac{5}{4}\right)^2 = \frac{9}{16}$$ — Factor the left side and simplify the right.

$$x + \frac{5}{4} = \pm\frac{3}{4}$$ — Solve.

$$x + \frac{5}{4} = \frac{3}{4} \quad \text{or} \quad x + \frac{5}{4} = -\frac{3}{4}$$

$$x = -\frac{1}{2} \qquad x = -2$$

Check: — We will use a calculator to check.

$$2(-2)^2 + 5(-2) + 2 = ?$$ — Check $x = -2$ by replacing x with -2.

    ENTER   **DISPLAY**

| 2 | × | 2 | +/− | x^2 | + | 5 |

| × | 2 | +/− | + | 2 | = | 0

$$2\left(-\frac{1}{2}\right)^2 + 5\left(-\frac{1}{2}\right) + 2 = ?$$

Check $x = -\frac{1}{2}$ by

replacing x with $-\frac{1}{2}$.

	ENTER	DISPLAY

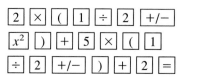

DISPLAY

0

The solution set is $\left\{-2, -\frac{1}{2}\right\}$.   ■

The communications industry provides us with an example of a quadratic equation that can be solved by completing the square.

EXAMPLE 11

The number of different two-station telephone connections (D) that can be made between N different stations is given by the formula

$$2(D - N) = N(N - 3).$$

If there are 45 possible connections on the operator's board, how many stations are in the system?

Formula:

$$2(D - N) = N(N - 3)$$

Substitute:

$$2(45 - N) = N(N - 3)$$

Substitute $D = 45$ in the formula and solve.

Solve:

$$90 - 2N = N^2 - 3N$$

$$N^2 - N - 90 = 0$$

$$N^2 - N = 90$$

The geometric formula for finding the number of diagonals in a polygon is the basis for the formula given here.

$$N^2 - N + \frac{1}{4} = 90 + \frac{1}{4}$$

$$\left(N - \frac{1}{2}\right)^2 = \frac{361}{4}$$

$$N - \frac{1}{2} = \pm\frac{19}{2}$$

$$N - \frac{1}{2} = \frac{19}{2} \quad \text{or} \quad N - \frac{1}{2} = -\frac{19}{2}$$
$$N = 10 \qquad\qquad N = -9$$

We reject -9 because the solution represents a number of stations.

Check:

$$2(45 - 10) = 10(10 - 3)$$
$$2(35) = 10(7)$$

Answer:

There are 10 stations in the system. ■

EXERCISE 5.1

A

Solve:

1. $x^2 = 1$ **2.** $z^2 = 64$ **3.** $a^2 = 0.16$ **4.** $b^2 = -0.09$

5. $a^2 = \dfrac{4}{9}$ **6.** $m^2 = 50$ **7.** $y^2 = 28$ **8.** $b^2 = -48$

9. $(x + 1)^2 = 1$ **10.** $(a + 3)^2 = 4$

Solve by completing the square; check by factoring when possible:

11. $x^2 - 2x - 3 = 0$ **12.** $x^2 + 4x + 3 = 0$ **13.** $x^2 - 10x - 24 = 0$

14. $x^2 + 14x - 32 = 0$ **15.** $x^2 - 2x - 8 = 0$ **16.** $y^2 - 4y - 5 = 0$

17. $z^2 + 2z - 3 = 0$ **18.** $a^2 + 8a - 20 = 0$ **19.** $x^2 - 7x + 12 = 0$

20. $x^2 - x - 12 = 0$

B

Solve:

21. $(y + 5)^2 = -1$ **22.** $(z - 2)^2 = -4$ **23.** $(x + 6)^2 = -5$

24. $(b - 4)^2 = -10$ **25.** $(x + 3)^2 = 2$ **26.** $(x - 1)^2 = 3$

27. $(x - 2)^2 = 48$ **28.** $(x + 3)^2 = 200$ **29.** $(c - 6)^2 = -18$

30. $(b + 4)^2 = -24$

Solve by completing the square.

31. $2x^2 + 24x + 8 = 0$ **32.** $3x^2 + 30x - 12 = 0$

33. $x^2 - x - 1 = 0$ **34.** $x^2 + x - 2 = 0$

35. $x^2 - 3x + 5 = 0$ **36.** $x^2 + 5x + 7 = 0$

37. $-2y^2 + 14y + 4 = 0$ **38.** $-5y^2 - 45y + 15 = 0$

39. $y^2 - \dfrac{3}{2}y - 5 = 0$ **40.** $y^2 - \dfrac{5}{3}y - 2 = 0$

C

41. $4x^2 - 12x + 9 = 25$ **42.** $25y^2 + 40y + 16 = 49$

43. $x^2 + 32x + 256 = -64$ **44.** $y^2 - 30y + 225 = -169$

45. $9x^2 - 12x + 4 = -81$ **46.** $16x^2 + 24x + 9 = -144$

47. $4x^2 - 28x + 49 = 40$ **48.** $9x^2 + 6x + 1 = 80$

49. $4x^2 - 20x + 25 = -24$ **50.** $9x^2 - 6x + 1 = -50$

51. $4x^2 + 4x + 7 = 0$ **52.** $4x^2 - 4x + 9 = 0$

53. $3x^2 + 2x = 6$ **54.** $2x^2 - 3x = 6$

55. $3x^2 - 8x + 2 = 0$ **56.** $5x^2 + 6x + 1 = 0$

57. $6x^2 = -3x - 1$ **58.** $8x^2 = 4x - 1$

59. $5y^2 - 6y - 2 = 0$ **60.** $5y^2 - 8y - 1 = 0$

61. $3x^2 - 4x + 5 = 0$ **62.** $3x^2 + 8x + 3 = 0$

D

63. The illumination (in foot candles) of a light source varies inversely as the square of the distance from the light source. The formula is $Id^2 = k$ (I is the illumination and d is the distance). If the constant of variation is 36, what is the distance from the light source if the illumination is 2 foot candles?

64. In Exercise 63, if the constant of variation is 25,000, what is the distance from the light source if the illumination is 100 foot candles?

Figure for Exercise 63

Solve using the Pythagorean theorem:

65. If $a = 3$ and $b = 4$, then $c = ?$

66. If $a = 6$ and $b = 8$, then $c = ?$

67. If $a = 5$ and $c = 13$, then $b = ?$

68. *If $a = 10$ and $c = 26$, then $b = ?$*

69. What is the rise of a rafter that is 20 feet in length and has a run of 16 feet?

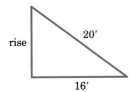

70. What is the run of a rafter that is 25 feet long and has a rise of 15 feet?

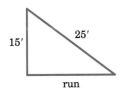

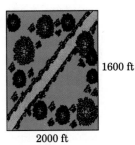

1600 ft

2000 ft

Figure for Exercise 72

71. Find the length of the third side of a garden plot that is in the shape of a right triangle if the two shorter sides are 12 ft and 16 ft in length.

72. Find the length, to the nearest tenth of a foot, of a path that is the diagonal of a rectangular park that is 1600 ft wide and 2000 ft long.

73. The number of two-station telephone connections (D) that can be made between N different stations is given by the formula

$$2(D - N) = N(N - 3).$$

If there are 66 possible connections on the operator's board, how many stations are in the system?

74. Using the formula in Exercise 73, find the number of stations in a system that has a total of 105 possible connections on the operator's board.

75. The number of diagonals D that a polygon of n sides has can be expressed in the formula

$$2D = n(n - 3).$$

If a polygon has 104 diagonals, how many sides does it have?

76. Using the formula in Exercise 75, how many sides does a polygon have if it has 44 diagonals?

77. The sum of two numbers is 21 and their product is 108. What are the numbers?

78. The sum of two numbers is 21 and their product is 90. What are the numbers?

79. The sum of two numbers is 6 and their product is 10. What are the numbers?

80. The sum of two numbers is 10 and their product is 29. What are the numbers?

In Exercises 81–88, assume $a \geq 0$, $b \geq 0$, and $c \geq 0$.

81. Solve $a^2 + b^2 = c^2$ for c.

82. Solve $a^2 + b^2 = c^2$ for a.

83. Solve $a^2 + b^2 = c^2$ for b.

84. Solve $x^2 = 9b$ for x.

85. Solve $x^2 = 16c$ for x.

86. Solve $(x + 2)^2 = b^2$ for x.

87. Solve $(x - 3)^2 = a^2$ for x.

88. Solve $(x - 5)^2 = c^2$ for x.

89. Solve the formula $V = \pi r^2 h$ for r.

90. Solve the formula $A = P(1 + r)^2$ for r.

91. The square of the sum of a number and 6 is 49. What is the number?

92. The square of the difference of a number and 8 is 100. What is the number?

STATE YOUR UNDERSTANDING

93. Explain how to use the completing the square method for solving a quadratic equation.

CHALLENGE EXERCISES

94. The formula for electrical power (P) measured in watts is given by the formula

$$P = I^2 \cdot R$$

where I is measured in amperes and R is measured in ohms. Find I when $R = 4$ ohms and $P = 2500$ watts. (Only positive values for I are acceptable.)

95. Use the formula in Exercise 93 to find I when $R = 3$ ohms and $P = 1200$ watts.

Solve each of the following for x by completing the square.

96. $x^2 + 2ax + b = 0$

97. $x^2 + 4ax + a = 0$

98. $2x^2 + 2bx + c$

99. $ax^2 + bx + c = 0$

MAINTAIN YOUR SKILLS (SECTIONS 4.2, 4.3)

Write using rational exponents:

100. $\sqrt{a^3b^5}$

101. $\sqrt[3]{a^2b}$

102. $3\sqrt{x^2 + 4}$

103. $\sqrt{3x^2 + 4}$

Simplify. Assume all variables represent positive numbers:

104. $\sqrt{108x^3y^5}$

105. $\sqrt{164c^4d^7}$

106. $\sqrt{360x^6y^9}$

107. $\sqrt[3]{360x^6y^9}$

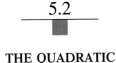

5.2

THE QUADRATIC FORMULA

OBJECTIVE

1. Solve quadratic equations using the quadratic formula.

■ **DEFINITION**

Quadratic Formula

When the literal equation $ax^2 + bx + c = 0$, $a > 0$ and a, b, and $c \in R$ is solved for x by completing the square, the result is a formula. The formula can be used to solve any quadratic equation. The *quadratic formula* is

$$x = \frac{-b \pm \sqrt{b^2 - 4ac}}{2a}.$$

Here is the derivation of the formula.

$ax^2 + bx + c = 0$, $a \neq 0$ Standard form.

$$x^2 + \frac{b}{a}x + \frac{c}{a} = 0$$ Divide both sides by a.

$$x^2 + \frac{b}{a}x = -\frac{c}{a}$$ Add $-\dfrac{c}{a}$ to both sides.

$$x^2 + \frac{b}{a}x + \frac{b^2}{4a^2} = -\frac{c}{a} + \frac{b^2}{4a^2}$$ Complete the square by adding $\left[\dfrac{1}{2} \cdot \dfrac{b}{a}\right]^2 = \dfrac{b^2}{4a^2}$ to both sides.

$$\left[x + \frac{b}{2a}\right]^2 = -\frac{c}{a} + \frac{b^2}{4a^2} \qquad \text{Factor the left side.}$$

$$\left[x + \frac{b}{2a}\right]^2 = -\frac{c}{a} \cdot \frac{4a}{4a} + \frac{b^2}{4a^2} \qquad \text{Add the fractions on the right. Their common denominator is } 4a^2.$$

$$\left[x + \frac{b}{2a}\right]^2 = \frac{b^2 - 4ac}{4a^2}$$

$$x + \frac{b}{2a} = \pm\sqrt{\frac{b^2 - 4ac}{4a^2}} \qquad \text{If } x^2 = k \text{ then } x = \pm\sqrt{k}.$$

$$x = -\frac{b}{2a} \pm \frac{\sqrt{b^2 - 4ac}}{2a}$$

$$x = \frac{-b \pm \sqrt{b^2 - 4ac}}{2a}$$

■ **PROCEDURE**

To solve a quadratic equation using the quadratic formula:

1. Write the equation in standard form: $ax^2 + bx + c = 0$.

2. Identify the values of a, b, and c:
   a is the coefficient of x^2.
   b is the coefficient of x.
   c is the constant term.

3. Substitute these values in the formula and simplify.

$$x = \frac{-b \pm \sqrt{b^2 - 4ac}}{2a}$$

EXAMPLE 1

Solve using the quadratic formula: $x^2 - 12x + 32 = 0$

$a = 1$, $b = -12$, $c = 32$ Identify a, b, and c.

$$x = \frac{-b \pm \sqrt{b^2 - 4ac}}{2a} \qquad \text{Quadratic formula.}$$

$$x = \frac{-(-12) \pm \sqrt{(-12)^2 - 4(1)(32)}}{2(1)} \qquad \text{Substitute.}$$

$$x = \frac{12 \pm \sqrt{144 - 128}}{2} \qquad \text{Simplify.}$$

$$x = \frac{12 \pm \sqrt{16}}{2}$$

$$x = \frac{12 \pm 4}{2}$$

$$x = \frac{12 + 4}{2} \quad \text{or} \quad x = \frac{12 - 4}{2}$$

$$x = 8 \qquad\qquad x = 4$$

■ CAUTION

When simplifying here, do not make the error of "canceling" the 2 and the 12. The expression $\dfrac{12 \pm \sqrt{16}}{2}$ cannot be reduced this way. That is,

$$\frac{12 \pm \sqrt{16}}{2} \neq 6 \pm \sqrt{16}.$$

Check:

$$(8)^2 - 12(8) + 32 = 0$$

$$64 - 96 + 32 = 0$$

$$(4)^2 - 12(4) + 32 = 0$$

$$16 - 48 + 32 = 0$$

Check in the original equation.

The solution set is $\{4, 8\}$.

■

EXAMPLE 2

Solve using the quadratic formula: $x^2 = x + 3$

$x^2 - x - 3 = 0$ Write in standard form.

$a = 1, b = -1, c = -3$ Identify a, b, and c. Recall that $-x = (-1)x$.

$$x = \frac{-b \pm \sqrt{b^2 - 4ac}}{2a}$$

Quadratic formula.

$$x = \frac{-(-1) \pm \sqrt{(-1)^2 - 4(1)(-3)}}{2(1)}$$

Substitute.

$$x = \frac{1 \pm \sqrt{1 + 12}}{2}$$

$$x = \frac{1 \pm \sqrt{13}}{2}$$

The solution set is $\left\{ \dfrac{1 \pm \sqrt{13}}{2} \right\}$.

■

If the lead coefficient is negative, we multiply each side by -1.

EXAMPLE 3

Solve using the quadratic formula: $-3x^2 + 8x + 12 = 0$

$3x^2 - 8x - 12 = 0$	Write in standard form by multiplying both sides by -1. We could also get the roots using $a = -3$, $b = 8$, and $c = 12$.
$a = 3$, $b = -8$, $c = -12$	Identify a, b, and c.
$x = \dfrac{-b \pm \sqrt{b^2 - 4ac}}{2a}$	Quadratic formula.
$x = \dfrac{-(-8) \pm \sqrt{(-8)^2 - 4(3)(-12)}}{2(3)}$	Substitute.
$x = \dfrac{8 \pm \sqrt{64 + 144}}{6}$	Simplify.
$x = \dfrac{8 \pm \sqrt{208}}{6}$	
$x = \dfrac{8 \pm 4\sqrt{13}}{6}$	$\sqrt{208} = \sqrt{16} \cdot \sqrt{13} = 4\sqrt{13}$
$x = \dfrac{2(4 \pm 2\sqrt{13})}{6}$	
$x = \dfrac{4 \pm 2\sqrt{13}}{3}$	Reduce.

The solution set is $\left\{ \dfrac{4 \pm 2\sqrt{13}}{3} \right\}$. ∎

EXAMPLE 4

Solve using the quadratic formula: $6x^2 + 7x = -3$

$6x^2 + 7x + 3 = 0$	Write in standard form.
$a = 6$, $b = 7$, $c = 3$	Identify a, b, and c.
$x = \dfrac{-b \pm \sqrt{b^2 - 4ac}}{2a}$	Quadratic formula.
$x = \dfrac{-(7) \pm \sqrt{(7)^2 - 4(6)(3)}}{2(6)}$	Substitute.

$$x = \frac{-7 \pm \sqrt{49 - 72}}{12} \qquad \text{Simplify.}$$

$$x = \frac{-7 \pm \sqrt{-23}}{12}$$

$$x = \frac{-7 \pm i\sqrt{23}}{12} \qquad \sqrt{-23} = i\sqrt{23}$$

$$x = -\frac{7}{12} \pm \frac{\sqrt{23}}{12}i \qquad \begin{array}{l}\text{Write the complex roots in the form} \\ a + bi \text{ (standard form).}\end{array}$$

The solution set is $\left\{-\dfrac{7}{12} \pm \dfrac{\sqrt{23}}{12}i\right\}$. ∎

The formula can be used in solving quadratic equations in which there is more than one variable.

EXAMPLE 5

Solve for r using the quadratic formula: $2r^2 - sr - s^2 = 0$, $s > 0$

$$2r^2 - sr - s^2 = 0$$

$$a = 2, \ b = -s, \ c = -s^2 \qquad \text{Identify } a, b, \text{ and } c.$$

$$r = \frac{-b \pm \sqrt{b^2 - 4ac}}{2a} \qquad \text{Quadratic formula.}$$

$$r = \frac{-(-s) \pm \sqrt{(-s)^2 - 4(2)(-s^2)}}{2(2)} \qquad \text{Substitute.}$$

$$r = \frac{s \pm \sqrt{s^2 + 8s^2}}{4} \qquad \text{Simplify.}$$

$$r = \frac{s \pm \sqrt{9s^2}}{4}$$

$$r = \frac{s \pm 3s}{4} \qquad \begin{array}{l}\sqrt{9s^2} = 3s, \ s > 0. \text{ If } b^2 - 4ac \text{ is a} \\ \text{perfect square the equation can be} \\ \text{solved by factoring.}\end{array}$$

$$r = s \quad \text{or} \quad r = -\frac{s}{2} \qquad\qquad\qquad\qquad ∎$$

The quadratic formula also works if the coefficients are not real numbers.

EXAMPLE 6

Solve: $x^2 + 3ix + 4 = 0$

$a = 1, b = 3i, c = 4$ The coefficient of x is a pure imaginary number.

$$x = \frac{-b \pm \sqrt{b^2 - 4ac}}{2a}$$ Quadratic formula.

$$x = \frac{-3i \pm \sqrt{(3i)^2 - 4 \cdot 1 \cdot 4}}{2 \cdot 1}$$ Substitute.

$$x = \frac{-3i \pm \sqrt{-9 - 16}}{2}$$ $(3i)^2 = -9$

$$x = \frac{-3i \pm \sqrt{-25}}{2}$$

$$x = \frac{-3i \pm 5i}{2}$$

$x = -4i$ or $x = i$

Check:

$(-4i)^2 + 3i(-4i) + 4 = 0$ Substitute $x = -4i$ in the original equation.

$16i^2 + (-12i^2) + 4 = 0$

$-16 + 12 + 4 = 0$

$(i)^2 + 3i(i) + 4 = 0$ Substitute $x = i$ in the original equation.

$-1 - 3 + 4 = 0$

The solution set is $\{-4i, i\}$.   ■

Compound interest can involve a quadratic equation (as shown in Example 7).

EXAMPLE 7

If a sum of I dollars is invested at r percent compounded annually, at the end of two years it will amount to $A = I(1 + r)^2$. What is the interest rate if $1000 grows to $1440 in two years?

$A = \$1440, I = \$1000, r = ?$ Identify the variables.

Formula:

$A = I(1 + r)^2$

Substitute:

$1440 = 1000(1 + r)^2$

Solve:

$$1440 = 1000(1 + 2r + r^2)$$ Square the binomial.

$$1440 = 1000 + 2000r + 1000r^2$$

$$1000r^2 + 2000r - 440 = 0$$ Write in standard form.

$$25r^2 + 50r - 11 = 0$$ Divide both sides by 40. The formula will still work if this step is omitted, but the numbers will be very large.

$a = 25$, $b = 50$, $c = -11$ Identify a, b, and c.

$$r = \frac{-b \pm \sqrt{b^2 - 4ac}}{2a}$$ Quadratic formula.

$$r = \frac{-(50) \pm \sqrt{(50)^2 - 4(25)(-11)}}{2(25)}$$ Substitute.

$$r = \frac{-50 \pm \sqrt{3600}}{50}$$

$$r = \frac{-50 \pm 60}{50}$$

■ **CAUTION**

When simplifying here, do not make the error of "canceling" the -50 and the 50. The expression
$$\frac{-50 \pm \sqrt{3600}}{50}$$
cannot be reduced this way. That is,
$$\frac{-50 \pm \sqrt{3600}}{50} \neq$$
$$-1 \pm \sqrt{3600}.$$

$$r = \frac{10}{50} \quad \text{or} \quad r = \frac{-110}{50}$$

$$r = 0.2 \qquad r = -2.2$$

We reject -2.2 as an answer since rate of interest is not negative.

Answer:

The interest rate is 0.2 or 20%. ■

Another application involving quadratic equations comes from the sheet metal industry.

EXAMPLE 8

A flat piece of aluminum that is 12 in. wide and 40 ft long is to be formed into a rectangular gutter by bending up an equal amount on each side. What will be the depth of the gutter if the area of the cross section is 18 in²?

$\underline{\hspace{3cm}}$ $\quad$ $x \lfloor \underline{\hspace{2cm}} \rfloor x$ 12 in. $\qquad$ $12 - 2x$	Sketch a diagram to show the cross section with an equal amount bent up on each side.

Formula:

$A = \ell \cdot w$ Formula for the area of a rectangle.

Substitute:

$18 = (12 - 2x)(x)$ $A = 18, \ell = 12 - 2x, w = x$

Solve:

$$18 = 12x - 2x^2$$

$2x^2 - 12x + 18 = 0$ Write in standard form.

$x^2 - 6x + 9 = 0$ Divide both sides by 2.

$a = 1, b = -6, c = 9$ Identify a, b, and c.

$x = \dfrac{-(-6) \pm \sqrt{(-6)^2 - 4(1)(9)}}{2(1)}$ Substitute in the formula.

$x = \dfrac{6 \pm \sqrt{36 - 36}}{2}$

$x = \dfrac{6 \pm 0}{2}$

$x = 3$ A single root, multiplicity two.

Answer:

The gutter is 3 in. deep. ■

Quadratic equations are solved by

1. Factoring (often the quickest method when it is possible to factor).

2. Using square roots (especially when $b = 0$ in the standard form).

3. Completing the square (a method that always works and has other applications in algebra and calculus).

4. The quadratic formula, which is discussed in this section.

EXERCISE 5.2

A

Solve for x using the quadratic formula. Assume all other variables represent positive numbers:

1. $x^2 + 9x + 14 = 0$

2. $x^2 - 9x + 14 = 0$

3. $x^2 + 5x - 24 = 0$

4. $x^2 - 5x - 24 = 0$

5. $2x^2 - 10x - 72 = 0$

6. $4x^2 + 20x - 144 = 0$

7. $x^2 + 10x + 29 = 0$

8. $x^2 - 10x + 34 = 0$

9. $x^2 - 20x = -96$

10. $x^2 + 20x = -96$

11. $3x^2 + 3 = -7x$

12. $3x^2 + 3 = 7x$

13. $4x^2 = -8x - 3$

14. $4x^2 = 8x - 3$

15. $5x^2 - 16x + 3 = 0$

16. $5x^2 + 16x + 3 = 0$

17. $3x^2 + 12x + 6 = 0$

18. $5x^2 + 30x + 20 = 0$

19. $x^2 - 4x + 5 = 0$

20. $x^2 - 10x + 29 = 0$

B

21. $x^2 - 4x - 7 = 0$

22. $x^2 + 5x - 2 = 0$

23. $x^2 + 8x - 2 = 0$

24. $x^2 - 7x - 3 = 0$

25. $3x^2 - 15x + 21 = 0$

26. $5x^2 + 15x + 35 = 0$

27. $x^2 - 9x = -4$

28. $x^2 + 10x = -2$

29. $-2x^2 + 3x + 7 = 0$

30. $-2x^2 - 4x + 9 = 0$

31. $3x^2 + 2 = 4x$

32. $3x^2 + 4 = -5x$

33. $6x^2 - 19ax + 15a^2 = 0$

34. $4x^2 + 25cx - 21c^2 = 0$

35. $x^2 - 3bx + b^2 = 0$

36. $x^2 + 6dx - d^2 = 0$

37. $3x^2 + 7x + 2 = 0$

38. $3x^2 - 7x + 2 = 0$

39. $x^2 - 2ix + 8 = 0$

40. $x^2 + 4ix + 32 = 0$

C

41. $6x^2 = -19x - 10$

42. $6x^2 = 19x - 10$

43. $8x^2 - 18x + 11 = 0$

44. $8x^2 + 18x + 11 = 0$

45. $-6x^2 = 13x - 5$

46. $-6x^2 = -13x - 5$

47. $8x^2 - 10x - 20 = 7x - 2x^2$

48. $12x^2 + 20x - 20 = 2x^2 + 3x$

49. $x^2 + 2x + 3 = 3x^2 + 7x - 3$

50. $x^2 - 3x - 2 = 3x^2 - 8x - 8$

51. $8x^2 + 5x + 10 = 3x^2 - 4x + 4$

52. $6x^2 - 5x + 1 = x^2 + 4x - 5$

53. $ax^2 - 3abx + 2b^2 = 0$

54. $cx^2 + 2cdx + d^2 = 0$

55. $2x^2 = -2d + 3cdx$

56. $5x^2 = -3ab - 4abx$

57. $2x^2 + \sqrt{2}x - 12 = 0$

58. $6x^2 + \sqrt{3}x - 6 = 0$

59. $ix^2 - 7x + 8i = 0$

60. $ix^2 - 7x - 10i = 0$

61. $6x^2 + 7ix + 3 = 0$

62. $8x^2 + 6ix + 5 = 0$

D

63. A piece of metal that is 16 in. wide is to be formed into a gutter of rectangular shape. Both sides are to be bent up an equal amount. What will be the depth of the gutter if the cross-sectional area is 32 in²?

64. What would be the depth of the gutter in Exercise 63 if the piece of metal is 20 in. wide and the cross-sectional area is 50 in²?

65. Elna wants to put her $1000 into an account that will be worth $1210 in two years. What rate of interest must she look for? (*Hint:* Use the formula in Example 7, with $A = 1210$ and $I = 1000$.)

66. Jimmy wants to buy a microcomputer that costs $1690. If he has $1000, what interest rate will he need to find so that he will have $1690 in two years? (Use the formula in Example 7.)

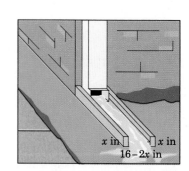

x in x in
16 – 2x in

Figure for Exercise 63

67. A baseball is thrown from the field into the stands. On its way down, it is caught by a fan in a seat that is 32 feet above the field. Its original speed is 48 feet per second. Ignoring friction, the laws of physics give the following formula:

$$h = -32 + 48t - 16t^2,$$

where h represents the height in feet measured from the seat in the stands and t represents the time in seconds. How long does it take the ball from the thrower to reach the person in the seat? ($h = 0$)

Figure for Exercise 67

68. In Exercise 67, how long will it take the ball if the original speed is 64 feet per second and the seat is 48 feet above the field? The formula is $h = -48 + 64t - 16t^2$.

69. The product of four more than a number and seven more than the same number is 108. What is the number?

70. The product of three more than a number and eight more than the same number is 150. What is the number?

STATE YOUR UNDERSTANDING

71. Explain how to use the quadratic formula.

CHALLENGE EXERCISES

72. The maximum distance an arrow will travel when shot into the air at an angle of 45° with the horizontal can be approximated by the formula

$$y = \frac{-32x^2}{v^2} + x$$

where x is the distance traveled in feet, y is the maximum height reached and v is the initial velocity of the arrow in feet per second. If an arrow leaves a bow with a speed of 192 feet per second and reaches a maximum height of 288 feet, how far will the arrow travel?

73. Using the formula of Exercise 72, how far will an arrow travel if the initial velocity is 144 feet per second and it reaches a maximum height of 162 feet?

MAINTAIN YOUR SKILLS (SECTIONS 3.2, 4.1)

Simplify using only positive exponents:

74. $\left(\dfrac{a^{5/8}}{a^{3/4}}\right)^{-1/2}$

75. $\left(\dfrac{b^{3/8}}{b^{-1/4}}\right)^{1/3}$

76. $(x^{-1/3}y^{-1/6})^2(x^2y^3)^{-1/6}$

77. $(p^{-1/2}q^{2/3})^{3/4}(pq^4)^{-1/4}$

78. $(a^{1/2} - 5)(a^{1/2} + 3)$

79. $(b^{2/5} + 15)(b^{2/5} - 15)$

Perform the indicated operations. Reduce if possible:

80. $\dfrac{2x^2 + 5x - 42}{12x^2 - 47x + 40} \cdot \dfrac{3x^2 - 11x + 8}{6x^2 - 19x - 7} \div \dfrac{2x^2 + 11x - 6}{12x^2 - 11x - 5}$

81. $\dfrac{a^2 + ac - 20c^2}{a^2 - ac - 6c^2} \cdot \dfrac{a^2 - 4c^2}{a^2 + 5ac}$

5.3

PROPERTIES OF QUADRATIC EQUATIONS

OBJECTIVES

1. Use the discriminant to describe the roots of a quadratic equation.

2. Given the roots, write a quadratic equation in standard form.

3. Check the solution of a quadratic equation using the sum and product of the roots.

We can use a portion of the quadratic formula to describe the roots of a quadratic equation. The portion we use is called the *discriminant*.

■ **DEFINITION**

Discriminant
$b^2 - 4ac$

In the quadratic formula, the expression under the radical, $b^2 - 4ac$, is called the *discriminant*.

The value of the discriminant yields two pieces of information.

1. The number of solutions.

2. The classification of solutions (real, complex, rational, irrational).

The following table shows some quadratic equations, their discriminants, and solutions.

Equation	Quadratic Formula	Discriminant $b^2 - 4ac$	Solution
$3x^2 - 5x + 8 = 0$	$x = \dfrac{5 \pm \sqrt{25 - 96}}{6}$	$25 - 96 = -71$	$x = \dfrac{5 \pm i\sqrt{71}}{6}$
$2x^2 - 7x - 5 = 0$	$x = \dfrac{7 \pm \sqrt{49 + 40}}{4}$	$49 + 40 = 89$	$x = \dfrac{7 \pm \sqrt{89}}{4}$

Equation	Quadratic Formula	Discriminant $b^2 - 4ac$	Solution
$x^2 - 8x + 16 = 0$	$x = \dfrac{8 \pm \sqrt{64 - 64}}{2}$	$64 - 64 = 0$	$x = 4$
$x^2 - 8x + 15 = 0$	$x = \dfrac{8 \pm \sqrt{64 - 60}}{2}$	$64 - 60 = 4$	$x = 3, 5$

From the table, we make the following observations:

If the Value of the Discriminant Is	Number and Description of Solutions
Negative	Two complex roots
Positive and not a perfect square	Two real roots that are irrational
Positive and a perfect square	Two real roots that are rational
Zero	One distinct real root that is rational. The root has multiplicity two.

EXAMPLE 1

Find the value of the discriminant and describe the roots of $x^2 - 8x - 9 = 0$.

$a = 1$, $b = -8$, and $c = -9$ Identify a, b, and c.

$b^2 - 4ac = (-8)^2 - 4(1)(-9)$ Substitute in the discriminant.

$\qquad = 64 + 36$

$\qquad = 100$ The discriminant is positive and a perfect square.

The discriminant is 100 and the equation has two rational roots. ■

EXAMPLE 2

Find the value of the discriminant and describe the roots of $2x^2 + 3x + 5 = 0$.

$a = 2$, $b = 3$, and $c = 5$ Identify a, b, and c.

$b^2 - 4ac = (3)^2 - 4(2)(5)$ Substitute in the discriminant.

$\qquad = 9 - 40$

$\qquad = -31$ The discriminant is negative.

The discriminant is -31 and the equation has two complex roots. ■

EXAMPLE 3

Find the value of the discriminant and describe the roots of $4x^2 + 12x + 9 = 0$.

$a = 4$, $b = 12$, and $c = 9$

$b^2 - 4ac = (12)^2 - 4(4)(9)$

$\qquad = 144 - 144$

$\qquad = 0$ The discriminant is zero. The root has multiplicity two.

The discriminant is 0 and the equation
has two equal rational roots.   ■

EXAMPLE 4

Find the value of the discriminant and describe the roots of $4x^2 - 9x - 15 = 0$.

$a = 4$, $b = -9$, and $c = -15$

$b^2 - 4ac = (-9)^2 - 4(4)(-15)$

$\qquad = 81 + 240$

$\qquad = 321$ The discriminant is positive and not a square.

The discriminant is 321 and there are
two real irrational roots.   ■

EXAMPLE 5

The number of units (N) produced each day at NCO Inc. is related to the number of employees. The relationship is given by

$N = x^2 - 14x + 59$,

where the number of employees is the larger of the two values of x found by solving the equation. What are the numbers of units for which the equation is valid (has real solutions)? Find the minimum number of employees for which the equation is valid.

$N = x^2 - 14x + 59$ For the equation to be valid (have real roots), the discriminant must be greater than or equal to zero.

$x^2 - 14x + (59 - N) = 0$ Write equation in standard form.

Formula:

$b^2 - 4ac \geq 0$

Substitute:

$(-14)^2 - 4(1)(59 - N) \geq 0$ $a = 1$, $b = -14$, $c = 59 - N$

Solve:

$$196 - 236 + 4N \geq 0$$

$$-40 + 4N \geq 0$$

$$4N \geq 40$$

$$N \geq 10$$

Answer:

The formula is valid for 10 or more units.

We now find the minimum number of employees.

Formula:

$$N = x^2 - 14x + 59$$

The minimum number of employees is found by substituting $N = 10$ in the formula.

Substitute:

$$10 = x^2 - 14x + 59$$

Solve:

$$x^2 - 14x + 49 = 0$$ Write in standard form.

$$(x - 7)^2 = 0$$ Factor the left side.

$$x = 7$$ So $x = 7$ when $N = 10$, and for any other value of $N > 10$, x will be greater than 7.

Answer:

The formula is valid for a minimum of 7 employees.   ■

Given the solution set of a quadratic equation, such as $\{-5, 4\}$, we can write a quadratic equation that has these roots. To do this, we reverse the steps used to solve by factoring.

If the solution set is $\{-5, 4\}$, then

$$x = -5 \quad \text{or} \quad x = 4.$$

When we reverse the steps, we have

$$x = -5 \quad \text{or} \quad x = 4$$

$$x + 5 = 0 \quad \text{or} \quad x - 4 = 0$$

$$(x + 5)(x - 4) = 0$$

$$x^2 + x - 20 = 0$$ Multiply.

The equation $x^2 + x - 20$ has solution set $\{-5, 4\}$.

In general,

$$x = r_1 \quad \text{or} \quad x = r_2$$

$$x - r_1 = 0 \quad \text{or} \quad x - r_2 = 0$$

$$(x - r_1)(x - r_2) = 0$$

$$x^2 - r_1 x - r_2 x + r_1 \cdot r_2 = 0$$

$$x^2 - (r_1 + r_2)x + r_1 \cdot r_2 = 0$$

■ FORMULA

If r_1 and r_2 are any real or complex numbers then r_1 and r_2 are roots of the quadratic equation

$$x^2 - (r_1 + r_2)x + r_1 \cdot r_2 = 0 \quad \text{General formula.}$$

EXAMPLE 6

Write a quadratic equation, in standard form, that has the solution set $\{-3, 8\}$.

$x = -3 \quad$ or $\quad x = 8$	The solution set $\{-3, 8\}$ allows us to write the two equations in the form $x = r_1$ or $x = r_2$.
$x + 3 = 0 \quad$ or $\quad x - 8 = 0$	Rewrite each equation so that the right side is zero.
$(x + 3)(x - 8) = 0$	
$x^2 - 5x - 24 = 0$	Write in standard form.

Alternatively, we can use the equation that shows the sum and product of the roots.

$\begin{aligned} r_1 + r_2 &= -3 + 8 \quad \text{or} \quad r_1 \cdot r_2 = (-3)(8) \\ &= 5 \qquad\qquad\qquad\quad\, = -24 \end{aligned}$	Find the sum and the product of the roots.
$x^2 - (r_1 + r_2)x + r_1 \cdot r_2 = 0$	Quadratic equation in terms of the roots.
$x^2 - (5)x + (-24) = 0$	Substitute in the formula.
$x^2 - 5x - 24 = 0$	

Example 7 shows an equation that has complex roots.

EXAMPLE 7

Write a quadratic equation, in standard form, that has the solution set $\{3 \pm \sqrt{2}i\}$.

$$x^2 - (r_1 + r_2)x + r_1 \cdot r_2 = 0 \qquad \text{Formula.}$$

$$r_1 + r_2 = (3 - \sqrt{2}i) + (3 + \sqrt{2}i)$$

$$= 6$$

$$r_1 \cdot r_2 = (3 - \sqrt{2}i)(3 + \sqrt{2}i)$$

$$= 9 - (2)i^2 \qquad \text{Product of conjugates.}$$

$$= 9 + 2$$

$$= 11$$

$$x^2 - (r_1 + r_2)x + r_1 \cdot r_2 = 0$$

$$x^2 - 6x + 11 = 0 \qquad \text{Substitute in the formula. When the roots contain radicals or are complex, this method is easier.} \quad \blacksquare$$

If we compare the general formula for the quadratic equation in terms of the roots with the standard form of the general quadratic equation, we can see that

$$x^2 - (r_1 + r_2)x + r_1 r_2 = 0 \quad \text{or} \quad x^2 + \frac{b}{a}x + \frac{c}{a} = 0$$

$$\text{Sum:} \qquad r_1 + r_2 = -\frac{b}{a}$$

$$\text{Product:} \qquad r_1 \cdot r_2 = \frac{c}{a}$$

For example, using the equation

$$3x^2 - 7x - 6 = 0 \qquad a^2 + bx + c = 0 \text{ form.}$$

$$x^2 - \frac{7}{3}x - 2 = 0 \qquad x^2 + \frac{b}{a}x + \frac{c}{a} = 0 \text{ form.}$$

Here $\dfrac{b}{a} = -\dfrac{7}{3}$ and $\dfrac{c}{a} = -2$.

Solving by factoring, we find

$$3x^2 - 7x - 6 = 0$$

$$(3x + 2)(x - 3) = 0$$

$$3x + 2 = 0 \qquad \text{or} \quad x - 3 = 0$$

$$x = -\frac{2}{3} \qquad\qquad x = 3$$

Let $r_1 = -\dfrac{2}{3}$ and $r_2 = 3$.

$$r_1 + r_2 = -\frac{2}{3} + 3$$

$$= -\frac{2}{3} + \frac{9}{3}$$

$$= \frac{7}{3} \qquad\qquad \text{From above } \frac{b}{a} = -\frac{7}{3}.$$

Thus, $r_1 + r_2 = \dfrac{7}{3}$, which is $-\left[-\dfrac{7}{3}\right]$.

$$(r_1)(r_2) = -\frac{2}{3}(3)$$

$$= -2 \qquad\qquad \text{Recall that } \frac{c}{a} = -2.$$

Thus $(r_1)(r_2) = -2$.

EXAMPLE 8

Solve $x^2 + 3x + 1 = 0$ and check the solution using the sum and product of the roots.

$a = 1$, $b = 3$, and $c = 1$ \qquad\qquad Solve the quadratic formula.

$$x = \frac{-3 \pm \sqrt{(3)^2 - 4(1)(1)}}{2(1)} \qquad\qquad \text{Substitute in the formula.}$$

$$x = \frac{-3 \pm \sqrt{5}}{2}$$

Check:

$$r_1 + r_2 = -\frac{3}{1} \qquad\qquad r_1 + r_2 = -\frac{b}{a}$$

$$\left[\frac{-3 + \sqrt{5}}{2}\right] + \left[\frac{-3 - \sqrt{5}}{2}\right] = -3$$

$$\frac{-6}{2} = -3$$

$$(r_1)(r_2) = \frac{1}{1} \qquad\qquad (r_1)(r_2) = \frac{c}{a}$$

$$\left[\frac{-3 + \sqrt{5}}{2}\right] \cdot \left[\frac{-3 - \sqrt{5}}{2}\right] = 1$$

$$\frac{9 - 5}{4} = 1$$

Therefore, the solution set is $\left\{\dfrac{-3 \pm \sqrt{5}}{2}\right\}$. ■

Exercise 5.3

A

Find the value of the discriminant and describe the roots of each of the following:

1. $x^2 + 10x + 16 = 0$ **2.** $x^2 - 12x + 20 = 0$ **3.** $x^2 + 3x + 9 = 0$

4. $x^2 + x + 6 = 0$ **5.** $x^2 + 16x + 64 = 0$ **6.** $x^2 - 14x + 49 = 0$

7. $x^2 - 3x = -36$ **8.** $x^2 + 5x = -15$ **9.** $x^2 = -x + 11$

10. $x^2 = 2x + 16$

Write a quadratic equation in standard form that has the following roots:

11. $x = -7$ or $x = -1$ **12.** $x = 8$ or $x = -12$ **13.** $x = 7$ or $x = -5$

14. $x = 8$ or $x = -3$ **15.** $x = i$ or $x = -i$ **16.** $x = 2i$ or $x = -2i$

Solve and check using the sum and the product of the roots:

17. $x^2 + 8x - 16 = 0$ **18.** $a^2 + 10x - 25 = 0$

19. $b^2 + 12b - 36 = 0$ **20.** $y^2 + 18y - 81 = 0$

B

Find the value of the discriminant and describe the roots of each of the following:

21. $3x^2 - 8x + 11 = 0$ **22.** $5x^2 - 6x + 9 = 0$ **23.** $2x^2 + 7x - 5 = 0$

24. $3x^2 - 8x - 4 = 0$ **25.** $7x^2 - 2x = -1$ **26.** $8x^2 = 3x - 2$

27. $5x^2 - 3x + 16 = 0$ **28.** $4x^2 + 7x + 20 = 0$ **29.** $7x^2 - 2x = x + 14$

30. $8x^2 + x = -2x + 2$

Write a quadratic equation in standard form that has the following roots:

31. $x = \dfrac{5}{7}$ or $x = -\dfrac{3}{2}$

32. $x = -\dfrac{3}{8}$ or $x = -\dfrac{5}{4}$

33. $x = 1 + \sqrt{5}$ or $x = 1 - \sqrt{5}$

34. $x = -2 + \sqrt{3}$ or $x = -2 - \sqrt{3}$

35. $x = 2 + i$ or $x = 2 - i$

36. $x = -3 + i$ or $x = -3 - i$

Solve and check using the sum and the product of the roots:

37. $5x^2 - 10x + 6 = 0$

38. $5x^2 - 6x + 9 = 0$

39. $12x^2 - 13x - 35 = 0$

40. $6x^2 - 17x - 28 = 0$

C

Find the value of the discriminant and describe the roots of each of the following:

41. $7x^2 + 4x = 3x - 10$

42. $5x^2 - 8x = 2x - 6$

43. $x^2 + 21 = 3x^2 + 2x + 12$

44. $15x - 8 = 3x^2 + x - 12$

45. $2x^2 - x + 2 = -2x^2 + 11x - 7$

46. $19x - 6 = -4x^2 - x - 31$

47. $15x^2 - 4x + 7 = (2x + 5)^2$

48. $(2x - 7)^2 - 9 = 2x(3x - 6)$

49. $(3x - 1)^2 - (x + 5)^2 = 6$

50. $(x - 5)(x + 7) - (3x + 2)(2x - 5) = 0$

Write an equation in standard form that has the following roots:

51. $x = 6 \pm \sqrt{5}$

52. $x = -3 \pm 3\sqrt{7}$

53. $x + \dfrac{2 \pm \sqrt{5}}{3}$

54. $x = \dfrac{-3 \pm \sqrt{2}}{2}$

55. $x = \dfrac{-5 \pm 2i}{3}$

56. $x = \dfrac{-2 \pm 3i}{5}$

57. $x = -3 \pm 2i\sqrt{5}$

58. $x = -4 \pm 3i\sqrt{2}$

Solve and check using the sum and the product of the roots:

59. $(2x - 7)^2 - 9 = 2x(3x - 6)$

60. $19x - 6 = -4x^2 - x - 31$

D

61. The number of units (N) produced each day at WHAT-CO, Inc. is related to the number of employees. The relationship can be expressed by the quadratic equation $N = x^2 - 20x + 121$, where the number of employees is the larger of the two values of x found by solving the equation. What are the numbers of units for which the equation is valid (has real solutions)? Find the minimum number of employees for which the equation is valid.

62. Solve Exercise 61 if the relationship is given by the equation $N = x^2 - 18x + 96$.

63. For what value of k will $x^2 + kx + 21 = 0$ have two equal real solutions?

64. For what value of k will $2x^2 - 3kx + 2 = 0$ have two equal real solutions?

65. Using the general roots of a quadratic equation, $x = \dfrac{-b \pm \sqrt{b^2 - 4ac}}{2a}$,

show that the product of the roots is $\dfrac{c}{a}$.

66. Given the equation $x^2 + bx - 24 = 0$, for what values of b will the equation have two real solutions?

67. Given the equation $x^2 + bx + 24 = 0$, for what values of b will the equation have one real solution?

68. Given the equation $x^2 + bx - 48 = 0$, for what values of b will the equation have no real solutions?

69. Given the equation $x^2 + bx + 48 = 0$, for what values of b will the equation have one real solution?

70. Given the equation $8x^2 + bx - 3 = 0$ with $\dfrac{1}{4}$ as one solution, find b and the other solution.

71. Given the equation $12x^2 + bx - 10 = 0$ with $-\dfrac{2}{3}$ as one solution, find b and the other solution.

STATE YOUR UNDERSTANDING

72. What information does the discriminant give us with respect to the solutions of the quadratic equation?

CHALLENGE EXERCISES

Find the value of b in each of the following equations.

73. $6bx^2 + 13x + 6b = 0$ if the sum of the roots is $-\dfrac{13}{6}$.

74. $5bx^2 - 11x - 3b = 0$ if the sum of the roots is $\dfrac{11}{10}$.

75. $bx^2 - 19x - (5b + 5) = 0$ if the product of the roots is -6.

76. $3bx^2 + 71x + 2b - 16 = 0$ if the product of the roots is -2.

MAINTAIN YOUR SKILLS (SECTIONS 4.2, 4.3, 4.4)

Write in radical form:

77. $(a^2 - 3)^{3/5}$

78. $(x^2 - x + 1)^{1/4}$

Simplify. Assume that variables represent positive numbers:

79. $3\sqrt{81a^4b^9}$

80. $4\sqrt{8x^4y^3}$

Simplify. Assume that variables represent real numbers:

81. $4\sqrt{256a^2b^4}$

82. $9\sqrt{27x^6y^8}$

Combine. Assume that variables represent positive numbers:

83. $\sqrt{80} - \sqrt{500} + \sqrt{125}$

84. $\sqrt{96} - \sqrt{600} + \sqrt{150}$

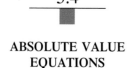

5.4

ABSOLUTE VALUE EQUATIONS

OBJECTIVE

1. Solve equations involving absolute value.

Recall that $|x|$ is the same as the distance between x and 0 on the number line and is never negative.

EXAMPLE 1

Solve: $|x| = 6$

$|x| = 6$ Original equation.

The graph of the solution is the set of all points on the number line that are 6 units from 0.

so $|x| = 6$ is equivalent to $x = \pm 6$.

Check:

$|6| = 6$ and $|-6| = 6$

The solution set is $\{\pm 6\}$.   ∎

More complicated equations can be solved in the same manner.

EXAMPLE 2

Solve: $|2x + 5| = 13$

$|2x + 5| = 13$

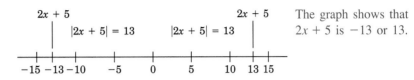

The graph shows that $2x + 5$ is -13 or 13.

$2x + 5 = 13$ or $2x + 5 = -13$

$\qquad 2x = 8 \qquad\qquad\quad 2x = -18$

$\qquad\quad x = 4 \qquad\qquad\quad\quad x = -9$

Check:

$\qquad |2x + 5| = 13$ or $\quad |2x + 5| = 13$

$\quad |2(4) + 5| = 13 \qquad |2(-9) + 5| = 13$

$\qquad\quad |13| = 13 \qquad\qquad\quad |-13| = 13$

The solution set is $\{-9, 4\}$.   ■

We saw in Examples 1 and 2 that the solution set was found by setting the expression in the absolute value bars equal to the coordinates of points that are equidistant from zero. This observation leads to the following method for solving absolute value equations.

■ **PROCEDURE**

To solve

$|ax + b| = c, \ c \geq 0$

1. Write and solve the two equations

   $ax + b = c$ or $ax + b = -c$.

2. Check the answers in the original equation.

EXAMPLE 3

Solve $|3x - 5| = 3$

$3x - 5 = 3$ or $3x - 5 = -3$ If $|ax + b| = c$ then
$ax + b = c$ or $ax + b = -c$.

     $3x = 8$ $3x = 2$ Solve each equation.

     $x = \dfrac{8}{3}$ $x = \dfrac{2}{3}$

Check:

$\left|3 \cdot \dfrac{8}{3} - 5\right| = 3$ $\left|3 \cdot \dfrac{2}{3} - 5\right| = 3$ Substitute $x = \dfrac{8}{3}$ and $\dfrac{2}{3}$ into the original

     $|3| = 3$ $|-3| = 3$ equation.

The solution set is $\left\{\dfrac{2}{3}, \dfrac{8}{3}\right\}$.   ■

If c is negative, the above procedure does not hold and the solution can be found by inspection. Since $|ax + b|$ cannot be negative, there is no solution. That is, the solution set is $\emptyset$.

EXAMPLE 4

Solve: $|2x + 9| = -11$

$|2x + 9| = -11$ Original equation.
The solution set is $\emptyset$.   The absolute value of a number cannot be negative.   ■

To solve an absolute value equation in which the variable appears both inside and outside the absolute value symbols, we use the same procedure. In this case, however, we must keep track of the restrictions on the variable (see Example 5) or check the answers (see Example 7). Checking the answers will reject restricted values and catch any computational errors.

■ **PROCEDURE**

> To solve
>
> $|ax + b| = cx + d, \ cx + d \geq 0$
>
> 1. Write and solve the two related equations
>
> $ax + b = cx + d$ or $ax + b = -(cx + d)$.
>
> 2. Check the answers.

EXAMPLE 5

Solve: $|x - 5| = 2x - 4$

By definition of absolute value, $2x - 4 \geq 0$. So $x \geq 2$ is a restriction on the variable x.

$x - 5 = 2x - 4$ or $x - 5 = -(2x - 4)$

$-1 = x$ $x - 5 = -2x + 4$

$3x = 9$

$x = 3$

If $|ax + b| = cx + d$ then $ax + b = cx + d$ or $ax + b = -(cx + d)$.

Check:

Since $-1 \neq 2$, we need to check only $x = 3$.

$|3 - 5| = 2 \cdot 3 - 4$

$|-2| = 2$

Substitute $x = 3$ in the original equation.

The solution set is $\{3\}$.

EXAMPLE 6

Solve: $|4x - 5| = 2x + 5$

$4x - 5 = 2x + 5$ or $4x - 5 = -(2x + 5)$

$2x = 10$ $4x - 5 = -2x - 5$

$x = 5$ $6x = 0$

$x = 0$

We will need to check since we did not restrict $2x + 5$ to nonnegative values.

Check:

Check by substituting each value for x.

$|4 \cdot 5 - 5| = 2 \cdot 5 + 5$ $|4 \cdot 0 + 5| = 2 \cdot 0 + 5$

$|20 - 5| = 10 + 5$ $|0 - 5| = 0 + 5$

$|15| = 15$ $|-5| = 5$

Substitute $x = 5$ and $x = 0$ into the original equation.

The solution set is $\{0, 5\}$.

EXAMPLE 7

Solve: $|x - 5| = 5 - 3x$

$x - 5 = 5 - 3x$ or $x - 5 = -(5 - 3x)$

$4x = 10$ $x - 5 = -5 + 3x$

$-2x = 0$

$x = 2.5$ or $x = 0$

Solve the two related equations.

Check:

$$|2.5 - 5| = 5 - 3(2.5) \qquad |0 - 5| = 5 - 3(0)$$

$$|-2.5| = 5 - 7.5 \qquad\qquad |-5| = 5$$

$$2.5 = -2.5 \qquad\qquad\qquad 5 = 5$$

Substitute each value for x in the original equation.

The value 2.5 does not check, so we reject it as a solution.

The solution set is $\{0\}$.   ∎

This type of absolute value equation can appear in different forms.

EXAMPLE 8

Solve: $|3x - 5| - 2x = 10$

$$|3x - 5| = 2x + 10$$

Write the equation in the form $|ax + b| = cx + d$.

$$3x - 5 = 2x + 10 \quad \text{or} \quad 3x - 5 = -(2x + 10)$$

$$5x = -5$$

$$x = 15 \qquad\qquad \text{or} \qquad\qquad x = -1$$

The check is left for the student.

The solution set is $\{-1, 15\}$.   ∎

Here is another variation of the absolute value equation.

EXAMPLE 9

Solve: $|2x + 3| = |x - 3|$

$$|2x + 3| = |x - 3|$$

For this equation to be true, the expressions $2x + 3$ and $x - 3$ must be either equal or opposites.

$$2x + 3 = x - 3 \quad \text{or} \quad 2x + 3 = -(x - 3)$$

$$2x + 3 = -x + 3$$

$$x = -6 \qquad \text{or} \qquad x = 0$$

Solve the related equations.

Check:

$$|2(-6) + 3| = |-6 - 3| \quad \text{or} \quad |12 \cdot 0 + 3| = |0 - 3|$$

$$|-12 + 3| = |-9| \qquad\qquad |0 + 3| = |-3|$$

$$|-9| = 9 \qquad\qquad\qquad |3| = 3$$

Substitute each value for x in the original equation.

The solution set is $\{-6, 0\}$.   ∎

EXAMPLE 10

The weekly cost (C) of producing crowbars at the Old Crow Manufacturing Company is given by

$$C = 10{,}000 + 2.5x,$$

where x is the number of crowbars produced. If the company keeps its weekly costs at \$30,000 plus or minus \$3000, what are the maximum and minimum number of crowbars that can be produced?

Simpler word form:

\$30,000 minus weekly cost = ±\$3000   Write a statement describing the problem.

In absolute value form:

$|\$30{,}000 - C| = \$3000$ Change to absolute value form: if $x = \pm c$, then $|x| = c$, $c \geq 0$.

Select a variable:

x represents the number of crowbars.

Translate to algebra:

$|30{,}000 - (10{,}000 + 2.5x)| = 3000$ Substitute $C = 10{,}000 + 2.5x$.

Solve:

$|20{,}000 - 2.5x| = 3000$

$20{,}000 - 2.5x = 3000$ or $20{,}000 - 2.5x = -3000$

$\qquad\qquad -2.5x = -17{,}000$ $\qquad\qquad\qquad -2.5x = -23{,}000$

$\qquad\qquad\qquad x = 6800$ or $\qquad\qquad\qquad x = 9200$

Check:

$\qquad x = 6800$ $\qquad\qquad\qquad\qquad x = 9200$

$C = 10{,}000 + 2.5(6800)$ $\qquad C = 10{,}000 + 2.5(9200)$

$\quad = 10{,}000 + 17{,}000$ $\qquad\qquad = 10{,}000 + 23{,}000$

$\quad = 27{,}000$ $\qquad\qquad\qquad\quad = 33{,}000$

$|30{,}000 - 27{,}000| = 3000$ $\qquad |30{,}000 - 33{,}000| = 3000$

$\qquad\quad |3000| = 3000$ $\qquad\qquad\qquad |-3000| = 3000$

Answer:

The minimum number of crowbars is 6800 and the maximum is 9200.   ∎

EXERCISE 5.4

A

Solve:

1. $|x| = 3$ 2. $|a| = 7$ 3. $|y| = 18$ 4. $|b| = 21$

5. $|x - 4| = 5$ 6. $|x + 3| = 6$ 7. $|x + 2| = -6$ 8. $|y - 3| = -3$

9. $|x + 5| = 0$ 10. $|a - 3| = 0$ 11. $|y + 13| = 20$ 12. $|b - 13| = 20$

13. $|x - 9| = 20$ 14. $|x + 7| = 14$ 15. $|x - 6| = -3$ 16. $|x + 3| = -10$

17. $|x - 5| = 12$ 18. $|x + 16| = 2$ 19. $|x + 15| = 1$ 20. $|x - 9| = 15$

B

21. $|x + 3| - 8 = 5$ 22. $|x - 2| + 6 = 12$

23. $2|x - 11| = 40$ 24. $3|x + 21| = 36$

25. $|3x + 6| = 9$ 26. $|5x - 10| = 25$

27. $|2x - 11| = 27$ 28. $|3x - 18| = 36$

29. $-2|2x - 14| = 40$ 30. $-2|5x + 6| = 14$

31. $|5x - 14| + 2 = 2$ 32. $|7x - 10| - 3 = -3$

33. $6 - |2x + 7| = -4$ 34. $5 - |5x + 6| = -9$

35. $4 + 2|4x - 2| = 16$ 36. $3 - 2|2x + 6| = -5$

37. $5|2x - 6| - 9 = 31$ 38. $8|3x - 7| + 11 = 51$

39. $-3|x - 9| - 7 = -25$ 40. $-4|x + 4| - 7 = -31$

C

41. $|x| = 2x - 3$ 42. $|x| = 3x + 2$

43. $2|x| = 2x - 6$ 44. $3x = 3|x| - 6$

45. $-3|x - 5| = 9 - 3x$ 46. $-2|x + 6| = 4 - 2x$

47. $|x + 5| = 2x - 8$ 48. $|x - 4| = 2x + 1$

49. $|2x + 3| = 6 - x$ 50. $|2x - 7| = 5 - x$

51. $|3x + 5| = 3x + 2$ 52. $|4x - 5| = 9 + 4x$

53. $|7x - 2| - 13 = 3x$ 54. $5 - |9x + 4| = 5x$

55. $3x - |11x + 3| = 14$ 56. $15 = 16x - |5x + 8|$

57. $|4x + 5| = |3x + 9|$ 58. $|8x - 1| = |6x + 29|$

59. $|2x - 7| = |3x - 8|$ 60. $|8x + 2| = |10x - 38|$

D

61. The weekly cost of producing crowbars at the Old Crow Manufacturing Company is given by

$$C = 10{,}000 + 2.5x,$$

where x is the number of crowbars produced. If the company keeps its weekly costs at \$25,000 plus or minus \$3000, what are the maximum and minimum number of crowbars that can be produced?

62. In Exercise 61, find the maximum and minimum number of crowbars if the cost is given by $C = 12{,}000 + 2.5x$ and the weekly costs are held the same.

63. During the past year, the price (p) of a share of Acme stock stayed within \$7 of the price it was on January 1. If its price on January 1 was \$38 what was the highest and lowest possible price for the year? (*Hint:* Solve by using absolute values, $|p - 38| = 7$.)

64. The Jimco Corporation stock did not vary by more than \$11 from its initial price during the year. What was its maximum and minimum price for the year if its initial price was \$110? (Solve by using absolute values.)

STATE YOUR UNDERSTANDING

65. Explain why some times $|x| = x$ and other times $|x| = -x$.

CHALLENGE EXERCISES

Solve:

66. $|x - 3| = x + 4$

67. $|x - 5| = x + 3$

68. $|x + 5| = x + 5$

69. $|x + 5| = -(x + 5)$

MAINTAIN YOUR SKILLS (SECTION 3.3)

Find the LCD of each of the following sets of rational expressions:

70. $\dfrac{2}{x - 3}, \dfrac{5x}{x^2 - 9}, \dfrac{4}{x + 3}$

71. $\dfrac{2}{3}, \dfrac{x + 2}{x - 3}, \dfrac{x - 1}{x + 5}$

72. $\dfrac{5}{x^2 + 2x + 1}, \dfrac{3}{x^2 + 5x + 4}$

73. $\dfrac{7}{x^2 + 6x + 8}, \dfrac{7}{x^2 - 2x - 8}, \dfrac{7}{x^2 - 8x + 16}$

74. $\dfrac{x}{x^2 + 5x + 6}, \dfrac{4}{x^2 + 4x + 3}, \dfrac{3}{x^2 + 3x + 2}$

75. $\dfrac{3}{x^2 - 4}, \dfrac{8}{x^2 - 4x + 4}, \dfrac{1}{x^2 + 4x + 4}$

76. $\dfrac{2x}{x^2 - 9}, \dfrac{4x}{2x^2 - 5x - 3}$

77. $\dfrac{4x}{x^2 + 8x + 15}, \dfrac{5x}{2x^2 + 7x - 15}, \dfrac{x + 2}{2x^2 + 3x - 9}$

5.5

**EQUATIONS
QUADRATIC IN FORM**

OBJECTIVES

1. Solve equations containing rational expressions that can be written as quadratic equations in standard form.

2. Solve equations that are quadratic in form.

It is now possible for us to solve some equations using the same procedures that are used for quadratic equations. The first we consider is an equation with rational expressions that is equivalent to a quadratic equation.

The equation

$$\frac{1}{x+1} - \frac{4}{x} = 1$$

is solved by multiplying each side by the LCD of the fractions. See Chapter 3 for the procedure for solving equations containing rational expressions.

EXAMPLE 1

Solve: $\dfrac{1}{x+1} - \dfrac{4}{x} = 1$

$$\frac{1}{x+1} - \frac{4}{x} = 1$$

$x \neq -1, 0$

Restrict the variable. Set each denominator equal to zero and solve.

$$x(x+1)\left[\frac{1}{x+1} - \frac{4}{x}\right] = x(x+1)1$$

Multiply both sides by the LCD of the fractions.

$$x(x+1)\left[\frac{1}{x+1}\right] - x(x+1)\left[\frac{4}{x}\right] = x(x+1)1$$

$$x - 4(x+1) = x^2 + x$$

$$x - 4x - 4 = x^2 + x$$

Simplify.

$$x^2 + 4x + 4 = 0$$

Write in standard form.

$$(x+2)^2 = 0$$

Factor and solve.

$$x = -2$$

Check:

$$\frac{1}{-2 + 1} - \frac{4}{-2} = 1$$

$$\frac{1}{-1} - (-2) = 1$$

The solution set is $\{-2\}$. ■

The procedure for solving these equations is the same as in Chapter 3. However, extraneous solutions can be introduced when solving fractional equations. We can identify these (and eliminate them) by

1. Checking each apparent solution or

2. Noting the necessary restrictions on the variable.

EXAMPLE 2

Solve: $\dfrac{1}{x - 1} + 3x = \dfrac{x}{x - 1}$

$\dfrac{1}{x - 1} + 3x = \dfrac{x}{x - 1}, \; x \neq 1$	Original equation. Restrict the variable.
$(x - 1)\left[\dfrac{1}{x - 1} + 3x\right] = (x - 1)\left[\dfrac{x}{x - 1}\right]$	Multiply both sides by the LCD, $(x - 1)$.
$1 + (x - 1)3x = x$	Reduce and simplify.
$1 + 3x^2 - 3x = x$	
$3x^2 - 4x + 1 = 0$	
$(3x - 1)(x - 1) = 0$	Write the quadratic equation in standard form and solve by factoring.
$3x - 1 = 0 \quad \text{or} \quad x - 1 = 0$	
$x = \dfrac{1}{3} \qquad\qquad x = 1$	$x = 1$ is the restricted value, so we reject it.
The solution set is $\left\{\dfrac{1}{3}\right\}$.	The check is left for the student. ■

It is possible for such an equation to have no solution.

EXAMPLE 3

Solve: $\dfrac{x}{x-1} + \dfrac{1}{x+1} = -\dfrac{2}{x^2-1}$

$$\dfrac{x}{x-1} + \dfrac{1}{x+1} = -\dfrac{2}{x^2-1}, \; x \neq -1, 1 \qquad \text{Restrict the variable.}$$

$$(x-1)(x+1)\left[\dfrac{x}{x-1} + \dfrac{1}{x+1}\right] = (x-1)(x+1)\left[-\dfrac{2}{x^2-1}\right] \qquad \text{Multiply by the LCD.}$$

$$x(x+1) + 1(x-1) = -2 \qquad \text{Simplify.}$$

$$x^2 + x + x - 1 = -2$$

$$x^2 + 2x + 1 = 0$$

$$(x+1)^2 = 0 \qquad \text{Solve by factoring.}$$

$$x + 1 = 0 \qquad \text{Take the square root of both sides.}$$

$$x = -1 \qquad \text{A restricted value, so it is rejected.}$$

The solution set is ∅.   ■

EXAMPLE 4

The Who Done It bookstore held its annual used book sale. On the first day, $1000 worth of books were sold at a fixed price per book. On the second day, the price per book was reduced by one dollar, and sales amounted to another $1000. If the total number of books sold was 1500, what was the price of a book each day?

	Price/ Book	Sales	No. of Books (Sales/Price)
First day	P	1000	$\dfrac{1000}{P}$
Second day	$P - 1$	1000	$\dfrac{1000}{P-1}$

Make a chart to organize the information.

Simpler word form:

$$\begin{pmatrix}\text{Number of}\\\text{books sold}\\\text{first day}\end{pmatrix} + \begin{pmatrix}\text{Number of}\\\text{books sold}\\\text{second day}\end{pmatrix} = \begin{array}{l}\text{Total number}\\\text{books sold}\end{array}$$

Translate to algebra:

$$\frac{1000}{P} + \frac{1000}{P-1} = 1500$$

Use the information from the chart.

Solve:

$$P(P-1)\left(\frac{1000}{P} + \frac{1000}{P-1}\right) = P(P-1)(1500)$$

Multiply by the LCD.

$$1000(P-1) + 1000P = 1500P(P-1)$$

Simplify.

$$1000P - 1000 + 1000P = 1500P^2 - 1500P$$

$$1500P^2 - 3500P + 1000 = 0$$

$$3P^2 - 7P + 2 = 0$$

Divide both sides by 500.

$$(3P - 1)(P - 2) = 0$$

$$3P - 1 = 0 \quad \text{or} \quad P - 2 = 0$$

$$P = \frac{1}{3} \qquad\qquad P = 2$$

The solution $P = \dfrac{1}{3}$ is rejected since the price on the second day would be negative.

Check:

	Price/ Book	Sales	No. of Books
First day	$2	1000	$\dfrac{1000}{2} = 500$
Second day	$1	1000	$\dfrac{1000}{1} = 1000$

Check in the original information chart.

The total number of books is 1500.

Answer:

The price per book was $2 on the first day and $1 on the second day. ■

Some equations that are not quadratic have the quadratic form

$$au^2 + bu + c = 0, \ a \neq 0$$

where u is an expression involving other variables. For instance,

$$x^4 + 3x^2 - 54 = 0$$

is quadratic in form. This equation can be written

$$(x^2)^2 + 3(x^2) - 54 = 0,$$

which is quadratic in x^2.

$(x^2)^2 + 3(x^2) - 54 = 0$	Let $u = x^2$. Then a quadratic
$u^2 + 3u - 54 = 0$	equation in u is the result.

This equation can be solved by factoring.

EXAMPLE 5

Solve: $x^4 + 3x^2 - 54 = 0$

Let $u = x^2$.

$u^2 + 3u - 54 = 0$	Substitute.
$(u + 9)(u - 6) = 0$	Factor.
$u + 9 = 0$ or $u - 6 = 0$	Solve.
$u = -9 \qquad\qquad u = 6$	
$x^2 = -9 \qquad\quad x^2 = 6$	Replace u by x^2 and solve.
$x = \pm\sqrt{-9} \qquad x = \pm\sqrt{6}$	
$x = \pm 3i$	

Check:

$x^4 + 3x^2 - 54 = 0$	Check in the original equation.
$(\pm 3i)^4 + 3(\pm 3i)^2 - 54 = 0$	Check $x = \pm 3i$. Since x is raised
$81i^4 + 27i^2 - 54 = 0$	only to even powers, we can
	check $3i$ and $-3i$ at the same
$81 - 27 - 54 = 0$	time. $i^4 = 1$ and $i^2 = -1$.
$(\pm\sqrt{6})^4 + 3(\pm\sqrt{6})^2 - 54 = 0$	Check $x = \pm\sqrt{6}$.
$36 + 18 - 54 = 0$	

The solution set is $\{\pm 3i, \pm\sqrt{6}\}$. ∎

In a similar manner, the equation

$$(x^2 + 2x)^2 - 5(x^2 + 2x) + 6 = 0$$

is quadratic in form.

$u^2 - 5u + 6 = 0$	Let $u = (x^2 + 2x)$. Then a quadratic equation in u is the result.

The solution is shown in Example 6.

EXAMPLE 6

Solve: $(x^2 + 2x)^2 - 5(x^2 + 2x) + 6 = 0$

$(x^2 + 2x)^2 - 5(x^2 + 2x) + 6 = 0$

Let $u = x^2 + 2x$.

Assign new variable and write as a quadratic equation.

$u^2 - 5u + 6 = 0$

Substitute.

$(u - 3)(x - 2) = 0$

Factor.

$u - 3 = 0$ or $u - 2 = 0$

$u = 3$ $u = 2$

Solution of the quadratic equation but not the equation involving x. We substitute $u = x^2 + 2x$ and continue.

$x^2 + 2x = 3$ $x^2 + 2x = 2$

$x^2 + 2x - 3 = 0$ $x^2 + 2x - 2 = 0$

Write in standard form.

$(x + 3)(x - 1) = 0$ $x = \dfrac{-2 \pm \sqrt{4 + 8}}{2}$

Solve one by factoring and the other by the quadratic formula.

$x = -3$ or $x = 1$ or $x = \dfrac{-2 \pm 2\sqrt{3}}{2}$

$x = -1 \pm \sqrt{3}$

The check is left for the student.

The solution set is $\{-3, 1, -1 \pm \sqrt{3}\}$.   ■

Our third example contains a square root radical. The equation

$3x - 31\sqrt{x} + 36 = 0$

is quadratic in $\sqrt{x}$.

$3(\sqrt{x})^2 - 31\sqrt{x} + 36 = 0$

$3u^2 - 31u + 36 = 0$

Let $u = \sqrt{x}$. Then a quadratic equation in u is the result.

The solution is shown in Example 7.

EXAMPLE 7

Solve: $3x - 31\sqrt{x} + 36 = 0$

$3x - 31\sqrt{x} + 36 = 0$

Let $u = \sqrt{x}$.

Replace $\sqrt{x}$ by u to write the equation as a quadratic equation in u.

$3u^2 - 31u + 36 = 0$

$u^2 = (\sqrt{x})^2 = x$

$(3u - 4)(u - 9) = 0$

Solve by factoring.

$3u - 4 = 0 \quad$ or $\quad u - 9 = 0$

$u = \dfrac{4}{3} \qquad\qquad u = 9$

Values of u.

$\sqrt{x} = \dfrac{4}{3} \qquad\qquad \sqrt{x} = 9$

Substitute $u = \sqrt{x}$ to solve for x.

$x = \dfrac{16}{9} \qquad\qquad x = 81$

Square both sides.

Check:

Check in the original equation. This check is essential since radical equations sometimes have extraneous roots.

$3\left(\dfrac{16}{9}\right) - 31\sqrt{\dfrac{16}{9}} + 36 = 0$

$\dfrac{16}{3} - 31\left(\dfrac{4}{3}\right) + 36 = 0$

$\dfrac{16 - 124}{3} + 36 = 0$

$-36 + 36 = 0$

$3(81) - 31(\sqrt{81}) + 36 = 0$

$243 - 279 + 36 = 0$

The solution set is $\left\{\dfrac{16}{9}, 81\right\}$.

Example 8 containing negative exponents, is also quadratic in form.

EXAMPLE 8

Solve: $y^{-2} + y^{-1} - 56 = 0$

$y^{-2} + y^{-1} - 56 = 0$

Let $u = y^{-1}$.

Replace y^{-1} by u to write the equation as a quadratic equation in u.

$u^2 + u - 56 = 0$

$u^2 = (y^{-1})^2 = y^{-2}$

$(u + 8)(u - 7) = 0$

Solve by factoring.

$$u + 8 = 0 \quad \text{or} \quad u - 7 = 0$$

$$u = -8 \qquad\qquad u = 7 \qquad\qquad \text{Values of } u.$$

$$y^{-1} = -8 \qquad\qquad y^{-1} = 7$$

$$\frac{1}{y} = -8 \qquad\qquad \frac{1}{y} = 7 \qquad\qquad y^{-1} = \frac{1}{y}$$

$$y = -\frac{1}{8} \qquad\qquad y = \frac{1}{7}$$

The solution set is $\left\{ -\dfrac{1}{8}, \dfrac{1}{7} \right\}$. ■

EXAMPLE 9

The sum of the areas of two squares is 272 square feet. If the number of feet in the side of the larger square is equal to the number of square feet in the area of the smaller square, what are the dimensions of the two squares?

Select variable:

Let x represent the length of a side of the smaller square.

Sketch a diagram to help organize the data. x^2 is the length of the side of the larger square.

Simpler word form:

$$\left(\begin{array}{c} \text{Area of} \\ \text{smaller} \\ \text{square} \end{array} \right) + \left(\begin{array}{c} \text{Area of} \\ \text{larger} \\ \text{square} \end{array} \right) = 272$$

Translate to algebra:

$$x^2 + x^4 = 272 \qquad\qquad \text{The area of a square is the square of the length of a side.}$$

Solve:

$$u + u^2 = 272 \qquad\qquad \text{Let } u = x^2.$$

$$u^2 + u - 272 = 0$$

$$(u + 17)(u - 16) = 0$$

$$u = -17 \quad \text{or} \quad u = 16$$

$$x^2 = 16$$

$$x = \pm 4$$

$u = -17$ is rejected because x^2 is nonnegative.

$x = -4$ is rejected because length is not negative.

Check:

	Side	Area
Smaller square	4	16
Larger square	16	256

Since the total area is 272, the solution 4 checks.

Answer:

The squares are 4 ft and 16 ft on a side.   ■

EXERCISE 5.5

A

Solve

1. $\dfrac{x^2 + 1}{x + 3} = \dfrac{11}{x + 3}$

2. $\dfrac{y^2 + 5}{y - 3} = \dfrac{13}{y - 3}$

3. $\dfrac{1}{x} - \dfrac{1}{3x} = \dfrac{2x}{3}$

4. $\dfrac{4}{y} - \dfrac{1}{4y} = \dfrac{3y}{4}$

5. $\dfrac{2}{x} = x - 1$

6. $\dfrac{3}{y} = y - 2$

7. $\dfrac{3}{x} = x + 2$

8. $\dfrac{7}{y} = y - 6$

9. $\dfrac{12}{y + 6} = y - 5$

10. $\dfrac{9}{x - 3} = x + 5$

11. $\dfrac{-17}{y + 8} = y - 10$

12. $\dfrac{-22}{x + 8} = x - 5$

13. $x^4 - 16 = 0$

14. $x^4 - 25 = 0$

15. $x^4 + 2x^2 - 15 = 0$

16. $x^4 + 2x^2 - 48 = 0$

17. $x^4 - 22x^2 = -120$

18. $x^4 - 13x^2 = -40$

19. $4x^4 - 25x^2 + 36 = 0$

20. $9x^4 - 16x^2 - 25 = 0$

21. $16x^4 - 8x^2 = -1$

22. $81x^4 + 72x^2 = -16$

23. $(x - 5)^2 - 3(x - 5) - 10 = 0$

24. $(x + 3)^2 - 7(x + 3) + 12 = 0$

B

25. $\dfrac{3}{x+2} + \dfrac{9x}{1} = \dfrac{6x}{x+2}$

26. $\dfrac{4}{x-2} + 3x = \dfrac{2x}{x-2}$

27. $\dfrac{2}{x-1} + \dfrac{x}{x-1} = 2x$

28. $\dfrac{x}{x+3} - \dfrac{3}{x+3} = 2x$

29. $\dfrac{1}{x+2} + \dfrac{1}{x-2} = \dfrac{x^2-8}{x^2-4}$

30. $\dfrac{2}{x-3} + \dfrac{2}{x+3} = \dfrac{x^2-21}{x^2-9}$

31. $\dfrac{3}{x-3} - \dfrac{x^2+7}{x^2-x-6} = \dfrac{-1}{x+2}$

32. $\dfrac{5}{x-5} = \dfrac{x^2-52}{x^2-x-20} - \dfrac{4}{x+4}$

33. $x - 3 = \dfrac{1}{x} + \dfrac{x^2+2}{x-2}$

34. $x - 4 - \dfrac{1}{x} = \dfrac{x^2+7}{x-3}$

35. $x + 7 = \dfrac{1}{x} + \dfrac{x^2-8}{x-3}$

36. $x - 5 = \dfrac{1}{x} + \dfrac{x^2-2}{x+1}$

37. $(x^2-1)^2 - 4(x^2-1) + 3 = 0$

38. $(x^2+1)^2 - 2(x^2+1) - 15 = 0$

39. $(x^2+8)^2 + 5(x^2+8) - 36 = 0$

40. $(x^2-11)^2 - 3(x^2-11) - 10 = 0$

41. $8(x^2-2)^2 - 10(x^2-2) - 63 = 0$

42. $4(x^2+2)^2 + 39(x^2+2) - 55 = 0$

43. $(x^2-2x)^2 - 14(x^2-2x) = 15$

44. $(x^2+5x)^2 - 2(x^2+5x) = 24$

45. $(x^2-10x)^2 = -11(x^2-10x) + 210$

46. $(4x^2-12)^2 = -17(4x^2-12) - 72$

C

47. $\dfrac{x-1}{x^2+x-6} + \dfrac{x-2}{x^2+4x+3} = \dfrac{x+1}{x^2-x-2}$

48. $\dfrac{x+1}{x^2+5x+6} + \dfrac{x-4}{x^2+2x-3} = \dfrac{x-3}{x^2+x-2}$

49. $\dfrac{1}{x^2-7x+12} - \dfrac{x}{x^2-2x-8} = \dfrac{2}{x^2-x-6}$

50. $\dfrac{x}{x^2+6x+5} - \dfrac{1}{x^2-x-2} = \dfrac{-3}{x^2+3x-10}$

51. $\dfrac{x}{x^2-9x+14} - \dfrac{2}{x^2-6x+8} = \dfrac{3}{x^2-11x+28}$

52. $\dfrac{x}{5x^2-4x-156} + \dfrac{3}{5x^2+76x+260} = \dfrac{2}{x^2+4x-60}$

53. $\dfrac{1}{2x^2+7x-4} - \dfrac{x-1}{2x^2+5x-3} = \dfrac{5}{x^2+7x+12}$

54. $\dfrac{1}{3x^2+13x-10} + \dfrac{4x}{3x^2+16x-12} = \dfrac{7}{x^2+11x+30}$

55. $\dfrac{3}{x^2+4x-21} - \dfrac{5x}{x^2-8x+15} = \dfrac{8}{x^2+2x-35}$

56. $\dfrac{6}{2x^2 + 13x - 24} + \dfrac{x}{3x^2 + 25x + 8} = \dfrac{3}{6x^2 - 7x - 3}$

57. $x^{-2} + 11x^{-1} + 30 = 0$

58. $x^{-2} + 8x^{-1} + 12 = 0$

59. $x^{-2} - 5x^{-1} - 84 = 0$

60. $x^{-2} - 12x^{-1} - 160 = 0$

61. $x - 4\sqrt{x} - 45 = 0$

62. $x + 7\sqrt{x} - 30 = 0$

63. $2x - 13\sqrt{x} - 99 = 0$

64. $3x - 4\sqrt{x} - 55 = 0$

65. $x^{-4} + 2x^{-2} - 24 = 0$

66. $x^{-4} - 3x^{-2} - 108 = 0$

67. $x^{-4} - 13x^{-2} + 36 = 0$

68. $x^{-4} - 10x^{-2} + 9 = 0$

69. $x^{-4} + 5x^{-2} + 4 = 0$

70. $x^{-4} + 17x^{-2} + 16 = 0$

D

71. The Who Done It bookstore held a used book sale. On the first day, $1000 worth of books were sold at a fixed price per book. On the second day, the price per book was reduced by one dollar, and sales amounted to another $1000. If the total number of books sold was 450, what would have been the price of a book each day?

72. If in Exercise 71 the price during the second day was reduced by $2 and the total number of books sold was 750, what was the price of a book each day?

73. Jane and Sally each earned $600 last month. Sally is paid one dollar an hour more than Jane. If together they worked a total of 220 hours, how much is each paid per hour?

74. Bob and Ray worked a total of 288 hours last month. Ray earns $2 an hour more than Bob. If each is paid $840, how much is each paid per hour?

75. The sum of the areas of two square parking lots is 4160 m². If the length of a side of the larger square is equal to the number of square units in the area of the smaller square, what is the length of a side of each square?

76. The sum of the areas of two circles is 2550π m². If the radius of the larger circle is the square of the radius of the smaller circle, what is the length of the radius of each circle?

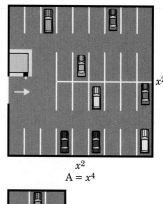

x^2

x^2

$A = x^4$

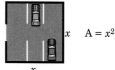

x $A = x^2$

x

Figures for Exercise 75

STATE YOUR UNDERSTANDING

77. What does it mean for an equation to be quadratic in form?

CHALLENGE EXERCISES

78. A plane completes a round trip of 960 miles in 7 hours while flying at a constant rate of speed. The first half of the trip was against a headwind of

20 miles per hour and the return trip was with a tail wind of 20 miles per hour. What was the speed of the plane in still air?

79. An airplane completes a round trip of 1080 miles in 5 hours while flying at a constant rate of speed. The first half of the trip was against a head wind of 60 miles per hour and the return trip was with a tail wind of 30 miles per hour. What was the speed of the plane in still air?

80. Three pipes together can fill a large tank in 80 minutes. One pipe can fill the tank in one-half the time of a second pipe, and in one less hour than the third pipe. How long would it take each to fill the tank alone?

MAINTAIN YOUR SKILLS (SECTION 4.5)

Multiply and simplify:

81. $(6\sqrt{14})(3\sqrt{21})$

82. $(14\sqrt{6})(21\sqrt{3})$

83. $(\sqrt{2} - \sqrt{3})^2$

84. $(\sqrt{2} - \sqrt{3})(\sqrt{3} + \sqrt{4})$

85. $\sqrt{6}(\sqrt{30} - \sqrt{54})$

86. $\sqrt{15}(\sqrt{30} - \sqrt{54})$

87. $(4s\sqrt{20s})(3s\sqrt{30s})$

88. $(2t\sqrt{20t})(4t\sqrt{32t})$

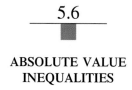

5.6

ABSOLUTE VALUE INEQUALITIES

OBJECTIVE

1. Solve absolute value inequalities.

To solve absolute value inequalities, we recall the definition of absolute value.

If $x \geq 0$, $|x| = x$.

If $x < 0$, $|x| = -x$.

■ DEFINITION

Compound Inequality

> A *compound inequality* is an inequality in which two inequalities are joined by the words "and" or "or."

For example, $x > 5$ or $x < 2$ is a compound inequality. The word *or* is the conjunction joining the two inequalities. For this compound inequality to be true, only one part of the compound inequality need be true for the entire statement to be true. When the conjunction "or" is used, the solution set of the compound inequality is the union of the solution sets of each part of the inequality.

In set builder notation, the solution set is

$\{x \mid x > 5\} \cup \{x \mid x < 2\}$.

In interval notation, the solution set is

$(-\infty, 2) \cup (5, +\infty)$.

Another compound inequality is

$x < 5$ and $x > 2$,

where the conjunction is the word *and*. For this compound inequality to be true, both of the inequality statements must be true. A shorter way of writing this compound inequality is $2 < x < 5$. This is read, "two is less than x and x is less than 5." When the conjunction "and" is used, the solution set of the compound inequality is the intersection of the solution sets of each part of the inequality.

Solution set $= \{x \mid x < 5\} \cap \{x \mid x > 2\}$ Set builder notation.

$= (-\infty, 5) \cap (2, +\infty) = (2, 5)$ Interval notation.

To solve an absolute value inequality we use the definition of absolute value to write an equivalent compound inequality.

EXAMPLE 1

Solve: $|x| > 5$

Case I	Case II

If $x \geq 0$, then $|x| = x$ or   If $x < 0$, then $|x| = -x$

$|x| > 5$ $|x| > 5$

$-x > 5$

$x > 5$ or $x < -5$ Multiply both sides by -1.

The graph of the solution set is

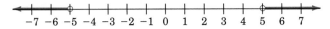

The graph of the solution consists of two intervals on the number line.

The numbers 5 and -5 are not included since neither is part of the solution.

Written in set builder notation, the solution set is

$\{x \mid x > 5\} \cup \{x \mid x < -5\}$,

and written in interval notation, the solution set is

$(-\infty, -5) \cup (5, +\infty)$. ∎

Example 1 suggests the rule.

■ **RULE**

> An inequality of the form
>
> $|x| > a, \, a \geq 0$
>
> is equivalent to $x > a$ or $x < -a$.

The solution can be shown in one of three ways:

1. Set builder notation: $\{x \mid x > a\} \cup \{x \mid x < -a\}$

2. Interval notation: $(-\infty, -a) \cup (a, +\infty)$

3. Graph:

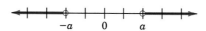

Because the absolute value of every number is nonnegative, the absolute value of any real number is always greater than any negative number. Therefore, the solution set of an absolute value inequality such as

$|x| > -2$

is immediately seen to be the set of all real numbers.

This can be shown in three ways:

1. Set builder notation: $\{x \mid x \in R\}$

2. Interval notation: $(-\infty, +\infty)$

3. Graph:

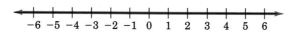

■ **RULE**

> The solution to the inequality $|x| > a, \, a < 0$, is the set of all real numbers since the absolute value of any real number is nonnegative.
>
> The solution to the inequality $|x| < a, \, a < 0$, is the empty set.

We use similar equivalent forms to solve other inequalities.

EXAMPLE 2

Solve: $|x + 4| > 6$

Case I	Case II

If $x + 4 \geq 0$, then $|x + 4| = x + 4$ or If $x + 4 < 0$, then $|x + 4| = -(x + 4)$

$$|x + 4| > 6 \qquad\qquad\qquad |x + 4| > 6$$
$$x + 4 > 6 \qquad\qquad\qquad -(x + 4) > 6$$
$$x > 6 - 4 \qquad\qquad\qquad x + 4 < -6$$
$$x > 2 \qquad\qquad \text{or} \qquad\qquad x < -10$$

The solution can be shown in one of three ways:

1. Set builder notation: $\{x \mid x > 2\} \cup \{x \mid x < -10\}$

2. Interval notation: $(-\infty, -10) \cup (2, +\infty)$

3. Graph:

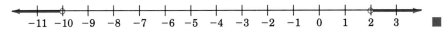

Example 2 suggests the rule.

■ **RULE**

> An inequality of the form
>
> $|ax + b| > c, \; c \geq 0$
>
> is equivalent to the compound inequality
>
> $ax + b > c$ or $ax + b < -c.$

The graph of the solution consists of two intervals on the number line.

If the absolute value inequality is "less than," we proceed in a similar manner.

EXAMPLE 3

Solve: $|x| < 5$

Case I	Case II

If $x \geq 0$, then $|x| = x$ or   If $x < 0$, then $|x| = -x$

$|x| < 5$ $|x| < 5$

                        $-x < 5$

$x < 5$ or $x > -5$

Graphing each of these on a number line, we have

Case I

$x < 5$ and $x \geq 0$

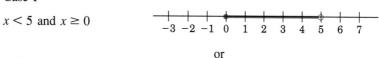

or

Case II

$x > -5$ and $x < 0$

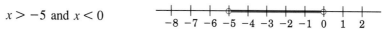

The word *or* indicates that both parts are in the solution set. Thus, we have

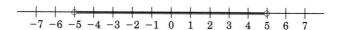

The solution written in set builder notation is

$\{x \mid -5 < x < 5\}$.

The solution written in interval notation is

$(-5, 5)$.   ■

Example 3 suggests the rule.

■ **RULE**

> An inequality of the form
>
> $|x| < a,\ a \geq 0$
>
> is equivalent to the statement
>
> $x < a$ and $x > -a$
>
> and can be written
>
> $-a < x < a$.

The graph of the solution set of such an absolute inequality is one interval on the number line.

We use a similar procedure to solve an inequality when there is an expression within the absolute value symbol.

EXAMPLE 4

Solve: $|x + 6| < 4$

Case I	Case II

If $x + 6 \geq 0$ then $x \geq -6$ or   If $x + 6 < 0$ then $x < -6$

$x + 6 < 4$ $-(x + 6) < 4$

                                       $x + 6 > -4$

$x < -2$ or $x > -10$

We graph each case.

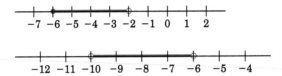

The graph of the solution is a single interval on the number line.

The solution written in set builder notation is

$\{x \mid -10 < x < -2\}$.

The solution written in interval notation is

$(-10, -2)$.

■ **RULE**

$|ax + b| < c$, $c \geq 0$, is equivalent to the compound inequality

$ax + b < c$ and $ax + b > -c$,

which can be written

$-c < ax + b < c$.

The graph of the solution consists of a single interval on the number line.

Summarizing the solution of absolute value inequalities, we have:

Type of Inequality	Equivalent to	Intervals
$\|x\| > a, a \geq 0$	$x > a$ or $x < -a$	2
$\|ax + b\| > c, c \geq 0$	$ax + b > c$ or $ax + b < -c$	2
$\|x\| < a, a \geq 0$	$x < a$ and $x > -a$ or $-a < x < a$	1
$\|ax + b\| < c, c \geq 0$	$ax + b < c$ and $ax + b > -c$ or $-c < ax + b < c$	1
Special Cases		
$\|x\| > a, a < 0$	$\{x \mid x \in R\}$	1
$\|ax + b\| > c, c < 0$	$\{x \mid x \in R\}$	1
$\|x\| < a, a < 0$	$\emptyset$	0
$\|ax + b\| < c, c < 0$	$\emptyset$	0

We now have two procedures for solving absolute value inequalities.

1. By cases.

2. By equivalent compound inequalities (rules above).

Example 5 illustrates the second method.

EXAMPLE 5

Solve: $|2x + 5| > x + 4$

$|2x + 5| > x + 4$ Original inequality. Since the relation is "greater than," $x + 4$ can be either positive or negative.

$2x + 5 > x + 4$ or $2x + 5 < -(x + 4)$ We write the equivalent compound statement making no restrictions on x.

$$2x + 5 < -x - 4$$

$$3x < -9$$

$x > -1$ or $x < -3$

In set builder notation:

$\{x \mid x > -1 \text{ or } x < -3\}$.

In interval notation:

$(-\infty, -3) \cup (-1, +\infty)$.   ■

EXAMPLE 6

Solve: $|2x + 5| < x + 4$

$\|2x + 5\| < x + 4$	Recall that the absolute value of a number is never negative. In this case, $x + 4 \geq 0$.

$x + 4 \geq 0$

$\qquad x \geq -4$ Any solution must satisfy this condition.

$2x + 5 < x + 4$ and $2x + 5 > -(x + 4)$ Write the equivalent compound statement.

$\qquad x < -1 \qquad\qquad 2x + 5 > -x - 4$

$\qquad\qquad\qquad\qquad\qquad 3x > -9$

$\qquad\qquad\qquad\qquad\qquad\quad x > -3$

$x < -1$ and $x > -3$ and $x \geq -4$ There are three conditions that must be satisfied.

The solution set is $\{x \mid -3 < x < -1\}$ or $(-3, -1)$.   ■

EXAMPLE 7

Solve and write the solution in interval notation:

$|2x - 5| < x - 10$

$2x - 5 < x - 10$ and $2x - 5 > -(x - 10)$ Write the absolute value inequality in its equivalent form. Recall that absolute value must not be negative, so that means that $x - 10 \geq 0$ or $x \geq 10$.

$\qquad x < -5 \qquad$ and $\quad 2x - 5 > -x + 10$ Solve the inequalities under that condition.

$\qquad\qquad\qquad\qquad\qquad\quad 3x > 15$

$\qquad x < -5 \qquad$ and $\qquad x > 5$

This says that the numbers in the solution must be less than -5 and at the same time be larger than 5. This is an impossibility, so the solution set is $\emptyset$.   ■

EXAMPLE 8

The storage of food in a home freezer is best accomplished if the temperature (T) in the freezer is maintained at 8°F with a variance of no more than 8° lower or higher. Find the range of permissible temperatures using absolute value inequalities.

$$|T - 8°| \leq 8$$

$$|T - 8| \leq 8$$

$$T - 8 \leq 8 \quad \text{and} \quad T - 8 \geq -8$$

$$T \leq 16 \qquad\qquad T \geq 0$$

If the temperature cannot vary more than 8° lower or higher, then the difference in the absolute value of the temperature in the freezer and 8° must be less than or equal to 8°. Write the absolute value inequality in its equivalent form and solve the inequalities. Since "and" joins the two inequalities, both must be satisfied.

For T to be less than or equal to 16 and at the same time be greater than or equal to 0, we have $0 \leq T \leq 16$. Thus, the inside temperature should be between 0° and 16°. ■

Exercise 5.6

A

Solve and graph the solution:

1. $|x| < 4$ **2.** $|x| < 7$ **3.** $|x| \geq 2$ **4.** $|x| \geq 6$

Solve and write the solution in set builder notation:

5. $|2x| \leq 10$ **6.** $|3x| \leq 15$ **7.** $|5x| \geq 15$

8. $|6x| \geq 12$ **9.** $|2x| - 5 < 3$ **10.** $|4x| - 6 > 6$

Solve and write the solution in interval notation:

11. $|2x - 5| \geq 3$ **12.** $|3x - 6| \geq 9$ **13.** $|x + 2| - 3 < 4$

14. $|x + 6| - 5 < 2$ **15.** $|x - 8| < -5$ **16.** $|x - 8| < -6$

17. $|x + 3| > -3$ **18.** $|x - 4| > -8$ **19.** $|x + 5| - 3 < -6$

20. $|x - 9| - 8 < -12$

B

21. $|2x + 3| \leq 5$ **22.** $|2x - 3| \leq 3$ **23.** $|3x - 6| \geq 6$

24. $|3x + 9| \geq 3$ **25.** $|3x + 5| < 2$ **26.** $|4x - 7| < 1$

27. $|5x - 4| > 1$ **28.** $|4x + 2| > 2$ **29.** $|2x + 4| > x - 5$

30. $|3x - 1| > 2x - 9$ **31.** $|x + 2| - 3 < x$ $\left(-\frac{5}{2}, \infty\right)$ **32.** $|2x - 3| + 3 \geq x$

Solve and write the solution in set builder notation:

33. $|4x - 7| \geq 3x - 12$ **34.** $|3x + 6| \geq 2x + 6$

35. $|2x - 9| < 5x + 3$ **36.** $|4x + 3| < 6x + 5$

37. $|4 - x| \geq 4$ **38.** $|9 - x| \geq 9$

39. $|8 - 2x| < 12$ **40.** $|15 - 5x| < 20$

C

41. $|12 - 4x| - 8 \leq 12$ **42.** $|15 - 3x| - 9 \leq 12$

43. $|x + 4| > 5x$ **44.** $|x - 5| > 6x$

45. $|2x + 1| > -x$ **46.** $|3x - 2| > -x$

47. $|3x - 2| < -x$ **48.** $|4x + 6| < -x$

49. $\left|\dfrac{2x - 3}{3}\right| < \dfrac{1}{3}$ **50.** $\left|\dfrac{4x + 2}{3}\right| \leq 1$

51. $\left|\dfrac{5a - 4}{2}\right| \geq \dfrac{1}{4}$ **52.** $\left|\dfrac{2a + 7}{3}\right| > \dfrac{1}{3}$

53. $\left|\dfrac{1}{2}x - 1\right| > 2$ **54.** $\left|\dfrac{1}{3}x + 4\right| < 1$

55. $\left|\dfrac{2}{3}x - 2\right| \leq 3$ **56.** $\left|\dfrac{3}{4}x + 1\right| \geq 4$

57. $|4x + 5| > x$ **58.** $|2x + 1| < x + 12$

59. $|3x + 4| < -x$ **60.** $|2x + 7| < -x - 5$

D

In Exercises 61–72, write each as an absolute value inequality and solve:

61. The distance from a number and 4 is less than 12.

62. The distance from a number and 8 is less than 8.

63. The distance from a number and 16 is more than 9.

64. The distance from a number and 12 is more than 11.

65. The distance from a number and 18 is at least 8.

66. The distance from a number and 11 is at least 9.

67. The distance from a number and 12 is no more than 7.

68. The distance from a number and 4 is no more than 2.

69. The storage of food in a home freezer is best accomplished if the temperature in the freezer is maintained at 6°F plus or minus 8°. Find the range of permissible temperatures using absolute value inequalities.

70. The temperature of a food dryer for which the dryer is most efficient is within 8 degrees of 150°F. Find the range of permissible temperatures using absolute value inequalities.

71. Erica's bathroom scale is off by no more than 2 pounds. If the scale measures her weight 115 pounds, what is the range for her true weight?

72. The scales at the produce section are off by no more than $\frac{1}{8}$ pound. If a bag of apples weighed on the scales indicates the weight is 4.5 pounds, what is the range for the true weight of the apples?

Figure for Exercise 72

73. Write $-5 \le x + 7 \le 5$ as an absolute value inequality.

74. Write $-6 < 3x + 1 < 6$ as an absolute value inequality.

75. Write $4x + 2 \ge 8$ or $4x + 2 \le -8$ as an absolute value inequality.

76. Write $8x + 6 \ge 4$ or $8x + 6 \le -4$ as an absolute value inequality.

77. Write $-8 \le 2x \le 2$ as an absolute value inequality.

78. Write $-7 < x < 17$ as an absolute value inequality.

STATE YOUR UNDERSTANDING

79. What is a compound inequality?

CHALLENGE EXERCISES

Solve:

80. $|x - a| < b, \ b > 0$

81. $|x + a| < b, \ b > 0$

82. $|x - a| > b, \ b > 0$

83. $|x + a| > b, \ b > 0$

84. $|x + a| \ge 0$

85. $|x - a| \ge 0$

86. $|x + a| \le 0$

87. $|x - a| \le 0$

MAINTAIN YOUR SKILLS (SECTIONS 2.8, 2.10)

Factor completely:

88. $10xy - 15wy - 2x + 3w$

89. $56a - 32b + 21ab - 12b^2$

90. $144t^2 - 120t + 25$

91. $121m^2 + 132mn + 36n^2$

92. $g^3 - 9h^2g$

93. $y^8 - 81$

94. Mary Jo can mow the lawn at the courthouse with an electric mower in three hours less time than Gil can do it with a push mower. If they both work together, they can mow the lawn in two hours. How long does it take each to mow it alone?

95. Mary Ann mows one-third of the lawn at the library in four hours. Then Maria joins her and they finish, together, in two more hours. How long would it have taken Maria to mow it alone?

5.7

QUADRATIC AND RATIONAL INEQUALITIES

OBJECTIVES

1. Solve quadratic inequalities.

2. Solve rational inequalities.

■ **DEFINITION**

Standard Form of a Quadratic Inequality

A *quadratic inequality in standard form* is an inequality that can be written in the form $ax^2 + bx + c < 0$, $a \neq 0$. The symbols $>$, $\leq$, and $\geq$ can also be used.

$x^2 - 7x + 22 \geq 0$ and $6x^2 - 4x + 13 < 0$ are quadratic inequalities in standard form.

$x^2 + 22x \geq 10$ and $13 < 4x - 6x^2$ are quadratic inequalities not in standard form.

■ **DEFINITION**

Critical Numbers

Critical numbers of a quadratic inequality written in standard form are values of the variable for which the left side is equal to zero.

The critical numbers of $(x + 7)(x - 4) > 0$ are -7 and 4.

To find the solution of a quadratic inequality, we write the inequality in standard form and use the real number line and critical numbers.

■ PROCEDURE

To solve a quadratic inequality:

1. Write in standard form.
2. Factor the left side.
3. Locate the critical numbers on the number line by setting the left side equal to 0 and solving.
4. Draw vertical lines through the critical numbers.
5. For each factor, write positive signs above the number line where the factor has positive values and negative signs where the factor has negative values.
6. Determine which combination of positive and negative factors will satisfy the inequality.
7. Draw the graph of the solution and write the solution. If equality is included ($\leq$ or $\geq$), the critical numbers that make the inequality equal to 0 are part of the solution.
8. If there are no critical numbers, the solution set is empty or the set of all real numbers. Use any value to check in the original inequality.

EXAMPLE 1

Solve, graph, and write the solution in interval notation: $(x + 5)(x - 6) < 0$

$(x + 5)(x - 6) < 0$ The left side is factored.

$x + 5 = 0$ or $x - 6 = 0$ Locate the critical numbers; that is, find where the left member is zero.

$x = -5$ $x = 6$

The critical numbers are -5 and 6.

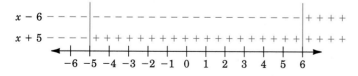

Draw vertical lines through the critical numbers. Determine whether each factor is positive or negative in the intervals.

The solution is the interval $(-5, 6)$.   The solution set contains those numbers for which one factor is positive and the other factor is negative.

The graph of the solution is

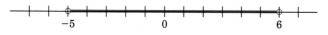

For those quadratic inequalities that are not written in factored form, we do the following.

EXAMPLE 2

Solve and write the solution in interval notation: $x^2 - x - 12 \geq 0$

$(x - 4)(x + 3) \geq 0$	Factor the left side.
The critical numbers are 4 and -3.	Locate the critical numbers.

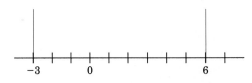

Draw vertical lines through the critical numbers. Determine whether each factor is positive or negative in those intervals.

Since the product is greater than or equal to zero, the intervals where both factors are positive or both are negative will make the product greater than zero. The critical numbers make the product zero.

The solution in interval notation is $(-\infty, -3] \cup [4, +\infty)$.   ■

If the inequality is not in standard form, then the first step of the procedure must be performed.

EXAMPLE 3

Solve, graph, and write the solution in interval notation: $2x^2 - 3x \leq x^2 + 18$

$2x^2 - 3x \leq x^2 + 18$	Write the inequality in standard form by collecting all terms on the left side with 0 on the right side.
$x^2 - 3x - 18 \leq 0$	
$(x - 6)(x + 3) \leq 0$	Find the critical numbers by factoring the left side.
The critical numbers are 6 and -3.	Both of these numbers make the left side equal to 0. Since the equality is part of the problem, each critical number is also a solution of the inequality.

On a number line, draw vertical lines above the critical numbers.

$x + 3$ $- -$ $+ + + + + + + + + + + + + + + +$ $+ + +$

$x - 6$ $- -$ $- - - - - - - - - - - - - - -$ $+ + +$

-3 0 6

Write positive or negative signs above the number line for each critical number showing where each factor is positive and where it is negative. The dots at -3 and 6 show that these numbers are part of the solution.

-3 0 6

The left side of the inequality in standard form is negative when one factor is positive and the other is negative. This occurs when x is between -3 and 6.

The solution in interval notation is $[-3, 6]$.

There are many applications that require the solution of quadratic inequalities.

EXAMPLE 4

The Fletco Corp. has found that it can make a profit if it can keep the cost (C) of materials under \$3000 in any given day. $(C < 3000)$. If the cost of producing n items (n must be positive) is given by $C = n^2 - 100n + 5400$, what is the number of items that can be produced in a day at a profit?

$C < 3000$ and $n > 0$

Cost restriction in order to make a profit with the provision n is positive.

$n^2 - 100n + 5400 < 3000$

Substitute $C = n^2 - 100n + 5400$.

$n^2 - 100n + 2400 < 0$

Write the inequality in standard form.

$(n - 40)(n - 60) < 0$

Find the critical numbers by factoring the left side.

The critical numbers are 40 and 60.

0 10 20 30 40 50 60 70 80

Draw vertical lines above the critical numbers. Here a vertical line is also drawn above zero since the number of items must be positive, not zero or negative.

$n - 60$ $- - - - - -$ $- - -$ $+ + +$

$n - 40$ $- - - - - -$ $+ + +$ $+ + +$

0 10 20 30 40 50 60 70 80

Write positive and negative signs above the number line for each factor.

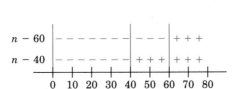

The number of items that can be produced at a profit must be between 40 and 60.

The inequality in standard form is true when one factor is positive and one is negative.

Graphically, the solution is

0 10 20 30 40 50 60 70 80

■

■ DEFINITION

Standard Form of a Rational Inequality

A *rational inequality in standard form* is an inequality whose left side is a single rational expression and whose right side is 0.

$\dfrac{x - 18}{x + 12} \geq 0$ and $\dfrac{(x + 1)(x - 3)}{x - 5} < 0$ are rational inequalities written in standard form.

$\dfrac{x}{x + 1} \leq \dfrac{18}{x + 1}$ and $\dfrac{x^2}{x - 10} + \dfrac{2x - 3}{x - 10} > 0$ are rational inequalities not written in standard form.

■ DEFINITION

Inequality Critical Numbers

Critical numbers of a rational inequality written in standard form are values of the variable for which the numerator or denominator is equal to 0.

The critical numbers for $\dfrac{x - 18}{x + 12} \leq 0$ are 18 and -12 since replacing x with -12 makes the denominator zero, and replacing x with 18 makes the numerator zero.

To find the solution of a rational inequality we write the inequality in standard form and use the real number line and critical numbers.

■ PROCEDURE

To solve a rational inequality:

1. Write in standard form.

2. If necessary, factor the left side.

3. Locate the critical numbers on the number line by setting the numerator and the denominator each equal to zero and solving.

4. Draw vertical lines through the critical numbers.

5. For each factor, write positive signs above the number line where the factor has positive values and negative signs where the factor has negative values.

6. Determine which combination of positive and negative factors will satisfy the inequality.

7. Draw the graph of the solution and write the solution. If equality is included ($\leq$ or $\geq$), the critical numbers that make the numerator zero are part of the solution.

8. If there are no critical numbers, the solution set is the empty set or the set of all real numbers. Use any value to check in the original inequality.

EXAMPLE 5

Solve, graph, and write the solution in set builder notation: $\dfrac{x + 1}{x - 3} \geq 0$

$\dfrac{x + 1}{x - 3} \geq 0$ The original inequality is written in standard form.

$x + 1 = 0 \qquad x - 3 = 0$ Find the critical numbers.

$\qquad x = -1 \qquad\quad x = 3$

The critical numbers are -1 and 3.

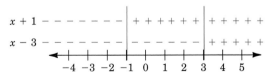

The graph of the solution is

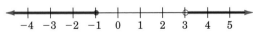

Write positive or negative signs above the number line for each factor. For the inequality to be greater than or equal to 0, both factors must be positive or both negative. Since equality is included, the critical number that makes the numerator 0 is a solution.

In set builder notation, the solution set is

$\{x \,|\, x \leq -1 \text{ or } x > 3\}$. ∎

EXAMPLE 6

Solve, graph, and write the solution in interval notation: $\dfrac{1}{x-1} < 2$

$\dfrac{1}{x-1} < 2$ is equivalent to

First write the inequality in standard form.

$\dfrac{1}{x-1} - 2 < 0$

$\dfrac{1}{x-1} - \dfrac{2(x-1)}{x-1} < 0$

$\dfrac{1-2x+2}{x-1} < 0$

Simplify the left side.

$\dfrac{3-2x}{x-1} < 0$

Standard form.

The critical numbers are 1 and $\dfrac{3}{2}$.

The denominator is 0 when $x = 1$, and the numerator is 0 when $x = \dfrac{3}{2}$.

On a number line, draw vertical lines above the critical numbers.

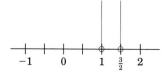

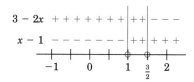

Write positive or negative signs above the number line between the critical numbers showing where each factor is positive and where it is negative.

The inequality $\dfrac{3-2x}{x-1} < 0$ is true (that is, the left side is negative) when x is less than 1 or when x is greater than $\dfrac{3}{2}$.

Determine which combination of positive and negative factors will satisfy the inequality.

So, the solution in interval notation is $(-\infty, 1) \cup \left(\dfrac{3}{2}, +\infty\right)$. The graph of the solution set is

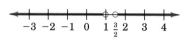

EXAMPLE 7

Solve and write the solution in interval notation: $\dfrac{1}{x^2 + 1} > 0$

$x^2 + 1 = 0$ Since $x^2 + 1 \neq 0$ there are no critical numbers.

Since there are no critical numbers, the solution set is either $\emptyset$ or the set of all real numbers. We select a test point. Let $x = -1$.

$$\dfrac{1}{(-1)^2 + 1} > 0$$

$$\dfrac{1}{1 + 1} > 0$$

$\dfrac{1}{2} > 0$ Since the statement is true, the solution is the set of all real numbers.

The solution set is $\{x \mid x \in R\}$.   ■

EXERCISE 5.7

A

Solve. Write the solution in interval notation:

1. $(x + 2)(x - 5) > 0$

2. $(x + 5)(x + 1) > 0$

3. $(x - 4)(x + 3) \leq 0$

4. $(x - 1)(x - 5) \leq 0$

5. $\dfrac{(5x + 3)}{(2x - 1)} \geq 0$

6. $\dfrac{(7x + 5)}{(3x - 2)} \geq 0$

7. $\dfrac{5(x - 2)}{2x + 1} < 0$

8. $\dfrac{2(2x - 3)}{x + 1} < 0$

9. $-5x(x - 3)(x + 2) \geq 0$

10. $-3x(x + 1)(x - 4) \geq 0$

11. $(x + 4)(x - 2)(x + 1) \leq 0$

12. $(x - 5)(x + 5)(x + 3) \geq 0$

13. $\dfrac{2x(2x + 3)}{(x - 5)} \geq 0$

14. $\dfrac{3x(4x + 3)}{(x + 4)} \leq 0$

15. $(3x + 2)(2x - 1)(4x - 5) \leq 0$

16. $(2x + 5)(3x - 4)(2x + 1) \geq 0$

17. $x(x + 2)(x - 3) < 0$

18. $x(x + 2)(x - 3) > 0$

19. $x(x + 1)(x - 2)(x - 3) \leq 0$

20. $x(x + 2)(x + 3)(x + 4) \geq 0$

B

21. $3x^2 + 5x + 4 > x^2 - 2x - 2$

22. $2x^2 - 8x + 5 > x - 5$

23. $\dfrac{5}{x} < 2$

24. $\dfrac{3}{x} < 3$

25. $\dfrac{2}{x} + 5 \geq 3$

26. $\dfrac{3}{x} - 4 \leq 2$

27. $4x^2 - 5x - 4 \leq 5x + 2$

28. $3x^2 - 15x + 15 \geq 2x - 5$

29. $x^2 + 5x + 10 < -5x - 15$

30. $x^2 - 10x + 10 < -4x + 1$

31. $6x^2 + 10x + 20 \geq 2x^2 - 10x - 5$

32. $6x^2 - 14x + 16 \geq -3x^2 + 10x$

33. $8x^2 + 20x + 15 \leq -8x^2 - 20x - 10$

34. $25x^2 - 10x + 2 \leq 10x - 2$

35. $2x^3 \leq 3x^2 + 2x$

36. $2x^3 > 4x^2 + 30x$

37. $\dfrac{-4}{x + 5} > 4$

38. $\dfrac{6}{x - 2} > 3$

39. $\dfrac{1}{x - 3} < -1$

40. $\dfrac{1}{x + 4} < -2$

C

41. $\dfrac{1}{x^2 + 1} < -1$

42. $\dfrac{3}{x^4 + 1} > -5$

43. $\dfrac{x}{x + 1} < 1$

44. $\dfrac{b}{b + 2} > 1$

45. $\dfrac{1}{x - 1} > \dfrac{3}{x - 2}$

46. $\dfrac{2}{x + 1} < \dfrac{3}{x + 2}$

47. $\dfrac{1}{x + 2} \leq \dfrac{2}{x - 3}$

48. $\dfrac{5}{x - 1} \leq \dfrac{1}{x + 1}$

49. $\dfrac{1}{x - 5} \geq \dfrac{1}{x + 6}$

50. $\dfrac{17}{x + 3} \geq \dfrac{-8}{x - 2}$

51. $\dfrac{x + 6}{(x + 2)(x + 3)} < 1$

52. $\dfrac{x - 16}{(x - 8)(x + 2)} < 1$

53. $\dfrac{x - 5}{(x - 1)(x + 4)} \leq 2$

54. $\dfrac{x + 11}{(x + 3)(x + 2)} \geq 1$

55. $\dfrac{5}{x^2 - 1} \geq -1$

56. $\dfrac{-13}{x^2 + 4} \leq -1$

57. $\dfrac{5}{x^2} + 4 \geq 0$

58. $\dfrac{-12}{-4 - x^2} \leq 0$

59. $\dfrac{x^2 + 4x + 4}{x^2 - 2x + 1} \geq 0$

60. $\dfrac{x^2 + 4x + 4}{x^2 + 6x + 9} < 0$

D

61. In an electronics factory, the cost (C) of producing x units of an item is given by $C = 8000 + 5x^2 - 350x$. If the cost is to be held at \$12,000 or below, what is the range of units that can be produced?

62. In Exercise 61, for what range of units will the cost exceed \$19,875?

63. In the equation $2x^2 - bx + 8 = 0$, for what values of b are there real solutions?

64. In the equation $-3x^2 + bx - 4 = 0$, for what values of b are there no real solutions?

STATE YOUR UNDERSTANDING

65. How are critical numbers determined?

CHALLENGE EXERCISES

Solve:

66. $x^3 - 8 \geq 0$ 67. $x^3 - 8 \leq 0$

68. $(x - 1)(x + 1)^2 > 0$ 69. $(x - 1)(x + 1)^2 < 0$

MAINTAIN YOUR SKILLS (SECTIONS 4.1, 4.6)

Simplify:

70. $25^{3/2}$ 71. $64^{1/3}$ 72. $4^{-5/2}$

73. $(-27)^{2/3}$ 74. $(81)^{1/4}$

Divide. Leave the answer in simplified form for radicals:

75. $\dfrac{2}{\sqrt{5}}$ 76. $\dfrac{1}{\sqrt[3]{4}}$ 77. $\dfrac{5}{1 + \sqrt{2}}$

CHAPTER 5
SUMMARY

Solution by Square Roots	If $x^2 = k$, then $x = \pm\sqrt{k}$.	(p. 325)
Completing the Square	1. Write in the form $x^2 + \dfrac{b}{a}x = k$.	(p. 329)
	2. Add $\left[\dfrac{1}{2} \cdot \dfrac{b}{a}\right]^2$ to both sides.	

3. Solve using square roots.

Quadratic Formula	$$x = \frac{-b \pm \sqrt{b^2 - 4ac}}{2a}$$	(pp. 337, 338)
Discriminant of the Quadratic Formula	$b^2 - 4ac$	(p. 348)

Properties of the Discriminant

If $b^2 - 4ac = 0$, the equation has one repeated solution. (p. 349)

If $b^2 - 4ac > 0$, the equation has two real solutions.

If $b^2 - 4ac < 0$, the equation has two complex solutions.

Sum and Product of the Roots of a Quadratic Equation

If r_1 and r_2 are the two roots of the quadratic equation (pp. 352, 353)

$$ax^2 + bx + c = 0,$$

then the following statements are true:

$$r_1 + r_2 = -\frac{b}{a} \quad \text{and} \quad (r_1)(r_2) = \frac{c}{a}$$

Absolute Value Equations and Their Equivalent

The equation $|ax + b| = cx + d$ is equivalent to $ax + b = cx + d$ or (p. 360) $ax + b = -(cx + d)$ if $cx + d \geq 0$.

To Solve Equations Quadratic in Form

1. Determine by inspection the expression for which the equation is (pp. 369, quadratic. 370)

2. Substitute a variable for that expression.

3. Solve the resulting quadratic equation.

4. Substitute the expression back into the problem.

5. Solve the resulting equations.

Absolute Value Inequalities and Their Equivalent

$|ax + b| \geq c$ is equivalent to $ax + b \geq c$ or $ax + b \leq -c$. (pp. 380, 382)

$|ax + b| \leq c$ is equivalent to $-c \leq ax + b \leq c$.

Critical Numbers of a Quadratic Inequality in Standard Form

Those that make the left side of the inequality equal to zero. (p. 388)

Critical Numbers of a Rational Inequality in Standard Form

Those numbers that make the numerator or the denominator of the rational expression equal to zero. (p. 392)

To Solve Quadratic Inequalities

1. Write in standard form. (p. 389)

2. Factor the left side.

3. Determine the critical numbers.

4. Make a number line drawing that shows where the factors are positive or negative.

5. From this drawing, determine from the original inequality those intervals of the real number line that are a part of the solution set.

To Solve an Inequality Containing Rational Expressions

1. Write in standard form. (pp. 392, 393)

2. Determine the critical numbers.

3. Make a number line drawing showing where the different factors are positive or negative.

4. From this drawing and the original inequality, determine the intervals of the real number line that make up the solution set of the inequality.

CHAPTER 5
REVIEW EXERCISES

SECTION 5.1   Objective 1

Solve using the square root method.

1. $x^2 = 256$ **2.** $a^2 = 361$ **3.** $(b + 4)^2 = 36$

4. $(y - 2)^2 = 64$ **5.** $x^2 = -81$ **6.** $y^2 = -100$

7. $a^2 - 8a + 16 = 48$ **8.** $b^2 + 16b + 64 = 80$ **9.** $x^2 - 12x + 36 = -9$

10. $x^2 + 6x + 9 = -16$

SECTION 5.1   Objective 2

Solve by completing the square.

11. $x^2 + 4x - 12 = 0$ **12.** $y^2 - 2y - 8 = 0$ **13.** $x^2 - 5x = 36$

14. $x^2 - 2x = 48$ **15.** $x^2 + 2x - 2 = 0$ **16.** $a^2 - 4a - 4 = 0$

17. $3x^2 + 4x - 15 = 0$ **18.** $4x^2 + 8x - 21 = 0$ **19.** $x^2 + 2x + 8 = 0$

20. $x^2 + 6x + 16 = 0$

SECTION 5.2   Objective 1

Solve using the quadratic formula.

21. $5x^2 + 13x - 6 = 0$ **22.** $4x^2 + 21x + 5 = 0$

23. $35x^2 + 24x - 35 = 0$ **24.** $32x^2 + 20x - 3 = 0$

25. $3x^2 - 6x - 5 = 0$ **26.** $3a^2 + 8a - 4 = 0$

27. $x^2 + 2x + 4 = 0$ **28.** $x^2 - 2x + 4 = 0$

29. $x^2 + 7ix - 12 = 0$ **30.** $x^2 - 8ix - 15 = 0$

SECTION 5.3   Objective 1

Use the discriminant to describe the roots of the following quadratic equations.

31. $x^2 - 8x + 12 = 0$

32. $x^2 + 9x + 20 = 0$

33. $3x^2 + 7x + 4 = 0$

34. $5x^2 - 4x - 12 = 0$

35. $9x^2 - 12x = -4$

36. $25a^2 + 16 = 40a$

37. $5x^2 + 8x - 8 = 0$

38. $4y^2 - 7y - 12 = 0$

39. $3x^2 + 2x + 5 = 0$

40. $4x^2 - 4x + 7 = 0$

SECTION 5.3   Objective 2

Write a quadratic equation in standard form that has the given solution set.

41. $\{1, 2\}$

42. $\{4, 5\}$

43. $\{-4, -2\}$

44. $\left\{\dfrac{1}{2}, \dfrac{2}{3}\right\}$

45. $\left\{-\dfrac{3}{4}, -\dfrac{4}{5}\right\}$

46. $\{1 \pm 2\sqrt{3}\}$

47. $\{2 \pm 3\sqrt{2}\}$

48. $\{-2 \pm 2i\}$

49. $\{-1 \pm 3i\}$

50. $\{2i, -5i\}$

SECTION 5.3   Objective 3

Use the sum and product of the roots to check the given solution set of each of the following quadratic equations.

51. $x^2 - 8x + 12 = 0$; $\{2, 6\}$

52. $x^2 + 9x + 20 = 0$; $\{-5, -4\}$

53. $3x^2 + 7x + 4 = 0$; $\left\{-1, -\dfrac{3}{4}\right\}$

54. $5x^2 - 4x - 12 = 0$; $\left\{-2, \dfrac{6}{5}\right\}$

55. $9x^2 - 12x = -4$; $\left\{\dfrac{2}{3}\right\}$

56. $25a^2 + 16 = 40a$; $\left\{\dfrac{4}{5}\right\}$

57. $5x^2 - 8x - 8 = 0$; $\left\{\dfrac{4 \pm 2\sqrt{14}}{5}\right\}$

58. $4y^2 - 7y - 12 = 0$; $\left\{\dfrac{7 \pm 4\sqrt{15}}{8}\right\}$

59. $3x^2 + 2x + 5 = 0$; $\left\{-\dfrac{1}{3} \pm \dfrac{\sqrt{14}}{3}i\right\}$

60. $4x^2 - 4x + 7 = 0$; $\left\{\dfrac{1}{2} \pm \dfrac{\sqrt{6}}{2}i\right\}$

SECTION 5.4   Objective 1

Solve the following absolute value equations.

61. $|x| = 8$

62. $|y| = 12$

63. $|x + 1| = 3$

64. $|a - 2| = 4$

65. $|2x + 3| = -1$

66. $|4x - 3| = -4$

67. $|2x + 1| = 3x - 6$

68. $|3x + 2| = x + 6$

69. $|5x + 2| - 3x = 6$

70. $|4x + 7| - 2 = 5x$

SECTION 5.5   Objective 1

Solve the following equations that contain rational expressions.

71. $\dfrac{1}{x+1} + \dfrac{1}{x} = \dfrac{5}{6}$

72. $\dfrac{1}{x-1} - \dfrac{1}{x} = \dfrac{1}{20}$

73. $\dfrac{1}{x+1} + \dfrac{2}{x+2} = \dfrac{5}{3x}$

74. $\dfrac{1}{x+2} + \dfrac{3}{x+3} = \dfrac{17}{3x+14}$

75. $\dfrac{2}{x+5} + \dfrac{1}{x-5} = \dfrac{1}{x-7}$

76. $\dfrac{x}{x+2} - \dfrac{4}{x-2} = \dfrac{8}{x^2-4}$

77. $\dfrac{3}{x-5} + \dfrac{x}{x+3} = \dfrac{12}{x^2-2x-15}$

78. $\dfrac{x}{2x+3} - \dfrac{1}{2x-5} = \dfrac{6}{4x^2-4x-15}$

79. $\dfrac{a}{a-4} - \dfrac{4}{a+4} = \dfrac{32}{a^2-16}$

80. $\dfrac{b}{b+3} - \dfrac{3}{b-3} = \dfrac{18}{b^2-9}$

SECTION 5.5   Objective 2

Solve the following equations that are quadratic in form.

81. $x^4 - 13x^2 + 36 = 0$

82. $x^4 - 10x^2 + 9 = 0$

83. $(x+1)^2 + 5(x+1) + 6 = 0$

84. $(x-5)^2 - 8(x-5) + 12 = 0$

85. $(x^2+3x)^2 - 8(x^2+3x) - 20 = 0$

86. $(x^2-x)^2 - 14(x^2-x) + 24 = 0$

87. $a^{-2} + 5a^{-1} - 6 = 0$

88. $x^{-4} - 2x^{-2} + 1 = 0$

89. $3x - 11\sqrt{x} + 6 = 0$

90. $6x + 7\sqrt{x} - 3 = 0$

SECTION 5.6   Objective 1

Solve the following absolute value inequalities. Write the solution in interval notation.

91. $|x| < 50$

92. $|x| > 50$

93. $|x+3| \geq 5$

94. $|x-5| \leq 8$

95. $|x-7| + 4 < 3$

96. $|x+2| + 2 < 3$

97. $|2x+7| > x+2$

98. $|3x-1| < x+5$

99. $|4x+2| < x-1$

100. $|3x+7| > x-5$

SECTION 5.7   Objective 1

Solve the following quadratic inequalities. Write the solution in set builder notation.

101. $(x+8)(x-2) < 0$

102. $(x-5)(x+3) > 0$

103. $x^2 + 12x + 36 \geq 0$

104. $x^2 - 8x + 16 < 0$

105. $x^2 - x - 30 \le 0$

106. $x + 2x - 48 \ge 0$

107. $3x^2 - 5x < 2x^2 - 6$

108. $4x^2 + 9x > 3x^2 - 20$

109. $12x^2 - 7x \le 12$

110. $6x^2 - 11x - 10 \ge 0$

SECTION 5.7   Objective 2

Solve the following rational inequalities. Write the solution in interval notation.

111. $\dfrac{x-3}{x-2} \ge 0$

112. $\dfrac{x+3}{x+2} \le 0$

113. $\dfrac{1}{x-3} < 2$

114. $\dfrac{1}{x+1} > 2$

115. $\dfrac{x^2-4}{x+3} \le 0$

116. $\dfrac{x^2-9}{x-2} \le 0$

117. $\dfrac{1}{x} < \dfrac{1}{x+2}$

118. $\dfrac{1}{a} > \dfrac{1}{a-1}$

119. $\dfrac{x+1}{x+2} < 1$

120. $\dfrac{x-3}{x-4} > 1$

CHAPTER 5
TRUE–FALSE CONCEPT REVIEW

Check your understanding of the language of algebra. Tell whether each of the following statements is true (always true) or false (not always true).

1. Every quadratic equation has two roots.

2. If $(x + 2)(x - 9) = 12$, then $x + 2 = 3$ and $x - 9 = 4$.

3. Every quadratic equation can be solved by completing the square.

4. The Pythagorean theorem applies to any triangle that has at least one right angle.

5. The easiest way to solve any quadratic equation is by using the quadratic formula.

6. Every quadratic equation has degree four.

7. In the quadratic formula $x = \dfrac{-b \pm \sqrt{b^2 - 4ac}}{2a}$, the expression $b^2 - 4ac$ is called the discriminant.

8. If a quadratic equation has two equal (double) roots, then the discriminant is equal to zero.

9. Every equation involving absolute value has two roots.

10. Every radical equation has either one or two roots.

11. If we multiply both sides of a fractional equation by the LCM and get a quadratic equation, the (original) fraction equation has two roots.

12. The quadratic formula can be used to solve some equations that are not quadratic equations.

13. The inequality, $|x| < a$, is equivalent to the compound inequality $-a < x < a$.

14. The same expression can be subtracted from both sides of any inequality and the result will be equivalent to the original inequality.

15. The same expression can be multiplied times both sides of any inequality and the result will be equivalent to the original inequality.

16. The critical numbers of an inequality are never solutions of the inequality.

17. It is possible to write a quadratic inequality that has no solution.

CHAPTER 5
TEST

1. Solve: $|x + 8| = 5$

2. Write a quadratic equation in standard form that has the roots $x = 15$ or $x = -11$.

3. Solve and graph the solution: $\dfrac{1}{x} \le \dfrac{1}{3}$

4. Solve, using the quadratic formula:
$x^2 + 18x + 70 = 0$

5. Solve and graph the solution: $|x + 1| \ge 1$

6. Solve: $3x^2 = 36$

7. Solve by completing the square:
$x^2 - 14x + 42 = 0$

8. Write a quadratic equation in standard form that has the roots $x = -\dfrac{1}{2}$ or $x = \dfrac{5}{2}$.

9. Solve: $4x^2 + 13x - 12 = 0$

10. Solve: $|2x + 3| = x - 12$

11. Solve, using the quadratic formula:
$12x^2 + 28x - 5 = 0$

12. Solve: $\dfrac{1}{5} - \dfrac{1}{x} = \dfrac{1}{x - 24}$

13. Solve: $(x + 5)^2 = 289$

14. Solve: $x + 3\sqrt{8x} - 14 = 0$

15. Solve: $\dfrac{6}{x - 2} - \dfrac{1}{x - 3} = \dfrac{5}{x - 1}$

16. Solve and graph the solution: $2x^2 + 3x - 9 < 0$

17. Find the value of the discriminant and describe the roots of $x^2 + 28x + 196 = 0$.

18. The product of two consecutive odd integers is 323. What are the two integers?

19. A technical artist can finish a certain job in 36 minutes. With the help of her apprentice, she can do the job in 20 minutes. How long would it take the apprentice alone to do the job?

CHAPTER

6

LINEAR EQUATIONS AND GRAPHING

Betting at horse and dog tracks continues to be a popular recreational activity for many people. Some race tracks take a small commission from all winnings to help offset total operational costs. In Exercise 74, Section 6.1, the track deducts $0.50 from every payout. The total amount received by the ticket holder is $2.50 for every dollar waged minus the commission. The payout for any number of dollars waged can be expressed as a linear equation. In Exercise 74 you are asked to write and graph this equation. *(Melchior DiGiacomo/The Image Bank)*

In Chapter 6 linear equations in two variables and their graphs are presented. There follows a discussion on the properties of graphs including slope, intercepts, distance formula, and useful forms of the linear equation. Identification of parallel and perpendicular lines is made using the slope is studied. Graphing inequalities flows from the prior work on equations. The chapter ends with the applications of variation, presenting direct, inverse, and joint variations. The ability to draw the graph a linear equation is used in chapters 7, 8 and 9. Graphing is a necessary tool for the further study of mathematics.

6.1

LINEAR EQUATIONS IN TWO VARIABLES

OBJECTIVES

1. Find ordered pairs that are solutions of a linear equation.

2. Draw the graph of a linear equation.

There are many ways in which two numbers can be related or paired. For instance, a number and its opposite

Number	Opposite Number
x	$-x$ (x is related to $-x$)

the diameter of a circle and its circumference

Diameter	Circumference
d	πd (d is related to circumference)

and the day of the year to day of the month

Day of the Year	Month and Day
61	3/2 (the 61st day of a nonleap year is March 2)

In this chapter, we study the relationships that can be expressed by a linear equation in two variables.

405

■ **DEFINITION**

Linear Equation
Standard Form

A linear equation in two variables is an equation that can be written in the form $Ax + By = C$ (standard form), A, B, and C real numbers, A and B not both zero.

In a linear equation, the exponent of each variable is one. The graph of a linear equation is a straight line. The equations, $x + y = 10$, $y = 4x - 7$, and $x = 2y - 3$ are linear equations.

A solution of a linear equation is an ordered pair of real numbers that satisfies the equation.

■ **DEFINITION**

Ordered Pair (x, y)

An ordered pair of real numbers (x, y) is a pair of real numbers where the x value is written first. The variable x is called the independent variable, and the variable y is called the dependent variable.

The ordered pair $(2, 3)$ is a solution of $2x - y = 1$ since the equation is true when x is replaced by 2 and y is replaced by 3.

$$2x - y = 1$$
$$2(2) - 3 = 1 \qquad \text{Replace } x \text{ with 2 and } y \text{ with 3.}$$
$$1 = 1 \qquad \text{True.}$$

A solution pair can be found by assigning a value to either variable and solving for the other.

EXAMPLE 1

Find the solution of $y = 4x - 7$ when x is 4.

$$y = 4x - 7 \qquad \text{Original equation.}$$
$$y = 4(4) - 7 \qquad \text{Replace } x \text{ with 4.}$$
$$= 9$$

The solution is $(4, 9)$.   The solution is written as an ordered pair. ■

■ **CAUTION**

Always write the value of x (the independent variable) first, (x, y).

EXAMPLE 2

Find the solutions of $4x - 3y = 15$ when $x = 3$, $x = -2$, $y = 0$, and $y = 0.5$.

$$4(3) - 3y = 15$$

Substitute each value of x or y in the equation and solve for the other variable.

$$12 - 3y = 15$$

$$y = -1$$

$$4(-2) - 3y = 15$$

$$-8 - 3y = 15$$

$$y = -\frac{23}{3}$$

$$4x - 3(0) = 15$$

$$4x = 15$$

$$x = \frac{15}{4}$$

$$4x - 3(0.5) = 15$$

$$4x - 1.5 = 15$$

$$x = 4.125$$

The solutions are $(3, -1)$, $\left(-2, -\frac{23}{3}\right)$, $\left(\frac{15}{4}, 0\right)$, and $(4.125, 0.5)$.

Write the solutions as ordered pairs.

◼ **PROCEDURE**

To find one solution of the equation $Ax + By = C$, $A \neq 0$, $B \neq 0$:

1. Assign a value to one of the variables.

2. Solve the resulting equation for the other variable.

3. Write the answer as an ordered pair, (x, y).

Equations of the form $x = k$ and $y = k$ have solutions of the form (k, y) and (x, k). That is, in the equation $x = 3$, only 3 can replace x and make the equation true. The value of y can be any real number. In standard form, the equation $x = 3$ is written $x + 0y = 3$.

EXAMPLE 3

Find three solutions of $y = -4$.

$(0, -4)$, $(3, -4)$, $(-18, -4)$ Any ordered pair with $y = -4$ is a solution. ◼

Ordered pair solutions can be graphed using a *rectangular coordinate system.* A rectangular coordinate system is formed by placing two number lines, called axes, so that one is horizontal (*x*-axis) and one is vertical (*y*-axis). The intersection of the two axes is the *origin.* The coordinates of the origin are the ordered pair (0, 0). The points that lie in the plane containing the two axes are associated with ordered pairs of real numbers (*x, y*). A rectangular coordinate system is shown in Figure 6.1 with its features labeled.

Ordered pairs of real numbers can be graphed (plotted) on the coordinate system by first locating the *x*-value (coordinate), called the *abscissa,* on the *x*-axis. The point is then found by counting up or down (positive or negative) the number of units indicated by the *y*-value (coordinate), which is called the *ordinate.* The graphs of (−4, −2) and (5, 3) are shown in Figure 6.2.

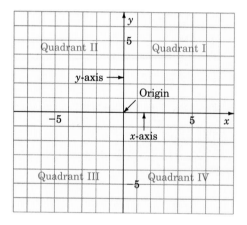

FIGURE 6.1

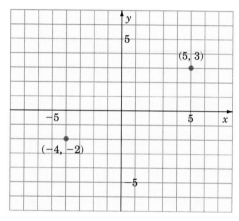

FIGURE 6.2

Using the rectangular coordinate system, we can graph the solution of an equation in two variables. To draw the graph of

$y = 3x - 5,$

find several ordered pairs that are solutions.

x	$y = 3x - 5$	(x, y)
0	$y = 3(0) - 5$	$(0, -5)$
2	$y = 3(2) - 5$	$(2, 1)$
3	$y = 3(3) - 5$	$(3, 4)$
4	$y = 3(4) - 5$	$(4, 7)$

Plot these points and connect them with a straight line (Figure 6.3). The line represents all ordered pairs in the solution set of $y = 3x - 5$.

FIGURE 6.3

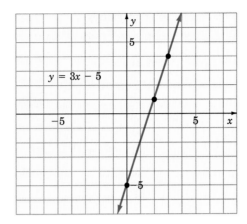

$y = 3x - 5$

■ PROPERTY

In the graph of a linear equation:

1. The coordinates of every point on the line are a solution of the equation.
2. Every solution of the equation is an ordered pair associated with a point on the line.

Since a straight line is determined by two points, we need to find only two ordered pairs to plot the graph of a linear equation. In practice, we use three points when graphing straight lines; the third point is used as a check.

■ PROCEDURE

To draw the graph of an equation that can be written in the form $Ax + By = C$:

1. Find three solutions (the coordinates of three points).
2. Plot the solutions on the coordinate system.
3. Draw the straight line passing through the points.

EXAMPLE 4

Draw the graph of $2x + y = 3$.

x	y
0	3
1	1
3	-3

Find three points by assigning values to x and solving for y.

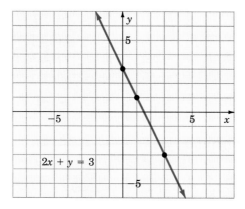

Plot each of the points on the coordinate system and draw the straight line passing through the points.

The arrows on the ends of the line indicate that the line has no endpoints.

If $B = 0$ in $Ax + By = C$, the graph is a vertical line, (parallel to the y-axis), and every ordered pair has the same abscissa.

EXAMPLE 5

Draw the graph of $3x = 9$.

$3x + 0y = 9$

When the equation is written in standard form zero is the coefficient of y.

x	y
3	-2
3	0
3	2

The equation can be simplified to $x = 3$. The equation is true for any value of y when $x = 3$.

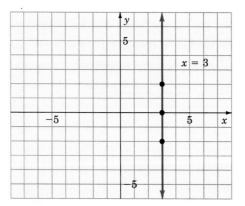

Plot the points and draw the graph. The graph is a line 3 units to the right of the y-axis. The coordinates of the points are all of the form $(3, y)$.

If $A = 0$, the graph is a horizontal line, and every ordered pair has the same y-value.

EXAMPLE 6

Draw the graph of $3y = 5$.

$3y = 5$ Solve for y. $y = \dfrac{5}{3}$. So the graph is $\dfrac{5}{3}$ units above the x-axis.

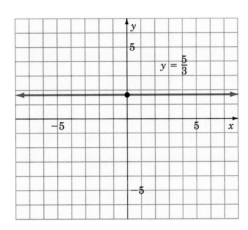

In order to conveniently show the graph of a linear equation, it is sometimes helpful to change the scale on one or both of the axes.

EXAMPLE 7

Draw the graph of $y = 250x + 120$.

x	y
-2	-380
0	120
1	370

Find the coordinates of three points.

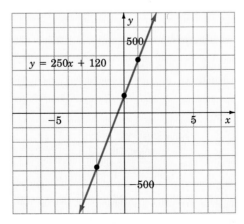

Because the values of y are so much larger than the x-values, we use a different scale (units) on the y-axis. Each unit on the x-axis is 1, and each unit on the y-axis is 100.

Graphs of physical information give the reader a "picture" of the trends that may affect conditions at any given time and therefore are useful in business presentations.

EXAMPLE 8

A strawberry grower advertises for pickers at wages (W) of $10 per day and 5 cents per pound (p). The total wages earned in a day is given by $W = 10 + 0.05p$. Draw the graph for $p \geq 0$.

p	W
50	12.5
100	15
200	20

Substitute the values of p in the equation and solve for W. The independent variable is p because the value of W depends on it. Wages are determined by the number of pounds picked.

We use different scales on the axes in order to show more of the graph.   ■

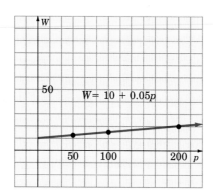

■ **CAUTION**

The graph does not continue to the left of the W-axis since the pounds picked can never be less than zero, $p \geq 0$.

■

EXERCISE   6.1

A

Find the ordered pairs that are solutions of the following equations that have the given values of x or y:

$2x + y = 5$

	x	y
1.	0	
2.	−2	
3.		0
4.		1

$-2x + y = 4$

	x	y
5.	0	
6.	1	
7.		−3
8.		1

$-3x + 2y = 6$

	x	y
9.		−3
10.	2	
11.		0
12.	0	

$5x - 2y = 10$

	x	y
13.		−5
14.	2	
15.	−2	
16.		5

Draw the graph of each equation:

17. $y = x - 2$ **18.** $y = x + 4$

19. $x + y = -5$ **20.** $x - y = 3$

B

Find the ordered pairs that are solutions of the following equations that have the given values of x or y:

$0.5x - 0.2y = 0.75$ $\dfrac{1}{3}x + \dfrac{3}{5}y = 1$

	x	y
21.	0.1	
22.		0.4
23.	4	
24.		-3

	x	y
25.	$\dfrac{1}{2}$	
26.		$-\dfrac{2}{3}$
27.	-3	
28.		$\dfrac{5}{2}$

Draw the graph of each equation:

29. $y = -2$ **30.** $y - 3 = 0$ **31.** $x = -4$

32. $x - 1 = 0$

33. $2x + y = 3$

34. $3x - y = 2$

35. $4x + 3y = 12$

36. $2x + 5y = 10$

37. $7y - 6x = 42$

38. $-5x + 7y = 35$

39. $2x + 7y = 14$

40. $-4x + 2y = 12$

41. $x - 5y = -6$

42. $5x - y = -6$

43. $x = 2y + 4$

44. $x = 3 - 5y$

45. $2x - 9y = 18$

46. $3x + 5y = 15$

47. $4x + 7y = 28$

48. $4x + 7y = -28$

49. $8x - y = 4$

50. $x + 8y = -4$

C

Draw the graph of each equation:

51. $y = \dfrac{2}{5}x + 2$

52. $y = -\dfrac{3}{5}x - 2$

53. $y = \dfrac{3}{8}x + 6$

54. $y = \dfrac{5}{7}x + 2$

55. $\dfrac{1}{5}x - y = 2$

56. $x + \dfrac{1}{3}y = 3$

57. $\dfrac{1}{7}x + \dfrac{1}{4}y + 1 = 0$

58. $-\dfrac{1}{8}x + \dfrac{1}{3}y = 1$

In Exercises 59–62 let one unit on the x-axis equal 5 and one unit on the y-axis equal 10 to draw the graph:

59. $10x - y = 50$

60. $50x + y = 200$

61. $25x - y = 125$

62. $30x + y = 70$

In Exercises 63–66 let one unit on the x-axis equal 50 and one unit on the y-axis equal 25 to draw the graph:

63. $y = 2.5x - 175$

64. $x - 2y = 25$

65. $x - y = 200$

66. $x + 2y = -300$

67. If the strawberry grower in Example 8 pays wages (W) of $5 per day and 10 cents per pound (p), draw the graph of $W = 5 + 0.10p$ for $p \geq 0$.

68. Peter Piper pays pickers to pick peppers in his field at the rate of $4 per day and 4 cents per pound. Draw the graph that represents the wages Peter pays for $p \geq 0$, where p represents the number of pounds picked.

69. A furniture refinishing outlet pays its employees $40 per day plus $10 per unit refinished. Write an equation to express an employee's daily wage (w) in terms of the units finished (x). Draw a graph for $x \geq 0$.

70. A minor league baseball pitcher is hired for $700 per month plus $100 per game won. Write an equation to express his monthly earnings in terms of games won (x) and graph the function for $x \geq 0$.

71. The cost of manufacturing a dinette set at the Walnut Grove Company is $75 per set. The company has a fixed overhead of $750. The formula for total cost (c) is $c = mn + p$, where m is the cost per unit, n is the number of units, and p is the fixed overhead cost. Draw the graph for the total cost of manufacturing 0 to 20 dinette sets.

72. Using the formula in Exercise 71, draw the graph of the cost of manufacturing 0 to 30 loveseats if each unit costs $60 to manufacture and the fixed overhead is $1000.

73. The height, H in feet, of an object dropped from a 100 ft high building after a time (t, in seconds) can be approximated by $H = -40t + 100$. Using H as the vertical axis and t as the horizontal axis, graph this relationship.

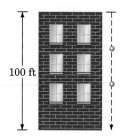

100 ft

Figure for Exercise 73

74. The payment (P) for a winning bet at the race track is $2.50 per dollar bet ($x$) less a one-time charge of $0.50 for the "take" of the track. Write the equation for the payment in terms of the amount bet and the track's share. Using P as the vertical axis and x as the horizontal axis, graph the equation for values of x greater than or equal to one.

75. In dealing with electricity, it was discovered that the resistance (y) of a wire is equal to a constant (depending on the material in the wire) times the length (x) of the wire. If the constant for a certain material is 0.20 ohms per foot, write the equation for the resistance in terms of the length. Graph the equation using values of x greater than or equal to zero.

76. "Straight-line" depreciation is given by $D = P - S$, where D is the depreciation, P is the purchase price, and S is the salvage value. If a corporation purchases a machine for $10,000, graph the equation using D as the vertical axis and S as the horizontal axis. (Note the "slant" of this line)

STATE YOUR UNDERSTANDING

77. Given the equation $2x - 3y = 6$, to graph the line we generally find three solutions by choosing any values for either variable. If, however, you find one of these solutions by letting $x = 0$ (and find the corresponding y) and another by letting $y = 0$ (and find x), where are these points on your graph?

78. The words "uphill" and "downhill" are relative words in the English language. They depend on where you are standing on a hill AND which way you are looking. How could we use these words to describe some of the lines we graphed in this section and communicate it to everyone in the class?

CHALLENGE EXERCISES

79. As discussed in this section, we frequently have to adjust our "scales" to fit the data on our graph. For the following equation, find three solutions, choose appropriate scales, and graph $y = -0.04x + 1000$.

80. Using "m" as the vertical axis and "n" as the horizontal axis, graph $2m + 3n = 6$. Then, on another graph, reverse the axes and graph the same line again. Are they the same line?

81. This time, using "m" as the vertical axis and "n" as the horizontal axis, graph $3m + 2n = 6$ and compare it to the two previous graphs. Is it the same as either?

82. If any two quantities are related to each other by the rule $y = kx$ (where "k" is some constant), the quantities are said to be "directly proportional" and k is called the "constant of proportionality." If y is known to be directly proportional to x and $y = 12$ when $x = 4$, find "k"; write the equation of the relationship; and graph the line.

MAINTAIN YOUR SKILLS (SECTIONS 2.9, 3.3)

Combine and reduce if possible:

83. $\dfrac{x + 1}{x - 6} + \dfrac{4}{x - 3} - \dfrac{2x - 4}{x^2 - 9x + 18}$

84. $\dfrac{1}{2x + 1} - \dfrac{3}{3x + 5} + \dfrac{5}{6x^2 + 13x + 5}$

Solve by factoring:

85. $4x^2 - 225 = 0$

86. $4x^2 + 12x + 9 = 0$

87. $w^2 + 7w - 18 = 0$

88. $8x^2 - 2x - 15 = 0$

89. $4(3 + 4x) = 3x^2$

90. $9y(3 - y) = 20$

6.2

SLOPE, INTERCEPTS, AND THE DISTANCE FORMULA

OBJECTIVES

1. Find the x- and y- intercepts given the equation of a line.

2. Find the slope of a line given two points on the line.

3. Determine whether or not two lines are parallel or perpendicular.

4. Find the distance between two points.

There are two points on the graph of $Ax + By = C$, $AB \neq 0$, for which their coordinates are easily computed. They are called the x-intercept and the y-intercept.

■ **DEFINITION**

x-intercept $(x, 0)$
y-intercept $(0, y)$

The x-intercept is the point where the line crosses the x-axis, $(x, 0)$; the y-intercept is the point where the line crosses the y axis, $(0, y)$.

■ **PROCEDURE**

1. To find the x-intercept of the graph whose equation is $Ax + By = C$, $A \neq 0$, substitute 0 for y in the equation and solve for x. The x-intercept is $\left(\dfrac{C}{A}, 0\right)$.

2. To find the y-intercept of the graph whose equation is $Ax + By = C$, $B \neq 0$, substitute 0 for x in the equation and solve for y. The y-intercept is $\left(0, \dfrac{C}{B}\right)$.

EXAMPLE 1

Find the x- and y-intercepts for the line $5x + 6y = 30$.

$$5x + 6(0) = 30 \qquad \text{Let } y = 0 \text{ to find the } x\text{-intercept.}$$

$$x = 6$$

$$5(0) + 6y = 30 \qquad \text{Let } x = 0 \text{ to find the } y\text{-intercept.}$$

$$y = 5$$

The x-intercept is $(6, 0)$, and the y-intercept is $(0, 5)$.   ■

Lines defined by equations written in the form $x = a$ or $y = b$ have only one intercept.

EXAMPLE 2

Find the x- and y-intercepts for the line given by $3x = 14$.

$$3x = 14 \qquad\qquad \text{The line is parallel to the } y\text{-axis and therefore has only an } x\text{-intercept.}$$

$$x = \frac{14}{3}$$

The x-intercept is $\left(\dfrac{14}{3}, 0\right)$.   ■

The intercepts are two of the easiest points to find and in many instances are useful in drawing the graph of a linear equation.

Another important property of a line is its slope.

■ DEFINITION

Slope of a Line

> The slope of a line is the ratio of the change in the vertical (rise) to the change in the horizontal (run).
>
> The letter m is used to denote slope, $m = \dfrac{\text{rise}}{\text{run}}$.

The line in Figure 6.4 passes through the points $A(x_1, y_1)$ and $B(x_2, y_2)$.

FIGURE 6.4

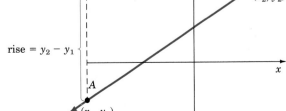

■ FORMULA

The slope of the line passing through (x_1, y_1) and (x_2, y_2) is

$$\text{Slope} = m = \frac{\text{rise}}{\text{run}} = \frac{\text{change in } y}{\text{change in } x} = \frac{y_2 - y_1}{x_2 - x_1} = \frac{y_1 - y_2}{x_1 - x_2}.$$

The slope of a line describes the slant or pitch of the line and is constant regardless of which pair of points is chosen on the line.

EXAMPLE 3

Find the slope of the line that passes through the points $(6, 3)$ and $(-1, 0)$.

$$m = \frac{y_2 - y_1}{x_2 - x_1} \qquad \text{Formula for the slope.}$$

$$m = \frac{3 - 0}{6 - (-1)} \qquad \text{Let } (x_2, y_2) = (6, 3) \text{ and } (x_1, y_1) = (-1, 0).$$

$$m = \frac{3}{7}$$

The slope is $\dfrac{3}{7}$. ■

A slope of $\dfrac{3}{7}$ indicates that for every 3 units of rise there is a corresponding 7 units of run as you move along the line from left to right.

EXAMPLE 4

Find the slope of the line joining the points $(-6, 10)$ and $(3, -8)$.

$$m = \frac{y_2 - y_1}{x_2 - x_1} \qquad \text{Substitute } (x_2, y_2) = (3, -8) \text{ and}$$
$$\qquad\qquad (x_1, y_1) = (-6, 10) \text{ in the slope formula.}$$

$$= \frac{-8 - 10}{3 - (-6)}$$

$$= \frac{-18}{9} = -2$$

The slope is -2. ■

There are many applications of slope and distance. One of the useful applications of slope is in predicting future events based on prior results.

EXAMPLE 5

The welding department of a local community college projects its enrollments on a straight line. If the first-year enrollment is 20, represented by $(1, 20)$, and the third-year enrollment is 27, represented by $(3, 27)$, draw the graph of the enrollment (y) for years (x), $x \geq 1$. What is the rate of growth (slope of the line)? What is the projected enrollment for year nine?

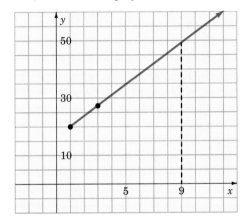

Plot the points and sketch the graph. Since the values of y are so much larger than the values of x, use a larger scale on the y-axis.

$$m = \frac{27 - 20}{3 - 1} = \frac{7}{2}$$

Calculate the slope (rate of growth).

The rate of growth is 3.5 enrollments per year. Projecting to the ninth year ($x = 9$), we predict an enrollment of 48.   ■

Using the concept of slope, we make four statements about lines:

1. A line that slopes *down to the right* has negative slope.

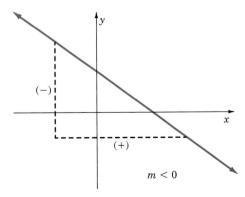

2. A line that slopes *up to the right* has positive slope.

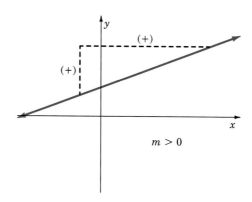

3. A horizontal line (parallel to the *x*-axis) has slope 0.

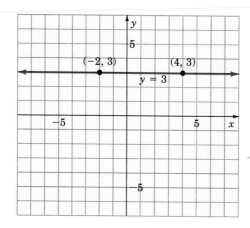

$$m = \frac{3 - 3}{4 - (-2)} = 0$$

4. A vertical line (parallel to the *y*-axis) has no slope, since the slope is undefined.

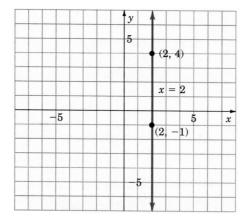

$$m = \frac{4 - (-1)}{2 - 2}$$

The slope is undefined since division by 0 is not defined.

■ DEFINITION

Two distinct nonvertical lines are parallel if and only if their slopes are the same (equal).

In Figure 6.5 we see that the line that passes through $(-1, 3)$ and $(2, 0)$ is parallel to the line that passes through the points $(1, 4)$ and $(3, 2)$ since their slopes are the same.

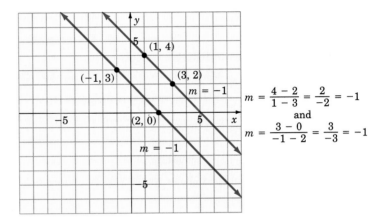

$$m = \frac{4 - 2}{1 - 3} = \frac{2}{-2} = -1$$
and
$$m = \frac{3 - 0}{-1 - 2} = \frac{3}{-3} = -1$$

FIGURE 6.5

Two distinct vertical lines are also parallel.

■ DEFINITION

Two nonvertical lines with slopes m_1 and m_2 are perpendicular if and only if the product of their slopes is -1.

$$m_1 m_2 = -1$$

The slope of one of two perpendicular lines is the opposite of the reciprocal of the slope of the other line. In Figure 6.6 we see that the line that passes through the points $(5, 5)$ and $(4, 2)$ is perpendicular to the line that passes through the points $(-3, 3)$ and $(0, 2)$ since the product of their slopes is -1.

FIGURE 6.6

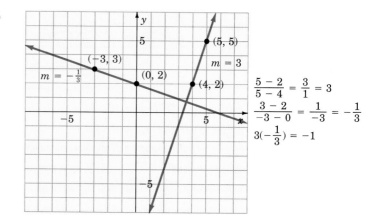

▪ PROCEDURE

To determine whether two lines are perpendicular or parallel:

1. Find the slope of each line.

2. If the slopes are equal and the intercepts are different, the lines are parallel.

3. If the product of the slopes is -1, the lines are perpendicular.

EXAMPLE 6

Are the graphs of the lines $3x + 2y = 6$ and $6x + 4y = 24$ parallel?

$3x + 2y = 6$		$6x + 4y = 24$		We find the intercepts of each line and the slopes.

x	y		x	y
0	3		0	6
2	0		4	0

Find the intercepts. The y-intercepts are different.

$$m_1 = \frac{3 - 0}{0 - 2} \qquad m_2 = \frac{6 - 0}{0 - 4}$$

Find the slopes using the intercepts.

$$= -\frac{3}{2} \qquad\qquad = -\frac{3}{2}$$

The slopes are the same.

The lines are parallel.

EXAMPLE 7

Are the graphs of the lines $3x + 2y = 1$ and $2x + 3y = 31$ perpendicular?

$3x + 2y = 1$ **$2x + 3y = 31$**   We need to find the product of the slopes of the lines.

x	y
1	−1
−1	2

x	y
5	7
−4	13

To find the slope, we first find two points on each line. Since the intercepts are not integers, we find other points.

$$m_1 = \frac{-1 - 2}{1 - (-1)} \qquad m_2 = \frac{7 - 13}{5 - (-4)}$$

$$= -\frac{3}{2} \qquad\qquad = -\frac{2}{3}$$

$$m_1 m_2 = \left(-\frac{3}{2}\right)\left(-\frac{2}{3}\right) = 1$$

If the lines are perpendicular, the product of the slopes is -1.

The lines are not perpendicular.   ■

Next we consider the distance between any two points in the coordinate system. We find the formula for the distance using the Pythagorean theorem from geometry (Figure 6.7).

The line segment joining the two points can be thought of as the hypotenuse of a right triangle (Figure 6.8).

By the Pythagorean theorem,

$$d^2 = (x_2 - x_1)^2 + (y_2 - y_1)^2.$$

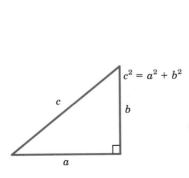

FIGURE 6.7

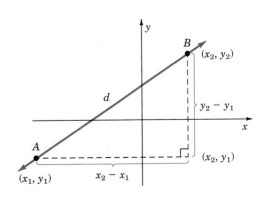

FIGURE 6.8

Solving for d, we have:

■ FORMULA

Distance Formula

$$d = \sqrt{(x_2 - x_1)^2 + (y_2 - y_1)^2}$$

The formula does not include the negative square root since negative numbers are not used to measure distance.

EXAMPLE 8

Find the distance between the points $(7, 3)$ and $(5, 8)$.

$$\begin{aligned}
d &= \sqrt{(x_2 - x_1)^2 + (y_2 - y_1)^2} \\
&= \sqrt{(5 - 7)^2 + (8 - 3)^2} \\
&= \sqrt{4 + 25} \\
&= \sqrt{29}
\end{aligned}$$

Distance formula.

Let $(x_2, y_2) = (5, 8)$ and $(x_1, y_1) = (7, 3)$.

The distance is $\sqrt{29}$ units.   ■

EXAMPLE 9

Show that the point $(0, 0)$, $(6, 0)$, and $(3, 3)$ are vertices of a right triangle.

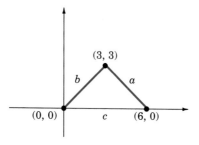

Sketch a graph of the triangle. If the triangle is a right triangle, then $a^2 + b^2 = c^2$. Find a, b, and c using the distance formula.

$$\begin{aligned}
d &= \sqrt{(x_2 - x_1)^2 + (y_2 - y_1)^2} \\
a &= \sqrt{(6 - 3)^2 + (0 - 3)^2} \\
&= \sqrt{18} \\
&= 3\sqrt{2} \\
b &= \sqrt{(3 - 0)^2 + (3 - 0)^2} \\
&= \sqrt{18} \\
&= 3\sqrt{2}
\end{aligned}$$

Distance formula.

$$c = \sqrt{(6-0)^2 + (0-0)^2}$$
$$= \sqrt{36+0}$$
$$= 6$$

$$a^2 + b^2 = c^2$$ Now verify that $a^2 + b^2 = c^2$.

$$(3\sqrt{2})^2 + (3\sqrt{2})^2 = (6)^2$$ Substitute $a = 3\sqrt{2}$, $b = 3\sqrt{2}$, and $c = 6$.

$$18 + 18 = 36$$ True.

The triangle is a right triangle.

Alternative solution:

$$m_a = \frac{0-3}{6-3} = -1$$ An alternative solution is to compute the slope of each side. If the product of the slopes of any two sides is -1, the two sides are perpendicular, so the triangle has a right angle.

$$m_b = \frac{3-0}{3-0} = 1$$

$$m_c = \frac{0-0}{0-6} = 0$$

$$(m_a)(m_b) = (-1)(1) = -1$$ Side a and side b are perpendicular.

The triangle is a right triangle.   Distance formula. ■

EXERCISE 6.2

A

Find the slope of the line joining the following pairs of points:

1. $(4, 0)$, $(0, -2)$ **2.** $(-5, -1)$, $(0, 0)$ **3.** $(0, 6)$, $(-4, 0)$ **4.** $(0, 0)$, $(-1, -3)$

5. $(1, -1)$, $(-1, 1)$ **6.** $(-2, -3)$, $(2, 3)$

Write the coordinates of the x- and y-intercepts:

7. $3x - 7y = 21$ **8.** $5x + 4y = 20$ **9.** $x - 7y = 14$ **10.** $8x + y = 8$

11. Is the graph of the line that passes through $(5, 8)$ and $(3, 10)$ parallel to the line that passes through $(9, 6)$ and $(4, 12)$?

12. Is the graph of the line that passes through $(-2, 3)$ and $(4, -1)$ parallel to the line that passes through $(-5, 6)$ and $(1, 2)$?

13. Determine whether or not the graph of the line that passes through $(3, 4)$ and $(6, 1)$ is perpendicular to the line that passes through $(4, 2)$ and $(7, 5)$.

14. Determine whether or not the graph of the line that passes through $(0, 5)$ and $(6, 3)$ is perpendicular to the line that passes through $(0, 2)$ and $(-4, 5)$.

Find the distance between the following pairs of points:

15. $(1, 5), (3, 5)$

16. $(4, 8), (6, 8)$

17. $(0, -2), (3, 0)$

18. $(3, -6), (0, 0)$

19. $(1, -1), (-1, 1)$

20. $(-2, -3), (2, 3)$

B

Write the coordinates of the x- and y-intercepts:

21. $2x - 3y = 11$

22. $7x - 4y = 15$

23. $x - 6y = 9$

24. $3x + y = -7$

25. $0.4x + 0.3y = 1.2$

26. $0.5x - 0.2y = 1$

Find the slope of the line joining the following points and find the distance between the points:

27. $(-6, 3), (8, -2)$

28. $(-11, -2), (-8, 1)$

29. $(3, -7), (8, -4)$

30. $(-1, -6), (-2, -5)$

31. $(2, -6), (5, -6)$

32. $(-5, 3), (5, 3)$

33. $(-2, 3), (-2, -3)$

34. $(4, 0), (4, -2)$

Are the line segments joining the following pairs of points perpendicular?

35. $(0, 3)$ and $(-2, 5)$
$(7, 6)$ and $(9, 8)$

36. $(8, 6)$ and $(5, 12)$
$(4, 9)$ and $(-4, 5)$

37. $(3, 5)$ and $(-4, 3)$
$(4, 1)$ and $(-3, 3)$

38. $(-6, 7)$ and $(-1, -3)$
$(8, -11)$ and $(4, -9)$

39. Determine if the line $y - 2x = 1$ is parallel or perpendicular to the line $y = 2x + 5$.

40. Determine if the line $x + 3y = 3$ is parallel or perpendicular to the line $y = 3x + 2$.

C

Write the coordinates of the x- and y-intercepts:

41. $\frac{2}{3}x + 6 = \frac{1}{2}y$

42. $-\frac{1}{2}x - \frac{1}{3}y = 8$

43. $2y - 7 = 0$

44. $3y - 2 = 0$

Find the slope of the line joining the following pairs of points and find the distance between the points:

45. $\left(\frac{1}{2}, 2\right), \left(3, \frac{1}{3}\right)$

46. $\left(-\frac{1}{4}, \frac{1}{2}\right), \left(\frac{3}{4}, -\frac{3}{2}\right)$

47. $\left(-\dfrac{3}{7}, 4\right), \left(-\dfrac{1}{14}, \dfrac{3}{2}\right)$

48. $\left(\dfrac{3}{5}, \dfrac{1}{3}\right), \left(-\dfrac{2}{5}, -\dfrac{2}{3}\right)$

49. $(0.39, 0.25), (-0.21, 0.40)$

50. $(-3.4, 1.29), (-1.88, -2.51)$

51. $(0.42, -0.35), (0.12, -0.25)$

52. $(1.37, 2.55), (-1.13, -1.45)$

53. Determine if the line $2y + x = 3$ is parallel or perpendicular to the line $y = 2x + 1$.

54. Determine if the line $10y - x = 20$ is parallel or perpendicular to the line $y = -10x + 5$.

Are the line segments joining the following pairs of points parallel, perpendicular, or neither?

55. $(-3, 5)$ and $(-7, 2)$
$(2, 6)$ and $(-2, 3)$

56. $(4, 8)$ and $(-2, -3)$
$(6, 2)$ and $(-5, 8)$

57. $(17, -42)$ and $(-18, 21)$
$(-67, 113)$ and $(133, -247)$

58. $(110, 57)$ and $(40, -43)$
$(-68, -53)$ and $(72, -151)$

59. $\left(\dfrac{2}{3}, -\dfrac{3}{4}\right)$ and $\left(\dfrac{1}{2}, -\dfrac{1}{4}\right)$
$\left(\dfrac{1}{6}, -\dfrac{1}{5}\right)$ and $\left(-\dfrac{31}{30}, -\dfrac{3}{5}\right)$

60. $\left(\dfrac{5}{7}, \dfrac{2}{3}\right)$ and $\left(-\dfrac{3}{7}, \dfrac{1}{2}\right)$
$\left(\dfrac{7}{12}, \dfrac{1}{6}\right)$ and $\left(\dfrac{5}{12}, -\dfrac{2}{3}\right)$

D

61. The enrollment in geography classes is projected on a straight line. If the enrollment in the first year is 50, represented by $(1, 50)$, and the enrollment in the second year is 60, represented by $(2, 60)$, find the rate of growth. Give the enrollment in the seventh year.

62. If the enrollment in literature is projected in the same manner as for geography in Exercise 61, and you are given $(1, 75)$ and $(4, 39)$, find the rate of growth and the enrollment in year five.

63. The opposite sides of a parallelogram are parallel. Show that the points $(-1, 5), (0, 9), (2, 2)$, and $(1, -2)$ are vertices of a parallelogram.

64. Show that $(-3, 7), (3, 13), (11, -3)$, and $(5, -9)$ are vertices of a parallelogram.

65. Show that the points $(2, -3), (8, -3)$, and $(5, 0)$ are vertices of an isosceles triangle (two equal sides).

66. Show that the points $(5, 5), (8, 4), (4, 2)$, and $(7, 1)$ are the vertices of a quadrilateral with all sides equal.

67. Use the Pythagorean theorem to show that the points $(-2, 3)$, $(-2, 7)$, and $(1, 3)$ are the vertices of a right triangle.

68. Use the Pythagorean theorem to show that the points $(0, 0)$, $(10, 0)$, and $(5, 5)$ are the vertices of a right triangle.

69. The points A $(2, 1)$, B $(5, 1)$, C $(5, 4)$, and D $(2, 4)$ are vertices of a square. Show that the diagonals of the square are perpendicular.

70. The points A $(-7, -1)$, B $(-2, -1)$, C $(-7, -5)$, and D $(-2, -5)$ form the vertices of a rectangle. Show that the diagonals of the rectangle have the same length.

71. Show that the points $(0, 0)$, $(10, 0)$, and $(5, 7)$ are not vertices of a right triangle.

72. The relationship between Celsius temperature (x) and Fahrenheit (y) on the two temperature scales is given by $9x - 5y + 160 = 0$. Find two solutions to the equation and use them to find the slope of the line.

STATE YOUR UNDERSTANDING

73. The slope formula can be given in two forms:

$$\frac{y_2 - y_1}{x_2 - x_1} \text{ or } \frac{y_1 - y_2}{x_1 - x_2}$$

Why, mathematically, are these two forms equivalent?

74. Given the equation $y = x^2 + 2$, complete the following solutions: $P_1(1, \ )$, $P_2(2, \ )$, and $P_3(3, \ )$
Calculate the slope using P_1 and P_2 and then using P_2 and P_3. Are they the same? Why?

CHALLENGE EXERCISES

75. When graphing it is customary to assign the independent variable to the horizontal axis and the dependent variable to the vertical axis. The slope can be thought of as

$$\frac{\text{change in the dependent variable}}{\text{change in the independent variable}}$$

Consequently, if you are given the equation $2m + 3n = 6$, the value of the slope will depend upon which variable is assigned as the independent variable. Let m be the independent variable and find the slope using two solutions. Now let n be the independent variable and find the slope. Compare the slopes.

76. Given the following equations, generate two solutions for each and find their slopes:

a) $y = 2x - 3$ b) $y = -3x + 1$ c) $y = \dfrac{2}{3}x + 2$

Any observations?

77. A college observes that in 1985, the average score on the verbal section of the SAT is 500, represented by the ordered pair (0, 500). Four years later, in 1989, the average score was 485, represented by (4, 485). Using the scores (S) as the dependent variable and the year (t) as the independent variable, locate the two points on a graph and connect them with a straight line. If the trend continues, what will the average score be in 1993?

78. Given the points (1, 3) and (4, k), find the value of k so that the slope of the line passing through them is -2.

MAINTAIN YOUR SKILLS (SECTIONS 4.3, 4.4, 5.1)

Simplify. Assume that all variables represent positive numbers:

79. $\sqrt{120x^5y^8}$

80. $\sqrt[3]{120x^5y^8}$

81. $\sqrt[3]{160a^7b^{10}}$

82. $\sqrt{160a^7b^{10}}$

83. $\sqrt{48s^3t^3} - 3st\sqrt{75st} + t\sqrt{12s^3t}$

84. $3\sqrt{180} - \sqrt{20} + 2\sqrt{80} - 5\sqrt{500}$

Solve by square roots:

85. $y^2 = 54$

86. $(a - 5)^2 = 40$

6.3

DIFFERENT FORMS OF THE LINEAR EQUATION

OBJECTIVES

1. Given the slope of a line and a point on the line, write the standard form of the equation of the line.

2. Given the equation of a line, find the slope and the y-intercept.

3. Draw the graph of a line using the slope and the y-intercept.

4. Given a linear equation, write the equation in both standard form and slope-intercept form.

The slope of a line is constant for any point (x, y) on the line and a given point, (x_1, y_1) (Figure 6.9).

FIGURE 6.9

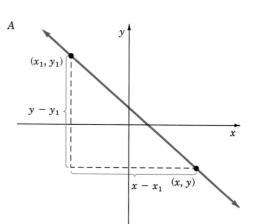

The slope (m) is

$$m = \frac{y - y_1}{x - x_1}, \quad x - x_1 \neq 0.$$

Multiplying by $(x - x_1)$ gives us

■ FORMULA

Linear Equation
Point-Slope Form

$y - y_1 = m(x - x_1)$	Point-slope form.

The equation is true for every point (x, y) on the line. The graph of every solution of the equation is a point on the line.

EXAMPLE 1

Given the line with slope $-\dfrac{2}{3}$ that passes through the point $(-4, 7)$, write the equation of the line in standard form.

$$y - y_1 = m(x - x_1)$$ Point-slope form.

$$y - 7 = -\frac{2}{3}[x - (-4)]$$ Substitute $m = -\dfrac{2}{3}$ and $(x_1, y_1) = (-4, 7)$.

$$3(y - 7) = -2(x + 4)$$

$$2x + 3y = 13$$ Standard form.

■

EXAMPLE 2

Write the equation, in standard form, of the line containing the points $(8, -5)$ and $(-6, 2)$.

$$m = \frac{2 - (-5)}{-6 - 8} = \frac{7}{-14} = -\frac{1}{2}$$ Find the slope.

$$y - (-5) = -\frac{1}{2}(x - 8)$$ Substitute $m = -\frac{1}{2}$ and $(x_1, y_1) = (8, -5)$ in $y - y_1 = m(x - x_1)$.

$$2y + 10 = -x + 8$$

$$x + 2y = -2$$ If you substitute $(x_1, y_1) = (-6, 2)$, you will get the same equation. ∎

Another useful form of the equation of a line is called the *slope-intercept form*. It is found by substituting the coordinates of the y-intercept into the point-slope form. Let $(0, b)$ represent the y-intercept. Then

$$y - y_1 = m(x - x_1)$$

$$y - b = m(x - 0)$$

$$y = mx + b.$$

■ **FORMULA**

**Linear Equation
Slope-Intercept Form**

$y = mx + b$	Slope-intercept form.

With the slope intercept form, we can solve a linear equation explicitly for y and immediately identify the slope and the y-intercept.

EXAMPLE 3

Find the slope and the y-intercept of the graph of the line $3x + 7y = 21$.

$$7y = -3x + 21$$

$$y = -\frac{3}{7}x + 3$$ Slope-intercept form.

$$m = -\frac{3}{7} \text{ and the } y\text{-intercept is } (0, 3).$$ ∎

EXAMPLE 4

Find the slope and the y-intercept of the graph of the line $9x - 5y = 15$.

$$9x - 5y = 15$$

$$-5y = -9x + 15$$

$$y = \frac{9}{5}x - 3 \qquad\qquad \text{Slope-intercept form.}$$

$m = \dfrac{9}{5}$ and the y-intercept is $(0, -3)$. ■

Using the slope and the y-intercept, we can draw the graph of the line. In Example 5, we use the information found in Example 4 to draw the graph.

EXAMPLE 5

Draw the graph of the line with slope $\dfrac{9}{5}$ and whose y-intercept is $(0, -3)$.

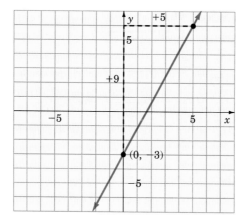

First plot the y-intercept, $(0, -3)$.

The slope, $\dfrac{9}{5}$, indicates that by moving up nine units, $+9$, and right 5 units, $+5$, we find a new point on the line. The new point is $(5, 6)$. Draw the line joining the points. ■

EXAMPLE 6

Draw the graph of $6x - 5y = -4$ using the slope and the y-intercept.

$$6x - 5y = -4 \qquad\qquad$$ Write the equation in the slope-intercept form.

$$y = \frac{6}{5}x + \frac{4}{5}$$

$m = \dfrac{6}{5}$ and the y-intercept is $\left(0, \dfrac{4}{5}\right)$.

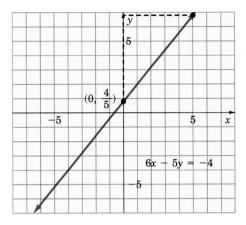

Plot the intercept. The slope tells us that another point can be found by moving up 6 units, +6, and right 5 units, +5. The new point is $\left(5, 6\frac{4}{5}\right)$. Draw the line passing through the points.

■

A given point and the slope can be used to predict probable happenings in the future.

EXAMPLE 7

The Acme Corp. projects that its profits will follow a straight line with an increase of $1000 per year (slope of line is 1000). If the profit at the end of year one is $9000, illustrated by the point $(1, 9000)$, draw the graph of the line and give the profit in year seven.

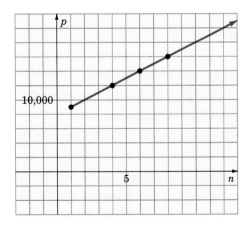

Use the point and the slope to sketch the graph. Note that the graph is valid only after year one. Now read the profit for year seven, $(7, 15{,}000)$.

The projected profit is $15,000 in the seventh year.   ■

Here is a list of the important formulas for linear equations.

▨ FORMULA

$Ax + By = C$	Standard form of a linear equation.
$y - y_1 = m(x - x_1)$	Point-slope form of a linear equation.
$y = mx + b$	Slope-intercept form of a linear equation.

You should memorize each of the above forms and be able to write a given linear equation in any one of the forms.

EXAMPLE 8

Write the following linear equation in both the standard and the slope-intercept form:

$$4(x - 2) + 6 = 3(y - 5) - 2y$$

$$4x - 8 + 6 = 3y - 15 - 2y$$

$$4x - y = -13 \qquad \text{Standard form. Now solve for } y.$$

$$-y = -4x - 13$$

$$y = 4x + 13 \qquad \text{Slope-intercept form.} \qquad ∎$$

EXERCISE   6.3

A

Write the equation of a line in standard form, given its slope and a point on the line:

1. $(3, 6)$, $m = 4$

2. $(-1, 2)$, $m = -1$

3. $(5, -4)$, $m = -2$

4. $(-4, 3)$, $m = \dfrac{1}{2}$

5. $(-7, 8)$, $m = -\dfrac{1}{3}$

6. $(4, 0)$, $m = -\dfrac{3}{4}$

7. $(3, -4)$, $m = 0$

8. $(-5, -2)$, $m = 0$

Find the slope and the y-intercept of each of the following lines:

9. $3x - 7y = 35$

10. $5x - 8y = 32$

11. $x + 9y = 36$

12. $x - 6y = -42$

Find the slope and the y-intercept of the graphs of the following equations.
Draw the graph using the slope and the y-intercept:

13. $y = -x + 3$

14. $y = x - 3$

15. $y = 2x - 4$

16. $y = -3x + 1$

17. $y = -\dfrac{2}{3}x + 5$

18. $y = \dfrac{3}{5}x - 2$

19. $y = -4$

20. $y = 3$

Write the equation of a line in standard form, given its slope and a point on the line:

21. $(-11, 14)$, $m = -3$

22. $(15, -20)$, $m = 5$

23. $\left(-\dfrac{1}{2}, -\dfrac{3}{5}\right)$, $m = -\dfrac{4}{5}$

24. $\left(\dfrac{7}{3}, -\dfrac{5}{4}\right)$, $m = \dfrac{3}{4}$

B

Write the equation of the line in standard form given two points on the line:

25. $(6, -5)$, $(8, -3)$

26. $(7, -2)$, $(9, 5)$

27. $(9, 0)$, $(1, 3)$

28. $(9, 0)$, $(-1, -4)$

29. $(-11, -5)$, $(3, -5)$

30. $(-2, 3)$, $(2, 3)$

31. $(4, -7)$, $(-5, -12)$

32. $(6, 13)$, $(-9, -3)$

Find the slope and the y-intercept of the graphs of the following equations.
Draw the graphs using the slope and the y-intercept:

33. $4x - y = 7$

34. $-3x + y = -2$

35. $-5x + 3y = 9$

36. $2x - 7y - 21 = 0$

37. $2x - 5y - 20 = 0$

38. $x + 3y - 12 = 0$

39. $x = -3$

40. $x = 5$

Write the equation of a line in slope-intercept form, given its slope and a point
on the line:

41. $(-34, -55)$, $m = -\dfrac{7}{12}$

42. $(-75, 45)$, $m = \dfrac{3}{20}$

43. $\left(\dfrac{12}{7}, \dfrac{15}{14}\right)$, $m = \dfrac{4}{21}$

44. $\left(-\dfrac{7}{15}, -\dfrac{13}{25}\right)$, $m = -\dfrac{7}{20}$

C

Write the equation of the line in standard form given two points on the line:

45. $\left(\dfrac{1}{2}, 4\right)$, $\left(-\dfrac{2}{3}, 4\right)$

46. $\left(-\dfrac{5}{6}, 1\right)$, $\left(\dfrac{4}{3}, 1\right)$

47. $(-2.5, 3)$, $(-2.5, 16)$

48. $(4.7, 3.5)$, $(4.7, -2.1)$

49. $\left(\dfrac{1}{3}, \dfrac{3}{4}\right)$, $\left(\dfrac{1}{2}, \dfrac{2}{3}\right)$

50. $\left(\dfrac{1}{2}, -\dfrac{4}{3}\right)$, $\left(-\dfrac{1}{4}, \dfrac{2}{3}\right)$

51. $\left(\dfrac{1}{4}, -\dfrac{1}{3}\right)$, $\left(-\dfrac{1}{3}, -\dfrac{3}{2}\right)$

52. $\left(-\dfrac{2}{5}, \dfrac{3}{2}\right)$, $\left(\dfrac{3}{5}, \dfrac{7}{8}\right)$

Write the following equations in standard form and slope-intercept form:

53. $3(x - 1) - 5 = 2(y - 1) + 7$

554. $2(x - y) + (y + 2) = 4 - 3x$

55. $5(x + 3) - 2(y + 5) - 17 = 0$

56. $-3(2x - 5) + 4(7 - 3y) - 12 = 0$

57. $4(x + 1) - 6 = 2(y + 4) + 5$

58. $2(x - 3y) + 6 = x - 3(y + 2)$

59. $5(2x - 6) + 7y = 3(y - 1) + 7$

60. $-2(10 - 3y) + 6x = 2(5 - 4x) + 3y$

D

61. The Uptown Corp. expects its profits to increase \$1500 per year. If the profit in year three was \$25,000, use a graph to predict what the profit will be in year 12.

62. In Exercise 61 if the increase is projected at \$2300 per year, what will the profit be in year 11?

63. Karla invests \$1000 at 10% simple interest for an indefinite period of time. The amount of interest (A) she will be paid on her investment (p) by the end of t years is given by

$A = prt,$

where r is the interest rate expressed as a decimal. (Assume that the interest is paid annually.) Write the equation showing the amount of interest paid for t years, $t \geq 0$.

64. In Exercise 63, if Karla had invested \$1500 at 12% interest, write the equation showing the interest earned in t years, $t \geq 0$.

65. Write the equation of a line that passes through the point $(-2, 5)$ and is parallel to the line whose equation is $2x - 3y = 4$.

66. Write the equation of a line that passes through the point $(4, -3)$ and is parallel to the line whose equation is $5x + 2y = 7$.

67. Write the equation of a line that passes through the point $(7, 2)$ and is perpendicular to the line whose equation is $4x + 3y = 5$.

68. Write the equation of a line that passes through the point $(-1, -4)$ and is perpendicular to the line whose equation is $5x - 7y = 12$.

69. Write the equation of the line that passes through the point $(2, -5)$ and is parallel to $y + 3 = 0$.

70. Write the equation of the line that passes through the point $(-1, -4)$ and is parallel to $x - 3 = 0$.

71. A police department finds that the number of crimes (y) committed in one week in a small city depends on the number of police officers (x) on special patrol. If there are 85 crimes committed when no police are on special patrol and the number drops to 75 when two police are on special patrol, write the equation for the relationship.

72. If velocity (V) of an object is used as the dependent variable and the time (t) in seconds is used as the independent variable, write the equation of the line that fits the following data: **(a)** $t = 2$ seconds; $V = 40$ feet/second   **(b)** $t = 4$ seconds; $V = 10$ feet/second.

73. The voltage (V) in a circuit is 40 volts when the resistance (R) is 8 ohms. The voltage is 80 volts when the resistance is 12 ohms. Using V as the dependent variable and the resistance (R) as the independent variable, write the equation of the voltage in terms of the resistance. What would the voltage be when $R = 20$ ohms?

STATE YOUR UNDERSTANDING

74. Graph the line $y = 4x + 20$ using the same scales on the x- and the y-axis. Then, using different scales on the two axes, graph the same line. Does the direction of the line change? Does the slope change? Explain.

75. We have been told that any equation in the form $y = mx + b$ is called the slope-intercept form of a straight line where "m" is the slope and "b" is the y-value where the line crosses the y-axis. Why is "b" always the y-intercept?

CHALLENGE EXERCISES

76. Given the points $(1, a)$ and $(3, 2a)$, find the value of a so that the slope of the line passing through the two points is $\dfrac{2}{3}$.

77. The mid-point of a line segment connecting two points $P_1(x_1, y_1)$ and $P_2(x_2, y_2)$ is given by the formula $\left(\dfrac{x_1 + x_2}{2}, \dfrac{y_1 + y_2}{2}\right)$. Find the midpoint of the line segment joining $(-3, 6)$ and $(5, 2)$.

78. Find the equation of the perpendicular bisector of the line segment joining the points $(-3, 6)$ and $(5, 2)$.

79. Given the equation $2ax + 3y + 6 = 0$, find the value of "a" so that the line has a slope of $\dfrac{5}{2}$.

80. If the payoff at the track for a winning horse is $7.20 for a $2.00 bet and $19.20 for a $5.00 bet, find the equation for this relationship using the variable y for the payoff and x for the amount that is bet.

MAINTAIN YOUR SKILLS (SECTIONS 3.2, 3.3, 3.4, 4.5, 5.1)

Perform the indicated operations and reduce if possible:

81. $\dfrac{2a^2 + 9a - 35}{18a^2 - 21a - 4} \cdot \dfrac{12a^2 - 13a - 4}{2a^2 + 19a + 35} \cdot \dfrac{12a^2 + 32a + 5}{8a^2 - 22a + 5}$

82. $\dfrac{1}{3 - y} - \dfrac{1}{3 + y} - \dfrac{1}{3 - y} + \dfrac{1}{3 + y}$

Multiply and simplify:

83. $(\sqrt{15} + \sqrt{24})(\sqrt{5} - \sqrt{6})$

84. $(\sqrt{6} + \sqrt{x - 4})(\sqrt{6} - \sqrt{x - 4})$

Simplify:

85. $\dfrac{x}{\dfrac{1}{3} + \dfrac{1}{y} + \dfrac{1}{6y}}$

86. $\dfrac{\dfrac{x}{3} + \dfrac{x}{y} + \dfrac{1}{6y}}{2}$

Solve by completing the square:

87. $x^2 + 2x = 7$

88. $3m^2 + 2m - 8 = 0$

6.4

■

LINEAR INEQUALITIES

OBJECTIVES

1. Draw the graph of a linear inequality.

2. Draw the graph of the intersection or union of two or more linear inequalities.

Every line divides a plane (rectangular coordinate system) into three distinct sets of points: the set of points on each side of the line, and the set of points on the line itself.

■ **DEFINITION**

Open Half Plane

An open half plane is the set of points on one side of the graph of a straight line.

Closed Half Plane

A closed half plane is the set of points formed by an open half plane and the line itself.

Table 6.1 below shows the standard forms of linear inequalities in two variables and describes the corresponding graph.

TABLE 6.1

Inequality	Graph
$Ax + By < C$	Open half plane
$Ax + By > C$	Open half plane
$Ax + By \leq C$	Closed half plane
$Ax + By \geq C$	Closed half plane

A linear inequality in two variables is graphed by shading all the points in the open half plane or closed half plane.

■ **PROCEDURE**

To draw the graph of a linear inequality in two variables:

1. Draw the graph of the corresponding equality. Use a solid line for a closed half plane and a broken line for an open half plane.

2. Test a point from one of the open half planes. If the coordinates of the point make the inequality true, all the points in the open half plane are in the solution set. If the coordinates of the point make it false, all the points in the other open half plane are in the solution set.

3. Shade the open half plane that represents the solutions.

EXAMPLE 1

Draw the graph of $4x + y \geq 0$.

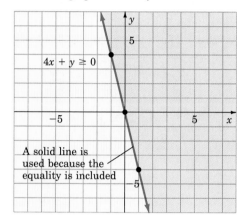

The graph is a closed half plane since the inequality symbol is $\geq$.

First, draw the graph of $4x + y = 0$. Use a solid line because equality is included.

Second, test any point in either open half plane. Here we test $(0, 1)$.

$$4x + y \geq 0$$
$$4(0) + 1 \geq 0$$
$$1 \geq 0 \quad \text{True}$$

Shade the open half plane containing $(0, 1)$.   ■

EXAMPLE 2

Draw the graph of $3x - 2y < 12$.

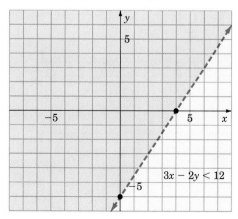

The graph is an open half plane.

First, draw the graph of $3x - 2y = 12$. Use a broken line because the inequality symbol is $<$.

Second, test any point in either open half plane. Here we test $(0, 0)$. (This is often the easiest point to test.)

$$3x - 2y < 12$$
$$3(0) - 2(0) < 12$$
$$0 < 12 \quad \text{True}$$

The graph is the open half plane that includes the origin.   ■

EXAMPLE 3

Draw the graph of $\frac{1}{3}x + \frac{1}{2}y < 2$.

$$6\left(\frac{1}{3}x + \frac{1}{2}y\right) = 6(2)$$

$$2x + 3y = 12$$

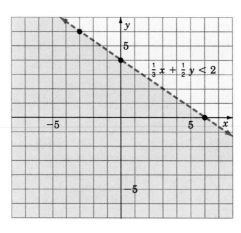

Simplify the corresponding equality.

First, draw the graph of $2x + 3y = 12$ using a broken line since the inequality symbol is $<$.

Second, test the point $(0, 0)$.

$$\frac{1}{3}(0) + \frac{1}{2}(0) < 2$$

$$0 < 2 \quad \text{True}$$

The graph is the open half plane that contains the origin.

EXAMPLE 4

Draw the graph of $y \le -5$.

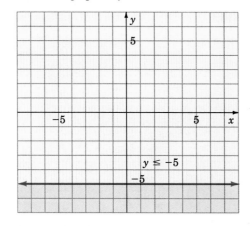

The graph is a closed half plane.

First, draw the graph of $y = -5$ using a solid line since the inequality symbol is $\le$.

Second, test the point $(0, 0)$.

$$0 \le -5 \quad \text{False}$$

The graph is the closed half plane that does not contain the origin.

Example 5 illustrates one application of linear inequalities.

EXAMPLE 5

A dump truck with a divided bed can haul two kinds of barkdust. The truck can carry up to 48 cubic yards, and the truck bed divider can be moved so that any amount (up to 48 yd³) of two kinds of barkdust, ground cover and decorator, can be hauled. Draw the graph that shows the possible loads the truck can haul. Let x represent the number of cubic yards of ground cover barkdust and y represent the number of cubic yards of decorator barkdust.

Simpler word form:

$$\begin{pmatrix} \text{Cubic yards} \\ \text{of ground} \\ \text{cover barkdust} \end{pmatrix} + \begin{pmatrix} \text{Cubic yards} \\ \text{of decorator} \\ \text{barkdust} \end{pmatrix} \leq 48$$

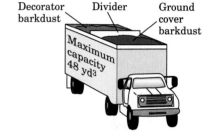

Translate to algebra:

$x + y \leq 48$

The total number of cubic yards of barkdust cannot exceed 48.

Draw the graph of $x + y = 48$.

Test the point (10, 10).

$10 + 10 \leq 48$
$\quad\quad 20 \leq 48$ True

■ **CAUTION**

> The graph is not valid for negative values of x or y.

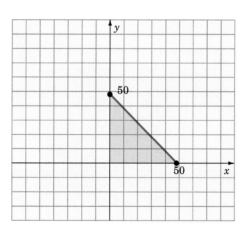

The shaded triangle represents the solution.

Any point on the line or in the shaded region has coordinates that represent a possible load for the truck. For instance, the point (15, 18) represents a load of 15 yd³ of ground cover barkdust and 18 yd³ of decorator barkdust.   ■

The intersection or union of two linear inequalities can be graphed by graphing the solution sets of both inequalities on the same coordinate system. Recall that the union of two solution sets is the set of all solutions of one inequality or the other. The intersection of two solution sets is the set of solutions that are common to both inequalities.

EXAMPLE 6

Draw the graph of the union of the solution sets of $x + 2y < 6$ and $2x - 5y > 10$.

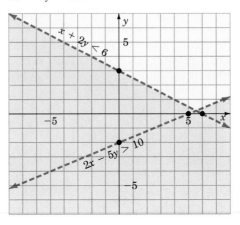

Draw the graph of each inequality separately. The graph of the union of the two inequalities is the entire shaded region.

EXAMPLE 7

Draw the graph of the intersection of $3x - y > 9$ and $2x + 3y \geq 12$.

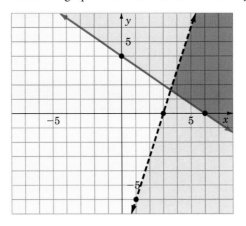

Draw the graph of each of the inequalities. The graph of the intersection. It is the region shaded green together with the lower part of the blue line.

If more than two inequalities are involved, you must identify the region on the graph that meets the given conditions.

EXAMPLE 8

Graph the intersection of $x + y < 4$, $x - y > 4$, and $y > -3$.

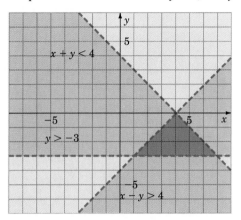

Draw the graph of each of the inequalities. The graph of the intersection is the region common to all three. It is shaded green and is triangular with vertices at $(4, 0)$, $(1, -3)$, and $(7, -3)$.

■ **CAUTION**

It is possible that the intersection of the solution sets of two linear inequalities is empty. The graphs have no points in common.

EXAMPLE 9

Draw the graph of the intersection of $4x + 2y < -5$ and $2x + y > 5$.

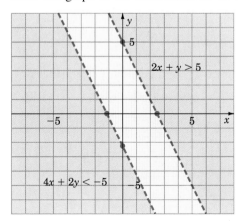

Draw the graph of each inequality.

The solution set is $\emptyset$.

The graphs have no points in common. ■

EXERCISE 6.4

A

Draw the graph of each inequality:

1. $y > x + 1$

2. $y < x + 1$

3. $y < 2x + 3$

4. $y > 2x + 3$

5. $y \leq -x + 5$

6. $y > -x - 1$

7. $y \geq x$

8. $y \leq x$

9. $x > -3$

10. $x \leq 4$

11. $y < 3$

12. $y \geq -2$

B

Draw the graph of each inequality:

13. $x - y < 4$ **14.** $y - x < 4$ **15.** $2x + y \geq 2$

16. $2x - y < 1$ **17.** $3x - 5y < 10$ **18.** $5x + 2y > 10$

19. $2x \geq 3y - 6$ **20.** $x > 3y - 4$

C

Draw the graph of each inequality:

21. $2x - 7y \leq 14$ **22.** $x + 12y > 0$ **23.** $x - 5 - y \geq 0$

24. $x + 5 - y \le 0$

25. $\dfrac{1}{2}y + \dfrac{1}{2} - x \le 0$

26. $\dfrac{1}{6}y + \dfrac{1}{3} - \dfrac{1}{2}x \ge 0$

27. $\dfrac{2}{3}y + \dfrac{5}{6}x \le \dfrac{5}{3}$

28. $-\dfrac{3}{2}x + \dfrac{1}{4}y \le \dfrac{1}{2}$

Draw the graph of the union of the solution sets of the given inequalities:

29. $x + y \le 1, x - y \le 1$

30. $x + y \ge 1, x - y \ge 1$

31. $3x + 2y > 6, 4x - 3y > -6$

32. $5x - 2y < 4, 3x + 4y > 8$

33. $\dfrac{1}{3}x - \dfrac{1}{4}y \le 1, \dfrac{1}{3}x + \dfrac{1}{4}y \ge 1$

34. $\dfrac{2}{3}x - \dfrac{3}{5}y \le 2, \dfrac{1}{4}x + \dfrac{1}{5}y \ge 1$

Draw the graph of the intersection of the solution sets of the given inequalities:

35. $x + y < 5, x - y > 6$

36. $x + y > 9, x - y < -2$

37. $3x + 2y \leq 12, 2x + 5y \geq 10$

38. $3x - 5y \geq 15, 2x + 3y \leq 6$

39. $x > y, 3x + 2y < 0$

40. $6x + 5y > 0, y > x$

41. $x > y + 1, x < y - 3$

42. $3x + 4y > 20, 3x + 4y < -4$

43. $x + 3y < 15, 2x - y > 3,$
  $x - 7y > -21$

44. $x - y > -4, x + 3y < 4,$
  $x + y > -2$

45. $2x - 3y \geq 6, 3x + 2y \leq 6,$
  $x - y \leq 6$

46. $2x + 5y \geq 10, x + 3y \leq 21,$
  $x - 3y \geq -21$

D

47. If the truck in Example 5 can haul 40 cubic yards at most, draw the graph that shows the possible loads the truck can haul.

48. A gravel truck has a divider fixed so that it can haul pea gravel and crushed rock. The divider is moveable so that the split between the types of gravel can vary. If the truck can hold at most 10 cubic yards, draw the graph that shows the possible loads this truck can haul.

49. The Miniature Collection Company manufactures tea sets at a cost of $8 per set and horse and buggy sets at a cost of $5 per set. If manufacturing costs for the week cannot exceed $1000, what are the possible combinations of tea sets (y) and horse and buggy sets (x) that can be made? Graph the solution. (*Note:* Negative values to represent sets have no meaning and are disregarded.)

50. At Van Pelt's Candy Company, fudge costs $3.90 per pound to make, and divinity costs $3.25 per pound to make. If production costs cannot exceed $234, what possible combinations of fudge (x) and divinity (y) can be made? Graph the solution. (Do not graph the negative values.)

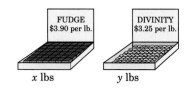

Total production costs of no more than $234

Figure for Exercise 50

51. John's math test consists of 50 true–false questions and 50 multiple-choice problems. Each true–false question is worth 3 points, and each multiple-choice problem is worth 5 points. Draw the graph that shows the combinations of multiple-choice and true–false questions he needs to answer correctly to receive a score of at least 75.

52. Referring to Exercise 51, draw the graph showing the combination John would have to answer correctly for a score of at least 90.

53. Given the same costs to manufacture the miniature sets in Exercise 49, the manufacturer wants to make at least 50 tea sets. Graph the solution of the original inequality with this new added condition.

54. In Exercise 50, the candy company wants to make at least 30 pounds of fudge. Graph your original inequality with this added condition.

55. Sue wants to invest up to $10,000 in stocks and bonds. What possible combinations of stocks (x) and bonds (y) can she have if she wishes to invest twice as much in bonds as she does stocks? (x and y are both greater than or equal to zero.)

56. In Exercise 55, if Sue wishes to invest at least $1,000 in stocks and at least $2,000 in bonds, graph the original inequality with these two new added conditions.

57. In Exercise 55, suppose Sue does not want to invest more than $2,000 in stocks because of the market at this time. Graph the original inequality with this added condition.

STATE YOUR UNDERSTANDING

58. For the graph of linear inequalities in two variables, why is the graph of the line broken when the original problem is $<$ or $>$? Why is the line solid when the original equation is $\leq$ or $\geq$?

59. If you were to know that the graph of $x^2 + y^2 = 9$ is a circle with its center at $(0, 0)$ and its radius is 3, what would the graph of $x^2 + y^2 < 9$ be?

CHALLENGE EXERCISES

60. A nutritionist finds that the total number of milligrams of iron (x) and calcium (y) in your diet should exceed 1000 mg but NOT exceed 2000 mg. Graph the solution to these inequalities. (x and y are both greater than or equal to zero).

61. If the nutritionist in Exercise 60 varies the amount of iron and calcium to be x mg of iron and $3y$ mg of calcium, rewrite the inequality and graph the solution.

62. A circuit calls for two resistors whose sum is not to exceed 500 ohms. If the ratio of resistor #1 (x) to resistor #2 (y) is 3 to 5, write the inequality and graph the solution.

63. From Exercise 62, if the resistance of resistor #1 is to be at least 50 ohm and the resistance of #2 is to be at most 150 ohms, graph the original solution WITH these added conditions.

MAINTAIN YOUR SKILLS (SECTIONS 3.2, 4.6, 4.8, 5.2)

Perform the indicated operations and reduce if possible:

64. $\left(\dfrac{1}{2+x} - \dfrac{1}{3+x}\right)\left(\dfrac{2}{x-2} + \dfrac{2}{x+3}\right)$

65. $\dfrac{x^2 - 15x + 54}{2x^2 + 13x + 21} \div \dfrac{x^2 - x - 72}{2x^2 - 3x - 35}$

Divide (rationalize the denominator) and simplify:

66. $\dfrac{\sqrt{3}}{\sqrt{6}} - \dfrac{\sqrt{3}}{\sqrt{15}}$

67. $\dfrac{\sqrt{3}}{\sqrt{6} - \sqrt{15}}$

Divide and write the answer in standard form:

68. $\dfrac{2}{3} + \dfrac{6}{i}$

69. $\dfrac{8}{3+i}$

Solve by the quadratic formula:

70. $y = 6 - \dfrac{8}{y}$

71. $\dfrac{3}{y+2} = \dfrac{y}{16}$

6.5

VARIATION

OBJECTIVES

1. Solve problems involving direct variation.

2. Solve problems involving joint variation.

3. Solve problems involving inverse variation.

The physical world is in a constant state of change or variation. Measurements of changes in temperature, rainfall, light, heat, speed, and fuel supplies are recorded. Changes in height, weight, and blood pressure can be important to a person's health. Many situations in business and industry will vary depending on cost, supplies, personnel, time, and space availability. Here we are concerned about direct, inverse, and joint variations.

■ **DEFINITION**

Direct Variation

> Two quantities vary directly when their quotient is a nonzero constant.

$\dfrac{y}{x} = k$

■ FORMULA

$\dfrac{y}{x} = k,\ k \neq 0$ or $y = kx,\ k \neq 0$

The nonzero constant, k, is called the constant of variation.

In direct variation, as x increases, y also increases.

Distance (D) traveled at a constant speed (r) varies directly as the time (t). Here the speed is the constant of variation.

$\dfrac{D}{t} = r$ or $D = rt$

The cost (C) of apples at a constant price per pound (p) varies directly as the weight (w).

$\dfrac{C}{w} = p$ or $C = pw$

In many applications of variation, you are given the data on one situation. You can apply this data to find the constant of variation.

EXAMPLE 1

The weight of a metal ingot varies directly as its volume. If an ingot containing $1\dfrac{2}{3}$ cubic feet weighs 350 pounds, what will an ingot of $2\dfrac{1}{3}$ cubic feet weigh?

Let w = weight and v = volume, then   Since this is direct variation, $\dfrac{w}{v}$ is
$w = kv$.

a nonzero constant.

First case:

$w = 350,\ v = 1\dfrac{2}{3}$ Identify w and v in the first case and solve for k.

$350 = \left(\dfrac{5}{3}\right)k$

$k = 210$

Second case:

$w = ?,\ v = 2\dfrac{1}{3}$ Identify v and w. Use the value of k found in the first case, $k = 210$.

$$w = \left(\frac{7}{3}\right)(210)$$

$$w = 490$$

The weight of a $2\frac{1}{3}$ cubic foot ingot is 490 pounds.   ■

EXAMPLE 2

A car salesman's salary varies directly as his total sales. If he receives $448 on sales of $6400, how much will he earn on sales of $16,350?

Let C = commission and s = sales, then $C = ks$.

The commission (C) varies directly as the total sales (s).

First case:

$C = \$448$, $s = \$6400$

$448 = 6400k$

$k = 0.07$

Substitute in the formula, $C = ks$, and solve for k.

Second case:

$C = ?$, $s = \$16,350$

$C = (16,350)(0.07)$

$\quad = 1144.5$

Use the constant of variation found in the first case to find the commission.

The salesman will earn a commission of $1144.50 on sales of $16,350.   ■

Another type of variation is joint variation.

■ DEFINITION

Joint Variation

> Three quantities vary jointly when one quantity varies directly as the product of two other quantities.

■ FORMULA

$\dfrac{z}{xy} = k$

$$\frac{z}{xy} = k,\ k \neq 0 \quad \text{or} \quad z = kxy,\ k \neq 0$$

The cost (C) of shipping goods varies jointly as the weight (w) and the distance shipped (d).

$C = kdw$

The temperature (T) of a gas varies jointly as the volume (V) and the pressure (P).

$T = kPV$

EXAMPLE 3

The interest paid on borrowed money varies jointly as the principal and the time. If $96 of interest is paid on a loan of $600 in two years, at the same rate, how much interest is paid on $1000 borrowed for three years?

Let i = interest, P = principal, and t = time.

$i = kPt$

This is an example of joint variation, so the quotient, $\dfrac{i}{Pt}$ is a constant (k).

Here the constant of variation is the interest rate.

First case:

$i = \$96$, $P = \$600$, $t = 2$ years

$96 = (k)(600)(2)$

$k = 0.08$

Use the values in the first case to solve for k.

The interest rate is 8%.

Second case:

$i = ?$, $P = \$1000$, $t = 3$ years

$i = (0.08)(1000)(3)$

$= 240$

Use the value of k from the first case to solve for i.

The interest paid is $240. ∎

The last type of variation we consider here is inverse variation.

■ DEFINITION

Inverse Variation

Two quantities vary inversely when their product is a nonzero constant.

■ FORMULA

$xy = k$

$xy = k$, $k \neq 0$ or $y = \dfrac{k}{x}$, $k \neq 0$

With inverse variation, as x increases, y decreases.

The number of donors (N) needed to raise a fixed amount of money (M) varies inversely as the average donation (D).

$ND = M$ Here, M is the constant of variation.

The gravitational attraction (F) between two bodies varies inversely with the square of the distance (d) separating them.

$Fd^2 = k$

EXAMPLE 4

The rate of vibration of a string under constant tension varies inversely as its length. If a string 42 cm long vibrates 512 times per second, what length of string will vibrate 768 times per second?

Let r = rate of vibration and L = length.

This is inverse variation, so the product, rL, is a nonzero constant, k.

$rL = k$

First case:

$r = 512, L = 42$

Use the values in the first case to solve for k.

$512(42) = k$
$21504 = k$

Second case:

$r = 768, L = ?$

Use the value of k found in the first case to solve for L.

$768L = 21504$
$L = 28$

Note that the length of the string decreased as the number of vibrations increased.

So, under similar conditions, a string 28 cm long will vibrate 768 times per second.

■ PROCEDURE

To solve a variation problem:

1. Identify the formula involved (the type of variation).

2. Using the data in the first case and the appropriate variation formula, solve for the constant of variation.

3. Substitute the constant of variation and the given data in the second case in the same formula and solve.

EXERCISE   6.5

A

1. If x varies directly as y, and if $x = 6$ when $y = 15$, find y when $x = 30$.

2. If d varies directly as t, and if $d = 6$ when $t = 45$, find t when $d = 8$.

3. If c varies directly as n, and if $c = 25$ when $n = 150$, find n when $c = 15$.

4. If w varies directly as v, and if $v = 35$ when $w = 8$, find v when $w = 28$.

5. If x varies inversely as y, and if $x = 24$ when $y = 6$, find x when $y = 15$.

6. If x varies inversely as y, and if $x = 54$ when $y = 120$, find y when $x = 135$.

7. If P varies inversely as T, and if $P = 75$ when $T = 65$, find P when $T = 78$.

8. If q varies inversely as m, and if $m = 36$ when $q = 15$, find q when $m = 45$.

9. If z varies jointly as x and y, and if $z = 36$ when $x = 4$ and $y = 9$, find z if $x = 28$ and $y = 7$.

10. If z varies jointly as x and y, and if $z = 72$ when $x = 15$ and $y = 9$, find z when $x = 12$ and $y = 10$.

11. If a varies jointly as b and c, and if $a = 100$ when $b = 5$ and $c = 25$, find a when $b = 12$ and $c = 60$.

12. If q varies jointly as r and t, and if $q = 42$ when $r = 10$ and $t = 35$, find q when $r = 6$ and $t = 21$.

B

13. The pressure P per square inch in water varies directly as the depth d. If $P = 5.77$ when $d = 13$, find P when $d = 32$ (to the nearest tenth).

14. Commission on sales varies directly as the total value of the sales. If a salesperson earns $653 on sales of $11,500, what would be the commission on sales of $41,000?

15. The time to drive from Portland to Reno varies inversely as the average speed driven. If it takes Mary 12 hours at 55 mph, how long would it take her at 60 mph?

16. The length of a rectangle with a constant area varies inversely as the width of the rectangle. If the length is 42 m when the width is 35 m, find the length when the width is 30 m.

17. The number of amperes varies directly as the number of watts. For a reading of 550 watts, the number of amperes is 5. What is the number of amperes when the reading of watts is 1600? (to the nearest tenth).

18. The force needed to raise an object with a lever varies inversely as the length of the lever. If it takes 45 lb of force to lift a certain object with a 2-ft-long lever, what force will be necessary if a 3-ft lever is used?

19. The intensity of illumination from a given source of light varies inversely as the square of the distance from the source in inches. If the intensity of illumination on a book 12 inches from the source is 360, what is the intensity of illumination if the book is placed 60 inches from the light source?

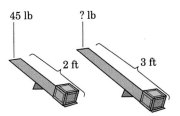

Figure for Exercise 18

20. The weight of wire varies directly as its length. If 3500 ft of wire weighs 150 lb, what will 14,250 ft of the wire weigh? (to the nearest pound).

C

21. At PPL Express, the cost of shipping goods (C) varies jointly as the distance shipped (d) and the weight (w). If it costs \$338 to ship 6.5 tons of goods 8 miles, how much will it cost to ship 24 tons of goods 1535 miles? (round to the nearest dollar).

22. The pressure in a liquid varies jointly with the depth and the density of the liquid. If the pressure is 320 when the depth is 240 and the density is 1.5, find the pressure when the depth is 360 and the density is 3.5 (to the nearest ten).

23. Simple interest earned during a fixed time varies jointly as the rate of interest and the principal invested. If \$1615 is earned when \$8500 is invested at 9.5%, how much interest will be earned if \$15,500 is invested at 11.5%?

24. The gravitational attraction between two bodies varies inversely as the square of the distance separating them. If the attraction measures 64 when the distance is 3 cm, what will it measure when the distance between them is 10 cm?

25. The volume of a box with a fixed depth varies jointly as the length and width of the bottom of the box. If the volume of the box is 1152 cm^3 when the dimensions of the bottom are 16 cm × 12 cm, find the volume when the bottom dimensions are 15 cm × 7 cm.

26. In electricity, the resistance (R) to the flow of current through a wire is directly proportional to the length (L) of the wire. If the resistance is $R = 100$ ohms when the length is 20 feet, write the equation for this relationship. What would be the resistance if the wire was 170 feet long?

27. In physics, when a wrench is attached to a bolt and a force is applied to the end of the wrench, the corresponding force applied to the bolt is called "torque." Torque is jointly proportional to the force (F) applied to the end of the wrench and the length (L) of the wrench. If a force of 40 pounds results in a torque of 200 foot-pounds, how much force would be required to produce 350 foot-pounds using a wrench of the same length?

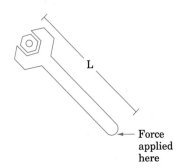

Figure for Exercise 27

28. In physics, Hooke's law states that the force (F) needed to stretch a spring is directly proportional to the distance (x) that it is stretched. Write the equation for this relationship using ''k'' as a constant (k is called the spring constant). If a force of 40 pounds is required to stretch the spring 1.2 feet, what is the value of k for this spring? How much force would be required to stretch the spring 3.6 feet?

Figure for Exercise 28

29. If the pressure (P) of a gas varies directly with the absolute temperature (T) in degrees Kelvin and inversely with the volume (V), write the equation of the relationship using the constant k. If $P = 15$ lbs/in² when the volume $V = 33$ in³ and the absolute temperature is $330°\,K$, what is the value of k (the constant of proportionality)? What will be the value of P when $V = 25$ in³ and $T = 350°\,K$?

STATE YOUR UNDERSTANDING

30. Write a sentence to describe the following direct, inverse and joint variations.

a. $y = \dfrac{kx}{t^2}$ **b.** $y = \dfrac{kx^3}{\sqrt{t}}$ **c.** $y = \dfrac{k}{x^3\sqrt{t}}$

CHALLENGE EXERCISES

31. L varies jointly with M and the square of N and inversely with the square root of P. Write the equation using a constant k.

32. y varies directly with the square root of r and inversely with the square of t. Write the equation using a constant k. If y is 5, when r is 25 and t is 6, find k (the constant).

33. Using the equation in Exercise 32, find the value of y when $r = 49$ and $t = 4$.

34. Given the following table for a direct proportion, fill in the missing values:

x	2	5	
y	10		65

35. Given the table for an inverse proportion, fill in the missing values:

x	3		4.5
y		21	8.4

36. Given the table for the joint variation where c varies jointly with a and b, fill in the missing values:

a	2		4
b	4	5	2
c		80	40

MAINTAIN YOUR SKILLS (SECTIONS 2.2, 2.3)

Simplify using only positive exponents:

37. $(w^{-1}y^{-1})^{-1}(wy)^{-1}$

38. $\left(\dfrac{a^{-2}}{b^{-2}}\right)\left(\dfrac{a^2}{b^2}\right)^{-2}$

39. $\dfrac{(3x^2y^4)^{-4}}{(2x^{-3}y^2)^{-2}}$

40. $(-3a^2)^2(2a^{-3})^2(-3a^4)^{-5}$

Write in scientific notation:

41. $7{,}843{,}000{,}000$

42. 0.0000007843

Write in place value form:

43. 7.843×10^{-1}

44. 7.843×10^5

CHAPTER 6

SUMMARY

$Ax + By = C$	Standard form of a linear equation. The graph is a straight line.	(pp. 406, 438)
(x, y)	Ordered pair of real numbers. The independent variable is x and the dependent variable is y. Used to write a solution of an equation in two variables. Also used as the coordinates of a point on a graph.	(p. 406)
x-intercept $(x, 0)$	The point where a line crosses the x-axis.	(p. 420)
y-intercept $(0, y)$	The point where a line crosses the y-axis.	(p. 420)
$m = \dfrac{\text{rise}}{\text{run}} = \dfrac{y_2 - y_1}{x_2 - y_2}$	Equation for the slope of a straight line.	(p. 422)
$d = \sqrt{(x_2 - x_1)^2 + (y_2 - y_1)^2}$	Formula for the distance between two points.	(p. 428)
Parallel Lines $m_1 = m_2$	Two lines are parallel if their slopes are equal and their y-intercepts are different.	(p. 425)
Perpendicular Lines $m_1m_2 = -1$	Two lines are perpendicular if the product of their slopes is -1.	(p. 425)
$y - y_1 = m(x_2 - x_1)$	Point-slope form of a linear equation.	(pp. 434, 438)
$y = mx + b$	Slope-intercept form of a linear equation.	(pp. 435, 438)

$Ax + By < C$ $Ax + By > C$ $Ax + By \leq C$ $Ax + By \geq C$	Standard forms of a linear inequality.	(p. 444)
Open-Half Plane	All the points on one side or the other of a straight line.	(p. 444)
Closed-Half Plane	All the points on one side or the other of a straight line, including the points on the line.	(p. 444)
Direct Variation $\dfrac{y}{x} = k$ or $y = kx$	The formula for direct variation, where y varies directly as x. The constant of variation is k.	(p. 458)
Joint Variation $\dfrac{z}{xy} = k$ or $z = kxy$	The formula for joint variation, where z varies directly as y and x.	(p. 459)
Inverse Variation $xy = k,\ k \neq 0$	The formula for inverse variation, where y varies inversely as x.	(p. 460)

CHAPTER 6
REVIEW EXERCISES

SECTION 6.1   Objective 1

Find the ordered pair that is a solution of $3x - 7y = 42$ given the value of x or y.

1. $x = 0$ **2.** $y = 0$ **3.** $x = 7$ **4.** $y = 6$

Find the ordered pair that is a solution of $9x - 2y = 36$ given the value of x or y.

5. $x = 5$ **6.** $x = -1$ **7.** $y = -\dfrac{9}{2}$ **8.** $y = -8$

SECTION 6.1   Objective 2

Draw the graph of each of the following.

9. $2x - y = 6$ **10.** $3x + 2y = 6$ **11.** $3x - 5 = 0$

12. $2x + 7 = 18$

13. $\dfrac{1}{4}x - \dfrac{1}{5}y = 20$

14. $\dfrac{2}{9}x + \dfrac{5}{6}y = \dfrac{1}{3}$

SECTION 6.2   Objective 1

Write the coordinates of the x- and y-intercepts.

15. $5x - 7y = 35$

16. $-3x + 2y = 18$

17. $2x + y = -8$

18. $3x - y = 15$

19. $\dfrac{1}{3}x + \dfrac{1}{7}y = 21$

20. $\dfrac{1}{5}x - \dfrac{1}{6}y = 5$

SECTION 6.2   Objective 2

Find the slope of the line joining the following points.

21. $(-3, 1), (4, 2)$

22. $(0, 5), (6, 1)$

23. $(-2, -5), (6, -1)$

24. $(3, -7), (0, 4)$

25. $(9, -7), (5, -6)$

26. $(8, -3), (-2, -4)$

27. $\left(\dfrac{1}{3}, -\dfrac{2}{5}\right), \left(\dfrac{3}{5}, \dfrac{2}{3}\right)$

28. $\left(-\dfrac{3}{4}, \dfrac{1}{2}\right), \left(\dfrac{3}{8}, -\dfrac{1}{2}\right)$

SECTION 6.2   Objective 3

Are the line segments joining the following pairs of points parallel, perpendicular, or neither?

29. $(4, -6), (5, 3)$
  $(7, 8), (10, 34)$

30. $(-3, -2), (5, 7)$
  $(-1, 6), (8, -2)$

31. $(12, -6), (8, -3)$
  $(2, -5), (-2, -2)$

32. $(11, 6), (5, 4)$
  $(6, -7), (5, -10)$

33. Do the coordinates $(6, 3)$, $(-2, 8)$, and $(-7, 16)$ form a right triangle?

34. Do the coordinates $(-7, -3)$, $(-5, 3)$, $(0, -1)$, and $(2, 5)$ form a parallelogram?

SECTION 6.2   Objective 4

Find the distance between the points.

35. $(8, -3), (5, 6)$

36. $(-1, -2), (-3, -7)$

37. $(-5, 0), (6, -4)$

38. $(3, -5), (-1, -1)$

39. $\left(\frac{1}{2}, -\frac{1}{5}\right), \left(\frac{1}{4}, -\frac{3}{5}\right)$

40. $\left(-\frac{3}{7}, -\frac{1}{2}\right), \left(-1, \frac{2}{7}\right)$

41. $(0.3, -0.2), (-0.1, 0.5)$

42. $(-0.15, 2), (0.55, -1.6)$

SECTION 6.3   Objective 1

Write the equation of the line, in standard form, given its slope and a point.

43. $(-5, -3), m = \frac{1}{3}$

44. $(7, -5), m = -\frac{2}{3}$

45. $(6, 0), m = -\frac{3}{8}$

46. $(-5, -7), m = \frac{5}{9}$

47. $\left(-\frac{1}{3}, -\frac{3}{4}\right), m = -\frac{2}{3}$

48. $\left(\frac{2}{7}, -\frac{3}{7}\right), m = -\frac{6}{7}$

49. $(0.15, -0.3), m = -0.5$

50. $(-0.35, 0.45), m = 0.4$

SECTION 6.3   Objective 2

State the slope and the y-intercept for the graph of each of the following.

51. $2x - 5y = 15$

52. $3x + 7y = -14$

53. $2y = 8x - 9$

54. $-2x = y + 12$

55. $\frac{1}{3}x + y = 7$

56. $-\frac{1}{2}y - x = 8$

57. $2x - \frac{1}{3}y + 8 = 0$

58. $\frac{1}{4}x - \frac{2}{5}y - 2 = 0$

SECTION 6.3   Objective 3

Draw the graph of the line given its slope and y-intercept.

59. $(0, -1), m = \frac{3}{4}$

60. $(0, -6), m = -\frac{3}{2}$

61. $(0, 5), m = -\dfrac{2}{3}$ **62.** $(0, 1), m = 4$

SECTION 6.3   Objective 4

Write each of the following equations in standard form and slope-intercept form.

63. $y - 5x + 7 = 0$ **64.** $2x = 5y + 9$ **65.** $y = 18 - \dfrac{2}{3}x$

66. $x = 12 - \dfrac{4}{5}y$ **67.** $-7y + 3x + 14 = 0$ **68.** $-\dfrac{1}{2}y + x - 8 = 0$

SECTION 6.4   Objective 1

Draw the graph of each of the following.

69. $2x - 3y > 0$ **70.** $2x + 3y \geq 0$ **71.** $-3x + 7y \leq 21$

72. $5x - 3y > 15$ **73.** $2x + 8 \geq 14$ **74.** $-3y + 9 \leq 6$

SECTION 6.4   Objective 2

Draw the graph of the union of the solution sets of the given inequalities.

75. $2x - y < 5, -3x + 2y \geq 6$

76. $y - 3x \leq 0, x - 3y \geq 0$

Draw the graph of the intersection of the solution sets of the given inequalities.

77. $5x - y \geq 10, 2x + y \leq 8$

78. $2x + 3y < 6, 3x - 2y > 6$

79. $3y + 5x < 15, 2x + 3y \geq 6$

80. $2x - 7y \leq 14, x \geq y$

81. $x > y, x \geq y + 1, x \geq y - 5$

82. $x + 2y < 6, x - 2y < 6,$
$x \geq -2$

SECTION 6.5   Objective 1

83. If d varies directly as t, and if $d = 30$ when $t = 45$, find d when $t = 36$.

84. If s varies directly as r, and if $s = 7.5$ when $r = 1.5$, find s when $r = 9$.

85. The taxes on a house vary directly as the value of the house. If the tax on an $80,000 home is $1750, what is the tax on a $125,000 home? (to the nearest dollar).

SECTION 6.5   Objective 2

86. If q varies jointly as s and t, and if $q = 42$ when $s = 3$ and $t = 7$, find q when $s = 2$ and $t = 20$.

87. If a varies jointly as b and c, and if $a = 3.6$ when $b = 5$ and $c = 2.4$, find a when $b = 30$ and $c = 1$.

88. Simple interest varies jointly as the rate of interest and the principle invested. If \$2500 interest is earned when \$15,000 is invested at 8%, find the interest earned when \$9000 is invested at 10.5%.

SECTION 6.5   Objective 3

89. If s varies inversely as t, and if $s = 14.5$ when $t = 6$, find s when $t = 10$.

90. If x varies inversely as y, and if $x = 24.6$ when $y = 50$, find x when $y = 20.5$.

91. The time it takes to travel from Seattle to San Francisco varies inversely with the average speed driven. If it takes Thong 14.5 hours at 55 mph, how long will it take him if he averages 65 mph? (to the nearest tenth of an hour).

CHAPTER 6
TRUE–FALSE CONCEPT REVIEW

Check your understanding of the language of algebra. Tell whether each of the following statements is true (always true) or false (not always true).

1. An equation with two variables can be a contradiction (inconsistent).

2. The graph of the point with coordinates $(2, -5)$ is in quadrant II.

3. The first number in an ordered pair is one of the values of the independent variable.

4. The value of the abscissa represents the distance from the y-axis.

5. All equations with two variables are linear equations.

6. A rectangular coordinate system contains two number lines.

7. A straight line always has an equation of the form $Ax + By = C$ if A and B are not both zero.

8. The graph of the equation $x = -12$ is a vertical line.

9. The x-intercept of a graph always occurs where $y = 0$.

10. A line that slopes downward toward the left has negative slope.

11. The slope of a vertical line is not defined.

12. There are lines that have no x-intercept.

13. The distance formula, $d = \sqrt{(x_2 - x_1)^2 + (y_2 - y_1)^2}$, is based on the slope of a line.

14. The point-slope form of the equation of a line is $A(y - y_1) = B(x - x_1) + C$.

15. The graph of a linear inequality is always an open half plane.

16. It is possible for a linear inequality to have no solution.

17. If $wv = a$, then w varies directly as v.

18. If $w = v + a$, then w varies directly as v.

CHAPTER 6

TEST

1. Find the slope of the line joining the points $(-8, 12)$ and $(-12, 8)$.

2. Find the distance between the points $(-12, 1)$ and $(0, -7)$.

3. Write the equation $\dfrac{7}{3}x = \dfrac{5}{6} - 5y$ in slope-intercept form.

4. Find the slope of the line joining the points $(18, 16)$ and $(21, 13)$.

5. Draw the graph of $x + 2y = 4$.

6. Is the graph of $y = -2x - 14$ perpendicular to the graph of $2x - y + 6 = 0$?

7. Find the ordered pair that is a solution of the equation $y = -9x + 34$ when $x = -4$.

8. Write the equation $\dfrac{7}{3}x = \dfrac{5}{6} - 5y$ in standard form.

9. Write the ordered pairs for the x- and y-intercepts of $7x - 6y = 56$.

10. Is the graph of $y = -3x - 4$ perpendicular to the graph of $x - 3y - 1 = 0$?

11. Draw the graph of $2x - 3y = 6$.

12. Find the slope of the line joining the points $(8, -12)$ and $(-8, 12)$.

13. Write the equation of the line, in standard form, that has a slope of $-\dfrac{2}{3}$ and passes through the point $(-3, 2)$.

14. Find the distance between the points $(-12, 15)$ and $(-6, 7)$.

15. State the slope and the y-intercept of the graph of $4x - 5y = 20$.

16. Draw the graph of $2x - 5y < 10$.

17. Write the ordered pairs for the x- and y-intercepts of $6x - 5y = 20$.

18. Find the slope of the line joining the points $(-13, 15)$ and $(17, 12)$.

19. Write the equation of the line, in standard form, that has a slope of $-\dfrac{2}{3}$ and that passes through the point $(2, 4)$.

20. Write the equation $\dfrac{1}{2}x = -\dfrac{3}{4}y - 3$ in slope-intercept form.

21. What is the slope of the graph of $\dfrac{2}{3}x + 2y = -5$?

22. State the slope and the y-intercept of the graph of $3x - 5y = 15$.

23. Draw the graph of $3x - 4y \geq 12$.

24. Draw the graph of the intersection of the solution sets of $5x - y < 5$ and $x + 2y \geq 6$.

25. Is the graph of $y = \dfrac{3}{4}x + 14$ parallel to the graph of $4x + 3y = -24$?

26. The time it takes to travel between two points varies inversely as the speed traveled. If it took 3.5 hours to cover the distance at 120 mph, how long would it take to cover the distance at 70 mph?

27. The circumference of a circle varies directly as its diameter. If the circumference of a circle with diameter 63.7 centimeters is 200.2 centimeters, what is the circumference of a circle with diameter 59.5 centimeters?

CHAPTER 4

Simplify and combine (when possible)

1. $(27)^{1/3}$

2. $\left(\dfrac{4}{9}\right)^{3/2}$

3. $\sqrt[3]{54}$

4. $\sqrt{50} - \sqrt{18} + \sqrt{2}$

5. $(2\sqrt{5} - \sqrt{2})(3\sqrt{5} + \sqrt{2})$

6. $\sqrt{\dfrac{3}{2}} + \sqrt{\dfrac{2}{3}}$

7. $\dfrac{2}{\sqrt{5} - 2}$

Simplify and combine (when possible)

8. $(8t^3)^{1/3}$

9. $(9x^{1/2}y^4)^{3/2}$

10. $\sqrt[3]{16a^3b^5}$

11. $\sqrt{8x^3y} - 2\sqrt{2x^3y} + x\sqrt{18xy}$

12. $\dfrac{a}{\sqrt{a} - b}$

13. Rationalize the numerator: $\dfrac{\sqrt{x} + \sqrt{y}}{\sqrt{x}}$

Write in standard form

14. $\sqrt{-64}$

15. $\sqrt{-50}$

16. $\sqrt{-18} + \sqrt{-8}$

Simplify

17. i^{33}

18. $i^{17} + i^{11}$

Perform the indicated operation

19. $(4 - 3i) + (2 - i)$

20. $(3 - 7i) - (-2 + 3i)$

21. $(3 - 4i)(2 + i)$

22. $\left(\dfrac{4}{1 - i}\right) + (3 - 2i)(1 - i)$

CHAPTER 5

Solve the following

23. $2x^2 + x - 6 = 0$

24. $5x^2 - 2x - 4 = x^2 - 2x + 1$

25. $3x^2 = 2x - 2$

26. $3x(x - 2) = 1 - x$

27. $|3x - 1| = 8$

28. $2|3 - 2x| - 3 = 7$

29. $\sqrt{7y + 4} - 2 = y$

30. $\sqrt{2y + 3} = 1 + \sqrt{y + 1}$

31. $4y^4 - 37y^2 + 9 = 0$

32. $y^3 = 2y^2 - 3y$

33. $8y^4 = y$

34. $\dfrac{x + 1}{x} + \dfrac{14}{x - 7} = \dfrac{3x - 7}{x^2 - 7x}$

35. $\dfrac{3a}{a + 1} + \dfrac{2}{a - 2} = 5$

36. $x^{-4} - 13x^{-2} + 36 = 0$

Solve the following inequalities and write the solution sets in interval form.

37. $|2x - 1| > 5$

38. $|2x - 3| \leq 3$

39. $|2x + 7| < x + 6$

40. $(x - 3)(2x + 1) \geq 0$

41. $2x(x + 4) < x + 15$

42. $y^3 + 2y^2 - 15y \leq 0$

43. $\dfrac{(x - 3)}{(x + 5)} \leq 0$

44. $\dfrac{x^2 - 9}{x^2 - 2x + 1} < 0$

45. $\dfrac{4}{x + 1} < \dfrac{3}{x + 2}$

CHAPTER 6

For the following linear equations, find: (a) the x-intercepts; (b) the y-intercepts; (c) a third solution; and (d) graph the lines

46. $3x - 2y = 6$

47. $y = -2x + 3$

48. $4x + 3y = 0$

49. $2x + 6 = 0$

50. $y = -3$

51. $\dfrac{x}{3} - \dfrac{y}{4} = 1$

Given the following pairs of points, find: (a) the slope of the line passing through them; (b) the distance between them; and (c) the equation of the line passing through them in standard form.

52. $(-2, 4)$ and $(1, 6)$

53. $(4, -2)$ and $(1, 7)$

54. $(-3, 4)$ and $(-3, -1)$

Determine if the lines passing through the given pairs of points are parallel, perpendicular, or neither.

55. $(2, 2)$ and $(-1, 4)$; $(1, -4)$ and $(3, -1)$

56. $(2, -2)$ and $(-1, 3)$; $(-3, -1)$ and $(0, 4)$

For the following linear inequalities, draw the graphs and shade the appropriate region.

57. $y \geq \dfrac{2}{3}x - 3$

58. $x + 2y < 0$

59. $y \leq -3$

60. The intersection of the following: $x + 2y \geq 3$, $y < 2x - 1$

61. The union of the following: $y > 2x - 3$, $y \geq \dfrac{1}{2}x$

For the following: (a) write the general form of the variation; (b) find the value of ''k''; and (c) solve for the unknown data

62. Given that ''a'' varies directly as the square of t, and $a = 20$ when $t = 2$, find a when $t = 5$.

63. Given that r varies directly as m and inversely as the square of n, and $r = 9$ when $m = 3$ and $n = 2$, find r when $m = 5.2$ and $n = 2.4$

64. The volume of a sphere varies directly as the cube of its radius. If the volume is 36π cubic inches when the radius is 3 inches, find the volume when the radius is 5 inches.

Solve

65. The square root of the sum of a number and seven is one more than the square root of that number. Find the number.

66. A ball is thrown vertically into the air from the top of a 100 ft building and its height (h) is given by $h = -16t^2 + 60t + 100$, where t is the time in seconds. How high is the ball after 1 second? How high is the ball after 3 seconds?

67. Jim finds a linear relationship between the profit, y, and the number of Christmas ornaments sold, x. He makes $14 when he sells 4 ornaments and $49 when he sells 14 ornaments. Write the equation relating profit and the number of ornaments sold. What would his profit be if he sold 40 ornaments?

68. Andy and Donna wish to invest up to $15,000 in real estate and stocks. If the amount invested in real estate is "x" and the amount invested in stocks is "y", write the inequality to show the relationship and graph it.

69. The number of cars (C) sold at a dealership varies directly as the number of customers (n) that enter the showroom and inversely as the day of the month (d). If 10 cars were sold when 100 people entered the showroom on the 20th of the month, how many cars can they anticipate selling if 70 people enter on the 10th day of the month.

CHAPTER

7

SYSTEMS OF
LINEAR
EQUATIONS

Systems of linear equations can be used to solve many types of mixture problems. In Exercise 63, Section 7.2, the Alaskan Fish Company is marketing a combination of shrimp and crab meat. The key points to remember in writing the system of equations is that the weight of the mixture must equal the sum of the original weights of the crab and shrimp. We assume no weight loss during the mixture process. Also, the income generated by the mixture is the same as that generated if the shrimp and crab are sold separately. (This is not always the case as many companies charge a premium to create the mixture.) *(Alvis Upitis/The Image Bank)*

PREVIEW

In chapter 7, linear systems of equations in two or three variables are solved. Solutions by substitution, linear combinations, Cramer's Rule, and matrices are illustrated. Determinants and matrices are introduced to support Cramer's Rule and matrix methods.

Similar techniques for solving systems involving quadratic equations are used in chapter 8. Chapter 10 contains applications that utilize the solution of systems.

7.1

SOLVING SYSTEMS BY GRAPHING AND SUBSTITUTION

OBJECTIVES

1. Solve a system of linear equations in two variables by graphing.

2. Solve a system of linear equations in two variables by substitution.

A *system of equations* consists of two or more equations. Systems of linear equations are discussed in this chapter.

■ DEFINITION

System of Linear Equations in Two Variables

A *system of linear equations in two variables* is a pair of linear equations that can be written in the form:

$$\begin{cases} A_1x + B_1y = C_1 \\ A_2x + B_2y = C_2 \end{cases}$$

The solution of the system is the intersection of the solution sets of the equations.

There are three possible outcomes when a system of two linear equations is graphed on a coordinate system. The graphs are the same line, parallel lines, or lines that intersect in one point. The following words are used to describe these systems.

■ DEFINITIONS

A system of linear equations in two variables is dependent if one equation is a multiple of the other. The system is independent if the equations are not multiples of each other.

(continued next page)

A system is consistent if the graphs of the equations intersect in at least one point. The system is inconsistent if the graphs do not intersect (parallel lines).

Examples and graphs of each of the three possibilities follow:

1. The same line, a dependent and consistent system (Figure 7.1).

$$\begin{cases} x + y = 8 \\ 2x + 2y = 16 \end{cases}$$

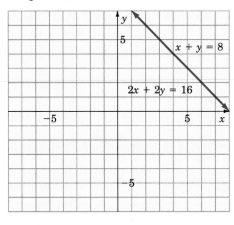

The second equation is a multiple of the first. $2x + 2y = 16$ is $2(x + y) = 2(8)$; the graphs are identical.

FIGURE 7.1

The solution set of the system can be written either $\{(x, y) \mid x + y = 8\}$ or $\{(x, y) \mid 2x + 2y = 16\}$.

2. The graphs are parallel lines, an independent and inconsistent system (Figure 7.2).

$$\begin{cases} 2x + 3y = 12 \\ 2x + 3y = 3 \end{cases}$$

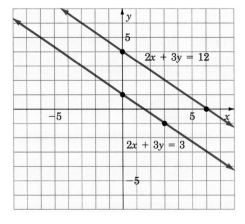

The lines are parallel. (Slopes are the same, and the y-intercepts are different.)

FIGURE 7.2

The solution set is $\emptyset$.

3. The graphs intersect in exactly one point, an independent and consistent system (Figure 7.3).

$$\begin{cases} 3x - 2y = 6 \\ 3x + 2y = 18 \end{cases}$$

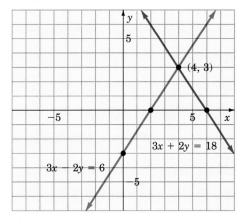

The lines intersect at the point $(4, 3)$.

FIGURE 7.3

The solution set is $\{(4, 3)\}$.

The three possibilities can be identified without graphing. Write the equations in the slope-intercept form, and compare their slopes and y-intercepts.

Type of System	Slopes of Lines	y-Intercepts	Solution
Dependent	Same	Same	Infinite number of solutions
Independent and inconsistent	Same	Different	$\emptyset$
Independent and consistent	Different	Same or different	One ordered pair in the solution

■ **PROCEDURE**

To solve a system of equations in two variables by graphing:

1. Write each equation in the slope-intercept form, $y = mx + b$.

2. Using the slope and the intercepts, classify the system:
   a. If the system is independent and inconsistent, there are no solutions; the solution set is $\emptyset$.
   b. If the system is dependent, there are an infinite number of solutions; the solution set is $\{(x, y)\,|\,A_1x + B_1y = C_1\}$ or $\{(x, y)\,|\,A_2x + B_2y = C_2\}$.   (continued next page)

c. If the system is independent and consistent, there is a unique solution. Draw the graphs of the lines to determine the coordinates of the point of intersection. The solution set is $\{(x, y)\}$.

EXAMPLE 1

Solve the system:

$$\begin{cases} x - 3y = 2 & (1) \\ 3x - 9y = 6 & (2) \end{cases}$$

(1)	(2)	
$x - 3y = 2$	$3x - 9y = 6$	Write each equation in slope-intercept form.
$-3y = -x + 2$	$-9y = -3x + 6$	
$y = \dfrac{1}{3}x - \dfrac{2}{3}$	$y = \dfrac{1}{3}x - \dfrac{2}{3}$	
$m = \dfrac{1}{3} \quad \left(0, -\dfrac{2}{3}\right)$	$m = \dfrac{1}{3} \quad \left(0, -\dfrac{2}{3}\right)$	The slopes and the y-intercepts are the same; the system is dependent.

The solution set is $\{(x, y) \mid x - 3y = 2\}$. ∎

EXAMPLE 2

Solve the system by graphing:

$$\begin{cases} 3x - y = 2 & (1) \\ x + 2y = 10 & (2) \end{cases}$$

(1)	(2)	
$3x - y = 2$	$x + 2y = 10$	Write each equation in the slope-intercept form.
$-y = -3x + 2$	$2y = -x + 10$	
$y = 3x - 2$	$y = -\dfrac{1}{2}x + 5$	
$m = 3 \quad (0, -2)$	$m = -\dfrac{1}{2} \quad (0, 5)$	The slopes are different, so the system is independent and consistent. There is a unique solution.

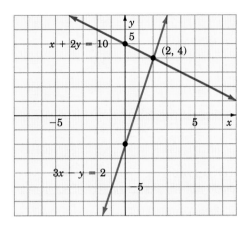

The intersection appears to be at $(2, 4)$. Check the pair $(2, 4)$.

Check:

$$3x - y = 2 \qquad x + 2y = 10$$
$$3(2) - 4 = 2 \qquad 2 + 2(4) = 10$$
$$6 - 4 = 2 \qquad 2 + 8 = 10$$
$$2 = 2 \qquad 10 = 10$$

The solution set is $\{(2, 4)\}$.

Solving systems of linear equations by graphing is not always the most accurate. It is sometimes difficult to identify the solution. This is particularly true if the ordered pair includes fractions or decimals. To find the exact solution, we use one of two other methods. Example 3 shows the method of substitution.

EXAMPLE 3

Solve the system by substitution:

$$\begin{cases} x + 4y = 6 & (1) \\ 3x + 5y = 11 & (2) \end{cases}$$

$$x + 4y = 6$$
$$\qquad x = -4y + 6$$

Solve equation (1) for x since it has a coefficient of one.

$$3(-4y + 6) + 5y = 11$$
$$-12y + 18 + 5y = 11$$
$$-7y = -7$$
$$y = 1$$

Substitute $-4y + 6$ for x in equation (2) and solve for y.

$$x = -4y + 6$$
$$\quad = -4(1) + 6$$
$$\quad = 2$$

To find x, substitute $y = 1$ in $x = -4y + 6$.

Check:

$$3x + 5y = 11$$

$$3(2) + 5(1) = 11$$

$$6 + 5 = 11$$

To check, substitute $x = 2$ and $y = 1$ in equation (2).

The solution set is $\{(2, 1)\}$.

■ **PROCEDURE**

Substitution

To solve a system of equations by substitution:

1. Solve one of the equations for y (or x).
2. Substitute that expression for y (or x) in the other equation.
3. Solve the resulting equation for x (or y). If both variables are eliminated and the new equation is
   a. an identity, the system is dependent.
   b. a contradiction, the system is inconsistent.
4. Substitute for x (or y) into the equation of step 1.
5. Solve for y (or x), and check the results in the equation not used in step 4.
6. The solution set is $\{(x, y)\}$.

EXAMPLE 4

Solve the system by substitution:

$$\begin{cases} 4x + 2y = 12 & (1) \\ 2x + \ y = 16 & (2) \end{cases}$$

$$2x + y = 16$$ Solve equation (2) for y.

$$y = -2x + 16$$

$$4x + 2(-2x + 16) = 12$$ Substitute $-2x + 16$ for y in equation (1). The variable x is eliminated, and the resulting equation is a contradiction.

$$4x - 4x + 32 = 12$$

$$32 = 12$$ The system is inconsistent.

The solution set is $\emptyset$.

EXAMPLE 5

Solve the system by substitution:

$$\begin{cases} x - \dfrac{1}{3}y = 2 & (1) \\ \dfrac{1}{2}x + \dfrac{1}{6}y = 3 & (2) \end{cases}$$

$$\begin{cases} 3x - y = 6 & (1') \\ 3x + y = 18 & (2') \end{cases}$$

Clear the fractions in both equations. Multiply equation (1) by 3, and relabel it (1'). Multiply equation (2) by 6, and relabel it (2').

$y = -3x + 18$

Solve equation (2') for y.

$3x - (-3x + 18) = 6$

Substitute $-3x + 18$ for y in (1') and solve for x.

$3x + 3x - 18 = 6$

$6x = 24$

$x = 4$

$y = -3x + 18$

To find y, substitute $x = 4$ in $y = -3x + 18$.

$= -3(4) + 18$

$= 6$

Check:

Since the solution was found using an equivalent system, check in both original equations.

$$x - \frac{1}{3}y = 2 \quad (1)$$

$$4 - \frac{1}{3}(6) = 2$$

$$4 - 2 = 2$$

$$\frac{1}{2}x + \frac{1}{6}y = 3 \quad (2)$$

$$\frac{1}{2}(4) + \frac{1}{6}(6) = 3$$

$$2 + 1 = 3$$

The solution set is $\{(4, 6)\}$.   ■

Systems of equations can be used to solve applications where there are two unknowns.

EXAMPLE 6

A retired postal worker has $20,000 to invest. She wants the annual income from the investment to be $2250. (The retiree is limited in the amount of money that can be earned.) The best long-term (5 years) investment rate available is 12%, and the best short-term (6 months) rate is 9%. How much can be invested at each rate to receive the $2250 maximum?

Simpler word form:

$$\begin{pmatrix} \text{Amount of money} \\ \text{invested at 12\%} \end{pmatrix} + \begin{pmatrix} \text{Amount of money} \\ \text{invested at 9\%} \end{pmatrix} = \$20,000$$

$$\begin{pmatrix} \text{Interest at} \\ 12\% \end{pmatrix} + \begin{pmatrix} \text{Interest at} \\ 9\% \end{pmatrix} = \$2250$$

To solve the application, we write two equations.

Translate to algebra:

$$\begin{cases} x + y = 20000 & (1) \\ 0.12x + 0.09y = 2250 & (2) \end{cases}$$

The interest earned at 12% is $0.12x$ and at 9% is $0.09y$.

$$\begin{cases} x + y = 20000 & (1) \\ 12x + 9y = 225000 & (2') \end{cases}$$

Clear the decimals in equation (2) by multiplying the equation by 100.

$$x + y = 20000 \qquad (1)$$

$$y = -x + 20000$$

Solve equation (1) for y.

$$12x + 9(-x + 20000) = 225000$$

$$12x - 9x + 180000 = 225000$$

$$3x = 45000$$

$$x = 15000$$

Substitute $-x + 20000$ for y in equation (2').

$$y = -x + 20000$$

$$y = -15000 + 20000$$

$$y = 5000$$

Substitute 15000 for y in $y = -x + 20000$.

Check:

$$0.12 + 0.09y = 2250$$

$$0.12(15000) + 0.09(5000) = 2250$$

$$1800 + 450 = 2250$$

Check in equation (2).

The retiree should invest $15,000 at 12% and $5000 at 9%.

■

EXERCISE 7.1

A

Solve by graphing:

1. $\begin{cases} x + y = -4 \\ x - y = 8 \end{cases}$

2. $\begin{cases} x - y = -2 \\ x + y = 8 \end{cases}$

3. $\begin{cases} 4x + y = 2 \\ x - y = 8 \end{cases}$

4. $\begin{cases} 3x - y = 4 \\ 2x + y = 1 \end{cases}$

5. $\begin{cases} y = x - 4 \\ y = -x - 2 \end{cases}$

6. $\begin{cases} y = -4x + 2 \\ y = 4x - 6 \end{cases}$

7. $\begin{cases} 6x - 3y = 15 \\ 2x - y = 5 \end{cases}$

8. $\begin{cases} 4x + 2y = 6 \\ 2x + y = 3 \end{cases}$

Solve the following systems by substitution:

9. $\begin{cases} y = x + 2 \\ x + y = 10 \end{cases}$

10. $\begin{cases} y = x + 1 \\ x + y = 3 \end{cases}$

11. $\begin{cases} x = y + 5 \\ 2x + 4y = 4 \end{cases}$

12. $\begin{cases} x = y - 3 \\ 4x - 3y = 2 \end{cases}$

13. $\begin{cases} x + y = 4 \\ 3x - 2y = 7 \end{cases}$

14. $\begin{cases} x + y = -2 \\ 2x + 3y = -9 \end{cases}$

15. $\begin{cases} 2x + y = 8 \\ 4x + 2y = 10 \end{cases}$

16. $\begin{cases} x = \dfrac{1}{3}y - 6 \\ 12x - 4y + 24 = 0 \end{cases}$

17. $\begin{cases} y = 2x - 3 \\ x = 2y - 9 \end{cases}$

B

18. $\begin{cases} x = 3y + 2 \\ y = 2x - 11 \end{cases}$

19. $\begin{cases} 3x + 4y = 1 \\ x - 3y = -4 \end{cases}$

20. $\begin{cases} 4x + 3y = 1 \\ x - 2y = 3 \end{cases}$

21. $\begin{cases} 4x - y = 16 \\ 3x + 2y = 1 \end{cases}$

22. $\begin{cases} 5x + 3y = -3 \\ 4x - y = -16 \end{cases}$

23. $\begin{cases} y = \dfrac{3}{2}x + 14 \\ -3x + 2y = 28 \end{cases}$

24. $\begin{cases} x = \dfrac{3}{4}y + 4 \\ 4x - 3y = 16 \end{cases}$

25. $\begin{cases} x = \dfrac{1}{2}y - 3 \\ 5x - 2y = -17 \end{cases}$

26. $\begin{cases} y = \dfrac{1}{3}x - 6 \\ 5x - 2y = -1 \end{cases}$

27. $\begin{cases} x + 4y = 0 \\ 2x + 7y = 1 \end{cases}$

28. $\begin{cases} 2x - y = -2 \\ 7x - 5y = 5 \end{cases}$

29. $\begin{cases} 6x + y = -5 \\ 3x + 5y = 2 \end{cases}$

30. $\begin{cases} y = -2x - 9 \\ 11x + 3y = -42 \end{cases}$

31. $\begin{cases} x = 5y + 4 \\ 9x - 16y = -22 \end{cases}$

32. $\begin{cases} x + y = -3 \\ -3x + 4y = 23 \end{cases}$

33. $\begin{cases} x = 2y + 6 \\ 3x - 4y = 12 \end{cases}$

34. $\begin{cases} x = -3y + 5 \\ 2x - 11y = 10 \end{cases}$

35. $\begin{cases} 4x + 6y = -1 \\ x + y = 0 \end{cases}$

36. $\begin{cases} 6x - 9y = -1 \\ 2x + y = 1 \end{cases}$

37. $\begin{cases} x + 3y = 5 \\ 3x - 5y = 12 \end{cases}$

38. $\begin{cases} 2x + y = 8 \\ 3x - 4y = 9 \end{cases}$

39. $\begin{cases} 4x + 3y = 0 \\ 2x + y = 2 \end{cases}$

40. $\begin{cases} 5x - 2y = 0 \\ x + 3y = 17 \end{cases}$

C

41. $\begin{cases} 8x - 4y = 11 \\ x + 3y = -3 \end{cases}$

42. $\begin{cases} 4x - y = 1 \\ 2x + 3y = -1 \end{cases}$

43. $\begin{cases} x + y = -1 \\ 5x - 5y = -1 \end{cases}$

44. $\begin{cases} x - 2y = 4 \\ 2x + y = 1 \end{cases}$

45. $\begin{cases} x - 2y = -5 \\ 6x + 3y = -10 \end{cases}$

46. $\begin{cases} x + 2y = -2 \\ x + y = -\dfrac{11}{4} \end{cases}$

47. $\begin{cases} x + y = -\dfrac{1}{2} \\ 2x - 3y = 0 \end{cases}$

48. $\begin{cases} x + y = \dfrac{1}{11} \\ -x + 2y = 1 \end{cases}$

49. $\begin{cases} x - 9y = -5 \\ -x - 3y = 1 \end{cases}$

50. $\begin{cases} 4x + y = 0 \\ 8x - 3y = 15 \end{cases}$

51. $\begin{cases} x - 3y = 7 \\ 6x + 4y = 9 \end{cases}$

52. $\begin{cases} 2x - y = 6 \\ 2x + 3y = 8 \end{cases}$

53. $\begin{cases} x - 2y = -1 \\ 4x + 4y = 5 \end{cases}$

54. $\begin{cases} 2x + 3y = -1 \\ x - y = \dfrac{7}{6} \end{cases}$

55. $\begin{cases} x - \dfrac{4}{3}y = -\dfrac{2}{3} \\ \dfrac{x}{2} + y = \dfrac{5}{2} \end{cases}$

56. $\begin{cases} \dfrac{x}{3} + \dfrac{y}{3} = 1 \\ \dfrac{x}{4} - y = 1 \end{cases}$

57. $\begin{cases} x + \dfrac{3}{4}y = 6 \\ \dfrac{2}{3}x + y = 6 \end{cases}$

58. $\begin{cases} \dfrac{1}{4}x + \dfrac{1}{3}y = 4 \\ x - \dfrac{1}{2}y = 5 \end{cases}$

59. $\begin{cases} \dfrac{1}{4}x + \dfrac{1}{5}y = -3 \\ \dfrac{2}{3}x - \dfrac{1}{3}y = -13 \end{cases}$

60. $\begin{cases} y = \dfrac{1}{6}x + 7 \\ \dfrac{1}{3}x + \dfrac{2}{3}y = 10 \end{cases}$

D

61. A total of \$40,000 is invested, part at 18% and part at 12%. The annual return is \$5220. How much is invested at each rate?

62. A total of \$100,000 is invested, part at 16% and part at 13%. The annual return is \$14,500. How much is invested at each rate?

63. Perry has 27 coins having a total value of \$4.95. If these coins are nickels and quarters, how many of each does he have?

64. Cynthia has 55 coins having a total value of \$21.50. If these coins are dimes and fifty-cent pieces, how many of each does she have?

65. The local Farmer's Feed Store in Ames, Iowa, has two grades of rabbit food. One grade sells for \$0.82 per pound and the other for \$0.62 per pound. How many pounds of each grade must be used to form 1000 pounds of a mixture that will sell for \$0.75 per pound?

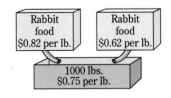

Figure for Exercise 65

66. The Blue Rose Theater sells tickets for \$4.50 and \$6.00. A total of 380 tickets were sold for their last performance of ''Mad Romance.'' If the sales for the performance totaled \$1972.50, how many tickets were sold at each price?

67. The Shocking Electronics Firm makes two kinds of resistors on the same assembly line. One resistor costs \$2 to make and takes 3 minutes to assemble. The other one costs \$1 to make and takes 2 minutes to assemble. If in one work week (2400 minutes) the cost of production was \$1375, how many of each type of resistor was produced?

68. The Fancy Pack Ice Cream Company can make a quart of vanilla in 4 minutes at a cost of $1. The company can make a quart of spumoni in 5 minutes at a cost of $1.45. If in a given eight-hour day they made only vanilla and spumoni and had production costs of $132, how many quarts of each did they make?

69. The current through two circuits on a television set is given by the equations $2I_1 + 3I_2 = 8$ and $3I_1 + I_2 = 5$. Using I_2 as the dependent variable and I_1 as the independent, graph both and determine the common solution. Would the common solution be different if you reversed the axes?

70. The sum of two numbers is 23. If you double the first and triple the second, the new sum is 54. Find the two numbers using the substitution method.

71. A rock group wants to buy amplifiers and speakers. Three amps and five speakers cost $700 while five amps and eight speakers would cost $1150. What is the cost of each amp and each speaker?

72. In a circuit, the sum of two capacitors is 55 microfarads (μF). The difference between them is 25 μF. What is the size of each capacitor? (solve by substitution)

STATE YOUR UNDERSTANDING

73. How do you know that a system is independent and consistent when you solve the system using substitution?

74. Write an example of a system that is independent and inconsistent (i.e. you create the example).

75. We know from the discussion in the text and previous exercises, that solving a system by graphing both lines frequently allows us only to estimate the solution to a system. The substitution technique allows us to find the exact solution. What is it about the way the equations appear in a system that makes it convenient to use the substitution method?

CHALLENGE EXERCISES

76. Solve $\begin{cases} 3x - 2y = 5 \\ x = 3 \end{cases}$ both graphically and by substitution.

77. Solve $\begin{cases} 2x - y = 4 \\ y + 3 = 0 \end{cases}$ both graphically and by substitution.

78. Solve $\begin{cases} x - y = a \\ x + 2y = 3 \end{cases}$ using substitution.

79. Solve the following system for x, y, and z using substitution.

$$\begin{cases} x + y - z = 5 \\ x + y \quad\;\; = 4 \\ \quad\;\; y \quad\;\; = -2 \end{cases}$$

80. The area of a rectangle is 35 square feet. The perimeter is 24 feet. Find the dimensions using substitution (Note: the system is not a linear system).

MAINTAIN YOUR SKILLS (SECTIONS 5.5, 6.1)

Solve:

81. $x^4 - 26x^2 + 25 = 0$ **82.** $x^4 - 12x^2 + 32 = 0$

83. $(w + 3)^4 - 12(w + 3)^2 + 32 = 0$

84. Find the ordered pair that is a solution of the equation $3y - |x - 7| = 0$ when $x = -38$.

Draw the graph of each equation:

85. $2x + 3y = 18$ **86.** $3x + 2y = 18$

87. $x = 5 - 2y$ **88.** $y = 5 - 2x$

7.2

**SOLVING SYSTEMS IN
TWO VARIABLES BY
LINEAR COMBINATIONS**

OBJECTIVE

1. Solve a system of linear equations in two variables by linear combinations.

A third method for solving a system is by linear combinations.

■ DEFINITION

Linear Combination

A *linear combination* of two equations is the sum of the two equations or the sum of multiples of the equations.

For a system of linear equations

$$\begin{cases} A_1x + B_1y = C_1 \\ A_2x + B_2y = C_2 \end{cases}$$

a combination is

$$a(A_1x + B_1y) + b(A_2x + B_2y) = aC_1 + bC_2, \qquad a \text{ and } b \in R.$$

For example, given the system

$$\begin{cases} 2x - 3y = 4 \\ x - 4y = 6 \end{cases}$$

two linear combinations are

$$3(2x - 3y) + 2(x - 4y) = 3(4) + 2(6) \quad \text{or}$$

$$8x - 17y = 24$$

and

$$-1(2x - 3y) - 3(x - 4y) = -1(4) - 3(6) \quad \text{or}$$

$$-5x + 15y = -22.$$

To solve a system of equations by linear combinations, we use a combination to eliminate one of the variables.

EXAMPLE 1

Solve using linear combinations:

$$\begin{cases} 3x + 2y = 5 & (1) \\ 4x - 3y = 1 & (2) \end{cases}$$

Since the terms involving y have opposite signs, we choose to eliminate y.

$$9x + 6y = 15 \quad (3)$$
$$\underline{8x - 6y = 2} \quad (4)$$
$$17x \quad\quad = 17$$

Multiply equation (1) by 3 and equation (2) by 2. This yields a new but equivalent system. Add equations (3) and (4).

$$x = 1$$

Solve for x.

$$3(1) + 2y = 5$$

Substitute 1 for x in equation (1).

$$2y = 2$$

$$y = 1$$

■ CAUTION

> Use one of the original equations in case an error was made in forming the equivalent system or be sure to check the solution in both original equations.

Check:

$$4x - 3y = 1 \quad (2)$$

$$4(1) - 3(1) = 1$$

Substitute 1 for x and 1 for y in equation (2).

$$4 - 3 = 1$$

The solution set is $\{(1, 1)\}$.   ■

■ PROCEDURE

> To solve a system of linear equations in two variables using linear combinations:
>
> 1. Write both equations in standard form, $Ax + By = C$.
>
> 2. Multiply one or both equations by factors so that the coefficients of y (or x) are opposites.
>
> 3. Add the two equations so that y (or x) is eliminated. If both x and y are eliminated and the resulting equation is
>   a. an identity, the system is dependent.
>   b. a contradiction, the system is inconsistent.
>
> 4. Solve.
>
> 5. Find the value of the other variable by substituting in one of the original equations.
>
> 6. Check in the equation not used in step 5.
>
> 7. The solution set is $\{(x, y)\}$.

EXAMPLE 2

Solve using linear combinations:

$$\begin{cases} 0.2x + 0.5y = 2.2 & (1) \\ 0.7x + 0.3y = -1 & (2) \end{cases}$$

$$\begin{cases} 2x + 5y = 22 & (3) \\ 7x + 3y = -10 & (4) \end{cases}$$

Clear the decimals by multiplying each equation by 10.

$$\begin{array}{r} -6x - 15y = -66 \\ 35x + 15y = -50 \\ \hline 29x \qquad\quad = -116 \\ x = -4 \end{array}$$

Form a third system to eliminate y by multiplying equation (3) by -3 and equation (4) by 5. Add.

$$0.2(-4) + 0.5y = 2.2$$

Substitute -4 for x in equation (1) to find y.

$$-0.8 + 0.5y = 2.2$$

$$0.5y = 3$$

$$y = 6$$

Check:

$$0.7x + 0.3y = -1$$

Check in equation (2).

$$0.7(-4) + 0.3(6) = -1$$

$$-2.8 + 1.8 = -1$$

The solution set is $\{(-4, 6)\}$.

EXAMPLE 3

Solve using linear combinations:

$$\begin{cases} 4x + 3y = 15 & (1) \\ 12x + 9y = 36 & (2) \end{cases}$$

$$\begin{array}{r} -12x - 9y = -45 \\ 12x + 9y = 36 \\ \hline 0 = -9 \end{array}$$

Form an equivalent system by multiplying equation (1) by -3. This makes the coefficients of x opposites. Add. Both variables are eliminated, and the result is a contradiction. The system is inconsistent.

The solution set is $\emptyset$.

EXAMPLE 4

Solve using linear combinations:

$$\begin{cases} 2x + y = 1 & (1) \\ 5x - 2y = 1 & (2) \end{cases}$$

$$4x + 2y = 2$$ Multiply equation (1) by 2.

$$\frac{5x - 2y = 1}{9x \quad\; = 3}$$ Add.

$$x = \frac{1}{3}$$

$$2\left(\frac{1}{3}\right) + y = 1$$ Use equation (1) to solve for y.

$$y = \frac{1}{3}$$

Check:

$$5\left(\frac{1}{3}\right) - 2\left(\frac{1}{3}\right) = 1$$ Check in equation (2).

$$1 = 1$$

The solution set is $\left\{\left(\frac{1}{3}, \frac{1}{3}\right)\right\}$. ∎

The next example is an application of a system of equations that might occur in a chemistry laboratory.

EXAMPLE 5

An amount of 40% acid solution is to be mixed with enough 10% acid solution to make a 25% acid solution. If there are to be 20 oz of the final mixture, how much of each solution should be mixed together?

Simpler word form:

$$\begin{pmatrix} \text{oz of 40\%} \\ \text{acid} \end{pmatrix} + \begin{pmatrix} \text{oz of 10\%} \\ \text{acid} \end{pmatrix} = \begin{pmatrix} \text{20 oz of} \\ \text{25\% acid} \end{pmatrix}$$

$$\begin{pmatrix} \text{Acid in} \\ \text{40\% mixture} \end{pmatrix} + \begin{pmatrix} \text{Acid in} \\ \text{10\% mixture} \end{pmatrix} = \begin{pmatrix} \text{Acid in} \\ \text{25\% mixture} \end{pmatrix}$$

Select variable:

Let x represent the number of ounces of the 40% acid solution and y represent the number of ounces of the 10% solution.

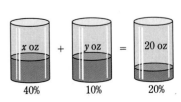

Translate to algebra:

$$\begin{cases} x + y = 20 & (1) \\ 0.40x + 0.10y = 0.25(20) & (2) \end{cases}$$

$0.40x$ represents the number of ounces of acid in the 40% solution, and $0.10y$ represents the number of ounces of acid in the 10% solution.

$$\begin{array}{r} x + y = 20 \\ -4x - y = -50 \\ \hline -3x = -30 \\ x = 10 \end{array}$$

Multiply equation (2) by -10 to clear the decimals and make the coefficients of y opposite.

$$10 + y = 20$$

$$y = 10$$

Use equation (1) to solve for y.

Check:

$$0.40(10) + 0.10(10) = 0.25(20)$$

Check in equation (2).

$$4 + 1 = 5$$

To get the 25% solution, 10 oz of the 40% solution and 10 oz of the 10% solution should be mixed together.   ■

EXERCISE 7.2

A

Solve using linear combinations:

1. $\begin{cases} x + y = 5 \\ x - y = 3 \end{cases}$
  2. $\begin{cases} x - y = -3 \\ x + y = 5 \end{cases}$
  3. $\begin{cases} 4x + y = 5 \\ 4x - y = 3 \end{cases}$

4. $\begin{cases} 2x - y = 8 \\ 2x + y = 4 \end{cases}$
  5. $\begin{cases} 4x - 3y = 8 \\ -4x + 3y = 7 \end{cases}$
  6. $\begin{cases} x + 7y = 13 \\ -x - 7y = -1 \end{cases}$

7. $\begin{cases} 2x - y = 18 \\ 10x - 5y = 90 \end{cases}$
  8. $\begin{cases} 3x + 5y = 1 \\ 18x + 30y = 6 \end{cases}$
  9. $\begin{cases} x + 3y = 6 \\ x - y = 2 \end{cases}$

10. $\begin{cases} x + 4y = 10 \\ x - y = 0 \end{cases}$
  11. $\begin{cases} 2x + 3y = -13 \\ 5x - y = 27 \end{cases}$
  12. $\begin{cases} 7x - 3y = 32 \\ 2x + y = 11 \end{cases}$

13. $\begin{cases} x + 4y = 8 \\ -2x + 5y = 23 \end{cases}$
  14. $\begin{cases} 2x - 5y = 11 \\ -x + 3y = -7 \end{cases}$
  15. $\begin{cases} x + y = -3 \\ 2x - 3y = 49 \end{cases}$

16. $\begin{cases} 2x + 3y = 3 \\ x - y = -11 \end{cases}$
  17. $\begin{cases} 4x - 5y = 15 \\ x + y = -3 \end{cases}$
  18. $\begin{cases} 2x + 7y = -4 \\ -x - 8y = 2 \end{cases}$

19. $\begin{cases} 5x - 3y = -2 \\ -x + 4y = -3 \end{cases}$

20. $\begin{cases} 2x - 5y = 9 \\ -x + 3y = -6 \end{cases}$

B

21. $\begin{cases} 3x + y = 17 \\ 4x + y = 22 \end{cases}$

22. $\begin{cases} x + 3y = 1 \\ x + 4y = 3 \end{cases}$

23. $\begin{cases} x + y = 1 \\ x - 5y = -23 \end{cases}$

24. $\begin{cases} x + y = 7 \\ x - 3y = -5 \end{cases}$

25. $\begin{cases} 3x + 2y = -4 \\ 4x + 3y = -5 \end{cases}$

26. $\begin{cases} 2x + 5y = -5 \\ 6x + 3y = 21 \end{cases}$

27. $\begin{cases} 5x + 4y = 48 \\ 3x - 7y = 10 \end{cases}$

28. $\begin{cases} 6x + 5y = -13 \\ 5x - 8y = 50 \end{cases}$

29. $\begin{cases} 2x - 7y = -1 \\ 6x - 21y = -3 \end{cases}$

30. $\begin{cases} -x + 3y = 5 \\ 3x - 9y = -15 \end{cases}$

31. $\begin{cases} 3x + 8y = -6 \\ 5x - 7y = -71 \end{cases}$

32. $\begin{cases} 4x - 7y = 3 \\ 5x + 6y = -70 \end{cases}$

33. $\begin{cases} 2x + 9y = -31 \\ 5x - 4y = 55 \end{cases}$

34. $\begin{cases} 7x - 5y = -12 \\ 3x + 4y = -42 \end{cases}$

35. $\begin{cases} x + \dfrac{5}{2}y = -\dfrac{29}{2} \\ \dfrac{3}{4}x - y = \dfrac{7}{2} \end{cases}$

36. $\begin{cases} \dfrac{3}{4}x + y = \dfrac{7}{2} \\ x + \dfrac{8}{5}y = \dfrac{22}{5} \end{cases}$

37. $\begin{cases} \dfrac{2}{3}x - y = \dfrac{13}{3} \\ \dfrac{1}{6}x + y = -\dfrac{8}{3} \end{cases}$

38. $\begin{cases} \dfrac{1}{5}x - 2y = \dfrac{34}{5} \\ \dfrac{1}{4}x + y = -2 \end{cases}$

39. $\begin{cases} 5x + 4y = 10 \\ \dfrac{5}{3}x + \dfrac{4}{3}y = 1 \end{cases}$

40. $\begin{cases} 2x - 3y = 18 \\ x - \dfrac{3}{2}y = 12 \end{cases}$

41. $\begin{cases} 20x + 25y = -7 \\ 10x - 15y = 13 \end{cases}$

42. $\begin{cases} 3x + 4y = 1 \\ 35x - 21y = 2 \end{cases}$

43. $\begin{cases} 27x + 63y = 13 \\ 12x - 3y = -8 \end{cases}$

44. $\begin{cases} 16x - 24y = 21 \\ -32x + 8y = -17 \end{cases}$

45. $\begin{cases} 3x + 3y = -2 \\ x - y = -1 \end{cases}$

46. $\begin{cases} 32x + 28y = 3 \\ 6x + 10y = -3 \end{cases}$

C

47. $\begin{cases} 4x + 3y = 5 \\ 12x + 6y = 5 \end{cases}$

48. $\begin{cases} 2x + 3y = 3 \\ 18x + 24y = 23 \end{cases}$

49. $\begin{cases} 4x - 6y = 5 \\ 10x - 8y = 9 \end{cases}$

50. $\begin{cases} 5x - 7y = -10 \\ 8x + 6y = 27 \end{cases}$

51. $\begin{cases} 5x + 3y = 3 \\ 7x - 11y = 8 \end{cases}$

52. $\begin{cases} 5x + 7y = 4 \\ 7x + 23y = 10 \end{cases}$

53. $\begin{cases} \dfrac{9}{4}x - \dfrac{15}{4}y = 2 \\ \dfrac{15}{8}x - \dfrac{9}{8}y = 1 \end{cases}$

54. $\begin{cases} \dfrac{3}{2}x + \dfrac{1}{4}y = \dfrac{13}{20} \\ \dfrac{1}{3}x - \dfrac{1}{2}y = \dfrac{1}{30} \end{cases}$

55. $\begin{cases} \dfrac{1}{4}x - \dfrac{1}{3}y = -\dfrac{5}{12} \\ \dfrac{1}{4}x + \dfrac{1}{3}y = \dfrac{1}{12} \end{cases}$

56. $\begin{cases} \dfrac{3}{2}x + \dfrac{4}{5}y = -\dfrac{13}{10} \\ \dfrac{7}{2}x - \dfrac{9}{5}y = -\dfrac{6}{5} \end{cases}$

57. $\begin{cases} 0.8x + 1.5y = 0.16 \\ 1.4x - 0.5y = -2.22 \end{cases}$

58. $\begin{cases} 1.3x - 0.9y = 0.08 \\ 2.1x + 0.7y = -2.24 \end{cases}$

59. $\begin{cases} 0.6x - 0.7y = 0.53 \\ 0.5x + 0.2y = 0.05 \end{cases}$

60. $\begin{cases} 0.6x + 0.5y = 0.69 \\ 0.3x - 0.2y = -0.33 \end{cases}$

D

61. How many cubic centimeters (cc) of a 50% saline solution must be mixed with how many cc of a 10% saline solution to yield 80 cc of a 25% saline solution?

62. How many oz of a 5% acid solution must be mixed with how many oz of a 30% acid solution to yield 100 oz of a 10% solution?

63. A 20-pound mixture of shrimp and crab meat is prepared by the Alaskan Fish Company and is sold for $76.95. If shrimp costs $3.10 a pound and crab costs $4.25 a pound, how many pounds of each were used?

64. A 300-pound mixture of two grades of coffee is packed by Pause That Refreshes Coffee Company and is sold for $1327. If one grade costs $4.15 per pound and the other grade costs $4.56 per pound, how many pounds of each grade were used in the mixture?

65. The WOW I Feel Good health food store is to fill a diet for a client that calls for two kinds of food. The first contains 40% of nutrient A and 16% of nutrient B. The second contains 35% of nutrient A and 20% of nutrient B. If the diet calls for exactly 4 ounces of nutrient A and 2 ounces of nutrient B, how many ounces of each food should be used?

66. The health food store in Exercise 65 has another order for the same two foods that calls for exactly 3 ounces of nutrient A and 1.6 ounces of nutrient B. How many ounces of each food should be used?

67. Statistics from the Girl Scout candy sale show that it takes 5 minutes to sell a $1.75 bar and 4 minutes to sell a $2.25 bar. If a Scout spent 120 minutes selling candy and raised $59.00, how many of each price bar did she sell?

68. A shoe salesman at Shoes Galore finds that it takes him 35 minutes to sell a $75 pair of Guchy shoes and 20 minutes to sell the latest style of Tenny Runners that cost $52.45 a pair. If in an eight-hour day he sells only these kinds of shoes and works the entire eight hours, how many pairs of each type of shoe does he sell if his total sales for the day is $1124.50?

69. In construction, one concrete mixture contains four times as much gravel as cement. If the total volume needed is 48 cubic yards, how much of each ingredient is needed?

70. The beam shown below rests on supports at points A and B.

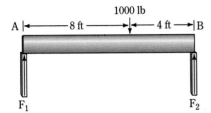

F₁   F₂   Figure for Exercise 70

The forces, F_1 and F_2, needed for equilibrium can be found by solving the equations $8F_1 - 4F_2 = 0$ and $F_1 + F_2 = 1000$. Find F_1 and F_2.

71. If the ⎰⎱ symbol is used to represent a resistor in a circuit, the equations for the following circuits:

Circuit number 1

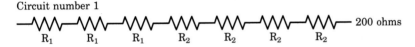

R_1 R_1 R_1 R_2 R_2 R_2 R_2

Circuit number 2

R_1 R_1 R_2 R_2 80 ohms

are $3R_1 + 4R_2 = 20$ and $2R_1 + 2R_2 = 12$, find R_1 and R_2.

STATE YOUR UNDERSTANDING

72. When solving systems algebraically (i.e. either substitution or linear combinations), what do you look for to determine which technique is best if you are given the option?

73. Given that the system $\begin{cases} 4x - 2y = 5 \\ 6x - 3y = 7 \end{cases}$ is independent and inconsistent, what is the "ratio" of the coefficients of x in the two equations? What is the ratio of the coefficients of y in the two equations? Conclusion? What is the ratio of the constants?

74. Given that the system $\begin{cases} 3x + 6y = -6 \\ -4x - 8y = 8 \end{cases}$ is dependent, what is the ratio of the coefficients of x and what is the ratio of the coefficients of y? What is the ratio of the two constants?

CHALLENGE EXERCISES

75. Solve $\begin{cases} ax + y = b \\ bx - y = a \end{cases}$ using linear combinations.

76. Solve $\begin{cases} ax + by = ab \\ bx - ay = b \end{cases}$ using linear combinations.

77. Replacing $\dfrac{1}{u}$ by x and $\dfrac{1}{v}$ by y, solve the following system for u and v.

$$\begin{cases} \dfrac{1}{u} + \dfrac{1}{v} = 4 \\[2mm] \dfrac{1}{u} - \dfrac{1}{v} = 2 \end{cases}$$

78. Using the same substitutions as in the previous problem, solve the following system for u and v.

$$\begin{cases} \dfrac{3}{u} - \dfrac{1}{v} = 2 \\[2mm] \dfrac{4}{u} + \dfrac{3}{v} = 1 \end{cases}$$

MAINTAIN YOUR SKILLS (SECTIONS 6.2, 6.3)

79. Write the coordinates of the x- and y-intercepts of the graph of $5x - 3y = 18$.

80. Write the coordinates of the x- and y-intercepts of the graph of $12y = x + 15$.

81. Write the coordinates of the x- and y-intercepts of the graph of $x = \dfrac{3}{5}y + \dfrac{1}{5}$.

82. Find the slope of the line through the points $(6, 3)$ and $(9, -5)$.

83. Find the distance between the points $(6, 3)$ and $(9, -5)$.

84. Write the equation, in standard form, of the line containing the points $(6, 3)$ and $(9, -5)$.

85. Is a line with slope $\dfrac{11}{2}$ perpendicular to the line containing the points $(7, 8)$ and $(5, -3)$?

86. Is a line with slope $\dfrac{11}{2}$ parallel to the line containing the points $(7, 8)$ and $(5, -3)$?

7.3

SOLVING SYSTEMS IN THREE VARIABLES BY LINEAR COMBINATIONS

OBJECTIVE

1. Solve a system of linear equations in three variables by linear combinations.

An equation containing two variables has an ordered pair of numbers for a solution. An equation that contains three variables has an ordered triple of numbers for a solution.

■ DEFINITION

Ordered Triple (x, y, z)

> An *ordered triple* is a solution of an equation containing three variables. The value of x is written first followed by the value of y and then the value of z. A solution of the linear equation, $Ax + By + Cz = D$, is written (x, y, z).

To find a solution of an equation in three variables, we can assign values to two of the variables and solve for the remaining one.

To find the solution of $3x - 2y + z = 10$ when $x = 2$ and $y = 3$, substitute 2 for x and 3 for y in the equation and solve for z.

$$3(2) - 2(3) + z = 10$$

$$6 - 6 + z = 10$$

$$z = 10$$

The solution is $(2, 3, 10)$.

In general the graph of a linear equation in three variables is a plane. To solve a system of three equations by graphing, we would need to graph the system in three dimensions, which is not practical. However, visualizing three planes, you can see that they can intersect in a single point, a line, a plane, or not at all (parallel planes). These possibilities are illustrated in Figure 7.4 on the following page.

There are other ways three planes can intersect that do not result in a common solution, but we will not consider them here.

To solve a system of three equations in three variables, we can use linear combinations similar to the way we solve a system of two equations in two variables.

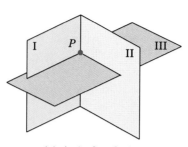

(a) A single solution

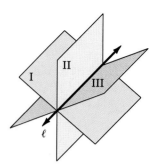

(b) Points of a line in common

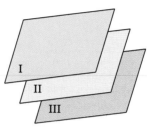

(c) No points in common

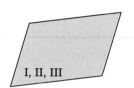

(d) All points in common

FIGURE 7.4

EXAMPLE 1

Solve using linear combinations:

$$\begin{cases} x + y + 2z = 0 & (1) \\ 2x - 2y + z = 8 & (2) \\ 3x + 2y + z = 2 & (3) \end{cases}$$

To solve the system, we first eliminate one of the variables using linear combinations of different pairs of equations.

$$\begin{array}{l} 2x + 2y + 4z = 0 \\ \underline{2x - 2y + z = 8} \\ 4x \qquad + 5z = 8 \qquad (4) \end{array}$$

Multiply equation (1) by 2 so that when it is added to equation (2), y is eliminated.
An equation in x and z.

$$\begin{array}{l} 2x - 2y + z = 8 \\ \underline{3x + 2y + z = 2} \\ 5x \qquad + 2z = 10 \qquad (5) \end{array}$$

Now add equations (2) and (3), which will also eliminate y.
A second equation in x and z.

$$\begin{cases} 4x + 5z = 8 & (4) \\ 5x + 2z = 10 & (5) \end{cases}$$

We now have a system of two equations in two variables. Solve as before.

$$\begin{array}{l} -8x - 10z = -16 \\ \underline{25x + 10z = 50} \\ 17x \qquad = 34 \end{array}$$

Multiply equation (4) by -2.
Multiply equation (5) by 5.

$$x = 2$$

$$4(2) + 5z = 8$$

Substitute 2 for x in equation (4) and solve for z.

$$5z = 0$$

$$z = 0$$

$$2 + y + 2(0) = 0$$

Substitute 2 for x and 0 for z in equation (1) and solve for y.

$$y = -2$$

Check:

$$2x - 2y + z = 8 \quad (2)$$

Check the solution in equations (2) and (3).

$$2(2) - 2(-2) + 0 = 8$$

$$4 + 4 + 0 = 8$$

$$3x + 2y + z = 2 \quad (3)$$

$$3(2) + 2(-2) + 0 = 2$$

$$6 - 4 + 0 = 2$$

The solution set is $\{(2, -2, 0)\}$.

■ **CAUTION**

Write the solution as an ordered triple (x, y, z).

■

■ **CAUTION**

When solving a system of three equations, take care to eliminate the same variable in the first two combinations. Otherwise, you may get a system of two equations in three variables, which does not have a unique solution.

If at any step of the solution process all the variables are eliminated and the result is

1. a contradiction.

The solution set is empty (inconsistent system).

2. an identity.

It is not possible, without further investigation, to determine if the solution set is empty or has an infinite number of solutions. Here we will say the system has no unique solution.

EXAMPLE 2

Solve using linear combinations:

$$\begin{cases} x + y - z = 8 & (1) \\ 2x + 2y - 2z = 2 & (2) \\ x + y + z = 9 & (3) \end{cases}$$

$$\begin{array}{r} x + y - z = 8 \\ \underline{x + y + z = 9} \\ 2x + 2y = 17 \quad (4) \end{array}$$

Add equations (1) and (3) to eliminate z.

$$\begin{array}{r} -2x - 2y + 2z = -16 \\ \underline{2x + 2y - 2z = 2} \\ 0 = -14 \end{array}$$

Multiply equation (1) by -2 so that the coefficient of z will be the opposite of the coefficient of z in (2). Add. A contradiction, so the solution set is empty.

The solution set is $\emptyset$. ∎

■ PROCEDURE

To solve a system of linear equations in three variables by linear combinations:

1. Write each of the three equations in the form $Ax + By + Cz = D$.

2. Reduce (using linear combinations) the system in three equations in three variables to a system of two equations in the same two variables.

3. If at any step all the variables are eliminated and the result is
   a. a contradiction, the solution set is $\emptyset$.
   b. an identity, there is no unique solution.

4. Solve the two-equation, two-variable system.

5. Substitute the values of the two known variables into one of the original equations to solve for the third variable.

6. Check the ordered triple in the two equations not used in step 5.

7. The solution set is $\{(x, y, z)\}$.

EXAMPLE 3

Solve using linear combinations:

$$\begin{cases} 3x + 4y - 5z = 4 & (1) \\ 4x + 3y + 2z = -16 & (2) \\ 5x - 2y + 4z = -13 & (3) \end{cases}$$

$$3x + 4y - 5z = 4$$
$$\underline{10x - 4y + 8z = -26}$$
$$13x \qquad + 3z = -22 \qquad (4)$$

To eliminate y, multiply equation (3) by 2 and add to equation (1). An equation in x and z.

$$8x + 6y + 4z = -32$$
$$\underline{15x - 6y + 12z = -39}$$

$$23x \qquad + 16z = -71 \qquad (5)$$

To eliminate y again, multiply equation (2) by 2, multiply equation (3) by 3, and add. An equation in x and z.

$$\begin{cases} 13x + 3z = -22 & (4) \\ 23x + 16z = -71 & (5) \end{cases}$$

The system of two equations in two variables.

$$208x + 48z = -352$$
$$\underline{-69x - 48z = 213}$$
$$139x \qquad = -139$$

Multiply equation (4) by 16, and equation (5) by -3, to eliminate z.

$$x = -1$$

$$13x + 3z = -22 \qquad (4)$$

$$13(-1) + 3z = -22$$

$$3z = -9$$

$$z = -3$$

Substitute -1 for x in equation (4) and solve for z.

$$3x + 4y - 5z = 4 \quad (1)$$

$$3(-1) + 4y - 5(-3) = 4$$

$$-3 + 4y + 15 = 4$$

$$4y = -8$$

$$y = -2$$

Substitute -1 for x and -3 for z in equation (1) and solve for y.

Check:

$$4x + 3y + 2z = -16 \quad (2)$$

$$4(-1) + 3(-2) + 2(-3) = -16$$

$$-4 - 6 - 6 = -16$$

$$5x - 2y + 4z = -13 \quad (3)$$

$$5(-1) - 2(-2) + 4(-3) = -13$$

$$-5 + 4 - 12 = -13$$

Check by substituting -1 for x, -2 for y, and -3 for z in equations (2) and (3).

The solution set is $\{(-1, -2, -3)\}$. ■

Finding the equation of a circle given three points on its circumference requires solving a system of equations with three variables.

EXAMPLE 4

The equation of a circle can be written in the form $x^2 + y^2 + ax + by + c = 0$. Find the equation of a circle that passes through the points $(2, 0)$, $(0, -2)$, and $(2, -4)$.

Since the points $(2, 0)$, $(0, -2)$, and $(2, -4)$ are on the circle, we can substitute these values for x and y in the equation, one pair at a time. We get a system in three variables.

$$x^2 + y^2 + ax + by + c = 0$$

$$2^2 + 0^2 + a(2) + b(0) + c = 0 \qquad \text{Substitute } (2, 0) \text{ in the equation.}$$

$$4 + 2a + c = 0$$

$$2a + c = -4 \quad (1)$$

$$0^2 + (-2)^2 + a(0) + b(-2) + c = 0 \qquad \text{Substitute } (0, -2) \text{ in the equation.}$$

$$4 - 2b + c = 0$$

$$-2b + c = -4 \quad (2)$$

$$2^2 + (-4)^2 + a(2) + b(-4) + c = 0 \qquad \text{Substitute } (2, -4) \text{ in the equation.}$$

$$4 + 16 + 2a - 4b + c = 0$$

$$2a - 4b + c = -20 \quad (3)$$

$$\begin{cases} 2a \qquad\; + c = -4 \quad (1) \\ \quad\; - 2b + c = -4 \quad (2) \\ 2a - 4b + c = -20 \quad (3) \end{cases} \qquad \text{We have a system in three variables to solve.}$$

$$\begin{aligned} 2a \qquad\; + c &= -4 \\ \underline{-2a + 4b - c} &= \underline{20} \\ 4b \qquad &= 16 \end{aligned} \qquad \text{To eliminate } c, \text{ multiply equation (3) by } -1 \text{ and add to equation (1).}$$

$$b = 4 \qquad \text{Solve for } b.$$

$$-2b + c = -4 \qquad \text{Substitute } b = 4 \text{ in equation (2) and solve for } c.$$

$$-2(4) + c = -4$$

$$c = 4$$

$$2a - 4b + c = -20 \qquad \text{Substitute } b = 4 \text{ and } c = 4 \text{ in equation (3) and solve for } a.$$

$$2a - 4(4) + 4 = -20$$

$$a = -4$$

The solution set of the system is $\{(-4, 4, 4)\}$.

The equation of the circle is $x^2 + y^2 - 4x + 4y + 4 = 0$. ■

EXERCISE 7.3

A

Solve using linear combinations:

1. $\begin{cases} x + y + z = 4 \\ x - y + 2z = 8 \\ 2x + y - z = 3 \end{cases}$

2. $\begin{cases} x + y + 2z = 5 \\ x + y - 2z = 1 \\ 2x + y + 2z = 6 \end{cases}$

3. $\begin{cases} x - y + z = 2 \\ x + y + z = 2 \\ 2x - y + z = 3 \end{cases}$

4. $\begin{cases} x + y + z = 6 \\ x + y - z = 2 \\ x + 2y + z = 8 \end{cases}$

5. $\begin{cases} x - y + z = 2 \\ 2x + y - 2z = -9 \\ 3x + y + 3z = -6 \end{cases}$

6. $\begin{cases} x + 4y - z = 6 \\ 2x - 5y - z = 11 \\ x + 7y + z = 4 \end{cases}$

7. $\begin{cases} x + y + z = -1 \\ 2x - y + 5z = 10 \\ 5x - y - z = 19 \end{cases}$

8. $\begin{cases} 2x + 3y - z = 20 \\ x - y + z = -7 \\ 3x + 2y - z = 16 \end{cases}$

9. $\begin{cases} x + 3y - 2z = 5 \\ -x + y + 3z = 20 \\ x - y - z = -10 \end{cases}$

10. $\begin{cases} 3x - 2y + z = 4 \\ x + y - z = 0 \\ x + 3y + z = 20 \end{cases}$

B

11. $\begin{cases} x + 2y + 3z = 4 \\ 2x + y + z = 0 \\ 3x + y + 4z = 2 \end{cases}$

12. $\begin{cases} 2x + 3y + z = 3 \\ x + 4y + 2z = 2 \\ x + y + 3z = 7 \end{cases}$

13. $\begin{cases} 3x + 4y + 2z = -13 \\ 2x - y + z = -1 \\ x - 2y + 3z = 0 \end{cases}$

14. $\begin{cases} 2x - 3y + 4z = -3 \\ 6x + 2y - z = -11 \\ 3x + 4y + 3z = 1 \end{cases}$

15. $\begin{cases} x + 2y + 3z = 2 \\ 2x + 5y - 2z = -3 \\ 3x + 4y + 4z = -1 \end{cases}$

16. $\begin{cases} 4x + 7y - 3z = 2 \\ 2x - 8y + z = 16 \\ 3x + 5y - 2z = 3 \end{cases}$

17. $\begin{cases} 2x - 3y + 5z = -3 \\ 2x + 5y + 6z = 28 \\ 3x - 2y + z = 12 \end{cases}$

18. $\begin{cases} 2x - 5y + 3z = 1 \\ 5x + 2y + 2z = -21 \\ 3x - 4y + 2z = -3 \end{cases}$

19. $\begin{cases} x + 2y - 4z = 2 \\ x + 3y - 6z = 1 \\ 2x + 3y + 4z = -15 \end{cases}$

20. $\begin{cases} 2x + 3y - 2z = -23 \\ 3x - 4y - 3z = -9 \\ 5x + 2y + 6z = 36 \end{cases}$

C

21. $\begin{cases} 9x + 3y + 2z = 3 \\ 4x + 2y + 3z = 9 \\ 3x + 5y + 4z = 19 \end{cases}$

22. $\begin{cases} 4x + 5y - 6z = -13 \\ 7x + 4y + 9z = 41 \\ 5x + 2y + 2z = 9 \end{cases}$

23. $\begin{cases} 2x - y - z = 0 \\ 2x - 3y - 3z = -2 \\ x + 2y - z = 1 \end{cases}$

24. $\begin{cases} 4x + 4y - 3z = 4 \\ 2x - y - 9z = -2 \\ 12x + 12y - 12z = 11 \end{cases}$

25. $\begin{cases} 3x - 7y + 5z = 52 \\ -2x + 6y - 3z = -41 \\ 4x - 3y + 2z = 8 \end{cases}$

26. $\begin{cases} 2x + 3y - 5z = 0 \\ 3x - 7y + 4z = 0 \\ 9x + 5y - 8z = -24 \end{cases}$

27. $\begin{cases} 4x + 5y - 2z = 1 \\ 3x - 2y + 4z = -3 \\ 6x + 5y - 4z = 1 \end{cases}$

28. $\begin{cases} 12x + 12y + 12z = 7 \\ 2x - y - 3z = 3 \\ 12x + 8y - 6z = 17 \end{cases}$

29. $\begin{cases} 10x + 15y + 10z = 1 \\ 15x - 20y + 15z = 27 \\ 2x + 5y - 3z = -5 \end{cases}$

30. $\begin{cases} 6x + 12y + 15z = -22 \\ 6x - 4y + 9z = 2 \\ 6x + 10y - 14z = -19 \end{cases}$

D

31. The general form of a parabola is $y = ax^2 + bx + c$. Find the equation of the parabola that passes through the points $(-2, 0)$, $(-5, 9)$, and $(1, 9)$.

32. The general equation of a circle is $x^2 + y^2 + ax + by + c = 0$. Find the equation of a circle that passes through the points $(7, 0)$, $(-3, 0)$, and $(2, 7)$.

33. The Civic Theater in Tucson, Arizona, sold 900 tickets to its production of "The Rainmaker." The tickets were priced at $8, $6.50, and $5. The total income from the sale of tickets was $5400. If twice as many $6.50 tickets were sold than $8 tickets, how many of each price tickets were sold?

34. The You Can Do It hardware store had a sale on three types of screwdrivers. Regular screwdrivers sold for $1.95 each, Phillips screwdrivers sold for $2.50 each, and rachet screwdrivers sold for $6.80 each. The total receipts from the sale of 128 screwdrivers were $421.50. If two more regular screwdrivers were sold than Phillips screwdrivers, how many of each type were sold?

35. The Alpine Fruit Company packs three different gift boxes of pears, apples, and oranges. Box A contains 5 pears, 3 apples, and 1 orange. Box B contains 2 pears, 5 apples, and 2 oranges. Box C contains 3 pears, 4 apples, and 4 oranges. The company has 84 pears, 94 apples, and 50 oranges. How many of each gift box can be prepared if all the fruit is to be used?

36. The Low Impedance Electronic Supplier sells three different packs of replacement resistors. Pack A contains 18 2-ohm resistors, 14 1-ohm resistors, and 20 3-ohm resistors. Pack B contains 20 2-ohm resistors, 10 1-ohm resistors, and 15 3-ohm resistors. Pack C contains 15 2-ohm resistors, 20 1-ohm resistors, and 18 3-ohm resistors. The Supplier has 1525 2-ohm resistors, 1450 1-ohm resistors, and 1610 3-ohm resistors. How many of each replacement pack can be made up if all the resistors are to be used?

37. The currents at various points in the circuit below are given by the following equations:

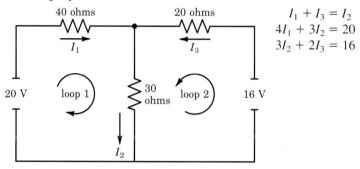

$$I_1 + I_3 = I_2$$
$$4I_1 + 3I_2 = 20$$
$$3I_2 + 2I_3 = 16$$

Figure for Exercise 37

Solve for the values of the currents I_1, I_2, and I_3 (the units of current is amps).

38. Three dowel rods are glued together as in the figure below:

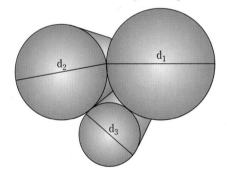

Figure for Exercise 38

By measuring, it is determined that the diameter of circle #1 (d_1) plus the diameter of circle #2 (d_2) is 12.5 inches. The diameter of circle #2 (d_2) plus the diameter of circle #3 (d_3) is 10.5 inches. Further, it is determined that the diameter of circle #1 (d_1) plus the diameter of circle #3 (d_3) is 11.0 inches. Find the diameters of the three rods.

39. Joan wants to invest $10,000 in stocks, bonds, and certificates of deposit (C.D.'s). Because of instability in the stock market, she wishes to invest four times as much in bonds as in stocks and $1,000 more in C.D.'s than in the stock market. How will the $10,000 be divided among the three investments?

STATE YOUR UNDERSTANDING

40. Could three planes have as their common solution (intersection) an entire line? Explain.

41. Could three planes have as their common solution (intersection) an entire plane? Explain.

42. What would the common solution be if two planes are identical and the third plane was parallel to the other two?

CHALLENGE EXERCISES

43. Using the following substitutions: $x = \dfrac{1}{u}$; $y = \dfrac{1}{v}$; and $z = \dfrac{1}{w}$, solve the following system for u, v, and w.

$$\begin{cases} \dfrac{1}{u} + \dfrac{1}{v} + \dfrac{1}{w} = 4 \\[2mm] \dfrac{1}{u} - \dfrac{1}{v} + \dfrac{1}{w} = 0 \\[2mm] \dfrac{1}{u} - \dfrac{1}{v} - \dfrac{1}{w} = 2 \end{cases}$$

44. Using the same substitutions as above, solve the following (careful):

$$\begin{cases} \dfrac{1}{u} + \dfrac{1}{v} + \dfrac{1}{w} = 1 \\[2mm] \dfrac{2}{u} + \dfrac{1}{v} - \dfrac{1}{w} = 8 \\[2mm] \dfrac{2}{v} + \dfrac{1}{w} = -2 \end{cases}$$

45. Even though the following system is not a three-by-three system, solve the system using linear combinations and notice the denominators for both the x- and y-values!

$$\begin{cases} ax + by = c \\ dx + ey = f \end{cases}$$

46. Solve using linear combinations:

$$\begin{cases} x + y + z = 4a - 1 \\ x - y + z = 5 \\ x - y - z = -3a + 1 \end{cases}$$

MAINTAIN YOUR SKILLS (SECTIONS 6.2, 6.3)

Write each of these equations in slope-intercept form:

47. $y - 18 = 3x + 7$

48. $\dfrac{x}{4} - \dfrac{y}{3} = 1$

Write each of these equations in standard form:

49. $3x - 18 = y + 7$

50. $\dfrac{x}{3} - \dfrac{y}{4} = 1$

Write the equation, in standard form, of each line described:

51. The line with slope $-\dfrac{3}{7}$ and containing the point $(10, -3)$.

52. The line with slope $-\dfrac{3}{7}$ and containing the point $(-3, 10)$.

Draw the graph of each line described:

53. The line with slope $-\dfrac{3}{5}$ and y-intercept $(0, 6)$.   **54.** The line with slope $-\dfrac{5}{3}$ and x-intercept $(6, 0)$.

7.4

EVALUATING DETERMINANTS

OBJECTIVES

1. Evaluate a determinant of order two.

2. Evaluate a determinant of order three.

3. Evaluate determinants of order greater than three.

4. Solve related equations.

Another way to solve a system of linear equations is to use Cramer's rule. In this section, we define and evaluate determinants. In the next section, we solve systems using Cramer's rule.

Determinants

■ DEFINITION

A *determinant* is a square array of numbers written between vertical lines.

$$\begin{vmatrix} a & b \\ c & d \end{vmatrix}$$
2×2

$$\begin{vmatrix} a_1 & b_1 & c_1 \\ a_2 & b_2 & c_2 \\ a_3 & b_3 & c_3 \end{vmatrix}$$
3×3

$$\begin{vmatrix} a_{11} & a_{12} & a_{13} \cdots a_{1n} \\ a_{21} & a_{22} & a_{23} \cdots a_{2n} \\ a_{31} & a_{32} & a_{33} \cdots a_{3n} \\ \vdots & \vdots & \vdots & \vdots \\ a_{n1} & a_{n2} & a_{n3} \cdots a_{nn} \end{vmatrix}$$

$n \times n$

The numbers in a determinant, a, a_1, and a_{11}, are called elements. Determinants are classified by the number of rows and columns.

Each determinant represents a real number. The formula for finding the value of the 2×2 determinant is:

■ FORMULA

Value of a 2 × 2 Determinant

$$\begin{vmatrix} a & b \\ c & d \end{vmatrix} = ad - cb$$

EXAMPLE 1

Evaluate the determinant:

$$\begin{vmatrix} 4 & -3 \\ 2 & 6 \end{vmatrix}$$

$a = 4, b = -3, c = 2, d = 6$ Identify a, b, c, and d.

$$\begin{vmatrix} 4 & -3 \\ 2 & 6 \end{vmatrix} = 4(6) - (2)(-3)$$ Substitute in the formula, $ad - cd$ and simplify.

$$= 24 + 6 = 30$$ ■

Note that we now have another symbol that represents the number 30, $\begin{vmatrix} 4 & -3 \\ 2 & 6 \end{vmatrix} = 30.$

EXAMPLE 2

Evaluate the determinant:

$$\begin{vmatrix} -1.5 & 0.5 \\ 0.3 & 0 \end{vmatrix}$$

$a = -1.5,\ b = 0.5,\ c = 0.3,\ d = 0$

$$\begin{vmatrix} -1.5 & 0.5 \\ 0.3 & 0 \end{vmatrix} = (-1.5)(0) - (0.3)(0.5)$$

$$= 0 - 0.15 = -0.15 \qquad\blacksquare$$

To evaluate a 3×3 determinant, we "expand by minors." Each entry of the determinant has a minor associated with it. The minor is the determinant formed by the entries of the determinant not in the row or column of the entry associated with the minor.

■ DEFINITION

Minor of a_1

For $\begin{vmatrix} a_1 & b_1 & c_1 \\ a_2 & b_2 & c_2 \\ a_3 & b_3 & c_3 \end{vmatrix}$ the determinant $\begin{vmatrix} b_2 & c_2 \\ b_3 & c_3 \end{vmatrix}$ is the minor of entry a_1.

Minors associated with other entries are found in a similar manner. For instance, the minor of b_3 is

$$\begin{vmatrix} a_1 & c_1 \\ a_2 & c_2 \end{vmatrix}.$$

To evaluate a 3×3 determinant by expanding by minors, we use an array of signs. Each minor has a sign associated with it as follows:

■ PROPERTY

Array of Signs

$$\begin{vmatrix} + & - & + \\ - & + & - \\ + & - & + \end{vmatrix} \qquad \text{Array of signs for a } 3 \times 3 \text{ determinant.}$$

Note that the signs alternate. The sign associated with the entry c_2 is negative. A minor, along with the sign associated with it, is called a *cofactor*.

■ **PROCEDURE**

To find the value of a 3×3 determinant:

1. Select a row or column to expand about.
2. Determine the minor of each entry of that row or column.
3. Determine the sign that goes with each minor.
4. Find the sum of the products for the row or column:

$$(\text{entry})(\pm 1)(\text{minor}).$$

The appropriate sign, ± 1, is found from the array of signs.

EXAMPLE 3

Evaluate the determinant:

$$\begin{vmatrix} 1 & -1 & -3 \\ 2 & 4 & -4 \\ 3 & -2 & 5 \end{vmatrix}$$

Expand about the first column. The entries are 1, 2, and 3. The signs associated with the entries are $+1$, -1, and $+1$, respectively.

$$= (1)(+1)\begin{vmatrix} 4 & -4 \\ -2 & 5 \end{vmatrix} + (2)(-1)\begin{vmatrix} -1 & -3 \\ -2 & 5 \end{vmatrix} + (3)(+1)\begin{vmatrix} -1 & -3 \\ 4 & -4 \end{vmatrix}$$

$$= 1[4(5) - (-2)(-4)] - 2[-1(5) - (-2)(-3)] \\ + 3[-1(-4) - 4(-3)]$$

Evaluate the 2×2 determinants.

$$= 1(20 - 8) - 2(-5 - 6) + 3(4 + 12)$$

$$= 1(12) - 2(-11) + 3(16)$$

$$= 12 + 22 + 48$$

$$= 82 \qquad ■$$

If a row or column of the determinant has 0 for an entry, it is usually easier to expand about that row or column since the product, $(\text{entry})(\pm 1)(\text{minor})$, will be 0.

EXAMPLE 4

Evaluate the determinant:

$$\begin{vmatrix} -1 & 0 & -4 \\ 5 & 7 & 0 \\ -4 & 0 & 6 \end{vmatrix}$$

$$= 0 + 7(+1) \begin{vmatrix} -1 & -4 \\ -4 & 6 \end{vmatrix} + 0$$

Expand about the second column since it contains two zeros.

$$= 0 + 7[-1(6) - (-4)(-4)] + 0$$

Evaluate the 2×2 determinant.

$$= 7(-6 - 16)$$

$$= 7(-22)$$

$$= -154 \qquad \blacksquare$$

There is a second method for finding the value of a 3×3 determinant.

First, copy columns one and two of the determinant on the right side of the determinant.

$$\begin{vmatrix} a_1 & b_1 & c_1 \\ a_2 & b_2 & c_2 \\ a_3 & b_3 & c_3 \end{vmatrix} \begin{matrix} a_1 & b_1 \\ a_2 & b_2 \\ a_3 & b_3 \end{matrix}$$

Second, find the sum, S_1, of the products along the diagonals starting with the upper left to lower right. Third, find the sum, S_2, of the products along the diagonals starting from lower left to upper right. These products are shown below.

$$a_3 b_2 c_1 + b_3 c_2 a_1 + c_3 a_2 b_1 = S_2$$

$$a_1 b_2 c_3 + b_1 c_2 a_3 + c_1 a_2 b_3 = S_1$$

Fourth, the value of the determinant is the difference of the sums:

$S_1 - S_2$.

■ CAUTION

This shortcut described above applies only to 3×3 determinants.

The proof that this gives the same result that is obtained by expanding by minors is left to the student.

EXAMPLE 5

Evaluate the determinant:

$$\begin{vmatrix} 1 & -1 & 1 \\ 0 & -3 & 2 \\ 4 & -2 & 2 \end{vmatrix}$$

$$S_2 = -12 + (-4) + 0 = -16$$

$$\begin{vmatrix} 1 & -1 & 1 \\ 0 & -3 & 2 \\ 4 & -2 & 2 \end{vmatrix} \begin{matrix} 1 & -1 \\ 0 & -3 \\ 4 & -2 \end{matrix}$$

Copy the first two columns on the right of the determinant. Now find the products along the diagonals. The products are found mentally.

$$S_1 = -6 + (-8) + 0 = -14$$

$$S_1 - S_2 = -14 - (-16) = 2$$ Subtract the sums of the products.   ■

A calculator is helpful in evaluating determinants with large entries or decimal entries.

A formula for the area of a triangle can be written as a determinant given the vertices of the triangle.

■ **FORMULA**

The area of a triangle with vertices at (x_1, y_1), (x_2, y_2), and (x_3, y_3) is $|A|$ where

$$A = \frac{1}{2} \begin{vmatrix} x_1 & y_1 & 1 \\ x_2 & y_2 & 1 \\ x_3 & y_3 & 1 \end{vmatrix}.$$

EXAMPLE 6

Find the area of a triangle with vertices at $(0, 4)$, $(0, -3)$, and $(6, 0)$.

$$A = \frac{1}{2} \begin{vmatrix} 0 & 4 & 1 \\ 0 & -3 & 1 \\ 6 & 0 & 1 \end{vmatrix}$$

To find the area of the triangle, substitute the coordinates of the points in the determinant and evaluate.

$$(x_1, y_1) = (0, 4)$$

$$(x_2, y_2) = (0, -3)$$

$$(x_3, y_3) = (6, 0)$$

$$A = \frac{1}{2} [0 + 0 + 6(+1)\begin{vmatrix} 4 & 1 \\ -3 & 1 \end{vmatrix}]$$ Expand about the first column.

$$= \frac{1}{2} [6(4 - (-3))]$$

$$= \frac{1}{2}(42)$$

$$= 21$$

The area of the triangle is 21 square units.   ■

Determinants of order greater than 3 are evaluated by expanding minors as was done for a 3×3 determinant. The expansion process is continued until the value is found. The array of signs can be extended to any order determinant. For example, the array of signs for a 5×5 determinant is

$$\begin{vmatrix} + & - & + & - & + \\ - & + & - & + & - \\ + & - & + & - & + \\ - & + & - & + & - \\ + & - & + & - & + \end{vmatrix}$$

EXAMPLE 7

Evaluate the determinant:

$$\begin{vmatrix} 3 & -1 & 4 & 0 & 5 \\ 0 & 1 & 0 & -2 & 0 \\ 0 & -3 & 0 & 1 & 2 \\ 6 & -3 & 1 & -4 & -3 \\ 0 & -1 & 0 & 0 & 4 \end{vmatrix}$$

Expand about the third column to take advantage of the zeros.

$$= 4(+1)\begin{vmatrix} 0 & 1 & -2 & 0 \\ 0 & -3 & 1 & 2 \\ 6 & -3 & -4 & -3 \\ 0 & -1 & 0 & 4 \end{vmatrix} + 0 + 0 + 1(-1)\begin{vmatrix} 3 & -1 & 0 & 5 \\ 0 & 1 & -2 & 0 \\ 0 & -3 & 1 & 2 \\ 0 & -1 & 0 & 4 \end{vmatrix} + 0$$

The value is now expressed in terms of 4×4 determinants.

Now expand each 4×4 about column one.

$$= 4[0 + 0 + 6(+1)\begin{vmatrix} 1 & -2 & 0 \\ -3 & 1 & 2 \\ -1 & 0 & 4 \end{vmatrix}] - 1[3(+1)\begin{vmatrix} 1 & -2 & 0 \\ -3 & 1 & 2 \\ -1 & 0 & 4 \end{vmatrix} + 0 + 0 + 0]$$

Now evaluate the 3×3 determinants.

$$= 24[(4 + 4 + 0) - (0 + 0 + 24)] - 3[(4 + 4 + 0) - (0 + 0 + 24)]$$

$$= 24(-16) - 3(-16) = -336$$ ∎

If an entry of a determinant is unknown and the value of the determinant is given, we can solve for the missing entry.

EXAMPLE 8

Solve for a:

$$\begin{vmatrix} a & 1 \\ -2 & 3 \end{vmatrix} = 14$$

$$3a - (-2)(1) = 14$$

$$3a + 2 = 14$$

$$3a = 12$$

$$a = 4$$

Expand the determinant and solve for a.

The solution set is {4}. ∎

EXAMPLE 9

Solve for b:

$$\begin{vmatrix} 0 & b & 2 \\ 2 & -1 & 1 \\ 3 & 2 & -1 \end{vmatrix} = 24$$

$$b(-1)\begin{vmatrix} 2 & 1 \\ 3 & -1 \end{vmatrix} + (-1)(1)\begin{vmatrix} 0 & 2 \\ 3 & -1 \end{vmatrix} + 2(-1)\begin{vmatrix} 0 & 2 \\ 2 & 1 \end{vmatrix} = 24$$

Expand about the column containing b.

$$-b(-2 - 3) - (0 - 6) - 2(0 - 4) = 24$$

$$5b + 6 + 8 = 24$$

$$5b = 10$$

$$b = 2$$

The solution set is {2}. ∎

EXERCISE 7.4

A

Evaluate the following determinants:

1. $\begin{vmatrix} 1 & -1 \\ -1 & 1 \end{vmatrix}$ **2.** $\begin{vmatrix} -1 & 1 \\ 1 & -1 \end{vmatrix}$ **3.** $\begin{vmatrix} 2 & -2 \\ -2 & 2 \end{vmatrix}$ **4.** $\begin{vmatrix} -2 & -2 \\ 2 & 2 \end{vmatrix}$

5. $\begin{vmatrix} 1 & -1 \\ 2 & 3 \end{vmatrix}$ **6.** $\begin{vmatrix} -1 & 2 \\ -3 & 1 \end{vmatrix}$ **7.** $\begin{vmatrix} -2 & -1 \\ -4 & 3 \end{vmatrix}$ **8.** $\begin{vmatrix} -4 & 1 \\ 2 & 3 \end{vmatrix}$

9. $\begin{vmatrix} 1 & 0 \\ 3 & -5 \end{vmatrix}$ **10.** $\begin{vmatrix} -6 & 4 \\ -3 & 0 \end{vmatrix}$ **11.** $\begin{vmatrix} -2 & -3 \\ -2 & 3 \end{vmatrix}$ **12.** $\begin{vmatrix} -4 & 1 \\ 4 & -1 \end{vmatrix}$

13. $\begin{vmatrix} 1.1 & -3.5 \\ 2.6 & -0.4 \end{vmatrix}$ **14.** $\begin{vmatrix} 0.35 & 2.1 \\ -0.65 & 1.31 \end{vmatrix}$

Solve for a:

15. $\begin{vmatrix} a & 1 \\ 0 & -1 \end{vmatrix} = 5$

16. $\begin{vmatrix} -2 & 0 \\ 5 & a \end{vmatrix} = 8$

17. $\begin{vmatrix} a & 1 \\ 1 & 1 \end{vmatrix} = 2$

18. $\begin{vmatrix} 1 & a \\ 1 & 1 \end{vmatrix} = 2$

19. $\begin{vmatrix} -1 & -1 \\ a & -1 \end{vmatrix} = 4$

20. $\begin{vmatrix} -1 & -1 \\ -1 & a \end{vmatrix} = 4$

B

Evaluate the following determinants:

21. $\begin{vmatrix} 4 & -7 \\ 3 & -6 \end{vmatrix}$

22. $\begin{vmatrix} 5 & -9 \\ 4 & -4 \end{vmatrix}$

23. $\begin{vmatrix} -8 & 9 \\ 9 & -8 \end{vmatrix}$

24. $\begin{vmatrix} 10 & -5 \\ -7 & 8 \end{vmatrix}$

25. $\begin{vmatrix} -13 & 11 \\ 5 & -4 \end{vmatrix}$

26. $\begin{vmatrix} -9 & 8 \\ 7 & -11 \end{vmatrix}$

27. $\begin{vmatrix} -20 & 5 \\ 30 & 10 \end{vmatrix}$

28. $\begin{vmatrix} 16 & -12 \\ 5 & 9 \end{vmatrix}$

29. $\begin{vmatrix} 3a & 2b \\ -4a & 5b \end{vmatrix}$

30. $\begin{vmatrix} 5a & 10 \\ -a & 6 \end{vmatrix}$

31. $\begin{vmatrix} 1 & 2 & -1 \\ 2 & 1 & 1 \\ 1 & 1 & 2 \end{vmatrix}$

32. $\begin{vmatrix} 2 & -1 & 1 \\ 1 & 2 & 1 \\ 2 & 1 & 1 \end{vmatrix}$

33. $\begin{vmatrix} -3 & -1 & 2 \\ 2 & 2 & -1 \\ -1 & 1 & 1 \end{vmatrix}$

34. $\begin{vmatrix} 4 & 1 & -1 \\ -2 & 2 & 1 \\ 2 & 2 & 1 \end{vmatrix}$

35. $\begin{vmatrix} 2 & 3 & 2 \\ 0 & -1 & 0 \\ 5 & 2 & 1 \end{vmatrix}$

36. $\begin{vmatrix} 1 & 0 & 4 \\ 2 & 4 & -1 \\ 3 & 0 & -2 \end{vmatrix}$

Solve for a:

37. $\begin{vmatrix} -9 & a \\ -6 & 7 \end{vmatrix} = 66$

38. $\begin{vmatrix} -11 & 5 \\ a & 7 \end{vmatrix} = 3$

39. $\begin{vmatrix} a & 12 \\ 3 & a \end{vmatrix} = 0$

40. $\begin{vmatrix} -a & 5 \\ 3 & a \end{vmatrix} = -40$

C

Evaluate the following determinants:

41. $\begin{vmatrix} -3 & -3 & 1 \\ 4 & 1 & 4 \\ 2 & 2 & -3 \end{vmatrix}$

42. $\begin{vmatrix} -4 & -6 & 2 \\ 5 & 2 & 1 \\ 1 & 3 & 1 \end{vmatrix}$

43. $\begin{vmatrix} \dfrac{3}{4} & \dfrac{1}{4} \\ -\dfrac{1}{2} & -\dfrac{2}{3} \end{vmatrix}$

44. $\begin{vmatrix} \dfrac{1}{2} & \dfrac{1}{4} \\ -\dfrac{2}{3} & -\dfrac{3}{4} \end{vmatrix}$

45. $\begin{vmatrix} 0 & 2 & -5 \\ 1 & 3 & 6 \\ 0 & 4 & 1 \end{vmatrix}$

46. $\begin{vmatrix} 1 & -3 & 2 \\ 0 & 1 & 1 \\ 0 & 4 & 1 \end{vmatrix}$

47. $\begin{vmatrix} 5 & 3 & -1 \\ 2 & -3 & 6 \\ 8 & -1 & -7 \end{vmatrix}$

48. $\begin{vmatrix} -2 & 6 & -5 \\ 3 & 7 & 2 \\ 4 & 4 & -2 \end{vmatrix}$

49. $\begin{vmatrix} 12 & 5 & 0 \\ -6 & -4 & 0 \\ 2 & 3 & -4 \end{vmatrix}$

50. $\begin{vmatrix} 0 & -2 & 1 \\ 3 & 13 & -4 \\ 0 & -5 & -6 \end{vmatrix}$

51. $\begin{vmatrix} 3 & -1 & 0 & 1 \\ 0 & 2 & 0 & 3 \\ 4 & -1 & 1 & 1 \\ 2 & 3 & 2 & 4 \end{vmatrix}$

52. $\begin{vmatrix} 0 & 1 & 0 & -2 \\ 0 & 3 & 4 & 0 \\ 5 & -1 & 1 & 2 \\ 3 & 1 & 4 & 0 \end{vmatrix}$

53.
$$\begin{vmatrix} 0 & 1 & 3 & 0 & 1 \\ 0 & 2 & 1 & 3 & 0 \\ 1 & 0 & 0 & -2 & 1 \\ 3 & 6 & -1 & 0 & 0 \\ 0 & 2 & 1 & 1 & 1 \end{vmatrix}$$

54.
$$\begin{vmatrix} 0 & 3 & -1 & 1 & 0 \\ 3 & 0 & 5 & 2 & 0 \\ 1 & 0 & 0 & 3 & 0 \\ -6 & 1 & 1 & 2 & -3 \\ 0 & 0 & 1 & 1 & 0 \end{vmatrix}$$

55.
$$\begin{vmatrix} -a & 2 & 1 \\ -a & 3 & 2 \\ a & 4 & -1 \end{vmatrix}$$

56.
$$\begin{vmatrix} b & 1 & -2 \\ b & 2 & 1 \\ -b & 3 & 1 \end{vmatrix}$$

Solve for a:

57.
$$\begin{vmatrix} -1 & -2 & 1 \\ a & 0 & 0 \\ -1 & 1 & 1 \end{vmatrix} = 3$$

58.
$$\begin{vmatrix} 2 & -1 & 0 \\ 1 & -2 & a \\ 3 & 1 & 0 \end{vmatrix} = 10$$

59.
$$\begin{vmatrix} a & -1 & 1 \\ 0 & 2 & -2 \\ 0 & 3 & a \end{vmatrix} = -4$$

60.
$$\begin{vmatrix} 1 & 0 & 3 \\ -1 & a & 0 \\ a & 0 & 2 \end{vmatrix} = -8$$

D

61. Find the area of a triangle with vertices at $(3, 4)$, $(6, 0)$, and $(2, 1)$.

62. Find the area of a triangle with vertices at $(-5, 5)$, $(5, -5)$, and $(5, 10)$.

STATE YOUR UNDERSTANDING

63. If a determinant contains either a row or a column of all zero's, what is the value of the determinant? Why?

64. Create any 3×3 determinant with two identical rows. Find the value of the determinant.

CHALLENGE EXERCISES

65. Evaluate the determinant:
$$\begin{vmatrix} 3 & -4 & 3 \\ 1 & 7 & 1 \\ 2 & 18 & 2 \end{vmatrix}$$

66. Evaluate the determinant:
$$\begin{vmatrix} 1 & -2 & 4 \\ -1 & -1 & 6 \\ -2 & 4 & -8 \end{vmatrix}$$

67. Evaluate the determinant: $\begin{vmatrix} 4 & -2 \\ 3 & 1 \end{vmatrix}$

Interchange the two rows and re-evaluate the determinant. Now, interchange the two columns of the original determinant and re-evaluate.

68. What value of a will make the determinant zero? $\begin{vmatrix} a & 3 \\ -4 & 2 \end{vmatrix}$

69. What value of b will make the determinant equal to 10? $\begin{vmatrix} b & -1 \\ 1 & b \end{vmatrix}$

70. Evaluate: $\begin{vmatrix} a & 5 & -3 \\ 0 & b & -4 \\ 0 & 0 & c \end{vmatrix}$

71. Evaluate: $\begin{vmatrix} a & 0 & 0 \\ -2 & b & 0 \\ -3 & 1 & c \end{vmatrix}$

MAINTAIN YOUR SKILLS (SECTIONS 4.7, 5.2, 5.5, 5.6, 5.7)

Solve:

72. $\dfrac{1}{x} - \dfrac{1}{x + 3} = 2$

73. $\dfrac{1}{x - 3} - \dfrac{1}{2} = \dfrac{1}{x}$

74. $\sqrt{x + 11} - 3 = \sqrt{x - 4}$

75. $2 - \sqrt{x + 2} = 3 - \sqrt{x + 3}$

76. $(x + 1)(x + 4) \le 4$

77. $x + 1 \le \dfrac{4}{x + 4}$

78. $|2x - 1| < 11$

79. $|2x - 11| < 1$

7.5

**SOLVING SYSTEMS
USING CRAMER'S RULE**

OBJECTIVES

1. Solve a system of linear equations in two variables using Cramer's rule.

2. Solve a system of linear equations in three variables using Cramer's rule.

Cramer's rule for finding the solution of a system of two linear equations in two variables can be derived by solving the general system using linear combinations.

$$\begin{cases} a_1x + b_1y = c_1 & (1) \\ a_2x + b_2y = c_2 & (2) \end{cases}$$

$$a_1b_2x + b_1b_2y = b_2c_1 \qquad \text{Multiply equation (1) by } b_2.$$

$$\underline{-a_2b_1x - b_1b_2y = -b_1c_2} \qquad \text{Multiply equation (2) by } -b_1.$$

$$a_1b_2x - a_2b_1x = b_2c_1 - b_1c_2 \qquad \text{Add.}$$

$$(a_1b_2 - a_2b_1)x = b_2c_1 - b_1c_2 \qquad \text{Solve for } x.$$

$$x = \frac{b_2c_1 - b_1c_2}{a_1b_2 - a_2b_1}$$

Both the numerator and the denominator can be written as determinants.

$$b_2c_1 - b_1c_2 = c_1b_2 - c_2b_1 = \begin{vmatrix} c_1 & b_1 \\ c_2 & b_2 \end{vmatrix}$$

$$a_1b_2 - a_2b_1 = \begin{vmatrix} a_1 & b_1 \\ a_2 & b_2 \end{vmatrix}$$

The values of x and y can be written as the quotient of determinants.

$$x = \frac{\begin{vmatrix} c_1 & b_1 \\ c_2 & b_2 \end{vmatrix}}{\begin{vmatrix} a_1 & b_1 \\ a_2 & b_2 \end{vmatrix}}$$

Similarly, when we solve for y, we have

$$y = \frac{\begin{vmatrix} a_1 & c_1 \\ a_2 & c_2 \end{vmatrix}}{\begin{vmatrix} a_1 & b_1 \\ a_2 & b_2 \end{vmatrix}}.$$

The determinant $\begin{vmatrix} a_1 & b_1 \\ a_2 & b_2 \end{vmatrix}$ is called D, the determinant $\begin{vmatrix} c_1 & b_1 \\ c_2 & b_2 \end{vmatrix}$ is called D_x, and the determinant $\begin{vmatrix} a_1 & c_1 \\ a_2 & c_2 \end{vmatrix}$ is called D_y. D is the determinant formed from the coefficients of the variables. We can remember D_x if we think of it as being formed by replacing the first column of D with the constant terms. D_y is formed by replacing the second column of D with the constant terms.

■ FORMULA

Cramer's rule

Cramer's rule for solving a linear system in two variables:

1. $x = \dfrac{D_x}{D}$ and $y = \dfrac{D_y}{D}$

2. The solution set is $\{(x, y)\}$.

EXAMPLE 1

Solve using Cramer's rule:

$$\begin{cases} x + y = 12 \\ 2x - 3y = -1 \end{cases}$$

$$D = \begin{vmatrix} 1 & 1 \\ 2 & -3 \end{vmatrix} = 1(-3) - 2(1) = -5$$

Find the value of D, D_x, and D_y.

$$D_x = \begin{vmatrix} 12 & 1 \\ -1 & -3 \end{vmatrix} = 12(-3) - (-1)(1) = -35$$

$$D_y = \begin{vmatrix} 1 & 12 \\ 2 & -1 \end{vmatrix} = 1(-1) - 2(12) = -25$$

$$x = \frac{D_x}{D} = \frac{-35}{-5} = 7$$

Use Cramer's rule to solve for x and y.

$$y = \frac{D_y}{D} = \frac{-25}{-5} = 5$$

Check:

Substitute 7 for x and 5 for y in the equations of the system.

$x + y = 12$	$2x - 3y = -1$
$7 + 5 = 12$	$2(7) - 3(5) = -1$
	$14 - 15 = -1$

The solution set is $\{(7, 5)\}$.   ■

Cramer's rule for a system of three equations in three variables can be developed along similar lines. Here we just show the determinants that are used in the solution. For the general 3×3 system:

$$\begin{cases} a_1x + b_1y + c_1z = d_1 \\ a_2x + b_2y + c_2z = d_2 \\ a_3x + b_3y + c_3z = d_3 \end{cases}$$

We define the following determinants:

$$D = \begin{vmatrix} a_1 & b_1 & c_1 \\ a_2 & b_2 & c_2 \\ a_3 & b_3 & c_3 \end{vmatrix} \qquad D_x = \begin{vmatrix} d_1 & b_1 & c_1 \\ d_2 & b_2 & c_2 \\ d_3 & b_3 & c_3 \end{vmatrix}$$

$$D_y = \begin{vmatrix} a_1 & d_1 & c_1 \\ a_2 & d_2 & c_2 \\ a_3 & d_3 & c_3 \end{vmatrix} \qquad D_z = \begin{vmatrix} a_1 & b_1 & d_1 \\ a_2 & b_2 & d_2 \\ a_3 & b_3 & d_3 \end{vmatrix}$$

As before, we can think of D_x as being formed by replacing column 1 of D (the coefficients of x) with the constants d_1, d_2, and d_3. In the same manner, D_y

and D_z are formed by replacing column 2 (the coefficients of y) and column 3 (the coefficients of z), respectively.

■ FORMULA

> Cramer's rule for solving a linear system in three variables:
>
> 1. $x = \dfrac{D_x}{D}$, $y = \dfrac{D_y}{D}$, and $z = \dfrac{D_z}{D}$
>
> 2. The solution set is $\{(x, y, z)\}$.

EXAMPLE 2

Solve using Cramer's rule:

$$\begin{cases} x + y + z = 6 \\ 2x - y + z = 3 \\ x + 2y - z = 2 \end{cases}$$

$$D = \begin{vmatrix} 1 & 1 & 1 \\ 2 & -1 & 1 \\ 1 & 2 & -1 \end{vmatrix} = 1(-1) - 1(-3) + 1(5) = 7$$

Find the value of D, D_x, D_y, and D_z.

Expand about the first row.

The value of each minor is in parentheses.

$$D_x = \begin{vmatrix} 6 & 1 & 1 \\ 3 & -1 & 1 \\ 2 & 2 & -1 \end{vmatrix} = 1(8) - 1(10) + 1(-1)(-9) = 7$$

Expand about the third column.

$$D_y = \begin{vmatrix} 1 & 6 & 1 \\ 2 & 3 & 1 \\ 1 & 2 & -1 \end{vmatrix} = 1(1) - 1(-4) + (-1)(-9) = 14$$

Expand about the third column.

$$D_z = \begin{vmatrix} 1 & 1 & 6 \\ 2 & -1 & 3 \\ 1 & 2 & 2 \end{vmatrix} = 1(-8) - 2(-10) + 1(9) = 21$$

Expand about the first column.

$$x = \frac{D_x}{D} = \frac{7}{7} = 1, \; y = \frac{D_y}{D} = \frac{14}{7} = 2,$$

Use Cramer's rule to find x, y, and z.

$$z = \frac{D_z}{D} = \frac{21}{7} = 3$$

The solution set is $\{(1, 2, 3)\}$.

The check is left to the student. ■

When using Cramer's rule, dependent and inconsistent systems are identified by the fact that $D = 0$. When $D = 0$, we will conclude that the system has no unique solution.

EXAMPLE 3

Solve using Cramer's rule:

$$\begin{cases} 2x + 6y + 4z = 12 \\ 4x - y - z = 8 \\ x + 3y + 2z = 6 \end{cases}$$

$$D = \begin{vmatrix} 2 & 6 & 4 \\ 4 & -1 & -1 \\ 1 & 3 & 2 \end{vmatrix} = 2(1) - 4(0) + 1(-2) = 0$$

Find D, D_x, D_y, and D_z.

$D = 0$, so there is no unique solution.

The system has no unique solution.   ■

Any application that can be expressed as a system of linear equations can be solved using Cramer's rule.

EXAMPLE 4

A total of 8736 ballots were cast in an election in which there were two candidates. The votes were cast in a ratio of 2 to 1. How many votes were cast for the winning candidate?

Simpler word form:

$$\left(\begin{array}{c} \text{Number of} \\ \text{votes received} \\ \text{by winner} \end{array} \right) + \left(\begin{array}{c} \text{Number of} \\ \text{votes received} \\ \text{by loser} \end{array} \right) = 8736 \text{ votes}$$

$$\left(\begin{array}{c} \text{Number of} \\ \text{votes received} \\ \text{by winner} \end{array} \right) = 2 \left(\begin{array}{c} \text{Number of} \\ \text{votes received} \\ \text{by loser} \end{array} \right)$$

To find the number of votes that the winning candidate received, we first write the simpler word form.

Select variables:

Let x represent the number of votes received by the winner.
Let y represent the number of votes received by the loser.

Translate to algebra:

$$\begin{cases} x + y = 8736 \\ x = 2y \end{cases}$$

Solve:

$$\begin{cases} x + y = 8736 & (1) \\ x - 2y = 0 & (2) \end{cases}$$

$$D = \begin{vmatrix} 1 & 1 \\ 1 & -2 \end{vmatrix}, \quad D_x = \begin{vmatrix} 8736 & 1 \\ 0 & -2 \end{vmatrix}, \quad D_y = \begin{vmatrix} 1 & 8736 \\ 1 & 0 \end{vmatrix}$$

Write D, D_x, and D_y.

$$x = \frac{D_x}{D} = \frac{\begin{vmatrix} 8736 & 1 \\ 0 & -2 \end{vmatrix}}{\begin{vmatrix} 1 & 1 \\ 1 & -2 \end{vmatrix}} = \frac{8736(-2) - 0(1)}{1(-2) - 1(1)} = \frac{-17472 - 0}{-2 - 1}$$

Use Cramer's rule to solve for x.

$$= \frac{-17472}{-3} = 5824$$

$$y = \frac{D_y}{D} = \frac{\begin{vmatrix} 1 & 8736 \\ 1 & 0 \end{vmatrix}}{-3} = \frac{1(0) - 1(8736)}{-3} = \frac{-8736}{-3} = 2912$$

Use Cramer's rule to solve for y.

Check:

Check $x = 5824$, $y = 2912$ in (1)

$5824 + 2912 = 8736$

Check $x = 5824$, $y = 2912$ in (2)

$5824 - 2(2912) = 0$

$5824 - 5824 = 0$

Answer:

The winner received 5824 votes.   ■

Cramer's rule also can be extended to a 4 × 4 or larger system. Answers are written as ordered quadruples, (x, y, z, w), etc.

EXERCISE   7.5

A

Solve using Cramer's rule:

1. $\begin{cases} 2x + 5y = 15 \\ 3x - 4y = 11 \end{cases}$

2. $\begin{cases} 6x - 5y = -31 \\ 3x + 4y = 17 \end{cases}$

3. $\begin{cases} 7x + y = 42 \\ 3x - y = 8 \end{cases}$

4. $\begin{cases} 8x - y = 34 \\ x + 8y = 53 \end{cases}$

5. $\begin{cases} 3x + 4y = 10 \\ 2x - 3y = 1 \end{cases}$

6. $\begin{cases} 5x - y = -29 \\ 2x + 3y = 2 \end{cases}$

7. $\begin{cases} 6x - 5y = 1 \\ 9x + 10y = 12 \end{cases}$

8. $\begin{cases} 5x - 2y = 5 \\ 20x + 10y = 11 \end{cases}$

9. $\begin{cases} x - 3y = 4 \\ 3x - 9y = 5 \end{cases}$

10. $\begin{cases} 3y - 2x = 2 \\ 12y - 8x = 8 \end{cases}$

11. $\begin{cases} 2.7x - 1.3y = 3.3 \\ 1.5x + 0.2y = 0.45 \end{cases}$

12. $\begin{cases} 0.6x + 1.4y = 2.46 \\ 2.3x - 0.4y = 2.51 \end{cases}$

B

13. $\begin{cases} x + y + z = 1 \\ x - y + z = 2 \\ x - y - z = 3 \end{cases}$

14. $\begin{cases} 3x - 2y + 4z = -8 \\ 5x + y - z = -8 \\ 3x - 3y + 2z = -13 \end{cases}$

C

15. $\begin{cases} 2x + 3y + z = 6 \\ x - 2y + 3z = -3 \\ 3x + y - z = 8 \end{cases}$

16. $\begin{cases} x + y + z = 6 \\ 3x - y + 2z = 7 \\ 2x + 3y - z = 5 \end{cases}$

17. $\begin{cases} 2x - 3y + z = -5 \\ 2x + 5y - 4z = -4 \\ 3x + 2y - 2z = -5 \end{cases}$

18. $\begin{cases} 5x - y + z = 12 \\ 3x - 4y + 3z = 0 \\ -2x + 3y + 2z = -12 \end{cases}$

19. $\begin{cases} 4x + 5y - 2z = -7 \\ 3x - 2y + 5z = -6 \\ 2x + 3y - 3z = -2 \end{cases}$

20. $\begin{cases} 5x + 6y - 3z = -8 \\ 3x - y + 2z = 9 \\ x - 2y - 3z = 8 \end{cases}$

21. $\begin{cases} 2x + 3y - 4z = 0 \\ x + 6y - 2z = 3 \\ 3x + 3y + 6z = 8 \end{cases}$

22. $\begin{cases} x + y + z = 0 \\ 2x - 3y - 3z = -1 \\ 3x + 3y - 7z = 4 \end{cases}$

23. $\begin{cases} 5x + 3y - z = -2 \\ 3x + y + 4z = 1 \\ 5x - 7y + z = 7 \end{cases}$

24. $\begin{cases} 2x + 3y + 10z = 4 \\ x - 4y + 20z = 10 \\ x - y + z = 1 \end{cases}$

25. $\begin{cases} x + y = 1 \\ y + z = 9 \\ x + z = -6 \end{cases}$

26. $\begin{cases} 2x + y - z = 7 \\ -x + y = 1 \\ -y + z = 1 \end{cases}$

27. $\begin{cases} x + 3z = 15 \\ 2x - 3y = -6 \\ 2y - 4z = -16 \end{cases}$

28. $\begin{cases} 3x + 4y = -1 \\ 2x - 3z = 8 \\ 2y - 5z = 8 \end{cases}$

29. $\begin{cases} 5x - 7y \quad\quad = -11 \\ \quad\quad 5y + 6z = -15 \\ 6x \quad\quad - 5z = 37 \end{cases}$

30. $\begin{cases} 9x \quad\quad - 2z = 19 \\ \quad\quad 3y - 5z = -29 \\ 7x + 4y \quad\quad = 9 \end{cases}$

31. $\begin{cases} 3x \quad\quad + 13z = -15 \\ 15x + 3y \quad\quad = 10 \\ \quad\quad 6y - 5z = 45 \end{cases}$

32. $\begin{cases} 12x - 75y \quad\quad = 9 \\ \quad\quad 15y - 4z = 16 \\ 4x \quad\quad - 12z = 31 \end{cases}$

33. $\begin{cases} x + y + z \quad\quad = 2 \\ \quad 2y - z - w = 2 \\ x \quad\quad - w = 1 \\ 2x \quad\quad + 3z + 2w = 0 \end{cases}$

34. $\begin{cases} 2x - y \quad\quad + 2w = 11 \\ \quad 2y - 3z - w = 2 \\ 3x \quad\quad - z + 2w = 12 \\ \quad y + 2z \quad\quad = -11 \end{cases}$

35. $\begin{cases} x - 3y + 2z \quad\quad = 17 \\ x \quad\quad - 4w = -1 \\ 2y - z + 3w = -6 \\ x \quad\quad + 3z \quad\quad = 6 \end{cases}$

36. $\begin{cases} 2x \quad\quad - 5z - 3w = -12 \\ \quad 2y + z + w = 8 \\ 3x + y \quad\quad - 2w = 4 \\ 2x - y + 3z - w = 6 \end{cases}$

Solve using Cramer's rule. Assume that a and b are nonzero constants, $a \neq b$:

37. $\begin{cases} ax + by = a \\ bx + ay = b \end{cases}$

38. $\begin{cases} ax + by = 1 \\ bx + ay = 1 \end{cases}$

39. $\begin{cases} ax - y = a^2 \\ bx - y = b^2 \end{cases}$

40. $\begin{cases} ax + by = 1 \\ b^2x + a^2y = b \end{cases}$

D

41. The sum of two numbers is 1459. The second number is 85 more than the first. What are the two numbers?

42. The sum of two numbers is 2091. The first number is 127 more than the second. What are the two numbers?

43. The Ace Fruit Company packs two kinds of mixed fruit. The first kind contains 2.5 ounces of grapes per quart, and the second kind contains 3.3 ounces of grapes per quart. How many quarts of each pack should they use to form 10 quarts of a new mixture that contains 3 ounces of grapes per quart?

44. At the One Price Suit Store, sport coats sell for $89 each, and suits sell for $132 each. In one day, the gross sales were $3582. If three more sport coats were sold than suits, how many of each were sold?

45. A stock broker buys two stocks, one at $37.50 a share, and the other at $14.75 a share. If the total purchase consisted of 650 shares at a cost of $14,137.50, how many shares were purchased at each price?

46. A buyer for the Beech Company bought two kinds of recliners. She paid $195 each for one type and $245 each for the other type. If the total purchase consisted of 24 recliners at a cost of $5130, how many of each type of reclincer did she buy?

47. A diet at the Reducing Salon calls for 10.5 ounces of nutrient U, 20 ounces of nutrient V, and 26 ounces of nutrient W. Three foods are available to fill the diet. Food A contains 30% nutrient U, 20% nutrient V, and no nutrient W. Food B contains 10% nutrient U, 30% nutrient V, and 30% nutrient W. Food C contains no nutrient U, 10% nutrient V, and 50% nutrient W. How many ounces of each food are needed to exactly meet the diet requirements?

48. The Nice-N-Pretty Cosmetic Firm finds that it takes $3 to produce, 10 minutes to sell, and $0.50 to deliver Brand A perfume; $5 to produce, 15 minutes to sell, and nothing to deliver Brand B perfume; and $4 to produce, 12 minutes to sell, and $1 to deliver Brand C perfume. The company sold perfume that cost $700 to produce, $125 to deliver, and 2200 minutes to sell. How many units of each brand of perfume were sold?

49. The following current loop diagram leads to the following equations:

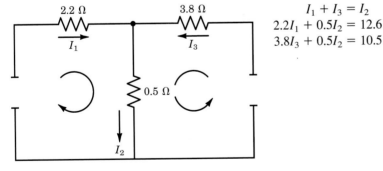

$$I_1 + I_3 = I_2$$
$$2.2I_1 + 0.5I_2 = 12.6$$
$$3.8I_3 + 0.5I_2 = 10.5$$

Figure for Exercise 49

Find, to the nearest tenth, I_1, I_2, and I_3 using Cramer's rule.

50. The federal business tax (x) on a small business is calculated on the gross receipts minus the amount of state business tax (y) that is paid. The state tax is calculated on the gross receipts less the amount of federal tax that is paid. The federal tax rate is 27% and the state tax rate is 11%, find, to the nearest cent, the amount of tax paid to both if the gross receipts are $150,000.

51. A teller is giving a customer $1000 in cash in $5, $10, and $20 bills. There are twice as many $10 bills as there are $5 bills. There are 15 more $10 bills than $20 bills. How many of each were given to the customer?

STATE YOUR UNDERSTANDING

52. Solve the following system using Cramer's rule. What conclusions can you draw from the solution?

$$\begin{cases} 3x - 2y + 5z = 0 \\ x - y + 3z = 0 \\ 2x - y + z = 0 \end{cases}$$

CHALLENGE EXERCISES

53. In a 2×2 system, if the value of the determinant D is zero, the system is either dependent or independent and inconsistent. Using Cramer's rule and the determinant D, what value for "a" would make one of these two occur?

$$\begin{cases} ax + 3y = 5 \\ 2x - y = -7 \end{cases}$$

54. In a 2×2 system, if the value of the determinant D and the value of either the determinants of D_x and D_y are zero, the system is dependent. Using the following example, show that these indeed occur.

$$\begin{cases} 5x - 10y = 15 \\ -3x + 6y = -9 \end{cases}$$

55. Write an example of a 2×2 system that is independent and inconsistent (but not dependent). Find the value of the determinants of D and D_x. Are they both zero?

56. Using Cramer's rule, determine what values of "a" and "b" would make the system dependent.

$$\begin{cases} ax + 4y = 4 \\ 2x - y = b \end{cases}$$

57. In a 3×3 system, it is mentioned in this section that if the value of the determinant D is zero, we say there is no unique solution. What value of "a" would make this happen?

$$\begin{cases} ax + 3y - z = 4 \\ 2y + 3z = 5 \\ ax - 3z = 1 \end{cases}$$

MAINTAIN YOUR SKILLS (SECTIONS 4.7, 5.2, 5.6, 6.4)

Solve:

58. $\sqrt{2x - 3} - \sqrt{x + 10} = -1$

59. $\sqrt{2x - 3} - \sqrt{4x - 13} = -1$

60. $|x + 3| - x > 7$

61. $|x - 7| + x < 3$

62. $(x + 5)^2 - (2x - 7)^2 = 3x^2 + 8$

63. $(x + 5)^2 + (2x - 7)^2 = 4x^2 - 6$

Draw the graph of the inequalities:

64. $x \geq 5 - 2y$

65. $y < -\dfrac{2}{3}x + 4$

7.6

SOLVING SYSTEMS OF LINEAR EQUATIONS BY MATRIX METHODS

OBJECTIVES

1. Solve a system of linear equations in two variables by matrix methods.

2. Solve a system of linear equations in three variables by matrix methods.

■ **DEFINITION**

Matrix

> A *matrix* is a rectangular array of numbers. Each number is said to be an entry of the matrix.

Matrices (plural of matrix) can be named using capital letters. Two examples of matrices are:

$$A = \begin{bmatrix} 2 & 3 & 4 \\ 0 & 1 & 5 \\ 6 & 9 & 1 \end{bmatrix} \qquad B = \begin{bmatrix} 3 & 7 & 4 & 1 \\ -1 & 0 & 0 & 3 \end{bmatrix}$$

Matrix A is 3×3 because it contains 3 rows and 3 columns and is said to be of order three. Matrix B is 2×4.

In this section, we present a method of solving linear systems using matrices. This method is important since it is easily adapted to computers.

Every system of linear equations has a matrix associated with it called the augmented matrix.

■ **DEFINITION**

The augmented matrix associated with a linear system in two variables

$$\begin{cases} ax + by = c \\ dx + ey = f \end{cases} \text{ is } \left[\begin{array}{cc|c} a & b & c \\ d & e & f \end{array}\right].$$

The augmented matrix is formed using the coefficients of the variables and the constant terms with a line separating them. This definition can be extended to systems involving more variables. So we have the following augmented matrix associated with the 3×3 system.

$$\begin{cases} 3x + 4y - 8z = 11 \\ 2x - 5y - 2z = -3 \\ x + 6y - z = 0 \end{cases} \qquad \left[\begin{array}{ccc|c} 3 & 4 & -8 & 11 \\ 2 & -5 & -2 & -3 \\ 1 & 6 & -1 & 0 \end{array}\right]$$

To solve a system using the augmented matrix, we use row operations to write a new matrix.

■ **DEFINITION**

Matrix Row Operations

The row operations are:

1. Any two rows of the matrix may be interchanged.

2. The numbers in any row may be multiplied by any nonzero real number.

3. Any row may be changed by adding to the numbers of the row the numbers of another row.

Using row operations we write a matrix of the form:

$$\left[\begin{array}{cc|c} 1 & m & n \\ 0 & 1 & p \end{array}\right] \quad \text{for a } 2 \times 2 \text{ system and}$$

$$\left[\begin{array}{ccc|c} 1 & m & n & p \\ 0 & 1 & q & r \\ 0 & 0 & 1 & s \end{array}\right] \quad \text{for a } 3 \times 3 \text{ system.}$$

The new matrix will yield a new system of equations that has the same solution as the original system (equivalent). The row operations parallel the operations used in solving a system of linear equations by linear combinations.

Now let's use row operations to solve a system.

EXAMPLE 1

Solve: $\begin{cases} x - 3y = 10 \\ 2x + y = -1 \end{cases}$

$$\begin{bmatrix} 1 & -3 & | & 10 \\ 2 & 1 & | & -1 \end{bmatrix}$$ First form the augmented matrix.

$$\begin{bmatrix} -2 & 6 & | & -20 \\ 2 & 1 & | & -1 \end{bmatrix}$$ Multiply the first row by -2 so that the numbers in the first column are opposites.

$$\begin{bmatrix} -2 & 6 & | & -20 \\ 0 & 7 & | & -21 \end{bmatrix}$$ Add the first row to the second row.

$$\begin{bmatrix} 1 & -3 & | & 10 \\ 0 & 1 & | & -3 \end{bmatrix}$$ Multiply the first row by $-\dfrac{1}{2}$.

Multiply the second row by $-\dfrac{1}{7}$.

$\begin{cases} x - 3y = 10 \\ y = -3 \end{cases}$ Write the corresponding system of equations. The value of y is now given, and we can solve for x using the other equation.

$x - 3(-3) = 10$ Solve for x.

$x + 9 = 10$

$x = 1$

The solution set is $\{(1, -3)\}$. ■

As you become more skilled with this method, you will be able to perform two row operations simultaneously. For instance, in Example 1 we could have multiplied the first row by -2 and added it to the second row at one time.

EXAMPLE 2

Solve the system: $\begin{cases} 4x - 5y = 1 \\ 2x + 3y = -5 \end{cases}$

$$\begin{bmatrix} 4 & -5 & | & 1 \\ 2 & 3 & | & -5 \end{bmatrix}$$ Form the augmented matrix.

$$\begin{bmatrix} 2 & 3 & | & -5 \\ 4 & -5 & | & 1 \end{bmatrix}$$ Exchange the rows of the matrix.

$$\begin{bmatrix} 2 & 3 & | & -5 \\ 0 & -11 & | & 11 \end{bmatrix}$$ Mentally multiply the first row by -2, and then add it to the second row.

$$\begin{bmatrix} 1 & \dfrac{3}{2} & | & \dfrac{-5}{2} \\ 0 & 1 & | & -1 \end{bmatrix}$$ Divide the first row by 2 and the second row by -11. This is the same as multiplying by $\dfrac{1}{2}$ and $-\dfrac{1}{11}$.

$$\begin{cases} x + \dfrac{3}{2}y = -\dfrac{5}{2} \\ \quad\quad y = -1 \end{cases}$$ Write the new system of equations.

$$x + \frac{3}{2}(-1) = -\frac{5}{2}$$ Solve for x.

$$x = -1$$

The solution set is $\{(-1, -1)\}$. ∎

When using this method to solve a 3×3 system, we start by getting the zeros in the first column.

EXAMPLE 3

Solve the system: $\begin{cases} 3x - 2y + z = 7 \\ 2x + 5y - z = 13 \\ x - y + 2z = -6 \end{cases}$

$$\begin{bmatrix} 3 & -2 & 1 & | & 7 \\ 2 & 5 & -1 & | & 13 \\ 1 & -1 & 2 & | & -6 \end{bmatrix}$$ Form the augmented matrix.

$$\begin{bmatrix} 1 & -1 & 2 & | & -6 \\ 2 & 5 & -1 & | & 13 \\ 3 & -2 & 1 & | & 7 \end{bmatrix}$$ Exchange rows one and three.

$$\begin{bmatrix} 1 & -1 & 2 & | & -6 \\ 2 & 5 & -1 & | & 13 \\ 0 & 1 & -5 & | & 25 \end{bmatrix}$$ Multiply row one by -3 and add to row three.

$$\begin{bmatrix} 1 & -1 & 2 & | & -6 \\ 0 & 7 & -5 & | & 25 \\ 0 & 1 & -5 & | & 25 \end{bmatrix}$$ Multiply row one by -2 and add to row two.

$$\begin{bmatrix} 1 & -1 & 2 & | & -6 \\ 0 & 1 & -5 & | & 25 \\ 0 & 7 & -5 & | & 25 \end{bmatrix}$$ Exchange rows two and three.

$$\begin{bmatrix} 1 & -1 & 2 & | & -6 \\ 0 & 1 & -5 & | & 25 \\ 0 & 0 & 30 & | & -150 \end{bmatrix}$$ Multiply row two by -7 and add to row three.

$$\begin{bmatrix} 1 & -1 & 2 & | & -6 \\ 0 & 1 & -5 & | & 25 \\ 0 & 0 & 1 & | & -5 \end{bmatrix}$$ Divide row three by 30.

$$\begin{cases} x - y + 2z = -6 \\ y - 5z = 25 \\ z = -5 \end{cases}$$ Write the corresponding system of equations.

$$y - 5(-5) = 25 \qquad \text{Use the second equation to solve for } y.$$

$$y = 0$$

$$x - 0 + 2(-5) = -6 \qquad \text{Use the first equation to solve for } x.$$

$$x - 10 = -6$$

$$x = 4$$

The solution set is $\{(4, 0, -5)\}$.   ■

If a 2×2 system is inconsistent, you will get a matrix of the form

$$\begin{bmatrix} a & b & | & c \\ 0 & 0 & | & d \end{bmatrix}, \quad d \neq 0.$$

$\begin{bmatrix} 3 & -5 & | & 7 \\ 0 & 0 & | & -2 \end{bmatrix}$, for example, gives the system

$$\begin{cases} 3x - 5y = 7 \\ \qquad\quad 0 = -2. \end{cases}$$

The contradiction $0 = -2$, indicates the system is inconsistent, and the solution set is $\emptyset$.

If a 2×2 system is dependent, you will get a matrix of the form

$$\begin{bmatrix} a & b & | & c \\ 0 & 0 & | & 0 \end{bmatrix}.$$

$\begin{bmatrix} 4 & -7 & | & 8 \\ 0 & 0 & | & 0 \end{bmatrix}$, for example, gives the system

$$\begin{cases} 4x - 7y = 8 \\ \qquad\quad 0 = 0. \end{cases}$$

The identity, $0 = 0$, indicates the system is dependent, and the solution set is $\{(x, y) \mid 4x - 7y = 8\}$.

Exercise 7.6

A

Solve each of the following systems using matrices:

1. $\begin{cases} x + y = -1 \\ 2x + 5y = -14 \end{cases}$

2. $\begin{cases} x - y = 12 \\ 3x + 2y = 1 \end{cases}$

3. $\begin{cases} x + 5y = -12 \\ 3x - 4y = 2 \end{cases}$

4. $\begin{cases} x - 3y = 8 \\ 5x + 2y = -11 \end{cases}$

5. $\begin{cases} 2x + 7y = 3 \\ x + y = 4 \end{cases}$

6. $\begin{cases} 3x - 5y = 12 \\ x + 2y = -7 \end{cases}$

7. $\begin{cases} 4x - y = -13 \\ x + 4y = 1 \end{cases}$

8. $\begin{cases} 5x - 2y = -24 \\ x + 3y = 19 \end{cases}$

9. $\begin{cases} 2x + 6y = 13 \\ x + 3y = 5 \end{cases}$

10. $\begin{cases} 4x - 2y = 12 \\ x - \dfrac{1}{2}x = 3 \end{cases}$

B

Solve each of the following systems using matrices:

11. $\begin{cases} 2x + 5y = 7 \\ 4x - 3y = 1 \end{cases}$

12. $\begin{cases} 3x - 2y = -1 \\ 6x - 7y = 1 \end{cases}$

13. $\begin{cases} 4x - y = 3 \\ 8x + 3y = -29 \end{cases}$

14. $\begin{cases} 10x + 7y = 8 \\ 5x + 3y = 2 \end{cases}$

15. $\begin{cases} 3x + y = -1 \\ 5x + 6y = 20 \end{cases}$

16. $\begin{cases} 4x + 5y = -9 \\ 3x - 2y = 22 \end{cases}$

17. $\begin{cases} \dfrac{1}{3}x - \dfrac{2}{3}y = 4 \\ \dfrac{2}{3}x + \dfrac{1}{3}y = 3 \end{cases}$

18. $\begin{cases} \dfrac{1}{4}x - \dfrac{1}{2}y = 1 \\ \dfrac{3}{4}x + \dfrac{3}{2}y = -15 \end{cases}$

19. $\begin{cases} \dfrac{1}{5}x - \dfrac{1}{2}y = \dfrac{5}{4} \\ \dfrac{3}{5}x + \dfrac{1}{2}y = \dfrac{3}{4} \end{cases}$

20. $\begin{cases} \dfrac{2}{3}x + \dfrac{1}{6}y = 0 \\ \dfrac{4}{3}x - \dfrac{2}{3}y = 3 \end{cases}$

C

Solve each of the following systems using matrices:

21. $\begin{cases} x + y + z = 3 \\ 2x + 3y + z = 6 \\ 2x - y - z = 0 \end{cases}$

22. $\begin{cases} x + 2y + z = -1 \\ 3x - y - z = 8 \\ 4x + y - 2z = 9 \end{cases}$

23. $\begin{cases} x - 3y + 2z = -7 \\ 2x + y - 3z = -7 \\ 3x + 2y - z = -2 \end{cases}$

24. $\begin{cases} x - 4y + 3z = -21 \\ 5x + 2y - z = 25 \\ 3x - y + 2z = -2 \end{cases}$

25. $\begin{cases} x + 2y = 6 \\ x + 3z = -7 \\ y + 3z = 2 \end{cases}$

26. $\begin{cases} x + 4z = 17 \\ y + 3z = 10 \\ 5x + 2y = -25 \end{cases}$

27. $\begin{cases} 4x + 7z = -1 \\ 3y + 2z = 5 \\ 3x - 5y = -11 \end{cases}$

28. $\begin{cases} 5y - 2z = 43 \\ 3x + z = -22 \\ 2x - y = -19 \end{cases}$

29. $\begin{cases} 2x - 5y - 7z = 1 \\ 3x - 2y + z = -10 \\ 4x + y + 5z = -17 \end{cases}$

30. $\begin{cases} x + 3y - 5z = 3 \\ 5x + 2y - 3z = -6 \\ 4x - 2y + 5z = -8 \end{cases}$

STATE YOUR UNDERSTANDING

31. Solve the following using the augmented matrix.

$$\begin{cases} 3x - 2y + 5z = 0 \\ x - y + 3z = 0 \\ 2x - y + z = 0 \end{cases}$$

When all of the constants are zero the system is called a homogeneous system. Could the solution by anything other than what you got when you solved the problem? Why?

32. Under what conditions would Cramer's rule be preferable over using the augmented matrix? Give an example of a problem that would be better solved by one method over the other.

CHALLENGE EXERCISES

33. Solve using the augmented matrix:

$$\begin{cases} ax + y = 3 \\ x - 2y = -4 \end{cases}$$

34. What value(s) for a will make the following system dependent? (Solve using the augmented matrix.)

$$\begin{cases} ax + 4y = 6 \\ -x - 2y = -3 \end{cases}$$

35. What value(s) for a will make the following system independent and inconsistent?

$$\begin{cases} ax + 3y = 5 \\ 2x - 4y = 3 \end{cases}$$

MAINTAIN YOUR SKILLS (SECTIONS 2.11, 6.1, 6.2, 6.3)

36. Find the ordered pair that is a solution of the equation $y = 2x + 7$ when $x = -37$.

37. Write the coordinates of the x- and y-intercepts of the graph of

$$y = \frac{1}{5}x - 9.$$

38. Is the graph of $y = \frac{1}{4}x - 18$ perpendicular to the graph of $x = -\frac{1}{4}y + 18$?

Write the equation, in standard form, of each line described:

39. The line parallel to $y = 2x - 3$ and having y-intercept $(0, 5)$.

40. The line perpendicular to $y = 2x - 3$ and having x-intercept $(5, 0)$.

Solve:

41. $(x + 3)^2 + 2(x + 23) = (3x + 5)^2$

42. $(x + 3)^2 - 2(x - 7)^2 = -11x - 183$

43. $(y - 2)^2 - 26(y - 2) + 25 = 0$

CHAPTER 7
SUMMARY

$\begin{cases} A_1x + B_1y = C_1 \\ A_2x + B_2y = C_2 \end{cases}$	A system of linear equations in two variables. The solution of the system is the intersection of the solutions of the two equations. The system can be solved by graphing, substitution, linear combinations, Cramer's rule, or matrices. (pp. 479, 481, 484, 493)
$\begin{cases} A_1x + B_1y + C_1z = D_1 \\ A_2x + B_2y + C_2z = D_2 \\ A_3x + B_3y + C_3z = D_3 \end{cases}$	A system of linear equations in three variables. The solution of the system is the intersection of the solutions of the three equations. The system can be solved by linear combinations, Cramer's rule, or matrices. (pp. 502, 504)
(x, y, z)	An ordered triple of numbers is used to write a solution of an equation in three variables. (p. 501)
$\begin{vmatrix} a & b \\ c & d \end{vmatrix} = ad - cb$	A 2×2 determinant and the formula for finding its value. (p. 512)
$\begin{vmatrix} a_1 & b_1 & c_1 \\ a_2 & b_2 & c_2 \\ a_3 & b_3 & c_3 \end{vmatrix}$ $= (a_1b_2c_3 + b_1c_2a_3 + c_1a_2b_3) - (a_3b_2c_1 + b_3c_2a_1 + c_3a_2b_1)$	A 3×3 determinant and a formula for findings its value. (pp. 514, 515)
Cramer's Rule for a 2×2 System	The solution of the system, $\begin{cases} a_1x + b_1y = c_1 \\ a_2x + b_2y = c_2 \end{cases}$, where (p. 522)

$$x = \frac{D_x}{D}, \; y = \frac{D_y}{D}$$

$$D_x = \begin{vmatrix} c_1 & b_1 \\ c_2 & b_2 \end{vmatrix}, \qquad D_y = \begin{vmatrix} a_1 & c_1 \\ a_1 & c_2 \end{vmatrix}, \quad \text{and}$$

$$D = \begin{vmatrix} a_1 & b_1 \\ a_2 & b_2 \end{vmatrix}, \qquad a_1 b_2 - a_2 b_1 \neq 0$$

Cramer's Rule for a 3×3
System

$$x = \frac{D_x}{D}, \; y = \frac{D_y}{D}, \; z = \frac{D_z}{D}$$

The solution of the system can be found on page 523.   (p. 524)

$$\begin{bmatrix} a_1 & b_1 & c_1 \\ a_2 & b_2 & c_2 \end{bmatrix}$$

The augmented matrix associated with the system of equations: (p. 532)
$$\begin{cases} a_1 x + b_1 y = c_1 \\ a_2 x + b_2 y = c_2 \end{cases}.$$

Matrix Row Operations

1. Any two rows of the matrix may be interchanged.   (p. 532)

2. The numbers in any row may be multiplied by any nonzero real number.

3. Any row may be changed by adding to the numbers of the row the numbers of another row.

CHAPTER 7
REVIEW EXERCISES

SECTION 7.1   Objective 1

Solve by graphing.

1. $\begin{cases} 2x - 5y = 10 \\ 3x - 2y = 4 \end{cases}$

2. $\begin{cases} 3x + y = 3 \\ 3x - y = 15 \end{cases}$

SECTION 7.1   Objective 2

Solve by substitution.

3. $\begin{cases} y = -4x + 1 \\ 2x - y = 5 \end{cases}$

4. $\begin{cases} y = 2x - 5 \\ 4x - 5y = 7 \end{cases}$

5. $\begin{cases} 7x - y = 9 \\ 5x + 6y = -7 \end{cases}$

6. $\begin{cases} 5x + y = 11 \\ 3x - 5y = 1 \end{cases}$

7. $\begin{cases} x - 4y = 3 \\ 2x - 5y = 4 \end{cases}$

8. $\begin{cases} x + 7y = 2 \\ 3x - 4y = 1 \end{cases}$

9. $\begin{cases} 6x - 2y = -1 \\ 9x - y = -2 \end{cases}$

10. $\begin{cases} x + y = 0 \\ 8x + 2y = 5 \end{cases}$

11. $\begin{cases} \dfrac{1}{3}x + y = -\dfrac{4}{5} \\ \dfrac{1}{6}x - y = \dfrac{11}{10} \end{cases}$

12. $\begin{cases} 9x - y = 2 \\ \dfrac{1}{5}x - 2y = -\dfrac{53}{9} \end{cases}$

SECTION 7.2 Objective 1

Solve using linear combinations.

13. $\begin{cases} 3x - y = 19 \\ 2x + 3y = -13 \end{cases}$

14. $\begin{cases} 5x - 2y = 5 \\ 7x + 4y = -61 \end{cases}$

15. $\begin{cases} x + 5y = -1 \\ 3x + 7y = 13 \end{cases}$

16. $\begin{cases} 5x - 2y = -31 \\ -3x + 4y = -15 \end{cases}$

17. $\begin{cases} x - 3y = -17 \\ 5x + 6y = -22 \end{cases}$

18. $\begin{cases} 5x + 7y = -38 \\ -2x + 3y = -37 \end{cases}$

19. $\begin{cases} 8x + 4y = -5 \\ 6x - 2y = 5 \end{cases}$

20. $\begin{cases} x + y = 1 \\ 4x - 6y = 39 \end{cases}$

21. $\begin{cases} \dfrac{1}{3}x - \dfrac{3}{7}y = 2 \\ \dfrac{1}{2}x + \dfrac{1}{7}y = 1 \end{cases}$

22. $\begin{cases} 0.3x + 2.4y = -1.8 \\ -1.2x - 0.8y = 1.92 \end{cases}$

SECTION 7.3 Objective 1

Solve using linear combinations.

23. $\begin{cases} 5x - 3y + z = 0 \\ 2x + 5y - 3z = 12 \\ 3x - 4y + 6z = -42 \end{cases}$

24. $\begin{cases} 7x - 4y - 3z = 9 \\ 5x + 2y - 2z = -26 \\ 6x + 5y + 2z = -1 \end{cases}$

25. $\begin{cases} 2x - 3y - 2z = 28 \\ 5x + 2y - 7z = 33 \\ 3x + 5y + 5z = -95 \end{cases}$

26. $\begin{cases} 6x - 5y - 3z = -21 \\ 2x + 7y - 7z = -1 \\ 5x - 11y + z = -21 \end{cases}$

27. $\begin{cases} 6x + y - z = 7 \\ 2x + 5y - 4z = 23 \\ 2x + 4y - 3z = 18 \end{cases}$

28. $\begin{cases} 2x + 3y - z = 9 \\ 3x + 5y + 7z = 10 \\ \dfrac{2}{3}x + y - \dfrac{1}{3}z = -4 \end{cases}$

SECTION 7.4 Objective 1

Evaluate.

29. $\begin{vmatrix} -8 & -5 \\ -6 & -9 \end{vmatrix}$

30. $\begin{vmatrix} 5 & -12 \\ -6 & 9 \end{vmatrix}$

31. $\begin{vmatrix} 5 & -7 \\ 12 & 15 \end{vmatrix}$

32. $\begin{vmatrix} 2 & -20 \\ 5 & -30 \end{vmatrix}$

33. $\begin{vmatrix} -8 & -9 \\ -6 & -5 \end{vmatrix}$

34. $\begin{vmatrix} 12 & -3 \\ -14 & -11 \end{vmatrix}$

35. $\begin{vmatrix} -3.5 & 0.75 \\ 2.3 & 0.34 \end{vmatrix}$

36. $\begin{vmatrix} 0.45 & 0.6 \\ -3.51 & -1.4 \end{vmatrix}$

37. $\begin{vmatrix} -\dfrac{1}{3} & \dfrac{1}{4} \\[2mm] \dfrac{1}{6} & -\dfrac{1}{2} \end{vmatrix}$

38. $\begin{vmatrix} -\dfrac{1}{5} & -\dfrac{1}{2} \\[2mm] \dfrac{4}{5} & \dfrac{3}{2} \end{vmatrix}$

SECTION 7.4   Objective 2

Evaluate.

39. $\begin{vmatrix} -3 & 2 & 7 \\ 5 & 8 & -1 \\ 0 & 3 & -5 \end{vmatrix}$

40. $\begin{vmatrix} -7 & 5 & 2 \\ -3 & -7 & -6 \\ 5 & 6 & 0 \end{vmatrix}$

41. $\begin{vmatrix} -10 & -3 & -1 \\ -3 & -2 & -1 \\ 1 & -3 & 2 \end{vmatrix}$

42. $\begin{vmatrix} 5 & -7 & -2 \\ -1 & 1 & -3 \\ -3 & 4 & -2 \end{vmatrix}$

43. $\begin{vmatrix} 3 & -14 & -4 \\ -15 & 2 & 8 \\ 4 & -5 & 4 \end{vmatrix}$

44. $\begin{vmatrix} -16 & 0 & 8 \\ -6 & -2 & -5 \\ 6 & -3 & 5 \end{vmatrix}$

SECTION 7.4   Objective 3

Evaluate.

45. $\begin{vmatrix} 0 & 1 & 1 & -3 \\ 0 & 2 & 3 & 0 \\ 2 & 0 & 1 & 1 \\ -1 & 1 & 0 & 1 \end{vmatrix}$

46. $\begin{vmatrix} -1 & 0 & 0 & -3 & 1 \\ 5 & 2 & 0 & 1 & 4 \\ -3 & 2 & 1 & 1 & 5 \\ -3 & 0 & 0 & 2 & 1 \\ 5 & 0 & 0 & 2 & 3 \end{vmatrix}$

SECTION 7.4   Objective 4

Solve.

47. $\begin{vmatrix} a & 9 \\ 3 & -4 \end{vmatrix} = 1$

48. $\begin{vmatrix} 2a & -3 \\ a & 1 \end{vmatrix} = 16$

49. $\begin{vmatrix} a & 6 \\ 2 & a \end{vmatrix} = 88$

50. $\begin{vmatrix} a & 3 & 1 \\ 0 & 2 & 1 \\ 1 & 5 & 1 \end{vmatrix} = 4$

51. $\begin{vmatrix} 0 & 3 & a \\ 2 & 1 & a \\ -1 & 3 & 2 \end{vmatrix} = 0$

SECTION 7.5   Objective 1

Solve using Cramer's rule.

52. $\begin{cases} 7x - y = -19 \\ -3x + 2y = 27 \end{cases}$

53. $\begin{cases} 6x - 3y = 18 \\ 2x - y = 6 \end{cases}$

54. $\begin{cases} 2x - y = 1 \\ 8x - 5y = 19 \end{cases}$

55. $\begin{cases} 4x - y = 25 \\ 2x - \dfrac{1}{2}y = -5 \end{cases}$

56. $\begin{cases} 5x - 3y = 6 \\ x + y = -1 \end{cases}$

57. $\begin{cases} x + y = 1 \\ 4x - y = -6 \end{cases}$

58. $\begin{cases} 0.3x - 1.1y = 0.52 \\ 2.8x + 0.9y = -4.08 \end{cases}$

59. $\begin{cases} 1.35x + 2.05y = 2.618 \\ 0.35x - 0.45y = 0.718 \end{cases}$

SECTION 7.5   Objective 2

Solve using Cramer's rule.

60. $\begin{cases} 3x - 2y = 1 \\ 5y + 3z = -7 \\ 2x + 5z = 45 \end{cases}$

61. $\begin{cases} 3x + 4y - 9z = -77 \\ 5x - 4y - 6z = 34 \\ 2x + 5y + 4z = -27 \end{cases}$

62. $\begin{cases} 3x + 3y + 3z = 4 \\ 2x + z = 3 \\ y + z = 1 \end{cases}$

63. $\begin{cases} x + 2y + z = -2 \\ 3x + 4y + 3z = -7 \\ 5x - 3z = -1 \end{cases}$

64. $\begin{cases} 5x - 5y - 6z = 116 \\ 15x + 25y + 4z = -214 \\ 3x - 30y - z = 257 \end{cases}$

65. $\begin{cases} 6x - 2y + 5z = 72 \\ 9x - 6y - 8z = -107 \\ 3x + 5y + 4z = -6 \end{cases}$

SECTION 7.6   Objective 1

Solve using matrix methods.

66. $\begin{cases} 4x - 9y = -10 \\ 2x - 5y = -6 \end{cases}$

67. $\begin{cases} 3x - 7y = 47 \\ x + 3y = -11 \end{cases}$

68. $\begin{cases} x - 6y = 35 \\ 3x + 5y = -10 \end{cases}$

69. $\begin{cases} 2x - y = -22 \\ 6x + 5y = -2 \end{cases}$

SECTION 7.6   Objective 2

Solve using matrix methods.

70. $\begin{cases} 3x - 5y + z = -49 \\ 2x - 3y - 5z = -2 \\ x + 2y - 3z = 26 \end{cases}$

71. $\begin{cases} 7x - 5y = -15 \\ 3y + 2z = 14 \\ 2x - 3z = 34 \end{cases}$

CHAPTER 7
TRUE–FALSE CONCEPT REVIEW

Check your understanding of the language of algebra. Tell whether each of the following statements is true (always true) or false (not always true).

1. If the graphs of two linear equations of a system are parallel, then the system is said to be independent and inconsistent.

2. When using Cramer's rule to solve systems of equations, if $D = 0$, the system has no unique solution.

3. When solving a system of equations, you arrive at an equation of the form $8 = 8$, an identity. The system has no solution.

4. If the graphs of the two linear equations of a system are identical, the system is said to be dependent.

5. $\begin{vmatrix} 0 & 0 & 3 \\ 1 & 0 & 2 \\ 0 & 3 & 4 \end{vmatrix} = 9$

6. Every linear system with three equations has an ordered triple for a solution.

7. If the graphs of the two linear equations of a system have a single point in common, the system is said to be independent and consistent.

8. Every matrix has the same number of rows and columns.

9. To solve a system of equations by substitution means to find the ordered pair by substituting trial coordinates in the equations.

10. The graph of the system $\begin{cases} x + y > 0 \\ 2x - y < 0 \end{cases}$ contains points in quadrant IV.

11. $\begin{vmatrix} a & b \\ c & d \end{vmatrix} = ad - bc$

12. When solving a system of equations, you arrive at an equation like $8 = 4$, a contradiction. The system has no solution.

13. $\begin{vmatrix} 3 & 5 \\ 6 & -1 \end{vmatrix} = -33$

14. In $\begin{vmatrix} a & b & c \\ d & e & f \\ g & h & i \end{vmatrix}$, the minor of g is $\begin{vmatrix} b & c \\ h & i \end{vmatrix}$.

15. In solving a system of equations by linear combinations, the equations or multiples of the equations are added to eliminate one of the variables in the sum.

16. The value of the minor of -5 in $\begin{vmatrix} 6 & 9 & -4 \\ 12 & -3 & 0 \\ -4 & -5 & 6 \end{vmatrix}$ is 0.

17. In $\begin{vmatrix} a & b & c \\ d & e & f \\ g & h & i \end{vmatrix}$, the sign associated with the minor of e is "+".

18. For the system $\begin{cases} a_1x + b_1y + c_1z = d_1 \\ a_2x + b_2y + c_2z = d_2, \\ a_3x + b_3y + c_3z = d_3 \end{cases}$

$$D_x = \begin{vmatrix} d_1 & d_2 & d_3 \\ a_2 & b_2 & c_2 \\ a_3 & b_3 & c_3 \end{vmatrix}.$$

19. In the system in problem 18, the solution is $\left(\dfrac{D_x}{D}, \dfrac{D_y}{D}, \dfrac{D_z}{D} \right)$.

20. A matrix row operation allows for adding the numbers of one row to the numbers of a different row.

CHAPTER 7
TEST

1. Solve the following system of equations by substitution:
$$\begin{cases} 3x - 3y = 7 \\ \quad\quad y = 1 - 4x \end{cases}$$

2. Solve using Cramer's rule:
$$\begin{cases} 3x + 2y = -2 \\ 4x - \quad y = -6 \end{cases}$$

3. Solve the following system of equations using linear combinations:
$$\begin{cases} 7x + 15y = 12 \\ 3x + 10y = \quad 8 \end{cases}$$

4. Evaluate: $\begin{vmatrix} 6 & -4 \\ -3 & 5 \end{vmatrix}$

5. Solve using Cramer's rule:
$$\begin{cases} x - \quad y + 2z = 3 \\ 2x - \quad y - \quad z = 2 \\ 2x - 2y - 2z = 3 \end{cases}$$

6. Solve the following system of equations using linear combinations:
$$\begin{cases} x + \quad y - \quad z = -1 \\ 2x + 3y + 3z = \quad 11 \\ 2x - 3y - 4z = \quad 2 \end{cases}$$

7. Solve the following system of equations using matrix methods:
$$\begin{cases} 2x - \quad y + \quad z = 3 \\ 4x + 2y - \quad z = 1 \\ 2x - 3y - 2z = 1 \end{cases}$$

8. Evaluate: $\begin{vmatrix} 5 & 2 & -1 \\ 1 & -2 & 3 \\ 3 & 4 & -1 \end{vmatrix}$

9. Solve the following system of equations by substitution:
$$\begin{cases} \quad\quad x = 2y - 7 \\ 2x - 2y = 11 \end{cases}$$

10. Evaluate: $\begin{vmatrix} 16 & -8 \\ -2 & 2 \end{vmatrix}$

11. Solve the following system of equations using linear combinations:
$$\begin{cases} x + 3y - \quad z = \quad 4 \\ 3x - 2y - 2z = \quad 9 \\ x + \quad y + 3z = -8 \end{cases}$$

12. Solve using Cramer's rule:
$$\begin{cases} 2x + 3y = -2 \\ \quad x - 4y = \quad 6 \end{cases}$$

13. Evaluate: $\begin{vmatrix} 2 & 4 & 0 \\ -1 & -1 & 2 \\ 3 & 2 & -1 \end{vmatrix}$

14. Solve the following system of equations using linear combinations:
$$\begin{cases} 7x + 5y = \quad 17 \\ 14x - 2y = -14 \end{cases}$$

15. A grocer bought two brands of frozen juice. She paid $4.70 per case for one brand and $5.20 per case for the other brand. If the total cost was $271 for 55 cases, how many cases of each brand of juice did she buy?

16. One alloy contains 60% copper, and another alloy contains 80% copper. How many pounds of each alloy are needed to make 30 pounds of a third alloy that is 72% copper?

8

SECOND-DEGREE EQUATIONS IN TWO VARIABLES

The paths of some comets can be shown to be a conic section. In Exercise 29, Section 8.5, one comet has a parabolic path while another orbiting body has an elliptical orbit. By plotting the two paths on a rectangular coordinate system, the points where the paths coincide can be determined. These points of intersection are the critical points to watch for a possible collision. *(NASA)*

PREVIEW

This chapter covers conic sections. The parabola is presented in detail with a complete discussion of vertices and maximum or minimum values. Other conic sections are treated in sections 8.2 and 8.3. The chapter concludes with the solutions of systems of equations involving quadratic equations. In 8.4 the systems are composed of one linear and one quadratic equation. In 8.5, the systems are composed of two equations that are quadratic. The graphs of the systems are used to illustrate the common solutions.

8.1

PARABOLAS

OBJECTIVES

1. Find the coordinates of the *x*-intercepts and *y*-intercepts of a parabola.

2. Find the vertex of a parabola.

3. Determine whether the vertex of a parabola is a maximum or minimum point of the graph.

4. Graph a parabola.

5. Write the equation of the axis of symmetry of a parabola.

Conic Section

■ DEFINITION

A *conic section* is a geometric figure formed by the intersection of a plane and a cone. Figure 8.1 shows the more important curves, the circle, ellipse, parabola, and hyperbola.

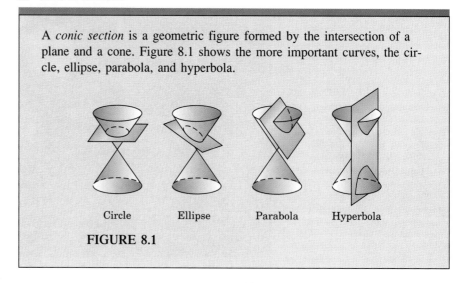

Circle   Ellipse   Parabola   Hyperbola

FIGURE 8.1

In this section, we study some of the properties of the parabola.

■ DEFINITION

Parabola

A *parabola* is the set of all points that are equidistant from a fixed point and a fixed line. The vertex of a parabola is its turning point. In some parabolas, the vertex is either the highest or lowest point of the graph. When the vertex is the lowest or highest point, the value of the dependent variable is called the minimum or maximum, respectively.

Vertex

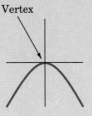

Our study of parabolas is limited to those having equations that can be written in the form

$$y = ax^2 + bx + c, \ a \neq 0 \quad \text{or} \quad x = ay^2 + by + c, \ a \neq 0.$$

The graphs of such equations are parabolas that open up, down, right, or left. The first property of parabolas we investigate is the x- and y-intercepts. As with linear equations, the x-intercepts are found by substituting 0 for y, and the y-intercepts are found by substituting 0 for x.

EXAMPLE 1

Find the intercepts of the graph of $y = x^2 - 4x - 5$.

$0 = x^2 - 4x - 5$ To find the x-intercepts, let $y = 0$.

$0 = (x + 1)(x - 5)$

$x = -1 \quad \text{or} \quad x = 5$

The x-intercepts are $(-1, 0)$ and $(5, 0)$.

$y = 0 - 0 - 5 = -5$ To find the y-intercept, let $x = 0$.

The y-intercept is $(0, -5)$.   ■

The graph of $y = ax^2 + bx + c$ has a vertex that is either the highest or the lowest point of the graph depending on whether the parabola opens up or down. To determine the coordinates of the vertex, we complete the square on the right side.

EXAMPLE 2

Find the coordinates of the vertex of the graph of $y = x^2 - 4x - 5$ by completing the square and graph the parabola.

$y = x^2 - 4x - 5$

$y = (x^2 - 4x \qquad) - 5$

$y = (x^2 - 4x + 4 - 4) - 5$

$y = (x^2 - 4x + 4) - 4 - 5$

$y = (x - 2)^2 - 9$

The vertex is $(2, -9)$.

To find the coordinates of the vertex, we complete the square on the right side by adding and subtracting the square of one-half of 4, the coefficient of x.

From this form, we can determine the vertex of the parabola.

Since $(x - 2)^2 \geq 0$ the smallest value of y is found when $(x - 2)^2 = 0$ or at $x = 2$. This is the turning point of the vertex. The y-value can be found by substituting $x = 2$ in the equation of the parabola.

To make a sketch of the graph, we make a table of values and draw a smooth curve through the points. Form a table of values for the equation by choosing values of x and calculating the values of y.

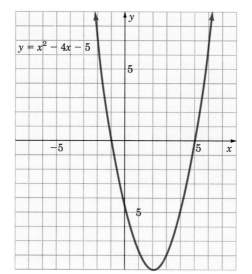

x	y
-3	16
-2	7
-1	0
0	-5
1	-8
2	-9
3	-8
4	-5
5	0
6	7
7	16

In Example 2 we see that the following pairs of points are points on the graph.

(x-values with equal y values)

$(-3, 16)$ and $(7, 16)$, $(-2, 7)$ and $(6, 7)$,

$(-1, 0)$ and $(5, 0)$, $(0, -5)$ and $(4, -5)$, and

$(1, -8)$ and $(3, -8)$.

These pairs of points are mirror images about the vertical line through the vertex and are said to be symmetric with respect to that vertical line. We conclude that the vertical line through the vertex of a parabola is the axis of symmetry of the parabola. The axis of symmetry for the parabola in Example 2 is $x = 2$.

To determine the formula for the coordinates of the vertex and the equation of the axis of symmetry we complete the square on the general equation.

$$y = ax^2 + bx + c$$

$$y = a\left(x^2 + \frac{b}{a}x\right) + c$$

$$y = a\left[x^2 + \frac{b}{a}x + \left(\frac{b}{2a}\right)^2\right] + c - a\left(\frac{b}{2a}\right)^2$$

$$y = a\left(x + \frac{b}{2a}\right)^2 + \frac{4ac - b^2}{4a}$$

If we let $h = -\dfrac{b}{2a}$ and $k = \dfrac{4ac - b^2}{4a}$, the equation can be written

$$y = a(x - h)^2 + k.$$

In the equation $y = a(x - h)^2 + k$, the value of a determines whether the vertex is a maximum point or a minimum point. Since $a \neq 0$, then $a > 0$ (positive) or $a < 0$ (negative).

If $a > 0$	If $a < 0$
$a(x - h)^2 \geq 0$, and the smallest value of y occurs when $x = h$. At $x = h$, $y = k$. The coordinates of the lowest or minimum point of the curve are (h, k).	$a(x - h)^2 \leq 0$, and the largest value of y occurs when $x = h$. At $x = h$, $y = k$. The coordinates of the highest or maximum point of the curve are (h, k).

If a is positive, the vertex of the parabola is the lowest point of the graph, and the parabola opens up. All other y-values are greater. If a is negative, the vertex of the parabola is the highest point of the graph, and the parabola opens down. All other y-values are smaller.

Using these facts we can find the vertex by using the general form.

■ PROCEDURE

Vertex of a Parabola

To find the vertex of a parabola (h, k):

1. Write the equation in the form $y = ax^2 + bx + c$.

2. Find the x-coordinate of the vertex using the formula $h = -\dfrac{b}{2a}$.

3. Find the y-coordinate of the vertex using the formula $k = \dfrac{4ac - b^2}{4a}$,

   or substitute the value of h in the equation of the parabola.

From the vertex we can write the general form for the axis of symmetry.

■ FORMULA

Axis of Symmetry

The equation of the axis of symmetry of the graph of
$y = ax^2 + bx + c$ is $x = h = -\dfrac{b}{2a}$.

Let us make use of this information to graph $y = -x^2 - 6x - 6$.

EXAMPLE 3

Graph $y = -x^2 - 6x - 6$. Write the coordinates of the vertex, the x-intercepts, and the y-intercepts, and the equation of the axis of symmetry.

The parabola opens downward.   Since $a = -1$.

The vertex is at $(-3, 3)$.   Vertex at (h, k). Since $h = -\dfrac{b}{2a}$

$$= -\dfrac{-6}{2(-1)} = -3$$

and
$k = -(-3)^2 - 6(-3) - 6 = 3$.

$y = -x^2 - 6x - 6$ The x-intercepts are found by
setting $y = 0$.

$0 = -x^2 - 6x - 6$

$x = \dfrac{6 \pm \sqrt{36 - 4(-1)(-6)}}{2(-1)}$

$$= \frac{6 \pm \sqrt{12}}{-1}$$

$$= \frac{-2(-3 \pm \sqrt{3})}{-2}$$

$$= -3 \pm \sqrt{3}$$

The x-intercepts are $(-3 + \sqrt{3}, 0)$ and $(-3 - \sqrt{3}, 0)$.

$y = -0^2 - 6(0) - 6$ 　　　　　　　　The y-intercept is found by setting
$\quad = -6$ 　　　　　　　　　　　　　$x = 0$.

The y-intercept is $(0, -6)$.

The equation of the axis of symmetry 　　The vertical line through the vertex
is $x = -3$. 　　　　　　　　　　　　$(-3, 3)$ is the axis of symmetry.

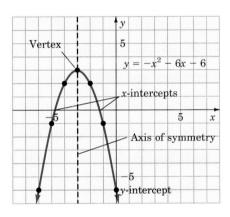

We can graph the parabola with this information and the calculated points $(-2, 2)$, $(-1, -1)$, $(0, -6)$ and $(1, -13)$. The corresponding points on the left side of the axis of symmetry are $(-4, 2)$, $(-5, -1)$, $(-6, -6)$ and $(-7, -13)$.

A summary of the properties of the parabola defined by $y = ax^2 + bx + c$ follows.

■ PROPERTY

The graph of $y = ax^2 + bx + c$ has these properties:

The vertex, (h, k) is $\left(-\dfrac{b}{2a}, \dfrac{4ac - b^2}{4a}\right)$.

If $a > 0$, the vertex is a minimum point with coordinates (h, k).

If $a < 0$, the vertex is a maximum point with coordinates (h, k).

The x-intercepts are found by setting $y = 0$.

The y-intercept is found by setting $x = 0$.

The equation of the axis of symmetry is $x = h$.

EXAMPLE 4

For the following equation:

1. Find the vertex.

2. Determine whether the vertex is a maximum or minimum point.

3. Find the x-intercepts.

4. Find the y-intercept.

5. Determine the equation of the axis of symmetry.

6. Graph.

$y = x^2 + 8x + 15$
$a = 1, b = 8, c = 15$

1. $h = -\dfrac{8}{2 \cdot 1} = -4$

1. The x-coordinate of the vertex (h) is found by letting $h = -\dfrac{b}{2a}$.

$$
\begin{aligned}
y = k &= (-4)^2 + 8(-4) + 15 \\
&= 16 - 32 + 15 \\
&= -1
\end{aligned}
$$

The vertex is at $(-4, -1)$.

Substitute -4 for x in the equation to determine the y-coordinate (k) of the vertex.

2. The vertex is a minimum point. The curve opens upward.

2. Since $a > 0$, the graph has a minimum point.

3. $0 = x^2 + 8x + 15$

$0 = (x + 5)(x + 3)$

$0 = x + 5$ or $0 = x + 3$
$-5 = x$ $-3 = x$

The x-intercepts are $(-5, 0)$ and $(-3, 0)$.

3. Substitute 0 for y, and solve the equation for x to find the x-intercepts.

4. $\begin{aligned} y &= (0)^2 + 8(0) + 15 \\ &= 0 + 0 + 15 \\ &= 15 \end{aligned}$

The y-intercept is $(0, 15)$.

4. Substitute 0 for x to find the y-intercept.

5. Since the vertex is at $(-4, -1)$, $x = -4$ is the equation of the axis of symmetry.

5. The vertical line that passes through the vertex is the axis of symmetry.

6.

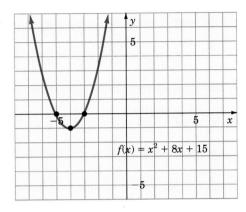

$$f(x) = x^2 + 8x + 15$$

EXAMPLE 5

If $y = 2x^2 + 4x$, write the coordinates of the vertex, and tell whether it is a maximum or minimum point. Write the coordinates of the x-intercepts, the y-intercepts, and the equation of the axis of symmetry and graph.

The vertex at $(-1, -2)$ is a minimum point.

Since $h = -\dfrac{b}{2a} = -\dfrac{4}{4} = -1$ and $k = 2(-1)^2 + 4(-1) = -2$. The parabola opens up and has a minimum point at $(-1, -2)$.

$0 = 2x^2 + 4x$

The x-intercepts, $y = 0$.

$2x(x + 2) = 0$

$x = 0$ or $x = -2$

$(-2, 0), (0, 0)$

$y = 2(0)^2 + 4(0) = 0$

The y-intercept, $x = 0$.

$(0, 0)$

The origin is both an x- and y-intercept.

The axis of symmetry is $x = -1$.

Since $x = h = -1$.

The coordinates of the symmetric pairs of points $(0, 0)$ and $(-2, 0)$, $(1, 6)$ and $(-3, 6)$ and $(2, 16)$ and $(-4, 16)$ can be used to draw the parabola more accurately.

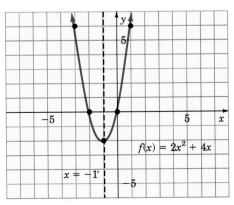

$$f(x) = 2x^2 + 4x$$

$x = -1$

EXAMPLE 6

For the following equation:

1. Find the vertex.

2. Determine whether the vertex is a maximum or minimum point.

3. Find the x-intercepts.

4. Find the y-intercept.

5. Determine the equation of the axis of symmetry.

6. Graph.

$$y = -\frac{1}{3}x^2 - \frac{4}{3}x - \frac{10}{3}$$

1. The vertex is at $(-2, -2)$.

1. Identify the vertex, (h, k).

$$h = -\frac{b}{2a} = -\frac{-\frac{4}{3}}{-\frac{2}{3}} = -2$$

$$k = -\frac{1}{3}(-2)^2 - \frac{4}{3}(-2) - \frac{10}{3}$$

$$= -2$$

2. The vertex is a maximum point. The curve opens downward.

2. Since $a < 0$, the graph has a maximum point.

3. $0 = -\frac{1}{3}x^2 - \frac{4}{3}x - \frac{10}{3}$

3. Substitute 0 for y to find the x-intercept(s).

$$0 = x^2 + 4x + 10$$

Multiply both sides by -3 to clear fractions.

$$x = \frac{-4 \pm \sqrt{16 - 40}}{2}$$

Use the quadratic formula to solve.

$$x = \frac{-4 \pm \sqrt{-24}}{2}$$

No x-intercept since $\sqrt{-24}$ is a complex number.

Since there is no x-intercept and the parabola has a maximum point, the entire curve is below the x-axis.

4. $y = -\frac{1}{3}(0)^2 - \frac{4}{3}(0) - \frac{10}{3}$

4. Substitute 0 for x to find the y-intercept.

$$y = -\frac{10}{3}$$

The y-intercept is $\left(0, -\frac{10}{3}\right)$.

5. Since the vertex is at $(-2, -2)$, the equation of the axis of symmetry is $x = -2$.

5. The vertical line that passes through the vertex is the axis of symmetry.

6.

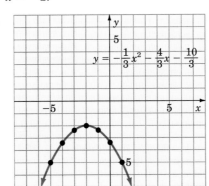

$$y = -\frac{1}{3}x^2 - \frac{4}{3}x - \frac{10}{3}$$

6. Plot the graph. Three points are known:

$(-2, -2)$, $\left(0, -\frac{10}{3}\right)$, and, by

symmetry, $\left(-4, -\frac{10}{3}\right)$. Find two

other points such as $(1, -5)$ and $(-5, -5)$.

■

EXAMPLE 7

A manufacturer estimated that the demand for a new T-shirt can be expressed as a quadratic equation during the first 12 weeks of sales. If y represents the demand (number of T-shirts) and x the number of weeks, the function is $y = -x^2 + 10x + 1000$. Find the week of highest demand and the number of T-shirts needed for that week to meet the demand.

$$y = -x^2 + 10x + 1000$$

$$h = -\frac{b}{2a} = -\frac{10}{2(-1)} = 5$$

$$k = -(5)^2 + 10(5) + 1000$$

$$= 1025$$

Vertex at $(5, 1025)$.

Since $a = -1$, the graph of the equation has a maximum point. To find the maximum point, find the vertex.

At the vertex, y has its maximum value, 1025.

Answer:

During the fifth week, the demand is at its peak and 1025 T-shirts are needed.   ■

Equations of the form

$$x = ay^2 + by + c$$

also have graphs that are parabolas. For example, if we interchange x and y in the equation $y = \frac{1}{2}x^2$, we have $x = \frac{1}{2}y^2$. The graphs of both parabolas are shown in Figure 8.2.

FIGURE 8.2

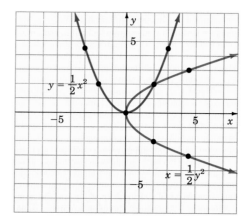

We see that the graph of $x = \dfrac{1}{2}y^2$ opens to the right. Equations of the form $x = ay^2 + by + c$ are equations of parabolas and have graphs that open to the left or right. Other properties of these relations are listed in the following table.

The Graph of $x = ay^2 + by + c$	
If $a > 0$	Parabola opens right
If $a < 0$	Parabola opens left
$(c, 0)$	Coordinates of the x-intercept
	(y-intercepts are the solution of the equation: $ay^2 + by + c = 0$)
$\left(\dfrac{4ac - b^2}{4a}, -\dfrac{b}{2a}\right)$	Coordinates of the vertex of the parabola
$y = -\dfrac{b}{2a}$	Equation of the axis of symmetry (horizontal line)

EXAMPLE 8

If $x = -y^2 + 2y + 3$, write the coordinates of the vertex, and tell whether the parabola opens right or left. Write the coordinates of the x-intercepts, the y-intercepts, and the equation of the axis of symmetry.

The parabola opens left.

Since $a = -1$, which is negative, the graph opens left.

The vertex is at $(4, 1)$.

$$-\frac{b}{2a} = -\frac{2}{2(-1)} = 1$$
$$x = -(1)^2 + 2(1) + 3$$
$$= 4$$

$$x = -(0)^2 + 2(0) + 3 \qquad\qquad \text{\textit{x}-intercept, } y = 0.$$

$$x = 3$$

The x-intercept is $(3, 0)$.

$$0 = -y^2 + 2y + 3 \qquad\qquad \text{\textit{y}-intercepts, } x = 0.$$

$$0 = (-y + 3)(y + 1)$$

$$y = 3 \quad \text{or} \quad y = -1$$

The y-intercepts are $(0, 3)$ and $(0, -1)$.

The axis of symmetry is the horizontal line $y = 1$.

With this information and a short table of values, we can sketch the graph.

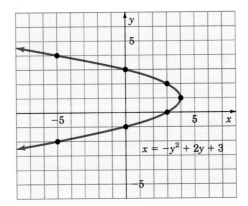

x	y
-12	-3 or 5
-5	-2 or 4
3	0 or 2

EXERCISE 8.1

A

Find the vertex, determine whether the vertex is a maximum or minimum point,
find the x- and y-intercepts, determine the equation of the axis of symmetry,
and graph:

1. $y = x^2 - 2$
2. $y = x^2 + 1$
3. $y = (x - 3)^2$

4. $y = (x + 2)^2$

5. $y = (x + 1)^2 - 2$

6. $y = (x - 3)^2 + 1$

7. $y = 2(x - 1)^2 - 2$

8. $y = \frac{1}{2}(x - 2)^2 + 2$

9. $y = -x^2 + 3$

10. $y = -2x^2 - 1$

11. $y = -2x^2 + 4x - 2$

12. $y = -\frac{1}{2}x^2 - 2x - 2$

Graph each relation:

13. $x = y^2 - 6$

14. $x = -y^2 + 3$

B

Find the vertex, determine whether the vertex is a maximum or minimum point, find the x- and y-intercepts, determine the equation of the axis of symmetry, and graph:

15. $y = -\dfrac{1}{3}x^2 + 2x$

16. $y = -\dfrac{1}{4}x^2 - x$

17. $y = -\dfrac{1}{3}x^2 + 2x - 5$

18. $y = -\dfrac{1}{9}x^2 - \dfrac{2}{3}x - 2$

19. $y = x^2 + 2x - 3$

20. $y = x^2 + 6x + 5$

21. $y = x^2 - 4x + 7$

22. $y = x^2 - 10x + 27$

Graph each relation:

23. $x = y^2 - 6y$

24. $x = -y^2 - 2y$

C

Find the vertex, determine whether the vertex is a maximum or minimum point, find the x- and y-intercepts, determine the equation of the axis of symmetry, and graph:

25. $y = x^2 - 2x - 5$

26. $y = x^2 + 4x - 3$

27. $y = 2x^2 + 4x - 2$

28. $y = 3x^2 - 12x + 13$

29. $y = -3x^2 - 6x - 3$

30. $y = -x^2 + 4x - 2$

31. $y = -2x^2 + 12x - 19$

32. $y = -\dfrac{1}{2}x^2 + x + \dfrac{1}{2}$

33. $y = \dfrac{1}{4}x^2 + \dfrac{1}{2}x - \dfrac{11}{4}$

34. $y = \dfrac{1}{5}x^2 - \dfrac{4}{5}x + \dfrac{9}{5}$

Graph each relation:

35. $x = y^2 - 6y + 1$

36. $x = -y^2 - 8y - 9$

D

37. If the demand curve of a manufacturer of T-shirts is
$y = -\dfrac{1}{2}x^2 + 4x + 1500$, find the week of highest demand and the number
of T-shirts needed to meet the demand.

38. The demand for the new video game Zam-Pow is estimated to be a
quadratic function during the first 52 weeks of sales. If the function is
defined by $y = -\dfrac{1}{2}x^2 + 36x + 2000$, find the week of highest demand and
the number of games needed that week to meet the demand.

39. The number of units (u) that a company can produce with n employees is
given by $u = -\dfrac{1}{3}n^2 + 12n$, with $0 < n \leq 30$. Find the maximum number
of units that can be produced and the number of employees needed to
produce them.

40. The number of units (u) that the Ajax Corp. can produce with n robots is
given by $u = -\dfrac{1}{4}n^2 + 16n$, with $0 \leq n \leq 50$. Find the maximum number
of units that can be produced and the number of robots needed to produce
them.

41. The height of an object thrown upward from a height of 6 ft at a velocity
of 96 ft/sec at any time (t) is given by $h = 96t - 16t^2 + 6$. What is the
maximum height reached by the object? In how many seconds will the
object reach its maximum height?

42. The height of an arrow shot straight up from the roof of a building 40 ft
tall at a speed of 128 ft/sec is given by $h = 128t - 16t^2 + 40$. What is the
maximum height reached by the arrow, and in how many seconds will it
reach its maximum height?

43. The work done in compressing a spring is given by $W = 50x^2$ where W is
the work in foot-pounds and x is the distance in feet that the spring is
compressed. Draw the graph of the curve using W as the dependent variable

(the vertical axis) for $0 \le x \le 10$. How much work is done if you compress the spring 2.7 feet? If the work done is 150 ft-lbs, how far is the spring compressed?

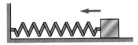

Figure for Exercise 43

44. The number of flu cases (during the flu season) reported by all of the hospitals in a large city is given by $C = -D^2 + 40D + 100$, where D is the number of days that have passed since the beginning of the report. How many cases are there 4 days after the report started? After how many days would there be 400 cases? What is the maximum number of cases?

45. The cost (in dollars) of producing N engine blocks is given by $C = N^2 + 135,000$. What is the cost of producing 15,000 engine blocks? For an investment of $550,000,000, how many blocks can be produced (to the nearest whole block)?

46. If the income generated by selling the engine blocks in problem (45) is given by $I = 1.2N^2 + 2400N$, what is the anticipated income if 15,000 engine blocks are produced? How many blocks must be produced for the company to break even?

47. John wants to fence in his back yard but he needs the fence only on 3 sides since he will use the back of his house as one side. If he has 200 feet of fence, what is the largest area he can enclose? (Hint: Let "x" be the two equal sides, then $200 - 2x$ will be the other.)

48. In economics, it is usual that the demand (q_d) for an object is dependent on the price for which it sells. If the relationship is given by $q_d = p^2 - 100p + 2500$, what would be the expected demand if the price is $25? What would be the expected price if the demand was for 900 items?

STATE YOUR UNDERSTANDING

49. If $y = ax^2 + bx + c$ is the general equation of a parabola that opens up or down, what can be said about the parabola if (a) $b^2 - 4ac > 0$? (b) If $b^2 - 4ac = 0$? (c) If $b^2 - 4ac < 0$?

50. Describe the equation of a parabola that opens to the right and is narrower than the standard parabola.

51. Explain why the parabola (a) $y = (x + 2)^2 + 1$, opens up and has vertex at $(-2, 1)$. (b) $y = -(x + 2)^2 + 1$ opens down and has vertex $(-2, 1)$.

CHALLENGE EXERCISES

52. Remember that $y = ax^2 + bx + c$ is the general equation for a parabola that opens up or down. Find the equation of the parabola (that opens up or down) that passes through $(4, -1)$, $(6, 2)$, and $(-2, 2)$. (Hint: substitute each of the ordered pairs in $y = ax^2 + bx + c$ and solve as a system of equations.)

53. Given $x = ay^2 + by + c$, find the equation of the parabola that opens left or right that passes through $(-2, -1)$, $(4, 5)$ and $(4, -3)$.

54. The cable of a suspension bridge which supports a roadway is attached to two 90 foot towers which are 300 ft. apart. If the cable "hangs" as a parabola and the shortest cable (in the middle) is 20 feet long, write an equation for the cable.

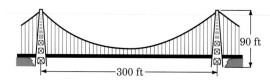

90 ft

—300 ft—

Figure for Exercise 54

55. Using the equation in Exercise 54, how long (to the nearest foot) are the support cables that are 40 feet from the middle of the bridge?

MAINTAIN YOUR SKILLS (SECTION 5.1)

Given $x^2 + y^2 = 400$. Find the real value(s) of y for each given value of x.

56. $x = 0$ **57.** $x = 12$

58. $x = 10$ **59.** $x = 25$

Given $2x^2 - y^2 = 8$. Find the real value(s) of y for each value of x.

60. $x = 6$ **61.** $x = 3$ **62.** $x = 2$ **63.** $x = 0$

8.2

■

CIRCLES

OBJECTIVES

1. Given the equation of a circle in standard form, write the coordinates of the center, the length of the radius, and draw the graph.

2. Given the coordinates of the center and the radius of a circle, write the equation in standard form.

3. Given an equation of a circle, write the equation in standard form, the coordinates of the center, the length of the radius, and draw the graph.

In this section, we study some of the properties of the circle.

■ **DEFINITION**

Circle, Center, Radius, Circumference

> A *circle* is a closed curve, all points of which are at a fixed distance from a fixed point called the *center*. The fixed distance is called the *radius*. The *circumference* of a circle is the distance around the circle.

To write an equation whose graph is a circle, we use the definition and the distance formula:

$$d = \sqrt{(x_2 - x_1)^2 + (y_2 - y_1)^2}.$$

Our goal is to write an equation for all points (x, y) that are a fixed distance, r, from (h, k) as shown in Figure 8.3.

FIGURE 8.3

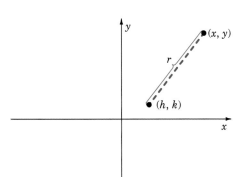

Using the distance formula, we have

$$\sqrt{(x - h)^2 + (y - k)^2} = r$$

$$(x - h)^2 + (y - k)^2 = r^2 \qquad \text{Square both sides.}$$

■ FORMULA

Circle with Center (h, k)

> If (x, y) is any point on the circle of radius r with center (h, k), the standard form of the equation is
>
> $$(x - h)^2 + (y - k)^2 = r^2, \; r \geq 0.$$

If $r = 0$, the circle is said to have degenerated to a single point. If the center of the circle is the origin, $(0, 0)$, the equation is

$$(x - 0)^2 + (y - 0)^2 = r^2.$$

■ FORMULA

Circle with Center $(0, 0)$

> If (x, y) is any point on the circle of radius r with center $(0, 0)$, the standard form of the equation is
>
> $$x^2 + y^2 = r^2, \; r \geq 0.$$

With the center at the origin, the x- and y-intercepts are readily determined. Substituting $y = 0$, the x-intercepts are $(\pm r, 0)$. Substituting 0 for x, the y-intercepts are $(0, \pm r)$, as shown in Figure 8.4. For instance,

$$x^2 + y^2 = 25$$

is the equation of a circle with center at the origin and radius 5, as shown in Figure 8.5. The x-intercepts are $(\pm 5, 0)$, and the y-intercepts are $(0, \pm 5)$.

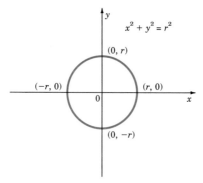

FIGURE 8.4

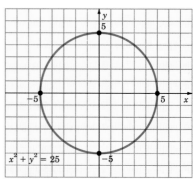

FIGURE 8.5

EXAMPLE 1

Write the coordinates of the center and the length of the radius, and graph the circle with equation $x^2 + y^2 = 36$.

Center $(0, 0)$
x-intercepts: $(\pm 6, 0)$
y-intercepts: $(0, \pm 6)$

This is the standard form of an equation whose graph is a circle with center at the origin.

$r = 6$

The radius is the square root of the constant term, $r = \sqrt{36}$.

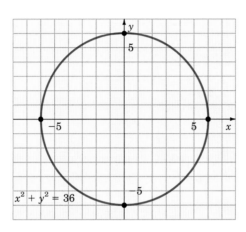

The graph of the equation

$$(x - 3)^2 + (y + 2)^2 = 16$$

is a circle with center $(3, -2)$ and radius 4. It is not obvious from the equation what the intercepts are, so we use the length of the radius to find the points at the extremities of the horizontal and vertical diameters of the circle. These points are 4 units from the center along the lines parallel to the x- and y-axes. The points are $(3, 2)$, $(3, -6)$, $(7, -2)$, and $(-1, -2)$, as shown in Figure 8.6.

FIGURE 8.6

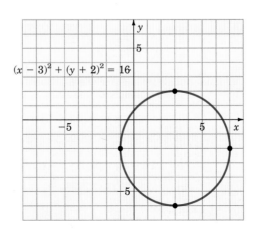

EXAMPLE 2

Write the coordinates of the center and the length of the radius, and graph the circle with equation $(x + 3)^2 + (y + 4)^2 = 9$.

Center $(-3, -4)$, $r = 3$

The equation is in standard form. The points $(-3, -1)$ and $(-3, -7)$ are found by plotting vertically 3 units (up and down) from the center. The points $(-6, 4)$ and $(0, -4)$ are found by plotting horizontally (right and left).

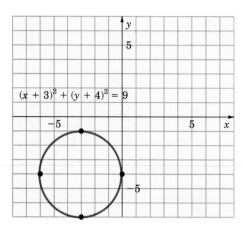

Given the center and radius of a circle, we can use the standard form to write its equation.

EXAMPLE 3

Write the equation of the circle with radius 8 and center at the origin.

The standard form of the equation is $x^2 + y^2 = 64$.

In general, the standard form of a circle at the origin is $x^2 + y^2 = r^2$. Since $r = 8$, $r^2 = 64$.

The graph of $x^2 + y^2 + 4x - 6y - 23 = 0$ is also a circle. To write the equation in standard form, we complete the square in both x and y.

EXAMPLE 4

Write the equation $x^2 + y^2 + 4x - 6y - 23 = 0$ in standard form, write the coordinates of the center and the length of the radius, and draw the graph.

$$x^2 + y^2 + 4x - 6y - 23 = 0$$

$$x^2 + 4x + \qquad + y^2 - 6y + \qquad = 23$$

$$x^2 + 4x + 4 + y^2 - 6y + 9 = 23 + 4 + 9$$

$$(x + 2)^2 + (y - 3)^2 = 36$$

First, write the equation in standard form by completing the squares in both x and y. Add 4 to both sides to complete the square in x and 9 to complete the square in y.

Center $(-2, 3)$, $r = 6$

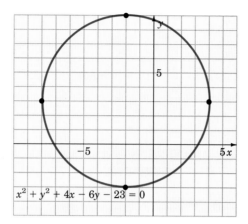

$x^2 + y^2 + 4x - 6y - 23 = 0$

We recognize the standard form of the equation of the circle with center at $(-2, 3)$ and radius 6.

■

The equation of a circle can be used to describe physical phenomena such as a circular track.

EXAMPLE 5

An architect is designing a building to enclose a circular track. The equation she is using for the track is

$$x^2 + y^2 = 44,100.$$

Draw the graph representing the track, and find the distance around the track using the formula for the circumference of a circle, $C = \pi d$. Round to the nearest foot.

$x^2 + y^2 = 44,100$

$r^2 = 44,100$, so $r = 210$.

This is the standard form of a circle with center at the origin and radius 210.

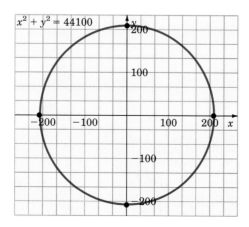

$x^2 + y^2 = 44100$

$$C = \pi d$$

$$C \approx (3.14)(420)$$

$$C \approx 1318.8$$

$$C \approx 1319$$

Use the formula for the circumference of a circle to find the distance around the track. Substitute $\pi \approx 3.14$ and $D = 2r = 420$.

It is approximately 1319 feet around the track, or approximately a quarter of a mile. ■

EXERCISE 8.2

A

Write the coordinates of the center and the length of the radius, and graph each of the following:

1. $x^2 + y^2 = 25$

2. $x^2 + y^2 = 64$

3. $x^2 + y^2 = 36$

4. $x^2 + y^2 = 1$

5. $2x^2 + 2y^2 = 8$

6. $3x^2 + 3y^2 = 27$

Write the equation in standard form of the circle with the given coordinates as the center and the given value of r as the radius:

7. $(0, 0), r = 9$

8. $(0, 0), r = 11$

9. $(2, 4), r = 7$

10. $(5, 6), r = 8$

11. $(0, 3), r = 2$

12. $(4, 0), r = 5$

13. $(1, 1), r = 6$

14. $(2, 2), r = 1$

B

Write the coordinates of the center and the length of the radius, and graph each of the following:

15. $(x - 5)^2 + (y + 4)^2 = 9$ **16.** $(x + 3)^2 + (y - 8)^2 = 16$ **17.** $(x + 5)^2 + (y - 3)^2 = 4$

18. $(x - 7)^2 + (y - 6)^2 = 1$ **19.** $(x - 5)^2 + (y - 4)^2 - 64 = 0$ **20.** $(x - 6)^2 + (y - 4)^2 - 25 = 0$

21. $(x - 1)^2 + (x - 2)^2 = \dfrac{9}{4}$ **22.** $(x + 2)^2 + (y - 2)^2 = \dfrac{25}{4}$

Write the equation in standard form of the circle with the given coordinates as the center and the given value of r as the radius:

23. $(-5, 2), r = 12$ **24.** $(6, -9), r = 11$

25. $(-3, -8), r = 16$ **26.** $(-9, -3), r = 15$

27. $(3, -1), r = \dfrac{2}{3}$ **28.** $(-1, 5), r = \dfrac{4}{5}$

C

Write the coordinates of the center and the length of the radius of each of the following:

29. $4x^2 + 4y^2 = 9$

30. $16x^2 + 16y^2 = 25$

31. $x^2 + y^2 - 6x + 2y + 6 = 0$

32. $x^2 + y^2 + 8x - 10y + 5 = 0$

Write the equation in standard form of the circle with the given coordinates as the center and the given value of r as the radius:

33. $(1, 1), r = \dfrac{7}{2}$

34. $(2, -2), r = \dfrac{3}{4}$

35. $(2, 2), r = \sqrt{2}$

36. $(4, 4), r = \sqrt{3}$

Each of the following is an equation whose graph is a circle. Write each in standard form, then write the coordinates of the center and the length of the radius:

37. $x^2 + y^2 - 10x + 4y + 17 = 0$

38. $x^2 + y^2 + 6x + 6y + 14 = 0$

39. $2x^2 + 2y^2 = 5$

40. $3x^2 + 3y^2 = 5$

Given the center and the radius of a circle, write an equation in the form $ax^2 + by^2 + cx + dy + e = 0$:

41. $(-2, -7), r = 1.2$

42. $(-8, 12), r = 3.5$

D

43. Draw the graph of a circular race track that has an equation $x^2 + y^2 = 3600$.

44. Draw the graph of a circular race track that has an equation $x^2 + y^2 = 8100$.

45. If $(x - h)^2 + (y - k)^2 = 0$ is the equation of the degenerate circle (i.e. the point (h, k)), what is the graph of $x^2 + y^2 - 4x + 8y + 20 = 0$?

46. Write the equation of the degenerate circle whose center is at $(-3, -2)$.

47. If we "solved" $x^2 + y^2 = 16$ for "y", we would get $y = \pm\sqrt{16 - x^2}$. What is the graph of $y = +\sqrt{16 - x^2}$? What about the graph of $y = -\sqrt{16 - x^2}$?

48. Graph the solution set to $x^2 + y^2 < 16$.

49. A tangent line is a line that intersects a circle at exactly one point. The tangent line is also perpendicular to the radius to the point where the tangent intersects the circle. Given the circle $x^2 + y^2 = 25$, find the equation of the tangent line at the point $(3, -4)$.

50. Determine the center, the radius, and graph: $x^2 + y^2 - 6x + 3y - 1 = 0$.

51. Determine the center, the radius, and graph: $4x^2 + 4y^2 - 8x + 16y + 11 = 0$.

52. The circumference of a circle is given by $C = \pi d$ (as discussed in this section) and the area is given by $A = \pi r^2$. Find the circumference and the area of the circle $x^2 + y^2 + 6x - 2y - 6 = 0$.

53. If you graph $x^2 + y^2 = 25$ and $x^2 + y^2 = 9$ on the same graph, what is the area of the region between the circles?

STATE YOUR UNDERSTANDING

54. If $x^2 + y^2 = r^2$ (where $r^2 > 0$) is the equation of a circle and $x^2 + y^2 = 0$ is the equation of a degenerate circle (i.e. a point), why are there no points on the Cartesian plane for $x^2 + y^2 = r^2$ if $r^2 < 0$?

55. How could you ''cut'' a cone and get the degenerate circle? How could you ''cut'' a cone and get no circle at all?

CHALLENGE EXERCISES

56. The height of a semicircular arch 1 foot from an end is 7 feet. What is the maximum height of the arch?

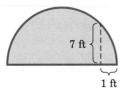

7 ft

1 ft

Figure for Exercise 56

57. Find the equation of the line tangent to the circle $(x - 3)^2 + (y + 1)^2 = 25$ at the point $(0, 3)$.

58. If $x^2 + y^2 + ax + by + c = 0$ is the general equation of a circle, using systems of equations in the previous chapter to find the equation of the circle passing through the points $(3, 1)$, $(0, 0)$, and $(8, 4)$.

59. For the circle, $x^2 + y^2 - 6x + 4y + 4 = 0$, find the center, the radius, the x-intercepts, and the y-intercepts.

MAINTAIN YOUR SKILLS (SECTIONS 2.11, 4.7, 5.5, 5.7)

Solve:

60. $4\sqrt{x - 1} + 2 = 4x - 1$

61. $4\sqrt{x - 1} + 2 = 1 - 4x$

62. $\dfrac{1}{x} - \dfrac{1}{x - 1} = 5$

63. $x - \dfrac{1}{x - 1} = \dfrac{1}{5}$

64. $7x^2 - 11x + 4 = 0$

65. $2y^2 - 47y + 66 = 0$

66. $2x^2 + 13x - 45 \leq 0$

67. $2x^2 - 13x - 45 > 0$

8.3

◼

ELLIPSES AND HYPERBOLAS

OBJECTIVES

◼

---

1. Graph an ellipse that is centered about the origin.

2. Graph a hyperbola that is centered about the origin.

3. Given a graph of an ellipse or hyperbola centered about the origin, write its equation in standard form.

In this section, we study some of the properties of ellipses and hyperbolas centered about the origin.

◼ **DEFINITION**

Ellipse

> An *ellipse* is the set of all points in a plane such that the sum of the distances from two fixed points is constant. Each fixed point is called a focus.

◼ **FORMULA**

Ellipse
Standard Form

> The standard form of the equation of an ellipse centered aobut the origin is
>
> $$\frac{x^2}{a^2} + \frac{y^2}{b^2} = 1, \text{ where } a > 0 \text{ and } b > 0.$$

The x-intercepts, $y = 0$, are $(\pm a, 0)$, and the y-intercepts, $x = 0$, are $(0, \pm b)$.

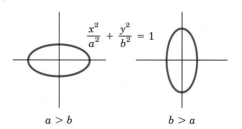

To sketch the graph of an ellipse, it is helpful to first form the rectangle that has vertices at (a, b), $(-a, b)$, $(a, -b)$, and $(-a, -b)$. The ellipse is bounded by the rectangle.

EXAMPLE 1

Sketch the graph of the ellipse: $\dfrac{x^2}{25} + \dfrac{y^2}{9} = 1$

$a = 5$ and $b = 3$

Since $a^2 = 25$, $b^2 = 9$, and a and b are both positive.

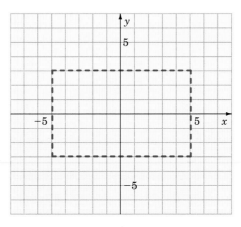

Draw the rectangle that is a boundary of the ellipse. The rectangle has vertices at $(5, 3)$, $(-5, 3)$, $(5, -3)$, and $(-5, -3)$.

The graph of the ellipse is inscribed (fitted within) the rectangle. In addition to the intercepts, calculate eight or ten pairs of coordinates to help sketch the graph more accurately. A calculator is helpful in making a table of values.

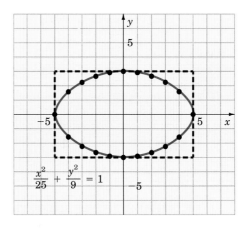

x	y
0	±3
±1	±2.94
±2	±2.75
±3	±2.4
±4	±1.8
±5	0

EXAMPLE 2

Sketch the graph of the ellipse: $9x^2 + y^2 = 36$

$\dfrac{9x^2}{36} + \dfrac{y^2}{36} = \dfrac{36}{36}$

Write the equation in standard form.

$\dfrac{x^2}{4} + \dfrac{y^2}{36} = 1$

Simplify. This is the standard form of an ellipse with intercepts at $(\pm 2, 0)$ and $(0, \pm 6)$.

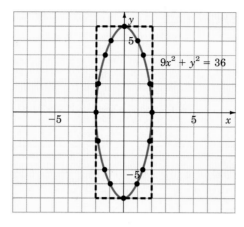

Draw the rectangle with vertices at $(2, 6)$, $(-2, 6)$, $(2, -6)$, and $(-2, -6)$. Inscribe the ellipse in the rectangle.

■

Elliptical shapes occur in the orbits of comets and planets, in the domes of some buildings, in coffee tables, and in race tracks.

EXAMPLE 3

A greyhound race track is in the shape of an ellipse that has the equation $\dfrac{x^2}{5625} + \dfrac{y^2}{2500} = 1$. Draw the graph representing the track.

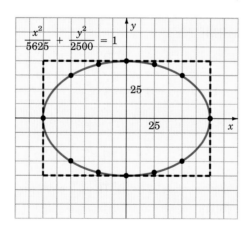

The graph of the track is an ellipse with vertices at $(\pm 75, 0)$ and $(0, \pm 50)$.

Table of values. (y-values are approximate.)

x	y
± 25	± 47
± 50	± 37

■

■ DEFINITION

Hyperbola

A *hyperbola* is the set of all points in a plane such that the difference of the distances from two fixed points is constant. Each fixed point is called a focus.

**Hyperbola
Standard Form**

■ FORMULA

The standard form of the equation of a hyperbola centered about the origin and opening horizontally is

$$\frac{x^2}{a^2} - \frac{y^2}{b^2} = 1, \text{ where } a > 0 \text{ and } b > 0.$$

The x-intercepts are $(\pm a, 0)$. There are no y-intercepts.

$$\frac{x^2}{a^2} - \frac{y^2}{b^2} = 1$$

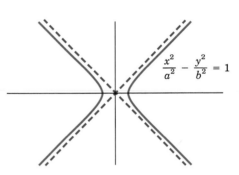

■ FORMULA

**Hyperbola
Standard Form**

The standard form of the equation of a hyperbola centered about the origin and opening vertically is

$$\frac{y^2}{b^2} - \frac{x^2}{a^2} = 1, \text{ where } a > 0 \text{ and } b > 0.$$

The y-intercepts are $(0, \pm b)$. There are no x-intercepts.

$$\frac{y^2}{b^2} - \frac{x^2}{a^2} = 1$$

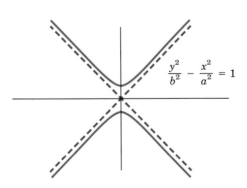

To sketch the graph of a hyperbola, we form the rectangle that has vertices at (a, b), $(a, -b)$, $(-a, b)$, and $(-a, -b)$, as we did for the ellipse. We then draw broken lines that contain the diagonals of the rectangle. These lines are called asymptotes (or guide lines) for the hyperbola. The hyperbola is contained between these lines either horizontally or vertically.

■ DEFINITION

Asymptote

An *asymptote* is a line that the graph of a relation approaches as a limit or boundary.

EXAMPLE 4

Sketch the graph of the hyperbola: $\dfrac{x^2}{4} - \dfrac{y^2}{9} = 1$

$a = 2$ and $b = 3$

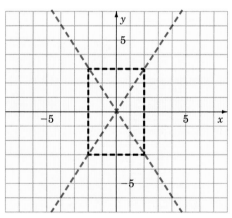

Since $a^2 = 4$ and $b^2 = 9$.

Draw the rectangle with vertices at $(2, 3)$, $(2, -3)$, $(-2, 3)$, and $(-2, -3)$, and extend the diagonals.

The graph is fitted to the extended diagonals by plotting the intercepts and about eight or ten additional points.

A calculator or square root table is helpful in making a table of values.

Table of values.

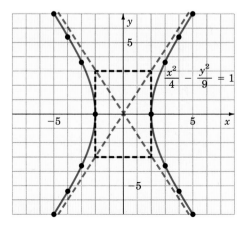

x	y
± 2	0
± 3	± 3.35
± 4	± 5.20
± 5	± 6.87
± 6	± 8.49

■

EXAMPLE 5

Sketch the graph of the hyperbola: $\dfrac{y^2}{16} - \dfrac{x^2}{1} = 1$

$b = 4$ and $a = 1$

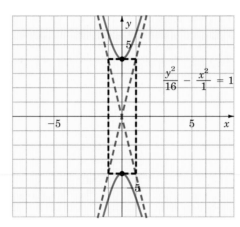

Since $b^2 = 16$ and $a^2 = 1$. Draw the rectangle with vertices at $(1, 4)$, $(-1, 4)$, $(1, -4)$, and $(-1, -4)$, and extend the diagonals.

The y-intercepts are $(0, \pm 4)$.

The graph is fitted to the extended diagonals by plotting the intercepts and a few additional points.

EXAMPLE 6

Sketch the graph of the hyperbola: $4x^2 - y^2 = 16$

$\dfrac{4x^2}{16} - \dfrac{y^2}{16} = \dfrac{16}{16}$

$\dfrac{x^2}{4} - \dfrac{y^2}{16} = 1$

Write the equation in standard form by first dividing both sides by 16.

This is the standard form of a hyperbola with intercepts at $(\pm 2, 0)$.

Draw the related rectangle, and extend the asymptotes.

Since $a^2 = 4$ and $b^2 = 16$, the vertices of the rectangle are $(2, 4)$, $(-2, 4)$, $(2, -4)$ and $(-2, -4)$.

Table of values.

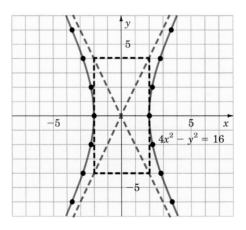

x	y
± 2	0
$\pm \sqrt{5}$	± 2
$\pm 2\sqrt{2}$	± 4
$\pm \sqrt{13}$	± 6

To write the equation of an ellipse or a hyperbola given its graph, use the denominators of x^2 and y^2 to identify the values of a and b.

EXAMPLE 7

Write the equation of the ellipse in standard form.

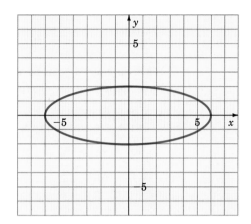

The *x*-intercepts are $(6, 0)$ and $(-6, 0)$.

The *y*-intercepts are $(0, 2)$ and $(0, -2)$.

Therefore, $a = 6$ and $b = 2$.

The equation is $\dfrac{x^2}{36} + \dfrac{y^2}{4} = 1$.

From the graph, write the coordinates of the *x*- and *y*-intercepts.

The *x*- and *y*-intercepts are at $(\pm a, 0)$ and $(0, \pm b)$.

Substitute the values of *a* and *b* in the standard form, $\dfrac{x^2}{a^2} + \dfrac{y^2}{b^2} = 1$.   ■

EXAMPLE 8

Write the equation of the hyperbola in standard form.

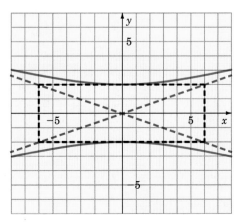

The *y*-intercepts are $(0, 2)$ and $(0, -2)$, so $b = 2$.

From the graph, write the coordinates of the *y*-intercepts.

There are no x-intercepts, but the rectangle extends 6 units to the right and left, so $a = 6$.

The equation is $\dfrac{y^2}{4} - \dfrac{x^2}{36} = 1$.

Substitute the values of a and b in the standard form, $\dfrac{y^2}{b^2} - \dfrac{x^2}{a^2} = 1$.

■

When an equation can be written in the form $Ax^2 + By^2 = C$, the graph or conic section can be identified by examining the values of A, B, and C.

1. If $A = B$ and A, B, and C have the same sign, the equation can be rewritten in the standard form for a circle. For instance:

$$8x^2 + 8y^2 = 32$$

$A = B$ and A, B, and C are positive.

$$\frac{8x^2}{8} + \frac{8y^2}{8} = \frac{32}{8}$$

Divide both sides by 8.

$$x^2 + y^2 = 4$$

This is the standard form for the equation of a circle with center at $(0, 0)$ and radius 2.

2. If $A \neq B$ and A, B, and C have the same sign, the equation can be rewritten in the standard form for an ellipse. For instance:

$$4x^2 + 9y^2 = 36$$

$A \neq B$ and A, B, and C are positive.

$$\frac{4x^2}{36} + \frac{9y^2}{36} = \frac{36}{36}$$

Divide both sides by 36.

$$\frac{x^2}{9} + \frac{y^2}{4} = 1$$

This is the standard form for the equation of an ellipse with center at $(0, 0)$ and intercepts $(\pm 3, 0)$ and $(0, \pm 2)$.

3. If A and B are opposite in sign, the equation can be rewritten in the standard form for a hyperbola. For instance:

$$25x^2 - 16y^2 = 400$$

A and B are opposite in sign.

$$\frac{25x^2}{400} - \frac{16y^2}{400} = \frac{400}{400}$$

Divide both sides by 400.

$$\frac{x^2}{16} - \frac{y^2}{25} = 1$$

This is the standard form for the equation of a hyperbola with center at $(0, 0)$ and intercepts $(\pm 4, 0)$.

EXERCISE 8.3

A

Draw the graphs of the following conic sections:

1. $x^2 - y^2 = 1$ **2.** $x^2 - y^2 = 9$ **3.** $\dfrac{x^2}{1} + \dfrac{y^2}{4} = 1$

4. $\dfrac{x^2}{4} + \dfrac{y^2}{1} = 1$ **5.** $\dfrac{x^2}{4} - \dfrac{y^2}{4} = 1$ **6.** $\dfrac{y^2}{1} - \dfrac{x^2}{1} = 1$

7. $\dfrac{x^2}{1} + \dfrac{y^2}{9} = 1$ **8.** $\dfrac{x^2}{9} + \dfrac{y^2}{1} = 1$

Write the equation of each conic section in standard form:

9.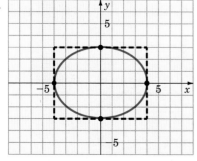

10.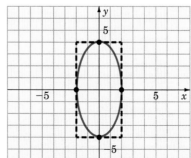

B

Draw the graphs of the following conic sections:

11. $\dfrac{x^2}{9} + \dfrac{y^2}{25} = 1$

12. $\dfrac{x^2}{25} + \dfrac{y^2}{9} = 1$

13. $\dfrac{x^2}{4} - \dfrac{y^2}{9} = 1$

14. $\dfrac{y^2}{9} - \dfrac{x^2}{4} = 1$

15. $\dfrac{x^2}{9} - \dfrac{y^2}{25} = 1$

16. $\dfrac{y^2}{25} - \dfrac{x^2}{9} = 1$

17. $\dfrac{x^2}{4} + \dfrac{y^2}{16} = 1$

18. $\dfrac{x^2}{16} + \dfrac{y^2}{4} = 1$

Write the equation of each conic section in standard form:

19.

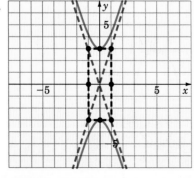

20.

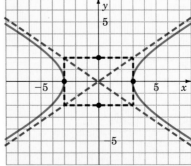

C

Draw the graphs of the following conic sections:

21. $\dfrac{x^2}{27} + \dfrac{y^2}{3} = 3$

22. $\dfrac{x^2}{16} + \dfrac{y^2}{4} = 4$

23. $10x^2 + 7y^2 = 140$

24. $8x^2 + 10y^2 = 80$

25. $5x^2 - 6y^2 = 60$

26. $10y^2 - 7x^2 = 140$

Write the equation of each conic section in standard form.

27.

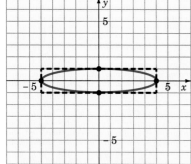

28.

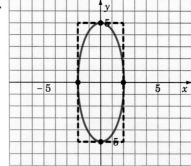

29.

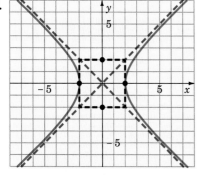

30.

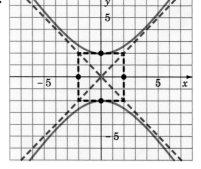

D

31. Draw the graph of an elliptical race track that has the equation
    $9x^2 + 25y^2 = 22,500$.

32. The plaza at the New Life Center has opposite sides that are hyperbolic.
    One set is contained in the graph described by the equation $x^2 - y^2 = 1600$,
    and the other pair is contained in the graph of $2y^2 - x^2 = 1225$. Draw the
    graph of the plaza.

33. If we solve $4x^2 + y^2 = 16$ (which is an ellipse) for y, we get
    $y = \pm\sqrt{16 - 4x^2}$ or $y = \pm2\sqrt{4 - x^2}$. What is the graph of
    $y = +\sqrt{16 - 4x^2}$? What is the graph of $y = -\sqrt{16 - 4x^2}$?

34. An object is moving through the air while it is spinning. It's total kinetic
    energy is 144 ft-lbs. If the velocity it is moving through the air is V and the
    spinning velocity is S, they are related by the equation $4V^2 + 9S^2 = 144$.
    Graph the relationship using S as the dependent variable and V as the
    independent variable.

35. A cam is in the shape of an ellipse. If the horizontal length of the cam is
    12 cm and the vertical length of the cam is 6 cm, write the equation of the
    cam.

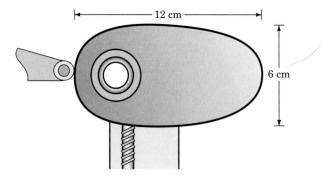

Figure for Exercise 35

36. By making a table of values and generating several solutions, graph the
    following: $xy = 4$. What conic section is it?

37. A circular plate has an elliptical hole in the center as in the diagram. Write the equation of the plate and the hole.

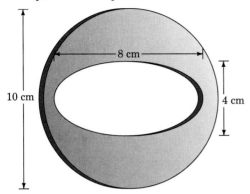

Figure for Exercise 37

38. A bridge span is in the shape of a semi-ellipse. Write the equation of the curve. How high is the span 10 feet from the center?

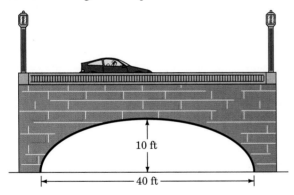

Figure for Exercise 38

39. Graph and shade the appropriate region for: $9x^2 + 4y^2 < 36$

40. Graph and shade the appropriate region for $x^2 - y^2 \leq 1$

STATE YOUR UNDERSTANDING

41. An ellipse is the set of all points such that the sum of the distances from two fixed points is a constant. Place two small nails in a board about 6 inches apart. Make a loop of string that will fit over the two nails with a little slack. Place a pencil inside the loop and move the pencil around the two nails keeping the loop taut. Why is the curve an ellipse?

42. What is the major difference between the equations of the ellipses and the hyperbolas that you found in this section?

43. Write the equation of a hyperbola with y-intercepts at $(0, \pm 5)$ and the other sides of the rectangle at $(\pm 3, 0)$.

CHALLENGE EXERCISES

44. If $x^2 + y^2 = 0$ is the degenerate circle (i.e. the point $(0, 0)$), what is the graph of $x^2 + 4y^2 = 0$?

45. Given $\dfrac{x^2}{4} - \dfrac{y^2}{9} = 1$, write the equations of the two asymptotes (the diagonals of the rectangle you use to graph the hyperbola). Are the diagonals perpendicular?

46. If $\dfrac{x^2}{9} + \dfrac{y^2}{16} = 1$ is graphed, what is the distance from the positive x-intercept to the positive y-intercept?

47. $x^2 - y^2 = 0$ is the degenerate hyperbola. What is the graph of this equation?

MAINTAIN YOUR SKILLS (SECTIONS 1.8, 3.5, 5.2, 6.4)

Solve:

48. $\dfrac{1}{a} + \dfrac{1}{a - 1} = \dfrac{2}{a^2 + 2a - 3}$

49. $\dfrac{b + 7}{2b^2 + 3b} = \dfrac{b + 1}{2b^2 - b - 6}$

50. $(x + 3)(x + 7) - 8 = 2(x + 6)^2$

Solve for w:

51. $\dfrac{1}{4}w - \dfrac{1}{5}a = \dfrac{1}{2}w - \dfrac{1}{3}a$

52. $3w^2 - 8cw - 3c^2 = 0$

53. $w + a(w + b) = w + 1$

54. $(w + a)(w + 6) = w(w + 1)$

Solve by graphing:

55. $\begin{cases} x - 5y \geq 0 \\ 3x + 4y < 6 \end{cases}$

8.4

SOLVING SYSTEMS OF EQUATIONS INVOLVING QUADRATIC EQUATIONS I

OBJECTIVE

1. Solve a system containing one linear and one quadratic equation by substitution.

Systems of equations in two variables can be solved by graphing or by substitution. The method of substitution is similar to that for systems of linear equations. The graph of the system is a visual illustration of the solution.

EXAMPLE 1

Solve the system: $\begin{cases} x^2 + y^2 = 16 & (1) \\ x + y = 4 & (2) \end{cases}$

$y = 4 - x$	To solve by substitution, solve equation (2) for y.
$x^2 + (4 - x)^2 = 16$	Substitute $y = 4 - x$ in equation (1).
$x^2 + 16 - 8x + x^2 = 16$	Solve for x.
$2x^2 - 8x = 0$	
$2x(x - 4) = 0$	
$2x = 0$ or $x - 4 = 0$	Zero-product property.
$x = 0$ or $x = 4$	
$0^2 + y^2 = 16$	To find the value of y, substitute $x = 0$ in equation (1).
$y^2 = 16$	
$y = \pm 4$	
$4^2 + y^2 = 16$	Substitute $x = 4$ in equation (1).
$y^2 = 0$	
$y = 0$	

The pairs $(0, 4)$, $(0, -4)$, and $(4, 0)$ are possible solutions.

Pairing the values of x and y yields the possible solutions. Since we used equation (1) to find the y-values, we can check the pairs using equation (2).

Check:

$0 + 4 = 4$ True	Substitute $(0, 4)$ in equation (2).
$0 + (-4) = 4$ False	Substitute $(0, -4)$ in equation (2).
$4 + 0 = 4$ True	Substitute $(4, 0)$ in equation (2).

The solution of the system is $\{(0, 4), (4, 0)\}$.

The check reveals that $(0, -4)$ is not a solution.

The graph illustrates the two pairs in the solution.

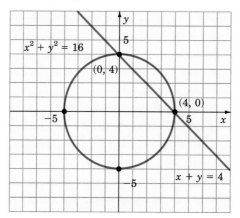

■ **PROCEDURE**

To solve a system containing a linear and a quadratic equation by substitution:

1. Solve the linear equation for x or y.

2. Substitute in the quadratic equation to eliminate one of the variables.

3. Solve the resulting equation.

4. To find the value(s) of the other variable, substitute the values found in step 3 in the quadratic equation and solve.

5. Check the pairs in the linear equation.

6. The solution contains the pairs that check.

Some systems have solutions that are complex numbers. The graph on a Cartesian plane does not show these solutions.

EXAMPLE 2

Solve the system: $\begin{cases} x = y^2 + 3y & (1) \\ x = y - 4 & (2) \end{cases}$

$y - 4 = y^2 + 3y$ Substitute $x = y - 4$ in equation (1).

$y^2 + 2y + 4 = 0$ Solve for y.

$y = \dfrac{-2 \pm \sqrt{4 - 4(1)(4)}}{2}$

$$y = \frac{-2 \pm \sqrt{-12}}{2}$$

$y = -1 \pm i\sqrt{3}$ The value of y is a complex number.

There are no real solutions.   The graph verifies that no real solution exists.

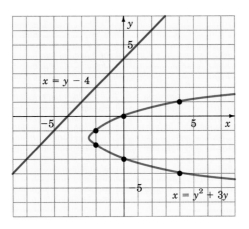

EXAMPLE 3

Solve the system: $\begin{cases} y = x^2 + 5 & (1) \\ y = 6x & (2) \end{cases}$

$6x = x^2 + 5$ Substitute $y = 6x$ in equation (1).

$0 = x^2 - 6x + 5$ Solve for x.

$0 = (x - 5)(x - 1)$

$x - 5 = 0$ or $x - 1 = 0$ Zero-product property.

$\quad x = 5$ or $\quad x = 1$

$y = 5^2 + 5 = 30$ Substitute $x = 5$ in equation (1).

$y = 1^2 + 5 = 6$ Substitute $x = 1$ in equation (1).

The pairs $(5, 30)$ and $(1, 6)$ are possible solutions.   Pairing the values of x and y yields the possible solutions.

Check:

$30 = 6(5)$ True   Substitute $(5, 30)$ in equation (2).

$\ \ 6 = 6(1)$ True   Substitute $(1, 6)$ in equation (2).

The solution set is $\{(5, 30), (1, 6)\}$.   The graph illustrates the solutions.

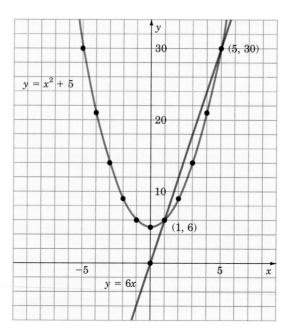

Some applications give rise to a system of equations in which one of the equations is quadratic.

EXAMPLE 4

The area of a rectangular plot of ground is 33,750 square feet. The perimeter of the plot is 750 feet. What are the dimensions of the plot of ground?

Select variables:

To determine the dimensions of the rectangle, we use the formulas for area, $A = \ell w$, and perimeter, $P = 2\ell + 2w$.

Let x represent the length and y represent the width.

System of equations:

$$\begin{cases} xy = A & (1) \\ 2x + 2y = P & (2) \end{cases}$$

Solve:

$$\begin{cases} xy = 33{,}750 & (1) \\ 2x + 2y = 750 & (2) \end{cases}$$

The area is 33,750, and the perimeter is 750.

$$2x = 750 - 2y$$

Solve equation (2) for x.

$$x = 375 - y$$

$(375 - y)y = 33{,}750$	Substitute $x = 375 - y$ in equation (1).
$375y - y^2 = 33{,}750$	
$y^2 - 375y + 33{,}750 = 0$	Solve for y.
$(y - 225)(y - 150) = 0$	
$y - 225 = 0 \quad$ or $\quad y - 150 = 0$	Zero-product property.
$\qquad y = 225 \quad$ or $\qquad\qquad y = 150$	
$x(225) = 33{,}750$	Substitute $y = 225$ in equation (1).
$\qquad x = 150$	
$x(150) = 33{,}750$	Substitute $y = 150$ in equation (1).
$\qquad x = 225$	
The solution set of the system is $\{(150, 225), (225, 150)\}$.	The check is left for the student.

Answer:

The rectangular plot is 225 feet $\times$ 150 feet.

EXERCISE 8.4

A

Solve:

1. $\begin{cases} y = x^2 \\ y = 4x \end{cases}$

2. $\begin{cases} y = x^2 \\ y = -5x \end{cases}$

3. $\begin{cases} y = x^2 + 3x \\ y = 2x + 6 \end{cases}$

4. $\begin{cases} y = x^2 - 5x \\ y = -3x + 3 \end{cases}$

5. $\begin{cases} x^2 + y^2 = 16 \\ x + y = 4 \end{cases}$

6. $\begin{cases} x^2 + y^2 = 25 \\ x + y = 5 \end{cases}$

7. $\begin{cases} 3x^2 + 3y^2 = 12 \\ \qquad\quad y = x + 2 \end{cases}$

8. $\begin{cases} 2x^2 + 2y^2 = 18 \\ x + y = 3 \end{cases}$

9. $\begin{cases} x^2 - y^2 = 1 \\ \qquad\quad y = 2 \end{cases}$

10. $\begin{cases} y^2 - x^2 = 1 \\ \qquad\quad x = 4 \end{cases}$

B

Solve:

11. $\begin{cases} x^2 + y^2 = 25 \\ 3y - 4x = 0 \end{cases}$

12. $\begin{cases} x^2 + y^2 = 100 \\ 4y - 3x = 0 \end{cases}$

13. $\begin{cases} x^2 - y^2 = 4 \\ x + y = 1 \end{cases}$

14. $\begin{cases} y^2 - x^2 = 4 \\ x + y = 1 \end{cases}$

15. $\begin{cases} 4x^2 + y^2 = 20 \\ 4x - y = 0 \end{cases}$

16. $\begin{cases} x^2 + 9y^2 = 82 \\ 3x + y = 0 \end{cases}$

17. $\begin{cases} y = x^2 - 4x + 2 \\ y - x = 2 \end{cases}$

18. $\begin{cases} y = x^2 + 4x + 1 \\ y - x = 5 \end{cases}$

19. $\begin{cases} y = x^2 - 2x + 4 \\ x - 3y = 3 \end{cases}$

20. $\begin{cases} y = x^2 + 4x + 5 \\ x - 2y = 4 \end{cases}$

C

Solve:

21. $\begin{cases} 3x^2 + 4y^2 = 12 \\ 3x + 2y = 0 \end{cases}$

22. $\begin{cases} 6x^2 + 5y^2 = 179 \\ 5x - 3y = 0 \end{cases}$

23. $\begin{cases} x^2 + y^2 = 6 \\ 2x - y = 6 \end{cases}$

24. $\begin{cases} x^2 + y^2 = 8 \\ 3x - y = 15 \end{cases}$

25. $\begin{cases} 4x^2 - y^2 = 15 \\ 2x - y + 3 = 0 \end{cases}$

26. $\begin{cases} y^2 - 4x^2 = 21 \\ 2x - y + 3 = 0 \end{cases}$

27. $\begin{cases} y = -x^2 + 10x - 24 \\ x - y = 4 \end{cases}$

28. $\begin{cases} y = -2x^2 + 12x - 14 \\ 2x + y = 6 \end{cases}$

29. $\begin{cases} x^2 + y^2 = 9 \\ y - x = 3\sqrt{2} \end{cases}$

30. $\begin{cases} 4x^2 - 9y^2 = 36 \\ -2x + y = 1 \end{cases}$

D

31. A rectangle has an area of 3200 cm^2 and a perimeter of 240 cm. What are its dimensions?

32. A rectangle has an area of 3000 m^2 and a perimeter of 220 m. What are its dimensions?

33. The cost of producing n units of steel at the Jimeo Corporation is given by $c = 10n$. The money received from the sale of n units of steel is given by $c = \dfrac{1}{10}n^2$. Find the number of units that Jimeo Corporation needs to produce to break even.

34. In Exercise 33, if the cost is given by $c = 15n + 64$ and the income is given by $c = \frac{1}{4}n^2$, when will the company break even?

35. The sum of two numbers is 12 and their product is 32. Find the two numbers.

36. The sum of two numbers is 1 and their product is -5, find them.

37. Is it possible to find two real numbers whose difference is 1 and whose product is 1. If so, find all the possibilities.

38. The product of the ages of two children, Jim and Sue, is 40. In 5 years, Jim will be 3 years older than twice Sue's age now. What are their ages now?

39. The position of a ball thrown in the air at night is given by $y = -x^2 + 10$. A beam of light is shining on the line $x - y + 4 = 0$. At what positons (points) will the light shine on the ball?

40. An object is moving in an elliptical path given by $4x^2 + y^2 = 16$. A laser beam is shining on the line $y = x - 4$. At what points will the laser beam hit the orbiting object?

STATE YOUR UNDERSTANDING

41. In a system containing two equations, one a hyperbola and the other a line, how many different possibilities are there for the number of points of intersection?

42. Given the system: $\begin{cases} 5x^2 + y^2 = 100 \\ x + y = 5 \end{cases}$

Would squaring both sides of the second equation allow us to solve the system using linear combinations? Explain.

CHALLENGE EXERCISES

43. The supply (q_s) of a certain product is dependent on the price (p) charged and given by $q_s = p^2 + 15p$. The demand (q_d) for the product is also dependent on the price and is given by $q_d = -5p + 300$. At what price is the supply equal to the demand?

44. Is the line $3x + 4y + 25 = 0$ tangent to the circle $x^2 + y^2 = 25$?

45. Graph the following system and estimate the solution:

$$\begin{cases} x^2 + y^2 + 4x - 6y = 7 \\ 2x + y = 5 \end{cases}$$

46. Solve the system in Exercise 45 algebraically.

47. The sum of the circumferences of two circles is 12π inches and the sum of their areas is 20π square inches. Find the radii of the two circles. (Hint: let "x" be the radius of one circle and "y" be the radius of the other. Also, remember that the circumference of a circle is given by $C = 2\pi r$ and the area is given by $A = \pi r^2$.)

48. The weekly demand for a certain item in a store is given by $q_d\, p = 1000$, where q_d is the demand and p is the selling price. The supply available to the store is given by $q_s = 5p - 50$. At what price is the supply equal to the demand?

MAINTAIN YOUR SKILLS (SECTIONS 7.2, 7.4)

Evaluate:

49. $\begin{vmatrix} 11 & -3 & 4 \\ 0 & -2 & 1 \\ 6 & 1 & 3 \end{vmatrix}$

50. $\begin{vmatrix} 11 & -3 & 4 \\ -2 & 0 & 1 \\ 6 & 1 & 3 \end{vmatrix}$

51. $\begin{vmatrix} 11 & -3 & 4 \\ 1 & -2 & 0 \\ 6 & 1 & 3 \end{vmatrix}$

52. $\begin{vmatrix} 11 & -3 & 4 \\ 6 & -2 & 1 \\ 0 & 1 & 3 \end{vmatrix}$

Solve by linear combinations:

53. $\begin{cases} 3x + 8y = 9 \\ 2x + 3y = 5 \end{cases}$

54. $\begin{cases} 5x - 2y = 12 \\ 3x + 5y = 1 \end{cases}$

55. Sonia takes 6 hours to row 12 miles upstream and 3 hours to row back down to her starting point. Find Sonia's average rowing rate in still water and the rate of the current of the stream.

56. How many quarts of distilled water must be added to five quarts of a 30% alcohol solution to produce a 20% alcohol solution?

8.5

SOLVING SYSTEMS OF EQUATIONS INVOLVING QUADRATIC EQUATIONS II

OBJECTIVE

1. Solve a system of quadratic equations of the form $\begin{cases} Ax^2 + By^2 = C \\ Dx^2 + Ey^2 = F \end{cases}$.

Systems of two quadratic equations in two variables where both variables have degree two, can be solved by combinations or by graphing. Solving by combinations is similar to that for systems of linear equations. We multiply one or both of the equations by numbers so that one of the variables is eliminated when the equations are added or subtracted. The graph of the system is a visual illustration of the solution.

EXAMPLE 1

Solve the system: $\begin{cases} x^2 + y^2 = 30 & (1) \\ x^2 - y^2 = 20 & (2) \end{cases}$

$$\begin{array}{ll} x^2 + y^2 = 30 & (1) \\ \underline{x^2 - y^2 = 20} & (2) \\ 2x^2 = 50 & \end{array}$$

To solve, we add equations (1) and (2).

$$x^2 = 25$$

$$x = \pm 5$$

$$(\pm 5)^2 + y^2 = 30$$

To find the value of y, substitute $x = \pm 5$ in equation (1).

$$25 + y^2 = 30$$

$$y^2 = 5$$

$$y = \pm\sqrt{5}$$

The pairs $(5, \sqrt{5})$, $(5, -\sqrt{5})$, $(-5, \sqrt{5})$, and $(-5, -\sqrt{5})$ are possible solutions.

The four pairs arise from the four values, two for $x = \pm 5$, and two for $y = \pm\sqrt{5}$.

Check:

$$5^2 - (\sqrt{5})^2 = 20$$

$$25 - 5 = 20 \quad \text{True}$$

Substitute $(5, \sqrt{5})$ in equation (2). The checks for the other 3 pairs are left for the student.

The solution set of the system is

$$\{(5, \sqrt{5}), (5, -\sqrt{5}), (-5, \sqrt{5}), (-5, -\sqrt{5})\}.$$

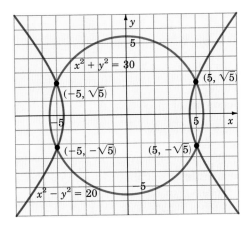

The graph illustrates the four solutions as points of intersection.

■ PROCEDURE

To solve a second-degree system in two variables where both variables have degree two:

1. Write each equation in the form $Ax^2 + By^2 = C$.

2. Multiply one or both equations by factors so that the coefficients of either variable have opposite values.

3. Add the two equations so that one of the variables is eliminated.

4. Solve for the remaining variable.

5. Find the value of the eliminated variable by substitution in one of the original equations, and check the solutions in the other.

6. Write the solutions as ordered pairs.

EXAMPLE 2

Solve the system: $\begin{cases} 2x^2 + y^2 = 6 & (1) \\ 5x^2 - 2y^2 = -3 & (2) \end{cases}$

$\begin{aligned} 4x^2 + 2y^2 &= 12 \\ \underline{5x^2 - 2y^2} &= \underline{-3} \\ 9x^2 &= 9 \end{aligned}$ Multiply equation (1) by 2, and add to equation (2) to eliminate the variable y.

$$x^2 = 1$$

$$x = \pm 1$$

To find the value of y:

$$2(\pm 1)^2 + y^2 = 6$$ Substitute $x = \pm 1$ in equation (1).

$$y^2 = 4$$

$$y = \pm 2$$

Pairing the values of x and y, we have $(1, 2)$, $(1, -2)$, $(-1, 2)$, and $(-1, -2)$.

The solution set of the system is   The check is left for the student.

$\{(1, 2), (-1, 2), (1, -2), (-1, -2)\}$.

■

Some systems have solutions that are complex numbers. The graph on a Cartesian plane does not show these solutions.

EXAMPLE 3

Solve the systems: $\begin{cases} x^2 + 4y^2 = 30 & (1) \\ x^2 + y^2 = 3 & (2) \end{cases}$

$$\begin{array}{rl} x^2 + 4y^2 = 30 & (1) \\ x^2 + y^2 = 3 & (2) \\ \hline 3y^2 = 27 & \end{array}$$

Subtract equation (2) from equation (1) to eliminate the variable x.

$$y^2 = 9$$

$$y = \pm 3$$

To find the value of x:

$$x^2 + 36 = 30$$

Substitute $y = \pm 3$ in equation (1).

$$x^2 = -6$$

$$x = \pm i\sqrt{6}$$

There are no real solutions.

The graph verifies that no real solution exists.

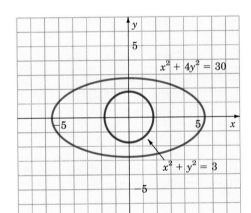

Conic sections can be found in some architectural forms.

EXAMPLE 4

A landscape architect wants to plant four pine trees 13 feet from the center of an ellipse but on the perimeter of the ellipse. If the elliptical garden is represented by the equation $4x^2 + 10y^2 = 1000$, where on the graph will he indicate the position of the trees? (*Hint:* If the trees are 13 feet from the center of the ellipse, they are on the circle whose equation is $x^2 + y^2 = 169$.)

$$\begin{cases} 4x^2 + 10y^2 = 1000 & (1) \\ x^2 + y^2 = 169 & (2) \end{cases}$$

To find the location of the pine trees, solve the system of equations. Multiply equation (2) by -4, and add to equation (1) to eliminate the variable x.

$$\begin{array}{r} 4x^2 + 10y^2 = 1000 \\ -4x^2 - 4y^2 = -676 \\ \hline 6y^2 = 324 \end{array}$$

$$y^2 = 54$$

Solve for y.

$$y = \pm 3\sqrt{6}$$

$$x^2 + (\pm 3\sqrt{6})^2 = 169$$

Substitute $y = \pm 3\sqrt{6}$ in equation (2).

$$x^2 + 54 = 169$$

$$x^2 = 115$$

$$x = \pm\sqrt{115}$$

The solution set of the system is

$$\{(\sqrt{115}, 3\sqrt{6}), (\sqrt{115}, -3\sqrt{6}), \\ (-\sqrt{115}, 3\sqrt{6}), (-\sqrt{115}, -3\sqrt{6})\}.$$

The check is left for the student.

The trees should be placed at the points whose coordinates are those given in the solution set.

The graph illustrates the solutions.

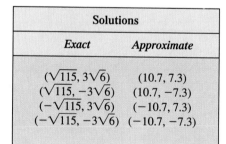

Solutions	
Exact	*Approximate*
$(\sqrt{115}, 3\sqrt{6})$	$(10.7, 7.3)$
$(\sqrt{115}, -3\sqrt{6})$	$(10.7, -7.3)$
$(-\sqrt{115}, 3\sqrt{6})$	$(-10.7, 7.3)$
$(-\sqrt{115}, -3\sqrt{6})$	$(-10.7, -7.3)$

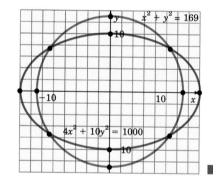

EXERCISE 8.5

A

Solve:

1. $\begin{cases} x^2 + y^2 = 12 \\ x^2 - y^2 = 4 \end{cases}$

2. $\begin{cases} x^2 + y^2 = 9 \\ x^2 - y^2 = 9 \end{cases}$

3. $\begin{cases} x^2 + 4y^2 = 16 \\ y^2 - x^2 = 4 \end{cases}$

4. $\begin{cases} 4x^2 + y^2 = 16 \\ x^2 - y^2 = 4 \end{cases}$

5. $\begin{cases} x^2 + y^2 = 4 \\ 4x^2 + y^2 = 4 \end{cases}$

6. $\begin{cases} x^2 + y^2 = 1 \\ x^2 + 9y^2 = 9 \end{cases}$

7. $\begin{cases} x^2 + y^2 = 4 \\ x^2 - 4y^2 = 4 \end{cases}$

8. $\begin{cases} x^2 + y^2 = 16 \\ y^2 - 25x^2 = 25 \end{cases}$

B

Solve:

9. $\begin{cases} 2x^2 + y^2 = 6 \\ 4x^2 + 3y^2 = 16 \end{cases}$

10. $\begin{cases} 3x^2 + y^2 = 12 \\ 4x^2 + 5y^2 = 49 \end{cases}$

11. $\begin{cases} x^2 + y^2 = 16 \\ x^2 - 4y^2 = 16 \end{cases}$

12. $\begin{cases} x^2 + y^2 = 25 \\ 25x^2 + y^2 = 25 \end{cases}$

13. $\begin{cases} x^2 + y^2 = 9 \\ y = x^2 + 3 \end{cases}$

14. $\begin{cases} x^2 + y^2 = 16 \\ y = -x^2 - 4 \end{cases}$

15. $\begin{cases} 3x^2 + 3y^2 = 36 \\ 2x^2 + 2y^2 = 4 \end{cases}$

16. $\begin{cases} x^2 + y^2 = 4 \\ 9x^2 + 25y^2 = 225 \end{cases}$

C

Solve:

17. $\begin{cases} 3x^2 + 4y^2 = 52 \\ 4x^2 - 5y^2 = 59 \end{cases}$

18. $\begin{cases} 2x^2 - 5y^2 = -37 \\ 4x^2 + 3y^2 = 43 \end{cases}$

19. $\begin{cases} x^2 - 4y^2 = 19 \\ x^2 - \dfrac{1}{2}y^2 = 26 \end{cases}$

20. $\begin{cases} x^2 - 9y^2 = 2 \\ x^2 - 4y^2 = 12 \end{cases}$

21. $\begin{cases} x^2 + y^2 = 2 \\ y = x^2 \end{cases}$

22. $\begin{cases} x^2 + y^2 = 6 \\ y = x^2 \end{cases}$

23. $\begin{cases} x^2 + y^2 = 10 \\ x^2 = 2y - 5 \end{cases}$

24. $\begin{cases} x^2 + y^2 = 32 \\ x^2 = 5y - 4 \end{cases}$

D

25. In Example 4, if the garden was represented by the equation $4x^2 + y^2 = 400$, where would the pine trees be placed?

26. An inner courtyard has two opposite sides that are hyperbolic.

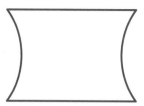

If benches are to be placed on the hyperbolic sides and exactly 12 ft from the center of the courtyard, designate where the benches should be placed if the sides are part of the hyperbola whose equation is $4x^2 - y^2 = 336$.

27. The sum of the squares of two numbers is 113. When 5 times the square of one is added to the square of the other, the sum is 309. What are the numbers?

28. The sum of the squares of two numbers is 225. The difference in these squares is 63. What are the numbers?

29. An orbiting body follows the path $4x^2 + y^2 = 16$ (where all the units are in 100,000 miles). A comet follows the path $y = x^2 - 4$. Will the two paths intersect? Where?

30. The position of one orbiting body moving in a circular path is given by $x^2 + y^2 = 25$. Another object is moving in an elliptical path given by $4x^2 + 3y^2 = 36$. Will the paths intersect? Where?

31. The sum of the squares of two positive numbers is 90. If the difference of their squares is 72, find the two positive numbers.

32. The logo for a corporation is designed to be the two curves $x^2 + y^2 = 25$ and $x^2 + 4y^2 = 100$. Do the two curves intersect at just two points as the designer wants? Where?

33. Two circular gears (one smaller than the other) are designed to touch at exactly one point. If their equations are $x^2 + y^2 = 25$ and $x^2 + y^2 - 14x + 45 = 0$, will they fit the requirement? Where?

STATE YOUR UNDERSTANDING

34. Explain the different ways that a circle and a hyperbola would not intersect at any points. Write the equations for such a system.

35. Explain the different ways that a circle and a parabola would intersect at just one point. Write the equations for such a system.

36. Can a circle and an ellipse intersect at just one point with one being inside the other? Explain how this would or could not happen.

CHALLENGE EXERCISES

37. Graph the "logo" in Exercise 32 to verify your points of intersection.

38. If concentric circles are circles with the same centers, is it possible for two concentric circles to intersect? Write a system to demonstrate your answer?

39. Solve: $\begin{cases} x^2 + y^2 = 17 \\ \quad\ xy = 4 \end{cases}$

40. Solve: $\begin{cases} x^2 \quad\ + y^2 = 25 \\ x^2 - xy + y^2 = 13 \end{cases}$

MAINTAIN YOUR SKILLS (SECTIONS 3.2, 3.3, 3.4, 7.3, 7.5, 7.6)

Combine. Reduce if possible:

41. $\dfrac{5}{x+3} - \dfrac{x}{x-5} + \dfrac{x^2-2}{x^2-2x-15}$

42. $\dfrac{y}{y+1} - \dfrac{y}{1-y} - \dfrac{y^2}{y^2-1}$

Simplify. Reduce if possible:

43. $\dfrac{x^2+x-30}{2x^2+13x-7} \cdot \dfrac{3x^2+22x+7}{5x^2-31x+30} \cdot \dfrac{10x^2+7x-6}{x^2+7x+6}$

44. $\dfrac{2x+4}{6x+2x^2} \cdot \dfrac{4x-x^2}{2x+x^2} \div \dfrac{16-4x}{18+6x}$

Simplify:

45. $\dfrac{\dfrac{1}{x-1} - \dfrac{1}{x+2}}{\dfrac{1}{x+2} + \dfrac{1}{x-1}}$

46. $\dfrac{\dfrac{x^2}{x^2-y^2} - 1}{\dfrac{xy}{y-x} + y}$

Solve by any method:

47. $\begin{cases} x + y + z = 8 \\ x - y + 2z = 12 \\ 2x + 4y - z = 6 \end{cases}$

48. $\begin{cases} 2x - y = 1 \\ 3y - 2z = 3 \\ 3x - z = 3 \end{cases}$

CHAPTER 8

SUMMARY

The properties of conic sections make them useful curves in a variety of applications, including applications in engineering, astronomy, architecture, and rocketry.

Conic Sections	A conic section is a geometric figure formed by the intersection of a plane and a cone. The more important curves are the circle, ellipse, parabola, and hyperbola.	(p. 546)

PARABOLA

Equations

$$y = ax^2 + bx + c = 0, a \neq 0 \text{ or } x = ay^2 + by + c = 0, a \neq 0 \qquad \text{(pp. 547, 556)}$$

Vertex

The vertex of a parabola with equation $y = ax^2 + bx + c = 0$ is at (h, k), where $h = -\dfrac{b}{2a}$ and $k = \dfrac{4ac - b^2}{4a}$. (pp. 550, 551)

The vertex is a maximum point if a is positive and a minimum point if a is negative.

Axis of Symmetry

The axis of symmetry of the parabola is the vertical line $x = h$. (pp. 550, 551)

CIRCLE

Equation with Center at the Origin

$$x^2 + y^2 = r^2, r \geq 0 \qquad \text{(p. 566)}$$

Center

$(0, 0)$

Radius

r

Equation with Center Not at Origin

$$(x - h)^2 + (y - k)^2 = r^2, r \geq 0 \qquad \text{(p. 566)}$$

Center

(h, k)

Radius

r

ELLIPSE

Equation

$$\frac{x^2}{a^2} + \frac{y^2}{b^2} = 1, \text{ where } a > 0 \text{ and } b > 0 \qquad \text{(p. 575)}$$

Boundary

The ellipse is inscribed in the rectangle with vertices at (a, b), $(-a, b)$, $(a, -b)$, and $(-a, -b)$. (pp. 575, 576)

HYPERBOLA

Equation

$$\frac{x^2}{a^2} - \frac{y^2}{b^2} = 1 \text{ or } \frac{y^2}{b^2} - \frac{x^2}{a^2} = 1, \text{ where } a > 0 \text{ and } b > 0 \qquad \text{(p. 578)}$$

Boundary

The hyperbola has two asymptotes that are the extended diagonals of the rectangle with vertices at (a, b), $(-a, b)$, $(a, -b)$, and $(-a, -b)$. (p. 579)

x- and *y*-intercepts	For all curves, the *x*-intercepts are the points that have *y*-coordinate zero, and the *y*-intercepts are the points that have *x*-coordinate zero.
Conic Section with Equation $Ax^2 + By^2 = C$	**1.** If $A = B$ and A, B, and C have the same sign, the equation can be rewritten in the standard form for a circle. (p. 582)
	2. If $A \neq B$ and A, B, and C have the same sign, the equation can be rewritten in the standard form for an ellipse.
	3. If A and B are opposite in sign, the equation can be rewritten in the standard form for a hyperbola.
Systems of Nonlinear Equations in Two Variables	Systems of nonlinear equations in two variables can be solved by graphing, by substitution, or by combinations. Solving by substitution or by combinations is similar to that for systems of linear equations. The graph of the system is a visual illustration of the solution. (pp. 590, 598)

CHAPTER 8
REVIEW EXERCISES

SECTION 8.1   Objectives 1–5

Find the vertex, determine whether the vertex is a maximum or a minimum point, find the *x*- and *y*-intercepts, determine the equation of the axis of symmetry, and graph.

1. $y = x^2 + 4$ **2.** $y = -x^2 + 4$ **3.** $y = x^2 - 2x + 4$

4. $y = x^2 + 2x - 2$

5. $y = -\dfrac{1}{4}x^2 - 2x - 6$

6. $y = \dfrac{1}{4}x^2 - 2x + 7$

7. $y = x^2 - 7x + 12$

8. $y = x^2 + 2x - 10$

9. $y = -2x^2 + 3x + 2$

10. $y = 3x^2 + 13x + 4$

SECTION 8.1   Objective 6

Sketch the graph of each of the following.

11. $x = -y^2 + 5$ **12.** $x = y^2 - 3$ **13.** $x = y^2 - 4y$

14. $x = -y^2 + 5y$ **15.** $x = y^2 - 3y - 10$ **16.** $x = -y^2 + 5y - 6$

SECTION 8.2   Objective 1

Write the coordinates of the center and the length of the radius, and graph each of the following equations whose graph is a circle.

17. $x^2 + y^2 = 16$ **18.** $x^2 + y^2 = 5$ **19.** $(x + 3)^2 + (y - 1)^2 = 4$

20. $(x - 5)^2 + (y - 2)^2 = 9$ **21.** $(x + 4)^2 + (y + 2)^2 = 16$ **22.** $(x - 1)^2 + (y + 3)^2 = 25$

23. $(x - 3)^2 + (y - 7)^2 = 4$ **24.** $(x - 5)^2 + (y - 1)^2 = 16$ **25.** $x^2 + (y - 4)^2 = 16$

26. $(x - 1)^2 + y^2 = 25$

SECTION 8.2   Objective 2

Write the equation in standard form of the circle with given coordinates as the center and the given value of r as the radius.

27. $(0, 0), r = \dfrac{1}{2}$ **28.** $(0, 0), r = \dfrac{4}{3}$

29. $(3, -4), r = 5$ **30.** $(2, 0), r = 5$

31. $(-6, -6), r = 4$ **32.** $(7, -8), r = 10$

33. $(-11, -12), r = 7$ **34.** $(-15, 15), r = 7$

35. $(-3, -8), r = \dfrac{2}{5}$ **36.** $(0, -8), r = \dfrac{3}{2}$

SECTION 8.2   Objective 3

Each of the following is the equation whose graph is a circle. Write each in standard form, and write the coordinates of the center and the length of the radius.

37. $4x^2 + 4y^2 = 144$ **38.** $3x^2 + 3y^2 = 90$

39. $x^2 + y^2 + 6y = 0$ **40.** $x^2 + y^2 + 14x = 0$

41. $x^2 + y^2 - 4y - 12 = 0$

42. $x^2 + y^2 - 6x - 7 = 0$

43. $x^2 + y^2 + 14x - 6y + 54 = 0$

44. $x^2 + y^2 - 10x - 2y - 10 = 0$

45. $x^2 + y^2 - 2x + 12y - 63 = 0$

46. $x^2 + y^2 - 8x - 14y + 64 = 0$

SECTION 8.3   Objective 1

Sketch the graph of each of the following.

47. $x^2 + 4y^2 = 16$

48. $9x^2 + y^2 = 36$

49. $\dfrac{x^2}{4} + \dfrac{y^2}{1} = 1$

50. $\dfrac{x^2}{1} + \dfrac{y^2}{9} = 1$

51. $\dfrac{x^2}{25} + \dfrac{y^2}{4} = 1$

52. $7x^2 + 28y^2 = 175$

SECTION 8.3   Objective 2

Sketch the graph of each of the following.

53. $x^2 - y^2 = 16$

54. $y^2 - x^2 = 36$

55. $\dfrac{x^2}{4} - \dfrac{y^2}{1} = 1$

56. $\dfrac{x^2}{1} - \dfrac{y^2}{9} = 1$

57. $\dfrac{y^2}{4} - \dfrac{x^2}{25} = 1$

58. $8y^2 - 32x^2 = 200$

SECTION 8.3   Objective 3

Write the equation in standard form of each ellipse and hyperbola.

59.

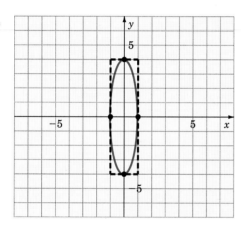

60.

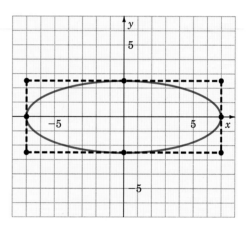

61.

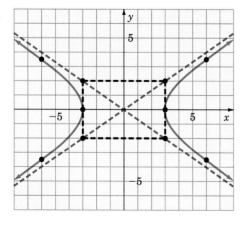

62.

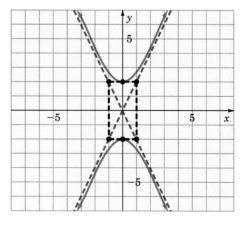

SECTION 8.4   Objective 1

Solve.

63. $\begin{cases} y = -x^2 \\ y = \dfrac{1}{2}x \end{cases}$

64. $\begin{cases} y = x^2 + 7 \\ y = -2x + 31 \end{cases}$

65. $\begin{cases} x^2 + y^2 = 16 \\ x + y = 2 \end{cases}$

66. $\begin{cases} x^2 - 9y^2 = 80 \\ 3x + y = 0 \end{cases}$

67. A rectangle has an area of 189 sq ft and a perimeter of 60 ft. What are its dimensions?

SECTION 8.5   Objective 1

Solve.

68. $\begin{cases} x^2 + y^2 = 35 \\ x^2 - y^2 = 15 \end{cases}$

69. $\begin{cases} 4x^2 + 5y^2 = 20 \\ 2x^2 + y^2 = 10 \end{cases}$

70. $\begin{cases} x^2 + 4y^2 = 144 \\ x^2 + y^2 = 3 \end{cases}$

71. $\begin{cases} x^2 + y^2 = 36 \\ x^2 - 9y^2 = 36 \end{cases}$

72. The sum of the squares of two numbers is 169. When eight times the square of one is added to the square of the other, the sum is 344. What are the numbers?

CHAPTER 8
TRUE–FALSE CONCEPT REVIEW

Check your understanding of the language of algebra. Tell whether each of the following statements is true (always true) or false (not always true).

1. A conic section is a geometric figure formed by the intersection of a plane with a cone.

2. An equation of the form $Ax^2 + By^2 = C$, $A \neq B$, has an ellipse for a graph.

3. The radius of a circle is two times the diameter.

4. A hyperbola has two maximum or minimum points.

5. The quadratic equation $y = ax^2 + bx + c$ has a parabola for a graph and defines a function.

6. The quadratic equation that has a circle for a graph defines a function.

7. The vertex of a parabola is also the maximum or minimum point of the graph.

8. $x^2 - y^2 = r^2$, $r \geq 0$, and $(x - h)^2 - (y - k)^2 = r^2$, $r \geq 0$, are the standard forms of the equation of a circle.

9. A system of equations that contains one quadratic and one linear will always have two solutions.

10. It is possible for a system of quadratic equations to have four solutions.

11. The equation of the axis of symmetry of the graph of a parabola is $x = -\dfrac{b}{a}$.

12. The graphs of all parabolas represent functions.

CHAPTER 8
TEST

1. Find the vertex of the graph of $y = x^2 - 12x - 8$, and determine if it is a maximum or a minimum point of the graph.

2. Sketch the graph of $x^2 + 4y^2 = 4$.

3. Find the x-intercepts of the parabola
$y = -2(x - 2)^2 + 2$.

4. Sketch the graph of $y = x^2 - 6x - 3$.

5. Solve the system by combinations:
$$\begin{cases} x^2 - y^2 = 4 \\ 4x^2 + y^2 = 16 \end{cases}$$

6. Find the equation of the axis of symmetry of the graph of $y = x^2 + 10x + 21$.

7. Write the equation in standard form of a circle with radius 7 and center at the origin.

8. Write the equation in standard form of a circle that has its center at $(2, -1)$ and a radius of 6.

9. Find the x-intercepts of the parabola
$y = x^2 - 2x - 2$.

10. Solve the system by substitution: $\begin{cases} x - y = 1 \\ x^2 - 8y = 8 \end{cases}$

11. Write the coordinates of the center and the length of the radius, and graph the circle whose equation is $x^2 + y^2 = 81$.

12. Find the equation of the axis of symmetry of the graph of $y = x^2 - 8x - 3$.

13. Sketch the graph of $x^2 - 4y^2 = 4$.

14. Find the vertex of the graph of
$y = -3(x - 7)^2 - 4$, and determine if it is a maximum or a minimum point of the graph.

15. Sketch the graph of $9x^2 + y^2 = 36$.

16. Write the equation in standard form of a circle whose center is at $(2, 1)$ and radius is 8.

17. Sketch the graph of $y = x^2 - 4x$.

19. Solve the system by substitution: $\begin{cases} x + y = 1 \\ x^2 - 4y = 8 \end{cases}$

20. Write the coordinates of the center and the length of the radius of the circle defined by $x^2 + 6x + y^2 - 10y + 25 = 0$.

18. Sketch the graph of $x = y^2 + 6y + 5$.

9

FUNCTIONS

An elephant-ear bamboo plant grows quite rapidly during the spring growing season. People often say they can see it grow. Scientists try to represent growing patterns mathematically so that the growth might be predicted. For this bamboo plant the growth in height was estimated by the formula $h = \sqrt{\dfrac{1}{2}d^2 + 4}$, where d represents the number of days of growth after the plant has reached a height of 4 ft. In Exercise 65, Section 9.1, you are asked to find the estimated height after 4, 6, 10, and 15 subsequent growth days. *(David R. Frazier/Photolibrary)*

PREVIEW

Relations and functions are defined and then symbolized using set notation with related ranges and domains. Functional notation is defined and is used throughout. Operations with functions, that is, addition, subtraction, multiplication, division, and composition of functions are discussed at great length. The inverse of a relation and one-to-one functions are defined. The inverse of one-to-one functions is applied in the study of exponential functions and logarithmic functions in Chapter 10.

9.1

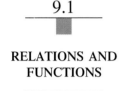

RELATIONS AND FUNCTIONS

OBJECTIVES

1. Write the ordered pairs of a relation, and state the domain and range.

2. Determine if a relation is a function, and state the domain and range.

3. Sketch the graph of a function.

Relationships between pairs of numbers are used to describe a variety of situations. Some examples are shown in the following chart.

Value	Related Value
Income	Tax based on income
Number of cars parked	Income from parking fee
Weight of item	Cost to mail an item
Hours worked	Wage
Dollars saved	Interest earned

In the last entry of the chart, if you have a $10,000 savings account at 7.5% annual interest the yearly interest earned is $750. This relationship can be written as an ordered pair (savings, interest) or ($10,000, $750). Another ordered pair in the same relationship, ($1500, $112.50), tells us that if $1500 is saved, the annual earnings will be $112.50. The interest depends on the amount of savings. The study of number pairs and how they are paired is the topic of this chapter.

Any pairing of two real numbers can be written as an ordered pair, (x, y).

Relation

■ DEFINITION

A *relation* is a set of ordered pairs of numbers.

For example, the set $\{(3, 4), (5, -2), (-6, 1), (0, 0)\}$ is a relation.

A relation can be defined or described by:

1. Listing the set of ordered pairs.

2. Writing an equation or rule that will yield each of the ordered pairs.

3. Drawing the graph that pictures the ordered pairs.

4. Writing a word description to tell how the numbers are paired.

Recall that the set of ordered pairs defined by the linear equation $y = 3x + 1$ is written $\{(x, y) \mid y = 3x + 1\}$.

■ DEFINITION

Domain

The set of first components (x-values) of the ordered pairs of a relation is called the *domain* of the relation.

Range

The set of second components (y-values) of the ordered pairs of a relation is called the *range* of the relation.

EXAMPLE 1

State the domain and the range of the relation,
$R = \{(2, 0), (3, -1), (-1, 6), (-2, 8)\}$.

Domain of R: $\{2, 3, -1, -2\}$ The set of first numbers (x-values) of the ordered pairs.

Range of R: $\{0, -1, 6, 8\}$ The set of second numbers (y-values) of the ordered pairs. ■

Frequently a relation is given in set builder notation with a specified domain. In these cases, we can find the ordered pairs of the relation by substituting values for x and solving for y. The range is then obtained from the ordered pairs.

EXAMPLE 2

Write the relation, $Q = \{(x, y) \mid y = |x|, x = 2, 1, 0, -1, -2\}$ as a set of ordered pairs, and state the range.

If $x = 2$, $y = |2| = 2$ Find a value of y for each value of x.

$x = 1$, $y = |1| = 1$

$x = 0$, $y = |0| = 0$

$x = -1$, $y = |-1| = 1$

$x = -2$, $y = |-2| = 2$

$Q = \{(2, 2),\ (1, 1),\ (0, 0),\ (-1, 1),\ (-2, 2)\}$

Range: $\{2, 1, 0\}$ Set of y-values   ■

A special type of relation that is very useful is a relation in which there is just one value in the range paired with each value in the domain. These relations are called functions.

■ DEFINITION

Function

> A *function* is a relation in which for each x-value there is exactly one y-value.

The equation $y = 3x$ defines a function because for each value of x there is only one value of y. The equation $y^2 = 3x$ does not define a function because when $x = 3$ then $y = \pm 3$, so the relation contains the ordered pairs $(3, -3)$ and $(3, 3)$, that is, two values of y when $x = 3$.

EXAMPLE 3

Does $y = |x^2 - 3|$ define a function?

Yes, a function   If x is a real number, then $x^2 - 3$ is a real number by the closure properties. Also, the absolute value of a real number is real and unique.   ■

EXAMPLE 4

Does $|y - 3| = x^2$ define a function?

No, not a function   If x is a real number, then x^2 is a real number. There are two numbers whose absolute value is a positive real number. For example, if $x = 2$, $x^2 = 4$, and $y = -1$ or $y = 7$.   ■

The graphs of the relations shown in Figure 9.1 can be identified as those that are graphs of functions and those that are not. Since a function can have only one ordinate value (y-value), for each abscissa value (x-value), we need to check whether some vertical line will intersect the graph at more than one

Vertical Line Test

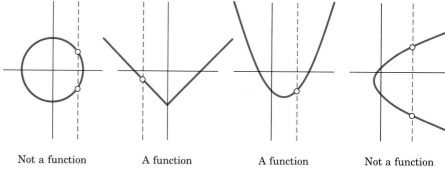

| Not a function | A function | A function | Not a function |

FIGURE 9.1

point. If it does, the relation has two y-values for the same x-value and therefore is not the graph of a function.

The following table shows different ways of describing a function.

Given	Conclusion
A value of x	There is exactly one value of y
Two different ordered pairs	x values are different
Graph of function	All vertical lines (that represent a value of x) intersect the graph in at most one point
A domain number	There is only one range number paired with it

■ PROCEDURE

Function Identification

To identify whether a relation is a function:

1. Solve for y. Each value assigned to x can yield only one value for y.

2. Draw the graph of the relation, and use the vertical line test on the graph. If it is a function, no vertical line can intersect the graph at more than one point.

If the domain of a function is not explicitly stated, it is determined by the rule (usually an equation). The domain is understood to consist of the largest set of real numbers for which y is also a real number. The range is then determined from the rule for pairing x and y.

For $\{(x, y)\,|\,y = |x|\}$, the domain is the set of all real numbers, $\{x\,|\,x$ is real$\}$. The range is the set of all real numbers greater than or equal to zero since $|x|$ cannot be negative, $\{y\,|\,y \geq 0\}$.

■ CAUTION

Unless explicitly stated, in this text, the domain of a function is the largest set of real numbers for which y is also a real number.

EXAMPLE 5

State the domain and the range of $\{(x, y)\,|\,y = \sqrt{3 - x}\}$.

Domain: $\{x\,|\,x \leq 3\}$ We restrict x to values less than or equal to 3 since the principal square root of a negative number is not real.

Range: $\{y\,|\,y \geq 0\}$ The range is nonnegative since a principal square root cannot be negative. ■

EXAMPLE 6

State the domain and range of $\left\{(x, y)\,\middle|\,y = \dfrac{1}{x - 2}\right\}$.

Domain: $\{x\,|\,x \neq 2\}$ If $x = 2$, the denominator is zero, and the division is not defined.

Range: $\{y\,|\,y \neq 0\}$ $\dfrac{1}{x - 2} \neq 0$ since a fraction cannot be zero unless the numerator is zero. ■

Graphs of many functions defined by equations can be sketched by listing a set of ordered pairs in the relation and then connecting them with a smooth curve.

EXAMPLE 7

Sketch the graph of $y = |x - 1| - 4$.

x	y
-4	1
-3	0
-2	-1
-1	-2
0	-3
1	-4
2	-3
3	-2
4	-1

First, find some of the ordered pairs in the relation.

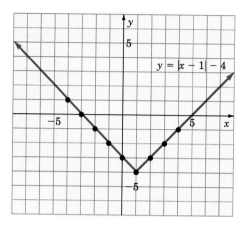

Now plot these points and draw the graph.

By the vertical line test, we see that the graph represents a function.

Domain: $\{x \mid x \in R\}$

There is a point corresponding to every point on the x-axis.

Range: $\{y \mid y \geq -4\}$

The lowest point of the graph is at $y = -4$. ∎

We can use any of the following names for the numbers in the ordered pair (x, y) of a function.

x(domain)	y(range)
Independent variable	Dependent variable
Abscissa	Ordinate
x-value	y-value
x-coordinate	y-coordinate
First component	Second component
First coordinate	Second coordinate
Directed distance from y-axis	Directed distance from x-axis

EXERCISE 9.1

A

State the range and domain of each relation:

1. $\{(1, 1), (2, 4), (3, 9), (4, 16)\}$

2. $\{(-4, 4), (-3, 4), (-2, 4), (-4, 4)\}$

3. $\{(-7, -10), (-3, 4), (0, 7), (3, 8)\}$

4. $\{(0, 0), (-1, 1), (-2, -8), (-3, -27), (-4, -64)\}$

5. $\{(-4, 8), (4, 8), (5, 8), (0, 8), (1, 8)\}$

6. $\{(5, 0), (5, -7), (5, 1), (5, 6), (5, -11)\}$

Write each of the following relations as a set of ordered pairs, and state the range:

7. $\{(x, y) \mid y = 2x^2 - 6x + 1, x = -4, -2, 0, 2, 4\}$

8. $\{(x, y) \mid y = x^3 - 3x, x = -2, 0, 2, 3\}$

9. $\{(x, y) \mid y = |5x - 6| + 4, x = -3, -1, 1, 3\}$

10. $\{(x, y) \mid y = 16 - x^2, x = -6, -5, -4, -3\}$

11. $\{(x, y) \mid y = |16 - 2x| + 1, x = 2, 3, 5, 8, 10\}$

12. $\{(x, y) \mid y = \sqrt{16 - 2x}, x = 2, 3, 5, 8\}$

From the graphs, state the range and domain of each relation. Identify those that are functions:

13.

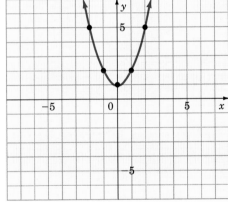

14.

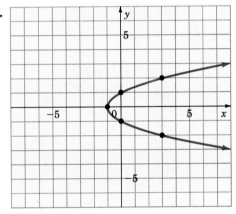

15.

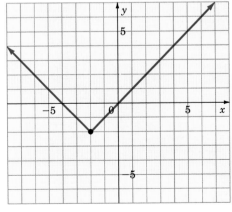

16.

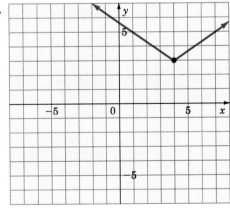

17.

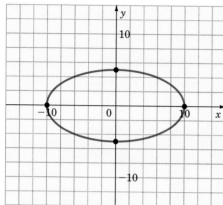

18.

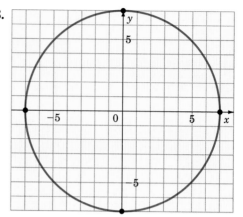

19.

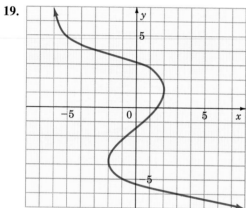

20.

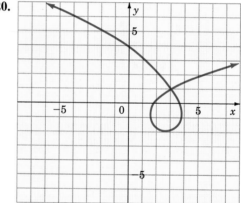

B

State the range and domain of the relation defined by each of the following, and identify those that are functions:

21. $y = \sqrt{100 - x^2}$

22. $y = \dfrac{x}{x - 3}$

23. $y^2 = 3x$

24. $y^2 = 3x - 2$

25. $y^2 = 3x^2 + 1$

26. $y = \sqrt{x^2 - 5}$

27. $y = |x|$

28. $y = |x + 3|$

29. $y = |x| - 7$

30. $y = |x + 3| - 7$

31. $y = \sqrt{x}$

32. $y = \sqrt{x} - 2$

33. $y = x^3$

34. $y = x^4$

35. $x = |y| + 1$

36. $x = \sqrt{y} - 2$

37. $y = \dfrac{1}{x}$

38. $x = \dfrac{1}{y}$

39. $y = \dfrac{2x}{x + 5}$

40. $y = \dfrac{x}{x - 6}$

C

State the range and domain of the relation defined by each of the following and identify those that are functions.

41. $y = \dfrac{x - 2}{x^2 - 9}$

42. $y = \dfrac{x + 4}{x^2 - 1}$

43. $y = \dfrac{3}{x^2 + 4}$

44. $y = \dfrac{-5}{x^2 + 1}$

45. $y = \dfrac{x - 7}{x^2 - x - 6}$

46. $y = \dfrac{x + 2}{x^2 + 5x + 4}$

47. $y < 2x - 5$

48. $y \geq x^2 + 1$

Draw the graph of each of the following. Use the graph to determine the range and domain. Use the vertical line test to identify functions:

49. $y = |x - 5|$

50. $y = |x| + 5$

51. $y = \sqrt{x} - 4$

52. $y = \sqrt{x - 4}$

53. $y = |3x - 6| - 1$

54. $y = |x - 4| + 3$

55. $y = \dfrac{1}{2}x^2 - 6$

56. $y = \dfrac{1}{2}x^2 - 3$

57. $x^2 + y^2 = 16$

58. $x^2 + y^2 = 25$

59. $x^2 - y^2 = 4$

60. $x^2 - 4y^2 = 16$

61. $4x^2 + y^2 = 16$

62. $x^2 + 9y^2 = 36$

D

63. The number of pairs of eyeglasses that the optical laboratory can produce is improved when it updates its equipment. The relationship between the number of pairs of glasses (G) and employees (n) is now $G = 2n^2 + 3n - 5$. Find the number of pairs of glasses produced by 5, 7, 8, or 10 employees. Express the answers as ordered pairs (n, G).

64. The number of articles (G) that the Big Top manufacturing firm can produce in a day is related to the number of employees (n) by the equation $G = n^3 - n^2 + 3$. Find the number of articles produced by 5, 6, 7, or 8 employees. Express the answers as ordered pairs (n, G).

65. During the growing season, the height (h) (in feet) of an elephant-ear bamboo is related to the number of good growing days (d) by the formula $h = \sqrt{\dfrac{1}{2}d^2 + 4}$. Find the height of the bamboo after 4, 6, 10, or 15 growing days. Express the answers as ordered pairs, rounding the height to the nearest tenth of a foot.

66. If fertilizer is applied, the bamboo in Exercise 65 grows by the formula $h = \sqrt{\dfrac{2}{3}d^2 + 5}$. Find the height of the bamboo after 4, 6, 10, and 15 growing days. Express the answers as ordered pairs, rounding the height to the nearest tenth of a foot.

67. The potential energy when a certain spring is compressed "x" millimeters is given by P.E. $= 5x^2$. What is the potential energy when the spring is compressed 20 millimeters? What is the P.E. when the spring is compressed 30.5 mm? Express answers as ordered pairs (x, P.E.).

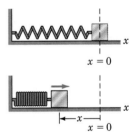

Figure for Exercise 67

68. The voltage (V) in a circuit is a function of the current (I) according to: $V = \dfrac{10}{I}$. Find V if $I = 2$ amps. Find V if $I = 0.5$ amps. Express answers as ordered pairs (I, V).

69. The height of a ball thrown from the top of a 100 ft building is given by $h = -16t^2 + 40t + 100$. What is the height 3 seconds after the ball is thrown? What is the height 4 seconds after the ball is thrown? Express answers as ordered pairs (t, h).

70. The demand (q_d) for a certain product is a function of the price at which it is sold given by: $q_d = -p^2 + 4p + 100$. What is the demand if the price is $10? What is the demand if the price is $12? Express answers as ordered pairs (p, q_d).

71. The number of cases (C) of mumps in a given school is given by $C = D^2 + 4D$, where D is the number of days that have passed since the first reported case. How many cases are there after 4 days? After 6 days? Express answers as ordered pairs (D, C).

72. The resistance (r) in a circuit after t milliseconds (ms) is given by $r = 8t^2 + 25$. Find r, in ohms, when t is 15 ms. Find r, in ohms, when t is 2.5 ms.

STATE YOUR UNDERSTANDING

73. What is the difference between a function and a relation? What is meant by the domain and the range of a function?

74. Is the graph of every straight line a function? Explain.

75. Is the graph of every circle not a function? Explain.

76. Is the graph of every hyperbola not a function? (Do you remember drawing the graph of $xy = 4$ in a previous chapter?) Explain.

CHALLENGE EXERCISES

77. Find two ordered pairs for the equation $x^2 + y^2 = 25$ to prove that the graph of this circle is not a function.

78. The length of time (T) that it takes a pendulum to complete one swing is a function of its length (L) of the pendulum and given by $T = 2\pi\sqrt{\dfrac{L}{10}}$. Find the time for one complete swing if $L = 2$ feet. (T is in seconds) Find the length of time if $L = 40$ feet. Round answer to the nearest tenth of a second.

79. Find the domain of the function $y = \sqrt{\dfrac{x}{x-1}}$. Careful!

MAINTAIN YOUR SKILLS (SECTIONS 6.2, 6.3, 7.1)

80. Find the slope of the line through the points $(6, -5)$ and $(-3, -2)$.

81. Find the equation, in standard form, of the line through the points $(6, -5)$ and $(-3, -2)$.

82. Find the distance between the points $(6, -5)$ and $(-3, -2)$.

83. Find the x-coordinate of a point, $(x, 4)$, that is 10 units' distance from the point $(-3, -2)$.

84. Find the equations of both lines that are perpendicular to the line segment joining $(2, 5)$ and $(4, -6)$ and that pass through the endpoints of the segment.

Solve:

85. $\begin{cases} y = x - 7 \\ 2x + 3y - 4 = 0 \end{cases}$

86. $\begin{cases} x = y - 7 \\ 2x + 3y - 4 = 0 \end{cases}$

87. $\begin{cases} 4x - 8y = -32 \\ x + 6y = -9 \end{cases}$

9.2

FUNCTION NOTATION

OBJECTIVES

1. Given a value of x, find $f(x)$.

2. Evaluate expressions containing function notation.

The functions we study in algebra are sets of ordered pairs of numbers. To claim that y is "a function of x" or in symbols "$f(x)$," we mean that for each value of x there is exactly one value of y. To communicate this in equation form, we write $y = f(x)$ so that we have two symbols for the same value: y and $f(x)$.

■ DEFINITION

> Function notation $y = f(x)$ is the algebraic form of the statement that y is a function of x (y depends on x). Consequently, x is the independent variable, and y[or $f(x)$] is the dependent variable.

■ CAUTION

> The symbol $f(x)$ does *not* mean f times x. This notation is used to refer to a functional relationship.

Here are some useful facts about this new symbolism:

1. The symbol $f(x)$ is often used in place of the variable y when writing the equation of a function.

2. The symbols $f(x)$, $g(x)$, $h(x)$, $F(x)$, $G(x)$, and $H(x)$ are the most common ones used to denote a function.

3. The symbol $f(x)$ is read "f at x" or "the value of f at x" or "f of x."

4. This symbol is used to denote that the second variable (y-value) is a function of x. The parentheses used for functions represent a new use of parentheses. It is a compound symbol for y that shows the x-value.

We now have two ways to write the ordered pairs of a function: $(x, f(x))$ or (x, y). Similarly, the function $\{(x, y) \,|\, y = x^2 - 4\}$ can also be written $\{(x, f(x)) \,|\, f(x) = x^2 - 4\}$. The y-value of the function associated with $x = 2$ is $f(2)$ and, since $2^2 - 4 = 0$, $f(2) = 0$.

The following table shows some other ordered pairs that belong to this function.

x	$f(x) = x^2 - 4$	$f(x)$	$(x, f(x))$
-2	$f(-2) = (-2)^2 - 4 = 0$	0	$(-2, 0)$
-1	$f(-1) = (-1)^2 - 4 = -3$	-3	$(-1, -3)$
0	$f(0) = (0)^2 - 4 = -4$	-4	$(0, -4)$
1	$f(1) = (1)^2 - 4 = -3$	-3	$(1, -3)$
2	$f(2) = (2)^2 - 4 = 0$	0	$(2, 0)$
3	$f(3) = (3)^2 - 4 = 5$	5	$(3, 5)$

The graph of the function (a parabola) is shown in Figure 9.2.

FIGURE 9.2

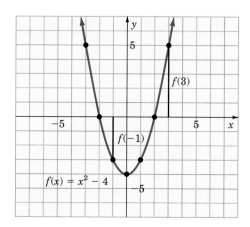

We see that $f(-1) = -3$; that is, the graph of the function is 3 units below the x-axis when $x = -1$. Furthermore, $f(3) = 5$; so the graph of the function is 5 units above the x-axis when $x = 3$.

EXAMPLE 1

Given $f(x) = \{(-3, 4), (-2, 4.2), (-1, -2), (0, 6), (1.2, 7), (2, -3), (3, -71)\}$, find $f(-1)$, $f(0)$, and $f(3)$.

$f(-1) = -2$ In the function, $x = -1$ is paired with $y = -2$.

$f(0) = 6$ In the function, $x = 0$ is paired with $y = 6$.

$f(3) = -71$ In the function, $x = 3$ is paired with $y = -71$. ■

EXAMPLE 2

Given $f(x) = 2x^2 - 3$, find $f(4)$, $f(-6)$, and $f(13)$.

$f(4) = 2(4)^2 - 3$ Replace x by 4.

    $= 29$

$f(-6) = 2(-6)^2 - 3$ Replace x by -6.

     $= 69$

$f(13) = 2(13)^2 - 3$ Replace x by h.

     $= 2(169) - 3$

     $= 335$ ■

EXAMPLE 3

Given $f(x) = x^2 - 3x + 4$, find $f(a)$ and $f(a + 2)$.

$f(a) = (a)^2 - 3(a) + 4$ Replace x by a.

    $= a^2 - 3a + 4$

$$f(a + 2) = (a + 2)^2 - 3(a + 2) + 4 \qquad \text{Replace } x \text{ by } a + 2.$$
$$= a^2 + 4a + 4 - 3a - 6 + 4$$
$$= a^2 + a + 2 \qquad \blacksquare$$

EXAMPLE 4

Given $f(x) = |x| - 8$ and $g(x) = 4 - x^2$, find $f(4) + g(11)$ and $f(-1) - g(2)$.

$$f(4) + g(11) = [|4| - 8] + [4 - 11^2]$$
$$= (-4) + (-117) = -121$$
$$f(-1) - g(2) = [|-1| - 8] - [4 - (2)^2]$$
$$= (-7) - (0) = -7 \qquad \blacksquare$$

EXAMPLE 5

Given $f(x) = 2x + 5$ and $g(x) = 3x + 2$, find $g[f(2)]$.

$f(2) = 2(2) + 5 = 9$	Working from inside the brackets find $f(2)$ by replacing x by 2 in $f(x)$.
$g[f(2)] = g(9)$	Substitute 9 for $f(2)$ in $g[f(2)]$.
$= 3(9) + 2$	Substitute 9 for x in $g(x)$.
$= 29$	Simplify. $\qquad \blacksquare$

EXERCISE   9.2

A

If $f(x) = \{(0.5, 2), (1, 3), (2, 5), (3, 7), (4, 9), (5, 11)\}$ and $g(x) = \{(0.5, 0.5), (1, 2), (2, 5), (3, 8), (4, 11), (5, 14)\}$, find the following:

1. $f(3)$ **2.** $g(3)$ **3.** $g(0.5)$ **4.** $f(4)$

5. $f(5)$ **6.** $f(0.5)$ **7.** $f(2) - f(1)$ **8.** $g(4) - f(5)$

9. $g(5) - f(4)$ **10.** $g(2) - f(5)$

Find the indicated values of $f(x)$:

11. $f(x) = 2x + 4$; $f(0), f(15)$ **12.** $f(x) = 6x - 1$; $f(-3), f(7)$

13. $f(x) = 16 - x$; $f(-6), f(6)$ **14.** $f(x) = 13 - x$; $f(-1), f(10)$

15. $f(x) = 5 - 3x$; $f(-1), f(2)$ **16.** $f(x) = 6 - 5x$; $f(-1), f(3)$

17. $f(x) = x^2 + x$; $f(-2), f(7)$ **18.** $f(x) = x^2 - 2x$; $f(-10), f(5)$

19. $f(x) = 2x^2 - x + 1$; $f(-3), f(0)$ **20.** $f(x) = 3x^2 + 2x - 5$; $f(1), f(0)$

B

If $f(x) = x^2 - 3x - 18$ and $g(x) = x - 6$, evaluate the following:

21. $f(6)$ **22.** $g(9)$ **23.** $g(6)$ **24.** $f(3)$

25. $f(g(9))$ **26.** $g(f(3))$ **27.** $f(f(-4))$ **28.** $f(g(0))$

29. $f(5) - g(4)$ **30.** $g(-1) - f(-3)$

If $f(x) = 7x + 13$, $g(x) = \sqrt{x + 9}$, and $h(x) = x^2 - 6x + 1$, evaluate the following:

31. $f(4) + g(7)$ **32.** $f(-5) + g(-5)$ **33.** $g(16) - h(-2)$ **34.** $f(12) - g(0)$

35. $g(9) + f(2)$ **36.** $g(18) + h(0)$ **37.** $f(2) - g(27)$ **38.** $h(5) - g(7)$

39. $f(g(0))$ **40.** $g(f(2))$

C

Given $f(x) = 6 - x$, $g(x) = x^2 - 4x + 7$, and $h(x) = |14 - 3x|$, evaluate the following:

41. $f(-27) + h(-23)$ **42.** $g(19) - h(18)$ **43.** $f(-12) - g(-4)$

44. $h(-11) - g(-8)$ **45.** $f(a)$ **46.** $g(a)$

47. $h(b)$ **48.**

50. $g(c) - h(b)$ **51.** $f(g(12))$ **52.** $g(f(12))$

53. $h(f(20))$ **54.** $h(g(10))$ **55.** $f(a^2 - 1)$

56. $g(a^2 - 1)$ **57.** $h(a^2 - 1)$ **58.** $f(a^2 + a)$

59. $g(a^2 + a)$ **60.** $h(a^2 + a)$

STATE YOUR UNDERSTANDING

61. What is the difference between $y = 2x$ and $f(x) = 2x$?

62. What is the advantage of the "$f(x)$" notation?

63. Are there any functions such that $f(a) + f(b) = f(a + b)$? Explain.

CHALLENGE EXERCISES

64. Given $f(x) = -x^3 + 3x^2 - 5x + 3$, find $f(0)$, $f(-1)$, and $f(2)$.

65. Given $f(x) = 2x - 3$, find $f(x + h)$. Substitute what you found into the expression $\dfrac{f(x + h) - f(x)}{h}$ and simplify.

66. Given $g(x) = -3x + 1$, find $g(x + h)$. Substitute what you found into the expression $\dfrac{g(x + h) - g(x)}{h}$ and simplify.

67. Given $f(x) = x^2 + 1$, find $f(x + h)$. Substitute what you found into the expression $\dfrac{f(x + h) - f(x)}{h}$ and simplify.

68. Given $F(x) = x^2 + 2x - 3$, find $F(x + h)$. Substitute what you found into the expression $\dfrac{F(x + h) - F(x)}{h}$ and simplify.

69. Given the two functions $f(x) = 2x - 3$ and $g(x) = 3x + 5$, we say that $(f + g)(x) = (2x - 3) + (3x + 5)$ which simplifies to $(f + g)(x) = 5x + 2$. Find $f(2)$, $g(2)$, then find $(f + g)(2)$. Is $f(2) + g(2) = (f + g)(2)$?

70. Given the two functions $f(x) = x^2 + 2x - 3$ and $g(x) = -3x - 5$, find $(f + g)(x)$ and simplify. Find $f(3)$, $g(3)$, then find $(f + g)(3)$. Is $f(3) + g(3) = (f + g)(3)$?

71. Given $h(x) = 3x$, find $h(2)$, $h(3)$, then find $h(2 + 3)$. (Note: this is not the same as problems 69 and 70!) Is $h(2) + h(3) = h(2 + 3)$?

72. Given $f(x) = 2x + 1$, find $f(2)$, $f(5)$, then find $f(2 + 5)$. Is $f(2) + f(5) = f(2 + 5)$?

73. Given $g(x) = x^2 + 1$, find $g(1)$, $g(3)$, then find $g(1 + 3)$. Is $g(1) + g(3) = g(1 + 3)$?

MAINTAIN YOUR SKILLS (SECTION 8.1)

Given: $y = -\dfrac{1}{2}x^2 + 6x - 16$

74. Find the vertex of the graph.   **75.** Find the axis of symmetry.

76. Find the x-intercepts.   **77.** Draw the graph.

Given: $y = 2x^2 + 8x - 3$

78. Find the vertex of the graph.   **79.** Find the axis of symmetry.

80. Find the x-intercepts.   **81.** Draw the graph.

9.3

CLASSIFICATION OF FUNCTIONS

OBJECTIVE

1. Classify a given function as a constant, linear, quadratic, cubic, polynomial, square root, absolute value, or reciprocal function, and graph the function.

In Chapters 6 and 8, the graphs of straight lines and parabolas are graphs of functions. The equations of these graphs are equations of functions since there is one value for y for each value of x. Here, some of these graphs are repeated and classified, and some new graphs are included.

■ **DEFINITION**

Linear Function

The function f is a linear function provided the equation of the function can be written in the form

$$f(x) = mx + b, \qquad m, b \text{ in } R \text{ and } m \neq 0.$$

EXAMPLE 1

Write $x + 2y = 6$ in the form $f(x) = mx + b$, and graph the function.

$$y = -\frac{1}{2}x + 3 \qquad\qquad \text{Solve for } y$$

$$f(x) = -\frac{1}{2}x + 3 \qquad\qquad \text{Replace } y \text{ by } f(x). \text{ This is a linear function with slope}$$
$$m = -\frac{1}{2} \text{ and } y\text{-intercept } b = 3.$$

The graph is drawn using the points $(0, 3)$, $(6, 0)$, and $(-6, 6)$.

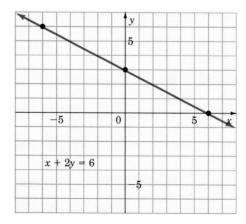

Functions of the form $y = k$ are called *constant functions* and have a graph that is a horizontal line. The linear function $y = x$ is called the *identity function*.

The equations of quadratic functions have the same standard form as the equations of parabolas that open up or down.

■ DEFINITION

Quadratic Function

> The function f is a quadratic function provided the equation of the function can be written in the form
>
> $f(x) = ax^2 + bx + c,$ a, b, c in R and $a \neq 0.$

EXAMPLE 2

Write $x^2 + y = 5$ in the form $f(x) = ax^2 + bx + c$, and graph the function.

$f(x) = -x^2 + 5$ Solve for y, and replace y by $f(x)$. This is a quadratic function with $a = -1$, $b = 0$, and $c = 5$.

The graph of this function is a parabola opening downward with vertex at $(0, 5)$.

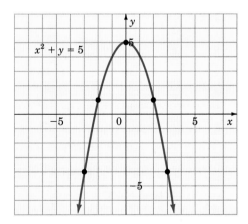

Linear, quadratic, and cubic functions are special cases of a set of functions called *polynomial functions*.

■ DEFINITION

Polynomial Function

> The function f is a polynomial function provided the equation of the function can be written in the form
>
> $f(x) = a_n x^n + a_{n-1} x^{n-1} + \cdots + a_1 x + a_0,$ $a_n, a_{n-1}, \ldots, a_1, a_0$ in R, $n \in N$ and $a_n \neq 0.$

An accurate graph of a polynomial function of degree three or more can be drawn with methods studied in calculus. However, a general idea of the graph can be obtained by plotting several points.

EXAMPLE 3

Graph the polynomial (cubic) function: $f(x) = x^3 + 3x^2 - 5$

x	-4	-3	-2	-1	0	1	2
$f(x)$	-21	-5	-1	-3	-5	-1	15

Choose enough values of x to get a general idea of the graph and find the corresponding values of $f(x)$.

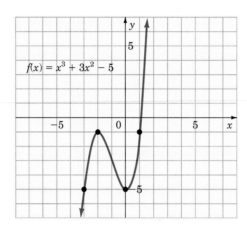

$$f(x) = x^3 + 3x^2 - 5$$

■ DEFINITION

Square Root Function

> The function f is a square root function provided the equation of the function can be written in the form
>
> $$f(x) = \sqrt{ax + b} + c, \qquad a \neq 0.$$

EXAMPLE 4

Graph the square root function: $f(x) = \sqrt{x + 4}$

x	-4	-3	-2	-1	0	1	2
$f(x)$	0	1	1.4	1.7	2	2.2	2.4

Choose enough values of x to get a general idea of the graph and find the corresponding values of $f(x)$. Some of the $f(x)$ values are approximations. For example, $f(1) = \sqrt{1 + 4} = \sqrt{5} \approx 2.2$.

■ **CAUTION**

In the formation of this table, we cannot use the values of x less than -4 since square roots of negative numbers are not real numbers.

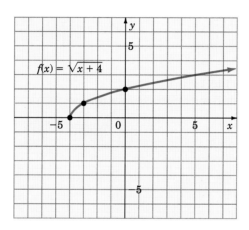

$f(x) = \sqrt{x} + 4$

Observe that the graph of a square root function is half the graph of a parabola that opens to the right or to the left.

■ **DEFINITION**

Absolute Value Function

The function f is an absolute value function provided the equation of the function can be written in the form

$$f(x) = |ax + b| + c, \qquad a \neq 0.$$

Graphs of absolute value functions are typically V-shaped.

EXAMPLE 5

Graph the absolute value function: $f(x) = |x + 4|$

x	-7	-6	-4	-2	0	2	3
$f(x)$	3	2	0	2	4	6	7

Choose enough values of x to get a general idea of the graph and find the corresponding values of $f(x)$.

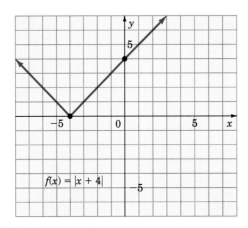

$f(x) = |x + 4|$

■ **DEFINITION**

Reciprocal Function

> The function f is a reciprocal function provided the equation of the function can be written in the form
>
> $$f(x) = \frac{1}{ax + b}, \qquad a \neq 0, \ ax + b \neq 0.$$

EXAMPLE 6

Graph the reciprocal function: $f(x) = \dfrac{1}{x - 5}$

x	1	2	3	4	6	7	8
$f(x)$	$-\dfrac{1}{4}$	$-\dfrac{1}{3}$	$-\dfrac{1}{2}$	-1	1	$\dfrac{1}{2}$	$\dfrac{1}{3}$

Choose enough values of x to get a general idea of the graph and find the corresponding values of $f(x)$.

■ **CAUTION**

In the formation of this table, we cannot use the value $x = 5$ since division by 0 is not defined.

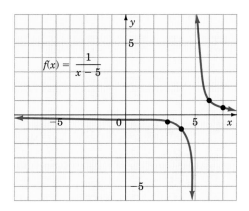

EXERCISE   9.3

A

Classify and graph each of the following functions:

1. $f(x) = 3x - 5$

2. $f(x) = 2x + 4$

3. $f(x) = x^2 + 2$

4. $f(x) = x^2 - 5$

5. $f(x) = \dfrac{1}{2}x - 2$

6. $f(x) = \dfrac{1}{4}x + 1$

7. $f(x) = \sqrt{x - 2}$

8. $f(x) = \sqrt{2 - x}$

9. $f(x) = 2x^2 + 3$

10. $f(x) = 3x^2 - 2$

B

Classify and graph each of the following functions:

11. $f(x) = x^2 - 3x - 4$ **12.** $f(x) = x^2 + 3x - 4$ **13.** $f(x) = x^2 + 5x + 6$

14. $f(x) = x^2 - 5x + 6$ **15.** $f(x) = \sqrt{2x - 4}$ **16.** $f(x) = \sqrt{4x + 8}$

17. $f(x) = |x - 4|$ **18.** $f(x) = |x + 4|$

19. $f(x) = |2x + 3|$

20. $f(x) = |2x - 3|$

C

Classify and graph each of the following functions:

21. $f(x) = \sqrt{x^2 + 4}$

22. $f(x) = \sqrt{x^2 - 4}$

23. $f(x) = |4x - 8|$

24. $f(x) = |3x + 6|$

25. $f(x) = \dfrac{x}{x + 1}$

26. $f(x) = \dfrac{x}{x - 1}$

27. $f(x) = \dfrac{2x}{x + 4}$

28. $f(x) = \dfrac{3x}{x - 4}$

29. $f(x) = |x - 4|$

30. $f(x) = |-x + 5|$

STATE YOUR UNDERSTANDING

31. What is the difference between an equation that defines a linear function and an equation that defines a quadratic function?

32. Why, arithmetically, do the y-values get very large as the x-values get close to 1 in Exercise 26?

33. Why, arithmetically, do the y-values get close to 1 as the x-values get very large in Exercise 26?

34. For the function $f(x) = \sqrt{x - 2}$ in problem 7, why are there no "points" to the left of $x = 2$? If the function had been $f(x) = |x - 2|$, would there be "points" to the left of $x = 2$? Explain.

CHALLENGE EXERCISES

For each of the following, make a table of values and graph.

35. $f(x) = -\sqrt{x - 2}$ (See Exercise 7 above)   **36.** $g(x) = \pm\sqrt{x - 2}$

37. $g(x) = -\sqrt{2 - x}$ (See Exercise 8 above)   **38.** $g(x) = -\sqrt{x^2 - 8x + 16}$ (See Exercise 17)

39. $g(x) = |2x + 4|$ (Use your table from Exercise 2 to help) Why is it not the same graph as in Exercise 2?

40. $g(x) = |x^2 + 2|$ (Use the values from Exercise 3 to help)

41. $g(x) = |x^2 - 4|$

42. $g(x) = |x| - 4$ Is the graph the same as Exercise 29? Explain.

43. $g(x) = \left| \dfrac{x}{x - 1} \right|$ (Use the values from Exercise 26 above)

44. $g(x) = \sqrt{-x^2 - 4}$ Careful!

MAINTAIN YOUR SKILLS (SECTIONS 8.2, 8.3)

45. Write the equation of the circle with the center at the origin and a radius of 25.

46. Write the standard form of the equation of the circle with a radius $\sqrt{17}$ and center of $(-3, -4)$.

47. Find the center of the circle whose equation is $x^2 + y^2 + 6x - 4y - 5 = 0$.

48. Find the length of the radius of the circle whose equation is $x^2 + y^2 + 6x - 4y - 5 = 0$.

Identify the conic section whose equation is:

49. $9x^2 - 25y^2 = 144$

50. $x^2 + y^2 - 49 = 0$

51. $6x^2 + 7y^2 = 42$

52. $x^2 - (2x^2 + 8) + y^2 = 18$

9.4

OPERATIONS WITH
FUNCTIONS

OBJECTIVES

1. Given two functions, find their sum, difference, product, or quotient, and state the domain of the result.

2. Evaluate the sum, difference, product, or quotient of two functions.

3. Find two functions whose sum, difference, product, or quotient is a given function.

4. Find the composite of two functions.

Operations that are similar to operations of arithmetic can be performed with functions. They are defined as follows:

■ DEFINITION

Operations with Functions

If the values of $f(x)$ and $g(x)$ are real numbers, then operations that define new functions are:

1. The sum of f and g, indicated by $f + g$:

$$(f + g)(x) = f(x) + g(x)$$

2. The difference of f and g, indicated by $f - g$:

$$(f - g)(x) = f(x) - g(x)$$

3. The product of f and g, indicated by $f \cdot g$:

$$(f \cdot g)(x) = f(x) \cdot g(x)$$

4. The quotient of f and g, indicated by $\dfrac{f}{g}$ or f/g:

$$\frac{f}{g}(x) = \frac{f(x)}{g(x)}, \text{ provided } g(x) \neq 0$$

The domain of each of these is the intersection of the domains of f and g. For the quotient of the two functions, there is the further restriction that $g(x) \neq 0$.

EXAMPLE 1

Let $f(x) = 3x - 5$ and $g(x) = 4x + 3$. Find $f + g$, $f - g$, $f \cdot g$, and f/g. State the domain of each.

$(f + g)(x) = f(x) + g(x)$

$= (3x - 5) + (4x + 3)$ Substitute $3x - 5$ for $f(x)$ and $4x + 3$ for $g(x)$.

$= 7x - 2$ Simplify.

Domain of $f = \{x \mid x \in R\}$

Domain of $g = \{x \mid x \in R\}$

Domain of $f + g = \{x \mid x \in R\} \cap \{x \mid x \in R\}$
$= \{x \mid x \in R\}$

The domain of $(f + g)(x)$ is the intersection of the domain of f and the domain of g. This is the set of real numbers.

$(f - g)(x) = f(x) - g(x)$

$= (3x - 5) - (4x + 3)$ Substitute $3x - 5$ for $f(x)$ and $4x + 3$ for $g(x)$

$= -x - 8$ Simplify.

Domain of $f - g = \{x \mid x \in R\}$

$(f \cdot g)(x) = (3x - 5)(4x + 3)$ Substitute $3x - 5$ for $f(x)$ and $4x + 3$ for $g(x)$.

$= 12x^2 - 11x - 15$ Multiply.

The domain of $f \cdot g = \{x \mid x \in R\}$.

$(f/g)(x) = \dfrac{f(x)}{g(x)}, \ g(x) \neq 0$

$= \dfrac{3x - 5}{4x + 3}$ Substitute $3x - 5$ for $f(x)$ and $4x + 3$ for $g(x)$.

The domain of $f/g = \{x \mid x \in R\} \cap \left\{ x \mid x \in R \text{ and } x \neq -\dfrac{3}{4} \right\}$.

$= \left\{ x \mid x \in R \text{ and } x \neq -\dfrac{3}{4} \right\}.$ ■

EXAMPLE 2

Let $f(x) = x^2 - 9$ and $g(x) = \sqrt[4]{x}$. The domain of g is restricted so that $g(x)$ is a real number. Find $f \cdot g$, f/g, and g/f. Give the domain of each.

$(f \cdot g)(x) = (x^2 - 9)(\sqrt[4]{x})$

$= x^2 \sqrt[4]{x} - 9\sqrt[4]{x}$ Perform the multiplication.

Domain of $f = \{x \mid x \in R\}$

Domain of $g = \{x \mid x \geq 0\}$

Domain of $f \cdot g = \{x \mid x \in R\} \cap \{x \mid x \geq 0\}$

$= \{x \mid x \geq 0\}$

$$(f/g)(x) = \frac{x^2 - 9}{\sqrt[4]{x}}$$

Domain of $f = \{x \mid x \in R\}$

Domain of $g = \{x \mid x > 0\}$

The domain of g is restricted since if $x = 0$, $\sqrt[4]{x} = 0$.

Domain of $f/g = \{x \mid x \in R\} \cap \{x \mid x > 0\}$

$$= \{x \mid x > 0\}$$

$$(g/f)(x) = \frac{\sqrt[4]{x}}{x^2 - 9}$$

Domain of $g = \{x \mid x \geq 0\}$

Domain of $f = \{x \mid x \neq \pm 3\}$

The domain of f is restricted to avoid division by zero.

Domain of $g/f = \{x \mid x \geq 0\} \cap \{x \mid x \in R\}$

$$= \{x \mid x \geq 0 \text{ and } x \neq 3\}$$ ∎

The evaluation of functions that are a result of one of the four operations just discussed can be done in two ways.

EXAMPLE 3

If $f(x) = 3x - 7$ and $g(x) = x^2 + 3x - 10$, find $(f + g)(-2)$.

$$(f + g)(-2) = f(-2) + g(-2)$$

Evaluate f, evaluate g, then find the sum.

$$= 3(-2) - 7 + (-2)^2 + 3(-2) - 10$$

$$= -6 - 7 + 4 - 6 - 10$$

$$= -25$$

An alternative method:

First find $(f + g)(x)$

$$(f + g)(x) = f(x) + g(x)$$

$$= (3x - 7) + (x^2 + 3x - 10)$$

$$= x^2 + 6x - 17$$

Now find $(f + g)(-2)$

$$(f + g)(-2) = (-2)^2 + 6(-2) - 17$$

Evaluate $f + g$.

$$= 4 - 12 - 17$$

$$= -25$$ ∎

EXAMPLE 4

If $f(x) = x + 4$ and $g(x) = 3x - 5$, find $(f \cdot g)(3)$.

$f(3) = 3 + 4 = 7$ Evaluate $f(x)$.

$g(3) = 3(3) - 5 = 9 - 5 = 4$ Evaluate $g(x)$.

$[f(3)][g(3)] = 7(4) = 28$ Find the product of $f(3)$ and $g(3)$.

An alternative method:
First find $(f \cdot g)(x)$

$(f \cdot g)(x) = (x + 4)(3x - 5)$ Find $(f \cdot g)(x)$.

$\qquad\quad\;\; = 3x^2 + 7x - 20$

Now find $(f \cdot g)(3)$
$(f \cdot g)(3) = 3(3)^2 + 7(3) - 20$ Evaluate $(f \cdot g)(x)$.

$\qquad\quad\;\; = 27 + 21 - 20$

$\qquad\quad\;\; = 28$ ■

In the study of higher mathematics, it is often useful to find two different functions whose sum, difference, product, or quotient results in the given function. That process is demonstrated in the following examples.

EXAMPLE 5

Find functions f and g such that $(f + g)(x) = x^2 - 3x + 2$.

There are many pairs of functions with this sum.

$f(x) = 2x^2 + x$ and $g(x) = -x^2 - 4x + 2$ or

$f(x) = x^2 + 2$ and $g(x) = -3x$ or

$f(x) = 3x^2 + x - 5$ and $g(x) = -2x^2 - 4x + 7$ ■

EXAMPLE 6

Find two functions f and g such that $(f \cdot g)(x) = 2x^2 - 7x - 15$.

Several such functions can be found. One pair can be found by factoring $2x^2 - 7x - 15$.

$(f \cdot g)(x) = 2x^2 - 7x - 15 = (2x + 3)(x - 5)$

Let $f(x) = 2x + 3$ and $g(x) = x - 5$, then

$(f \cdot g)(x) = (2x + 3)(x - 5) = 2x^2 - 7x - 15$.

Another pair is $f(x) = 4x^2 - 14x - 30$ and $g(x) = 0.5$.   ■

The function $G(x) = (3x + 6)^{10}$ can be thought of as a combination of two other functions. The first is the function $(3x + 6)$, and the second is the power of 10 function. This combination is called the composite of the two functions.

The composite of two functions is defined as follows:

■ DEFINITION

Composite Function

> The *composite function* $f \circ g$ is
>
> $(f \circ g)(x) = f(g(x))$.
>
> The domain of $f \circ g$ contains all those numbers in the domain of g for which $g(x)$ is a number in the domain of f.
>
> The composite function $g \circ f$ is
>
> $(g \circ f)(x) = g(f(x))$.
>
> The domain of $g \circ f$ contains all those numbers in the domain of f for which $f(x)$ is a number in the domain of g.

EXAMPLE 7

If $f(x) = 3x - 5$ and $g(x) = x^2$, find $(f \circ g)(3)$.

$(f \circ g)(x) = f(g(x))$	Definition of composition.
$(f \circ g)(3) = f(g(3))$	We need to evaluate f at $g(3)$.
$g(3) = (3)^2$	First evaluate $g(3)$.
$g(3) = 9$	$g(3)$ is in the domain of f.
$f(g(3)) = f(9)$	Second evaluate f at $g(3) = 9$.
$\quad = 3(9) - 5$	
$\quad = 27 - 5$	
$\quad = 22$	

Therefore, $(f \circ g)(3) = 22$.   ■

There is a pictorial view of the composite function $f \circ g$, in the following diagram. Note that the resulting function is defined only for those members of the domain of g for which $g(x)$ is a member of the domain of f.

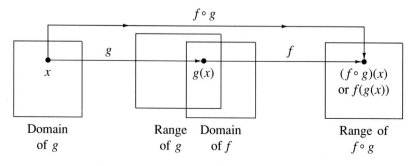

EXAMPLE 8

If $f(x) = 3x - 5$ and $g(x) = x^2$, find $(g \circ f)(3)$.

$(g \circ f)(x) = g(f(x))$ Definition of composition. Note the

$(g \circ f)(3) = g(f(3))$ functions f and g are the same as in Example 7, but the composition is commuted.

$f(3) = 3(3) - 5$ Evaluate f at 3.

$\quad\quad = 9 - 5$

$\quad\quad = 4$

$(g \circ f)(3) = g(4)$

$\quad g(4) = 4^2$

$\quad\quad = 16$ ■

The composition of two functions defines a new function, so we can write $h(x) = (f \circ g)(x)$.

EXAMPLE 9

If $f(x) = 2x - 3$ and $g(x) = x + 3$, find $h(x) = (f \circ g)(x)$.

$h(x) = (f \circ g)(x) = f(g(x))$ Definition of composition.

$\quad\quad\quad = f(x + 3)$ $g(x) = x + 3$

$\quad\quad\quad = 2(x + 3) - 3$ Evaluate f at $x + 3$.

$\quad\quad\quad = 2x + 6 - 3$

$\quad\quad\quad = 2x + 3$

The composite function is $h(x) = 2x + 3$.   ■

EXAMPLE 10

If $f(x) = 2x - 3$ and $g(x) = x + 3$, find $h(x) = (g \circ f)(x)$.

$h(x) = (g \circ f)(x) = g(f(x))$ Definition of composition.

$\quad\quad\quad = g(2x - 3)$ $f(x) = 2x - 3$

$\quad\quad\quad = 2x - 3 + 3$ Evaluate g at $2x - 3$.

$\quad\quad\quad = 2x$

We note that in this case the composition of functions is not commutative.   ■

■ **CAUTION**

The composition of functions is not commutative.

In general, $(f \circ g)(x) \neq (g \circ f)(x)$

Those functions for which $f \circ g = g \circ f$ are called inverse functions. They are discussed in Section 9.5.

The value of a composite function for a specific value of x can be computed in either of one of two ways.

EXAMPLE 11

If $f(x) = x^2 - 5$ and $g(x) = 3x$, find $(f \circ g)(-2)$.

$(f \circ g)(-2) = f(g(-2))$

$\qquad = f(-6)$

$\qquad g(x) = 3x$ so $g(-2) = 3(-2) = -6$
Since -6 is in the domain of f, the composite function is defined at -2.

$\qquad = (-6)^2 - 5 = 31.$

$\qquad f(x) = x^2 - 5$

An alternative method:

Domain of $g = \{x \mid x \in R\}$

Range of $g = \{x \mid x \in R\}$

Domain of $f = \{x \mid x \in R\}$

To use the alternative method it is necessary to verify that $g(-2)$ is in the domain of f before evaluating the composition.

Therefore, $g(-2)$ is in the domain of f.

$(f \circ g)(x) = f(g(x))$

$\qquad = f(3x)$

$\qquad = (3x)^2 - 5$

$\qquad f(x) = x^2 - 5$

$\qquad = 9x^2 - 5$

$(f \circ g)(-2) = 9(-2)^2 - 5$

$\qquad = 9(4) - 5$

$\qquad = 36 - 5$

$\qquad = 31$ $\blacksquare$

EXERCISE 9.4

A

Find $(f + g)(x)$, $(f - g)(x)$, $(f \cdot g)(x)$, and $\left(\dfrac{f}{g}\right)(x)$. State the domain of each result:

1. $f(x) = x + 4$; $g(x) = x - 4$

2. $f(x) = 2x + 5$; $g(x) = x - 3$

3. $f(x) = 3x + 7$; $g(x) = 4x + 3$

4. $f(x) = 4x - 5$; $g(x) = 2x + 1$

5. $f(x) = 2x - 9$; $g(x) = x^2$

6. $f(x) = 5x - 1$; $g(x) = 3x^2$

7. $f(x) = x^2 + 2x - 8$; $g(x) = x + 4$

8. $f(x) = 2x^2 - 8$; $g(x) = x^2 - 4x + 4$

9. $f(x) = x^3 - 3x^2 + 4x - 7$; $g(x) = x^3$

10. $f(x) = 4x^5$; $g(x) = x^2 - 1$

Let $f(x) = 3x + 4$ and $g(x) = x^2 - 5x + 4$. Find the following:

11. $(f + g)(-1)$

12. $(f - g)(3)$

13. $(f \cdot g)(1)$

14. $(f/g)(-2)$

15. $(g - f)(1)$

16. $(g + f)(4)$

17. $(g/f)(1)$

18. $(g \cdot f)(4)$

19. $(f \cdot f)(-4)$

20. $(g \cdot g)(-5)$

B

Find two functions f and g such that the given operation on f and g will yield the given function:

21. $h(x) = 4x^2 - 5x + 2$; $f + g$

22. $h(x) = x^2 - 3x + 5$; $f + g$

23. $h(x) = x^2 - 3x + 4$; $f - g$

24. $h(x) = 6x^2 + 13x + 6$; $f - g$

25. $h(x) = x^2 + 8; f + g$

26. $h(x) = 3x^3 - 4x + 9; f + g$

27. $h(x) = x^3 - 8; f \cdot g$

28. $h(x) = x^3 + 27; f \cdot g$

29. $h(x) = 4x - 3; f/g$

30. $h(x) = 3x + 7; f/g$

31. $h(x) = x^2 - 3x + 1; f/g$

32. $h(x) = x^2 + 4x + 2; f/g$

Evaluate each of the following given that $f(x) = 2x - 3$ and $g(x) = 6x^2 - 13x + 6$:

33. $f(g(-1.5))$

34. $g(f(1.5))$

35. $g(f(-1.5))$

36. $f(g(1.5))$

37. $f(g(0))$

38. $g(f(0))$

C

Find $(f \circ g)(x)$ and $(g \circ f)(x)$ for each of the following:

39. $f(x) = 3x + 7; g(x) = 2x - 9$

40. $f(x) = 5x + 9; g(x) = 2x + 3$

41. $f(x) = 2x + 4; g(x) = x^2$

42. $f(x) = 3x - 9; g(x) = x^3$

43. $f(x) = 2x^2 - 5; g(x) = x^2 + 3$

44. $f(x) = x^2 - 4x; g(x) = 3x^2 + 1$

45. $f(x) = x + 2; g(x) = x - 2$

46. $f(x) = x - 8; g(x) = x + 8$

47. $f(x) = 2x; g(x) = 0.5x$

48. $f(x) = 4x; g(x) = 0.25x$

Let $f(x) = 3x$ and $g(x) = x - 1$. Evaluate each of the following:

49. $(f \circ g)(-1)$

50. $(f \circ g)(-4)$

51. $(f \circ g)(1)$

52. $(f \circ g)(2)$

53. $(f \circ f)(1)$

54. $(g \circ g)(1)$

55. $(g \circ f)(-1)$

56. $(g \circ f)(1)$

57. $(g \circ f)(2)$

58. $(g \circ g)(-4)$

Find (a) $f(x + h)$, (b) $f(x + h) - f(x)$, (c) $\dfrac{f(x + h) - f(x)}{h}$. Item (c) is important in the study of calculus:

59. $f(x) = 2x - 7$

60. $f(x) = 3x - 5$

61. $f(x) = x^2$

62. $f(x) = 3x^2$

63. $f(x) = -2x - 9$

64. $f(x) = -4x + 7$

65. $f(x) = 2x^2 - 4x$

66. $f(x) = -x^2 - 4x$

STATE YOUR UNDERSTANDING

67. Describe the procedure for finding the composition of two functions.

68. Given $g(x) = x - 2$ and $f(x) = \sqrt{x}$, why, mathematically, is it that $f(g(0))$ is not defined?

69. If $f(x)$ and $g(x)$ are two functions, why is it true that $(f + g)(a) = f(a) + g(a)$ for any "a" in the domain of both but that $f/g(a) = \dfrac{f(a)}{g(a)}$ may not be true?

CHALLENGE EXERCISES

Given $g(x) = x - 3$ and $f(x) = \dfrac{1}{x}$, find:

70. $f(g(2))$ **71.** $g(f(4))$ **72.** $f(g(1.5))$ and $g(f(1.5))$

73. $f(g(0))$ **74.** $g(f(0))$

Given $g(x) = 2x - 3$ and $f(x) = \sqrt{x + 1}$ find:

75. $f(g(1))$ **76.** $g(f(1))$ **77.** $g(f(5))$

78. $f(g(0))$ **79.** $g(f(0))$

Given $g(x) = x + 2$ and $f(x) = \dfrac{x}{x - 1}$, find:

80. $f(g(3))$ **81.** $g(f(3))$ **82.** $f(g(0))$

83. $g(f(0))$ **84.** $f(g(-1))$ **85.** $g(f(-1))$

MAINTAIN YOUR SKILLS (SECTIONS 3.4, 3.5, 4.7)

Simplify:

86. $\dfrac{\dfrac{1}{x} - \dfrac{2}{y}}{\dfrac{1}{2x} + \dfrac{3}{y}}$ **87.** $\dfrac{\dfrac{x}{x - 1} + 3}{\dfrac{1}{x - 1} - \dfrac{1}{x}}$

Solve:

88. $\dfrac{2}{3x} - \dfrac{3}{4x} - \dfrac{4}{5x} - \dfrac{5}{6x} = \dfrac{1}{30}$ **89.** $\dfrac{3y - 1}{4} - 4 = \dfrac{4y - 5}{5} + \dfrac{7y + 5}{10}$

90. $\sqrt{x + 4} = 7$ **91.** $\sqrt{x + 4} = \sqrt{7}$

92. $\sqrt{x} + \sqrt{4} = 7$ **93.** $x + 4 = \sqrt{7}$

9.5

INVERSE FUNCTIONS

1. Find the inverse of a function.

2. Write the equation of the inverse of a function.

3. Determine from the graph of a function if it is a one-to-one function.

4. Graph a function and its inverse on the same set of axes.

Quite often, it useful to interchange the components of the ordered pairs in a function. For example:

$F = \{(1, 2), (3, 4), (5, 6)\}$
$G = \{(2, 1), (4, 3), (6, 5)\}$

The components of F have been interchanged to form G. As a result, G is said to be the inverse of F. Also, the function F is the inverse of the function G.

Domain of $F = \{1, 3, 5\}$

Domain of $G = \{2, 4, 6\}$

Range of $F = \{2, 4, 6\}$

Range of $G = \{1, 3, 5\}$

■ DEFINITION

Inverse of a Function

> Associated with every function is a unique relation called the *inverse of the function*. The inverse is formed by interchanging the x- and y-values of the ordered pairs.

The symbol F^{-1} is used to name the inverse of the function F. In this case, the symbol -1 is not an exponent; it means the inverse of the function F. We read F^{-1} "F inverse."

■ CAUTION

> Do not misread the symbol F^{-1} to mean the reciprocal of the real number F.

If

$F = \{(2, 1), (-3, 5), (4, 6)\},$

then

$$F^{-1} = \{(1, 2), (5, -3), (6, 4)\},$$

and if

	f		then		f^{-1}	

x	y
1	3
7	21
−5	−15

x	y
3	1
21	7
−15	−5

f^{-1} is read "f inverse" or "the inverse of f," and $f^{-1}(x)$ is read "f inverse at x."

EXAMPLE 1

Given the function $f = \{(2, -1), (5, 3), (7, 2), (8, -6)\}$, find f^{-1}.

$f^{-1} = \{(-1, 2), (3, 5), (2, 7), (-6, 8)\}$ To find the inverse of a function, interchange the x and y values of the ordered pairs. ■

Given the equation that defines a function, the equation that defines the inverse is found by interchanging the variables in the rule that defines the function. So

if $f = \{(x, y) \mid y = x^2\}$, then $f^{-1} = \{(x, y) \mid x = y^2\}$.

If we solve for y in terms of x, we find

$$f^{-1} = \{(x, y) \mid y = \pm\sqrt{x}\}.$$

Note that x and y are interchanged only in the rule $y = x^2$ and not within the pair (x, y). Also this example points out that not all functions have inverses that are functions. Clearly, $y = \pm\sqrt{x}$ does not define a function since there are two values of y for most values of x.

EXAMPLE 2

Given $F = \{(x, y) \mid y = 2x + 1\}$, find the equation that defines F^{-1}.

$y = 2x + 1$ The equation that defines F.

$x = 2y + 1$ Interchange x and y.

$x - 1 = 2y$ Solve for y.

$y = \dfrac{1}{2}x - \dfrac{1}{2}$

So $F^{-1} = \left\{(x, y) \mid y = \dfrac{1}{2}x - \dfrac{1}{2}\right\}$. ■

■ **PROCEDURE**

To find the inverse of a function

$F = \{(x, y) \mid y = F(x)\}$:

If F is in ordered pair form, interchange the x- and y-values.

If F is given in equation form:

1. Interchange the variables in the equation that defines F.
2. Solve explicitly for y.
3. Write the inverse of F in the form

$$F^{-1} = \{(x, y) \mid y = F^{-1}(x)\}.$$

Consider the quadratic function

$f(x) = \{(x, y) \mid y = 4x^2 + 1\}.$

To find f^{-1}, interchange y and x in the equation that defines $f(x)$ and solve for y.

$y = 4x^2 + 1$

$x = 4y^2 + 1$ Step 1.

$y^2 = \dfrac{x - 1}{4}$ Step 2.

$y = \pm \dfrac{1}{2}\sqrt{x - 1}$

$f^{-1}(x) = \{(x, y) \mid y = \pm \dfrac{1}{2}\sqrt{x - 1}\}$ Step 3.

f^{-1} is not a function since at least one value of x is paired with two values of y: $(5, 1)$, $(5, -1)$.

■ **DEFINITION**

One-to-One Function

A function is said to be a *one-to-one function* when for each x in the domain there is only one y in the range and, conversely, for each y in the range there is only one x in the domain.

From the definition, we see that no two different ordered pairs of a one-to-one function will have the same second component.

An alternative definition of, f, is: a one-to-one function if $f(a) = f(b)$, then $a = b$. One-to-one functions are of special interest because their inverses are also functions.

The graph of a one-to-one function can be recognized using the *horizontal line test*. If every horizontal line intersects the given function in at most one point, then the function is one-to-one and its inverse is a function (Figure 9.3).

Horizontal Line Test

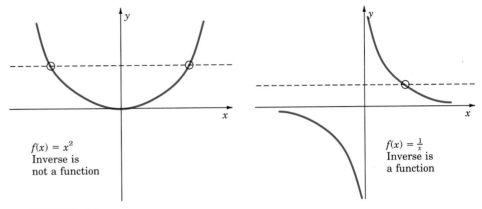

$f(x) = x^2$
Inverse is
not a function

$f(x) = \frac{1}{x}$
Inverse is
a function

FIGURE 9.3

In Figure 9.3, the graph of $f(x) = y = x^2$ does not pass the horizontal line test. That is, a horizontal line intersects the graph in more than one point; therefore, this function is not one-to-one. Because of this, the inverse of $f(x) = x^2$ is not a function.

The graph of $f(x) = \dfrac{1}{x}$ does pass the horizontal line test.

No horizontal line intersects the graph in more than one point. Therefore, $f(x)$ is one-to-one, and consequently, $f^{-1}(x)$ is a function.

EXAMPLE 3

Use the horizontal line test to determine whether or not the function that is graphed is one-to-one.

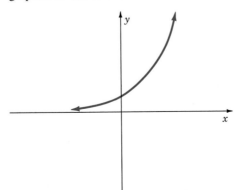

The graph is the graph of a one-to-one function since no horizontal line will intersect it more than one time.

■

If $(a, b) \in f$, then $(b, a) \in f^{-1}$. The ordered pairs (a, b) and (b, a) are "mirror images" or "reflections" about the graph of the identity function. Therefore, the graphs of a function and its inverse are symmetric with respect to the line $y = x$. See Figure 9.4.

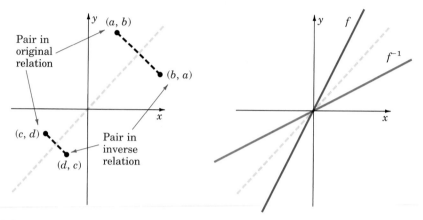

FIGURE 9.4

EXAMPLE 4

Given $f(x) = \left\{ (x, y) \mid y = -\dfrac{1}{2}x + 6 \right\}$, write the equation defining $f^{-1}(x)$, and draw the graphs of f and f^{-1} on the same axes. State whether f^{-1} is a function.

$f(x) = y = -\dfrac{1}{2}x + 6$ Definition of f.

$x = -\dfrac{1}{2}y + 6$ To write the equation of the inverse, interchange the variables in the equation of f.

$2x = -y + 12$ Multiply both sides by 2.

$y = -2x + 12$ Solve for y.

So $f^{-1}(x) = -2x + 12$.

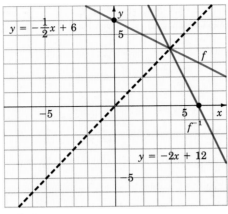

From the graph, we see that $f(x)$ is a one-to-one function. We see that the graphs are symmetric with respect to the line whose equation is $y = x$.

f^{-1} is a function   ■

EXAMPLE 5

Given $g(x) = \frac{1}{2}(x + 1)^2 - 1$, write the equation that defines g^{-1}, and draw the graphs of g and g^{-1} on the same axes. State whether g^{-1} is a function.

$$g(x) = y = \frac{1}{2}(x + 1)^2 - 1$$ Definition of $g(x)$.

$$x = \frac{1}{2}(y + 1)^2 - 1$$ To write the definition of the inverse, interchange the variables in the definition of g.

$$2x = (y + 1)^2 - 2$$ Multiply both sides by 2.

$$2x + 2 = (y + 1)^2$$

$$\pm\sqrt{2x + 2} = y + 1$$

$$g^{-1}(x) = -1 \pm \sqrt{2x + 2}$$

g^{-1} is not a function

Since two values of y are obtained for some values of x, g^{-1} is not a function.

Note the symmetry with respect to the line whose equation is $y = x$. Also note from the graph that $g(x)$ is a function but not a one-to-one function.

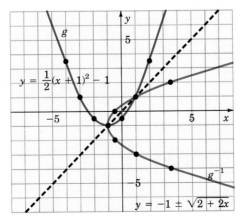

                                                                                                ■

EXERCISE   9.5

A

Find the inverse of the given function:

1. $(3, 2), (4, 3), (5, 4), (6, 5)\}$

2. $\{(-2, 2), (-1, 1), (0, 0), (1, 1)\}$

3. $\{(9, 4.5), (8, 4), (7, 3.5)\}$

4. $\{(5, 1), (10, 2), (15, 3), (20, 4)\}$

5. $\{(9, 3), (4, 2), (1, 1), (0, 0), (-1, -1), (-4, -2), (-9, -3)\}$

6. $\{(2, 2), (3, 3), (4, 4), (5, 5), (6, 6), (7, 7)\}$

Write the equation of the inverse of the following functions:

7. $f(x) = 3x - 5$

8. $f(x) = 4x - 3$

9. $f(x) = -x + 3$

10. $f(x) = -2x + 1$

11. $f(x) = 3x - 12$

12. $f(x) = 4x + 6$

13. $f(x) = 5x + 4$

14. $f(x) = 6x + 12$

B

15. $f(x) = \dfrac{1}{2}x + 2$

16. $f(x) = \dfrac{1}{4}x + 4$

17. $f(x) = \dfrac{2}{3}x - 5$

18. $f(x) = \dfrac{2}{5}x - 8$

19. $f(x) = \dfrac{1}{2}x + \dfrac{2}{3}$

20. $f(x) = \dfrac{1}{3}x + \dfrac{3}{4}$

21. $f(x) = x^2$

22. $f(x) = x^3$

23. $f(x) = x^3 + 1$

24. $f(x) = x^2 - 1$

Determine whether or not each function graphed is a one-to-one function:

25.

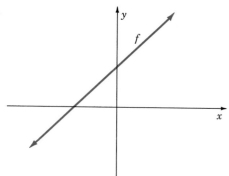

26.

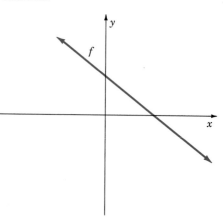

27.

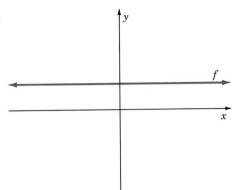

28.

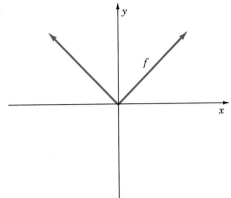

29.

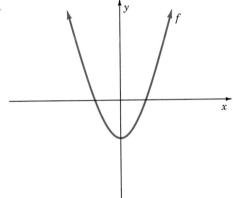

30.

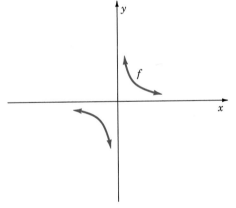

31.

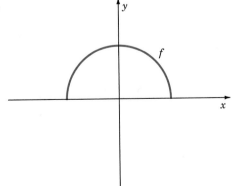

32.

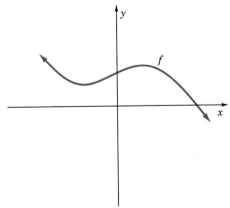

C

Each equation defines a function f. Write the equation of f^{-1}, sketch the graph of f and f^{-1} on the same set of axes, and state whether or not f^{-1} is a function:

33. $f(x) = 4x - 3$

34. $f(x) = 3x - 5$

35. $f(x) = 5$

36. $f(x) = -3$

37. $f(x) = x^2 + 2$

38. $f(x) = x^2 - 4$

39. $f(x) = |x|$

40. $f(x) = |-x|$

Find the composite function $(f \circ f^{-1})$:

41. $f(x) = x - 5$

42. $f(x) = x + 4$

43. $f(x) = -3x - 5$

44. $f(x) = -2x + 1$

STATE YOUR UNDERSTANDING

45. What is the inverse of a function, and when will the inverse be a function?

46. The graph of the curve $x^2 + y^2 = 25$ is a circle with a center at $(0, 0)$ and a radius of 5. What is/are the largest portions of the curve you could use so that the curve is a function? What is/are the largest portions of the curve that you could use so that the curve is one-to-one? What is/are the largest portions of the curve that you could use that are both a function and 1 to 1?

47. Given $y = x^2 - 2x + 3$ is the equation of a parabola that is a function but not one-to-one, what portion of the domain (which is all reals) could you use so that the curve is both a function and one-to-one?

CHALLENGE EXERCISES

48. In Exercise 7, you were asked to find the equation for f^{-1}. Using the result, find $f(f^{-1}(3))$ and $f^{-1}(f(3))$.

49. In Exercise 13, you were asked to find the equation for f^{-1}. Using the result, find $f(f^{-1}(1))$ and $f^{-1}(f(1))$.

50. In Exercise 29, the curve is not one-to-one. What "portion" of the curve could you use so the curve is one-to-one?

Using your best artistic skills, draw the inverse of the following curves:

51.

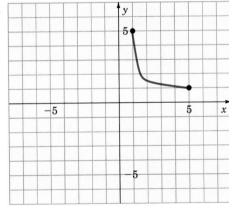

52.

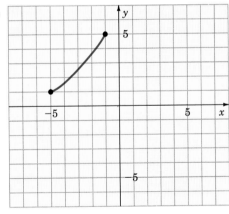

53.

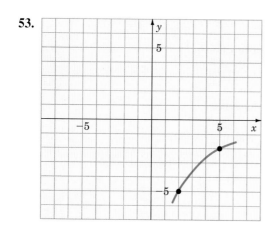

54. Graph $f(x) = \sqrt{x} - 2$ and using the fact that f^{-1} is symmetric with respect to the line $y = x$, graph f^{-1}. (Do not write the equation)

55. Graph $g(x) = \sqrt{x+2}$ and using the symmetry with respect to $y = x$, graph g^{-1}. (Do not write the equation of the inverse)

56. Sketch the graph of any curve that will pass the horizontal line test, but is not a function. (keep it simple)

57. Using the symmetry with respect to $y = x$, draw the inverse of the curve you just drew in exercise 56. Is it a function? Is it one-to-one?

MAINTAIN YOUR SKILLS (SECTIONS 5.1, 5.2)

Solve:

58. $(x - 4)^2 = 4$

59. $t^2 - 9t + \dfrac{81}{4} = \dfrac{121}{4}$

60. $m^2 + \dfrac{2}{5}m = \dfrac{7}{5}$

61. $p^2 + \dfrac{1}{2}p = \dfrac{1}{8}$

62. $2 + \dfrac{4}{s - 2} = \dfrac{3}{s + 8}$

63. $3 + \dfrac{2t + 1}{t - 1} = \dfrac{2t - 1}{t + 1}$

Solve for x:

64. $x^2 + dx + f = 0$

65. $dx^2 - fx + 1 = 0$

CHAPTER 9
SUMMARY

Relation	Any set of ordered pairs.	(p. 616)
Domain of a Relation	The set of first components (x-values) of the ordered pairs of a relation.	(p. 616)
Range of a Relation	The set of second components (y-values) of the ordered pairs of a relation.	(p. 616)
Function	A relation in which for each x-value there is exactly one y-value.	(p. 617)
Vertical Line Test	Used to determine whether or not a graph is the graph of a function. If a vertical line intersects the graph of a relation in more than one point, then the graph is not that of a function.	(p. 618)
Function Notation	The notation $f(x)$ is used to denote a function. It is read ''f at x'' or ''f of x.'' The notations $f(x)$ and y have the same meaning. That is, $y = f(x)$.	(p. 627)
Independent Variable	The first component of an ordered pair, (x), is called the independent variable.	(p. 620)
Dependent Variable	The second component of an ordered pair, (y), is called the dependent variable.	(p. 620)
Linear Function	Defined by $f(x) = y = mx + b$, $m \neq 0$.	(p. 632)
Quadratic Function	Defined by $f(x) = y = ax^2 + bx + c$, $a \neq 0$.	(p. 633)
Polynomial Function	Defined by $f(x) = y = a_n x^n + a_{n-1} x^{n-1} + \cdots + a_1 x + a_0$, where $a_n, a_{n-1}, \ldots, a_1, a_0$ are real numbers and $a_n \neq 0$.	(p. 633)
Square Root Function	Defined by $f(x) = y = \sqrt{ax + b} + c$, where a is any algebraic expression in which $a \geq 0$.	(p. 634)
The Sum of Two Functions	The sum of f and g, indicated by $f + g$, is defined by $(f + g)(x) = f(x) + g(x)$.	(p. 642)
The Difference of Two Functions	The difference of f and g, indicated by $f - g$, is defined by $(f - g)(x) = f(x) - g(x)$.	(p. 642)
The Product of Two Functions	The product of two functions f and g, indicated by $f \cdot g$, is defined by $(f \cdot g)(x) = f(x) \cdot g(x)$.	(p. 642)
The Quotient of Two Functions	The quotient of two functions f and g, indicated by $\dfrac{f}{g}$ or f/g, is defined by $\dfrac{f}{g}(x) = \dfrac{f(x)}{g(x)}$, provided $g(x) \neq 0$.	(p. 642)
Composition Notation	The notation $(f \circ g)(x)$ indicates the composite of two functions.	(p. 646)

Composite of Two Functions	Defined by $(f \circ g)(x) = f(g(x))$.	(p. 646)
Inverse of a Function	Associated with every function is a unique relation called the inverse of the function. To find the inverse of a function:	(pp. 652, 654)
	If the function is in ordered pair form, interchange the x- and y-values.	
	If the function is defined by an equation:	
	1. Interchange the variables in the equation.	
	2. Solve explicitly for y.	
	3. Write the inverse using function notation.	
One-to-One Function	A function is said to be a one-to-one function when for each x in the domain there is only one y in the range and, conversely, for each y in the range there is only one x in the domain.	(p. 654)
Inverse of a One-to-One Function	If a function is one-to-one, its inverse is a function.	(p. 655)
Horizontal Line Test for One-to-One Function	The graph of a one-to-one function can be recognized using the horizontal line test. If every horizontal line intersects the given function in at most one point, the function is one-to-one and its inverse is also a function.	(p. 655)

CHAPTER 9
REVIEW EXERCISES

SECTION 9.1 Objective 1

Write each of the following relations as a set of ordered pairs, and state the range.

1. $\{(x, y) \mid y = x^2 + 2x + 1, x = -2, -1, 0, 1, 2\}$

2. $\{(x, y) \mid y = x^3 - 3x^2 - 5, x = -2, -1, 0, 1, 2\}$

3. $\{(x, y) \mid y = \sqrt{x + 2}, x = 0, 1, 2, 3\}$

4. $\{(x, y) \mid y = |4x - 5|, x = -4, -1, 0, 1, 2\}$

State the domain and range of each of the following relations.

5. $\{(2, 1), (3, 2), (4, 6), (9, 2)\}$

6. $\{(0, 0), (1, 1), (2, 2), (3, 3), (4, 4)\}$

7. $\{(1, -1), (2, -2), (3, -3), (4, -4), (5, -5)\}$

8. $\{(5, 1), (6, 1), (7, 1), (8, 1), (9, 1), (10, 1)\}$

SECTION 9.1   Objective 2

State the domain and range of the relation defined by each of the following, and identify those that are functions.

9. $y = x^2 - 8x - 5$ **10.** $y = 2x^2 + 4x - 7$

11. $y = \sqrt{x^2 - 5}$ **12.** $y = \sqrt{x^2 + 9}$

13. $y = \sqrt{15 - x^2}$ **14.** $y = \sqrt{20 + x^2}$

15. $y = |3x - 5|$ **16.** $y = |8x + 4|$

17. $y = \dfrac{x}{x + 2}$ **18.** $y = \dfrac{x}{x - 5}$

SECTION 9.1   Objective 3

Draw the graph of each of the following. Use the graph to determine the range and domain. Use the vertical line test to identify functions.

19. $y = 4x - 7$ **20.** $y = 5x + 2$

21. $y = \sqrt{x} + 2$ **22.** $y = \sqrt{x} - 3$

23. $y = |x + 1|$ **24.** $y = |x + 2|$

25. $x^2 + y^2 = \dfrac{49}{4}$

26. $x^2 + y^2 = \dfrac{81}{4}$

SECTION 9.2   Objective 1

Find the indicated values of $f(x)$.

27. $f(x) = 6x - 9$; $f(12)$, $f(-4)$

28. $f(x) = -5x - 9$; $f(-2)$, $f(4)$

29. $f(x) = x^2 - 8x + 12$; $f(6)$, $f(2)$

30. $f(x) = x^2 + 9x + 20$; $f(-4)$, $f(-5)$

31. $f(x) = |2x + 7|$; $f(0)$, $f(-7)$

32. $f(x) = |4x + 9|$; $f(0)$, $f(-4)$

SECTION 9.2   Objective 2

If $f(x) = x^2 - 3x - 54$ and $g(x) = 2x + 3$, evaluate the following.

33. $f(9)$ **34.** $g(3)$ **35.** $g(0)$ **36.** $f(-6)$ **37.** $f(g(3))$ **38.** $g(f(-6))$

SECTION 9.3   Objective 1

Classify each of the following functions, and draw their graph.

39. $f(x) = |3x - 6|$

40. $f(x) = \sqrt{x^2 - 9}$

41. $f(x) = 2x^2 + x - 6$

42. $f(x) = \dfrac{3x}{2x - 6}$

SECTION 9.4   Objective 1

Given $f(x) = x + 5$, $g(x) = x - 2$, and $h(x) = x^2 + 2x + 1$, find the following.

43. $(f + g)(x)$

44. $(f + h)(x)$

45. $(h - f)(x)$

46. $(h - g)(x)$

47. $(f - h)(x)$

48. $(g - f)(x)$

49. $(g + h)(x)$

50. $(g + f)(x)$

51. $(f \cdot g)(x)$

52. $(f \cdot h)(x)$

53. $(g \cdot h)(x)$

54. $(f \cdot f)(x)$

55. $\left(\dfrac{f}{g}\right)(x)$

56. $\left(\dfrac{f}{h}\right)(x)$

57. $\left(\dfrac{g}{h}\right)(x)$

58. $\left(\dfrac{h}{f}\right)(x)$

SECTION 9.4   Objective 2

Given $f(x) = 2x - 3$ and $g(x) = x^2 - 3x + 2$, find the following.

59. $(f + g)(-1)$

60. $(f - g)(3)$

61. $(f \cdot g)(1)$

62. $\left(\dfrac{g}{f}\right)(-1)$

63. $(g \cdot f)(2)$

64. $(f \cdot f)(-3)$

SECTION 9.4   Objective 3

Find two functions f and g such that the given operation on f and g will yield the given function.

65. $h(x) = 4x^2 - 8x - 12$, $f + g$

66. $h(x) = x^2 - 9x + 5$, $f - g$

67. $h(x) = x^2 - 3x - 18$, $f \cdot g$

68. $h(x) = x + 5$, $\dfrac{f}{g}$

SECTION 9.4   Objective 4

Find $f \circ g(x)$ and $g \circ f(x)$ for each of the following.

69. $f(x) = 2x - 1$; $g(x) = 3x + 2$

70. $f(x) = x + 9$; $g(x) = -x - 8$

71. $f(x) = x^2 + 3$; $g(x) = x + 1$

72. $f(x) = 5x$; $g(x) = 0.2x$

SECTION 9.5   Objective 1

Find the inverse of the given function.

73. $\{(1, 2), (2, 3), (3, 4), (4, 5)\}$

74. $\{(3, 1), (4, 1), (5, 1), (6, 1)\}$

75. $\{(8, 2), (9, 3), (-8, 1), (-6, 7)\}$

76. $\{(-1, 1), (0, 0), (1, 1), (2, 2), (3, 3)\}$

SECTION 9.5   Objective 2

Write the equation of the inverse of the following.

77. $f(x) = 2x - 3$ **78.** $f(x) = 3x + 2$

79. $f(x) = 2x + 3$ **80.** $f(x) = 3x - 2$

81. $f(x) = \dfrac{1}{2}x + 4$ **82.** $f(x) = \dfrac{1}{4}x - 1$

83. $f(x) = 2x^2$ **84.** $f(x) = 3x^2 - 4$

SECTION 9.5   Objective 3

Determine whether or not each function graphed is a one-to-one function.

85. 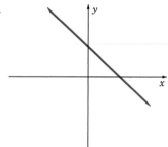   **86.** 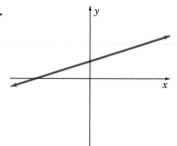   **87.**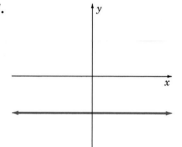

88. 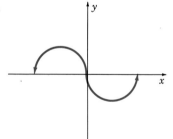   **89.** 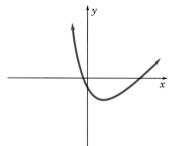   **90.**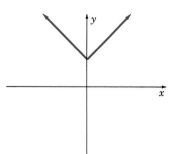

91. 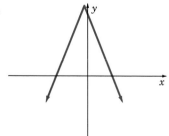   **92.**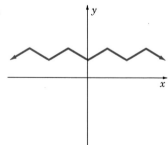

SECTION 9.5   Objective 4

Each equation defines a function f. Write the equation f^{-1}, sketch the graph of f and f^1 on the same set of axes, and state whether or not f^{-1} is a function.

93. $f(x) = 3x + 3$

94. $f(x) = -5x - 15$

95. $f(x) = x^2$

96. $f(x) = 3x^2$

97. $f(x) = \dfrac{2}{3}x - \dfrac{4}{3}$

98. $f(x) = -\dfrac{2}{5}x + 4$

CHAPTER 9
TRUE–FALSE CONCEPT REVIEW

Check your understanding of the language of algebra. Tell whether each of the following statements is true (always true) or false (not always true).

1. Every set of ordered pairs is called a relation.

2. Every set of points on a rectangular coordinate system is the graph of some relation.

3. It is possible for the domain of a relation to contain a single number.

4. It is possible for the range of a relation to contain a single number.

5. If the ordered pairs (a, b) and (c, b) are members of a function, then a and c must be equal.

6. The symbol $f(x)$ represents a value in the range of a function.

7. Every function is a relation.

8. A function is said to be one-to-one if and only if whenever $f(a) = f(b)$, then $a = b$.

9. If the graph of a function will pass the horizontal line test, then the inverse of that function is also a function.

10. $(f + g)(x) = f(x) + g(x)$

11. $(f - g)(x) = f(x) + g^{-1}(x)$

12. $(f \cdot g)(x) = [f(x)][g(x)]$

13. $(f/g)(x) = [f(x)][g^{-1}(x)]$

14. $(g \circ f)(x) = f(g(x))$

15. $(f \circ g)(0) = 0$

CHAPTER 9
TEST

1. Given $f(x) = 3x + 4$ and $g(x) = x^2 + 4x + 3$,
   a. State the domain of $f(x)$.
   b. State the range of $g(x)$.

2. Given $f(x) = x^2 + 2$,
   a. Write the equation that defines $f^{-1}(x)$.
   b. Graph $f(x)$ and $f^{-1}(x)$ on the same set of axes.
   c. Does $f^{-1}(x)$ define a function?

c.

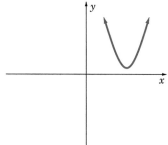

3. Given $f(x) = 3x - 5$ and $g(x) = 2x + 7$,
   a. Find $(f \circ g)(x)$.
   b. Find $(g \circ f)(x)$.

4. Which of the following (if any) is the graph of a one-to-one function?
   a.

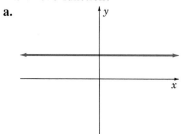

   b.

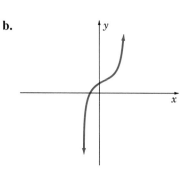

5. Given $f(x) = 2x + 5$ and $g(x) = 2x^2 - x - 15$,
   a. Find $(f + g)(x)$.
   b. Find $(g - f)(x)$.
   c. Find $(f \cdot g)(x)$.
   d. Find $(g/f)(x)$.

6. Given the relation $\{(-7, 4), (-5, 2), (-3, 0), (-1, -2), (1, -4), (3, -6)\}$,
   a. State the domain of the relation.

   b. State the range of the relation.

   c. Is the relation a function?

7. If $f(x) = 6x - 9$ and $g(x) = 5x + 1$,
   a. Find $f(g(2))$.
   b. Find $g(f(2))$.

8. Classify the following functions.
   a. $f(x) = 2$
   b. $f(x) = 2x$
   c. $f(x) = 2x^2$

9. Given $f(x) = \dfrac{1}{4}x - 2$, write the equation that defines $f^{-1}(x)$.

10. Given $f(x) = \{(-7, 4), (-5, 2), (-3, 0), (-1, -2), (1, -4), (3, -6)\}$, find $f^{-1}(x)$.

11. Given $h(x) = 3x^2 + 4x - 12$,
    a. Find $f(x)$ and $g(x)$ such that $(f + g)(x) = h(x)$.

    b. Find $f(x)$ and $g(x)$ such that $(f - g)(x) = h(x)$.

c. Find $f(x)$ and $g(x)$ such that $(f \cdot g)(x) = h(x)$.

d. Find $f(x)$ and $g(x)$ such that $(f/g)(x) = h(x)$.

12. Given $f(x) = x^2 - 5x - 6$,
  a. Find $f(-2)$
  b. Find $f(2)$.

13. Given $f(x) = \{(-7, 4), (-5, 2), (-3, 0), (-1, -2), (1, -4), (3, -6)\}$,
  a. Find $f(-5)$.
  b. Find $f(3)$.

14. Given $f(x) = 2x + 5$ and $g(x) = x^2 - x - 6$,
  a. Find $(f + g)(3)$.
  b. Find $f(g(3))$.

CHAPTER 7

Solve the following systems by substitution

1. $\begin{cases} 4x + y = 16 \\ y = 3x - 5 \end{cases}$

2. $\begin{cases} 2x + 3y = 9 \\ x - y = -3 \end{cases}$

3. $\begin{cases} 4x + 3y = -1 \\ y = -4 \end{cases}$

4. $\begin{cases} x = -4y - 3 \\ 2x + 8y = -6 \end{cases}$

Solve the following using linear combinations

5. $\begin{cases} 2x + y = 1 \\ -2x + y = 1 \end{cases}$

6. $\begin{cases} 3x - 2y = 6 \\ 4x - 3y = 7 \end{cases}$

7. $\begin{cases} 10x - 5y = 7 \\ 2x - y = 4 \end{cases}$

8. $\begin{cases} 9x - 3y = 4 \\ 2x + y = 5 \end{cases}$

Evaluate the following determinants

9. $\begin{vmatrix} 5 & 1 \\ -2 & 3 \end{vmatrix}$

10. $\begin{vmatrix} -1 & 7 \\ 4 & 0 \end{vmatrix}$

11. $\begin{vmatrix} -3 & 0 & 4 \\ 5 & 2 & -3 \\ 7 & 0 & 6 \end{vmatrix}$

12. $\begin{vmatrix} 1 & 1 & 1 \\ 3 & 2 & -6 \\ 9 & -4 & 12 \end{vmatrix}$

Solve the following systems using Cramer's Rule

13. $\begin{cases} 6x - 5y = -7 \\ 3x - 4y = 5 \end{cases}$

14. $\begin{cases} 2x - 4y = 5 \\ -4x + 2y = 7 \end{cases}$

15. $\begin{cases} 2x + 3y + z = 1 \\ 6x + y + 5z = 1 \\ 3x + 4y + 2z = 1 \end{cases}$

16. $\begin{cases} x + y + z = 4 \\ 2x + 4y + 6z = -3 \\ x + 3y + 5z = -1 \end{cases}$

Solve the following using matrix methods.

17. $\begin{cases} 5x - 3y = 8 \\ 9x + 2y = 7 \end{cases}$

18. $\begin{cases} 2x + y + z = 3 \\ x + 4y - 2z = -3 \\ -5x - 2y + 3z = -14 \end{cases}$

CHAPTER 8

For the following parabolas, find: (a) the vertex; (b) the x-intercepts; (c) the y-intercepts; (d) the equation of the axis of symmetry; and (e) graph it.

19. $y = 2x^2 - 8$

20. $y = -2x^2 + 4x + 6$

21. $y = x^2 - x - 3$

22. $x = -y^2 + 2y - 1$

For the following circles, find: (a) the center; (b) the radius; and (c) the graph.

23. $x^2 + y^2 = 9$

24. $(x + 2)^2 + (y - 4)^2 = 25$

25. $x^2 + y^2 - 8y = 0$

26. $x^2 + y^2 - 6x + 2y + 6 = 0$

For the following ellipses, find: (a) the x-intercepts; (b) the y-intercepts; and (c) the graph

27. $\dfrac{x^2}{4} + \dfrac{y^2}{9} = 1$

28. $\dfrac{x^2}{5} + \dfrac{y^2}{12} = 1$

29. $25x^2 + y^2 - 25 = 0$

30. $9x^2 + 36y^2 = 324$

For the following hyperbolas, find: (a) the x-intercepts; (b) the y-intercepts; and (c) the graph.

31. $\dfrac{x^2}{4} - \dfrac{y^2}{9} = 1$

32. $\dfrac{y^2}{9} - \dfrac{x^2}{16} = 1$

33. $x^2 - 4y^2 - 100 = 0$

34. $x^2 - y^2 + 4 = 0$

Solve the following systems

35. $\begin{cases} y = x^2 - 5x - 1 \\ 2x + y = 3 \end{cases}$

36. $\begin{cases} x^2 + y^2 = 9 \\ x - y = 3 \end{cases}$

37. $\begin{cases} x^2 + y^2 = 9 \\ x + y = 7 \end{cases}$

38. $\begin{cases} x^2 + y^2 = 5 \\ 4x^2 + 9y^2 = 25 \end{cases}$

39. $\begin{cases} x^2 + y^2 = 9 \\ x^2 - y^2 = 5 \end{cases}$

CHAPTER 9

For the following relations, determine: (a) the domain; (b) the range; (c) is it a function?; and (d) is it one-to-one?

40. $R = \{(2, 0), (-1, 3), (4, 0), (5, 3)\}$

41. $R = \{(-1, -4), (5, 4), (-1, 3), (4, 2)\}$

42. R:

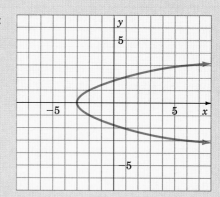

43. R:

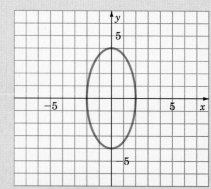

44. R:

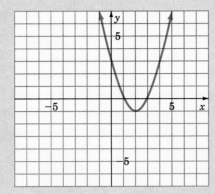

Determine the domain.

45. $y = x^2 - 3x + 1$

46. $y = \dfrac{x}{x + 2}$

47. $y = \sqrt{5 - x}$

48. $y = \dfrac{x + 2}{x^2 - 9}$

49. $y = \dfrac{1}{x^2 + 16}$

Given $f(x) = \dfrac{x}{x - 3}$, $g(x) = \sqrt{x - 2}$, and

$h(x) = x^2 - 3x + 1$, find:

50. $f(-2)$

51. $g(10)$

52. $h(2)$

53. $h(3)$

54. $g(0)$

55. $f(4) + g(6)$

56. $g(f(4))$

57. $f(h(4))$

58. $f(g(3))$

59. $f(a + b)$

60. $h(a - b)$

61. $(f + g)(4)$

Classify the following functions and graph

62. $f(x) = |2x - 4|$

63. $f(x) = x^2 - x - 6$

64. $f(x) = \sqrt{x + 3}$

65. $f(x) = \dfrac{1}{x + 2}$

Write the equation of the inverse (f^{-1}) of each of the following and sketch the graphs of f and f^{-1} on the same axes.

66. $f(x) = 2x - 3$

67. $f(x) = \dfrac{2}{3}x + 1$

68. $f(x) = \dfrac{1}{5}x + 1$

69. $f(x) = x^2 - 1$

Solve

70. The sum of two resistors is 140 ohms and their difference is 60 ohms. Find the resistance of each resistor.

71. The sum of three numbers is 3. The sum of the first two numbers is 6. If you double the first and add it to the third, the sum would be -1. Find the three numbers.

72. The number of cases (C) of mumps in a school is given by $C = -\dfrac{1}{4}d^2 + 6d + 3$, where "$d$" is the number of days that have passed since the epidemic started. How many cases are there 10 days after the epidemic started? What is the maximum number of cases that will occur and on what day will they occur?

73. The area of a rectangle is 65 square feet and the perimeter is 36 feet. Find the dimensions of the rectangle.

CHAPTER

10

EXPONENTIAL AND LOGARITHMIC FUNCTIONS

The intensity of an earthquake is measured by the Richter scale. The value on the Richter scale is given by the formula $R = \log \dfrac{E}{E_0}$, where E is the energy released by the current quake and E_0 is the energy released by the earthquake used for comparison, called the zero quake. Each quake that measures above 4 on the scale can cause severe damage, such as those seen recently in San Francisco and Mexico City. In Exercise 66, Section 10.2, you are given the value of the ratio $\dfrac{E}{E_0}$ and are asked to find the intensity of the earthquake. *(Roberto Valladares/The Image Bank)*

677

PREVIEW

Exponential and logarithmic functions are presented in this chapter. Properties of logarithms are given. Logarithms are found using a calculator. Base e and base ten logs are found and you are given a method of finding the logarithm of a number with any positive base. Equations involving variable exponents and equations involving logarithms are solved.

10.1

EXPONENTIAL FUNCTIONS

OBJECTIVES

1. Draw the graph and state the domain and range of an exponential function.

2. Solve an exponential equation.

In this section, we define *exponential functions,* list their properties, and look at some of their useful applications.

■ DEFINITION

Exponential Function

> An *exponential function* is a function defined by an equation of the form $f(x) = b^x$, $b \neq 1$, and $b > 0$. The number b is called the base of the exponential function.

In Chapters 2 and 4, you studied the meaning of integer and rational exponents such as

$$4^3, \quad 6^{1/2}, \quad 8^{-1/4}, \quad 15^{1.8}, \quad \text{and} \quad 1^{-3.46}.$$

In this section, we make the assumption that the expression b^x has exactly one value for each real number x when b is a positive number ($b > 0$). Thus, expressions with irrational exponents such as

$$3^{\sqrt{2}}, \quad 7^{\pi}, \quad 15^{\sqrt[3]{2}}, \quad \text{and} \quad 8^{-\sqrt{7}}$$

have one and only one real number value. It is shown in an advanced mathematics course that this assumption is reasonable. With this property established, the exponential function is a one-to-one function.

When a positive number b has been chosen, the equation $f(x) = b^x$ defines a function whose domain is the set of real numbers and whose range is the set of positive real numbers.

Once we have defined and graphed the exponential function for all real number values of x, we can see from the graph that y is never negative or zero. As a matter of fact, no exponential function defined by

$$y = b^x, \qquad b > 0, \qquad b \neq 1$$

has negative or zero values.

■ DEFINITION

Domain and Range of an Exponential Function

For any exponential function defined by

$$y = b^x, \qquad b > 0, \qquad b \neq 1$$

the *domain* is the set of real numbers.

Domain $= \{x \mid x \in R\}$

The *range* is the set of positive real numbers.

Range $= \{y \mid y > 0\} = R^+$

For example, if $b = 2$, then we have the function with equation:

$$y = 2^x.$$

Using some rational values of x, we can make the following table of values.

x	-3	-2	-1	0	0.5	1	1.5	2	3	4
y or $f(x)$	0.125	0.25	0.5	1	$(\sqrt{2})$ 1.414	2	$(\sqrt{8})$ 2.828	4	8	16

By our assumption that for every real number x there is a unique real number y, we sketch the graph of this function by plotting the pairs from the table and connecting them with a smooth curve (Figure 10.1).

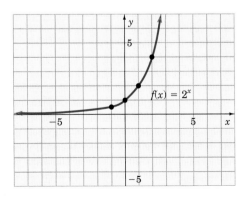

FIGURE 10.1

Note that the curve approaches the x-axis but does not cross it. The x-axis is an asymptote of the curve.

If the base, b, is greater than one, then the values of b^x increase as x increases.

Values of x	-3	-2	-1	0	1	2	3
Values of 3^x	$\dfrac{1}{27}$	$\dfrac{1}{9}$	$\dfrac{1}{3}$	1	3	9	27
Values of 10^x	$\dfrac{1}{1000}$	$\dfrac{1}{100}$	$\dfrac{1}{10}$	1	10	100	1000

These results are characteristic of *increasing functions*.

■ DEFINITION

Increasing Function

> An *increasing function* is a function for which the values of the function (y-values) increase as x increases. That is, if $x_2 > x_1$, then $f(x_2) > f(x_1)$.

EXAMPLE 1

Make a table of values, draw the graph, and write the domain and range of the function defined by $y = f(x) = 4^x$.

x	-2	-1	0	0.5	1	1.5	2	3
y	0.0625	0.25	1	2	4	8	16	64

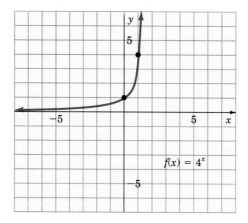

$f(x) = 4^x$

To calculate the values on a calculator, use the y^x key.

Plot the points and draw the curve.

The x-axis is an asymptote.

The function is increasing.

Domain: $\{x \mid x \in R\}$
Range: $\{y \mid y > 0\}$ ■

We now sketch the graph of an exponential function in which $0 < b < 1$. See Figure 10.2. The function is defined by

$$f(x) = \left[\frac{1}{2}\right]^x.$$

Using rational values of x, we can make the following table of values for the function.

x	-3	-2	-1	0	0.5	1	2	3
y	8	4	2	1	0.71	0.5	0.25	0.125

Substitute values for x in the equation, and solve for y. Recall that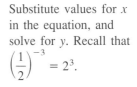

Sketch the graph by connecting the points with a smooth curve. From the graph, we see that it is a decreasing function. The graph approaches the x-axis but does not cross it.

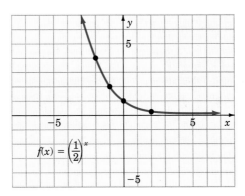

FIGURE 10.2

EXAMPLE 2

Make a table of values, draw the graph, and write the domain and range of the function defined by $y = f(x) = \left[\dfrac{1}{4}\right]^x$.

x	-3	-2	-1	0	1	2	3
y	64	16	4	1	$\dfrac{1}{4}$	$\dfrac{1}{16}$	$\dfrac{1}{64}$

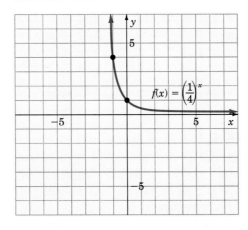

Domain: $\{x \mid x \in R\}$
Range: $\{y \mid y > 0\}$

If the base, b, is between one and zero, then the values of b^x decrease as x increases.

Values of x	-3	-2	-1	0	1	2	3
Values of $\left[\dfrac{1}{5}\right]^x$	125	25	5	1	$\dfrac{1}{5}$	$\dfrac{1}{25}$	$\dfrac{1}{125}$
Values of $\left[\dfrac{1}{9}\right]^x$	729	81	9	1	$\dfrac{1}{9}$	$\dfrac{1}{81}$	$\dfrac{1}{729}$

These results are characteristic of *decreasing functions*.

■ DEFINITION

Decreasing Function

A *decreasing function* is a function for which the values of the function (y-values) decrease as x increases. That is, if $x_2 > x_1$, then $f(x_2) < f(x_1)$.

Some general properties of exponential functions are listed in the following chart.

$f(x) = b^x,\ b \neq 1,\ b > 0$	$0 < b < 1$	$b > 1$
Domain	All reals	All reals
Range	$y > 0$	$y > 0$
Asymptote	$x = 0$	$x = 0$
Increasing or decreasing?	Decreasing	Increasing
Graph [Every graph contains the point (0, 1)]		

Example 3 shows an application of exponential functions from physics.

EXAMPLE 3

A certain isotope of radium has a half-life of 1600 years. This means that after 1600 years, half the radioactive radium has disintegrated. If the original amount of radium is 1 mg (milligram), the formula for the amount A that is left after t years is approximately

$A \approx (0.999567)^t$.

How much of the 1 mg of radium is left after 10 years?

Formula:

$A \approx (0.999567)^t$ The formula defines an exponential
                                                        function.

Substitute:

$A \approx (0.999567)^{10}$ Substitute $t = 10$.

Solve:

ENTER	DISPLAY
0.999567 y^x 10 $=$	0.99567843

A calculator is used to perform the
calculation. Otherwise, we would
need to multiply 10 factors of
0.999567.

Answer:

Approximately 0.9957 mg of radium is left.   ■

The exponential function is a one-to-one function.

■ **PROPERTY**

> If $b^x = b^y$
>
> then $x = y$.

We use this property to solve exponential equations.

EXAMPLE 4

Solve for x: $16^x = 256$

$16 = 2^4$, $256 = 2^8$ 16 and 256 can each be written as
                                                        powers of the same base, 2.

$16^x = 256$ Original equation.

$(2^4)^x = 2^8$ Substitute 2^4 for 16 and 2^8 for 256.

$2^{4x} = 2^8$ Property of exponents for raising a
                                                        power to a power.

$4x = 8$ If $b^x = b^y$, then $x = y$.

$x = 2$

Check:

$$16^2 = 256$$ Substitute $x = 2$ in the original equation.

The solution set is $\{2\}$.   ■

EXAMPLE 5

Solve for x: $25^x = 3125$

$25 = 5^2$, $3125 = 5^5$ Write each base as a power of the same base. Each is a power of 5.

$25^x = 3125$

$(5^2)^x = 5^5$ Substitute 5^2 for 25 and 5^5 for 3125.

$5^{2x} = 5^5$ Property of exponents for raising a power to a power.

$2x = 5$ If $b^x = b^y$, then $x = y$.

$$x = \frac{5}{2}$$ The check is left for the student.

The solution set is $\left\{ \dfrac{5}{2} \right\}$.   ■

The procedure in Examples 4 and 5 works only when both numbers can be written as a power of the same base by inspection. If it can't be done by inspection, more "tools" are needed. See Section 10.5.

EXERCISE 10.1

A

Complete the following tables of values:

1. $f(x) = 3^x$

x	-3	-2	-1	0	1	2	3
$f(x)$							

2. $f(x) = \left(\dfrac{1}{3} \right)^x$

x	-3	-2	-1	0	1	2	3
$f(x)$							

3. $f(x) = 5^x$

x	-3	-2	-1	0	1	2	3
$f(x)$							

4. $f(x) = \left(\dfrac{1}{5}\right)^x$

x	-3	-2	-1	0	1	2	3
$f(x)$							

5. $f(x) = \left(\dfrac{2}{3}\right)^x$

x	-3	-2	-1	0	1	2	3
$f(x)$							

6. $f(x) = \left(\dfrac{3}{2}\right)^x$

x	-3	-2	-1	0	1	2	3
$f(x)$							

7. $f(x) = \left(\dfrac{3}{4}\right)^x$

x	-3	-2	-1	0	1	2	3
$f(x)$							

8. $f(x) = \left(\dfrac{4}{3}\right)^x$

x	-3	-2	-1	0	1	2	3
$f(x)$							

9. $f(x) = (10)^x$

x	-3	-2	-1	0	1	2	3
$f(x)$							

10. $f(x) = \left(\dfrac{1}{10}\right)^x$

x	-3	-2	-1	0	1	2	3
$f(x)$							

11. $f(x) = \left(\dfrac{5}{2}\right)^x$

x	-3	-2	-1	0	1	2	3
$f(x)$							

12. $f(x) = \left(\dfrac{2}{5}\right)^x$

x	-3	-2	-1	0	1	2	3
$f(x)$							

Solve:

13. $2^x = 32$ **14.** $3^x = 81$ **15.** $4^x = 64$ **16.** $5^x = 125$

B

Use the appropriate tables in Exercises 1 to 12 to graph the function defined by each of these equations. Write the domain and range for each function:

17. $f(x) = 3^x$

18. $f(x) = \left(\dfrac{1}{3}\right)^x$

19. $f(x) = 5^x$

20. $f(x) = \left(\dfrac{1}{5}\right)^x$

21. $f(x) = \left(\dfrac{3}{4}\right)^x$

22. $f(x) = \left(\dfrac{4}{3}\right)^x$

Solve:

23. $\left[\dfrac{1}{2}\right]^x = \dfrac{1}{8}$

24. $\left[\dfrac{1}{3}\right]^x = \dfrac{1}{27}$

25. $\left[\dfrac{2}{3}\right]^x = \dfrac{16}{81}$

26. $\left[\dfrac{4}{5}\right]^x = \dfrac{256}{625}$

C

Graph the function defined by each of these equations. Write the domain and range of each:

27. $f(x) = 10^x$

28. $f(x) = (0.1)^x$

29. $f(x) = \left(\dfrac{3}{2}\right)^x$

30. $f(x) = \left(\dfrac{2}{3}\right)^x$

31. $f(x) = \left(\dfrac{5}{2}\right)^x$

32. $f(x) = \left(\dfrac{2}{5}\right)^x$

33. $f(x) = (1.2)^x$

34. $f(x) = (0.3)^x$

35. $f(x) = 2^x + 2$

36. $f(x) = 3^x - 1$

Solve:

37. $2^{-x} = \left[\dfrac{1}{16}\right]$

38. $3^{-x} = \left[\dfrac{1}{81}\right]$

39. $\left[\dfrac{1}{2}\right]^{-x} = 16$

40. $\left[\dfrac{1}{3}\right]^{-x} = 81$

41. $\left[\dfrac{3}{4}\right]^{-x} = \dfrac{64}{27}$

42. $\left[\dfrac{2}{3}\right]^x = \dfrac{81}{16}$

D

43. An isotope of radium has a half-life of 1600 years. The formula for the amount left after t years is $A = (0.999567)^t$. How much of 1 mg of radium is left after 5 years?

44. In Exercise 43, how much of the 1 mg of radium is left after 25 years?

45. Based on the growth of the population for the past 20 years, the formula for the future world population is (approximately) $P = 4e^{0.2t} \times 10^9$, where t is the number of years and $e \approx 2.718$. What is the population now $(t = 0)$? What will the population be (approximately) 10 years from now?

46. If the world population continues to grow according to the formula in Exercise 45, what will the population be (approximately) 20 years from now? Do you think the population will continue to grow at this rate?

47. The rate for radioactive decay can be expressed by the exponential function $y = y_0 e^{bt}$, where y_0 is the amount originally present, t measures time in years, y is the amount present after an interval t, and b is constant.

   The decay rate of a certain radioactive substance is expressed by $y = y_0 e^{-0.0125t}$, where t is measured in years. If you begin with 100 mg of the substance, how much is left after 10 years? ($e \approx 2.718$) (Round to the nearest tenth.)

48. The number of bacteria in a given culture is given by the exponential function $y = y_0 2^t$, where y_0 is the amount originally present, t is the amount of time (in hours) after the original count, and y is the amount currently present. If there were 5000 bacteria initially present, how many are present after 6 hours?

49. The formula for interest compounded quarterly is $A = P\left(1 + \dfrac{r}{4}\right)^n$, where A is the total amount on deposit after n quarter-year periods if P dollars are invested at r percent per year.

   Ms. Skinflint invests $1000 at 8% interest compounded quarterly, which is to be given to her first great-grandchild upon its twenty-first birthday. If the great-granddaughter is 21 years of age 40 years after the money is invested (160 periods), how much does she receive? (Round to the nearest dollar.)

50. If $10,000 is invested at 6% compounded monthly, how much will this amount to at the end of 20 years? The formula is $A = P\left(1 + \dfrac{r}{12}\right)^n$ (see Exercise 49), where n represents the number of monthly periods.

51. If $1000 is invested at 8% compounded semiannually, how much will this amount to at the end of 10 years? The formula is $A = P\left(1 + \dfrac{r}{2}\right)^n$, where n represents the number of semiannual periods.

52. If $5000 is invested at 10% compounded semiannually, how much will this amount to at the end of 5 years? (Use the formula in Exercise 51.)

53. The amount of bacteria in a culture is given by $N = N_0 e^{0.06t}$, where N_0 is the initial amount of bacteria, t is the time in hours, and N is the amount present after "t" hours. If the initial bacterial count is 3500, how many are there after 5 hours? Let $e \approx 2.718$. Round your answer to the nearest whole number.

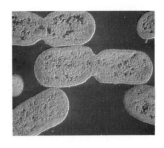

Figure for Exercise 53

54. The amount of current (in amps) in a certain circuit drops according to $I = I_0(1 - e^{-t/2})$, where I_0 is the initial current, t is the time in seconds, and I is the current after t seconds. If the initial current is 2.5 amps, what is the current after 2.4 seconds? (answer to the nearest hundredth) Let $e \approx 2.718$

55. The approximate density of sea water at a depth of "h" miles is given by $d = 64e^{0.0068\,h}$. What is the density (in pounds per cubic foot) at a depth of 2.2 miles? (nearest hundredth) Let $e \approx 2.718$

56. The value of any piece of property that depreciates over time is given by $V = V_0\,e^{-kt}$, where V_0 is the original price, k is the percentage that the item depreciates per year (as a decimal) and t is the number of years since the item was purchased. Find the present value (V), if the original price was $100,000, the rate of depreciation is 10%, and the item is 10 years old. (nearest hundredth) Let $e \approx 2.718$

57. Over several years of testing, it has been found that the percentage (P) of a city's population that contributes to a charity fund-raiser is given by $P = 70(1 - e^{-0.04\,t})$ where t is the time in days that the campaign has run. What percentage of the population of the city will have contributed after 20 days? (nearest tenth) Let $e \approx 2.718$

STATE YOUR UNDERSTANDING

58. Define an exponential function.

59. Why, arithmetically, is the curve $f(x) = 2^x$ asymptotic to the x-axis. In other words, why does the curve get closer to the x-axis for negative values of x?

60. To solve an exponential equation, what do you have to do to both sides of the equation?

61. The graph of $f(x) = 3^x$ and the graph of $g(x) = -3^x$ are symmetric to each other. What is the "line of symmetry"? Are they inverses of each other?

CHALLENGE EXERCISES

Solve:

62. $2^{x+1} = 8$

63. $3^{x-1} = \dfrac{1}{27}$

64. $\left(\dfrac{1}{2}\right)^{x+1} = 4$

65. $2^{x-2} = 4^x$

66. $2^{x-3} = 8^{x+2}$

67. $4^{x+1} = 16^{x-3}$

68. $3^{x+1} = 1$

69. $\left(\dfrac{1}{2}\right)^{x+1} = 4^x$

70. $2^{2x-3} = 4^{x+1}$

71. Using the table of values in Exercise 1 as a basis, generate a table of values for $g(x) = -3^x$ and graph the curve.

72. Create a table of values for the function and graph: $f(x) = 3^{-x}$ Does it look like any curve you have previously drawn?

73. Create two tables of values and graph both curves on the same graph:
(a) $f(x) = 3^x + 1$ and (b) $g(x) = 3^{x+1}$

74. Do the curves $y = 2^{x+1}$ and $y = 4^x$ intersect? Where? (Solve the system algebraically)

75. Do the curves $y = 4^{3x-1}$ and $y = 8^{2x-3}$ intersect? Where? (Solve the system algebraically)

MAINTAIN YOUR SKILLS (SECTION 9.5)

Write the equation of the inverse of the following functions:

76. $f(x) = x - 3$

77. $f(x) = x + 4$

78. $f(x) = 2x + 3$

79. $f(x) = 3x - 4$

80. $f(x) = \dfrac{1}{2}x + 2$

81. $f(x) = \dfrac{1}{3}x - 4$

82. $f(x) = x^2 + 2$

83. $f(x) = x^3 - 1$

10.2

■

LOGARITHMIC FUNCTIONS

_____

1. Write exponential equations in logarithmic form.

2. Write logarithmic equations in exponential form.

3. Draw the graph of a logarithmic function.

4. Solve a logarithmic equation.

_____

Since an exponential function is one-to-one, the inverse is also a function. The equation that defines the inverse is found by interchanging the variables x and y. Thus the inverse of

$$y = b^x, \quad b > 0 \quad \text{and} \quad b \neq 1$$

is

$$x = b^y, \quad b > 0 \quad \text{and} \quad b \neq 1.$$

The inverse is not written explicitly in terms of y, so we define a new symbol. This special symbol, "log" (abbreviation of "logarithm"), is used to denote the inverse of an exponential function.

■ FORMULA

Logarithmic Function

> The equivalent equations
>
> $$x = b^y \quad \text{and} \quad y = \log_b x, \qquad x > 0, \qquad b > 0, \qquad b \neq 1$$
>
> both define the *logarithmic function* base b.

Since the logarithmic function and exponential function are inverses of each other, x and y are interchanged. Therefore, the domain and range are interchanged.

Equation	Domain	Range
$y = 2^x$	All reals	Positive reals
$x = 2^y$ (inverse)	Positive reals	All reals

The usual step is to solve the inverse equation for y. In this case, the conventional methods of solving for y cannot be used. So we use the new term "$\log_b$" to write the inverse. We write

if $x = 2^y$ then $y = \log_2 x$.

This is read, "if $x = 2$ to the y, then y equals log x, base 2." The following table shows the relation between the exponential and log forms.

Exponential Form	Log Form	Vocabulary
$2^3 = 8$	$\log_2 8 = 3$	2 is the base 8 is the power 3 is the exponent or logarithm
$10^2 = 100$	$\log_{10} 100 = 2$	10 is the base 100 is the power 2 is the exponent or logarithm
$5^{-2} = \dfrac{1}{25}$	$\log_5 \dfrac{1}{25} = -2$	5 is the base $\dfrac{1}{25}$ is the power -2 is the exponent or logarithm
$\left(\dfrac{1}{2}\right)^{-3} = 8$	$\log_{1/2} 8 = -3$	$\dfrac{1}{2}$ is the base 8 is the power -3 is the exponent or logarithm

The two equations $x = b^y$ and $y = \log_b x$ are equivalent, so they refer to the same set of ordered pairs. These ordered pairs constitute a function. Just as each different base defines a different exponential function, so each different base defines a different logarithmic function. Note that by this definition a logarithm is an exponent. In both equations

$y = \log_b x$ and $x = b^y$

y is the exponent of the base b that will yield the power x.

EXAMPLE 1

Write $3^4 = 81$ in logarithmic form.

$3^4 = 81$ 3 is the base, 4 is the logarithm or exponent, and 81 is the power.

$\log_3 81 = 4$ ∎

Examples 2 and 3 show that even if the exponent is not a whole number, the process is still the same.

EXAMPLE 2

Write $16^{1/2} = 4$ in logarithmic form.

$16^{1/2} = 4$ 16 is the base, $\dfrac{1}{2}$ is the exponent or logarithm, and

$\log_{16} 4 = \dfrac{1}{2}$ 4 is the power.   ∎

EXAMPLE 3

Write $3^{-2} = \dfrac{1}{9}$ in logarithmic form.

$3^{-2} = \dfrac{1}{9}$ 3 is the base, -2 is the exponent or logarithm, and

$\log_3 \left[\dfrac{1}{9}\right] = -2$ $\dfrac{1}{9}$ is the power.   ∎

Changing from logarithmic form to exponential form is just a matter of reversing the process.

EXAMPLE 4

Write $\log_3 9 = 2$ in exponential form.

$\log_3 9 = 2$ 3 is the base, 2 is the exponent, and 9 is the

$\qquad 3^2 = 9$ power.   ∎

EXAMPLE 5

Write $\log_7 \sqrt{7} = \dfrac{1}{2}$ in exponential form.

$\log_7 \sqrt{7} = \dfrac{1}{2}$ 7 is the base, $\dfrac{1}{2}$ is the exponent or logarithm, and

$\qquad 7^{1/2} = \sqrt{7}$ $\sqrt{7}$ is the power.   ∎

EXAMPLE 6

Write $\log_5 \left[\dfrac{1}{25}\right] = -2$ in exponential form.

$\log_5 \left[\dfrac{1}{25}\right] = -2$ 5 is the base, -2 is the exponent or logarithm, and $\dfrac{1}{25}$

$\qquad (5)^{-2} = \dfrac{1}{25}$ is the power.   ∎

The inverse relation between logarithmic and exponential functions gives us a convenient way to make tables of values for the logarithmic functions. We make a table for the logarithmic form and use our knowledge of the exponential form to find y.

x	Log Form	Exponential Form		y or $\text{Log}_2 x$
x	$y = \log_2 x$	$x = 2^y$		y
$\dfrac{1}{8}$	$y = \log_2 \dfrac{1}{8}$	$\dfrac{1}{8} = 2^y$ or	$2^{-3} = 2^y$	-3
$\dfrac{1}{4}$	$y = \log_2 \dfrac{1}{4}$	$\dfrac{1}{4} = 2^y$ or	$2^{-2} = 2^y$	-2
$\dfrac{1}{2}$	$y = \log_2 \dfrac{1}{2}$	$\dfrac{1}{2} = 2^y$ or	$2^{-1} = 2^y$	-1
1	$y = \log_2 1$	$1 = 2^y$ or	$2^0 = 2^y$	0
2	$y = \log_2 2$	$2 = 2^y$ or	$2^1 = 2^y$	1
4	$y = \log_2 4$	$4 = 2^y$ or	$2^2 = 2^y$	2

We can use this information to graph a logarithmic function.

EXAMPLE 7

Graph the function defined by $y = \log_2 x$, using the above table of values.

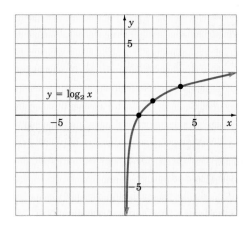

Using these pairs of numbers and connecting the points with a smooth curve, we get the graph.

When we graph both the logarithmic curve of Example 7 and the exponential curve with base two, we can see the reflection of the graphs about the line $y = x$ (Figure 10.3).

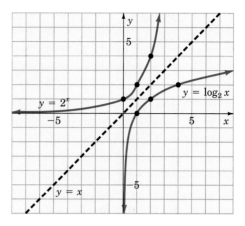

FIGURE 10.3

EXAMPLE 8

Complete a table of values, and graph $y = \log_3 x$.

x	$\dfrac{1}{9}$	$\dfrac{1}{3}$	1	3	9
y	-2	-1	0	1	2

Change $y = \log_3 x$ to exponential form to make a table of values.

$x = 3^y$

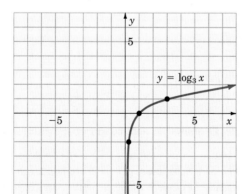

Plot the points and connect with a smooth curve.

Note: The y-axis is a vertical asymptote.

Domain: $\{x \mid x > 0\}$
Range: $\{y \mid y \in R\}$

3^y cannot be zero or negative. From the graph, we see that there are no restrictions on y. ■

To solve a logarithmic equation, we first change from logarithmic form to exponential form. Then solve the exponential equation.

EXAMPLE 9

Solve for w: $w = \log_4 16$

$4^w = 16$	Change from log form to exponential form. 4 is the base.
$4^w = 4^2$	w is the exponent. 16 is the power.
$w = 2$	If $b^x = b^y$, then $x = y$.

The solution set is $\{2\}$.   ■

EXAMPLE 10

Solve for w: $w = \log_{1/16} 64$

$$\left[\frac{1}{16}\right]^w = 64 \qquad \text{Change to exponential form.}$$

$$\left[\frac{1}{2^4}\right]^w = 2^6 \qquad \text{Rewrite both using base 2.}$$

$$[2^{-4}]^w = 2^6 \qquad\qquad \frac{1}{2^4} = 2^{-4}$$

$$2^{-4w} = 2^6$$

$$-4w = 6 \qquad \text{If } b^x = b^y, \text{ then } x = y.$$

$$w = -\frac{3}{2}$$

The solution set is $\left\{-\dfrac{3}{2}\right\}$.   ■

Example 11 shows an application of logarithms in chemistry.

Chemists use a measurement called pH to describe the degree of acidity or alkalinity of a substance. pH measurements run from 0 (strong acid such as battery acid) through 7 (neutral; pure water has pH 7) to 14 (strong alkali such as lye). The pH of most food and drink is less than 7 (lemonade may be about 3, coffee about 5), whereas the pH of most cleaning agents is more than 7 (soap about 7.5, detergents about 9.5, and ammonia about 12). Human blood has a pH of about 7.4.

EXAMPLE 11

The pH of a chemical solution is given by the formula

$$pH = \log_{10}\left(\frac{1}{H^+}\right),$$

where H^+ represents the hydrogen ion concentration. Find the pH of a solution with $H^+ = 1 \times 10^{-3}$.

Formula:

$$pH = \log_{10}\left(\frac{1}{H^+}\right)$$

To find the pH, substitute the hydrogen ion concentration in the formula.

Substitute:

$$pH = \log_{10}\left(\frac{1}{1 \times 10^{-3}}\right) \qquad H^+ = 1 \times 10^{-3}$$

Solve:

$$pH = \log_{10} 10^3 \qquad\qquad \frac{1}{1 \times 10^{-3}} = \frac{1}{10^{-3}} = 10^3$$

$$10^{pH} = 10^3 \qquad\qquad \text{Write in exponential form.}$$

$$pH = 3 \qquad\qquad \text{(3 is approximately the pH of carbonated water.)}$$

Answer:

The pH of the chemical solution is 3. ■

EXERCISE   10.2

A

Write each logarithmic equation in exponential form:

1. $\log_3 27 = 3$ **2.** $\log_2 8 = 3$ **3.** $\log_{10} 1 = 0$

4. $\log_{10} 10 = 1$ **5.** $\log_x x^2 = 2$ **6.** $\log_y y^5 = 5$

7. $\log_5 \sqrt{5} = \frac{1}{2}$ **8.** $\log_6 \sqrt[4]{6} = \frac{1}{4}$ **9.** $\log_{1/2} \frac{1}{8} = 3$

10. $\log_{1/3} \frac{1}{9} = 2$

Write each exponential equation in logarithmic form:

11. $4^2 = 16$ **12.** $8^2 = 64$ **13.** $2^6 = 64$

14. $3^4 = 81$ **15.** $\left(\frac{1}{2}\right)^4 = \frac{1}{16}$ **16.** $\left(\frac{1}{3}\right)^3 = \frac{1}{27}$

17. $8^{1/2} = \sqrt{8}$ 　　　　　　**18.** $10^{1/5} = \sqrt[5]{10}$ 　　　　　**19.** $5^{-2} = \dfrac{1}{25}$

20. $3^{-4} = \dfrac{1}{81}$

B

Find w:

21. $w = \log_6 36$ 　　　　　**22.** $w = \log_{12} 144$ 　　　　　**23.** $w = \log_{11} \sqrt{11}$

24. $w = \log_6 \sqrt{6}$ 　　　　　**25.** $w = \log_{10} \dfrac{1}{10}$ 　　　　　**26.** $w = \log_{10} 10^{-2}$

27. $w = \log_7 49$ 　　　　　**28.** $w = \log_8 512$ 　　　　　**29.** $w = \log_5 \dfrac{1}{625}$

30. $w = \log_4 \dfrac{1}{1024}$ 　　　　**31.** $w = \log_{81} 9$ 　　　　　**32.** $w = \log_{121} \dfrac{1}{11}$

33. $w = \log_{1/4} 16$ 　　　　**34.** $w = \log_{1/2} 8$

Complete the table of values:

35. $y = \log_4 x$

x	$\dfrac{1}{16}$	$\dfrac{1}{4}$	1	4	16	64
y						

36. $y = \log_5 x$

x	$\dfrac{1}{25}$	$\dfrac{1}{5}$	1	5	25	125
y						

37. $y = \log_{10} x$

x	$\dfrac{1}{100}$	$\dfrac{1}{10}$	1	100	1000
y					

38. $y = \log_8 x$

x	$\dfrac{1}{64}$	$\dfrac{1}{8}$	1	8	64
y					

39. $y = \log_{1/2} x$

x	4	2	1	$\dfrac{1}{2}$	$\dfrac{1}{4}$
y					

40. $y = \log_{1/3} x$

x	9	3	1	$\dfrac{1}{3}$	$\dfrac{1}{9}$
y					

C

Find z:

41. $z = \log_4 8$ 　　　　　**42.** $z = \log_9 27$ 　　　　　**43.** $z = \log_{25} \dfrac{1}{5}$

44. $z = \log_{64} \dfrac{1}{16}$ 　　　　**45.** $z = \log_{32} 16$ 　　　　　**46.** $z = \log_{64} 32$

47. $z = \log_{2/5} \left(\dfrac{25}{4}\right)$ 　　　**48.** $z = \log_{1/9} 27$ 　　　　**49.** $z = \log_{3/2} \left(\dfrac{8}{27}\right)$

50. $z = \log_{1/16} 2$

51. $z = \log_{1/25} 125$

52. $z = \log_{1/64} 16$

Use the appropriate tables in Exercises 35 to 40 to sketch the graphs of the following; state the domain and range:

53. $y = \log_4 x$

54. $y = \log_5 x$

55. $y = \log_{10} x$

56. $y = \log_8 x$

57. $y = \log_{1/2} x$

58. $y = \log_{1/3} x$

D

59. The pH of a chemical solution is given by the formula

$$pH = \log_{10}\left[\frac{1}{H^+}\right],$$

where H^+ represents the hydrogen ion concentration. Find the pH of a solution with an H^+ value of 1×10^{-2}.

60. Find the pH of a solution with an H^+ value of 1×10^{-7}.

61. Find the pH of a solution with an H^+ value of 1×10^5.

62. Find the pH of a solution with an H^+ value of $1 \times 10^{2.2}$.

63. After a test in math, it is determined that the rate of remembering can be expressed by the function defined by $F(t) = 78 - 12\log_{10}(1 + t)$, where 78 is the average score on the original test, t is the time in weeks after the test is given, and $F(t)$ is the average score obtained on the retest. What average score can be expected after a period of 9 weeks has elapsed?

64. In the function of Exercise 63, what average score can be expected after 99 weeks have passed?

65. The age (t, in years) of an object is given by approximately $t = -18000 \log N_0$, where N_0 is the percentage (as a decimal) of the total carbon remaining that is Carbon 14. If 1% of the total carbon left in a bone is Carbon 14, what is the approximate age? What if the percentage is 0.1%, what is the approximate age?

66. The rating (R) of an earthquake on the Richter Scale is given by the formula $R = \log \dfrac{E}{E_0}$, where E is the energy released by the current quake and E_0 is the energy released by the earthquake used for comparison called the zero quake. If the ratio, $\dfrac{E}{E_0}$ is given as 1,000,000, what is the rating on the Richter Scale?

67. The loudness of sound in decibels (dB) is given by $dB = 10 \log \dfrac{P}{P_0}$, where P is the power, in amps, of the measured sound and P_0 is the power, in amps, of the lowest sound you can hear. If the ratio, $\dfrac{P}{P_0}$, for a jet plane is given as 1,000,000,000,000, what is the dB rating?

68. What is the decibel rating of a classical music selection if the ratio of $\dfrac{P}{P_0}$ is 100,000,000?

STATE YOUR UNDERSTANDING

69. What is the relationship between an exponential function and a logarithmic function?

70. Given the exponential equation $2^{x+1} = 8$, rewrite it in logarithmic form. Do not solve.

71. $y = b^x$ (for $b > 1$) is always an increasing function; what about $y = \log_b x$ (for $b > 1$)? Explain.

72. $y = b^x$ (for $0 < b < 1$) is always a decreasing function; what about $y = \log_b x$ (for $0 < b < 1$)? Explain.

CHALLENGE EXERCISES

Solve:

73. $3 = \log_2 x$

74. $\log_9 x = \dfrac{1}{2}$

75. $\log_4 x = -3$

76. $x + 1 = \log_4 16$

77. $\log_5 5 = 2x - 4$

78. $\log_9 27 = x + 1$

79. $\log_4 8 = 2x + 1$ **80.** $\log_x 9 = 2$ (Careful)   **81.** $\log_x 1 = 0$ (Careful)

82. Create a table of values and graph $y = -\log_2 x$ **83.** Create a table of values and graph $y = \log_2 (x + 1)$

84. Create a table of values and graph $y = (\log_2 x) + 1$

MAINTAIN YOUR SKILLS (SECTIONS 2.1, 2.2, 2.3, 2.6)

Perform the indicated operations:

85. $(3a^2b^3c)(-15a^6b^5c^3)$ **86.** $(-3c^4d^5)^3$ **87.** $\left(\dfrac{3x}{4y}\right)^{-2}$

88. $\dfrac{(5x^2y^3)^{-3}}{(20xy^4)^{-2}}$ **89.** $(x + y)^2(x + y)^4$ **90.** Write 8.2×10^{-5} in decimal form.

91. Write 0.00000271 in scientific notation.   **92.** Write 3.5×10^8 in decimal form.

10.3

PROPERTIES OF LOGARITHMS

OBJECTIVE

1. Write logarithmic expressions in different forms using the basic properties of logarithms.

By definition, the equations $y = \log_b x$ and $b^y = x$ are equivalent. Therefore, in the exponential form of the statement when y is replaced with $\log_b x$, we have

$b^{\log_b x} = x$

$$b^{\log_b x} = x.$$

We expect this since the exponential and logarithmic functions are inverse functions.

The logarithm of an exponential expression (using the same base) is the exponent.

$\log_b b^x = x$

$\log_b b^x = x$

Since exponential and logarithmic functions are inverse functions, they are each one-to-one functions. From this fact, we conclude that if $x = y$, then $\log_b x = \log_b y$ and vice-versa.

▦ PROPERTY

1. $b^{\log_b x} = x$
2. $\log_b b^x = x$
3. If $x = y$, then $\log_b x = \log_b y$.
4. If $\log_b x = \log_b y$, then $x = y$.

Other properties of logarithms are based on the laws of exponents.

▦ PROPERTY

Property (in Symbols)	Property (in Words)
5. $\log_b PQ = \log_b P + \log_b Q$	The logarithm of a product is the sum of the logarithms of the factors.
6. $\log_b \dfrac{P}{Q} = \log_b P - \log_b Q$	The logarithm of a quotient is the difference of the logarithm of the dividend and the logarithm of the divisor.
7. $\log_b P^n = n \log_b P$	The logarithm of a power is the exponent times the logarithm of the base.
8. $\log_b \sqrt[n]{P} = \dfrac{1}{n} \log_b P$	The logarithm of the root of a number is the reciprocal of the index number times the logarithm of the number.

We will derive Property Five and Property Seven; Properties Six and Eight, left as exercises, can be proved in a similar manner.

Property Five

$\log_b PQ = \log_b P + \log_b Q$

Let $s = \log_b P$ and $t = \log_b Q$

then

$b^s = P$ and $b^t = Q$	Definition of logarithm.
$PQ = b^s \cdot b^t$	Multiply P and Q.
$PQ = b^{s+t}$	Property of exponents.
$\log_b PQ = \log_b b^{s+t}$	Take the log of both sides. Property Three.
$\quad = s + t$	$\log_b b^x = x$; logarithms and exponentials are inverse functions. Property Two.
$\quad = \log_b P + \log_b Q$	Substitute: $s = \log_b P$, $t = \log_b Q$.

Property Seven

$\log_b P^n = n \log_b P$

Let $P = b^s$ so $s = \log_b P$	Logarithm form.

then

$P^n = (b^s)^n$	Raise both sides to the nth power.
$P^n = b^{sn}$	Property of exponents.
$\log_b P^n = \log_b b^{sn}$	Take the log of both sides.
$\quad = sn$	$\log_b b^x = x$. Property Two.
$\quad = n \log_b P$	Substitute: $s = \log_b P$.

Note that the properties of logarithms apply only to logs of products (multiplication), quotients (division), powers, and roots. The properties *do not* apply to addition or subtraction. There is no property for the expression $\log_b (P + Q)$.

For instance, we can see that

$$\log_{10}(10^2 + 10^3) \neq \log_{10} 10^2 + \log_{10} 10^3$$
$$\log_{10} (100 + 1000) \neq 2 + 3$$
$$\log_{10} 1100 \neq 5$$
$$10^5 \neq 1100$$

Since $b^0 = 1$ and $b^1 = b$, we have two more properties of logarithms:

■ PROPERTY

9. $\log_b 1 = 0$ The logarithm of 1 to any base b is 0.

10. $\log_b b = 1$ The logarithm of the base number to any base b is 1.

Before the days of the hand-held electronic calculators, these properties (together with a table of values of logarithms) were used to perform calculations involving multiplication, division, powers, and roots. Now these properties are used primarily to write log expressions in different forms and to solve logarithmic and exponential equations and formulas. For example,

$$\log_b x + \log_b y - \log_b w$$

can be written as a single logarithm using Properties Five and Six.

$$\log_b x + \log_b y - \log_b w = \log_b xy - \log_b w \qquad \text{Log of a product.}$$

$$= \log_b \frac{xy}{w} \qquad \text{Log of a quotient.}$$

Using the properties in the other direction, the expression

$$\log_b \left(\frac{x^3 \sqrt{y}}{w^2} \right)$$

can be written as the combination of logs of single variables as follows:

$$\log_b \left(\frac{x^3 \sqrt{y}}{w^2} \right) = \log_b x^3 y^{1/2} - \log_b w^2 \qquad \text{Log of a quotient.}$$

$$= \log_b x^3 + \log_b y^{1/2} - \log_b w^2 \qquad \text{Log of a product.}$$

$$= 3 \log_b x + \frac{1}{2} \log_b y - 2 \log_b w \qquad \text{Log of a base to a power.}$$

EXAMPLE 1

Write $\log_b m - \log_b n - \log_b r$ as a single logarithm.

$$\log_b m - \log_b n - \log_b r \qquad \text{Original expression.}$$

$$\log_b \frac{m}{n} - \log_b r \qquad \text{The difference of two logs is the log of a quotient.}$$

$$\log_b \frac{m}{nr} \qquad \text{The difference of two logs is the log of a quotient.} \quad \blacksquare$$

EXAMPLE 2

Write $3 \log_b a + \frac{1}{2} \log_b c$ as a single logarithm.

$$3 \log_b a + \frac{1}{2} \log_b c \qquad \text{Original expression.}$$

$$\log_b a^3 + \log_b c^{1/2} \qquad \text{Log of a base to a power.}$$

$$\log_b a^3 \sqrt{c} \qquad \text{The sum of two logs is the log of a product. Also } c^{1/2} = \sqrt{c}. \quad \blacksquare$$

EXAMPLE 3

Write $\dfrac{1}{3}(\log_b 5 - 2\log_b 7)$ as a single logarithm.

$\dfrac{1}{3}(\log_b 5 - 2\log_b 7)$ Original expression.

$\dfrac{1}{3}(\log_b 5 - \log_b 7^2)$ Log of a base raised to a power.

$\dfrac{1}{3}\log_b \dfrac{5}{49}$ The difference of two logs is the log of a quotient.

$\log_b \left(\dfrac{5}{49}\right)^{1/3}$ or $\log_b \sqrt[3]{\dfrac{5}{49}}$ Log of a base raised to a power.   ■

EXAMPLE 4

Write $\log_b \dfrac{99(574)}{237}$ as a combination of logarithms of single numbers.

$\log_b \dfrac{99(574)}{237}$ Original expression.

■ **CAUTION**

$$\log_b \frac{p}{q} \neq \frac{\log_b p}{\log_b q}$$

$\log_b 99(574) - \log_b 237$ Log of a quotient.

■ **CAUTION**

$$\log_b pq \neq (\log_b p)(\log_b q)$$

$\log_b 99 + \log_b 574 - \log_b 237$ Log of a product.   ■

EXAMPLE 5

Write $\log_b \sqrt{\dfrac{x}{y}}$ as a combination of logarithms of single variables.

$\log_b \sqrt{\dfrac{x}{y}}$ Original expression.

$\log_b \left(\dfrac{x}{y}\right)^{1/2}$ $\sqrt{m} = m^{1/2}$

$\dfrac{1}{2}\log_b \dfrac{x}{y}$	Log of a base raised to a power.
$\dfrac{1}{2}(\log_b x - \log_b y)$	Log of a quotient.
$\dfrac{1}{2}\log_b x - \dfrac{1}{2}\log_b y$	Distributive property. ∎

EXAMPLE 6

Write $\log_b \dfrac{1}{w^4 z^7}$ as a combination of logarithms of single variables.

$\log_b \dfrac{1}{w^4 z^7}$	Original expression.
$\log_b 1 - \log_b w^4 z^7$	Log of a quotient.
$0 - \log_b w^4 z^7$	$\log_b 1 = 0 \ (b^0 = 1)$
$-(\log_b w^4 + \log_b z^7)$	Log of a product.
$-(4 \log_b w + 7 \log_b z)$	Log of a base raised to a power.
$-4 \log_b w - 7 \log_b z$	∎

Example 7 is another application of logarithms from chemistry. A scientific calculator is useful to find the pH of a solution.

EXAMPLE 7

The hydrogen ion concentration of a chemical solution is $H^+ = 0.000005$. Find the pH of the solution.

Formula:

$$pH = \log_{10} \frac{1}{H^+}$$

Substitute:

$$pH = \log_{10} \frac{1}{5 \times 10^{-6}} \qquad H^+ = 0.000005 = 5 \times 10^{-6}$$

Solve:

$$= \log_{10} 200{,}000$$

Use the reciprocal key to find the reciprocal of 5×10^{-6}.

$$\approx 5.3010$$

Use the log key.   ∎

The pH of the solution is approximately 5.3.

The Richter scale is a measure of the intensity of an earthquake. The scale uses base 10 logarithms.

EXAMPLE 8

The magnitude $M(i)$ on the Richter scale for a given earthquake is defined to be

$$M(i) = \log_{10} \frac{i}{i_0},$$

where i is the amplitude of the ground motion of the quake, and i_0 is the amplitude of the ground motion of a "zero" quake. What is the magnitude on the Richter scale of an earthquake that is 1000 times stronger than the zero earthquake?

Formula:

$$M(i) = \log_{10} \frac{i}{i_0}$$

Substitute:

$$M(1000i_0) = \log_{10} \frac{1000i_0}{i_0} \qquad i = 1000i_0$$

Solve:

$$= \log_{10} 1000 \qquad \text{Reduce.}$$
$$= \log_{10} 10^3 \qquad \text{Write 1000 as a power of 10.}$$
$$= 3 \log_{10} 10 \qquad \text{Log of a base raised to a power.}$$
$$= 3(1) \qquad \log_{10} 10 = 1$$
$$= 3$$

Answer:

The Richter scale reads 3. ■

Examples 9, 10, and 11 show how to use the properties of logarithms to find other logarithm values.

EXAMPLE 9

Given that $\log_{10} 2 = 0.30103$, find $\log_{10} 16$.

$16 = 2^4$	Write 16 as a power of 2.
$\log_{10} 16 = \log 2^4$	Substitute 2^4 for 16.
$= 4(\log 2)$	Log of a base raised to a power.
$= 4(0.30103)$	
$= 1.20412$	■

EXAMPLE 10

Given that $\log_{10} 2 = 0.30103$ and $\log_{10} 3 = 0.47712$, find $\log_{10} 36$.

$36 = 2^2 \cdot 3^2$	Prime factor 36.
$\log_{10} 36 = \log_{10} (2^2 \cdot 3^2)$	Substitute $2^2 \cdot 3^2$ for 36.
$= (\log_{10} 2^2) + (\log_{10} 3^2)$	Log of a product.
$= 2(\log_{10} 2) + 2(\log_{10} 3)$	Log of a base raised to a power.
$= 2(0.30103) + 2(0.47712)$	
$= 1.5563$	■

EXAMPLE 11

Given that $\log_{10} 2 = 0.30103$ and $\log_{10} 3 = 0.47712$, find $\log_{10} \sqrt{180}$.

$180 = 2 \cdot 3^2 \cdot 10$	Factor 180 into factors of 2, 3, and 10.
$\log_{10} \sqrt{180} = \log_{10} (180)^{1/2}$	Change the radical to an exponent.
$= 0.5(\log_{10} 180)$	$\dfrac{1}{2} = 0.5$
$= 0.5[\log_{10} (2 \cdot 3^2 \cdot 10)]$	
$= 0.5[\log_{10} 2 + 2(\log_{10} 3) + \log_{10} 10]$	
$= 0.5[0.30103 + 2(0.47712) + 1]$	
$= 0.5[2.25527]$	
$= 1.127635$	■

Exercise 10.3

A

Write the following as logarithms of a single number:

1. $\log_b 6 + \log_b 8$

2. $\log_b 10 - \log_b 2$

3. $\log_{10} 250 - \log_{10} 50$

4. $\log_2 5 + \log_2 3$

5. $2 \log_5 8$

6. $3 \log_5 2$

7. $\dfrac{1}{2} \log_3 25$

8. $\dfrac{1}{3} \log_5 8$

9. $-2 \log_4 3$

10. $-3 \log_5 2$

Write the following as a combination of logarithms of a single number:

11. $\log_b 34(65)$

12. $\log_b \dfrac{12}{17}$

13. $\log_b (132)^4$

14. $\log_b (44)^{1/3}$

15. $\log_b 13(21)^2$

16. $\log_b 16\sqrt{15}$

Given $\log_{10} 5 = 0.69897$ and $\log_{10} 7 = 0.84510$, find the following:

17. $\log_{10} 25$

18. $\log_{10} 49$

19. $\log_{10} 35$

20. $\log_{10} 125$

B

Write the following as logarithms of single numbers:

21. $3 \log_b 4 + \log_b 3$

22. $2 \log_b 6 - \log_b 9$

23. $4 \log_b 2 - \dfrac{1}{2} \log_b 16$

24. $\dfrac{1}{3} \log_b 27 + 2 \log_b 5$

25. $2 \log_b 5 + \log_b 2 - \log_b 10$

26. $3 \log_b 2 + \log_b 9 - \log_b 6$

27. $\dfrac{1}{2} (\log_{10} 2 + \log_{10} 32)$

28. $\dfrac{1}{2} (\log_{10} 88 - \log_{10} 8)$

Write the following as combinations of logarithms of single variables:

29. $\log_b \left(\dfrac{z}{x} \right)$

30. $\log_4 \dfrac{ab}{c}$

31. $\log_{10} \dfrac{w^2}{z}$

32. $\log_{10} \dfrac{r}{s^3}$

33. $\log_5 \dfrac{x^2}{y^2}$

34. $\log_5 a^2 b^2 c$

35. $\log_b \dfrac{x^3}{yz^2}$

36. $\log_b \dfrac{x^2 y}{z^3}$

Given $\log_{10} 2 = 0.30103$, $\log_{10} 5 = 0.69897$ and $\log_{10} 3 = 0.47712$, find the following:

37. $\log_{10} 40$ **38.** $\log_{10} 90$ **39.** $\log_{10} 120$ **40.** $\log_{10} 180$

C

Write the following as logarithms of single numbers:

41. $2 \log_b 4 + 3 \log_b 2 - 3 \log_b 4$

42. $5(\log_b 2 + \log_b 3 - \log_b 6)$

43. $5(\log_b 2 + \log_b 3) - \log_b 6$

44. $3(\log_b 5 + \log_b 4) - 3 \log_b 10$

45. $\dfrac{1}{3} \log_b 8 + \dfrac{1}{2} \log_b 49$

46. $3 \log_c 4 + \dfrac{1}{4} \log_c 16$

47. $\dfrac{1}{3} \log_b 64 - \dfrac{1}{2} \log_b 4$

48. $\dfrac{1}{2} \log_c 121 - \dfrac{1}{2} \log_c 4 - \log_c 3$

Write the following as combinations of logarithms of single variables:

49. $\log_c z\sqrt{xy}$

50. $\log_x c\sqrt[3]{\dfrac{a}{b}}$

51. $\log_7 \dfrac{\sqrt{mn}}{p^2}$

52. $\log_7 \dfrac{st}{\sqrt[3]{r}}$

53. $\log_x \sqrt[3]{(a + b)^2}$

54. $\log_x \sqrt[4]{(x^2 + y^2)^3}$

55. $\log_b (x^2 y^3)^4$

56. $\log_b \left(\dfrac{x^4 y^3}{z^2}\right)^5$

Given $\log_{10} 2 = 0.30103$, $\log_{10} 3 = 0.47712$, $\log_{10} 5 = 0.69897$, and $\log_{10} 7 = 0.84510$, find the following:

57. $\log_{10} 84$ **58.** $\log_{10} 105$ **59.** $\log_{10} 315$ **60.** $\log_{10} 252$

D

61. The magnitude on the Richter scale for a given earthquake is defined by
$M(i) = \log_{10} \dfrac{i}{i_0}$, where i is the amplitude of the ground motion of the
quake, and i_0 is the amplitude of the ground motion of a "zero" quake.
What is the magnitude on the Richter scale of an earthquake that is $10^{2.5}$
times stronger than the zero quake?

62. What is the magnitude on the Richter scale of an earthquake that is $10^{6.2}$
times stronger than the zero quake? (use the formula from Exercise 61.)

63. The law of forgetting for a certain test is $f(t) = 82 - 16 \log_{10} (1 + t)$,
where t is the time in months after the test is given. How many months can
be expected to elapse before the average score will be $66 (f(t) = 66)$?

64. The law of forgetting for a test is $f(t) = 88 - 12 \log_{10} (1 + t)$, where t is the time in weeks after the test is given. How many weeks can be expected to elapse before the average score will be $64(f(t) = 64)$?

65. Find the pH of a solution of hydrochloric acid (a strong acid) if its hydrogen ion concentration (H^+) is 5×10^{-2}. (Use $\log_{10} 0.2 \approx -0.6990$, and round to the nearest tenth.)

66. Find the pH of a caustic soda solution (strong alkalinity) for which $H^+ = 3.2 \times 10^{-13}$. (Use $\log_{10} 0.3215 \approx -0.5051$, and round to the nearest tenth.)

67. Using $P = b^s$ and $Q = b^t$, show that $\log_b \dfrac{P}{Q} = \log_b P - \log_b Q$.

68. Using $P = b^s$, show that $\log_b \sqrt[n]{P} = \dfrac{1}{n} \log_b P$.

69. Show that $\log_{10} (\log_2 (\log_3 9)) = 0$.

70. Show that $\log_4 (\log_3 (\log_2 512)) = 0.5$.

71. What is the magnitude on the Richter scale of a quake that is 1,500,000 times the zero quake? (Note: $\log_{10} 1.5 \approx 0.17609$) Use the formula from Exercise 61. Round answer to the nearest tenth.

72. If the sound intensity in decibels (dB) is given by $dB = 10 \log \dfrac{i}{i_0}$, what is the sound intensity at a fraternity party where $\dfrac{i}{i_0} = 360{,}000{,}000$? (i.e. the sound intensity if 360,000,000 times the lowest threshold of hearing) [Use $\log_{10} 3.6$ is approx. 0.55630]. Round answer to the nearest tenth.

73. The age of a fossil, in years, is given by $t = -18{,}000 \log_{10} R$ where R is the percentage of the remaining carbon in the sample that is Carbon 14. (Note: $\log_{10} 1.5 \approx 0.17609$) What is the age of the fossil if 15% of the remaining carbon is Carbon 14. (Remember to convert 15% to a decimal in the formula)

74. The difference in altitude, A (in feet), between two levels having barometric readings of B_1 and B_2 inches of mercury is approximately $A = 60{,}500 \log_{10} \left(\dfrac{B_2}{B_1}\right)$. If the barometric pressure at B_1 is 24.00 inches of mercury and at B_2 it is 30.00 inches, what is the difference in altitude of the two locations? ($\log_{10} 1.25 \approx 0.09691$)

STATE YOUR UNDERSTANDING

75. Why is the $\log_b 1 = 0$ true for any value of b, if $b > 0$ and $b \neq 1$?

76. Why is the following incorrect? $\log_b (x^2 + y^2) = 2 \log_b x + 2 \log_b y$

77. Why is the following incorrect? $\dfrac{\log_b x^4}{\log_b y^3} = 4 \log_b x - 3 \log_b y$

CHALLENGE EXERCISES

If $\log_b 3 = 0.7925$, $\log_b 5 = 1.1610$, and $\log_b 7 = 1.4037$, where ''b'' is some unknown base, find:

78. $\log_b 21$

79. $\log_b\left(\dfrac{3}{7}\right)$

80. $\log_b\left(\dfrac{5}{3}\right)$

81. $\log_b 45$

82. $\log_b 125$

83. $\log_b \sqrt{5}$

84. $\log_b \sqrt{35}$

85. $\log_b \sqrt[3]{9}$

86. $\log_b\left(\dfrac{1}{5}\right)$

87. $\log_b\left(\dfrac{1}{\sqrt{7}}\right)$

MAINTAIN YOUR SKILLS (SECTIONS 7.3, 8.2, 8.3)

88. Find the radius of the circle $x^2 + 6x + y^2 - 1 = 0$.

89. Find the coordinates of the center of the circle $x^2 + 6x + y^2 - 1 = 0$.

90. Write the standard form of the equation of the circle with radius $\sqrt{17}$ and center at the origin.

91. Write the standard form of the equation of the circle with radius $\sqrt{17}$ and center at $(-3, -4)$.

92. Identify the conic section whose equation is $36x^2 + 49y^2 = 1764$.

Solve:

93. $\begin{cases} x + 3y - 2z = 8 \\ x + y - z = 4 \\ 2x - y + z = -3 \end{cases}$

94. Joe has 80 coins, all nickels and dimes. The value of the coins is $6.75. How many of each coin does Joe have?

95. The Colonial Jewelry Store has some gold that is 85% pure and some that is 70% pure. How many ounces of each should they use to make 15 ounces of 75% pure gold?

10.4

LOGARITHMS AND ANTILOGARITHMS

OBJECTIVES

1. Find the base 10 or base e logarithm of a number.

2. Find a number (antilogarithm) given its logarithm base 10 or base e.

There are two bases for logarithms that are used in mathematics and science. The one that is a result of our base ten number system is called a *common logarithm*.

■ DEFINITION

Common Logarithm

> *Common logarithms* are logarithms with base 10. The common abbreviation for $\log_{10} x$ is $\log x$.

Common logarithms of powers of 10 are easy to determine. Recalling that a logarithm is defined as an exponent, when we change from exponent form to logarithmic form we have

Exponential Form	**Logarithmic Form**
$1000 = 10^3$	$\log 1000 = 3$
$100 = 10^2$	$\log 100 = 2$
$10 = 10^1$	$\log 10 = 1$
$1 = 10^0$	$\log 1 = 0$
$0.1 = 10^{-1}$	$\log 0.1 = -1$
$0.01 = 10^{-2}$	$\log 0.01 = -2$

EXAMPLE 1

Find the log 1,000,000.

$1,000,000 = 10^6$ Change to exponential form.

Therefore, log 1,000,000 = 6.   ■

EXAMPLE 2

Find the log 0.0001.

$0.0001 = 10^{-4}$ Change to exponential form.

Therefore, log 0.0001 = −4.   ■

To find the common logarithm of a number that is not a power of 10, a scientific calculator or a table of logarithms is used. In the following example, a calculator is used. The calculator has a $\boxed{\log}$ key to find these logarithms.

EXAMPLE 3

 Use a calculator to find log 93 correct to four decimal places.

$\log 93 \approx 1.9685$ Enter 93, press $\boxed{\log}$ key.   ■

EXAMPLE 4

Use a calculator to find log 0.00251 correct to four decimal places.

$\log 0.00251 \approx -2.6003$ Calculator reading.   ■

■ DEFINITION

Natural Logarithm

> *Natural logarithms* are logarithms with the base *e*. The common abbreviation for $\log_e x$ is ln x.

The number *e* is an irrational number that is used extensively in the study of calculus. An approximation of its value is 2.718281828.

Based on the logarithm properties of Section 10.3, we can verify the following statements:

$\ln e^3 = 3$

$\ln e^2 = 2$

$\ln e^1 = 1$

$\ln e^0 = \ln 1 = 0$

$\ln e^{-1} = -1$

$\ln e^{-2} = -2$

The natural logarithm of *e* to any power is the exponent of *e*. Hence, powers of *e* are the only numbers for which natural logarithms are integers.

The calculator has a $\boxed{\ln}$ key to find these natural logarithms.

EXAMPLE 5

Find ln 5.38 correct to four decimal places.

$\ln 5.38 \approx 1.6827$ Enter 5.38, press $\boxed{\ln}$.   ■

EXAMPLE 6

Find ln 0.00123 correct to four decimal places.

ln 0.00123 ≈ −6.7007   Enter .00123, press $\boxed{\text{ln}}$.   ■

■ DEFINITION

Antilogarithm or Antilog

> Given a logarithm of a number, such as log $x = y$, the number (x) is called the *antilogarithm* or *antilog* of y, base 10.

 To find the antilog given a common logarithm of the number, we use the $\boxed{10^x}$ key. On some calculators, the two keys $\boxed{\text{INV}}$ $\boxed{\text{log}}$ or $\boxed{\text{2nd}}$ $\boxed{\text{log}}$ give the same value.

EXAMPLE 7

Find x to the nearest whole number given log $x = 2.8451$.

$x \approx 700$ Enter 2.8451, press $\boxed{10^x}$ key.   ■

EXAMPLE 8

Find x to the nearest thousandth given log $x = -1.6382$.

$x \approx 0.023$ Enter −1.6382, press $\boxed{10^x}$ key.   ■

Some of the applications of logarithms and antilogarithms are given in the next two examples.

EXAMPLE 9

The pH of a chemical solution is given by the formula

$$pH = \log \frac{1}{H^+},$$

where H^+ is the concentration of hydrogen ions. What is the pH of a chemical solution with a hydrogen ion concentration of 4.7×10^{-7}?

Formula:

$$pH = \log \frac{1}{H^+}$$

Substitute:

$$pH = \log \frac{1}{4.7 \times 10^{-7}}$$ Substitute 4.7×10^{-7} for H^+ and solve.

Solve:

pH $\approx$ log 2127659.57

$\dfrac{1}{4.7 \times 10^{-7}} \approx 2127659.57$

≈ 6.32790214 By calculator.

≈ 6.3 Rounded to the nearest tenth.

Answer:

The pH of the chemical solution is approximately 6.3.   ■

EXAMPLE 10

The intensity i of a sound is given by the function defined by

$$S(i) = 10 \log \dfrac{i}{i_0},$$

where i_0 is the least intensity that can be heard by the human ear, and $S(i)$ is measured in decibels. What is the decibel level of soft music that is 10,000 times the least intensity?

Formula:

$S(i) = 10 \log \dfrac{i}{i_0}$ To find the decibel reading for the music, substitute in the formula and solve.

Substitute:

$S(i) = 10 \log \dfrac{10,000 i_0}{i_0}$

Solve:

$S(i) = 10 \log 10,000$

$ = 10(4)$

$ = 40$

Answer:

The soft music measures 40 decibels.   ■

EXERCISE   **10.4**

A

Find these logarithms without using a table or calculator:

1. $\log_{10} \dfrac{1}{10}$ **2.** $\log_{10} 10$ **3.** $\log_{10} 100$ **4.** $\log_{10} 1$

5. $\log_{10} \dfrac{1}{1000}$ **6.** $\log_{10} 10{,}000$ **7.** $\log_e e^5$ **8.** $\log_e e^{1/3}$

9. $\ln \sqrt{e}$ **10.** $\ln e^{-1}$ **11.** $\ln \dfrac{1}{e^2}$ **12.** $\ln \sqrt{\dfrac{1}{e}}$

Find the antilogarithm without using a table or calculator:

13. $\log x = 4$ **14.** $\log x = -2$ **15.** $\log x = -5$ **16.** $\log x = 3$

17. $\log x = 6$ **18.** $\log x = -4$ **19.** $\log x = 8$ **20.** $\log x = -8$

21. $\ln x = 0$ **22.** $\ln x = 1$

B

Find the logarithms:

23. $\log 7$ **24.** $\log 7.8$ **25.** $\log 2.3$ **26.** $\log 2.36$

27. $\log 6.9$ **28.** $\log 6.92$ **29.** $\log 8.8$ **30.** $\log 8.85$

Using a calculator, find the logarithms:

31. $\ln 4.32$ **32.** $\ln 0.24$ **33.** $\ln 16.2$ **34.** $\ln 0.59$

Find the antilogarithms:

35. $\log x = 0.6232$ **36.** $\log x = 0.8692$ **37.** $\log x = 2.2304$

38. $\log x = 1.7324$ **39.** $\log x = -0.1308$ **40.** $\log x = -1.4248$

41. $\log x = -4.2733$ **42.** $\log x = -3.0301$

C

Find the logarithms:

43. $\log 70$ **44.** $\log 78$ **45.** $\log 230$

46. $\log 0.236$ **47.** $\log 0.069$ **48.** $\log 6920$

Find the antilogarithms:

49. $\log x = 2.6522$ **50.** $\log x = 3.9469$ **51.** $\log x = -1.1481$

52. $\log x = -0.4989$ **53.** $\log x = 4.7777$ **54.** $\log x = 7.3355$

55. $\log x = -2.7891$ **56.** $\log x = -3.5555$

Using a calculator, find the antilogarithms:

57. $\ln x = 2$ **58.** $\ln x = -3$

59. $\ln x = 3.42$ **60.** $\ln x = -2.5$

D

61. The Slippery Rock band plays with an intensity that is 10^{12} times the least intensity that can be heard. What is the decibel level? (Use the formula in Example 10.)

62. After a three-hour concert, the intensity of the sound created by the Burning Zephyrs is only 10^{10} times the least intensity that can be heard. What is the decibel level?

63. A normal jet engine can create a sound of 120 decibels. If a new jet engine creates a sound whose intensity is one billion times the least intensity, what is the decibel level? Is this more or less than the normal?

64. The Plan-On Background Music Company adjusts their speakers to produce a sound whose intensity is 1000 times the least intensity that can be heard. What is the decibel level?

65. Find the pH of a mixture with a hydrogen ion concentration of 6.2×10^{-6}. (Use the formula in Example 9.)

66. Find the pH of a mixture with a hydrogen ion concentration of 8.7×10^{-5}.

67. Find the pH of a mixture with a hydrogen ion concentration of 2.2×10^{-2}.

68. Find the pH of a mixture with a hydrogen ion concentration of 1.7×10^{-9}.

69. The heat loss (L) from a four inch pipe with one inch of insulation that is carrying hot water is approximately $L = \dfrac{50}{\ln\left(\dfrac{5}{4}\right)}$. Find the heat loss (to the nearest tenth) for the given pipe. (The units are BTU per hour where a "BTU" is a British Thermal Unit) (nearest tenth)

70. Under special circumstances, the pressure (P) and the volume (V) of a certain gas are related by $\ln(P) = C - 1.6 \ln(V)$ where C is a constant for that certain gas. Find C if $P = 8.4$ and $V = 2.5$. (nearest tenth)

71. 5 milligrams (mg) of a 100 mg sample of radium are still radioactive after a period of "t" years. Find t (to the nearest tenth) if $t = 2500(\ln 100 - \ln 5)$.

72. The time, in hours, that it takes a certain virus to double is given by $t = \dfrac{\ln 2}{k}$, where "k" is a certain constant for the particular virus. Find the time it takes the number of virus to double if $k = 0.3$ for this particular strain. (nearest tenth)

73. The price (p) of a certain item depend on the quantity demanded (q_d), and is given by $p = \$150 - \ln (q_d)$. Find the anticipated price if the demand is for 500,000 of the items. (To the nearest cent.)

STATE YOUR UNDERSTANDING

74. Why are ln 50 and log 50 not equal?

75. Without using your calculator, the log 35 is between what two whole numbers? Why?

76. Without using your calculator, the log 5,374 is between what two whole numbers? Why?

77. Use your calculator to find the y-values for the following x-values using $y = \ln x$. (3,), (4,), and (5,). Then, using the first two ordered pairs, calculate the slope. Using the second two ordered pairs, calculate the slope. Are the slopes the same? What does this tell you about the graph of $y = \ln x$?

CHALLENGE EXERCISES

Using the properties of logarithms, rewrite the following expressions so that the log of just one number need be taken. Then find the log of that number.

78. $\log 5 + \log 2 =$

79. $\log 10 - \log 2 =$

80. $\log 7 + \log 4 - \log 2 =$

81. $2 \log 5 + \log 2 - \log 10 =$

82. $\ln 12 + 2 \ln 3 - \ln 6 =$

83. $5 \ln 2 - \ln 2 - \ln 4 =$

MAINTAIN YOUR SKILLS (SECTIONS 7.4, 8.1, 9.5)

84. Find the x-intercepts of the graph of $y = (x - 9)^2 - 121$.

85. Find the x-intercepts of the graph of $y = x^2 - 9x - 111$.

Write the equation for $G^{-1}(x)$ if:

86. $G(x) = -\dfrac{3}{5}x + \dfrac{7}{5}$

87. $G(x) = 2x^2 - 3x$

88. $G(x) = -3x + 6$

89. $G(x) = x^3 + 2$

Evaluate:

90. $\begin{vmatrix} 6 & 8 \\ 9 & -2 \end{vmatrix}$

91. $\begin{vmatrix} 6 & 8 & 1 \\ 9 & -2 & 0 \\ 1 & 1 & -1 \end{vmatrix}$

10.5

EXPONENTIAL EQUATIONS, LOGARITHMIC EQUATIONS, AND FINDING LOGARITHMS, ANY BASE

OBJECTIVES

1. Solve equations that are in exponential form or logarithmic form.

2. Find the logarithm of a number, any positive base.

■ **DEFINITION**

Exponential Equation

Logarithmic Equation

> An *exponential equation* is an equation in which a variable occurs in an exponent.
>
> A *logarithmic equation* is an equation that has a variable as part of a logarithmic expression.

The equation $5^x = 25$ can be solved by inspection to get $x = 2$. However, the equation $5^x = 12$ does not have a solution that we readily recognize. We first find the logarithms of both sides of the equation. Since the two numbers 5^x and 12 are equal, their logarithms will be equal.

EXAMPLE 1

Solve for x (round answer to three decimal places): $5^x = 12$

$\log 5^x = \log 12$ Take the logarithm, base 10, of each sides.

$x(\log 5) = \log 12$ Log of a base raised to a power.

$x = \dfrac{\log 12}{\log 5}$

$x \approx \dfrac{1.0792}{0.6990}$

■ **CAUTION**

> $\dfrac{\log 12}{\log 5} \neq \log 12 - \log 5.$

$x \approx 1.544$ Simplify.

The solution set is $\{1.544\}$.

EXAMPLE 2

Solve for x (round answer to four decimal places): $27^x = 243$

$\log 27^x = \log 243$ Take the logarithm, base 10, of each side.

$x \log 27 = \log 243$ Log of a base raised to a power.

$$x = \frac{\log 243}{\log 27}$$ Solve for x.

$$x \approx 1.6666667$$

The solution set is $\{1.6667\}$. ∎

EXAMPLE 3

Solve for x (round answer to four decimal places): $7^x = 128$

$\log 7^x = \log 128$	Take the logarithm of each side.
$x \log 7 = \log 128$	Log of a power raised to a power.
$x = \dfrac{\log 128}{\log 7}$	Solve for x.
≈ 2.4934503	Simplify.

The solution set is $\{2.4935\}$. ∎

To solve a logarithmic equation, we write the equation in exponential form and solve using the laws of equations.

EXAMPLE 4

Solve for x: $\log_x 9 = 2$

$x^2 = 9$	Write in exponential form.
$x = \pm 3$	Take the square root; the root, -3, is rejected since the base is restricted to positive values.

The solution set is $\{3\}$. ∎

EXAMPLE 5

Solve for x: $\log_2 (x - 1) + \log_2 x = 1$

$\log_2 x(x - 1) = 1$	The sum of two logs is the log of their product.
$x(x - 1) = 2^1$	Change to exponential form.
$x^2 - x - 2 = 0$	Rewrite as a quadratic equation in standard form.
$(x - 2)(x + 1) = 0$	Factor.
$x = 2 \text{ or } x = -1$	Reject -1 because $\log_2 (-1)$ is not defined.
Check:	Check in the original equation.

$$\log_2 (2 - 1) + \log_2 2 = 1$$
$$\log_2 1 + \log_2 2 = 1$$
$$0 + 1 = 1 \qquad \text{True.}$$

The solution set is $\{2\}$. ∎

EXAMPLE 6

Solve for x: $\log_5 (x - 4) = -1$

$(x - 4) = 5^{-1}$ Change to exponential form.

$x - 4 = \dfrac{1}{5}$

$x = 4\dfrac{1}{5} = 4.2$ The check is left for the student.

The solution set is {4.2}. ■

EXAMPLE 7

Solve for x: $\log (x - 9) - \log (x - 2) = 1$

$\log \dfrac{x - 9}{x - 2} = 1$ The difference of two logs is the log of a quotient.

$\dfrac{x - 9}{x - 2} = 10$ Change to exponential form.

$x - 9 = 10(x - 2)$

$x - 9 = 10x - 20$

$-9x = -11$

$x = \dfrac{11}{9}$

■ **CAUTION**

$\log (x - 9)$ is defined only for $x - 9 > 0$ or $x > 9$, so we reject this value of x.

The solution set is $\emptyset$. ■

The formula for compound interest is an application of an exponential equation.

EXAMPLE 8

The formula for the value of an investment or a savings account that is compounded 4 times a year is

$$V = P\left(1 + \frac{r}{4}\right)^{4y},$$

where V represents the value of the account, P represents the amount invested, r represents the annual rate of interest, and y represents the number of years.

How many years will it take for a \$50 investment to grow to a value of \$74.30 at a rate of 8% annually?

Formula:

$$V = P\left(1 + \frac{r}{4}\right)^{4y}$$

Substitute:

$$74.30 = 50\left(1 + \frac{0.08}{4}\right)^{4y}$$

Substitute the given values in the formula and solve: $V = \$74.30$, $P = \$50$, $r = 8\% = 0.08$.

Solve:

$$1.486 = (1 + 0.02)^{4y}$$

Divide both sides by 50.

$$\log 1.486 = 4y(\log 1.02)$$

Take the log of each side.

$$4y = \frac{\log 1.486}{\log 1.02}$$

$$y = \frac{\log 1.486}{4(\log 1.02)}$$

Solve for y.

$$\approx 5.00044691$$

Simplify.

It will take approximately 5 years.   ■

EXAMPLE 9

The amount (A) of radioactive material in an object at any time t is given by the formula

$$A = A_o 2^{-t/k},$$

where A_o is the original amount present, and k is the half-life of the material.

How many years will it take 100 mg of a radioactive substance to decay to 20 mg if its half-life is 1000 years?

$A = 20$ mg, $A_o = 100$, $k = 1000$ Identify the variables.

Formula:

$$A = A_o 2^{-t/k}$$

Substitute:

$$20 = 100(2^{-t/1000})$$

Solve:

$$0.2 = 2^{-t/1000}$$

$$\log 0.2 = -\frac{t}{1000}(\log 2)$$

$$\frac{1000(\log 0.2)}{\log 2} = -t$$

$$-\frac{1000(\log 0.2)}{\log 2} = t$$

$$2321.928 \approx t \qquad\qquad \text{Approximately 2322 years.}$$

It will take approximately 2322 years.   ∎

If we have a calculator with a logarithm key, we can find the logarithm of a number to any other (positive) base. For instance, we can find

$$\ln 2 \text{ or } \log_5 2$$

by using the base 10 logarithm key on the calculator.

The formula for changing the base of any logarithm is

$$\log_a x = \frac{\log_b x}{\log_b a}$$

In particular, to change from base 10 to base e, use

$$\ln x = \frac{\log x}{\log e},$$

and to change from base e to base 10, use

$$\log x = \frac{\ln x}{\ln 10}.$$

For example,

$$\ln 2 = \frac{\log 2}{\log e} \approx \frac{\log 2}{\log 2.72}$$

$$\approx \frac{0.30103}{0.43429} \approx 0.69315$$

The accuracy is limited since the display on the calculator is rounded. A calculator that has an $\boxed{\ln}$ key shows the same approximation that is obtained using the formula and base 10 logarithms.

EXAMPLE 10

Find $\log_5 2$ using base 10 logarithms.

$$\log_a x = \frac{\log_b x}{\log_b a}$$

Formula for changing the base of a logarithm. The formula does not need to be memorized. We could also write

$$x = \log_5 2$$

$$5^x = 2$$

$$\log 5^x = \log 2$$

$$x \log 5 = \log 2$$

and solve for x.

$$\log_5 2 = \frac{\log_{10} 2}{\log_{10} 5}$$

Substitute in the formula: $x = 2$, $b = 10$, $a = 5$.

$$\log_5 2 \approx 0.4307$$

Rounded to four decimal places.

EXAMPLE 11

Find $\log_{11} 22$.

$$\log_{11} 22 = \frac{\log 22}{\log 11}$$

Substitute in the formula for changing the base of a logarithm: $x = 22$, $b = 10$, $a = 11$.

$$\approx 1.2891$$

Rounded to four decimal places.

Using base e (natural) logarithms

$$\log_{11} 22 = \frac{\ln 22}{\ln 11}$$

Formula for changing the base of a logarithm using natural logs.

$$\approx 1.2891$$

So $\log_{11} 22 \approx 1.2891$.

EXERCISE   10.5

A

Solve these exponential and logarithmic equations without using a calculator:

1. $4^x = 32$

2. $9^x = 27$

3. $2^{x+1} = 8^x$

4. $3^{x-2} = 9^x$

5. $\log_3 2 + \log_3 x = 2$

6. $\log_2 3 + \log_2 x = 3$

7. $\log_4 x - \log_4 3 = 2$

8. $\log_5 x - \log_5 2 = 2$

9. $5^{x+1} = 125^x$

10. $6^{2x-2} = 36^{x+1}$

11. $\log_x 81 = 4$

12. $\log_x 125 = 2$

13. $4^{2x+1} = 32^{x+5}$
14. $49^{x+2} = 7^{3x+1}$
15. $\log_2 25 - \log_2 (x - 1) = 3$

16. $\log_3 16 - \log_3 (2x + 1) = 3$

B

Use the formulas of this section and a calculator to find each of the logarithms correct to four decimal places:

17. $\log_7 25$
18. $\log_8 25$
19. $\log_5 48$

20. $\log_4 110$
21. $\log_{13} 6$
22. $\log_{13} 16$

Use a calculator to solve these exponential equations (round to four decimal places):

23. $4^x = 30$
24. $9^x = 30$

25. $25^x = 50$
26. $6^x = 50$

Solve these logarithmic equations:

27. $\log_4 x + \log_4 x = 3$
28. $\log_3 x + \log_3 x = 4$

29. $\log_2 x - \log_2 (x - 2) = 2$
30. $\log_5 (x + 2) - \log_5 x = 1$

C

Use the formulas of this section and a calculator to find each of the logarithms correct to four decimal places:

31. $\log_{1.5} 10$
32. $\log_{1.5} 16$
33. $\log_{2.5} 45$
34. $\log_{4.3} 100$

35. $\log_{1/4} 48$
36. $\log_{3/4} 0.55$
37. $\log_{1/5} 0.33$
38. $\log_{7/4} 101$

Use a calculator to solve these exponential equations (round to four decimal places):

39. $6^{x+1} = 3$
40. $4^{x+2} = 5$
41. $3^{x+1} = 4$

42. $2^{x-1} = 7$
43. $5^{x+1} = 3^x$
44. $3^{x-1} = 2^x$

Solve these logarithmic equations:

45. $\log_3 (x + 6) + \log_3 x = 3$
46. $\log_4 (x - 12) + \log_4 x = 3$
47. $\log_2 (x + 1) - \log_2 x = 3$

48. $\log_5 (x^2 + 6) - \log_5 x = 1$
49. $\log_2 x + \log_2 (x - 2) = 3$
50. $\log_5 x + \log_5 (x + 20) = 3$

D

51. The formula for the value of an investment or a savings account that is compounded 4 times a year is

$$V = P\left(1 + \frac{r}{4}\right)^{4y},$$

where V represents the value of the account, P represents the amount invested, r represents the annual rate of interest, and y represents the number of years. How many years will it take for the $50 to reach a value of $66 if it is invested at 8% compounded quarterly?

52. How long will it take for the $50 in Exercise 51 to double in value?

53. How long will it take for the $50 in Exercise 51 to triple in value?

54. Using the formula and interest rate in Exercise 51, how long will it take $100 to amount to $400?

55. The value of P dollars invested at 5% with interest compounded continuously is $V = Pe^{0.05t}$, where V represents the final value, P represents the amount invested, t represents the number of years, and $e \approx 2.718$. Find the value of $1 compounded continuously at 5% for 40 years.

56. Use the formula in Exercise 55 to find the value of $5 if you put it in a savings account that pays 5% interest compounded continuously and your great-grandchild takes the money out in 100 years.

57. The amount (A) of radioactive material present at time t is given by the formula

$$A = A_o 2^{-t/k},$$

where A_o is the initial amount present, and k is the materials half-life. How many years will it take 100 mg of a radioactive substance to decay to 25 mg if its half-life is 1000 years?

58. Using the formula and half-life in Exercise 57, how many years will it take 100 mg to decay to 10 mg?

59. The population (A) of a certain bacteria present in a culture at some time t is given by

$$A = A_o e^{0.05t},$$

where A_o is the population at time $t = 0$. Let the number of bacteria in the culture be 5000. How many days will it take for the bacteria to increase to 10,000?

60. Using the formula in Exercise 59, how many days will it take the 5000 bacteria to increase to 50,000?

61. Show that $\ln x \approx 2.3026(\log x)$.

62. A certain bank finds that the percentage (P) of credit cards used in a given month depends on the time, t (in months), that has passed since those cards were issued. For a certain bank, it is found that $P = 100(1 - e^{-0.07t})$. For those cards issued 6 months ago, what percentage will be used? (nearest tenth)

63. A heated steel ingot whose initial temperature is 1500°C is cooled for a period of time, T, until its temperature is 1000°C. The cooling of this type of steel is given by $t = t_0 e^{-0.7T}$, where t_0 is the initial temperature and t is the cooler temperature. If T is the time in hours, how long did it take to cool this steel? (To the nearest hundredth of an hour.)

STATE YOUR UNDERSTANDING

64. Define a logarithm.

65. What are the differences in the approaches for solving the following exponential and logarithmic equations: $3^x = 5$ and $\log_x 9 = 2$?

66. For many of the "decay" problems, the basic formula is $A = A_0 e^{-kt}$, and for many of the "growth" equations, the formula is $A = A_0 e^{+kt}$. Why is the exponent "negative" for the decay problems and "positive" for the growth problems?

CHALLENGE EXERCISES

Solve: (answers to the nearest hundredth)

67. $\log_4 x + \log_4 (x + 1) = \log_4 12$

68. $\log_7 x - \log_7 (x - 2) = \log_7 3$

69. $\log_5 x - \log_5 (x - 4) = \log_5 (x - 6)$

70. $\log_3 (2x + 7) - \log_3 (x - 1) = \log_3 (x - 7)$

71. $6^{2x+1} = 5^{x+2}$

72. $5^{2x-3} = 4^{x-2}$

73. $5^x = (4)2^x$

74. $10^{2x-5} = 5^{x+2}$

75. $\log_3 (x - 4) = \dfrac{1}{2} \log_3 2x$

76. $2^x 5^x = 4$

MAINTAIN YOUR SKILLS (SECTIONS 7.3, 7.4, 8.4, 8.5)

Evaluate:

77. $\begin{vmatrix} 6 & 0 & 2 \\ 0 & 4 & 8 \\ 1 & 2 & 5 \end{vmatrix}$

78. $\begin{vmatrix} 1 & -1 & -2 \\ -3 & -1 & 2 \\ 1 & 1 & 0 \end{vmatrix}$

Solve:

79. $\begin{cases} 6x + 2z = 3 \\ 4y + 8z = 0 \\ x + 2y + 5z = 4 \end{cases}$

80. $\begin{cases} x - y - 2z = 1 \\ -3x - y + 2z = 4 \\ x + y = 0 \end{cases}$

Solve by any method:

81. $\begin{cases} 4y = 3x \\ 2x^2 - 3xy + 4 = 0 \end{cases}$

82. $\begin{cases} x^2 + y^2 = 25 \\ x + y - 1 = 0 \end{cases}$

83. $\begin{cases} x^2 - y^2 + 9 = 0 \\ 3x^2 - 2y^2 + 2 = 0 \end{cases}$

84. $\begin{cases} 4x - y^2 = 15 \\ 2x + 3y^2 = 11 \end{cases}$

CHAPTER 10
SUMMARY

Exponential Function	An exponential function is a function defined by an equation of the form $f(x) = y = b^x$, where b is a constant with $b \neq 1$ and $b > 0$.	(p. 678)
Domain of an Exponential Function	The set of real numbers.	(p. 679)
Range of an Exponential Function	The set of positive real numbers.	(p. 679)
Increasing Function	A function for which the value of the function (y-value) increases as x increases. That is, for all x in the domain when $x_2 > x_1$, then $f(x_2) > f(x_1)$.	(p. 680)
Decreasing Function	A function for which the value of the function (y-value) decreases as x increases. That is, for all x in the domain when $x_2 > x_1$, then $f(x_2) < f(x_1)$.	(p. 682)
Logarithmic Function	The equivalent equations $x = b^y$ and $y = \log_b x$, $x > 0$, $b > 0$, $b \neq 1$ define a logarithmic function.	(p. 691)
Logarithm Comparison	A logarithm is sometimes referred to as an exponent.	(p. 692)
Properties of Logarithms	$b^{\log_b x} = x$	(pp. 702, 703)
	$\log_b b^x = x$	
	If $x = y$, then $\log_b x = \log_b y$.	
	If $\log_b x = \log_b y$, then $x = y$.	
	$\log_b (PQ) = \log_b P + \log_b Q$	
	$\log_b \dfrac{P}{Q} = \log_b P - \log_b Q$	
	$\log_b P^n = n(\log_b P)$	
	$\log_b \sqrt[n]{P} = \dfrac{1}{n}(\log_b P)$	
	$\log_b 1 = 0$	
	$\log_b b = 1$	
Common Logarithms	Logarithms with a base of 10. The common abbreviation for $\log_{10} x$ is $\log x$.	(p. 713)
Natural Logarithms	Logarithms with the base of e. The common abbreviation for $\log_e x$ is $\ln x$.	(p. 714)

Antilogarithm (Antilog)	When given the logarithm of a number, such as $\log x = 1.2345$, or $\ln x = 1.2345$, the number (x) is called the antilogarithm or antilog of 1.2345 base 10 or base e.	(p. 715)
Exponential Equation	An equation in which a variable is an exponent.	(p. 720)
Logarithmic Equation	An equation that has a variable as part of the logarithmic expression.	(p. 720)
Formula for Changing the Base of a Logarithm	$\log_a x = \dfrac{\log_b x}{\log_b a}$	(p. 724)

CHAPTER 10
REVIEW EXERCISES

SECTION 10.1   Objective 1

Draw the graphs of the following exponential functions.

1. $f(x) = 5^x$

2. $f(x) = (0.5)^x$

SECTION 10.1   Objective 2

Solve the following exponential equations.

3. $4^x = 256$ **4.** $3^x = 243$ **5.** $8^x = 256$

6. $9^x = 243$ **7.** $16^x = 4096$ **8.** $25^x = 15625$

SECTION 10.2   Objective 1

Write the following exponential equations in logarithmic form.

9. $4^5 = 1024$ **10.** $5^4 = 625$ **11.** $36^{1/2} = 6$

12. $343^{1/3} = 7$ **13.** $27^{2/3} = 9$ **14.** $16^{3/4} = 8$

SECTION 10.2   Objective 2

Write the following logarithmic equations in exponential form.

15. $\log_5 125 = 3$ **16.** $\log_3 243 = 5$ **17.** $\log_{32} 2 = \dfrac{1}{5}$

18. $\log_{27} 3 = \dfrac{1}{3}$ **19.** $\log_{16} 64 = \dfrac{3}{2}$ **20.** $\log_{125} 625 = \dfrac{4}{3}$

SECTION 10.2   Objective 3

Draw the graph of each of the following.

21. $f(x) = \log_3 x$ **22.** $f(x) = \log_{1/3} x$

SECTION 10.2   Objective 4

Solve the following equations.

23. $a = \log_5 25$ **24.** $b = \log_8 64$ **25.** $c = \log_{1/2} 128$ **26.** $d = \log_{1/3} 729$

SECTION 10.3   Objective 1

Write each of the following expressions as a single logarithm.

27. $\log_b x + \log_b y + \log_b z$ **28.** $\log_b x - \log_b y - \log_b z$

29. $\dfrac{1}{2} \log_b x + \dfrac{1}{2} \log_b y - \log_b z$ **30.** $\dfrac{1}{5}(\log_b x + 2 \log_b y - 3 \log_b z)$

Write each of the following as a combination of logs of single variables.

31. $\log_b x^2 y^3$ **32.** $\log_b \dfrac{3x^3}{y^4}$

33. $\log_b \dfrac{\sqrt{xy^2}}{z}$ **34.** $\log_b x^{1/2} y^{2/3} z^{1/4}$

Given $\log_{10} 2 = 0.30103$, $\log_{10} 3 = 0.47712$, $\log_{10} 5 = 0.69897$, and $\log_{10} 7 = 0.84509$, find the following.

35. $\log_{10} 42$ **36.** $\log_{10} 210$ **37.** $\log_{10} 1260$ **38.** $\log_{10} 1050$

SECTION 10.4　Objective 1

Use a calculator to find the following logarithms correct to four decimal places.

39. log 36　　　　　　**40.** ln 36　　　　　　**41.** ln 0.58　　　　　　**42.** log 0.58

43. log 156　　　　　**44.** ln 156　　　　　**45.** ln 0.014　　　　　**46.** log 0.014

SECTION 10.4　Objective 2

Find the antilogs correct to two decimal places.

47. $\log x = 1.3578$　　　　**48.** $\ln x = 1.3578$　　　　**49.** $\ln x = -0.2143$

50. $\log x = -0.2143$　　　**51.** $\log x = 3.2154$　　　**52.** $\ln x = 3.2154$

53. $\ln x = -1.2357$　　　　**54.** $\log x = -1.2357$

SECTION 10.5　Objective 1

Solve the following.

55. $\log x + \log (x + 3) = 1$　　　　　　**56.** $\log x + \log (x - 3) = 1$

57. $\log x + \log (3x + 5) = 2$　　　　　**58.** $\log x + \log (6x + 1) = 2$

59. $\log_2 x + \log_2 (x + 2) = 3$　　　　　**60.** $\log_3 x + \log_3 (x - 6) = 3$

61. $3^{2x+3} = 729$　　　　**62.** $2^{2a-3} = 16$　　　　**63.** $4^{x+2} = 32$

64. $9^{2x+1} = 27$　　　　**65.** $2^{2x} = 5$　　　　**66.** $3^{3x} = 6$

SECTION 10.5　Objective 2

Find each of the following logarithms using base 10 logarithms.

67. $\log_5 10$　　　　　　**68.** $\log_3 8$　　　　　　**69.** $\log_2 17$

70. $\log_2 168$　　　　　**71.** $\log_5 100$　　　　　**72.** $\log_{50} 100$

CHAPTER 10
TRUE–FALSE CONCEPT REVIEW

Check your understanding of the language of algebra. Tell whether each of the following statements is true (always true) or false (not always true).

1. An exponential function is defined by an equation of the form $x = b^y$.

2. An increasing function is a function for which the value of the function increases as x increases.

3. An asymptote is a line that the graph of a relation gets very close to or approaches as a limit or boundary.

4. In the exponential function $f(x) = b^x$, the value of b can be any real number.

5. A logarithmic function is the inverse of an exponential function.

6. The domain of the logarithmic function is the set of real numbers.

7. $\log_b x + \log_b y = \log_b (x + y)$

8. $\log_b \sqrt[n]{a} = n(\log_b a)$

9. $\log_b b^x = x$

10. Common logarithms have e for a base.

11. log is the abbreviation used to indicate $\log_{10}$.

12. ln is the abbreviation used to indicate $\log_e$.

13. Finding a number when the logarithm of the number is known is called finding the antilogarithm of the number.

14. The exponential equation $a^2 = 35$ is equivalent to the logarithmic equation $\log_a 35 = 2$.

15. The inverse of a logarithmic function is an exponential function.

CHAPTER 10
TEST

1. Write the following exponential equation in logarithmic form: $a^3 = 15$

2. Solve for x: $\log_3 x + \log_3 9 = 3$

3. Complete the table of values, and sketch the graph of the function $f(x) = 2^x$.

x	-3	-2	-1	0	1	2	3
2^x							

4. Find the domain and range of the function defined by $y = \log_4 x$.

5. Write the following logarithmic equation in exponential form: $\log_5 125 = 3$

6. Write the following as a single logarithm:
$$\frac{1}{3} \log_b 2 + \frac{1}{3} \log_b x - \log_b y$$

7. Solve for x: $9^x = 243$

8. Write the following exponential equation in logarithmic form: $8^{1/3} = 2$

9. Write the following expression as a combination of logarithms of single numerals or variables:
$$\log_a \frac{x^3 y^2}{z}$$

10. Use a calculator or table of logarithms to find log 23,100.

11. Find the domain and the range of the function defined by $f(x) = 2^x$.

12. Use a calculator or table of logarithms to find x (antilog) if log $x = 1.9069$.

13. Use a calculator or table of logarithms to find log 0.0123.

14. Complete the table of values, and sketch the graph of $y = \log_4 x$.

x	1	4	16	$\dfrac{1}{4}$	$\dfrac{1}{16}$	$\dfrac{1}{64}$
$\log_4 x$						

15. Write the following logarithmic equation in exponential form: $\log_b c = d$

16. Use a calculator or table of logarithms to find $\log_5 30$.

17. Given $\log_{10} 2 = 0.30103$ and $\log_{10} 3 = 0.47712$, find $\log_{10} 18$.

18. Solve for x: $27^x = \dfrac{1}{243}$

19. Given $\log_{10} 5 = 0.69897$ and $\log_{10} 3 = 0.47712$, find $\log_{10} 45$.

CHAPTER

11

FUNCTIONS OF COUNTING NUMBERS

SECTIONS

11.1
Sequences and Series

11.2
Arithmetic Progressions
(Sequences)

11.3
Geometric Progressions
(Sequences)

11.4
Infinite Geometric
Progressions

11.5
Binomial Expansion

Often salaries are negotiated on a long term basis using a fixed annual raise. Initial placement is usually based on experience or unusual talent. In Exercise 65, Section 6.2, two basketball players are hired with different initial salaries and scheduled pay increases. The pay agreement follows an arithmetic progression so the wage for any year can be easily computed. Using the given information determine the arithmetic progression involved. By writing an equation showing the nth terms of each progression equal, you can determine the year when they will be paid the same. *(Walter Ioosa Jr./The Image Bank)*

735

PREVIEW

In this chapter functions of counting numbers are presented. Emphasis is placed on progressions, both arithmetic and geometric. The general binomial expansion and finding a specified term in the expansion concludes the coverage.

11.1

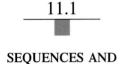

SEQUENCES AND SERIES

OBJECTIVES

1. Find a specified term of a sequence given the rule that defines the sequence function.

2. Find a rule that defines a sequence function given the first five terms of a sequence.

3. Expand summation notation.

4. Write a series in summation notation.

In this chapter, we present functions that involve sequences of numbers. You are familiar with sequences such as

1, 3, 5, 7, . . .   Odd numbers.

0, 2, 4, 6, . . .   Even numbers.

We now examine sequences that are the ranges of functions, which are called *sequence functions*.

■ DEFINITION

Sequence

> A *sequence* is a function whose domain is a set of consecutive positive integers.

The *range* values of the function are called the terms of the sequence.

For the sequence defined by

$c(n) = 2n + 3, \quad \text{Domain} = \{n \mid n \in J^+\}$

the sequence is

5, 7, 9, 11, 13, ...

This sequence is an *infinite sequence* since the domain is the set of positive integers.

A sequence defined by

$$c(n) = 2n + 3, \quad \text{with } n = 1, 2, 3, 4, 5, 6$$

is a *finite sequence*. It contains a finite number of terms since its domain is
$\{1, 2, 3, 4, 5, 6\}$. The range is computed to be $\{5, 7, 9, 11, 13, 15\}$, as illustrated
below.

$$c(1) = 2 \cdot 1 + 3 = 5$$

$$c(2) = 2 \cdot 2 + 3 = 7$$

$$c(3) = 2 \cdot 3 + 3 = 9$$

$$c(4) = 2 \cdot 4 + 3 = 11$$

$$c(5) = 2 \cdot 5 + 3 = 13$$

$$c(6) = 2 \cdot 6 + 3 = 15$$

To find a particular term of a sequence given the sequence rule $c(n)$, substitute
the number of the specified term in the rule for the sequence.

Sequence Rule	General Term (nth Term)	Specified Term(s)	Substitute	Desired Term(s) of the Sequence
$c(n) = n + 2$	$n + 2$	Second	$2 + 2$	4
$c(n) = 3n - 1$	$3n - 1$	Fourth	$3 \cdot 4 - 1$	11
$c(n) = n^2 + 2$	$n^2 + 2$	Third	$3^2 + 2$	11
$c(n) = \dfrac{1}{n}$	$\dfrac{1}{n}$	First 3	$\dfrac{1}{1}, \dfrac{1}{2}, \dfrac{1}{3}$	$1, \dfrac{1}{2}, \dfrac{1}{3}$
$c(n) = n^2$	n^2	First 4	$1^2, 2^2, 3^2, 4^2$	1, 4, 9, 16

EXAMPLE 1

Given the sequence defined by $c(n) = 5n - 6$, find $c(7)$, $c(11)$, and $c(21)$.

$c(n) = 5n - 6$ Replace n with 7, 11, and 21, respectively, and simplify.

$$\begin{aligned} c(7) &= 5(7) - 6 \\ &= 35 - 6 \\ &= 29 \end{aligned}$$ Seventh term.

$$\begin{aligned} c(11) &= 5(11) - 6 \\ &= 55 - 6 \\ &= 49 \end{aligned}$$ Eleventh term.

$$\begin{aligned} c(21) &= 5(21) - 6 \\ &= 105 - 6 \\ &= 99 \end{aligned}$$ Twenty-first term.

So $c(7) = 29$, $c(11) = 49$, and $c(21) = 99$.   ∎

EXAMPLE 2

For every year that wheat is planted on the same acreage, the yield is reduced by 11 bushels per acre. If the yield in the first year was 110 bushels per acre, what will the yield be in the fifth year?

110, 99, 88, 77, 66   The yield drops 11 bushels per acre each year, so we can write the first five terms.

The yield in the fifth year is 66 bushels.   ■

To find the sequence rule given the terms of the sequence, you must look for a pattern. For example, suppose you are given the sequence of six terms 1, 4, 9, 16, 25, 36.

Note the following:

Number of Terms	Term	
1	$1 = 1^2$	Number of term squared
2	$4 = 2^2$	Number of term squared
3	$9 = 3^2$	Number of term squared
4	$16 = 4^2$	Number of term squared
5	$25 = 5^2$	Number of term squared
6	$36 = 6^2$	Number of term squared
n	n^2	Number of term squared

The sequence rule is $c(n) = n^2$.

EXAMPLE 3

Write a rule for a sequence function whose first five terms are 5, 8, 11, 14, and 17.

$3n$

$3(1) = 3$

$3(2) = 6$

$3(3) = 9$

$3(4) = 12$

$3(5) = 15$

Note that the difference between consecutive terms is 3. Any time two consecutive integers are multiplied by 3, the difference between the products is 3. This indicates that $3n$ is in the rule. Finding the rule for a sequence is a guessing process based on observation. There are no general rules. If we add 2 to each value $(3n + 2)$, we get the sequence.

So the sequence rule is $c(n) = 3n + 2$.   ■

Associated with any sequence is an indicated sum called a *series*. The indicated sum of a given sequence is a series.

For example, given the finite sequence 5, 7, 9, 11, 13, 15, there is an indicated sum $5 + 7 + 9 + 11 + 13 + 15$ that is the series associated with the sequence. Algebraically, the terms in the series are the same as those in the corresponding sequence, so we can refer to the first term or the second term or the general term of a series in the same manner as we do for a sequence.

■ DEFINITION

Sigma or Summation Notation

The indicated sum of a sequence is called a *series*. The *sigma* or *summation notation* (Σ) is used with the general term or *n*th term of the sequence to represent a series.

For example,

$$\sum_{n=1}^{4} (2n - 1)$$

represents the indicated sum of the sequence whose rule is $c(n) = 2n - 1$, with $n = 1, 2, 3, 4$. The variable n is called the *index* of summation. The *expanded form* of a series is the indicated sum from one to n. That is:

$$\sum_{n=1}^{4} (2n - 1) = 1 + 3 + 5 + 7$$

SUMMATION NOTATION   EXPANDED FORM

In a finite sequence with n terms, the rule for the last term is the same as the rule for the general term.

The following table summarizes the vocabulary associated with a sequence function:

Sequence Function	$\{(1, 5), (2, 7), (3, 9), (4, 11), (5, 13), (6, 15)\}$
Domain	$\{1, 2, 3, 4, 5, 6\}$
Range	$\{5, 7, 9, 11, 13, 15\}$
Sequence of elements or terms	5, 7, 9, 11, 13, 15
Rule (equation) of the sequence	$c(n) = 2n + 3$
Series (expanded form)	$5 + 7 + 9 + 11 + 13 + 15$
Series (summation notation)	$\displaystyle\sum_{n=1}^{6} (2n + 3)$

EXAMPLE 4

Write $\displaystyle\sum_{n=1}^{6} \frac{n}{4}$ in expanded form.

$$\sum_{n=1}^{6} \frac{n}{4} = \frac{1}{4} + \frac{2}{4} + \frac{3}{4} + \frac{4}{4} + \frac{5}{4} + \frac{6}{4}$$

Expand by replacing n with 1, 2, 3, 4, 5, and 6. Simplify, and write the indicated sum of the terms.

$$= \frac{1}{4} + \frac{1}{2} + \frac{3}{4} + 1 + \frac{5}{4} + \frac{3}{2}$$

∎

The problem of writing a series in summation notation given the expanded notation is similar to the problem of finding the rule (equation) for a sequence function given the first five terms of the sequence.

For example, to write the finite series

$$1 + \frac{1}{2} + \frac{1}{3} + \frac{1}{4} + \frac{1}{5} \text{ in summation notation,}$$

inspect the series term by term.

Number of Term	Term	
1	$1 = \dfrac{1}{1}$	$\dfrac{1}{\text{Number of term}}$
2	$\dfrac{1}{2} = \dfrac{1}{2}$	$\dfrac{1}{\text{Number of term}}$
3	$\dfrac{1}{3} = \dfrac{1}{3}$	$\dfrac{1}{\text{Number of term}}$
4	$\dfrac{1}{4} = \dfrac{1}{4}$	$\dfrac{1}{\text{Number of term}}$
5	$\dfrac{1}{5} = \dfrac{1}{5}$	$\dfrac{1}{\text{Number of term}}$
n	$\dfrac{1}{n}$	

The sigma notation for the above series of five terms is

$$\sum_{n=1}^{5} \frac{1}{n}.$$

The sigma notation for the infinite series whose general term is also $\dfrac{1}{n}$ is:

$$\sum_{n=1}^{\infty} \frac{1}{n} = 1 + \frac{1}{2} + \frac{1}{3} + \frac{1}{4} + \frac{1}{5} \cdots .$$

EXAMPLE 5

Write the following series in summation notation: $-1 + 2 + 7 + 14 + 23 + \cdots$

n^2	There is no common difference between the terms. This tells us that multiples of the counting numbers do not help make up the terms. We now try other alternatives such as powers (squares, cubes, roots) of counting numbers.
$(1)^2 = 1$	
$(2)^2 = 4$	
$(3)^2 = 9$	Note that each term of the sequence is 2 less than the square of n, $(n^2 - 2)$. If this observation had not been useful, we could have looked at cubes or other powers of n.
$(4)^2 = 16$	
$c(n) = n^2 - 2$	Rule or nth term.
$\displaystyle\sum_{n=1}^{\infty} (n^2 - 2)$	The ellipses (three dots) at the end of the series indicates that it is an infinite series. ■

EXAMPLE 6

Write a rule for a sequence function defined by the five terms

$$\frac{x^2}{3}, \frac{x^4}{5}, \frac{x^6}{7}, \frac{x^8}{9}, \frac{x^{10}}{11}.$$

1. $\dfrac{x^2}{3}$ or $\dfrac{x^{2\cdot 1}}{2 \cdot 1 + 1}$ We compare the value of each term with the *number* of the term. Note that the exponents as well as the denominators differ by 2, so $2n$ is involved in each.

2. $\dfrac{x^4}{5}$ or $\dfrac{x^{2\cdot 2}}{2 \cdot 2 + 1}$

3. $\dfrac{x^6}{7}$ or $\dfrac{x^{2\cdot 3}}{2 \cdot 3 + 1}$

4. $\dfrac{x^8}{9}$ or $\dfrac{x^{2\cdot 4}}{2 \cdot 4 + 1}$

5. $\dfrac{x^{10}}{11}$ or $\dfrac{x^{2\cdot 5}}{2 \cdot 5 + 1}$

So the rule is

$$c(n) = \frac{x^{2n}}{2n + 1}$$ ■

EXAMPLE 7

For the sequence function in Example 6, find $c(9)$ and write the summation notation for the first 10 terms.

$$c(9) = \frac{x^{2 \cdot 9}}{2 \cdot 9 + 1} \qquad \text{Find } c(9) \text{ by substituting } n = 9 \text{ in the rule.}$$

$$= \frac{x^{18}}{19}$$

$$\sum_{n=1}^{10} \frac{x^{2n}}{2n + 1} \qquad \text{To write the summation notation, use the } n\text{th term, and let } n \text{ range from 1 to 10.} \qquad \blacksquare$$

EXERCISE   11.1

A

Write the first five terms of the sequence defined by each of the following:

1. $c(n) = n + 6$ **2.** $c(n) = n^3$

3. $c(n) = 6n - 4$ **4.** $c(n) = 10 - 3n$

Write a rule for a sequence given the first five terms:

5. 3, 5, 7, 9, 11   **6.** 4, 5, 6, 7, 8   **7.** 2, 4, 6, 8, 10

8. 5, 10, 15, 20, 25   **9.** 3, 6, 9, 12, 15   **10.** 5, 9, 13, 17, 21

11. −1, 0, 1, 2, 3   **12.** −2, −4, −6, −8, −10

Write in expanded form, and find the sum:

13. $\displaystyle\sum_{n=1}^{3} (2n + 7)$ **14.** $\displaystyle\sum_{n=1}^{4} (3n - 3)$ **15.** $\displaystyle\sum_{n=1}^{6} \frac{n}{2}$

16. $\displaystyle\sum_{n=1}^{5} \frac{n}{5}$ **17.** $\displaystyle\sum_{n=1}^{5} (-2n - 5)$ **18.** $\displaystyle\sum_{n=1}^{5} (-3n - 4)$

B

Write a rule for a sequence given the first five terms, and find $c(8)$:

19. $\dfrac{2}{3}, \dfrac{4}{3}, 2, \dfrac{8}{3}, \dfrac{10}{3}$ **20.** $\dfrac{10}{7}, \dfrac{20}{7}, \dfrac{30}{7}, \dfrac{40}{7}, \dfrac{50}{7}$

21. $1, \dfrac{1}{4}, \dfrac{1}{9}, \dfrac{1}{16}, \dfrac{1}{25}$ **22.** $\sqrt{2}, \sqrt{3}, 2, \sqrt{5}, \sqrt{6}$

23. $-1, -4, -9, -16, -25$ **24.** $-2, -4, -6, -8, -10$

Write in expanded form, and find the sum:

25. $\displaystyle\sum_{n=1}^{6} (4n - n^2)$ **26.** $\displaystyle\sum_{n=1}^{4} \sqrt{4n}$ **27.** $\displaystyle\sum_{n=1}^{7} (5n - 9)$

28. $\displaystyle\sum_{n=1}^{6} (n^2 + 2n - 3)$ **29.** $\displaystyle\sum_{n=1}^{9} (n^3 - n^2)$ **30.** $\displaystyle\sum_{n=1}^{11} (n^2 - 5n + 1)$

Write the following series in summation notation:

31. $4 + 8 + 12 + 16 + 20 + \cdots$ **32.** $-6 + (-12) + (-18) + (-24) + (-30) + \cdots$

33. $1 + \dfrac{1}{8} + \dfrac{1}{27} + \dfrac{1}{64} + \dfrac{1}{125} + \cdots$ **34.** $\dfrac{1}{4} + \dfrac{1}{2} + \dfrac{3}{4} + 1 + \dfrac{5}{4} + \cdots$

35. $0 + 1 + \sqrt{2} + \sqrt{3} + 2 + \cdots$ **36.** $\dfrac{1}{2} + \dfrac{2}{3} + \dfrac{3}{4} + \dfrac{4}{5} + \dfrac{5}{6} + \cdots$

C

Write a rule for a sequence function given the first five terms, and find $c(9)$;
write the summation notation for the first 10 terms:

37. $\dfrac{1}{7}, \dfrac{2}{8}, \dfrac{3}{9}, \dfrac{4}{10}, \dfrac{5}{11}$ **38.** $\dfrac{1}{\sqrt{2}}, \dfrac{2}{\sqrt{3}}, \dfrac{3}{2}, \dfrac{4}{\sqrt{5}}, \dfrac{5}{\sqrt{6}}$

39. $x, \dfrac{x^2}{2}, \dfrac{x^3}{3}, \dfrac{x^4}{4}, \dfrac{x^5}{5}$ **40.** $-x^2, 2x^2, -3x^2, 4x^2, -5x^2$

41. $0, 2\log 2, 3\log 3, 4\log 4, 5\log 5$ **42.** $1 \cdot 4, 2 \cdot 9, 3 \cdot 16, 4 \cdot 25, 5 \cdot 36$

43. $1, \dfrac{2y}{5}, \dfrac{4y^2}{25}, \dfrac{8y^3}{125}, \dfrac{16y^4}{625}$ **44.** $\dfrac{3}{4}, \dfrac{9x}{8}, \dfrac{27x^2}{16}, \dfrac{81x^3}{32}, \dfrac{243x^4}{64}$

45. $5x^3, 4x^6, 3x^9, 2x^{12}, x^{15}$ **46.** $\dfrac{3}{x}, 5, 7x, 9x^2, 11x^3$

47. $-0.01, 0.1, -1, 10, -100$ **48.** $-1, 4, -9, 16, -25$

Write in expanded form, and find the sum:

49. $\displaystyle\sum_{n=4}^{7} (n^2 - 5n)$

50. $\displaystyle\sum_{n=4}^{8} (3n - 21)^2$

51. $\displaystyle\sum_{n=16}^{21} (3n + 14)$

52. $\displaystyle\sum_{n=28}^{34} (100 - 2n)$

53. $\displaystyle\sum_{n=10}^{14} (-2n + 5)$

54. $\displaystyle\sum_{n=50}^{55} (-5n + 25)$

55. $\displaystyle\sum_{n=5}^{9} (-2)^n$

56. $\displaystyle\sum_{n=4}^{6} \left(-\frac{1}{3}\right)^n$

D

57. For every year that strawberries are planted in the same field, the yield is reduced by 8 tons. If the yield the first year is 91 tons, what is the yield in the fourth year?

58. If the strawberry field in Exercise 57 is fertilized, the yield can be held to a reduction of 6 tons annually. If the yield is 91 tons the first year and if the field is fertilized, what is the yield in the fourth year?

59. Mary starts a savings plan in which she agrees to put a penny in the bank on the first day, 3 pennies on the second day, 5 pennies on the third day, and so on, adding 2¢ more to the deposit each day. How much will she need to deposit on the tenth day?

60. Marcia starts a savings plan in which she agrees to put a penny in the bank on the first day, 2¢ on the second day, 4¢ on the third day, and so on, doubling the amount of the deposit each day. How much will she need to deposit on the tenth day?

61. A basketball dropped from 10 ft rebounds 5 ft on the first bounce and 2.5 ft on the second bounce. If it continues to rebound each time to half the previous height, how high will it rebound on the eighth bounce?

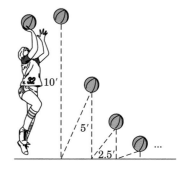

Figure for Exercise 61

62. If the ball in Exercise 61 is dropped from 12 ft, how high will it rebound on the eighth bounce?

63. At a discount clothing store, the price of a suit is reduced by $15 at the end of each week that the suit remains in the store. If the original price was $139, what will the price be after seven weeks if the suit does not sell?

64. The value of a certain baseball card was $70 in the first year. The second year, its value was $92 and the third year, it was worth $114. What will be its value in the fifth year, if the present trend continues?

65. The value of a sheet of first issue postage stamps was $7 in the first year it was issued. The value increased to $14 the second year and $28 the third year. What is the expected value in the sixth year?

Figure for Exercise 63

66. Gambling at the dice table in Atlantic City, you start with a bet of $10 and win $10, you bet it all on the next roll. If you win 6 times in a row and quit, how much will you leave with?

STATE YOUR UNDERSTANDING

67. How does one recognize a sequence?

68. What is the main difference between a sequence and a series?

69. What ''trick'' can be used to get the signs to alternate in a sequence?

CHALLENGE EXERCISES

In the following problems, you are given the first four terms of a sequence; (a) write the rule for the sequence; and (b) find the seventh term of the sequence.

70. $-2, 0, 2, 4, \ldots$

71. $\dfrac{3}{2}, \dfrac{9}{4}, \dfrac{27}{8}, \dfrac{81}{16}, \ldots$

72. $1, -1, -3, -5, \ldots$

73. $-3, 1, 5, 9, \ldots$

74. $0, 7, 26, 63, 124, \ldots$
(hint: see exercise #2, in this section)

75. $\sqrt{2}, \sqrt{5}, \sqrt{10}, \sqrt{17}, \ldots$

76. $3, 6, 11, 18, \ldots$

77. $\dfrac{4}{1}, \dfrac{9}{2}, \dfrac{16}{3}, \dfrac{25}{4}, \ldots$

78. $0, 3, 8, 15, \ldots$

MAINTAIN YOUR SKILLS (SECTIONS 6.2, 6.3, 8.1)

79. Write the coordinates of the x-intercept of the graph of $3x + 5y = 8$.

80. Write the coordinates of the y-intercept of the graph of $3x + 5y = 8$.

81. Write the slope of the line that is perpendicular to the graph of $3x + 5y = 8$.

82. Write the equation of the line parallel to $3x + 5y = 8$ that contains the point $(-3, 5)$.

83. Find the coordinates of the vertex of the parabola described by $y = 2x^2 - 28x + 108$.

84. Write the coordinates of the x-intercepts of the graph of $y = 2x^2 - 28x + 108$.

85. Write the coordinates of the y-intercepts of the graph of $y = 2x^2 - 28x + 108$.

86. What is the minimum value of $y = 2x^2 - 28x + 108$?

11.2

ARITHMETIC PROGRESSIONS (SEQUENCES)

OBJECTIVES

1. Write an arithmetic progression (sequence) given the first term and the common difference of any two consecutive terms.

2. Find a specified term of an arithmetic progression (sequence).

3. Find all the terms between two given terms of an arithmetic progression (sequence).

4. Find the number of terms in a finite arithmetic progression (sequence).

5. Find the sum of a finite arithmetic progression (sequence).

■ **DEFINITION**

Arithmetic Progression

An *arithmetic progression* is a sequence in which there is a common difference (d) between any two consecutive terms.

The sequence

1, 3, 5, 7, 9

is an arithmetic progression because there is a difference of 2, ($d = 2$), between any two consecutive terms.

Consecutive Terms	Difference
1, 3	$3 - 1 = 2$
3, 5	$5 - 3 = 2$
5, 7	$7 - 5 = 2$
7, 9	$9 - 7 = 2$

EXAMPLE 1

Write the first five terms of the arithmetic progression which has first term 12 and common difference 6.

12	First term.
$12 + 6 = 18$	Second term.
$18 + 6 = 24$	Third term.
$24 + 6 = 30$	Fourth term.
$30 + 6 = 36$	Fifth term.

The first five terms are 12, 18, 24, 30, and 36.   ■

In general, an arithmetic progression can be described in the following way:

Number of the Term	Term
1	a_1
2	$a_1 + d$
3	$(a_1 + d) + d = a_1 + 2d$
4	$(a_1 + 2d) + d = a_1 + 3d$
5	$(a_1 + 3d) + d = a_1 + 4d$
6	$(a_1 + 4d) + d = a_1 + 5d$
⋮	
n	$= a_1 + (n - 1)d$

Notice that for each term the coefficient of d is one less than the number of the term.

■ FORMULA

In an arithmetic progression if a_1 represents the first term and d the common difference, then the nth term, a_n, is

$$a_n = a_1 + (n - 1)d.$$

To find the twelfth term of the progression

6, 9, 12, . . .

we first identify, a_1 and d.

$a_1 = 6$	First term.
$d = 9 - 6 = 3$	Difference between first term and second term.
$a_n = a_1 + (n - 1)d$	Formula for nth term.
$a_{12} = 6 + (12 - 1)(3)$	$n = 12$ (twelfth term).
$= 6 + 33$	
$= 39$	

The twelfth term is 39.

EXAMPLE 2

Write the next four terms of the arithmetic progression 3, 8, 13, . . . and find the twentieth term.

$a_1 = 3$, $a_1 + d = 8$

$d = 8 - 3 = 5$ The difference between two consecutive terms is $d = 5$.

3, 8, 13, 18, 23, 28, 33   Add 5 to each succeeding term to get the next four terms.

$a_n = a_1 + (n - 1)d$ Formula for general term.

$a_{20} = 3 + (20 - 1)5$ Find twentieth term by letting $a_1 = 3$, $n = 20$, and $d = 5$.

$= 3 + 95$

$= 98$

The next four terms are 18, 23, 28, and 33. The twentieth term is 98.   ■

An arithmetic progression can be identified given

1. a_1 and d (see Example 1), or

2. a_1 and a specific term, or

3. d and a specific term, or

4. Any two specific terms

For example,

2. **Given a_1 and a specific term** If $a_1 = 20$ and the fifteenth term is 62, we first find d.

$a_n = a_1 + (n - 1)d$

$62 = 20 + (15 - 1)d$ Substitute $a_n = 62$, $a_1 = 20$, $n = 15$.

$62 = 20 + 14d$

$42 = 14d$

$3 = d$

So the progression is

20, 23, 26, 29, ...

3. **Given d and a specific term** The progression for which $d = -5$ and whose eleventh term is 92 is found by using the same formula.

$a_n = a_1 + (n - 1)d$

$92 = a_1 + (11 - 1)(-5)$ Substitute $d = -5$, $a_n = 92$, $n = 11$.

$92 = a_1 - 50$

$142 = a_1$

So the progression is

142, 137, 132, 127, 122, ...

4. Given any two specific terms   If $a_3 = 7$ and $a_8 = 17$, we substitute into the general term formula and then solve the resulting system of equations for a_1 and d.

$a_n = a_1 + (n - 1)d$

$\begin{cases} 7 = a_1 + (3 - 1)d = a_1 + 2d \\ 17 = a_1 + (8 - 1)d = a_1 + 7d \end{cases}$ Substitute $a_n = 7$, $n = 3$.
    Substitute $a_n = 17$, $n = 8$.

$\begin{cases} a_1 + 2d = 7 \quad (1) \\ a_1 + 7d = 17 \quad (2) \end{cases}$

$-5d = -10$ Subtract (2) from (1).

$d = 2$

$a_1 + 14 = 17$ Substitute $d = 2$ in (2).

$a_1 = 3$

The arithmetic progression is

3, 5, 7, 9, 11, 13, 15, 17, ...

EXAMPLE 3

If the eighth term of an arithmetic progression is 210 and the thirteenth term is 275, find all the terms between them.

$n = 8$, $a_8 = 210$	Given information.
$n = 13$, $a_{13} = 275$	Given information.
$a_n = a_1 + (n - 1)d$	Formula for general term.
	Write two equations with two variables to find a_1 and d.
$\begin{cases} 210 = a_1 + 7d \quad (1) \\ 275 = a_1 + 12d \quad (2) \end{cases}$	Substitute $n = 8$, $a_8 = 210$ to get equation (1). Substitute $n = 13$, $a_{13} = 275$ to get equation (2).
$65 = 5d$	Solve the system of equations by subtracting (1) from (2).
$d = 13$	
210, 223, 236, 249, 262, 275	Start with the smaller term and add 13 (the common difference) to get the terms that are in between.

The missing terms are 223, 236, 249, and 262.   ■

Given the finite arithmetic progression

$-6, -1, 4, \ldots, 84,$

we can find the number of terms (n) in the progression using the same formula. Here

$a_1 = -6$ First term.

$d = (-1) - (-6) = 5$ Difference between second term and first term.

$a_n = 84$ The nth or last term.

$a_n = a_1 + (n - 1)d$

$84 = -6 + (n - 1)5$ Substitute $a_n = 84$, $d = 5$, $a_1 = 6$.

$84 = -6 + 5n - 5$

$95 = 5n$

$19 = n$

So the progression has 19 terms.

EXAMPLE 4

Find the number of terms of the finite arithmetic progression
$12, 25, 38, \ldots, 129$.

$a_1 = 12$, $d = 25 - 12 = 13$ $a_1 = 12$ is the first term, $d = 13$ is the difference between the first two terms and hence the common difference. The last term, a_n, is 129.

$a_n = 129$

$a_n = a_1 + (n - 1)d$ Formula for nth term.

$129 = 12 + (n - 1)13$ Substitute $a_n = 129$, $a_1 = 12$, and $d = 13$ in the formula, and solve for n.

$129 = 12 + 13n - 13$

$130 = 13n$

$n = 10$

There are 10 terms in the progression. ■

EXAMPLE 5

John sets a goal of increasing the weight he can bench-press by 12 pounds per week. If initially he can press 125 pounds, how many weeks will it take for him to press 281 pounds?

$125, 137, 149, \ldots, 281$ Since John is adding 12 lb per week, we have an arithmetic progression with $d = 12$, $a_1 = 125$, and $a_n = 281$.

Formula:

$$a_n = a_1 + (n-1)d$$

This example is similar to Example 4. We are looking for the number of terms, n.

Substitute:

$$281 = 125 + (n-1)12$$

Solve:

$$281 = 113 + 12n$$
$$168 = 12n$$
$$n = 14$$

Answer:

The 14th term of the progression is 281. Since John can press 125 pounds the first week, it will take him 13 more weeks to reach his goal. ■

If an arithmetic progression is finite (has n terms), it is possible to find a formula for its sum. (In this text, the "sum of the progression" is defined as the "sum of the terms of the progression.") If S_n represents the sum of an arithmetic progression of n terms, then we can write

$$S_n = a_1 + (a_1 + d) + (a_1 + 2d) + (a_1 + 3d) + \cdots + (a_1 + (n-1)d).$$

Also,

$$S_n = a_n + (a_n - d) + (a_n - 2d) + (a_n - 3d) + \cdots + (a_n - (n-1)d).$$

Adding these together gives us

$$2S_n = (a_1 + a_n) + (a_1 + a_n) + (a_1 + a_n) + (a_1 + a_n) + \cdots + (a_1 + a_n)$$
$$2S_n = n(a_1 + a_n).$$

■ FORMULA

$$S_n = \frac{n(a_1 + a_n)}{2} \qquad \text{Sum of an arithmetic progression.}$$

In words, this formula says that the sum of an arithmetic progression of n terms is the product of the number of terms in the progression and the average of the first and last terms. There is another formula for finding the sum. Consider the following:

Since $a_n = a_1 + (n - 1)d$,

$$S_n = \frac{n(a_1 + a_n)}{2} = \frac{n}{2}(a_1 + a_n)$$

$$S_n = \frac{n}{2}[a_1 + (a_1 + (n - 1)d)].$$

■ FORMULA

$$S_n = \frac{n}{2}[2a_1 + (n - 1)d] \qquad \text{Sum of an arithmetic progression.}$$

EXAMPLE 6

Find the sum of the first 15 terms of the arithmetic progression 6, 14, 22, ...

$a_1 = 6, d = 14 - 6 = 8$

$\quad n = 15$ Specified number of terms.

$a_n = a_1 + (n - 1)d$

$a_{15} = 6 + (15 - 1)8$ To find the sum, we first find the fifteenth

$a_{15} = 6 + 112$ term.

$\quad\;\; = 118$

$\quad S_n = \dfrac{n(a_1 + a_n)}{2}$ Formula for the sum of n terms of an
           arithmetic progression.

$\quad S_{15} = \dfrac{15(6 + 118)}{2}$ Substitute and simplify.

$\qquad = \dfrac{15(124)}{2}$

$\qquad = 15(62)$

$\qquad = 930$

The sum of the first 15 terms is 930.   ■

EXAMPLE 7

Find the sum of the first twenty terms of the arithmetic progression
$6 + 11 + 16 + \cdots$.

$a_1 = 6, d = 5$ First we compute d. $d = 11 - 6 = 5$

$$a_{20} = 6 + (20 - 1)5$$

$$= 6 + 19 \cdot 5$$

$$= 101$$

$$S_n = \frac{n(a_1 + a_n)}{2}$$

$$S_{20} = \frac{20(6 + 101)}{2}$$

$$= 10(107)$$

$$= 1070$$

The sum is 1070.

Using the formula for the general term of an arithmetic progression

$$a_n = a_1 + (n - 1)d,$$

we can write the sum of the first k terms of a progression in summation notation.

$$S_n = \sum_{n=1}^{k} a_n = \sum_{n=1}^{k} [a_1 + (n - 1)d]$$

EXAMPLE 8

Find the sum: $\displaystyle\sum_{n=1}^{23} (4n - 5)$

$$-1, 3, 7, 11, \ldots, a_{23}$$

$$a_1 = -1, \ d = 4, \ n = 23$$

$$a_{23} = -1 + (23 - 1)4$$

$$= -1 + 88$$

$$= 87$$

$$S_n = \frac{n(a_1 + a_n)}{2}$$

$$S_{23} = \frac{23(-1 + 87)}{2}$$

$$= 989$$

The sum is 989.

EXERCISE 11.2

A

Write the first five terms of the arithmetic progression, and find the indicated term:

1. $a_1 = 7, d = 3$ Find a_{10}

2. $a_1 = 4, d = -6$ Find a_{15}

3. $a_1 = 21, d = 7$ Find a_{13}

4. $a_1 = 50, d = 25$ Find a_{20}

5. $a_1 = 7, a_2 = 19$ Find a_{11}

6. $a_1 = 15, a_2 = 46$ Find a_{18}

7. $a_1 = 28, a_2 = 20$ Find a_{10}

8. $a_1 = 45, a_2 = 27$ Find a_{14}

9. $a_1 = 2, a_4 = 14$ Find a_6

10. $a_1 = 1, a_5 = -11$ Find a_3

Find the number of terms in each of the following finite arithmetic progressions:

11. 6, 9, . . . , 108

12. 11, 16, . . . , 66

13. 17, 13, . . . , -23

14. 36, 30, . . . , -30

15. 42, 55, . . . , 341

16. 64, 81, . . . , 421

B

Find all the terms between the given terms of the following arithmetic progressions:

17. $a_1 = 10, a_6 = 0$

18. $a_1 = -14, a_7 = 16$

19. $a_7 = 80, a_{11} = 56$

20. $a_5 = 37, a_{11} = -5$

21. $a_{10} = 22, a_{15} = 37$

22. $a_7 = 44, a_{11} = 76$

Find the sum of the indicated number of terms of the following arithmetic progressions:

23. 5, 8, 11, . . .   20 terms

24. 11, 17, 23, . . .   17 terms

25. 15, 9, 3, . . .   10 terms

26. 18, 14, 10, . . .   8 terms

27. 2.73, 3.01, 3.29, . . .   61 terms

28. 9.73, 9.15, 8.57, . . .   100 terms

29. $-2, -\dfrac{3}{2}, -1, \ldots$ 9 terms

30. $3, \dfrac{8}{3}, \dfrac{7}{3}, \ldots$ 10 terms

Write the first five terms of the arithmetic progression, and find the indicated term:

31. $a_1 = 5, d = \dfrac{1}{2}$ Find a_{10}

32. $a_1 = \dfrac{3}{4}, d = \dfrac{2}{3}$ Find a_{12}

33. $a_1 = 5, a_2 = 3.5$ Find a_{15}

34. $a_1 = 0, a_2 = -1.05$ Find a_{10}

35. $a_3 = 21, a_8 = 31$ Find a_{20}

36. $a_{14} = 36, a_{18} = 53$ Find a_{25}

37. $a_5 = 4, a_{15} = -11$ Find a_{31}

38. $a_4 = -6.4, a_{11} = -12$ Find a_{21}

C

Find the number of terms and the sum of the following finite arithmetic progressions:

39. 5.5, 6.25, . . . , 13.75

40. $\dfrac{4}{3}$, 2, . . . , $\dfrac{34}{3}$

41. 14.2, 12.9, . . . , -14.4

42. 17, $16\dfrac{3}{8}$, . . . , -18

43. 6.95, 5.82, 4.69, . . . , -23.56

44. -75.03, -62.45, -49.87, . . . , 327.53

Find the indicated sum:

45. $\displaystyle\sum_{n=1}^{18} (3n + 9)$

46. $\displaystyle\sum_{n=1}^{40} (4n + 3)$

47. $\displaystyle\sum_{n=1}^{20} \left(\dfrac{1}{2}n - 2\right)$

48. $\displaystyle\sum_{n=1}^{35} \dfrac{1}{2}(6 - n)$

49. $\displaystyle\sum_{n=1}^{25} (-0.4n - 10)$

50. $\displaystyle\sum_{n=1}^{30} \left(-\dfrac{2}{5}n + 5\right)$

Find the sum of the indicated number of terms of the following progressions:

51. $a_{10} = 41$, $a_{15} = 66$ Find S_{18}

52. $a_{11} = 8$, $a_{20} = -10$ Find S_{25}

53. $a_5 = -5$, $a_{15} = -30$ Find S_{36}

54. $a_8 = 6$, $a_{14} = 9$ Find S_{40}

55. $a_5 = 29.48$, $a_{11} = 47.66$ Find S_{20}

56. $a_{13} = \dfrac{26}{3}$, $a_{20} = 25$ Find S_{50}

D

57. Pete intends to increase the distance he runs each week by two miles. If he now runs 15 miles per week, in how many weeks will he be running 63 miles per week?

58. Mary starts a diet that will cause her to lose $1\dfrac{1}{2}$ lb per week. If she now weighs 168 lb, how many weeks will it take her to meet her goal of 135 lb?

59. Joanna and Phil are hired by the Rey Taco Diner. Joanna is hired with the understanding she will get $220 per week with an increase of $5 per week every week. Phil's salary will be $220 per week with a raise of $20 per week at the end of every four weeks. After 12 weeks who has earned more money? How much more?

60. Ms. Bond and Mr. Stox are hired by a brokerage firm. Both start at $300 per week. Ms. Bond is to receive a weekly raise of $10. Mr. Stox's weekly salary will be increased by $40 at the end of every four weeks. After 52 weeks, whose total income is the larger? How much larger?

61. Willy is stacking blocks. How many blocks are there in his stack if there are 18 blocks in the bottom row, 17 in the second row, 16 in the third row, and so on, until there is 1 block in the top row?

16 blocks
17 blocks
18 blocks

Figure for Exercise 61

62. Use the formula for the sum of an arithmetic series to show that the sum of the first n odd numbers.

$$1 + 3 + 5 + 7 + 9 + \cdots + (2n - 1)$$

is n^2.

63. At a mill that makes pipe, the pipe is stacked with 35 pieces of pipe on the bottom row, 34 pieces of pipe in the second row, 33 in the third row, and etc. If there are 17 rows of pipe, (a) how many are in the top row? (b) how many pieces of pipe are there in all 17 rows?

64. An equilateral triangle has three equal sides and a sequence of triangles is given as follows:

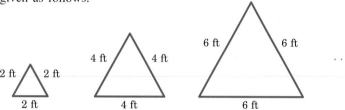

    a. What is the length of each side of the 14th triangle?
    b. If the perimeter of each triangle is 3 times the length of one side, what would be the sum of all the perimeters of the 14 triangles?

65. Two basketball players, Davis and Mooney, sign long-term contracts with a team. Davis is more experienced, so his salary at the beginning is higher. Mooney shows a lot of promise so his increases are larger. Their salaries for the first three years of their contracts are as follows:

	Davis	Mooney
1st year	$150,000	$73,000
2nd year	$155,000	$85,000
3rd year	$160,000	$97,000

    a. In what year will their salaries be the same?
    b. How much will each earn in that year?
    c. How much will each player earn from the first year to, and including, the year that their salaries are the same?

66. A wooden ladder gets narrower as you approach the top rung. If the bottom rung is 36 inches wide, the top rung is 20 inches wide, and there are 20 rungs on the ladder, how many inches of wood are needed to make the rungs (assuming no waste)?

STATE YOUR UNDERSTANDING

67. What is an arithmetic progression?

68. To determine the general term of an arithmetic sequence, what two pieces of information are critical?

20″

36″

Figure for Exercise 66

CHALLENGE EXERCISES

For the following arithmetic sequences, find: (a) the tenth term of the sequence; and (b) the sum of the first ten terms.

69. 7, 7, 7, 7, . . .

70. −8, −2, 4, 10, . . .

71. $4, \dfrac{13}{3}, \dfrac{14}{3}, 5, \ldots$

72. $6\dfrac{1}{3}, 6\dfrac{5}{6}, 7\dfrac{1}{3}, 7\dfrac{5}{6}, \ldots$

73. $\sqrt{3}, \sqrt{12}, \sqrt{27}, \sqrt{48}, \ldots$ (Hint: simplify all radicals)

MAINTAIN YOUR SKILLS (SECTIONS 10.1, 10.2)

Write in exponential form:

74. $\log_5 125 = 3$

75. $\log_7 49 = 2$

76. $\log_{13} 2197 = 3$

77. $\log_{1/2} 0.015625 = 6$

Write in logarithmic form:

78. $3^{11} = 177147$

79. $4096^{1/6} = 4$

80. $(x + 3)^m = 5$

81. $(a + b)^9 = y$

11.3

GEOMETRIC PROGRESSIONS (SEQUENCES)

OBJECTIVES

1. Write a geometric progression (sequence) given the first term and the common ratio.

2. Find a specified term of a geometric progression (sequence).

3. Find all the terms between two given terms of a geometric progression (sequence).

4. Find the number of terms in a finite geometric progression (sequence).

5. Find the sum of a finite geometric progression (sequence).

■ DEFINITION

Geometric Progression

A *geometric progression* is a sequence in which there is a *common ratio* (*r*) between any two consecutive terms.

The sequence

2, 4, 8, 16, 32

is a geometric progression because there is a ratio of 2, $(r = 2)$, between any two consecutive terms.

Consecutive Terms	Ratio (r)	Consecutive Terms	Ratio (r)
2, 4	$\dfrac{4}{2} = 2$	8, 16	$\dfrac{16}{8} = 2$
4, 8	$\dfrac{8}{4} = 2$	16, 32	$\dfrac{32}{16} = 2$

Here is another way of describing the preceding geometric progression. *Each term is the product of the preceding term and a common factor* (the common ratio).

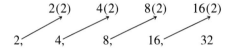

$$2(2) \quad 4(2) \quad 8(2) \quad 16(2)$$
$$2, \quad 4, \quad 8, \quad 16, \quad 32$$

EXAMPLE 1

Write the first five terms of the geometric progression if $a_1 = 4$ and $r = 3$.

$a_1 = 4$ The first term is a_1. Each term thereafter is found
                              by multiplying the preceding term by $r = 3$.
$a_2 = 4(3) = 12$

$a_3 = 12(3) = 36$

$a_4 = 36(3) = 108$

$a_5 = 108(3) = 324$

So the first five terms are 4, 12, 36, 108, and 324.   ■

In general, a geometric progression can be described in the following way:

Number of the Term	Term	Example
1	a_1	7
2	$(a_1) \cdot r = a_1 r$	$(7)3 = (7)3^1 = 21$
3	$(a_1 r) \cdot r = a_1 r^2$	$(21)3 = (7)3^2 = 63$
4	$(a_1 r^2) \cdot r = a_1 r^3$	$(63)3 = (7)3^3 = 189$
5	$(a_1 r^3) \cdot r = a_1 r^4$	$(189)3 = (7)3^4 = 567$
$\vdots$	$\vdots$	$\vdots$
n	$(a_1 r^{n-2}) \cdot r = a_1 r^{n-1}$	$(7)3^{n-1}$

Notice that for each term the exponent of r is one less than the number of the term.

■ FORMULA

In a geometric progression if a_1 represents the first term and r is the common ratio, then the nth term, a_n, is

$$a_n = a_1 r^{n-1}.$$

To find the eighth term of the progression

$3, 9, 27, \ldots ,$

we first identify a_1 and r, then substitute in the formula for the nth term.

$a_1 = 3, \; r = \dfrac{9}{3} = 3$

$a_n = a_1 r^{n-1}$

$a_8 = 3(3)^{8-1} = 3(3^7)$

$\quad = 3(2187)$

$\quad = 6561$

The eighth term is 6561.

EXAMPLE 2

Write the next four terms of the geometric progression 7, 14, 28, . . . and find the fifteenth term.

$a_1 = 7, \; r = \dfrac{14}{7} = \dfrac{28}{14} = 2$
The first term is $a_1 = 7$. The common ratio is $r = 2$.

7, 14, 28, 56, 112, 224, 448
Find each of the next four terms by multiplying the preceding term by $r = 2$.

$a_n = a_1 r^{n-1}$
Formula for the general or nth term of a geometric progression.

$a_{15} = 7(2)^{15-1}$
Substitute $a_1 = 7$, $r = 2$, and $n = 15$ in the formula.

$a_{15} = 7(16{,}384)$
$2^{14} = 16{,}384$

$a_{15} = 114{,}688$

The next four terms are 56, 112, 224, and 448, and the fifteenth term is 114,688. ■

EXAMPLE 3

The next time he goes to Nevada, Sam Slick decides, he will double his bet each time he loses. If he initially bets $5, how much will he bet after eight consecutive losses?

	Sam is following a geometric progression in his betting.
5, 10, 20, . . .	Write the first three bets to identify the geometric progression.
$a_1 = 5$, $r = 2$	Since Sam had eight straight losses, he will be making his ninth bet ($n = 9$).
$n = 9$	

Formula:

$a_n = a_1 r^{n-1}$ $a_1 = 5$, $r = 2$, $n = 9$

Substitute:

$a_9 = 5(2)^{9-1}$

Solve:

$a_9 = 5(256)$ $2^8 = 256$

$ = 1280$

Answer:

Sam's ninth bet must be $1280.   ∎

We can identify a geometric progression given

1. a_1 and r_1 (see Example 1), or

2. a_1 and a specific term, or

3. Any two specific terms

For example,

2. Given a_1 and a specific term. If $a_1 = 5$ and the fifth term is 1280, we first find the common ratio. Substitute the values for a_n, n, and a_1 in the formula.

$a_n = a_1 r^{n-1}$ where $a_1 = 5$, $n = 5$

So $a_n = a_5 = 1280$

$1280 = 5 \cdot r^{5-1}$

$1280 = 5r^4$

$256 = r^4$

$r = \pm 4$

There are two such progressions.
Using $r = 4$, the progression is

5, 20, 80, 320, 1280, ...

Using $r = -4$, the progression is

5, -20, 80, -320, 1280, ...

3. Given any two specific terms. If $a_5 = 4802$ and $a_8 = 1,647,086$, we substitute into the general term formula and then solve the resulting system of equations for a_1 and r.

$$a_n = a_1 r^{n-1}$$

$$4802 = a_1 r^{5-1} \qquad\qquad a_5 = 4802, n = 5$$

$$1,647,086 = a_1 r^{8-1} \qquad\qquad a_8 = 1,647,086, n = 8$$

$$\begin{cases} 4802 = a_1 r^4 & (1) \\ 1,647,086 = a_1 r^7 & (2) \end{cases}$$

$$4802 = a_1 r^4 \qquad\qquad \text{Solve equation (1) for } a_1.$$

$$a_1 = \frac{4802}{r^4}$$

$$1,647,086 = \frac{4802}{r^4} \cdot r^7 \qquad\qquad \text{Substitute the expression } a_1 \text{ in equation (2).}$$

$$1,647,086 = 4802 r^3$$

$$343 = r^3$$

$$7 = r$$

Now substitute $r = 7$ in

$$a_1 = \frac{4802}{r^4}$$

$$= \frac{4802}{7^4}$$

$$= \frac{4802}{2401}$$

$$= 2$$

The progression is

2, 14, 98, 686, 4802, 33,614, 235,298, 1,647,086, ...

EXAMPLE 4

Find all the terms of the geometric progression(s) between the terms $a_5 = 176$ and $a_9 = 2816$.

$a_n = a_1 r^{n-1}$	Formula for the nth term of a geometric progression.
$a_5 = a_1 r^4, \; a_9 = a_1 r^8$	

So,

$\begin{cases} 176 = a_1 r^4 & (1) \\ 2816 = a_1 r^8 & (2) \end{cases}$	Write two equations with two unknowns using the given data to find a_1 and r.
$a_1 = \dfrac{176}{r^4}$	Solve (1) for a_1.
$2816 = \left(\dfrac{176}{r^4}\right) r^8$	Substitute in (2).
$2816 = 176 r^4$	Simplify.
$r^4 = 16$	
$r = \pm 2$	There are two values of r, so there are two sets of terms that will satisfy the conditions.
$176, \; 352, \; 704, \; 1408, \; 2816$	Using 2 as the common ratio.
$176, \; -352, \; 704, \; -1408, \; 2816$	Using -2 as the common ratio.

The missing terms are either 352, 704, and 1408 or -352, 704, and -1408.   ∎

Given the finite geometric progression

5, 3, 1.8, ... 0.648,

we can find the number of terms (n) in the progression using the same formula. Here

$$a_1 = 5, \; r = \frac{3}{5} = 0.6, \; a_n = 0.648$$

$a_n = a_1 r^{n-1}$	
$0.648 = 5(0.6)^{n-1}$	
$0.1296 = (0.6)^{n-1}$	
$(0.6)^4 = (0.6)^{n-1}$	Substitute $(0.6)^4$ for 0.1296
$4 = n - 1$	If $x^a = x^b$ then $a = b$.
$5 = n$	

So the progression has 5 terms.

EXAMPLE 5

Find the number of terms in the finite geometric progression
46,875, 9375, 1875, . . . , 3.

$a_1 = 46{,}875, \; r = \dfrac{9375}{46{,}875} = \dfrac{1}{5}$ *Method I:* Find the common ratio, and use it to find the missing terms. Then count the number of terms.

46,875, 9375, 1875, 375, 75, 15, 3

There are 7 terms.

$a_n = a_1 r^{n-1}$ *Method II:* Use the formula to solve for n, where $a_1 = 46{,}875$, $r = \dfrac{1}{5}$, and $a_n = 3$.

$3 = 46{,}875 \left(\dfrac{1}{5} \right)^{n-1}$

$\dfrac{1}{15{,}625} = \left(\dfrac{1}{5} \right)^{n-1}$ $15{,}625 = 5^6$

$\left(\dfrac{1}{5} \right)^6 = \left(\dfrac{1}{5} \right)^{n-1}$

$6 = n - 1$ If $x^a = x^b$, then $a = b$.

$n = 7$

There are 7 terms in the progression.   ■

If a geometric progression is finite (has n terms), we can find the sum of those n terms. If S_n is the sum of a geometric progression of n terms, then

$$S_n = a_1 + a_1 r + a_1 r^2 + a_1 r^3 + \cdots + a_1 r^{n-1}$$

and

$$rS_n = a_1 r + a_1 r^2 + a_1 r^3 + \cdots + a_1 r^{n-1} + a_1 r^n.$$

Subtracting, we get

$S_n - rS_n = a_1 - a_1 r^n$ All intermediate terms drop out.

$S_n(1 - r) = a_1 - a_1 r^n \quad r \neq 1$ Factor the left side.

■ FORMULA

$$S_n = \dfrac{a_1 - a_1 r^n}{1 - r} \quad r \neq 1$$ Sum of a geometric progression.

Or since $a_1 r^{n-1} = a_n$, we can write

$$S_n = \dfrac{a_1 - (a_1 r^{n-1}) r}{1 - r} \quad r \neq 1.$$

■ FORMULA

$$S_n = \frac{a_1 - ra_n}{1 - r} \quad r \neq 1 \qquad \text{Sum of a geometric progression.}$$

The sum can be found using whichever formula is the most convenient. For example, the sum of the first eight terms of

2, 6, 18, . . .

can be found by using $S_n = \dfrac{a_1 - a_1 r^n}{1 - r}$, where $a_1 = 2$, $r = 3$, $n = 8$.

$$S_8 = \frac{2 - 2 \cdot 3^8}{1 - 3}$$

$$S_8 = 6560$$

Or first solve for a_n, and use the formula $S_n = \dfrac{a_1 - ra_n}{1 - r}$.

$$a_8 = 2(3^7) = 4374$$

$$S_8 = \frac{2 - 3(4374)}{1 - 3}$$

$$S_8 = 6560$$

EXAMPLE 6

Find the sum of the first ten terms of the geometric progression 9, 18, 36, ...

$a_1 = 9$, $r = \dfrac{18}{9} = 2$, $n = 10$ Find a_1, r, and n.

$S_n = \dfrac{a_1 - a_1 r^n}{1 - r}$ Substitute in the formula for the sum of n terms and simplify.

$S_{10} = \dfrac{9 - 9(2)^{10}}{1 - 2}$

$S_{10} = \dfrac{9 - 9(1024)}{-1}$ $2^{10} = 1024$

$S_{10} = \dfrac{9 - 9216}{-1}$

$S_{10} = 9207$

The sum of the first ten terms is 9207.   ■

The formula for finding the nth term of a geometric progression is

$$a_n = a_1 r^{n-1},$$

so we can write the sum of the progression in summation notation.

$$S_n = \sum_{n=1}^{k} a_n = \sum_{n=1}^{k} a_1 r^{n-1}$$

EXAMPLE 7

Find the sum: $\displaystyle\sum_{n=1}^{6} 2(-4)^n$

$$\sum_{n=1}^{6} 2(-4)^n = 2(-4)^1 + 2(-4)^2 + 2(-4)^3$$

$$+ \; 2(-4)^4 + 2(-4)^5 + 2(-4)^6$$

$$= (-8) + (32) + (-128)$$
$$+ \, (512) + (-2048) + (8192)$$

$$= 6552$$

Method I: Expand and add.

$$S_n = \frac{a_1 - a_1 r^n}{1 - r}$$

$$S_6 = \frac{-8 - (-8)(-4)^6}{1 - (-4)}$$

$$= \frac{-8 + 32768}{5}$$

$$= 6552$$

Method II: Use the formula for the sum. Substitute $a_1 = -8$, $r = -4$, and $n = 6$.

The sum is 6552.   ■

Exercise   11.3

A

Write the first four terms of the geometric progression, and find the indicated term:

1. $a_1 = 1$, $r = 3$ Find a_6

2. $a_1 = 4$, $r = \dfrac{1}{2}$ Find a_7

3. $a_1 = 8$, $a_2 = 40$ Find a_8

4. $a_1 = 12$, $a_2 = 3$ Find a_7

Find the number of terms in each finite geometric progression:

5. 6, 12, . . . , 1536

6. 1, 10, . . . , 100,000

7. $\frac{1}{9}, \frac{1}{3}, \ldots, 243$

8. $\frac{1}{16}, \frac{1}{4}, \ldots, 4096$

Find all the terms between the given terms of the following geometric progressions:

9. $a_1 = 2, a_5 = 162$

10. $a_1 = \frac{1}{2}, a_6 = 16$

11. $a_4 = 14, a_{10} = 896$

12. $a_{18} = 10, a_{22} = 6250$

13. $a_2 = -1, a_7 = 243$

14. $a_3 = \frac{8}{5}, a_8 = -\frac{256}{5}$

Find the sum of the indicated number of terms of the following geometric progressions:

15. 10, 20, 40, . . .   11 terms

16. 6, 18, 54, . . .   7 terms

17. 2, −6, 18, . . .   6 terms

18. 21, −42, 84, . . .   8 terms

B

Write the first four terms of the geometric progression, and find the indicated term:

19. $a_1 = 60, r = 0.1$ Find a_8

20. $a_1 = 30, r = \frac{1}{3}$ Find a_7

21. $a_1 = 9, a_2 = 6$ Find a_8

22. $a_1 = 8, a_2 = -6$ Find a_7

23. $a_3 = 20, a_6 = 0.16$ Find a_{10}

24. $a_5 = 100, a_9 = 0.81$ Find a_{12}

25. $a_2 = \frac{1}{6}, a_5 = \frac{1}{162}$ Find a_7

26. $a_2 = -\frac{1}{6}, a_5 = \frac{1}{48}$ Find a_8

Find the number of terms and the sum of the following finite geometric progressions:

27. $\frac{1}{6}, \frac{1}{3}, \ldots, 42\frac{2}{3}$

28. $\frac{1}{10}, -\frac{1}{2}, \ldots, 1562\frac{1}{2}$

29. 80, 20, . . . , $\frac{5}{64}$

30. 45, −15, . . . , $-\frac{5}{243}$

31. 0.2, 0.06, . . . , 0.000486

32. 1.2, 0.24, . . . , 0.000384

33. −2.1, 1.05, . . . , −0.13125

34. $-\frac{1}{4}, -\frac{1}{2}, \ldots, -32$

C

Find the indicated sum:

35. $\displaystyle\sum_{n=1}^{5} 3 \cdot 2^n$

36. $\displaystyle\sum_{n=1}^{6} 4(-3)^n$

37. $\displaystyle\sum_{n=1}^{10} 10(-1)^n$

38. $\displaystyle\sum_{n=1}^{8} 36\left(\frac{1}{2}\right)^n$

39. $\displaystyle\sum_{n=1}^{7} 4(3)^n$

40. $\displaystyle\sum_{n=1}^{5} 5(-1)^n$

41. $\displaystyle\sum_{n=1}^{11} 2187(3)^{-n}$

42. $\displaystyle\sum_{n=1}^{9} 5764801(7)^{-n}$

43. $\displaystyle\sum_{n=1}^{7} 11(-5)^n$

44. $\displaystyle\sum_{n=1}^{10} (-4)(-6)^n$

45. $\displaystyle\sum_{n=1}^{5} (3)\left(-\frac{1}{3}\right)^n$

46. $\displaystyle\sum_{n=1}^{6} (5.4)(-0.2)^n$

47. $\displaystyle\sum_{n=1}^{10} (2)\left(\frac{3}{2}\right)^n$

48. $\displaystyle\sum_{n=1}^{8} \left(\frac{5}{4}\right)\left(-\frac{1}{2}\right)^n$

49. $\displaystyle\sum_{n=1}^{5} \left(\frac{3}{2}\right)\left(\frac{1}{3}\right)^n$

50. $\displaystyle\sum_{n=1}^{4} (-2.5)\left(\frac{1}{5}\right)^n$

D

51. Sam Slick's brother Joe triples his bets to cover losses. How much will Joe bet after eight straight losses if his initial bet was $10?

52. A certain casino sets its slot machines so that the average gambler loses 20% of his investment per hour. If a gambler starts out with $100, how much would he have left after five hours?

53. The price of a can of juice increases at a rate of 5% per year because of inflation. If the price of the juice is now 73¢, what will the price be four years from now? Twenty years from now? (*Hint:* Use $r = 1.05$ since the price each year is 105% of last year's price.)

54. At an inflation rate of 4%, what will the price of a gallon of gas be five years from now if the price now is $1.45 per gallon?

55. The number of bacteria in a culture doubles every day. If the original culture contains 50 bacteria, how many will there be after 10 days?

56. A person vows to put 5 cents in a piggy bank on the first day of the month, 15 cents on the second, and 45 cents on the third. If they continue the pattern, how much will they put into the bank on the tenth day? How much will they put in the bank on the fifteenth day? What will be the total in the bank after fifteen days?

57. If you start with a single square and draw a vertical and a horizontal line through the square as shown, you will now have four smaller squares. If you divide each of those, you will have sixteen still smaller squares. If this process is continued six times, how many very small squares will there be?

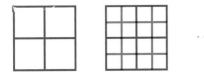

Figure for Exercise 57

58. A teacher receives a starting salary of $20,000 and a raise of 5% per year. The salary in the second year is now $20,000 + (0.05)($20,000), which could be written $20,000(1 + 0.05)^1$. How would you write a similar expression for the teachers salary in the third year? What will the salary be for the eleventh year? What will be the total earnings of the teacher for those eleven years? (all answers to the nearest cent)

59. A student stands 6 feet from a classroom wall and steps off one-half of the distance from her original position to the wall (i.e. 3 feet). She then steps one-half of the remaining distance with each successive step. After eleven steps, how far is the student from the wall? What distance has she stepped-off so far? (to the nearest hundredth of an inch)

├──── 6 ft ────┤

Figure for Exercise 59

60. A young kitten splashes his paw in a bowl containing 6 ounces of milk and one-tenth of the milk splashes out of the bowl. If each time the kitten splashes in the milk, one-tenth splashes out of the bowl, how many ounces of milk remain after five splashes? (to the nearest hundredth of an ounce)

61. A ball is dropped from a height of 40 feet and it bounces up to 70% of the previous height. How high will it bounce on the seventh bounce? What will be the total "up and down" distance that the ball has travelled after its seventh bounce (Careful)? (all answers to the nearest hundredth)

STATE YOUR UNDERSTANDING

62. What is a geometric progression?

63. What is the difference between an arithmetic and a geometric progression?

64. In exercise 59, where the student steps one-half of the distance, will she ever reach the wall? Explain your answer.

CHALLENGE EXERCISES

For each of the following geometric sequences, find: (a) the seventh term of the sequence; and (b) the sum of the first seven terms.

65. $4, -2, 1, -\dfrac{1}{2}, \ldots$

66. $\dfrac{1}{6}, -\dfrac{1}{3}, \dfrac{2}{3}, -\dfrac{4}{3}, \ldots$

67. $\sqrt{3}, \sqrt{6}, 2\sqrt{3}, 2\sqrt{6}, \ldots$

68. $0.1, 0.01, 0.001, 0.0001, \ldots$

69. $1, \sqrt{5}, 5, \sqrt{125}, \ldots$

MAINTAIN YOUR SKILLS (SECTIONS 10.2, 10.5)

Solve without a calculator:

70. $q = \log_4 64$

71. $r = \log_{512} 8$

72. $s = \log_{10} \sqrt[3]{10}$

73. $5^{x-1} = 125$

Solve with the help of a calculator or logarithm tables; write the answer to the nearest ten-thousandth:

74. $6^x = 42$

75. $5^{x-1} = 15$

76. $\log_5 17 = x$

77. $\log_{17} 5 = x$

11.4

INFINITE GEOMETRIC PROGRESSIONS

OBJECTIVE

1. Find the sum of an infinite geometric progression for which $|r| < 1$.

■ **DEFINITION**

Infinite Geometric
Progression

> An *infinite geometric progression* is a geometric progression in which the number of terms is infinite.

In this section, the expressions "the sum of the terms of an infinite geometric progression" and "the sum of an infinite geometric series" have the same meaning.

Consider the infinite geometric progression

$$4, 2, 1, \frac{1}{2}, \frac{1}{4}, \ldots$$

The common ratio is $\frac{1}{2}$. Each succeeding term is getting smaller and closer to zero. In fact, we can find a term that is as close to zero as we want. We call the progression infinite if the terms continue without end. $a_n = 4\left(\frac{1}{2}\right)^{n-1}$ is the nth term of the progression $4, 2, 1, \frac{1}{2}, \frac{1}{4}, \ldots$. The tenth and twentieth terms are

$$n = 10 \qquad a_{10} = \frac{1}{128} = 0.0078125$$

$$n = 20 \qquad a_{20} = \frac{1}{131072} \approx 0.00000763$$

Note that the terms are getting smaller and close to zero. Now look at the sum of n terms of this progression.

$$S_n = \frac{a_1(1 - r^n)}{1 - r} \qquad \text{where } a_1 = 4, \ r = \frac{1}{2}$$

$$S_n = \frac{4\left(1 - \left(\frac{1}{2}\right)^n\right)}{1 - \frac{1}{2}}$$

$$= \frac{4\left(1 - \frac{1}{2^n}\right)}{\frac{1}{2}}$$

$$= 8\left(1 - \frac{1}{2^n}\right)$$

$$= 8 - \frac{8}{2^n}$$

The larger the value of n, the closer the sum is to 8. If $n = 10$,

$$S_{10} = 8 - 0.0078125$$

$$S_{10} = 7.9921875.$$

If $n = 20$,

$$S_{20} \approx 8 - 0.00000763$$

$$S_{20} \approx 7.99999237.$$

We define the sum of the terms of this infinite geometric series to be 8. The symbol* S_∞ is used to denote the sum of an infinite geometric series:

$$S_\infty = 4 + 2 + 1 + \frac{1}{2} + \cdots = 8$$

Look again at the formula for the sum of a geometric progression.

$$S_n = \frac{a_1(1 - r^n)}{1 - r}$$

If $|r| < 1$, then as n gets very large, r^n gets close to zero, and the sum becomes

$$S_n \approx \frac{a_1(1 - 0)}{1 - r} = \frac{a_1}{1 - r}.$$

■ **FORMULA**

The sum of an infinite geometric series where $|r| < 1$ is defined as

$$S_\infty = \frac{a_1}{1 - r}.$$

If $r \geq 1$ or $r \leq -1$, the sum is not defined.

To find the sum of the infinite geometric series

$$28 + \frac{56}{3} + \frac{112}{9} + \cdots,$$

first identify, a_1 and r, and then substitute in the formula

$$a_1 = 28, \ r = \frac{\dfrac{56}{3}}{28} = \frac{56}{3} \cdot \frac{1}{28} = \frac{2}{3}$$

$$S_n = \frac{a_1}{1 - r}$$

$$= \frac{28}{1 - \dfrac{2}{3}}$$

$$= \frac{28}{\dfrac{1}{3}}$$

$$= 84$$

*The symbol ∞ is read "infinity." Here it shows that the progression continues forever without end. The symbol ∞ does not represent a number.

EXAMPLE 1

Find the sum of the infinite geometric series $60 + 30 + 15 + \cdots$.

$a_1 = 60,\ r = \dfrac{30}{60} = \dfrac{1}{2}$ Identify a_1 and r.

$S_\infty = \dfrac{a_1}{1-r}$ Use the formula to find the sum.

$S_\infty = \dfrac{60}{1 - \left(\dfrac{1}{2}\right)}$ Substitute.

$\quad = 120$

The sum is 120. ■

EXAMPLE 2

Find the sum of the infinite geometric series $6 - 3 + \dfrac{3}{2} - \cdots$.

$a_1 = 6,\ r = \dfrac{-3}{6} = -\dfrac{1}{2}$ Identify a_1 and r.

$S_\infty = \dfrac{a_1}{1-r}$ Use the formula to find the sum.

$S_\infty = \dfrac{6}{1 - \left(-\dfrac{1}{2}\right)}$ Substitute.

$\quad = 4$

The sum is 4. ■

EXAMPLE 3

A repeating decimal is an infinite geometric series. The sum is the common fraction name for the decimal. Find the sum (common fraction name) of $0.151515\ldots$.

$0.151515\cdots = 0.15 + 0.0015 + 0.000015 + \cdots$ Write as a series to help identify a_1 and r.

$a_1 = 0.15,\ r = \dfrac{0.0015}{0.15} = 0.01$

$S_\infty = \dfrac{a_1}{1-r}$ Use the formula for the sum of an infinite geometric progression.

$$S_\infty = \frac{0.15}{1 - 0.01}$$

$$= \frac{5}{33}$$

So the sum (common fraction name) of $0.151515\ldots$ is $\frac{5}{33}$.   ∎

Sigma notation is also used to indicate the sum of an infinite geometric progression.

EXAMPLE 4

Find the sum: $\displaystyle\sum_{n=1}^{\infty} \left[\frac{1}{2}\right]^n$

$$\sum_{n=1}^{\infty} \left[\frac{1}{2}\right]^n = \frac{1}{2} + \frac{1}{4} + \frac{1}{8} + \cdots$$

$$a_1 = \frac{1}{2}, \; r = \frac{1}{2}$$

$$S_\infty = \frac{\dfrac{1}{2}}{1 - \dfrac{1}{2}}$$

$$= \frac{\dfrac{1}{2}}{\dfrac{1}{2}}$$

$$= 1$$ ∎

EXAMPLE 5

A silver mine yields 30 tons of high-grade ore in the first year. If each year thereafter it will yield two-thirds of the amount of the previous year, find the number of tons of high-grade ore that was in the mine.

First year = 30 tons

Second year = $\dfrac{2}{3}$(30 tons) = 20 tons

Third year = $\dfrac{2}{3}$(20 tons) = $\dfrac{40}{3}$ tons

$$30 + 20 + \frac{40}{3} + \cdots$$ Write the series, and identify a_1 and r.

$$a_1 = 30, \ r = \frac{2}{3}$$ Since $|r| < 1$, the sum is finite.

$$S_\infty = \frac{a_1}{1 - r}$$

$$= \frac{30}{1 - \left(\frac{2}{3}\right)}$$

$$= 90$$

The mine contained 90 tons of high-grade ore.   ■

Exercise 11.4

A

Find the sum of each of the following infinite geometric series:

1. $8 + 4 + 2 + \cdots$

2. $10 + 2 + \frac{2}{5} + \cdots$

3. $15 + 3 + \frac{3}{5} + \cdots$

4. $50 + 5 + \frac{1}{2} + \cdots$

5. $21 + 7 + \frac{7}{3} + \cdots$

6. $20 + 5 + \frac{5}{4} + \cdots$

7. $36 + 12 + 4 + \cdots$

8. $80 + 20 + 5 + \cdots$

9. $30 + 15 + \frac{15}{2} + \cdots$

10. $100 + 10 + 1 + \cdots$

11. $80 + 40 + 20 + \cdots$

12. $60 + 12 + \frac{12}{5} + \cdots$

13. $-100 + (-50) + (-25) + \cdots$

14. $-49 + (-7) + (-1) + \cdots$

15. $\sum_{n=1}^{\infty} \left[\frac{1}{3}\right]^n$

16. $\sum_{n=1}^{\infty} \left[\frac{1}{4}\right]^n$

B

Find the sum of each of the following infinite geometric series:

17. $15 - 3 + \frac{3}{5} - \cdots$

18. $250 - 25 + 2.5 - \cdots$

19. $12 - 6 + 3 - \cdots$

20. $18 - 6 + 2 - \cdots$

21. $\frac{2}{3} + \frac{1}{3} + \frac{1}{6} + \cdots$

22. $\frac{5}{8} + \frac{1}{8} + \frac{1}{40} + \cdots$

23. $\frac{2}{7} - \frac{1}{7} + \frac{1}{14} - \cdots$

24. $\frac{15}{8} - \frac{5}{8} + \frac{5}{24} - \cdots$

25. $-22 - 8.8 - 3.52 - \cdots$

26. $-30 - 21 - 14.7 - \cdots$

27. $28 - 12.6 + 5.67 - \cdots$

28. $-19 + 6.08 - 1.9456 + \cdots$

29. $5 - \dfrac{15}{2} + \dfrac{45}{4} - \cdots$

30. $2 - \dfrac{10}{3} + \dfrac{50}{9} - \cdots$

31. $\displaystyle\sum_{n=1}^{\infty} \left[\dfrac{2}{3}\right]^n$

32. $\displaystyle\sum_{n=1}^{\infty} \left[\dfrac{3}{4}\right]^n$

33. $\displaystyle\sum_{n=1}^{\infty} \left[\dfrac{7}{8}\right]^n$

34. $\displaystyle\sum_{n=1}^{\infty} \left[\dfrac{9}{10}\right]^n$

C

Find the common fraction name for the following repeating decimals:

35. $0.55555\ldots$

36. $0.7777\ldots$

37. $0.151515\ldots$

38. $0.828282\ldots$

39. $0.016016016\ldots$

40. $0.315315315\ldots$

41. $0.181818\ldots$

42. $0.242424\ldots$

43. $0.108108108\ldots$

44. $0.369369369\ldots$

45. $0.027027027\ldots$

46. $0.072072072\ldots$

47. $\displaystyle\sum_{n=1}^{\infty} \dfrac{1}{2}\cdot\left[\dfrac{5}{9}\right]^n$

48. $\displaystyle\sum_{n=1}^{\infty} \dfrac{4}{7}\cdot\left[\dfrac{7}{11}\right]^n$

49. $\displaystyle\sum_{n=1}^{\infty} \dfrac{3}{5}\cdot\left[\dfrac{4}{7}\right]^n$

50. $\displaystyle\sum_{n=1}^{\infty} \dfrac{2}{3}\cdot\left[\dfrac{5}{7}\right]^n$

D

51. A mine in Idaho yields 150 tons of high-grade gold ore in the first year, and each year thereafter yields $\dfrac{3}{4}$ the tonnage of the previous year. Find the amount of high-grade ore that was in the mine.

52. An oil well in Oklahoma yields 5000 barrels of oil in the first year. Each year thereafter it yields $\dfrac{7}{10}$ of the previous year's output. How many barrels of oil were in the well?

53. A ball rebounds to $\dfrac{1}{3}$ of its previous height with each bounce. How far has the ball traveled if it is dropped from a height of 10 feet? (Assume the ball bounces an infinite number of times.)

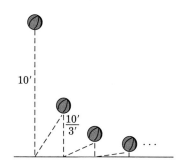

10'

10'
3'

Figure for Exercise 53

54. How far does the ball in Exercise 53 travel if it rebounds $\dfrac{3}{4}$ of its previous height with each bounce?

55. Find the sum of $\dfrac{1}{x} + \dfrac{1}{x^2} + \dfrac{1}{x^3} + \dfrac{1}{x^4} + \cdots,\ |x| > 1.$

56. Find the sum of $\dfrac{x}{y} + \dfrac{x^2}{y^2} + \dfrac{x^3}{y^3} + \dfrac{x^4}{y^4} + \cdots,\ |x| < |y|.$

57. An equilateral triangle has three equal sides and the dimensions of three triangles in a sequence are given below:

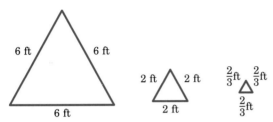

If the perimeter of each triangle is three times the length of one side, find sum of the perimeters of all of the triangles in the sequence if the sequence w were to continue indefinitely.

58. The yield, in tons of wheat, from a certain field starts at 1000 tons in the first year. Due to soil depletion, the yield the next year is 800 tons and the year after, it is only 640 tons. If the trend continues, what will be the TOTAL yield of the field over a long period.

59. Suppose the federal reserve releases $10,000,000,000 into the economy and they find that people save 10% of this new money and spend 90%. Then, of the 90%, they save 10% and spend 90%. If this trend continues, how much spending is generated by the original 10 billion dollars?

60. A fund is designed to pay the holder most of its return "up-front" and it pays $10,000 in the first year. In the second year, it pays $6000 and in the third year, it pays $3,600. If the process continues, what will be paid to the holder and his/her survivors over a long period of time?

61. A diamond (a square which is resting on a corner) is 4 feet by 4 feet. By connecting the midpoints of the sides of the diamond, a square is formed inside. What is the length of each side of the new square? Then, another diamond is formed by connecting the midpoints of the sides of the square, and so on. If the area of each is found by squaring one of its sides, what would be the sum of the areas of all of the diamonds and squares if the process were to continue forever?

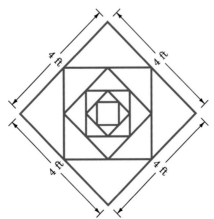

If the perimeter of each figure is found by multiplying each side by four, what would be the sum of all of the perimeters of the diamonds and squares?

STATE YOUR UNDERSTANDING

62. What two quantities are needed to find the sum of an infinite geometric series?

63. Why, mathematically, does the formula for the sum of an infinite geometric series FAIL when the common ratio is 1?

64. What is a necessary condition for finding the sum of an infinite geometric progression?

CHALLENGE EXERCISES

65. If the sum of the following geometric series is 4, find "r."

$$3 + 3r + 3r^2 + 3r^3 + \ldots$$

66. Find the sum: $\displaystyle\sum_{n=1}^{\infty} \left(\frac{1}{2}\right)^{2n}$

67. Convert each of the following to fractions using the method in this section.

    a. 1.17171717 . . .

    b. 0.9999999 . . .

    c. 1.7999999 . . .

MAINTAIN YOUR SKILLS (SECTIONS 10.3, 10.5)

Write as a combination of logarithms of single variables:

68. $\log \dfrac{r^3}{s}$

69. $\log \dfrac{2a}{b}$

70. $\ln \dfrac{\sqrt{x}}{y}$

71. $\ln \dfrac{2\sqrt{w}}{z^3}$

72. $\ln p + \ln q - \ln \dfrac{1}{r}$

73. $\ln x^{-1} - \ln y^{-1} + \ln \dfrac{1}{z}$

Solve:

74. $\log_5 (x + 3) - \log_5 x = 2$

75. $\log (x + 3) - \log x = 2$

11.5

BINOMIAL EXPANSION

1. Evaluate an expression written in factorial notation.

2. Write a binomial of the form $(a + b)^n$ in expanded form.

3. Find a specified or particular term of a binomial expansion.

When writing the general binomial $(a + b)^n$ in expanded form, it is useful to have a short way of writing the product of consecutive positive integers. To do this, we use the symbol $n!$. This symbol is read "n factorial" or "factorial n."

■ FORMULA

Factorial Notation

$$n! = n(n - 1)(n - 2)(n - 3) \ldots (2)(1)$$

EXAMPLE 1

Write $10!$ in expanded form and simplify.

$$10! = 10 \cdot 9 \cdot 8 \cdot 7 \cdot 6 \cdot 5 \cdot 4 \cdot 3 \cdot 2 \cdot 1$$

$$= 3,628,800$$ ■

Products of consecutive integers such as $9 \cdot 8 \cdot 7 \cdot 6$ can be written using quotients of factorials.

$$9 \cdot 8 \cdot 7 \cdot 6 = \frac{9 \cdot 8 \cdot 7 \cdot 6 \cdot 5 \cdot 4 \cdot 3 \cdot 2 \cdot 1}{5 \cdot 4 \cdot 3 \cdot 2 \cdot 1}$$

$$= \frac{9!}{5!}$$

EXAMPLE 2

Simplify: $\dfrac{18!}{15! \, 3!}$

Note that $16! = 16 \cdot 15!$ $22! = 22 \cdot 21!$ and in general $n! = n(n - 1)!$.

$$\frac{18!}{15! \, 3!} = \frac{18 \cdot 17 \cdot 16 \cdot \cancel{15!}}{\cancel{15!} \, 3!}$$

We use this to rewrite $18!$ as $18 \cdot 17 \cdot 16 \cdot 15!$ so we can divide out $15!$

$$= \frac{18 \cdot 17 \cdot 16}{3 \cdot 2 \cdot 1}$$

$$= 816$$ ■

For convenience in the binomial expansion, we define 0! as follows:

■ DEFINITION

$0! = 1$

$0! = 1$

The expanded form of the binomial $(a + b)^n$ is an interesting and useful series.

$(a + b)^1 = a + b$

$(a + b)^2 = a^2 + 2ab + b^2$

$(a + b)^3 = a^3 + 3a^2b + 3ab^2 + b^3$

$(a + b)^4 = a^4 + 4a^3b + 6a^2b^2 + 4ab^3 + b^4$

$(a + b)^5 = a^5 + 5a^4b + 10a^3b^2 + 10a^2b^3 + 5ab^4 + b^5$

If b^0 is inserted as a factor of the first term and a^0 is inserted as a factor of the last term, we can verify the following statements about the expansion.

The exponent of a	Starts with a^n in the first term and descends consecutively to a^0 in the last term
The exponent of b	Starts with b^0 in the first term and increases consecutively to b^n in the last term
First term	$\dfrac{n!}{n!0!}a^nb^0 = a^n$
Second term	$\dfrac{n!}{(n-1)!1!}a^{n-1}b^1 = na^{n-1}b$
Third term	$\dfrac{n!}{(n-2)!2!}a^{n-2}b^2 = \dfrac{n(n-1)}{2\cdot 1}a^{n-2}b^2$
Fourth term	$\dfrac{n!}{(n-3)!3!}a^{n-3}b^3 = \dfrac{n(n-1)(n-2)}{3\cdot 2\cdot 1}a^{n-3}b^3$

■ FORMULA

The formula for a binomial expansion:

$$(a + b)^n = \frac{n!}{n!0!}a^nb^0 + \frac{n!}{(n-1)!1!}a^{n-1}b^1 + \frac{n!}{(n-2)!2!}a^{n-2}b^2 + \cdots$$

$$+ \frac{n!}{(n-k)!k!}a^{n-k}b^k + \cdots + \frac{n!}{(n-n)!n!}a^0b^n,$$

where k is one less than the number of the term.

EXAMPLE 3

Write $(y - 2)^5$ in expanded form.

$n = 5$, $a = y$, $b = -2$ Identify a, b, and n.

$\frac{5!}{5!0!}(y)^5(-2)^0 + \frac{5!}{4!1!}(y)^4(-2)^1 + \frac{5!}{3!2!}(y)^3(-2)^2 +$ Use the formula to expand $(y - 2)^5$.

$\frac{5!}{2!3!}(y)^2(-2)^3 + \frac{5!}{1!4!}(y)^1(-2)^4 + \frac{5!}{0!5!}(y)^0(-2)^5$

$= (1)(y^5)(1) + 5(y^4)(-2) + 10(y^3)(4) +$ Simplify.
$\quad 10(y^2)(-8) + 5(y)(16) + 1(1)(-32)$

$= y^5 - 10y^4 + 40y^3 - 80y^2 + 80y - 32$ ∎

EXAMPLE 4

Write the first four terms of the expansion of $(x - 5y^2)^{18}$.

$a = x$, $b = -5y^2$, $n = 18$ Identify a, b, and n.

$\frac{18!}{18!0!}(x)^{18}(-5y^2)^0 + \frac{18!}{17!1!}(x)^{17}(-5y^2)^1 +$ Use the formula to expand.

$\frac{18!}{16!2!}(x)^{16}(-5y^2)^2 + \frac{18!}{15!3!}(x)^{15}(-5y^2)^3 + \cdots$

$= (x^{18})(1) + 18(x^{17})(-5y^2) + 153(x^{16})(25y^4) +$ Simplify.
$\quad 816(x^{15})(-125y^6) + \cdots$

$= x^{18} - 90x^{17}y^2 + 3825x^{16}y^4 - 102000x^{15}y^6 + \cdots$ ∎

A pattern that can be used as a shortcut for expanding a binomial is called Pascal's triangle. The triangle is formed by using the coefficients of the expanded form of the binomial $(a + b)^n$.

		Pascal's Triangle
$(a + b)^0$	1	
$(a + b)^1$	$a + b$	
$(a + b)^2$	$a^2 + 2ab + b^2$	
$(a + b)^3$	$a^3 + 3a^2b + 3ab^2 + b^3$	

Pascal's triangle can now be extended by continuing to write "1" along the left and right side of the triangle and then adding the pairs of terms in the last line to find the next line.

The triangle can then be used to expand a binomial by using the appropriate line for the coefficients. For example, to expand $(y - z)^6$:

From the triangle above, we see that the coefficients in the expansion will be in the seventh row.

$$(y - z)^6 = \boxed{1}\ y^6 + \boxed{6}\ y^5(-z) + \boxed{15}\ y^4(-z)^2 + \boxed{20}\ y^3(-z)^3$$
$$+ \boxed{15}\ y^2(-z)^4 + \boxed{6}\ y(-z)^5 + \boxed{1}\ (-z)^6$$
$$= y^6 - 6y^5z + 15y^4z^2 - 20y^3z^3 + 15y^2z^4 - 6yz^5 + z^6$$

EXAMPLE 5

Write $(2y - 3)^4$ in expanded form.

Use the fifth row of Pascal's triangle to determine the coefficients:

$$\begin{array}{ccccccccc}
 & & & & 1 & & & & \\
 & & & 1 & & 1 & & & \\
 & & 1 & & 2 & & 1 & & \\
 & 1 & & 3 & & 3 & & 1 & \\
1 & & 4 & & 6 & & 4 & & 1
\end{array}$$

$$\boxed{1}\ (2y)^4(-3)^0 + \boxed{4}\ (2y)^3(-3)^1 + \boxed{6}\ (2y)^2(-3)^2$$
$$+ \boxed{4}\ (2y)^1(-3)^3 + \boxed{1}\ (2y)^0(-3)^4$$

$16y^4 + 4(8y^3)(-3) + 6(4y^2)(9) + 4(2y)(-27) + 81$ Simplify.

$16y^4 - 96y^3 + 216y^2 - 216y + 81$ ■

■ FORMULA

To find a specified term of an expression, use the formula for a specified term in the general expansion:

$$\frac{n!}{(n-k)!k!}a^{n-k}b^k, \quad k \text{ is one less than the number of the term.}$$

For instance, to find the twelfth term of the expansion of $(3x + y)^{15}$, identify n, a, b, and k and substitute.

$a = 3x,\ b = y,\ n = 15,\ k = 12 - 1 = 11$

$$\frac{15!}{(15-11)!11!}(3x)^{15-11}(y)^{11}$$

$$\frac{15 \cdot 14 \cdot 13 \cdot 12}{4 \cdot 3 \cdot 2 \cdot 1}(81x^4)(y^{11})$$

$1365(81x^4)(y^{11})$

$110565x^4y^{11}$

EXAMPLE 6

Find the eighth term of the expanded form of $(4x^2 - y^3)^{10}$.

$a = 4x^2$, $b = -y^3$, $n = 10$, $k = 7$ Identify a, b, n, and k. Since we want the 8th term, use

$k = 8 - 1 = 7.$

$$\frac{n!}{(n-k)!k!}a^{n-k}b^k$$ Formula for a specific term of an expansion.

$$\frac{10!}{(10-7)!7!}(4x^2)^{10-7}(-y^3)^7$$ Substitute.

$120(64x^6)(-y^{21})$ Simplify.

$-7680x^6y^{21}$ ■

EXAMPLE 7

If a coin is tossed four times in succession, the number of ways that exactly two heads and two tails will show is the coefficient of the third term of the expansion $(H + T)^4$. Find the number of ways that you can get exactly two heads and two tails.

$a = H$, $b = T$, $n = 4$, $k = 3 - 1 = 2$ Find a, b, n, and k.

Formula:

$$\frac{n!}{(n-k)!k!}a^{n-k}b^k$$ Formula for a specific term of an expansion.

Substitute:

$$\frac{4!}{(4-2)!2!}(H)^{4-2}(T)^2$$

Simplify:

$6H^2T^2$

Answer:

There are 6 ways you can get exactly two heads and two tails when you toss a coin four times.

     ■

EXERCISE   11.5

A

Evaluate:

1. $3!$ **2.** $8!$ **3.** $7!$ **4.** $0!$

5. $\dfrac{9!}{6!}$ **6.** $\dfrac{18!}{15!}$ **7.** $\dfrac{20!}{15!}$ **8.** $\dfrac{30!}{28!}$

Write in expanded form:

9. $(x + 1)^4$ **10.** $(y - 3)^3$ **11.** $(x - 1)^5$

12. $(y + 2)^4$ **13.** $(x + 2)^8$ **14.** $(x - 4)^7$

15. $(2x + 1)^5$ **16.** $(3x - 1)^4$

B

Evaluate:

17. $\dfrac{18!}{16!\,2!}$ **18.** $\dfrac{21!}{18!\,3!}$ **19.** $\dfrac{30!}{25!\,5!}$ **20.** $\dfrac{40!}{36!\,4!}$

21. $\left(3x - \dfrac{y}{3}\right)^4$ **22.** $\left(\dfrac{x}{2} + 2y\right)^5$

23. $(x^2 - 2y^3)^6$ **24.** $(x^4 + 3y^2)^4$

25. $(2x - 3)^5$ **26.** $(3x - 1)^6$

27. $(2x^2 + 3y^2)^5$ **28.** $(2x^3 - y^4)^6$

Find the specified term of the expanded form:

29. $(x + 1)^{10}$, seventh term   **30.** $(y - 1)^{14}$, twelfth term

31. $(2x + 1)^9$, sixth term   **32.** $(3x - y^2)^{12}$, tenth term

C

Write the first four terms of the expanded form:

33. $(2x + y)^{24}$ **34.** $(3x - y)^{17}$

35. $(x - 5y)^{20}$ **36.** $(y + 3z)^{14}$

37. $(x^2 - 2y^3)^{12}$

38. $(x^3 + 3y^2)^{15}$

39. $(xy - 3)^{23}$

40. $(x^2y^2 - 2z^2)^{21}$

Find the specified term of the expanded form:

41. $(x + 1)^{22}$, eleventh term

42. $(y - 1)^{18}$, tenth term

43. $\left(x + \dfrac{1}{2}\right)^{17}$, ninth term

44. $\left(2x - \dfrac{y}{2}\right)^{24}$, thirteenth term

45. $\left(\dfrac{x}{2} + 2\right)^{15}$, eighth term

46. $\left(\dfrac{y}{3} - 3\right)^{20}$, tenth term

47. $(3x^2 - y^3)^{23}$, twentieth term

48. $\left(6x - \dfrac{1}{2}y\right)^{18}$, fifteenth term

D

49. If the coin in Example 7 is tossed eight times, the number of ways you can get exactly four heads and four tails is the coefficient of the fifth term of the expansion of $(H + T)^8$. Find the number of ways.

50. In Exercise 49, the number of ways you can get exactly two heads and six tails is the coefficient of the seventh term of the expanded form of the same binomial. Find the number of ways.

51. If an initial investment, A_0, is invested at "r" percent (expressed as a decimal) is compounded once at the end of the year, then the return, A, after "n" years is given by $A = A_0(1 + r)^n$. Use binomial expansion to find the algebraic expression that would show the value of your investment after 5 years.

52. Use your expansion in problem (51) to find the return A, if $A_0 = \$5,000$ and the rate is 7% (for the same 5-year time period).

53. If the number of bacteria in a culture initially containing 1000 bacteria, is given by $N = 1000(1.02)^t$ (where t is the time in hours), use the binomial expansion of $(1 + 0.02)^6$ to find the number present after 6 hours.

54. Remembering how to deal with powers of i, use binomial expansion to expand $(1 + i)^7$, then simplify the expression.

55. Use the binomial expansion of $((1) + (-1))^8$, to verify that the result is zero.

The coefficient, $\dfrac{n!}{(n - k)!k!}$, is also used to calculate the number of ways of selecting k-elements from a set containing n-elements, if the order of the elements you select does not matter. For example, if set $A = \{2, 4, 6, 8, 10\}$, and you are asked, "How many subsets can be composed containing three of the members of set A without considering order?", the calculation is given by:

$\dfrac{5!}{(5 - 3)!3!}$ (which is 10).

56. A committee of 6 people is to be selected from a group of 10 accountants. How many different committees can be formed?

57. Of the 12 member nations of an international treaty group, how many subgroups can be formed containing representatives from 4 nations?

58. The mathematics department contains 7 men and 4 women. How many committees can be formed containing 3 people?

59. Using the information from exercise 58, how many three person committees can be formed if exactly one of the members has to be a man?

STATE YOUR UNDERSTANDING

60. What is Pascal's triangle?

61. Why is the sum of the coefficients of the expansion of $(a - b)^5$ equal to zero?

CHALLENGE EXERCISES

62. If $f(x) = x^3$, find $f(x + h)$ in expanded form. Use the result to find $\dfrac{f(x + h) - f(x)}{h}$ (Simplify).

63. If $f(x) = x^5$, find $f(h + h)$ in expanded form. Use the result to find $\dfrac{f(x + h) - f(x)}{h}$ (Simplify).

64. If $g(x) = x^4 - x^2$, find $g(x + h)$ in expanded form. Use the result to find $\dfrac{g(x + h) - g(x)}{h}$ (Simplify).

65. Use binomial expansion to find the sixth term of $\left(x + \dfrac{1}{2y}\right)^9$.

66. Use binomial expansion to find the fourth term of $\left(\dfrac{1}{x} - \dfrac{1}{y}\right)^7$.

MAINTAIN YOUR SKILLS (SECTIONS 10.4, 10.5)

Find the logarithm to four decimal places:

67. log 5950

68. ln 0.0378

69. ln 5950

Find the antilogarithm to four decimal places:

70. $\log x = 2.7586$

71. $\log x = -2.3112$

72. $\ln x = 2.7586$

73. $\ln x = -2.3112$

Solve:

74. $\log_6 (x - 2) - \log_6 (x - 1) = 1$

CHAPTER 11
SUMMARY

Sequence Function	A function whose domain is the set of counting numbers.	(p. 736)
Range of a Sequence Function	Called a sequence when the elements are placed in order.	(p. 736)
Elements (Terms) of a Sequence	Those numbers that make up the sequence are called the elements or terms of the sequence.	(p. 736)
Series	The indicated sum of any sequence is called a series.	(p. 739)
Summation Notation	Sigma or summation notation (Σ) is used with the general term of the sequence to represent the series.	(p. 739)
Arithmetic Progression	A sequence in which there is a common difference (d) between any two consecutive terms.	(p. 746)
General (nth) Term of an Arithmetic Progression	If a_1 represents the first term of an arithmetic progression and d the common difference, then the nth term is $a_n = a_1 + (n - 1)d$.	(p. 747)
Sum of a Finite Arithmetic Progression of n Terms	$$S_n = \frac{n(a_1 + a_n)}{2} \quad \text{or} \quad S_n = \frac{n}{2}[2a_1 + (n - 1)d]$$	(pp. 751, 752)
Geometric Progression	A sequence function in which there is a common ratio (r) between any two consecutive terms.	(p. 757)
General (nth) Term of a Geometric Progression	If a_1 represents the first term of a geometric progression and r the common ratio, then the nth term is $a_n = a_1 r^{n-1}$.	(p. 759)
Sum of a Finite Geometric Progression of n Terms	$$S_n = \frac{a_1 - ra_n}{1 - r}, \; r \neq 1$$	(p. 763)
Infinite Geometric Progression	A geometric progression in which the terms of the sequence have no end.	(p. 769)
Equivalent Expressions	The expressions "the sum of the terms of an infinite geometric progression" and "the sum of an infinite geometric series" have the same meaning.	(p. 769)
Sum of an Infinite Geometric Progression	$$S_\infty = \frac{a_1}{1 - r}, \; |r| < 1$$	(p. 771)
Sum of an Infinite Geometric Progression if $|r| \geq 1$	Not defined.	(p. 771)
Factorial Notation	The symbol $n!$ is read "n factorial" or "factorial n."	(p. 778)
Definition of Factorial Notation	$n! = n(n - 1)(n - 2)(n - 3) \ldots (2)(1)$ $0! = 1$	(pp. 778, 779)

Binomial Expansion Formula

$$(a + b)^n = \frac{n!}{n!0!}a^nb^0 + \frac{n!}{(n-1)!1!}a^{n-1}b^1 + \frac{n!}{(n-2)!2!}a^{n-2}b^2 + \cdots \quad \text{(p. 779)}$$

$$+ \frac{n!}{(n-k)!k!}a^{n-k}b^k + \cdots + \frac{n!}{(n-n)!n!}a^0b^n,$$

where k is one less than the number of the term.

Pascal's Triangle

A pattern formed by using the coefficients of the expanded form of the (p. 780) binomial $(a + b)^n$.

```
                    1
                1       1
            1       2       1
        1       3       3       1
      1     4       6       4     1
    1     5     10      10      5     1
  1     6     15      20      15     6     1
```

Specified Term of a Binomial Expansion Formula

To find a specified term of an expansion, use the formula for a speci- (p. 781) fied term in the general expansion:

$$\frac{n!}{(n-k)!k!}a^{n-k}b^k, \quad k \text{ is one less than the number of the term.}$$

CHAPTER 11
REVIEW EXERCISES

SECTION 11.1 Objective 1

Write the first five terms of the sequence defined by each of the following.

1. $c(n) = 2n + 1$ **2.** $c(n) = 3n + 4$ **3.** $c(n) = 8 - 2n$

4. $c(n) = 12 - 5n$ **5.** $c(n) = n^2 + 1$ **6.** $c(n) = n^3 - 1$

SECTION 11.1 Objective 2

Write a rule for a sequence function given the first five terms. Find $c(8)$.

7. $\dfrac{3}{5}, \dfrac{6}{5}, \dfrac{9}{5}, \dfrac{12}{5}, 3$ **8.** $\dfrac{1}{2}, \dfrac{1}{4}, \dfrac{1}{6}, \dfrac{1}{8}, \dfrac{1}{10}$

9. $\dfrac{1}{2}, \dfrac{2}{3}, \dfrac{3}{4}, \dfrac{4}{5}, \dfrac{5}{6}$ **10.** $\dfrac{1}{3}, \dfrac{1}{2}, \dfrac{3}{5}, \dfrac{2}{3}, \dfrac{5}{7}$

11. $\dfrac{2}{1}$, 1, $\dfrac{2}{3}$, $\dfrac{1}{2}$, $\dfrac{2}{5}$

12. $\dfrac{5}{2}$, $\dfrac{5}{3}$, $\dfrac{5}{4}$, 1, $\dfrac{5}{6}$

SECTION 11.1   Objective 3

Write in expanded form, and find the sum.

13. $\displaystyle\sum_{n=1}^{5} (2n + 4)$

14. $\displaystyle\sum_{n=1}^{6} (n + 1)^2$

15. $\displaystyle\sum_{n=1}^{4} \sqrt{9n}$

16. $\displaystyle\sum_{n=1}^{7} (n^2 + 3n + 2)$

17. $\displaystyle\sum_{n=1}^{4} (n^2 + 5n + 5)$

18. $\displaystyle\sum_{n=1}^{5} (n^2 - 1)$

SECTION 11.1   Objective 4

Write the following in summation notation.

19. $5 + 10 + 15 + 20 + 25$

20. $-2 - 4 - 6 - 8 - 10$

21. $1 + 4 + 9 + 16 + 25$

22. $1 + \dfrac{1}{8} + \dfrac{1}{27} + \dfrac{1}{64} + \dfrac{1}{125}$

23. $\sqrt{2} + \sqrt{3} + 2 + \sqrt{5} + \sqrt{6}$

24. $\dfrac{2}{3} + \dfrac{3}{4} + \dfrac{4}{5} + \dfrac{5}{6} + \dfrac{6}{7}$

SECTION 11.2   Objectives 1 and 2

Write the first five terms of the arithmetic progressions, and find the indicated term.

25. $a_1 = 10, d = 5$ Find a_{10}

26. $a_1 = 7, d = -4$ Find a_{11}

27. $a_1 = 2, d = 16$ Find a_{14}

28. $a_1 = 12, d = 14$ Find a_{20}

29. $a_1 = 4, d = 0.5$ Find a_8

30. $a_1 = -12, d = 0.9$ Find a_{12}

31. $a_1 = -0.8, d = -0.01$ Find a_{12}

32. $a_1 = 0.25, d = 0.1$ Find a_{12}

SECTION 11.2   Objective 3

Find all the terms between the given terms of the following arithmetic progressions.

33. $a_1 = 15, a_8 = 99$

34. $a_1 = -8, a_7 = -5$

35. $a_1 = -12, a_7 = 36$

36. $a_3 = 1.5, a_6 = 0$

37. $a_{12} = -19, a_{20} = -67$

38. $a_5 = 8.2, a_9 = -5$

SECTION 11.2   Objective 4

Find the number of terms in each of the following arithmetic progressions.

39. 8, 12, . . . , 56

40. $-12, -14, \ldots , -42$

41. 14, 2, . . . , -142

42. 15.8, 17.3, . . . , 29.3

43. $-3.5, 4.2, \ldots , 88.9$

44. $-1.25, 2, \ldots , 28$

SECTION 11.2   Objective 5

Find the sum of the indicated number of terms of the following arithmetic progressions.

45. 15, 18, 21, . . .   15 terms

46. 12, 9, 6, . . .   20 terms

47. 3.51, 2.14, 0.77, . . .   12 terms

48. $a_5 = 16$, $a_{12} = 44$ Find S_{15}

49. $a_{19} = 50$, $a_{15} = 20$ Find S_{40}

50. $a_{14} = 3.2$, $a_{20} = 6.2$ Find S_{50}

SECTION 11.3   Objectives 1 and 2

Write the first four terms of the geometric progression, and find the indicated term.

51. $a_1 = 5$, $r = 2$ Find a_6

52. $a_1 = 2$, $r = 0.25$ Find a_5

53. $a_1 = 8$, $a_2 = 4$ Find a_8

54. $a_1 = 27$, $a_2 = 9$ Find a_5

55. $a_2 = 40$, $a_6 = 2.5$ Find a_8

56. $a_2 = 40$, $a_4 = 1000$ Find a_6

57. $a_2 = 40$, $a_5 = 0.32$ Find a_7

58. $a_3 = -\dfrac{1}{12}$, $a_6 = \dfrac{1}{96}$ Find a_8

SECTION 11.3   Objective 3

Find all the terms between the given terms of the following geometric progressions.

59. $a_1 = 5$, $a_5 = 1280$

60. $a_1 = 0.5$, $a_7 = 2048$

61. $a_3 = -72$, $a_7 = -5832$

62. $a_2 = \dfrac{8}{5}$, $a_7 = -\dfrac{256}{5}$

SECTION 11.3   Objective 4

Find the number of terms in each finite geometric progression.

63. $\dfrac{1}{8}, \dfrac{1}{2}, 2, \ldots , 512$

64. 75, 15, 3, . . . , $\dfrac{3}{625}$

65. 48, 72, 108, . . . , 820.125

66. 6, 8, . . . , $25\dfrac{23}{81}$

SECTION 11.3   Objective 5

Find the sum of each of the following finite geometric progressions.

67. 15, 30, 60, . . . 8 terms

68. $\dfrac{1}{32}, \dfrac{1}{8}, \dfrac{1}{2}, \ldots$ 10 terms

69. $\displaystyle\sum_{n=1}^{4} 4 \cdot 2^n$

70. $\displaystyle\sum_{n=1}^{5} 5 \cdot \left[\dfrac{1}{5}\right]^n$

SECTION 11.4   Objective 1

Find the sum of each of the following infinite geometric progressions.

71. $80 + 40 + 20 + \cdots$

72. $90 + 30 + 10 + \cdots$

73. $12 - 6 + 3 - \cdots$

74. $54 - 18 + 6 - \cdots$

75. $0.121212 \ldots$

76. $0.102102 \ldots$

77. $0.540540 \ldots$

78. $0.801801 \ldots$

79. $\displaystyle\sum_{n=1}^{\infty} \left[\dfrac{3}{4}\right]^n$

80. $\displaystyle\sum_{n=1}^{\infty} \left[\dfrac{7}{8}\right]^n$

SECTION 11.5   Objective 1

Evaluate the following.

81. $12!$

82. $10!$

83. $\dfrac{16!}{14!}$

84. $\dfrac{25!}{8! \, 17!}$

85. $\dfrac{16!}{13! \, 3!}$

86. $\dfrac{19!}{8! \, 11!}$

SECTION 11.5   Objective 2

Write the first four terms of the expanded form.

87. $(a + b)^{12}$

88. $(a - b)^{14}$

89. $(x + 2y)^7$

90. $(2x - y)^8$

91. $(2x - 3y)^{10}$

92. $(3x + 2y)^9$

SECTION 11.5   Objective 3

Find the specified term of the expanded form in each of the following.

93. $(x + 5)^6$, fifth term

94. $(x - 3)^{12}$, eleventh term

95. $(a - 3b)^8$, fourth term

96. $(2a + b)^{10}$, eighth term

97. $(3x + 2y)^9$, sixth term

98. $(2x - 3y)^7$, seventh term

CHAPTER 11
TRUE–FALSE CONCEPT REVIEW

Check your understanding of the language of algebra. Tell whether each of the following statements is true (always true) or false (not always true).

1. The domain of a sequence function is the set of counting numbers.

2. The range of a sequence function is the set of integers.

3. The Greek letter Σ (sigma) is used to represent a sum.

4. It is possible to find the one hundred thirty-first term of a sequence without knowing all one hundred thirty of the preceding terms.

5. It is possible to find the sum of one hundred thirty terms of a sequence without knowing the first term.

6. In any arithmetic sequence, you can write as many terms as you like by adding the same number repeatedly.

7. Every sequence function is either an arithmetic sequence or a geometric sequence.

8. In any geometric sequence, you can write as many terms as you like by multiplying by the same number repeatedly.

9. Every finite arithmetic sequence has a finite sum.

10. Any given infinite geometric sequence has a finite sum.

11. The factorial symbol, !, as in $n!$, is a way of writing special multiplication problems.

12. The FOIL shortcut for squaring a binomials is a special case of binomial expansion.

CHAPTER 11
TEST

1. Write the first five terms of the arithmetic progression in which $a_1 = -8$ and $d = 6$.

2. Write the first five terms of the geometric progression in which $a_1 = -2$ and $r = 3$.

3. Write the first five terms of the sequence defined by $c(n) = 10n - 7$.

4. Find the value of $\dfrac{11!}{5! \ 6!}$.

5. Write a rule for a sequence function whose first six terms are 22, 18, 14, 10, 6, 2.

6. Write $\displaystyle\sum_{n=1}^{5} (45 - 5n)$ in expanded form.

7. Find the twenty-first term of the arithmetic progression in which $a_1 = -8$ and $d = 6$.

8. Find the seventh term of the geometric progression 40, 20, 10, 5, ...

9. Write the first three terms of the expanded form of $(x + 3)^7$.

10. Find the number of terms in the finite arithmetic progression $-1, 5, 11, 17, \ldots, 59$.

11. Find all the terms between the terms of the geometric progression in which $a_3 = 8$ and $a_6 = 27$.

12. Find the sum of the terms of the finite arithmetic progression 3, 8, 13, 18, . . . , 53

13. Find the sum of the terms of the finite geometric progression 1, 2, 4, 8, . . . , 256.

14. Write the following series in summation notation:

6 + 13 + 20 + 27 + 34

15. Find the sum of the infinite geometric series

$$9 + 3 + 1 + \frac{1}{3} + \frac{1}{9} + \cdots.$$

Prime Factors of Numbers 1 Through 100

	PRIME FACTORS		PRIME FACTORS		PRIME FACTORS		PRIME FACTORS
1	none	26	$2 \cdot 13$	51	$3 \cdot 17$	76	$2^2 \cdot 19$
2	2	27	3^3	52	$2^2 \cdot 13$	77	$7 \cdot 11$
3	3	28	$2^2 \cdot 7$	53	53	78	$2 \cdot 3 \cdot 13$
4	2^2	29	29	54	$2 \cdot 3^3$	79	79
5	5	30	$2 \cdot 3 \cdot 5$	55	$5 \cdot 11$	80	$2^4 \cdot 5$
6	$2 \cdot 3$	31	31	56	$2^3 \cdot 7$	81	3^4
7	7	32	2^5	57	$3 \cdot 19$	82	$2 \cdot 41$
8	2^3	33	$3 \cdot 11$	58	$2 \cdot 29$	83	83
9	3^2	34	$2 \cdot 17$	59	59	84	$2^2 \cdot 3 \cdot 7$
10	$2 \cdot 5$	35	$5 \cdot 7$	60	$2^2 \cdot 3 \cdot 5$	85	$5 \cdot 17$
11	11	36	$2^2 \cdot 3^2$	61	61	86	$2 \cdot 43$
12	$2^2 \cdot 3$	37	37	62	$2 \cdot 31$	87	$3 \cdot 29$
13	13	38	$2 \cdot 19$	63	$3^2 \cdot 7$	88	$2^3 \cdot 11$
14	$2 \cdot 7$	39	$3 \cdot 13$	64	2^6	89	89
15	$3 \cdot 5$	40	$2^3 \cdot 5$	65	$5 \cdot 13$	90	$2 \cdot 3^2 \cdot 5$
16	2^4	41	41	66	$2 \cdot 3 \cdot 11$	91	$7 \cdot 13$
17	17	42	$2 \cdot 3 \cdot 7$	67	67	92	$2^2 \cdot 23$
18	$2 \cdot 3^2$	43	43	68	$2^2 \cdot 17$	93	$3 \cdot 31$
19	19	44	$2^2 \cdot 11$	69	$3 \cdot 23$	94	$2 \cdot 47$
20	$2^2 \cdot 5$	45	$3^2 \cdot 5$	70	$2 \cdot 5 \cdot 7$	95	$5 \cdot 19$
21	$3 \cdot 7$	46	$2 \cdot 23$	71	71	96	$2^5 \cdot 3$
22	$2 \cdot 11$	47	47	72	$2^3 \cdot 3^2$	97	97
23	23	48	$2^4 \cdot 3$	73	73	98	$2 \cdot 7^2$
24	$2^3 \cdot 3$	49	7^2	74	$2 \cdot 37$	99	$3^2 \cdot 11$
25	5^2	50	$2 \cdot 5^2$	75	$3 \cdot 5^2$	100	$2^2 \cdot 5^2$

APPENDIX II
Formulas

PERIMETER AND AREA

Square

Perimeter: $P = 4s$

Area: $A = s^2$

Rectangle

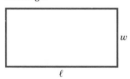

Perimeter: $P = 2\ell + 2w$

Area: $A = \ell w$

Triangle

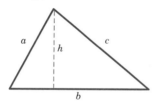

Perimeter: $P = a + b + c$

Area: $A = \dfrac{bh}{2}$

Parallelogram

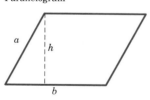

Perimeter: $P = 2a + 2b$

Area: $A = bh$

Trapezoid

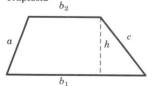

Perimeter: $P = a + b_1 + c + b_2$

Area: $A = \dfrac{1}{2}(b_1 + b_2) \cdot h$

Circle

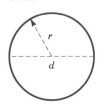

Circumference: $C = 2\pi r$

$C = \pi d$

Area: $A = \pi r^2$

$A = \dfrac{\pi d^2}{4}$

Volume

Rectangular solid

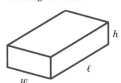

$V = \ell w h$

Cube

$V = e^3$

Sphere

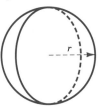

$V = \dfrac{4}{3}\pi r^3$

Cylinder

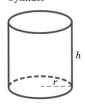

$V = \pi r^2 h$

Cone

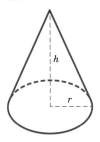

$V = \dfrac{1}{3}\pi r^2 h$

Miscellaneous

FORMULA	DESCRIPTION
$D = rt$	Distance equals rate (speed) times time.
$I = prt$	Interest equals principal times rate (%) times time.
$C = np$	Cost equals number (of items) times price (per item).
$S = c + m$	Selling price equals cost plus markup.
$F = \dfrac{9}{5}C + 32$	Fahrenheit temperature equals $\dfrac{9}{5}$ times Celsius temperature plus 32.
$S = v + gt$	Velocity (of object falling) equals initial velocity plus gravitational force times time.
$R \cdot B = A$	Rate (%) times base equals amount (percentage).
$a^2 + b^2 = c^2$	The square of the length of one leg of a right triangle plus the square of the length of the other leg is equal to the square of the length of the hypotenuse.

Many other formulas appear within problems in the text.

Squares and Square Roots (0 to 199)

n	n^2	$\sqrt{n}$	n	n^2	$\sqrt{n}$	n	n^2	$\sqrt{n}$	n	n^2	$\sqrt{n}$
0	0	0.000	50	2,500	7.071	100	10,000	10.000	150	22,500	12.247
1	1	1.000	51	2,601	7.141	101	10,201	10.050	151	22,801	12.288
2	4	1.414	52	2,704	7.211	102	10,404	10.100	152	23,104	12.329
3	9	1.732	53	2,809	7.280	103	10,609	10.149	153	23,409	12.369
4	16	2.000	54	2,916	7.348	104	10,816	10.198	154	23,716	12.410
5	25	2.236	55	3,025	7.416	105	11,025	10.247	155	24,025	12.450
6	36	2.449	56	3,136	7.483	106	11,236	10.296	156	24,336	12.490
7	49	2.646	57	3,249	7.550	107	11,449	10.344	157	24,649	12.530
8	64	2.828	58	3,346	7.616	108	11,664	10.392	158	24,964	12.570
9	81	3.000	59	3,481	7.681	109	11,881	10.440	159	25,281	12.610
10	100	3.162	60	3,600	7.746	110	12,100	10.488	160	25,600	12.649
11	121	3.317	61	3,721	7.810	111	12,321	10.536	161	25,921	12.689
12	144	3.464	62	3,844	7.874	112	12,544	10.583	162	26,244	12.728
13	169	3.606	63	3,969	7.937	113	12,769	10.630	163	26,569	12.767
14	196	3.742	64	4,096	8.000	114	12,996	10.677	164	26,896	12.806
15	225	3.873	65	4,225	8.062	115	13,225	10.724	165	27,225	12.845
16	256	4.000	66	4,356	8.124	116	13,456	10.770	166	27,556	12.884
17	289	4.123	67	4,489	8.185	117	13,689	10.817	167	27,889	12.923
18	324	4.243	68	4,624	8.246	118	13,924	10.863	168	28,224	12.961
19	361	4.359	69	4,761	8.307	119	14,161	10.909	169	28,561	13.000
20	400	4.472	70	4,900	8.367	120	14,400	10.954	170	28,900	13.038
21	441	4.583	71	5,041	8.426	121	14,641	11.000	171	29,241	13.077
22	484	4.690	72	5,184	8.485	122	14,884	11.045	172	29,584	13.115
23	529	4.796	73	5,329	8.544	123	15,129	11.091	173	29,929	13.153
24	576	4.899	74	5,476	8.602	124	15,376	11.136	174	30,276	13.191
25	625	5.000	75	5,625	8.660	125	15,625	11.180	175	30,625	13.229
26	676	5.099	76	5,776	8.718	126	15,876	11.225	176	30,976	13.266
27	729	5.196	77	5,929	8.775	127	16,129	11.269	177	31,329	13.304
28	784	5.292	78	6,084	8.832	128	16,384	11.314	178	31,684	13.342
29	841	5.385	79	6,241	8.888	129	16,641	11.358	179	32,041	13.379
30	900	5.477	80	6,400	8.944	130	16,900	11.402	180	32,400	13.416
31	961	5.568	81	6,561	9.000	131	17,161	11.446	181	32,761	13.454
32	1,024	5.657	82	6,724	9.055	132	17,424	11.489	182	33,124	13.491
33	1,089	5.745	83	6,889	9.110	133	17,689	11.533	183	33,489	13.528
34	1,156	5.831	84	7,056	9.165	134	17,956	11.576	184	33,856	13.565
35	1,225	5.916	85	7,225	9.220	135	18,225	11.619	185	34,225	13.601
36	1,296	6.000	86	7,396	9.274	136	18,496	11.662	186	34,596	13.638
37	1,369	6.083	87	7,569	9.327	137	18,769	11.705	187	34,969	13.675
38	1,444	6.164	88	7,744	9.381	138	19,044	11.747	188	35,344	13.711
39	1,521	6.245	89	7,921	9.434	139	19,321	11.790	189	35,721	13.748
40	1,600	6.325	90	8,100	9.487	140	19,600	11.832	190	36,100	13.784
41	1,681	6.403	91	8,281	9.539	141	19,881	11.874	191	36,481	13.820
42	1,764	6.481	92	8,464	9.592	142	20,164	11.916	192	36,864	13.856
43	1,849	6.557	93	8,649	9.644	143	20,449	11.958	193	37,249	13.892
44	1,936	6.633	94	8,836	9.659	144	20,736	12.000	194	37,636	13.928
45	2,025	6.708	95	9,025	9.747	145	21,025	12.042	195	38,025	13.964
46	2,116	6.782	96	9,216	9.798	146	21,316	12.083	196	38,416	14.000
47	2,209	6.856	97	9,409	9.849	147	21,609	12.124	197	38,809	14.036
48	2,304	6.928	98	9,604	9.899	148	21,904	12.166	198	39,204	14.071
49	2,401	7.000	99	9,801	9.950	149	22,201	12.207	199	39,601	14.107
n	n^2	$\sqrt{n}$	n	n^2	$\sqrt{n}$	n	n^2	$\sqrt{n}$	n	n^2	$\sqrt{n}$

APPENDIX IV
Common Logarithms

n	0	1	2	3	4	5	6	7	8	9
1.0	0.0000	0.004321	0.008600	0.01284	0.01703	0.02119	0.02531	0.02938	0.03342	0.03743
1.1	0.04139	0.04532	0.04922	0.05308	0.05690	0.06070	0.06446	0.06819	0.07188	0.07555
1.2	0.07918	0.08279	0.08636	0.08991	0.09342	0.09691	0.1004	0.1038	0.1072	0.1106
1.3	0.1139	0.1173	0.1206	0.1239	0.1271	0.1303	0.1335	0.1367	0.1399	0.1430
1.4	0.1461	0.1492	0.1523	0.1553	0.1584	0.1614	0.1644	0.1673	0.1703	0.1732
1.5	0.1761	0.1790	0.1818	0.1847	0.1875	0.1903	0.1931	0.1959	0.1987	0.2014
1.6	0.2041	0.2068	0.2095	0.2122	0.2148	0.2175	0.2201	0.2227	0.2253	0.2279
1.7	0.2304	0.2330	0.2355	0.2380	0.2405	0.2430	0.2455	0.2480	0.2504	0.2529
1.8	0.2553	0.2577	0.2601	0.2625	0.2648	0.2672	0.2695	0.2718	0.2742	0.2765
1.9	0.2788	0.2810	0.2833	0.2856	0.2878	0.2900	0.2923	0.2945	0.2967	0.2989
2.0	0.3010	0.3032	0.3054	0.3075	0.3096	0.3118	0.3139	0.3160	0.3181	0.3201
2.1	0.3222	0.3243	0.3263	0.3284	0.3304	0.3324	0.3345	0.3365	0.3385	0.3404
2.2	0.3424	0.3444	0.3464	0.3483	0.3502	0.3522	0.3541	0.3560	0.3579	0.3598
2.3	0.3617	0.3636	0.3655	0.3674	0.3692	0.3711	0.3729	0.3747	0.3766	0.3784
2.4	0.3802	0.3820	0.3838	0.3856	0.3874	0.3892	0.3909	0.3927	0.3945	0.3962
2.5	0.3979	0.3997	0.4014	0.4031	0.4048	0.4065	0.4082	0.4099	0.4116	0.4133
2.6	0.4150	0.4166	0.4183	0.4200	0.4216	0.4232	0.4249	0.4265	0.4281	0.4298
2.7	0.4314	0.4330	0.4346	0.4362	0.4378	0.4393	0.4409	0.4425	0.4440	0.4456
2.8	0.4472	0.4487	0.4502	0.4518	0.4533	0.4548	0.4564	0.4579	0.4594	0.4609
2.9	0.4624	0.4639	0.4654	0.4669	0.4683	0.4698	0.4713	0.4728	0.4742	0.4757
3.0	0.4771	0.4786	0.4800	0.4814	0.4829	0.4843	0.4857	0.4871	0.4886	0.4900
3.1	0.4914	0.4928	0.4942	0.4955	0.4969	0.4983	0.4997	0.5011	0.5024	0.5038
3.2	0.5051	0.5065	0.5079	0.5092	0.5105	0.5119	0.5132	0.5145	0.5159	0.5172
3.3	0.5185	0.5198	0.5211	0.5224	0.5237	0.5250	0.5263	0.5276	0.5289	0.5302
3.4	0.5315	0.5328	0.5340	0.5353	0.5366	0.5378	0.5391	0.5403	0.5416	0.5428
3.5	0.5441	0.5453	0.5465	0.5478	0.5490	0.5502	0.5514	0.5527	0.5539	0.5551
3.6	0.5563	0.5575	0.5587	0.5599	0.5611	0.5623	0.5635	0.5647	0.5658	0.5670
3.7	0.5682	0.5694	0.5705	0.5717	0.5729	0.5740	0.5752	0.5763	0.5775	0.5786
3.8	0.5798	0.5809	0.5821	0.5832	0.5843	0.5855	0.5866	0.5877	0.5888	0.5899
3.9	0.5911	0.5922	0.5933	0.5944	0.5955	0.5966	0.5977	0.5988	0.5999	0.6010
4.0	0.6021	0.6031	0.6042	0.6053	0.6064	0.6075	0.6085	0.6096	0.6107	0.6117
4.1	0.6128	0.6138	0.6149	0.6160	0.6170	0.6108	0.6191	0.6201	0.6212	0.6222
4.2	0.6232	0.6243	0.6253	0.6263	0.6274	0.6284	0.6294	0.6304	0.6314	0.6325
4.3	0.6335	0.6345	0.6355	0.6365	0.6375	0.6385	0.6395	0.6405	0.6415	0.6425
4.4	0.6435	0.6444	0.6454	0.6464	0.6474	0.6484	0.6493	0.6503	0.6513	0.6522
4.5	0.6532	0.6542	0.6551	0.6561	0.6571	0.6580	0.6590	0.6599	0.6609	0.6618
4.6	0.6628	0.6637	0.6646	0.6656	0.6665	0.6675	0.6684	0.6693	0.6702	0.6712
4.7	0.6721	0.6730	0.6739	0.6749	0.6758	0.6767	0.6776	0.6785	0.6794	0.6803
4.8	0.6812	0.6821	0.6830	0.6839	0.6848	0.6857	0.6866	0.6875	0.6884	0.6893
4.9	0.6902	0.6911	0.6920	0.6928	0.6937	0.6946	0.6955	0.6964	0.6972	0.6981
5.0	0.6990	0.6998	0.7007	0.7016	0.7024	0.7033	0.7042	0.7050	0.7059	0.7067
5.1	0.7076	0.7084	0.7093	0.7101	0.7110	0.7118	0.7126	0.7135	0.7143	0.7152
5.2	0.7160	0.7168	0.7177	0.7185	0.7193	0.7202	0.7210	0.7218	0.7226	0.7235
5.3	0.7243	0.7251	0.7259	0.7267	0.7275	0.7284	0.7292	0.7300	0.7308	0.7316
5.4	0.7324	0.7332	0.7340	0.7348	0.7356	0.7364	0.7372	0.7380	0.7388	0.7396

(continued)

n	0	1	2	3	4	5	6	7	8	9
5.5	0.7404	0.7412	0.7419	0.7427	0.7435	0.7443	0.7451	0.7459	0.7466	0.7474
5.6	0.7482	0.7490	0.7497	0.7505	0.7513	0.7520	0.7528	0.7536	0.7543	0.7551
5.7	0.7559	0.7566	0.7574	0.7582	0.7589	0.7597	0.7604	0.7612	0.7619	0.7627
5.8	0.7634	0.7642	0.7649	0.7657	0.7664	0.7672	0.7679	0.7686	0.7694	0.7701
5.9	0.7709	0.7716	0.7723	0.7731	0.7738	0.7745	0.7752	0.7760	0.7767	0.7774
6.0	0.7782	0.7789	0.7796	0.7803	0.7810	0.7818	0.7825	0.7832	0.7839	0.7846
6.1	0.7853	0.7860	0.7868	0.7875	0.7882	0.7889	0.7896	0.7903	0.7910	0.7917
6.2	0.7924	0.7931	0.7938	0.7945	0.7952	0.7959	0.7966	0.7973	0.7980	0.7987
6.3	0.7993	0.8000	0.8007	0.8014	0.8021	0.8028	0.8035	0.8041	0.8048	0.8055
6.4	0.8062	0.8069	0.8075	0.8082	0.8089	0.8096	0.8102	0.8109	0.8116	0.8122
6.5	0.8129	0.8136	0.8142	0.8149	0.8156	0.8162	0.8169	0.8176	0.8182	0.8189
6.6	0.8195	0.8202	0.8209	0.8215	0.8222	0.8228	0.8235	0.8241	0.8248	0.8254
6.7	0.8261	0.8267	0.8274	0.8280	0.8287	0.8293	0.8299	0.8306	0.8312	0.8319
6.8	0.8325	0.8331	0.8338	0.8344	0.8351	0.8357	0.8363	0.8370	0.8376	0.8382
6.9	0.8388	0.8395	0.8401	0.8407	0.8414	0.8420	0.8426	0.8432	0.8439	0.8445
7.0	0.8451	0.8457	0.8463	0.8470	0.8476	0.8482	0.8488	0.8494	0.8500	0.8506
7.1	0.8513	0.8519	0.8525	0.8531	0.8537	0.8543	0.8549	0.8555	0.8561	0.8567
7.2	0.8573	0.8579	0.8585	0.8591	0.8597	0.8603	0.8609	0.8615	0.8621	0.8627
7.3	0.8633	0.8639	0.8645	0.8651	0.8657	0.8663	0.8669	0.8675	0.8681	0.8686
7.4	0.8692	0.8698	0.8704	0.8710	0.8716	0.8722	0.8727	0.8733	0.8739	0.8745
7.5	0.8751	0.8756	0.8762	0.8768	0.8774	0.8779	0.8785	0.8791	0.8797	0.8802
7.6	0.8808	0.8814	0.8820	0.8825	0.8831	0.8837	0.8842	0.8848	0.8854	0.8859
7.7	0.8865	0.8871	0.8876	0.8882	0.8887	0.8893	0.8899	0.8904	0.8910	0.8915
7.8	0.8921	0.8927	0.8932	0.8938	0.8943	0.8949	0.8954	0.8960	0.8965	0.8971
7.9	0.8976	0.8982	0.8987	0.8993	0.8998	0.9004	0.9009	0.9015	0.9020	0.9025
8.0	0.9031	0.9036	0.9042	0.9047	0.9053	0.9058	0.9063	0.9069	0.9074	0.9079
8.1	0.9085	0.9090	0.9096	0.9101	0.9106	0.9112	0.9117	0.9122	0.9128	0.9133
8.2	0.9138	0.9143	0.9149	0.9154	0.9159	0.9165	0.9170	0.9175	0.9180	0.9186
8.3	0.9191	0.9196	0.9201	0.9206	0.9212	0.9217	0.9222	0.9227	0.9232	0.9238
8.4	0.9243	0.9248	0.9253	0.9258	0.9263	0.9269	0.9274	0.9279	0.9284	0.9289
8.5	0.9294	0.9299	0.9304	0.9309	0.9315	0.9320	0.9325	0.9330	0.9335	0.9340
8.6	0.9345	0.9350	0.9355	0.9360	0.9365	0.9370	0.9375	0.9380	0.9385	0.9390
8.7	0.9395	0.9400	0.9405	0.9410	0.9415	0.9420	0.9425	0.9430	0.9435	0.9440
8.8	0.9445	0.9450	0.9455	0.9460	0.9465	0.9469	0.9474	0.9479	0.9484	0.9489
8.9	0.9494	0.9499	0.9504	0.9509	0.9513	0.9518	0.9523	0.9528	0.9533	0.9538
9.0	0.9542	0.9547	0.9552	0.9557	0.9562	0.9566	0.9571	0.9576	0.9581	0.9586
9.1	0.9590	0.9595	0.9600	0.9605	0.9609	0.9614	0.9619	0.9624	0.9628	0.9633
9.2	0.9638	0.9643	0.9647	0.9652	0.9657	0.9661	0.9666	0.9671	0.9675	0.9680
9.3	0.9685	0.9689	0.9694	0.9699	0.9703	0.9708	0.9713	0.9717	0.9722	0.9727
9.4	0.9731	0.9736	0.9741	0.9745	0.9750	0.9754	0.9759	0.9763	0.9768	0.9773
9.5	0.9777	0.9782	0.9786	0.9791	0.9795	0.9800	0.9805	0.9809	0.9814	0.9818
9.6	0.9823	0.9827	0.9832	0.9836	0.9841	0.9845	0.9850	0.9854	0.9859	0.9863
9.7	0.9868	0.9872	0.9877	0.9881	0.9886	0.9890	0.9894	0.9899	0.9903	0.9908
9.8	0.9912	0.9917	0.9921	0.9926	0.9930	0.9934	0.9939	0.9943	0.9948	0.9952
9.9	0.9956	0.9961	0.9965	0.9969	0.9974	0.9978	0.9983	0.9987	0.9991	0.9996

APPENDIX V
Calculators

The wide availability and economical price of current hand-held calculators make them ideal for doing time-consuming arithmetic operations. You are encouraged to use a calculator as you work through this text. Calculator examples throughout the text show where the use of a calculator is appropriate. Your calculator will be especially useful for

1. doing the fundamental operations of arithmetic (add, subtract, multiply, and divide),
2. checking solutions to equations,
3. checking solutions to problems,
4. finding square roots of numbers, and
5. finding powers of numbers.

To practice solving the problems with your calculator, you will need to know whether the calculator is a basic calculator or a scientific calculator. Here are examples of each.

Basic calculator · · · · · · · · · · · · Scientific calculator

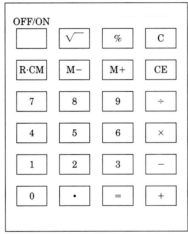

All computers, many scientific calculators, and some other calculators have the fundamental order of operations built into their circuitry. To tell if your calculator has this feature, do the following exercise:

$6 + 4(9)$

ENTER	DISPLAY
6	6.
+	6.
4	4.
×	4.
9	9.
=	42.

If the display reads "42," your calculator finds values of combinations according to the agreed-upon order. If the display reads "90," you can get the correct result, 42, by entering the problem in the calculator following the rules for the order of operations.

ENTER	DISPLAY
4	4.
×	4.
9	9.
+	36.
6	6.
=	42.

A calculator that has parentheses keys can also override the order of operations.

ENTER 6 + (4 × 9) =   **DISPLAY** 42.

If yours is a scientific calculator, it has other keys that you may find useful in later mathematics courses. Recall that parentheses and fraction bars are used to group operations to show which one is done first.

EXPRESSION	BASIC CALCULATOR Enter	SCIENTIFIC CALCULATOR Enter	DISPLAY
$144 \div 3 - 7$	144 ÷ 3 − 7 =	144 ÷ 3 − 7 =	41.
$\dfrac{28 + 42}{10}$	28 + 42 = ÷ 10 =	28 + 42 = ÷ 10 =	7.
$\dfrac{288}{6 + 12}$	6 + 12 = **STEP 1**		18.
	288 ÷ 18 = **STEP 2**		16.
		6 + 12 = $\frac{1}{x}$ × 288 =	16.
$13^2 + 4(17)$	13 × 13 = **STEP 1**		169.
	4 × 17 = **STEP 2**		68.
	169 + 68 = **STEP 3**		237.
		13 × 13 + 4 × 17 =	237.
		or	
		13 x^2 + 4 × 17 =	237.
$\sqrt{5184}$		5184 $\sqrt{\ }$	72.

APPENDIX VI

Sets

Sets.1

SETS, SUBSETS, AND THEIR SYMBOLS

OBJECTIVES

1. Describe a set in words.

2. Describe a set using a roster method.

3. Describe a set using set builder notation.

4. Identify the members of a set.

5. Identify a subset of a given set.

The idea of a *set* is so basic that it is difficult to define in simpler terms. The following group of synonyms will help give meaning to the word set: clump, family, grouping, collection, flock, cluster, and assembly. In mathematics, sets typically contain numbers, geometric figures, and other mathematical objects.

There are several ways to designate a set. Three ways are explained here. First, a set may be described in words. We may talk about "the set of whole numbers between 0 and 6." The same set may be described in words by writing "the set whose elements are one, two, three, four, and five."

The *elements* of a set are the objects in the set. They are also called *members* of the set.

A second choice is to make a roster of the set by listing all the elements or members between a pair of braces. In such a case we often use a capital letter to name the set. The set described above in words can also be described by writing

$A = \{1, 2, 3, 4, 5\}$.

EXAMPLE 1

Use a roster to describe the set of all whole numbers greater than 4 and less than 10. Call this set G.

$G = \{5, 6, 7, 8, 9\}$ The set consists of five members. ■

EXAMPLE 2

Describe the set $\{2, 4, 6, 8\}$ in words.

The set of even whole numbers   The elements of the set are whole
greater than one and less than nine.   numbers and are even.   ■

The roster method is easy to use when the sets contain only a few elements as in Examples 1 and 2. It would be quite tedious, however, to make a roster of all of the whole numbers between 5,748,919 and 5,901,337. Such rosters appear in telephone directories. In mathematics sets with many elements are described using a third method that is called *set builder notation.*

$$A = \qquad \{x \qquad | \qquad x \text{ is a whole number and } 0 < x < 6\}$$

A is the set   of all x such that x is a whole number and $0 < x < 6\}$

EXAMPLE 3

Use set builder notation to indicate B, the set of whole numbers larger than 15 and less than 40.

$B = \{x \mid 15 < x < 40 \text{ and } x \text{ is a whole number}\}$ ■

EXAMPLE 4

Use set builder notation to write all natural numbers less than 57 that are multiples of 4.

$F = \{x \mid x \in N, x < 57 \text{ and } x = 4n\}$ We name the set F. We first
indicate that x is a natural
number, then x is smaller than 57,
and finally that x is four times
some natural number, n.   ■

The symbol $\in$ is used to abbreviate the phrase "is an element of" or "is a member of" so from $G = \{5, 6, 7, 8, 9\}$, we can make the following statements:

The number 7 is a member of set G,   or $7 \in G$

The number 9 is an element of set G,   or $9 \in G$

The number 5 is an element of set G,   or $5 \in G$

The symbol $\notin$ is used to abbreviate the phrase "is not an element of" or "is not a member of," so we make the following statements:

The number 4 is not a member of set G,   or $4 \notin G$

The number 10 is not a member of set G,   or $10 \notin G$

EXAMPLE 5

True or false. If $A = \{x \mid x$ is a whole number and $15 < x < 68\}$, then $42 \in A$.

The statement is true.

The symbol $\in$ means "is a member of," and 42 is a number that satisfies the conditions. ∎

EXAMPLE 6

True or false. If $A = \{x \mid x$ is a whole number and $15 < x < 68\}$, then $36 \notin A$.

The statement is false.

The symbol $\notin$ means "is not an element of," and 36 is larger than 15 and less than 68. ∎

Some of the sets of symbols and numbers, together with the letter that identifies each set, that are used in mathematics are

Digits $= \{0, 1, 2, 3, 4, 5, 6, 7, 8, 9\}$

The set of digits is a group of symbols with which we write the place value notation for the different sets of numbers.

Natural or counting numbers $N = \{1, 2, 3, 4, 5, 6, \ldots\}$

Whole numbers $W = \{0, 1, 2, 3, 4, 5, 6, \ldots\}$

Integers $J = \{\ldots, -5, -4, -3, -2, -1, 0, 1, 2, 3, 4, 5, \ldots\}$

In these descriptions we used the roster method. We did this despite the fact that not all the members can be listed. Enough members of the set are listed to establish a pattern. The three dots (ellipses) mean that the listing continues without end.

Study the following sets, which are described in words.

The set of all whole numbers less than zero.

The set of all even numbers that end in 7.

The set of all multiples of 10 that are not divisible by 10.

A set that contains no elements is called the *empty* set or *null* set. We can write the empty set with an empty pair of braces.

Empty set $= \{\ \}$

Another symbol used to indicate the empty set is $\emptyset$.

Empty set $= \emptyset$

When a symbol is needed to indicate an empty set, either symbol can be used. However, in mathematics $\emptyset$ is normally used to indicate an empty set.

■ **CAUTION**

> The symbol {0} does not name the empty set nor does the symbol {∅} name the empty set. In each case, the set contains a member.

Let $A = \{0, 1, 2, 3, 4, 5, 6, 7, 8, 9\}$ and $B = \{0, 2, 4, 6, 8\}$. We note that every element of B is also a member of A. In such a case, we say that "B is a subset of A" and we write

$$B \subseteq A$$

Let $C = \{1, 3, 5, 7, 9\}$ and $D = \{1, 3, 5, 7, 8\}$. Notice that not every member of D is a member of C ($8 \notin C$). Therefore D is not a subset of C and we write

$$D \nsubseteq C$$

We also make the observation that A is a subset of A since every member of A is contained in A. This satisfies the definition of subset. Since this is true for every set, we make the general statement that every set is a subset of itself. Also ∅ will be considered to be a subset of every set. (Since the empty set has no elements the definition of subset is not violated because every element of ∅ is in every set.)

Two sets are *equal* if they contain exactly the same elements.

A set F will be called a *proper subset* of E if $F \subseteq E$ and $F \neq E$. To indicate that proper subset relationship, we write $F \subset E$. The symbol for subset ($\subseteq$) and the symbol for proper subset ($\subset$) are similar to the symbols $\leq$ and $<$. For a first set to be a subset of a second, the first set (subset) can contain fewer or exactly the same number of members as the second set. For a first member to be a proper subset of a second set, the first set (proper subset) must contain fewer members than the second set.

EXAMPLE 7

If $A = \{1, 2, 3, 4, 5, 6, 7, 8, 9, 10\}$, $B = \{1, 2, 3\}$, $C = \{0, 2, 4, 6\}$, $D = \{3, 1, 2\}$, and $E = \{2, 4\}$, are the following true or false?

a. $B \subseteq A$ True   All of the members of B are contained in A.

b. $C \subseteq A$ False $0 \notin A$

c. $C \subseteq D$ False   Not all of the members of C are contained in D, i.e., $0 \notin D$ and $6 \notin D$.

d. $B \subset A$ True $B \subseteq A$ and $B \neq A$

e. $B = D$ True   They each contain the same members. D is stated in a different order, but that makes no difference. ■

A

Write each of the following sets using a roster.

1. The set of even whole numbers less than 17.

2. The set of even whole numbers less than 22.

3. The set of all whole numbers from 9 to 15 inclusive.

4. The set of all whole numbers from 11 to 21 inclusive.

Describe each set in words.

5. {11, 12, 13, 14, 15}   **6.** {18, 20, 22, 24, 26, 28}   **7.** {1}   **8.** {2}

Write each set using set builder notation.

9. The set of whole numbers less than 84.

10. The set of whole numbers greater than 48.

11. The set of even numbers between 51 and 133.

12. The set of odd numbers between 42 and 92.

Tell whether each statement is true or false.

13. If $B = \{x \mid x$ is a whole number and $6 < x < 106\}$, then $3 \in B$.

14. If $B = \{x \mid x$ is a whole number and $6 < x < 106\}$, then $10 \in B$.

15. If $C = \{x \mid x$ is a whole number and $x > 19\}$, then $19.6 \in B$.

16. If $C = \{x \mid x$ is a whole number and $x > 19\}$, then $300 \in B$.

17. If $S = \{x \mid x \in W\}$, and $T = \{0\}$, then $T \subseteq S$.

18. If $S = \{x \mid x \in W\}$, and $T = \{100, 203, 305\}$, then $T \subseteq S$.

19. If $A = \{x \mid x \in J\}$, and $B = \left\{\dfrac{2}{3}\right\}$, then $B \subseteq A$.

20. If $A = \{x \mid x \in J\}$, and $B = \{-31, -28, -15\}$, then $B \subseteq A$.

B

Write each of the following sets using a roster.

21. The set of all multiples of 4 between 15 and 38.

22. The set of all multiples of 3 between 16 and 35.

23. The set of all fractions with numerators that are counting numbers less than 4 and with denominators that are counting numbers less than 5.

24. The set of all fractions with numerators that are counting numbers less than 5 and with denominators that are counting numbers less than 4.

Describe each set in words.

25. {9, 12, 15, 18, 21}

26. {18, 12, 16, 20}

27. {88, 99, 110, 121, 132}

28. {96, 108, 120, 132}

Write each set using set builder notation.

29. The set of the first five consecutive counting numbers larger than 23.

30. The set of the largest five consecutive counting numbers less than 96.

31. The set of counting numbers that are divisors of 18.

32. The set of counting numbers that are divisors of 20.

Tell whether each statement is true or false.

33. If $J = \{x \mid x$ is an integer$\}$, then $-2 \in J$.

34. If $J = \{x \mid x$ is an integer$\}$, then $\dfrac{1}{2} \in J$.

35. If $K = \left\{\dfrac{x}{y} \,\middle|\, x$ is an integer and y is a counting number$\right\}$, then $\dfrac{3}{4} \in K$.

36. If $K = \left\{\dfrac{x}{y} \,\middle|\, x$ is an integer and y is a counting number$\right\}$, then $\dfrac{4}{0} \in K$.

37. If $T = \{x \mid x$ is a multiple of 3$\}$, $V = \{x \mid x$ is a multiple of 6$\}$, then $T \subseteq V$.

38. If $T = \{x \mid x$ is a multiple of 3$\}$, $V = \{x \mid x$ is a multiple of 6$\}$, then $V \subseteq T$.

39. If $E = \{x \mid x$ is an even number$\}$, and $F = \{x \mid x$ is a multiple of 10$\}$, then $E \subseteq F$.

40. If $E = \{x \mid x$ is an even number$\}$, and $F = \{x \mid x$ is a multiple of 10$\}$, then $F \subseteq E$.

Write each of the following sets using a roster.

41. The set of all multiples of 13 between 200 and 240.

42. The set of all multiples of 13 between 300 and 340.

43. The set of all multiples of 14 between 200 and 240.

44. The set of all multiples of 14 between 300 and 340.

Describe each set in words.

45. {128, 132, 136, 140}

46. {132, 140, 148, 156}

47. {32, 34, 35, 36, 38, 40}

48. {35, 36, 38, 40, 42, 44, 46, 49}

Write each of the following sets using a roster.

49. $\{t \mid t \in W \text{ and } 36 < t \le 40\}$

50. $\{t \mid t \in W \text{ and } 36 \le t < 40\}$

51. $\{s \mid s \in J \text{ and } -4 \le s < 5\}$

52. $\{s \mid s \in J \text{ and } -5 \le s \le 4\}$

Given that $A = \{100, 101, 102, 103, 104, 105, 106, 107, 108, 109, 110\}$ and $B = \{x \mid x \text{ is a multiple of 3 and } 100 < x < 110\}$, and $C = \{x \mid x \in W \text{ and } x < 111\}$, tell whether each statement is true or false.

53. $100 \in A$

54. $100 \in B$

55. $33 \in B$

56. $33 \in C$

57. $109 \in B$

58. $109 \in C$

59. $B \subseteq A$

60. $A \subseteq B$

61. $C \subseteq A$

62. $C \subseteq B$

63. $B \subseteq C$

64. $A \subseteq C$

STATE YOUR UNDERSTANDING

65. Describe the empty set.

66. What is the difference between a subset and a proper subset?

CHALLENGE EXERCISES

Given $A = \{0, 1, 2, 3, \ldots, 50\}$ and $B = \{2, 4, 6, 8, \ldots, 60\}$, write the following using the roster method:

67. $C = \{s \mid s \in A \text{ and } s \text{ is both a multiple of 3 and a multiple of 5}\}$.

68. $D = \{s \mid s \in A, s \in B, \text{ and } s \text{ is both a multiple of 3 and } 9 < s < 60\}$.

69. $E = \{s \mid s \in A, s \in B, \text{ and } s \text{ is a multiple of 7}\}$.

Sets.2

IDENTIFYING TYPES OF NUMBERS, SET OPERATIONS, AND VENN DIAGRAMS

OBJECTIVES

1. Identify specified subsets of a given set.

2. Find the union of two sets.

3. Find the intersection of two sets.

4. Draw a Venn diagram showing the relationship between two sets.

5. Find the complement of a set.

In the study of mathematics it is often necessary to be able to identify different types of numbers. In addition to the sets of numbers previously defined, we define and give examples of three more sets.

Rational numbers $\qquad Q = \left\{ \dfrac{p}{q} \,\middle|\, p \text{ and } q \in J, \, q \neq 0 \right\}$

Rational numbers are referred to as fractions, both positive and negative, such as $\dfrac{3}{4}, \dfrac{7}{8}, \dfrac{11}{15}, \dfrac{9}{4}, -\dfrac{15}{16}, -\dfrac{8}{3}, \dfrac{5}{1}, -\dfrac{4}{2}$.

An alternate definition of the set of rational numbers is

$Q = \{x \mid x \text{ is a terminating or a nonterminating repeating decimal}\}$

Numbers such as $0.333\ldots$, $0.234234\ldots$, and $3.121212\ldots$ are rational numbers since they are nonterminating repeating decimals. A nonterminating repeating decimal can also be indicated by drawing a line over the repeating group of digits to indicate the repetition. So, $0.234234\ldots = 0.\overline{234234}$ and $3.121212\ldots = 3.1\overline{212}$.

Irrational numbers $\qquad I = \{x \mid x \text{ is a nonterminating and nonrepeating decimal}\}$

Thus an irrational number is a number that cannot be written as a fraction or as a nonterminating repeating decimal. Numbers such as $0.1011011101111011111\ldots$, π, $\sqrt{2}$, $\sqrt{5}$, and $\sqrt[5]{5}$ are irrational numbers.

Real numbers $\qquad R = \{x \mid x \in Q \text{ or } x \in I\}$

As a result of this definition, we can say that real numbers are either rational or irrational numbers.

We make the following observations about the six sets of numbers we have discussed:

1. The set of whole numbers contains all of the numbers that are in the set of natural numbers.

2. The set of integers contains all of the numbers that are in the set of whole numbers.

3. The set of rational numbers contains all of the numbers that are in the set of integers. (Write the integer as a fraction with a denominator of 1.)

4. The set of real numbers contains all of the numbers that are in the set of rational numbers and the set of irrational numbers.

As a result of those observations, we make the following statements.

1. $N \subset W$, $N \subset J$, $N \subset Q$, and $N \subset R$
   The set of natural numbers is a proper subset of the sets of whole numbers, integers, rational numbers, and real numbers.

2. $W \subset J$, $W \subset Q$, and $W \subset R$
   The set of whole numbers is a proper subset of the sets of integers, rational numbers, and real numbers.

3. $J \subset Q$ and $J \subset R$
   The set of integers is a proper subset of the sets of rational numbers and real numbers.

4. $Q \subset R$
   The set of rational numbers is a proper subset of the set of real numbers.

5. $I \subset R$
   The set of irrational numbers is a proper subset of the set of real numbers.

In Example 1, this information is used to write subsets of given sets.

EXAMPLE 1

Given $A = \left\{ -34.5, -21, -\dfrac{43}{10}, -\pi, -\sqrt{5}, 0, \dfrac{3}{4}, \pi, 14, 19.2\overline{2} \right\}$, list using the roster method (a) $\{x \mid x \in N \text{ and } x \in A\}$, (b) $\{x \mid x \in W \text{ and } x \in A\}$, (c) $\{x \mid x \in J \text{ and } x \in A\}$, (d) $\{x \mid x \in Q \text{ and } x \in A\}$, (e) $\{x \mid x \in I \text{ and } x \in A\}$, and (f) $\{x \mid x \in R \text{ and } x \in A\}$.

a. $\{14\}$

b. $\{0, 14\}$

c. $\{-21, 0, 14\}$

d. $\left\{ -34.5, -21, -\dfrac{43}{10}, 0, \dfrac{3}{4}, 14, 19.2\overline{2} \right\}$

e. $\{-\pi, -\sqrt{5}, \pi\}$

f. $\left\{ -34.5, -21, -\dfrac{43}{10}, -\pi, -\sqrt{5}, 0, \dfrac{3}{4}, \pi, 14, 19.2\overline{2} \right\}$

■

There are two operations with sets that we will discuss. They are the operations union and intersection of two or more sets.

The *union* of sets A and B, denoted by $A \cup B$, is a third set containing all of the elements of A and all of the elements of B.

EXAMPLE 2

Given $A = \{-5, -1, 3, 7, 9\}$ and $B = \{-2, -1, 4, 8\}$, find $A \cup B$.

$A \cup B = \{-5, -2, -1, 3, 4, 7, 8, 9\}$

Note the elements of the union are in either or both of the sets. The element -1 was contained in both sets but was listed only one time in the union. ∎

The *intersection* of sets A and B, denoted by $A \cap B$, is a third set containing all of the elements that are in both set A and set B. The intersection contains those elements that are common to both sets.

EXAMPLE 3

Using sets A and B of Example 2, find $A \cap B$.

$A \cap B = \{-1\}$

There was only one element, -1, contained in both sets A and B. ∎

EXAMPLE 4

Given $A = \{2, 5, 9\}$, $B = \{3, 5, 9, 11\}$, $C = \{4, 6, 11, 12\}$, and $D = \emptyset$, find
(a) $A \cup B$, (b) $A \cup B \cup C$, (c) $A \cap B$, (d) $A \cap C$, (e) $B \cap C$, (f) $A \cup D$, (g) $A \cap D$.

a. $A \cup B = \{2, 3, 5, 9, 11\}$

List the elements of A and B.

b. $A \cup B \cup C = \{2, 3, 4, 5, 6, 9, 11, 12\}$

List the elements contained in all three sets.

c. $A \cap B = \{5, 9\}$

5 and 9 are elements of both A and B.

d. $A \cap C = \emptyset$

A and C have no elements in common, therefore the intersection is empty. Two sets are said to be disjoint when their intersection is the empty set.

e. $B \cap C = \{11\}$

11 is common to both B and C.

f. $A \cup D = \{2, 5, 9\} = A$

The union of a set and $\emptyset$ is the set itself. The empty set does not add any new members.

g. $A \cap D = \emptyset$

The empty set contains no elements, therefore the two sets have nothing in common. ■

Venn diagrams, named for the English mathematician John Venn, are pictures that can be used to represent the union or intersection of sets as well as showing subset relationships. Figures 1 and 2 each show a Venn diagram. Figure 1 depicts the union of sets A and B since the circle representing set A and the circle representing set B are each totally shaded. Figure 2 depicts the intersection of sets A and B since only that area which they have in common is shaded. The rectangle containing the two circles is referred to as the *universe* or the *universal set* and is labeled U. The universal set is a set that contains all elements of the type being discussed in the problem.

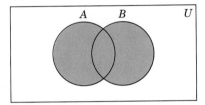

FIGURE 1

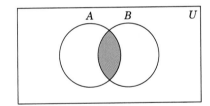

FIGURE 2

Figure 3 is a Venn diagram illustrating two disjoint sets, that is, their intersection is empty. (The two sets, A and B, do not intersect.) Figure 4 illustrates that set A is a proper subset of set B. From the diagram we can see that $A \cap B = A$.

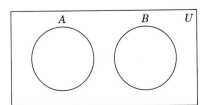

FIGURE 3

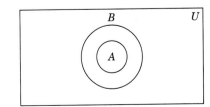

FIGURE 4

EXAMPLE 5

Use a Venn diagram to illustrate the set of all men that are bald.

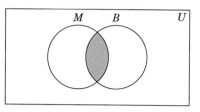

Set M is the set of all men. Set B is the set of all bald people. The universe is the set of all people. Thus all people not in set M are women. The set $M \cap B$ is the set of all men that are bald. ■

COMPLEMENT

Another important set that is often used is called the *complement* of a set. Suppose our universal set is the set of all women, and B is the set of all women with blonde hair. Then the complement of B is the set of all women that do not have blonde hair. The complement of B is written $\sim B$. We can show the meaning of the complement of a set, using a Venn diagram.

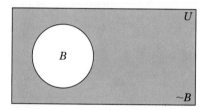

FIGURE 5

EXAMPLE 6

If the universe is the set $U = \{0, 1, 2, 3, 4, 5, 6, 7, 8, 9\}$ and $E = \{2, 4, 6, 8\}$, what are the members of $\sim E$?

$\sim E = \{1, 3, 5, 7, 9\}$

$\sim E$ contains all members of U that are not in E. ■

EXAMPLE 7

If the universe is the set of all automobiles and C is the set of all Chevrolets, describe $\sim C$.

$\sim C$ is the set of all automobiles that are not Chevrolets. ■

EXERCISES

A

Given: $T = \left\{ -15.7\overline{7}, -\dfrac{19}{4}, -4, -\dfrac{7}{6}, -\pi, 0, 2, 3.121121112\ldots, 10, \dfrac{42}{3} \right\}$

If $x \in T$, list the following subsets of T.

1. $\{x \mid x \in N\}$ **2.** $\{x \mid x \in J\}$ **3.** $\{x \mid x \in Q\}$ **4.** $\{x \mid x \in I\}$

Given: $A = \{3, 6, 9, 12, 15\}$, $B = \{2, 4, 6, 8, 10, 12\}$, and $C = \{0, 5, 10, 15, 20\}$
Find:

5. $A \cup B$ **6.** $B \cap C$ **7.** $A \cup B \cup C$ **8.** $A \cap B \cap C$

Given: $U = \{0, 1, 2, 3, \ldots, 15\}$, $A = \{0, 5, 10, 15\}$, and $B = \{1, 3, 5, 7, 9, 11, 13, 15\}$
Find:

9. $\sim A$ **10.** $\sim B$

Given: $A = \{0, 1, 2, 3, 4, 5, 6, 7, 8, 9\}$ $B = \{2, 4, 6, 8\}$, and $C = \{3, 6, 9\}$
True or false:

11. $B \subset A$ **12.** $C \subset B$ **13.** $B \subseteq B$ **14.** $A \subset C$

Draw a Venn diagram to show the relationship between the following sets and describe the intersection of the sets.

15. $A = \{x \,|\, x$ is an odd whole number$\}$
$B = \{x \,|\, x$ is an even whole number$\}$

16. $C = \{x \,|\, x$ is a multiple of 3 and $x \in W\}$
$D = \{x \,|\, x$ is a multiple of 2 and $x \in W\}$

B

Given: $M = \left\{-27, -\sqrt{13}, -\dfrac{9}{4}, -3, -1.01, 0, \dfrac{5}{11}, 3, 4.57, 9, 16.13\overline{13}\right\}$

If $x \in M$, list the following subsets of M.

17. $\{x \,|\, x \in W\}$ **18.** $\{x \,|\, x \in Q\}$ **19.** $\{x \,|\, x \in I\}$ **20.** $\{x \,|\, x \in R\}$

Given: $S = \left\{-4, -3, -\dfrac{1}{2}, 0, \dfrac{1}{2}, 3, 4\right\}$, $T = \left\{-\dfrac{1}{2}, -\dfrac{1}{4}, 0, \dfrac{1}{4}, \dfrac{1}{2}\right\}$, and
$V = \{-4, -2, 0, 2, 4\}$
Find:

21. $S \cap V$ **22.** $S \cup T$ **23.** $S \cap T \cap V$ **24.** $S \cup T \cup V$

Given: $U = W = \{$all whole numbers$\}$, $A = \{$all even whole numbers$\}$, and
$B = \{$all whole numbers that are a multiple of 5$\}$
Find:

25. $\sim A$ **26.** $\sim B$ **27.** $\sim (A \cap B)$

28. $\sim (W \cap A)$ **29.** $\sim (W \cap B)$ **30.** $\sim (A \cup B)$

Given: $H = \{x \,|\, x \in J\}$, $K = \{-5, -4, -3, -2, -1, 0, 1, 2, 3, 4, 5\}$, and
$L = \left\{\dfrac{1}{4}, \dfrac{2}{4}, \dfrac{3}{4}, \dfrac{4}{4}, \dfrac{5}{4}, \dfrac{6}{4}, \dfrac{7}{4}\right\}$

True or false:

31. $K \subset H$ **32.** $L \subset H$ **33.** $L \subseteq L$ **34.** $K \subseteq H$

Draw a Venn diagram to show the relationship between the given sets and describe the union of the sets.

35. $A = \{$all left-handed males$\}$ **36.** $C = \{$all ducks$\}$
   $B = \{$all left-handed females$\}$ $D = \{$all mallard ducks$\}$

Given: $A = \{x \mid x < 20$ and $x \in N\}$, $B = \{x \mid x > 10$ and $x \in W\}$, and $C = \{x \mid x <$
   18 and $x \in J\}$
List the following sets:

37. $\{x \mid x \in A$ and $x \in Q\}$ **38.** $\{x \mid x \in C$ and $x \in N\}$

39. $\{x \mid x \in B$ and $x \in J\}$ **40.** $\{x \mid x \in C$ and $x \in W\}$

41. $A \cap B$ **42.** $B \cap C$ **43.** $A \cap C$ **44.** $A \cap B \cap C$

45. $B \cup C$ **46.** $A \cup C$ **47.** $A \cup B$ **48.** $A \cup B \cup C$

Given: $U = \{$all people$\}$, $A = \{$all males under 25 years old$\}$, and $B = \{$all people with red hair$\}$
Find:

49. $\sim A$ **50.** $\sim B$ **51.** $\sim(A \cap B)$

52. $\sim(A \cup B)$ **53.** $\sim(U \cup B)$ **54.** $\sim[(U \cup A) \cap B]$

Draw a Venn diagram to show the relationship between the following sets and describe the intersection of the sets.

55. $A = \{$state capitals$\}$ **56.** $A = \{x \mid x \in N$ and x is a multiple of 5$\}$
   $B = \{$the largest city in each state$\}$ $B = \{x \mid x \in J$ and $x < 10\}$

STATE YOUR UNDERSTANDING

57. What is the meaning of $A \cup B$?

58. What can be said about the set A if $A \cap B = B$?

CHALLENGE EXERCISES

Given: $A = \{x \mid x > 25$ and $x \in J\}$, $B = \{x \mid x < 40$ and $x \in W\}$, and $C = \{x \mid x$ is
   a multiple of 5 and $x \in N\}$
Find:

59. $(A \cup B) \cap C$ **60.** $(B \cap C) \cup A$ **61.** $(A \cap C) \cup B$ **62.** $(B \cup C) \cap A$

ANSWERS TO SELECTED EXERCISES

CHAPTER 1

Exercise 1.1

1. T **3.** T **5.** F **7.** T **9.** T **11.** F
13. T **15.** F **17.** T **19.** F **21.** No
23. $\{6, 91\}$ **25.** Yes **27.** $\{-8, 0, 5, 42\}$
29. $\{-\sqrt[3]{3}, \sqrt{39}\}$ **31.** $\{0, 5, 42\}$ **33.** $\{-29, 62\}$
35. $\{\sqrt[3]{9}\}$ **37.** $<$ **39.** $>$
41. $\{-3, -2, -1, 0, 1, 2, 3\}$ **43.** $\{-2\}$
45. $\{-3, -2, -1, 0, 1, 2, 3, 5, 7\}$ **47.** $\emptyset$ **49.** Yes
51. F **53.** T **55.** T **57.** T **59.** T **61.** In
words **63.** $\emptyset$ **65.** B

Exercise 1.2

1. Commutative property of addition **3.** Associative
property of multiplication **5.** Multiplication property of
one **7.** Closure property of addition **9.** Commutative
property of addition **11.** Commutative property of
multiplication **13.** Addition property of zero
15. Distributive property **17.** True **19.** False
21. False **23.** True **25.** True **27.** True
29. False **31.** True **33.** False **35.** Multiplication
property of one **37.** Distributive property
39. Multiplication property of zero **41.** Multiplication
property of reciprocals **43.** Associative property of
addition **45.** Addition property of opposites
47. Addition property of zero **49.** Multiplication
property of one **51.** Addition property of opposites
53. Distributive property **55.** Addition property of zero
57. Distributive property **59.** Multiplication property of
zero **61.** Addition property of opposites **63.** Addition
property of zero **65.** Closure property of addition
67. No **69.** No **71.** Yes

Exercise 1.3

1. 6 **3.** 6 **5.** -4 **7.** 6 **9.** -10 **11.** 0
13. -5 **15.** 12 **17.** 15 **19.** 6 **21.** -17
23. 2 **25.** -6 **27.** 0 **29.** -26 **31.** $-x - 2$
33. $-x + 2$ or $2 - x$ **35.** $a + b$ **37.** $-2x$
39. $-a - b$ **41.** -63 **43.** -25 **45.** -8.8
47. 3.59 **49.** 8 **51.** 7 **53.** 11 **55.** -5
57. 9 **59.** 9.8 **61.** \$220 increase **63.** \$17,179
65. \$605.35 **67.** 7640 ft **69.** \$267.25 **71.** No; If
both a and b are positive or both are negative **73.** Yes
75. Yes

Exercise 1.4

1. 6 **3.** -17 **5.** -4 **7.** 2 **9.** $11a$ **11.** $2b$
13. $-17abc$ **15.** $35xy$ **17.** -176 **19.** -181

21. $\dfrac{3}{8}$ **23.** -1 **25.** -0.1 **27.** $16xy$ **29.** $4x$
31. $7a$ **33.** $18x + 8$ **35.** $-19x$ **37.** -23.3
39. 19.22 **41.** 38.8 **43.** $6x + 8y$ **45.** $-4x + 3y$
47. $\dfrac{5}{4}x - \dfrac{5}{8}y$ **49.** $-\dfrac{5}{12}a - \dfrac{3}{4}b$ **51.** $-0.92x - y$
53. $3.52y + 6.3z + 2$ **55.** $-2.8x - 3$ **57.** $-1.2x$
59. $x + 5y + 5$ **61.** $48°F$ **63.** $-60°$ **65.** 850 ft
67. \$53.96 (overdrawn) **69.** \$11.15 **71.** In words
73. Yes; If $c = 0$

Exercise 1.5

1. $\dfrac{1}{6}$ **3.** 3 **5.** 35 **7.** -70 **9.** 0 **11.** -22
13. -8 **15.** 4 **17.** $5x$ **19.** -400 **21.** $-\dfrac{12}{11}$
23. $\dfrac{2}{3}$ **25.** 24 **27.** -24 **29.** 10 **31.** -7
33. $24a$ **35.** $96b$ **37.** $\dfrac{3}{5}x$ **39.** -60 **41.** 90
43. 30 **45.** 4 **47.** $-\dfrac{20}{11}$ or $-1.\overline{81}$ **49.** $-15x$
51. $36y$ **53.** $-\dfrac{10}{7}z$ **55.** $8a + 4$ **57.** $-12x + 24$
59. $-16x - 40$ **61.** $25a - 10b + 15$
63. $18w + 24y + 54z$ **65.** $-4ax - 8ay + 6a$
67. $10°C$ **69.** $-10°C$ **71.** -3.75 **73.** $-\$625.00$
75. -300 ft **77.** In words **79.** Yes; if x is a
negative number **81.** Yes; if the number is 1

Exercise 1.6

1. 5 **3.** -36 **5.** 25 **7.** $15x - 4$ **9.** $2x$
11. 22 **13.** -17 **15.** 17 **17.** -6 **19.** -84
21. $7x + 6$ **23.** $2x - 8$ **25.** $-4a - 9b$
27. $8a - 8$ **29.** $-10x - 45$ **31.** 40 **33.** 7
35. -10 **37.** -14 **39.** -2 **41.** -19 **43.** 3
45. -42 **47.** 1 **49.** $-15x + 6$ **51.** $9x + 41$
53. 57 **55.** 121 **57.** 144 **59.** 704 **61.** $122°F$
63. $-31°F$ **65.** \$750.32 **67.** 26 yd **69.** 144 ft
71. In words **73.** Yes, if $x = 1$

Exercise 1.7

1. $\{3\}$ **3.** $\{12\}$ **5.** $\{2\}$ **7.** $\{-5\}$ **9.** $\{-2\}$
11. $\{4\}$ **13.** $\{-21\}$ **15.** R, identity **17.** $\emptyset$
19. $\{0\}$ **21.** R, identity **23.** $\{3\}$ **25.** $\emptyset$
27. $\left\{\dfrac{28}{9}\right\}$ **29.** $\left\{\dfrac{53}{8}\right\}$ **31.** $\left\{\dfrac{5}{4}\right\}$ **33.** $\left\{-\dfrac{3}{5}\right\}$

35. $\{2\}$ **37.** $\{2\}$ **39.** $\{-2\}$ **41.** $\left\{-\dfrac{5}{7}\right\}$

43. $\left\{\dfrac{9}{7}\right\}$ **45.** $\{15\}$ **47.** $\{-3\}$ **49.** $\left\{\dfrac{4}{3}\right\}$

51. $\{-34\}$ **53.** $\{1\}$ **55.** $\left\{-\dfrac{11}{8}\right\}$ **57.** $\left\{\dfrac{11}{16}\right\}$

59. $\left\{-\dfrac{71}{4}\right\}$ **61.** TV \$700, radio \$150   **63.** The numbers are $-3, -7$ **65.** \$200   **67.** 2.5 hr   **69.** The numbers are 5, 17   **71.** In words   **73.** $b = -6$ **75.** $-218.°C$

Exercise 1.8

1. $t = \dfrac{d}{r}$ **3.** $r = \dfrac{i}{pt}$ **5.** $h = \dfrac{V}{\pi r^2}$ **7.** $b = \dfrac{2A}{h}$

9. $m = \dfrac{Fr^2}{GM}$ **11.** $y = -x + 12$ **13.** $x = \dfrac{y+3}{m}$

15. $b = 2a - 3$ **17.** $t = \dfrac{2s-8}{3}$ **19.** $C = \dfrac{-7D-5}{6}$

21. $x = \dfrac{5y+2}{a}$ **23.** $t = \dfrac{(A-P)}{Pr}$ **25.** $n = \dfrac{PV}{RT}$

27. $h = \dfrac{S - 2\pi r}{2\pi r^2}$ **29.** $B = An + C$

31. $C = B - An$ **33.** $d = \dfrac{(L-f)}{(n-1)}$

35. $h = \dfrac{A - \pi r^2}{2\pi r}$ **37.** $P = \dfrac{3At + d}{A}$

39. $C = \dfrac{8d+3}{14}$ **41.** $L = \dfrac{P - 2W}{2}$

43. $x = \dfrac{F - ma}{k}$ **45.** $g = \dfrac{2s - 2vt - 2x}{t^2}$

47. $x = \dfrac{W - mad}{kd}$ **49.** $x = \dfrac{(5y + 25)}{4}$

51. $R = \dfrac{(E - rI)}{I}$ **53.** $B = \dfrac{(2A - bh)}{h}$

55. $c = 2S - a - b$ **57.** $a = 2S - c$

59. $S = \dfrac{11L + 1}{3}$ **61.** In words   **63.** \$260

65. 233.15K

Exercise 1.9

1.

3.

5.

7.

9.

11. $(12, +\infty)$ **13.** $(-\infty, 4)$ **15.** $(-4, +\infty)$
17. $\{x \mid x < 5 \text{ and } x \in R\}$ **19.** $\{x \mid x \le -2\}$
21. $\{x \mid x \le -2 \text{ and } x \in R\}$
23.

25.

27.

29. $\left\{x \mid x \le -\dfrac{1}{2}\right\}$ **31.** $\varnothing$ **33.** $\left\{x \mid x > \dfrac{15}{8}\right\}$

35. $\left(-\infty, \dfrac{1}{2}\right)$ **37.** $\left(-\infty, -\dfrac{1}{2}\right]$ **39.** $(-\infty, 17)$

41.

43.

45. $\varnothing$ **47.** $(2, 6)$ **49.** $\left[-\dfrac{9}{5}, 5\right)$ **51.** $[-4, 1]$

53. $\left[\dfrac{22}{7}, 9\right]$ **55.** $[-1, 1]$

57. $\{x \mid x < 5\} \cup \{x \mid x > 6\}$ **59.** $\left\{x \mid -\dfrac{7}{2} \le x \le \dfrac{3}{2}\right\}$

61. Yes, the trip can be made   **63.** 29 sacks or less   **65.** 7 cans   **67.** 91%   **69.** No   **71.** In words   **73.** $-25 < -x < -9$ **75.** The number can be between -24 and -6

Exercise 1.10

1. Length 31.2 ft, width 7.8 ft   **3.** 276.125 sq cm

5. 3 hr 15 min or $3\dfrac{1}{4}$ hour   **7.** 21% decrease

9. 183% increase   **11.** \$318.75   **13.** 13 in, 23 in
15. 41°

Chapter 1 Review Exercises

1. T   **3.** F   **5.** T
7. $\{-5, -4, -3, -2, -1, 0, 1, 2, 3, 4, 5\}$
9. $\{-4, -2, 0, 2, 4\}$ **11.** $\{0, 15\}$ **13.** $\{15\}$
15. $\{-\sqrt{3}, \sqrt[3]{6}\}$ **17.** Conditional   **19.** Identity
21. T   **23.** F   **25.** F   **27.** Distributive property
29. Multiplication property of zero   **31.** 7   **33.** 0.67
35. $-2x - 3$ or $-(2x + 3)$ **37.** 0.75   **39.** $3x$
41. 6   **43.** 12   **45.** 78   **47.** -61 **49.** -91

51. $-x$ **53.** $13ab$ **55.** $-\dfrac{2}{5}x$ **57.** -68

59. -0.075 **61.** $-\dfrac{1}{5}$ **63.** $\dfrac{5}{4}$ **65.** $\dfrac{20}{7}$ **67.** -3

69. -5 **71.** $-36x$ **73.** $31ab$ **75.** $3a$
77. -147 **79.** 174   **81.** 12   **83.** -2 **85.** -8

87. $\{5\}$ **89.** $\{5\}$ **91.** $t = \dfrac{i}{pr}$ **93.** $G = \dfrac{Fr^2}{Mm}$

95. $a = \dfrac{W - kdx}{dm}$ **97.** $[-3, +\infty)$ **99.** $\left\{x \mid x \geq -\dfrac{5}{2}\right\}$

101. $\{x \mid x < 6\} \cup \{x \mid x > 10\}$ **103.** $\{x \mid x < -18\}$

105. $\left(\dfrac{11}{2}, \dfrac{19}{2}\right]$ **107.** 3.75 hr   **109.** 214, 216, 218, 220

Chapter 1 True–False Concept Review

1. False   **3.** True   **5.** False   **7.** True   **9.** True
11. False   **13.** False   **15.** False   **17.** False
19. True   **21.** True   **23.** False   **25.** True

Chapter 1 Test

1. $-\dfrac{23}{200}$ **2.** Multiplication property of one
3. $-15a + 15b$ **4.** -420 **5. a.** F **b.** T **c.** T
6. $\{x \mid x \geq 3\}$ or $[3, +\infty)$ **7.** $\{a, b, c, d, e, f, g\}$ **8.** 321
9. $-4b$ **10.** Addition property of opposites
11. a. $\{-7, 1, 9\}$ **b.** $\{\sqrt{19}\}$

c. $\left\{-7, -3.6, -\dfrac{3}{5}, 1, 9, 12.7, \sqrt{19}\right\}$ **12.** $\{-1\}$

13. $-x + 10y$ **14.** $-5y - 9z$ **15.** Commutative

property of addition   **16.** $-\dfrac{19}{24}$ **17.** -3.7

18. Multiplication property of zero   **19.** 16.1

20. $\left\{\dfrac{17}{15}\right\}$ **21.** $n = \dfrac{L + d - f}{d}$ **22.** Commutative

property of multiplication   **23.** $10a - 2b$ **24.** -33.4
25. $\{-2, 0, 2\}$ **26.** $-30x$ **27.** Distributive property
28. 208   **29.** $\{-5\}$ **30.** -17.03 **31.** -44
32. -108 **33.** $0.3x - 10.4y$ **34.** $-5.6x + 19.7y$
35. $\{x \mid -9 \leq x < 1\}$ or $[-9, 1)$ **36.** $y = 6 - 2x$

37. $\left\{x \mid x \leq \dfrac{1}{2}\right\}$ or $\left(-\infty, \dfrac{1}{2}\right]$ **38.** $\{1\}$ **39.** 24 ft by

51 ft   **40.** 4 hr

CHAPTER 2

Exercise 2.1

1. r^{12} **3.** $60a^6$ **5.** a^{16} **7.** $9t^6$ **9.** $9x^4y^6$
11. $20z^8$ **13.** x^6y^6 **15.** $16x^8$ **17.** $64t^{12}$
19. $121s^8$ **21.** $-30x^{38}$ **23.** $15x^{11}y^8z^{14}$
25. $196p^4q^6$ **27.** $-512p^{18}q^{12}r^9$ **29.** t^{12}
31. $108x^5$ **33.** $-32x^5$ **35.** c^5y^{13} **37.** x^6y^6
39. $a^{12}b^8$ **41.** $-32x^5y^6$ **43.** $12a^8b^8c^5$
45. $-108t^{19}$ **47.** $-768x^7y^{18}z^{14}$ **49.** $(x + y)^6$
51. 2^5 or 32   **53.** $2t^{6n-2}$ **55.** $5y^{7n+4}$ **57.** $4x^{3m+3}$

59. w^{3t-1} **61.** $V = \pi r^7$ **63.** $V = \dfrac{1}{2}a^7$

65. $A = \dfrac{1}{4}\pi d^2$ **67.** $A = b^4$ **69.** $V = 3\pi h^7$

71. In words   **73.** 8^{x+y+z} **75.** $5^{a(b+c)}$ or 5^{ab+ac}

77. -32.42 **79.** $\dfrac{37}{120}$ **81.** $-\dfrac{3}{10}x - \dfrac{5}{4}y$

83. -3318 ft

Exercise 2.2

1. $\dfrac{1}{64}$ **3.** $\dfrac{1}{s}$ **5.** $\dfrac{b^2}{a}$ **7.** $\dfrac{r^4}{s^2}$ **9.** a^4b^3

11. $\dfrac{1}{36}$ **13.** $\dfrac{a^2}{b^4}$ **15.** $\dfrac{1}{xy}$ **17.** $\dfrac{w}{xy}$ **19.** y^2

21. $-\dfrac{12}{xy^2}$ **23.** $\dfrac{y^6}{25x^4}$ **25.** $-\dfrac{2x^8}{27y^6}$ **27.** $\dfrac{5x}{y^8}$

29. mn **31.** $\dfrac{x}{y}$ **33.** $\dfrac{b^8}{a^{20}}$ **35.** $\dfrac{1}{8x^9y^6}$ **37.** $\dfrac{x^4y^6}{9}$

39. $\dfrac{9x^4}{y^6}$ **41.** $\dfrac{1}{64x^{12}}$ **43.** $\dfrac{9z^8}{25x^2y^{12}}$ **45.** 2500

47. $4x^{14}$ **49.** $\dfrac{9x^3}{2y^3}$ **51.** p^3q^2 **53.** $(x + a)^3$

55. $(x + y)^{m+2}$ **57.** $(2r - 3)^7$ **59.** $(x + 3)(x - 4)^3$
61. 0.46   **63.** 0.0000049   **65.** 0.0073
67. 0.000001   **69.** In words   **71.** True for any value

of x and y such that $xy \neq 0$ or $x + y \neq 0$ **73.** $\left[\dfrac{7}{3}, +\infty\right)$

75. $3x^2 - 13x - 12$ **77.** $-34y + 309$ **79.** \$296

Exercise 2.3

1. 5×10^4 **3.** 4×10^{-5} **5.** 4.3×10^5
7. 9×10^{-7} **9.** 8.25×10^5 **11.** 930   **13.** 89,100
15. 0.00232   **17.** 0.0000067   **19.** 314,200
21. 3.77×10^5 **23.** 7.01×10^{-5} **25.** 6.11×10^8
27. 0.321   **29.** 68,900   **31.** 2.38×10^9
33. 1.08×10^{-6} **35.** 1.44×10^4 **37.** 2.7×10^{-7}
39. 3.0×10^8 **41.** 3.784×10^3 **43.** 3.448×10^4
45. 4.84×10^{-4} **47.** 38,400,000   **49.** 0.00000000236
51. 2.9×10^4; 29,000   **53.** 1.2×10^{-5}; 0.000012
55. 2.10×10^4; 21,000   **57.** 2.5×10^4; 25,000
59. 7.3×10^0; 7.3   **61.** 5.87×10^{12} mi
63. 5×10^4 cycles/sec   **65.** 1×10^{-6} cm
67. 0.00004 cm   **69.** 25,500,000,000,000 miles
71. 3.00×10^8 m/sec   **73.** In words
75. 0.000000000001   **77.** 2.3901×10^{-1}
79. -171.21 **81.** $0.8p + 8.5pq + 3.1q - 2.4$

83. $\left(-\infty, \dfrac{20}{9}\right)$ **85.** 19%

Exercise 2.4

1. 2—binomial; 1 (linear)   **3.** 3—trinomial;
2 (quadratic)   **5.** 1—monomial; 0 (zero degree)   **7.** 3—
trinomial; 2 (quadratic)   **9.** 2—binomial; 3 (cubic)
11. 4—polynomial; 1 (linear)   **13.** 3—trinomial;
1 (linear)   **15.** 4—polynomial; 5 (fifth degree)   **17.** 3—
trinomial; 2 (quadratic)   **19.** 4—polynomial; 3 (cubic)

21. In words **23.** In words **25.** No **27.** True
29. True **31.** Commutative property of multiplication
33. $\dfrac{4}{7}$

Exercise 2.5

1. $10x + 15y$ **3.** $4x^2 + 2x + 4$ **5.** $8x - 36$
7. $15x - y + 6$ **9.** $11x^2 + 3x + 14$
11. $\dfrac{1}{6}x^2 + \dfrac{2}{3}x - \dfrac{1}{2}y$ **13.** $-1.1p + 0.7q + 5.7$
15. $5x - 7w$ **17.** $15r - 9s + 6$ **19.** $-4x^2 - x + 9$
21. $5x^3 + 7x^2 - 11x + 6$ **23.** $4x^2y - 3xy + 22xy^2 - 2$
25. $12a^2 + 3ab + 8b^2 + 3$ **27.** $6a^2 - 4b^2$
29. $3a^4 - 2a^3 + a^2 + a$ **31.** $0.2x + 1.5y - 0.1$
33. $3x - 13$ **35.** $\left\{-\dfrac{10}{3}\right\}$ **37.** $\left\{\dfrac{8}{9}\right\}$ **39.** $\{1\}$
41. $\dfrac{11}{5}p + \dfrac{7}{4}q - 1$ **43.** $-0.29r + 0.33s - 0.71t$
45. $4x^2 + 7x - 5$ **47.** $6y - 10$ **49.** $11x - 2y$
51. $10r^2 - 6r + 3$ **53.** $-6x + 12y$ **55.** $\{-2\}$
57. $\{6\}$ **59.** $\{4\}$ **61.** $C = \dfrac{26}{5}n^2 - 44n + 70;$
$10,870$ **63.** $P = x^2 + 2x - 1500$
65. $V = 3x^3 - 4x + 50; 66$ **67.** $B = 4t^2 - 3t + 350;$
351 bolts **69.** $T = \dfrac{7}{2}t^2 - 6t + 30; 32$ **71.** In words
73. $2x^{2n} + 10x^n - 8$ **75.** $6x^n + 4x^2 - 4x - 2$
77. -119 **79.** 70 **81.** -66 **83.** -158

Exercise 2.6

1. $2x + 6$ **3.** $x^2 - x$ **5.** $6x + 14$ **7.** $-12a + 18b$
9. $y^3 + 3y^2 - 8y$ **11.** $2a^3 - 6a^2 + 12a$
13. $3x^3y - 18x^2y^2 + 3xy^3$
15. $-6a^2bc + 9ab^2c - 3abc^2 + 3abc$
17. $2a^7 - 5a^6 + 3a^4 + a^3$
19. $156x^4y^2 - 143x^3y^3 + 104x^3y^2$ **21.** $x^2 + x - 12$
23. $4x^2 - 8xy + 3y^2$ **25.** $-12m^2 + 31mt - 7t^2$
27. $a^2 + 2ab + b^2 + 3a + 3b$
29. $2x^4 - 7x^3 + 7x^2 - 7x + 5$
31. $4y^5 + y^4 + 12y^3 - 9$ **33.** $\{-4\}$ **35.** $\{-1\}$
37. $\left\{\dfrac{3}{13}\right\}$ **39.** $\left\{\dfrac{11}{4}\right\}$ **41.** $a^2 - b^2 + 2bx - x^2$
43. $c^2 + 2cd + d^2 + 9c + 9d + 20$
45. $2x^3 - 3x^2y - 7xy^2 + 12x + 3y^3 + 18y$
47. $x^2 + 2xy + y^2 + 2yz + 2xz + z^2$
49. $x^3 + 2x^2 - 5x - 6$
51. $5x^5 + 3x^4 + 8x^3 + 5x^2 - 4x - 2$ **53.** $\left\{-\dfrac{1}{4}\right\}$
55. $\{-9\}$ **57.** $\{1\}$ **59.** $\left\{\dfrac{1}{4}\right\}$ **61.** $6\dfrac{3}{4}$ hours going,
$8\dfrac{3}{4}$ hours returning **63.** 54 qt at $1.30, 36 qt at $1.60

65. 6 lb of hard candy, 9 lb of caramels **67.** $\dfrac{3}{5}$ hr or
36 min **69.** 1 hr **71.** In words
73. $x^{2n} + 2x^ny^n + y^{2n}$ **75.** $x^n - 1 + 2x^{-n}$
77. 5.12 or $5\dfrac{3}{25}$ **79.** $-420y$ **81.** $34x - 136w$
83. $-\$31$

Exercise 2.7

1. $x^2 + 7x + 12$ **3.** $y^2 - 2y - 35$ **5.** $b^2 - 5b + 4$
7. $y^2 - 11y + 24$ **9.** $x^2 - x - 42$ **11.** $a^2 - 100$
13. $x^2 - 18x + 81$ **15.** $w^2 - 6w + 9$ **17.** $4x^2 - 121$
19. $p^2 + 26p + 169$ **21.** $3x^2 - 23x + 30$
23. $2a^2 - a - 15$ **25.** $15x^2 + 17x - 42$
27. $4a^2 - 25$ **29.** $4a^2 + 20a + 25$
31. $9x^2 + 30xy + 25y^2$ **33.** $64w^2 - 208w + 169$
35. $\{-9\}$ **37.** $\{-2\}$ **39.** $\left\{-\dfrac{1}{10}\right\}$
41. $15x^2 + 12x - 99$ **43.** $0.18x^2 - 0.27x + 0.1$
45. $\dfrac{1}{6}x^2 + \dfrac{13}{36}x + \dfrac{1}{6}$ **47.** $a^2 + 2ab + b^2 - 64$
49. $4x^2 + 4xy + y^2 - 20x - 10y + 25$
51. $9x^2 - 6xy + y^2 + 21x - 7y + 12$
53. $a^2 + 2ab + b^2 - c^2 - 2dc - d^2$ **55.** $\{1\}$
57. $\{-2\}$ **59.** $\left\{\dfrac{2}{3}\right\}$ **61.** 50 ft **63.** John's age 26,
Mary's age 21 **65.** length—19 ft, width—16 ft
67. Jim—16 years, Pam—6 years **69.** Johnny—
9 years, Jason—3 years **71.** In words **73.** Select any
value for y except 0 and select $x = 0$ or select any non-zero
values for x and y such that $x \neq y$. **75.** $4x^{2a} - 9$
77. $a^3 - b^3$ **79.** $262x - 8$ **81.** 267 **83.** 321
85. $294

Exercise 2.8

1. $12(m - n)$ **3.** $16(a - 2)$ **5.** $6a(b - 2c + 3d)$
7. $(y + 4)(c + d)$ **9.** $(3y + z)(x + 2)$
11. $(2y + 3z)(3x + 4)$ **13.** $(x + y)(a + b + c)$
15. $\{5, 9\}$ **17.** $\left\{\dfrac{1}{2}, \dfrac{1}{4}\right\}$ **19.** $\left\{-\dfrac{3}{4}, \dfrac{4}{3}\right\}$
21. $6xy^2(3x - 5)$ **23.** $2a(a + b - 7)$
25. $(5x + 2y)(4x - 3)$ **27.** $(x + 2)(5x + 6)$
29. $(2x^2 + 1)(y - 3)$ **31.** Prime polynomial
33. $\{0, -6\}$ **35.** $\left\{-\dfrac{3}{5}, -\dfrac{5}{4}\right\}$ **37.** $\left\{-\dfrac{3}{7}, \dfrac{5}{2}\right\}$
39. $\left\{-7, \dfrac{7}{4}\right\}$ **41.** $3x^2y^2z(2x - 3y - 8z)$
43. $12x^5y^6(3x^2 - 7y^3 + 6)$ **45.** $(6x + 5y)(7x - 3y)$
47. $(5x + 8y)(x^2 + 5x + 2)$ **49.** $(2x - 3)(a^2 + b + c)$
51. $\{-3, 5\}$ **53.** $\{5, 11\}$ **55.** $\{0, a\}$ **57.** $\{-3a, 2a\}$
59. $\{-5, 2b\}$ **61.** 144 sets **63.** 256 ft, 5 sec
65. 4 sec **67.** 3 sec **69.** 800 toy soldiers

71. In words **73.** $x^{2n}(x^n - 1)$ **75.** $x^{n+3}(x^n + 1)$
77. 208 **79.** $\{-8\}$ **81.** $\{-5\}$ **83.** Betty is 9 years old

Exercise 2.9

1. $(x + 7)(x + 5)$ **3.** $(x - 7)(x - 3)$
5. $(a - 9)(a + 2)$ **7.** $(a + 7)(a - 3)$ **9.** Prime polynomial **11.** $(5y + 1)(y + 3)$ **13.** $\{1, 2\}$
15. $\{-2, 4\}$ **17.** $\{-9, -7\}$ **19.** $\{-9, 7\}$
21. $(x + 11)(x - 10)$ **23.** $(a - 12)(a - 8)$
25. $(2a - 3)(a + 6)$ **27.** $(5x + 2)(x - 3)$

29. $(2t + 5)(t - 7)$ **31.** $\{-6, 12\}$ **33.** $\left\{-\dfrac{4}{9}, 1\right\}$

35. $\left\{-\dfrac{2}{3}, 1\right\}$ **37.** $\left\{-\dfrac{4}{3}, \dfrac{3}{4}\right\}$ **39.** $\left\{\dfrac{1}{2}, 11\right\}$

41. $(x + 12)(x - 18)$ **43.** $(x - 13)(x - 15)$
45. Prime polynomial **47.** $(3b + 2)(5b - 9)$
49. $(x + 2y - 7)(x + 2y + 6)$ **51.** $2(2a + 7)(a - 1)$

53. $2(3t - 10)(6t + 1)$ **55.** $\{7, 16\}$ **57.** $\left\{\dfrac{9}{2}, -\dfrac{7}{5}\right\}$

59. $\left\{-\dfrac{7}{4}, \dfrac{8}{5}\right\}$ **61.** 12 in. by 6 in. **63.** 65 ft by 80 ft

65. 5 in. by 10 in. **67.** $(x^n - 1)(x^n - 2)$
69. $(x^2 + 6)(x^2 + 2)$ **71.** $\{-4\}$ **73.** R **75.** $\{-21\}$

77. $x = \dfrac{5t + 45}{3}$ or $x = \dfrac{5}{3}t + 15$

Exercise 2.10

1. $(a + 1)(a - 1)$ **3.** $(x + 3)(x - 3)$ **5.** Prime polynomial **7.** $(x + 4)^2$ **9.** $(c + 3)^2$ **11.** $(y - 14)^2$
13. $(x - 1)(x^2 + x + 1)$ **15.** $(y - 5)(y^2 + 5y + 25)$
17. $\{\pm 7\}$ **19.** $\{\pm 16\}$ **21.** Prime polynomial
23. $(3ab + 8c)(3ab - 8c)$ **25.** $(4y + 3)^2$
27. $(2a - 7)^2$ **29.** $(ab - c)(a^2b^2 + abc + c^2)$

31. $(2w + 1)(4w^2 - 2w + 1)$ **33.** $\left\{\pm\dfrac{5}{2}\right\}$ **35.** $\{-3\}$

37. $\left\{\dfrac{9}{2}\right\}$ **39.** $\{-16\}$

41. $(3x - 2y)(3x + 2y)(9x^2 + 4y^2)$
43. $(a + b + c)(a + b - c)$ **45.** $(6ab - 11c)^2$
47. $(10y - 13)^2$ **49.** $(4a - 3b)(16a^2 + 12ab + 9b^2)$
51. $(10cd - 7)(100c^2d^2 + 70cd + 49)$
53. $(5x + y + 5)(x + y - 5)$
55. $(a - b + x - y)(a - b - x + y)$
57. $(x - 5 + y)(x - 5 - y)$

59. $(12 + 2t + s)(12 - 2t - s)$ **61.** $\left\{\pm\dfrac{13}{2}\right\}$

63. $\left\{\dfrac{7}{2}\right\}$ **65.** 8 sec **67.** 21 years **69.** 9th hour

71. True **73.** In words **75.** $(x^n + 1)(x^{2n} - x^n + 1)$

77. $(x^n - 1)(x^{2n} + x^n + 1)$ **79.** $-8a + 18$ **81.** $\emptyset$

83. $\left\{y \mid y > \dfrac{2}{5}, y \in R\right\}$ **85.** 18 crates

Exercise 2.11

1. $3(3c + 2b - 5)$ **3.** $x(x - 8)$ **5.** $(x - 6)(x - 2)$
7. $(x + 10)(x - 10)$ **9.** Prime polynomial
11. $(a + 3)(a + 4)$ **13.** $\{0, 2\}$ **15.** $\{-4, 4\}$
17. $\{3, 5\}$ **19.** $4xy(4x - 2y + 3)$
21. $(x - 11)(x - 5)$ **23.** $(c - 9)(c + 5)$
25. $3(x + 5)(x - 2)$ **27.** $(7y + 6)(7y - 6)$
29. Prime polynomial **31.** $(x + 4)(x^2 - 4x + 16)$
33. $(10c - d)(100c^2 + 10cd + d^2)$
35. $(a^2 + 1)(4a - b)$ **37.** $(2x + 7)(x + 3)$
39. $(3x - 2)(2x + 5)$ **41.** $\{3, 8\}$ **43.** $\{\pm 9\}$
45. $\{-2, 1\}$ **47.** $(4c - 3d)(3c + 5d)$ **49.** $(6w + 5)^2$
51. $5(x + 6)(x - 6)$ **53.** $2(5x - 2)(25x^2 + 10x + 4)$
55. $3(x + 7)(x - 10)$ **57.** $2a(2a + 5)(7a - 3)$
59. $2(x + 9)(x + 1)$ **61.** $(x + 2)(x - 2)(x^2 + 3)$
63. $(w + 6)(w - 6)(w^2 + 1)$
65. $(x + 1)(x^2 - x + 1)(x - 1)(x^2 + x + 1)$
67. $(t + 2)(t^2 - 2t + 4)(t - 2)(t^2 + 2t + 4)$

69. $\left\{\dfrac{7}{2}, \dfrac{10}{3}\right\}$ **71.** $\left\{\dfrac{1}{3}, -\dfrac{5}{4}\right\}$ **73.** $\left\{\dfrac{1}{4}, \dfrac{1}{24}\right\}$

75. 8 ft by 13 ft **77.** In words
79. $x^{3n}(x^n - 3)(x^{2n} + 3x^n + 9)$
81. $(x + 1 + y)(x + 1 - y)$ **83.** $5a + 50$

85. $a = -\dfrac{54}{7}$ **87.** $a \le 4$ **89.** 12 hr

Chapter 2 Review Exercises

1. t^{15} **3.** $(-6)^{27}$ **5.** $-99r^5s^5t^4$ **7.** $196x^{10}y^6$

9. 7^{2w} **11.** $1024t^{18}$ **13.** $\dfrac{1}{169}$ **15.** 196 **17.** $\dfrac{c^2}{d^4}$

19. xy **21.** $\dfrac{81xy^2}{5}$ **23.** $(x + 7)^{a+2}$ **25.** 3.6×10^7

27. 3.6×10^{-4} **29.** 7.322×10^6 **31.** 83,400,000
33. 0.000834 **35.** 0.000834 **37.** 2.66×10^2
39. 7.6×10^{-8} **41.** binomial, degree 2
43. trinomial, degree 6 **45.** monomial, degree 0
47. $24a^2 - 5a + 16$ **49.** $6a^2 + 3a + 22$
51. $6 - 10w$ **53.** $2 + 2w$ **55.** $\{1\}$ **57.** $\{7\}$
59. $\{-2\}$ **61.** $-39r^2s^2t^3 + 91rs^3t^3 - 117rs^2t^4 + 13rs^2t^3$
63. $v^2 + 23v + 126$ **65.** $w^6 - 3w^3b - 108b^2$

67. $x^3 - 2x^2y + 3x^2 - 2xy + 4y^2 - 6y$ **69.** $\left\{-\dfrac{9}{4}\right\}$

71. $\left\{\dfrac{7}{10}\right\}$ **73.** $\{4\}$ **75.** $y^2 + 18y + 72$

77. $4w^2 - 25w - 56$ **79.** $28t^2 - 57t + 14$
81. $y^2 - 196$ **83.** $t^2 - 441$ **85.** $16b^2 - 225$
87. $y^2 + 22y + 121$ **89.** $49t^2 - 112t + 64$

91. $169a^2 - 52a + 4$ **93.** $\left\{\dfrac{5}{2}\right\}$ **95.** $\left\{\dfrac{15}{2}\right\}$

97. $\left\{-\dfrac{3}{5}\right\}$ **99.** $17wx^2(w + 3)$

101. $6m^2n^2(5mn + 12m - 1)$ **103.** $(x - 3y)(2x - 5y)$
105. $(s - 3)(t + m)$ **107.** $(b - p)(b - q)$
109. $(1 - 8s)(1 + 9t)$ **111.** $\{-7, 0\}$
113. $\{-12, -11\}$ **115.** $\{-5, 13\}$ **117.** $(z - 6)(z + 4)$
119. $(t + 11)(t - 10)$ **121.** $(b - 25)(b - 3)$
123. $(5y - 1)(y - 5)$ **125.** $(4w + 9)(2w + 9)$
127. $(3a - 7)(7a - 6)$ **129.** $\{-40, 25\}$
131. $\left\{-12, \dfrac{1}{6}\right\}$ **133.** $\left\{-16, -\dfrac{1}{3}\right\}$
135. $(5pq + 14)(5pq - 14)$ **137.** $(2 + 11t)(2 - 11t)$
139. $(13 + 2a - 7b)(13 - 2a + 7b)$ **141.** $(y - 13)^2$
143. $(5p + 11)^2$ **145.** $(6 + 7w)^2$
147. $(y + 8)(y^2 - 8y + 64)$
149. $(1 - 7m)(1 + 7m + 49m^2)$
151. $(3 + 10t)(9 - 30t + 100t^2)$ **153.** $\left\{\pm\dfrac{4}{5}\right\}$
155. $\left\{-\dfrac{3}{13}\right\}$ **157.** $5(w^2 + 5y)(w^2 - 5y)$
159. $5(2p + 3)(4p^2 - 6p + 9)$ **161.** $4x^2(s - 6)(t + 1)$
163. $2y(9y - 5)(2y + 7)$ **165.** $\left\{\pm\dfrac{10}{3}\right\}$
167. $\{-14, 12\}$ **169.** $\{-7, 9\}$

Chapter 2 True–False Concept Review

1. False   **2.** True   **3.** False   **4.** True   **5.** True
6. False   **7.** True   **8.** False   **9.** False   **10.** True
11. True   **12.** True   **13.** False   **14.** True
15. False   **16.** True   **17.** False   **18.** False
19. True   **20.** True   **21.** False   **22.** False
23. False   **24.** False   **25.** True

Chapter 2 Test

1. $81x^{12}y^{16}$ **2.** $6a^2 + 19a - 36$ **3.** $\left\{\dfrac{7}{2}, 4\right\}$
4. $(4c - 3)^2$ **5.** 4.78×10^7 **6.** $10a^2 - 12ab + 4b^2$
7. $-\dfrac{b^9}{8a^{12}c^6}$ **8.** $4q^2 - 2qr - 2r^2$ **9.**   True
10. $-36x^8y^3z^8$ **11.** $(a + 6)(a - 13)$
12. $5cd(2c - 3cd + d)$ **13.** $\left\{-\dfrac{1}{2}\right\}$
14. $-20s^2 - 30st + 28t^2$ **15.** $\dfrac{x^4z^6}{9y^4}$ **16.** $\left\{-5, \dfrac{1}{3}\right\}$
17. $(3c + 5)(9c^2 - 15c + 25)$ **18.** $(7y - 5)(x + 3)$
19.   0.0062   **20.** $(5b - 2)(3b + 7)$ **21.** $\left\{\dfrac{5}{3}, 3\right\}$
22. $\left\{\pm\dfrac{9}{2}\right\}$ **23.** $-\dfrac{17a}{y^5}$ **24.** $y^3 - 8y^2 + 11y + 20$
25. 6   **26.** $25a^2 - 16b^2$ **27.** $(11xy - 9)(11xy + 9)$
28. $49m^2 + 56m + 16$ **29.**   6 in. $\times$ 8 in.   **30.**   45 feet
wide; 30 feet deep

CHAPTER 3

Exercise 3.1

1. $x \neq 3$ **3.** $x \neq -6$ **5.** $x \neq -5, 2$ **7.** $x \neq -9, 1$
9. $x \neq -2, 10$ **11.** $x \neq -\dfrac{2}{3}, 1$ **13.** mx **15.**   9
17. $10w$ **19.** $9x + 6$ **21.** $\dfrac{c}{6x}$ **23.** $\dfrac{4y}{5x}$ **25.** $\dfrac{5}{6}$
27. $\dfrac{4}{3b}$ **29.** $6m$ **31.** $12xy$ **33.** $3x + 3$
35. $2x^2 - 2x$ **37.** $\dfrac{3bc}{4a^2}, abc \neq 0$
39. $\dfrac{3}{x - 2}, x \neq 2, 4$ **41.** $\dfrac{x - 1}{4}, x \neq -1$
43. $x - 3, x \neq -3$ **45.** $2m^2 - 3m - 14, m \neq \dfrac{7}{2}, -\dfrac{7}{2}$
47. $-7y - 14, y \neq -2$ **49.** $21x^2 - x - 2, x \neq \dfrac{1}{2}, \dfrac{1}{3}$
51. $2y^3 - y^2 + y + 1, y \neq -1$ **53.** $\dfrac{x + 4}{x + 3}$
55. $\dfrac{3x + 1}{x - 5}$ **57.** $\dfrac{x - 3}{x + 1}$ **59.** $\dfrac{a - 3}{a^2 - 3a + 9}$
61. $\dfrac{x - 3}{3(x^2 - x + 1)}$ **63.** In words   **65.** In words
67. $\dfrac{2x + 3}{x + 3}$ **69.** $\dfrac{x^2 - 2x - 8}{x^2 + 1}$ **71.** $-\dfrac{45}{209}$
73. $-\dfrac{3}{10}$ **75.** $-\dfrac{112}{165}$ **77.** $-\dfrac{2}{3}$

Exercise 3.2

1. $\dfrac{2y}{w}$ **3.** $\dfrac{3x - 3}{7x + 14}$ **5.** $\dfrac{-a}{2a - 6}$ **7.** $\dfrac{3}{4}$ **9.** 3
11. $\dfrac{30}{x}, x \neq 0$ **13.** $\dfrac{28y}{9z}, z \neq 0$ **15.** $\dfrac{2x - 8}{x - 3}, x \neq 3$
17. $\dfrac{t - 4}{30}$ **19.** $\dfrac{1}{2}, x \neq -\dfrac{3y}{2}$ **21.** $\dfrac{2a}{3b}$ **23.** $\dfrac{1}{10}$
25. $\dfrac{1}{6}$ **27.** $-\dfrac{1}{2}$ **29.** $4(x + 1)$ or $4x + 4$
31. $\dfrac{5}{4}; x \neq \pm 2$ **33.** $-\dfrac{8}{y}; y \neq 0, 1, x \neq \dfrac{3}{5}$
35. $\dfrac{2z + 6}{3z - 15}; z \neq 0, 5, -3, 13$ **37.** $\dfrac{2(x - 5)}{x - 3};$
$x \neq \pm 3, 5$ **39.** $3(a + 3)$ or $3a + 9, a \neq 4$ **41.** $\dfrac{5}{8}$
43. $\dfrac{49 - x^2}{5(x + 3)}$ **45.** $\dfrac{3x + 9}{2x - 6}$ **47.** $\dfrac{x + 9}{3x - 1}$

49. $\dfrac{4x^2 + 28x + 49}{4x^2 + 4x - 15}$ **51.** $\dfrac{x^3 - 5x^2 + 4x - 20}{x^2 + 3x + 9}$

53. $\dfrac{3(y^2 + 9)}{(y - 2)(3 - y)(3 + y)}, \; y \neq -3, 0, 2, 3$

55. $x + 3, \; x \neq -7, -5, -2, 0, 2$ **57.** $\dfrac{b + c}{a - b}$

59. $\dfrac{2c + 3d}{c + d}$ **61.** $\dfrac{3x - 2}{x - 4}$ **63.** $8(x^2 - 4) = 8x^2 - 32$

65. In words   **67.** $\dfrac{(x^{2n} + 1)(x^n + 2)}{x^n - 5}$ **69.** $\dfrac{x^n - 2}{x^n - 4}$

71. $\dfrac{47}{36}$ **73.** $\dfrac{5}{63}$ **75.** $-\dfrac{97}{10}$ **77.** $-\dfrac{17}{21}$

Exercise 3.3

1. $5abc$ **3.** $6(y - 2)$ **5.** $\dfrac{4}{x}$ **7.** $\dfrac{3w}{14}$

9. $\dfrac{-a - 4}{6}$ **11.** $\dfrac{1}{6(a + 1)}$ **13.** $\dfrac{b + c}{5abc}$

15. $\dfrac{-3y}{2(y - 2)}$ **17.** $\dfrac{23}{6(y - 1)}$ **19.** $\dfrac{-6x + 9}{3x - 2}$

21. $x - 2$ or $2 - x$ **23.** $3x(x - 1)$ **25.** $\dfrac{6}{w - 4}$

27. $\dfrac{41}{42x}$ **29.** $\dfrac{8t - 1}{5}$ **31.** $\dfrac{5x - 6}{x - 2}$

33. $\dfrac{11x + 21}{(x + 6)(2x - 3)}$ **35.** $\dfrac{x^2 + 11x - 3}{3x(x - 1)}, \; x \neq 0, 1$

37. $\dfrac{-2x^2 - 8x + 21}{(2x - 3)(x + 3)}, \; x \neq \dfrac{3}{2}, -3$

39. $\dfrac{5y + 20}{y(y - 2)(y + 2)}, \; y \neq 0, 2, -2$

41. $(x - 1)(x - 3)(x + 5)$ **43.** $(x + 3)(x - 2)(x + 4)$

45. $\dfrac{-19}{(x + 3)(x - 5)}, \; x \neq -3, 5$

47. $\dfrac{2x^2 - 20}{(x - 2)(x + 2)(x - 5)}, \; x \neq 2, -2, 5$

49. $\dfrac{5x + 1}{(x + 3)(x - 2)(x + 4)}, \; x \neq -3, 2, -4$

51. $\dfrac{-2x^2 + 6x + 12}{(2x + 1)(x - 5)(x + 1)}, \; x \neq -\dfrac{1}{2}, -1, 5$

53. $\dfrac{-x + 1}{x(x + 1)}, \; x \neq 0, -1, -2$

55. $\dfrac{-3y^2 - 3y - 2}{(y + 1)(y - 1)}, \; y \neq \pm 1$

57. $\dfrac{p^2 q^2 - 1}{pq}, \; p \neq 0, q \neq 0$

59. $\dfrac{x^3 + 2x^2 + 1}{x(x + 2)}, \; x \neq 0, -2$ **61.** $\dfrac{2x^2 - 1}{x^2(x + 1)}$

63. $\dfrac{2x}{x^2 - 4}$ **65.** $\dfrac{2x - 10}{(x - 2)(x + 2)(x + 3)}$ **67.** $\dfrac{r_1 + r_2}{r_1 r_2}$

69. In words   **71.** $\dfrac{x^2 + 2x + 3}{(x + 2)(x - 2)}$ **73.** $\{8\}$

75. $\left\{-\dfrac{7}{5}\right\}$ **77.** $\left\{-\dfrac{41}{17}\right\}$ **79.** $\left\{\dfrac{19}{15}\right\}$

Exercise 3.4

1. $\dfrac{y}{2x}$ **3.** $\dfrac{19}{11}$ **5.** $\dfrac{x - 1}{x + 1}$ **7.** $\dfrac{m + 6}{3n}$

9. $\dfrac{3q - 1}{3p + 1}$ **11.** $\dfrac{3w + 9}{2w + 10}$ **13.** $\dfrac{2a - 1}{3a + 1}; \; a \neq 0, -\dfrac{1}{3}$

15. $\dfrac{1 + 2x}{1 - 3x}; \; x \neq 0, \dfrac{1}{3}$ **17.** $\dfrac{x + 6}{x + 1}$ **19.** $\dfrac{x}{x + 5y + 10}$

21. $\dfrac{w^2 - tw}{w^2 + 1}$ **23.** $\dfrac{11x + 8}{x - 1}$ **25.** $\dfrac{20y - 32}{y + 5}$

27. $\dfrac{12a + 2b}{3a - 7b}$ **29.** $\dfrac{2b - 3a}{ab(b - a)}, \; a \neq b, a \neq 0, b \neq 0$

31. $\dfrac{3x + 7}{4x + 9}, \; x \neq -2, \dfrac{-9}{4}$ **33.** $-\dfrac{b}{a}$

35. $\dfrac{y^2 + 4y - 11}{y^2 - 5y + 7}$ **37.** $\dfrac{3a + 1}{6}$ **39.** $\dfrac{2y + 2}{4y + 1}$

41. $\dfrac{x^2 - 4}{-4}$ or $\dfrac{4 - x^2}{4}$ **43.** $\dfrac{-1 - x}{2x^2 - x - 1}$ or

$\dfrac{x + 1}{1 + x - 2x^2}$ **45.** $\dfrac{(12a - 3)(6a + 1)}{(6a - 1)(18a + 5)}$

47. $\dfrac{(x^2 - 2)(x - 3)}{x^2 - 3x + 1(x^2 - 1)}$ **49.** $\dfrac{9}{10}$ **51.** Take the

reciprocal of each side.   **53.** Replace x by a number and

evaluate each side.   **55.** $\dfrac{1}{x^{2n} - 1}$ **57.** $\{-19\}$

59. $\left\{-\dfrac{5}{36}\right\}$ **61.** $\{9\}$ **63.** $\left\{-\dfrac{3}{82}\right\}$

Exercise 3.5

1. $\{-2\}$ **3.** $\{-26\}$ **5.** $\left\{\dfrac{13}{4}\right\}$ **7.** $\{-17\}$

9. $\left\{-\dfrac{13}{3}\right\}$ **11.** $\left\{\dfrac{49}{20}\right\}$ **13.** $\left\{\dfrac{22}{5}\right\}$ **15.** $\{3\}$

17. $\left\{-\dfrac{7}{2}\right\}$ **19.** $\{12\}$ **21.** $\{-10\}$ **23.** $\emptyset$

25. $\{24\}$ **27.** $\left\{-\dfrac{1}{2}\right\}$ **29.** $\left\{-\dfrac{13}{7}\right\}$

31. $c = \dfrac{ab}{a + b}$ **33.** $a = \dfrac{b - 5b^2}{5b + 2}$ **35.** $b = c - 5x$

37. $\{-5\}$ **39.** $\{-2\}$ **41.** $\{-10\}$ **43.** $\left\{-\dfrac{13}{5}\right\}$

45. $\left\{\dfrac{5}{8}\right\}$ **47.** $\left\{\dfrac{3}{10}\right\}$ **49.** $\left\{-\dfrac{2}{7}\right\}$ **51.** $\{16\}$

53. $\left\{-\dfrac{5}{2}\right\}$ **55.** $\{1\}$ **57.** $\{-3\}$ **59.** $\{16\}$

61. Freda, 18 min; Frank, 36 min   **63.** 6 mph
65. Charlie—65 mph; Jim—60 mph   **67.** 2, 10

69. 12 days   **71.** $R = \dfrac{r_1 r_2}{r_1 + r_2}$ **73.** In words

75. $\{9\}$ **77.** $\{16\}$ **79.** $w = \dfrac{x + 4}{a - b}$

81. $\left(-\dfrac{5}{3}, +\infty\right)$ **83.** $2304a^{40}b^{26}$

85. $2b^3 - 14b^2 + 25b - 25$

Exercise 3.6

1. $x^2 + 3x + 2$ **3.** $2a^3 - 3a + 1$ **5.** $2x^2 - 4x - \dfrac{6}{7}$

7. $-10xy + 9y^2 - 7yz$ **9.** $2xy - 3y^2 - \dfrac{4}{x}$

11. $\dfrac{7x^2y^2}{z} + \dfrac{8xz^2}{y^2} + \dfrac{10}{xy^2z}$ **13.** a **15.** a **17.** $3x$

19. $3y^2$ **21.** $x + 3$ **23.** $x + 2$ **25.** $x - 2$

27. $x - 25$ **29.** $2x + 7$ **31.** $x - 10 + \dfrac{17}{x + 2}$

33. $3x + 7$ **35.** $x + 6 + \dfrac{4}{4x - 9}$

37. $x - 3 + \dfrac{19}{2x + 1}$ **39.** $3x - 6 + \dfrac{13}{2x + 3}$

41. $x^2 + 2x + 1$ **43.** $x^2 + 4x + 4$ **45.** $x^2 + 3$

47. $x^4 + x - 3 + \dfrac{27}{x + 3}$ **49.** $2x^2 + x - 1 + \dfrac{2}{3x + 1}$

51. $x^4 - x^3 + x^2 - x + 1$

53. $x^5 - x^4 + x^3 - x^2 + x - 1 + \dfrac{2}{x + 1}$

55. $(7x + 9)(x + 2)$ **57.** $(x + 5)(x^2 + 5x + 11)$
59. No   **61.** $(x + 2)(x + 3)(x + 4)$
63. $(x - 2)(3x + 1)(2x + 3)$
65. $(x + 3)(x - 2)(x + 2)(x^2 + 4)$ **67.** In words

69. $x^n + 3$ **71.** $2x^n + \dfrac{9}{x^n + 3}$ **73.** $k = 29$

75. $216x^3 - 1$ **77.** $16y^4 - 1$
79. $p^6 - 4p^2 + 4p - 1$ **81.** $27y^3 + 54y^2 + 36y + 8$

Exercise 3.7

1. $x + 3$ **3.** $2x - 9$ **5.** $4x - 11$ **7.** $3x + 13$
9. $x^2 + 3x - 5$ **11.** $x^2 - 10x + 36$

13. $x - 5 + \dfrac{17}{x + 2}$ **15.** $x - 12 + \dfrac{57}{x + 4}$

17. $x^2 - 2x + 6 + \dfrac{3}{x - 3}$ **19.** $x^2 - 4x + 1 - \dfrac{4}{x + 5}$

21. $2x^2 - 3x + 1 + \dfrac{4}{x - 1}$ **23.** $2x^2 - 6x + 5$

25. $x^2 - 2x + 1 - \dfrac{9}{x + 2}$ **27.** $x^3 - 5x^2 + 7x + 5$
29. $2x^3 - 5x^2 - 3x + 6$ **31.** $x^5 - 4x^4 - 2x^2 + 6$
33. $x^3 - 2x^2 + 4x - 8$

35. $x^6 + x^5 + x^4 + x^3 + x^2 + x + 1 - \dfrac{1}{x - 1}$ **37.** No

39. No   **41.** No   **43.** $(x + 1)^3$ **45.** Yes   **47.** No
49. Yes   **51.** In words   **53.** $x^{3n} - 4x^{2n} - 14x^n - 15$
55. $x^2 + 9x + 18$ **57.** $(x^2 + 3)(x + 2)(x - 2)$
59. $(x + 5)(x + 1)(x - 1)$ **61.** $(m + n)(x + 3)(x - 3)$

63. $\dfrac{x - 1}{7x}$

Chapter 3 Review Exercises

1. $a \neq 4$ **3.** $x \neq 0$ **5.** None   **7.** $a \neq -2, 3$
9. $x \neq -5, -1$ **11.** $x \neq 2$ **13.** $12x^2$ **15.** $3a + 3b$
17. $3x^2$ **19.** $(2a + b)(a + b)$
21. $(x + y)(x^2 + xy + y^2)$ **23.** $(5x + 2)(2x + 3)$

25. $\dfrac{a^3}{2}$ **27.** $\dfrac{1}{4a^3}$ **29.** $\dfrac{2x + 3}{5x - 4}$ **31.** $\dfrac{7a + 4}{5a + 6b}$

33. $\dfrac{x + 1}{x - 4}$ **35.** $\dfrac{a + b}{a^2 + ab + b^2}$ **37.** $\dfrac{y + 4}{y + 6}$

39. -1 **41.** $\dfrac{x + y}{x - y}$ **43.** $\dfrac{x + 3}{x - b}$ **45.** $\dfrac{2x}{y}$ **47.** $\dfrac{5}{3}$

49. $\dfrac{ab}{4}$ **51.** $-\dfrac{2}{3}$ **53.** $\dfrac{5}{6}$ **55.** $\dfrac{8a + 32}{a(a + 2)}$

57. $\dfrac{5}{a + 4}$ **59.** -1 **61.** $\dfrac{2x + 1}{x + 3}$

63. $\dfrac{x^2 - 6x + 5}{x + 1}$ **65.** 1   **67.** $48abc$ **69.** $2y - 3$
or $3 - 2y$ **71.** $5z(z + 2)$ **73.** $(a + 2)(a - 3)$

75. $(x - 2)(x + 3)(x + 4)$ **77.** $\dfrac{7y^2 + y - 10}{5y(y - 2)}$

79. $\dfrac{-2x^2 - 7x - 11}{(2x + 3)(x - 1)}$ **81.** $\dfrac{4a - 3b}{a(a - b)(a + b)}$

83. $\dfrac{x^2 - x + 6}{(x - 2)(x + 3)(x + 1)}$ **85.** $\dfrac{y^2 + 8y + 44}{(y - 1)(y - 3)(y + 6)}$

87. $-\dfrac{a + 7}{3a}$ **89.** $\dfrac{2x}{(x + y)(x - y)}$ **91.** $\dfrac{6}{7}$ **93.** $\dfrac{1}{x}$

95. $\dfrac{4}{3xy}$ **97.** $\dfrac{x - y}{2}$ **99.** $\dfrac{1}{2c + 3b}$ **101.** $\{3\}$

103. $\{-14\}$ **105.** $\left\{\dfrac{5}{6}\right\}$ **107.** $\left\{-\dfrac{3}{2}, \dfrac{2}{3}\right\}$

109. $\left\{-\dfrac{8}{5}, 1\right\}$ **111.** $5a^2 - 3b - 4$

113. $5x + 3 + \dfrac{2}{x - 2}$ **115.** $2x - 1$

117. $a^2 + 5a + 25$ **119.** $(2a + 5)(a + 3)(a - 2)$

121. $x^2 + 3x - 14 + \dfrac{16}{x + 2}$ **123.** $x^3 - 7x + 6$

125. $x^3 - 5$ **127.** Yes

Chapter 3 True–False Concept Review

1. False   **2.** False   **3.** False   **4.** True   **5.** True
6. False   **7.** True   **8.** False   **9.** True   **10.** True
11. False   **12.** False   **13.** False   **14.** False
15. True   **16.** False   **17.** True   **18.** True
19. True   **20.** True

Chapter 3 Test

1. $\dfrac{5a^3}{6b^2}$ **2.** $4t^2 - 2t$ **3.** $18a^3b^2$ **4.** $\dfrac{3 - 2c}{29 - 3c}$

5. $\dfrac{4x^2 - 14x + 15}{2(x - 1)(3 - 2x)}$ or $\dfrac{-4x^2 + 14x - 15}{2(x - 1)(2x - 3)}$ **6.** $\dfrac{y - 2}{y - 4}$

7. $\left\{-\dfrac{6}{11}\right\}$ **8.** $3x^2 + 4x + 1$ **9.** $-\dfrac{2s^4t^3}{3}$

10. $42x^4y^2$ **11.** $(x + 2)(x + 6)(x - 8)$

12. $\dfrac{y^2 + 7y - 10}{(y + 5)(y - 9)(y + 9)}$ **13.** $\dfrac{3x + 1}{x + 7}$

14. $\dfrac{3x^2 + x + 4}{(x - 1)(x + 1)}$ **15.** $\dfrac{4a^3}{3}$ **16.** $\{-2\}$

17. $\dfrac{c - 6}{2c}$ **18.** $x^3 - 3x + 1 + \dfrac{2}{x + 3}$

19. 29 minutes   **20.** \$12, \$10

Chapters 1–3 Cumulative Review

1. 52   **2.** 22   **3.** -55 **4.** 62   **5.** $\{-4\}$

6. $\left\{\dfrac{11}{2}\right\}$ **7.** $\emptyset$ **8.** $\{0\}$ **9.** $a = 4 - 2b$

10. $x = 5 + 4y$ **11.** $a = x + 3b$ **12.** $b_2 = \dfrac{2A}{h} - b_1$,

or $b_2 = \dfrac{2A - b_1h}{h}$ **13.** $(-4, 2)$ **14.** $[-5, 2]$

15. $\left(-\dfrac{5}{2}, \infty\right)$ **16.** $(-\infty, -22)$ **17.** The number is 8.

18. The number is 48   **19.** 12 in. by 21 in.
20. 3.75 hr   **21.** $8a - 9b + 10$ **22.** $6x^2 + 2x - 8$
23. $-6s^2 - 9t + 16$ **24.** $7a - 4b$ **25.** $-2w - 18$
26. $-10x^2 - 5xy - 15x$ **27.** $16a^2 + 24ab + 32ac$
28. $12y^2 - 5y - 77$ **29.** $2x^2 + 5xy + x + 2y^2 - y - 1$
30. $3t^4 - 33t^3 + 11t^2 + 44t - 20$ **31.** $3x^2 - 2x - 17$
32. $169t^2 - 196m^2$ **33.** $3x^2 - 14x + 8$
34. $2x(4x - 2y - 3)$ **35.** $(x - 7)(x + 2)$
36. $(x^2 + 2)(a - 4)$ **37.** $(2w + 17)(2w - 3)$
38. $(7x - 1)(8x - 15)$ **39.** $3(3w + 2)(9w^2 - 6w + 4)$
40. $4(x + 4)(x + 13)$ **41.** prime polynomial
42. $25(a - 2)(a + 2)(a^2 + 4)$ **43.** $\{-22, 1\}$
44. $\{-0.5, 2\}$ **45.** $\{5, 13\}$ **46.** $\{-8, -4\}$

47. $\dfrac{x - 7}{4}$ **48.** $\dfrac{3a^3}{7}$ **49.** $\dfrac{4x + 1}{x + 1}$ **50.** $-x^2(x - 1)$

51. $\dfrac{-6a + 5b}{4(a - 3)}$ **52.** $\dfrac{2x^2 + 2x + 11}{2(x + 1)}$

53. $\dfrac{2x - 14}{(x + 2)(x + 4)(x - 4)}$ **54.** $\dfrac{-5}{(x - 2)(x + 2)(x - 5)}$

55. $\dfrac{x + 20}{5x^2}$ **56.** $\dfrac{a^2 + 2a + 6}{2a + 1}$ **57.** $\dfrac{4a^2 - 1}{3a^2 + 1}$

58. $\dfrac{2x + 2}{4x + 3}$ **59.** $\left\{\dfrac{31}{8}\right\}$ **60.** $\emptyset$ **61.** $\left\{-\dfrac{4}{3}\right\}$

62. $b = 2a - c$ **63.** $5x - 8y - 12 + \dfrac{16}{xy}$ **64.** $x + 4$

65. $x^2 + x + 1$ **66.** Yes; $(x + 5)(x^2 + 3x - 10)$
67. $3x + 5$ **68.** $x^4 - 2x + 1$

69. $x^2 + 5x + 5 + \dfrac{11}{x - 2}$ **70.** No

CHAPTER 4

Exercise 4.1

1. 12   **3.** 8   **5.** $5^{4/5}$ **7.** $c^{9/10}$ **9.** $x^{1/3}$
11. $3^{1/2}$ **13.** $10^{3/2}$ **15.** $7w^{3/4}$ **17.** x^6 **19.** w^6
21. $15x^{3/4}$ **23.** $36b^{1/8}$ **25.** $4x^{6/7}$ **27.** $4x^{1/6}$
29. $-8d^{1/2}$ **31.** $125p^{1/12}$ **33.** $9x^{1/2}$ **35.** $x^{3/2}y^2$

37. $-64c^{3/4}d^{3/2}$ **39.** $\dfrac{a^{3/2}}{b^{1/2}}$ **41.** $x^{7/6}y^{13/5}$

43. $d^{7/3}e^{11/12}f^4$ **45.** $9a^{6/5}b^{7/4}$ **47.** $\dfrac{x^{3/8}}{2}$ **49.** $\dfrac{x^{1/3}}{y^{4/9}}$

51. $-9x^{11/6}$ **53.** $\dfrac{x^{5/12}}{z^{1/3}}$ **55.** $a - b$

57. $a - 2a^{2/3}b^{2/3} + 3a^{1/3}b^{1/3} - 6b$ **59.** $w - 64$
61. 64   **63.** 13   **65.** 1000   **67.** In words

69. x^{2n} **71.** $\dfrac{x^{1/n}y^{1/n}}{z^2}$ **73.** $x^{1/2}(x + 1)$

75. 0.0000913   **77.** 8.46×10^3 **79.** 6.37×10^6
81. $(x - y - 1)(x + y - 7)$

Exercise 4.2

1. 2   **3.** 8   **5.** $\sqrt{47a}$ **7.** $\sqrt[3]{4w^2}$ **9.** $(xy)^{1/2}$
11. $3(ab)^{1/2}$ **13.** $8\sqrt{s}$ **15.** $2\sqrt[3]{2t}$ **17.** $|w|$

19. $10|m^3|$ **21.** $6\sqrt{x}$ **23.** $\dfrac{2}{\sqrt{a}}$ **25.** $(5xy)^{1/2}$

27. $(x + 1)^{1/2}$ **29.** 19   **31.** 31   **33.** -3
35. Not a real number   **37.** $2|t|$ **39.** $8p^2$ **41.** $11x^2$
43. $3x^2$ **45.** $\sqrt[3]{w^2 + 3}$ **47.** $\sqrt[3]{(a^2 + a)^2}$

49. $\dfrac{1}{\sqrt{p - q}}$ **51.** $m^{4/3}n^{2/3}$ **53.** $5^{1/2}x^{3/2}y^{5/2}$

55. $(7 - x)^{1/2}$ **57.** $|9 - x|$ **59.** $|a - b|$ **61.** 14 m
63. 111 ft   **65.** 45   **67.** In words   **69.** 20 miles per

hour   **71.** 24 sq ft   **73.** $\dfrac{49a^2c^2}{20b^2}$ **75.** $-\dfrac{y + 3}{(y - 3)^2}$

77. $\dfrac{3s}{t}$ **79.** $\dfrac{2q}{5}$

Exercise 4.3

1. $10\sqrt{2}$ **3.** $3\sqrt{10}$ **5.** $3\sqrt{3}$ **7.** $3\sqrt{7}$
9. $2y\sqrt{10}$ **11.** $3p\sqrt{6p}$ **13.** $2t\sqrt{6s}$ **15.** $5\sqrt{ab}$
17. $7x^2$ **19.** $3p\sqrt{2q}$ **21.** 14.142 **23.** 9.487
25. 66.543 **27.** $4x^2\sqrt{2}$ **29.** $5m^6\sqrt{6}$ **31.** $5a\sqrt{5a}$
33. $4y^3\sqrt{5y}$ **35.** $13|y|$ **37.** $2|a^3|\sqrt{7}$
39. $3x^2|y^3|\sqrt{5}$ **41.** $5abc^2\sqrt{7b}$ **43.** $2y\sqrt[3]{2}$
45. $2xy^2\sqrt[3]{x}$ **47.** $2a^2$ **49.** $7\ell^5m^5\sqrt{3m}$
51. $2t^4w^2\sqrt[4]{w}$ **53.** $4c^4|d^5|\sqrt{3}$ **55.** $3x$ **57.** $2|s|t^2$
59. $\sqrt{5w}$ **61.** $\sqrt{5mn}$ **63.** $\sqrt{3c}$
65. a. $8\sqrt{3}$ yd **b.** 13.9 yd **67. a.** $10\sqrt{5}$ cm
b. 22.4 cm **69.** 2.18 m **71.** 3.5 inches
73. In words **75.** $x^{2n}y^{3n}$ **77.** y^{3n} **79.** y^9
81. 88 ft/sec **83.** $6x^2 - 17x + 5$ **85.** $\dfrac{4}{2x - 5}$

87. $\dfrac{14y^2 + 21y + 12}{2y + 3}$ **89.** $\dfrac{1}{3x - 8}$

Exercise 4.4

1. $2\sqrt{3}$ **3.** $3\sqrt{7}$ **5.** $\sqrt{2}$ **7.** $6\sqrt{5}$ **9.** $-3\sqrt{35}$
11. $6\sqrt{2}$ **13.** $-\sqrt{2}$ **15.** $6\sqrt{3}$ **17.** 0
19. $-4\sqrt{5}$ **21.** $-\sqrt{6}$ **23.** $2\sqrt{2x}$ **25.** $9\sqrt{3y}$
27. $2\sqrt{2x}$ **29.** $10\sqrt{x}$ **31.** $8y\sqrt{22y}$ **33.** 0
35. $5x\sqrt[3]{3}$ **37.** $\sqrt[3]{4}$ **39.** $11\sqrt{2}$ **41.** $28x\sqrt{2x}$
43. $28y^2\sqrt{11y}$ **45.** $29y^2\sqrt{7y}$ **47.** $5a\sqrt{5a}$
49. $-5m\sqrt{10n}$ **51.** $39rs\sqrt{2}$ **53.** $4x$ **55.** $3a\sqrt[4]{2}$
57. $x^2\sqrt[4]{x^2}$ **59.** $(q + r - 2)\sqrt[3]{p}$
61. a. $12\sqrt{5}$ in. **b.** 26.8 in. **63.** $16\sqrt{2}$
65. In words **67.** $2\sqrt[3]{2}$ **69.** $12x^2$
71. $\dfrac{17a^2 + 3b^2}{ab}$ **73.** $\dfrac{5x^2 + 2x + 8}{4x(x - 2)}$ **75.** $\dfrac{18}{x + 1}$
77. $\dfrac{x^2 + x + 15}{(x - 5)(x + 4)}$

Exercise 4.5

1. 4 **3.** $2\sqrt{3}$ **5.** $4\sqrt{15}$ **7.** $-12\sqrt{3}$
9. $\sqrt[3]{10x^2}$ **11.** 20 **13.** $5x\sqrt{5}$ **15.** $4a\sqrt{a}$
17. $15\sqrt{15} - 45$ **19.** $11\sqrt{6}$ **21.** $6\sqrt{10y}$
23. $400\sqrt{3}$ **25.** $30s^2\sqrt{2t}$ **27.** $-12\sqrt[3]{6}$
29. $10t^2\sqrt{15t}$ **31.** $2\sqrt{3} + 2$ **33.** 3
35. $10 - 10\sqrt{3}$ **37.** $21 - 21\sqrt{2}$
39. $6\sqrt{2} + 6\sqrt{3} - 6\sqrt{10}$ **41.** $12 + 4\sqrt{5}$ **43.** 4
45. $2x - 7\sqrt{x} - 4$ **47.** $25a - 60\sqrt{a} + 36$
49. $4b - c$ **51.** $\sqrt[6]{12{,}500}$ **53.** $10\sqrt[4]{18}$
55. $18\sqrt[12]{x^7}$ **57.** $2x\sqrt[12]{81x}$ **59.** $2\sqrt{2}$
61. a. $5\sqrt{3}$ ft^2 **b.** 8.7 ft^2 **63. a.** $18\sqrt{3}$ cm^2
b. 31.2 cm^2 **65.** $50\sqrt{2}$ m^3 **67. a.** $\dfrac{35\sqrt{3}}{2}$ dm^2
b. 30.3 dm^2 **69.** In words **71.** x **73.** $2x^2$

75. 2 **77.** $a + b$ **79.** $\dfrac{1}{(a - 2)(a + 5)}$
81. $\dfrac{3y - 1}{4y + 1}$ **83.** $\dfrac{2y^2 + 6y + 1}{-3y - 8}$ **85.** $\left\{-\dfrac{58}{15}\right\}$

Exercise 4.6

1. $\dfrac{2\sqrt{5}}{5}$ **3.** $\dfrac{\sqrt{7}}{7}$ **5.** $\dfrac{\sqrt{2}}{4}$ **7.** $\dfrac{3\sqrt{2}}{2}$ **9.** $-\dfrac{\sqrt{6}}{2}$
11. $\sqrt{7}$ **13.** $\dfrac{\sqrt{6}}{4}$ **15.** $\dfrac{\sqrt{xy}}{y}$ **17.** $\dfrac{\sqrt{14}}{2}$
19. $\dfrac{\sqrt{2}}{2}$ **21.** $\dfrac{\sqrt{2a}}{2}$ **23.** $\dfrac{\sqrt{3}}{3}$ **25.** $\dfrac{\sqrt{10}}{5}$
27. $\sqrt[3]{x}$ **29.** $-\dfrac{4\sqrt{3}}{3}$ **31.** $\dfrac{\sqrt[3]{ab}}{b}$ **33.** $\dfrac{\sqrt[3]{6}}{2}$
35. $\dfrac{\sqrt[4]{12}}{2}$ **37.** $\sqrt{3}$ **39.** $\dfrac{25\sqrt{2}}{12}$ **41.** $\dfrac{3}{\sqrt{6}}$
43. $\dfrac{ab}{\sqrt{3ab}}$ **45.** $\dfrac{3 + 2\sqrt{3}}{3}$ **47.** $\dfrac{7 - 2\sqrt{10}}{3}$
49. $\dfrac{\sqrt{15} + 5 + \sqrt{6} + \sqrt{10}}{2}$ **51.** $\dfrac{2x + 2\sqrt{xy}}{x - y}$
53. $\dfrac{a\sqrt{2} - \sqrt{ab}}{2a - b}$ **55.** $\dfrac{13 - 5\sqrt{5}}{12}$
57. $\dfrac{y - 2\sqrt{y} + 1}{y - 1}$ **59.** $\dfrac{15x - 11\sqrt{xy} + 2y}{25x - 4y}$
61. $\dfrac{1}{2\sqrt{3} - 3}$ **63.** $\dfrac{1}{-1 + \sqrt{3}}$
65. a. $\dfrac{\pi\sqrt{2}}{4}$ sec **b.** 1.1 sec **67. a.** $\dfrac{\pi\sqrt{6}}{12}$ sec
b. 0.6 sec **69.** In words **71.** $2\sqrt[3]{2} + 2 + \sqrt[3]{4}$
73. $\dfrac{2 + \sqrt[3]{4} + \sqrt[3]{2}}{2}$ **75.** $\dfrac{9b + 3}{(1 - b)(1 + b)}$
77. $\dfrac{3b - 2a}{ab(b + a)}$ **79.** $\dfrac{2x + 4}{3x + 4}$ **81.** $\left\{\dfrac{18}{17}\right\}$

Exercise 4.7

1. $\{121\}$ **3.** $\emptyset$ **5.** $\{1\}$ **7.** $\{38\}$ **9.** $\{4\}$
11. $\left\{\dfrac{27}{2}\right\}$ **13.** $\{128\}$ **15.** $\emptyset$ **17.** $\{125\}$
19. $\{-29\}$ **21.** $\left\{\dfrac{1}{2}\right\}$ **23.** $\{7\}$ **25.** $\{-2, -1\}$
27. $\{11\}$ **29.** $\{3\}$ **31.** $\left\{-\dfrac{4}{9}, 4\right\}$ **33.** $\{10\}$
35. $\left\{\dfrac{3}{2}\right\}$ **37.** $\{9\}$ **39.** $\{6\}$ **41.** $\{16\}$ **43.** $\{7\}$
45. $\emptyset$ **47.** $\left\{\dfrac{49}{36}\right\}$ **49.** $\emptyset$ **51.** $\{30\}$ **53.** $\left\{0, \dfrac{9}{4}\right\}$
55. $\{-2\}$ **57.** $\{5, 13\}$ **59.** $\{-4, 9\}$
61. $p = 2.8$ psi **63.** 64 ft^2, 576 ft^2 **65.** $\$2420$

67. In words   **69.** 7 ft   **71.** $y = \pm\sqrt{x^2 + z^2}$
73. $(-4, 7)$ **75.** $(2, 6)$ **77.** $[-4, 1\}$ **79.** $[-1, 1]$

Exercise 4.8

1. $0 + 4i$ **3.** $0 + 2i\sqrt{2}$ **5.** $0 + 3i\sqrt{2}$
7. $2\sqrt{2} + 2i\sqrt{3}$ **9.** $0 + 4i\sqrt{3}$ **11.** i **13.** -1
15. i **17.** $7 + 6i$ **19.** $2 - 3i$ **21.** $6 + 4i$
23. $-1 - 2i$ **25.** $15 + i$ **27.** $7\sqrt{3} + 2i\sqrt{2}$
29. $11 + 2i$ **31.** -28 **33.** $-15 + 6i$ **35.** i
37. $-i$ **39.** 1 **41.** $-i$ **43.** $16 - 2i$
45. $6 - 43i$ **47.** $-5 + 12i$ **49.** $-\dfrac{1}{2}i$ **51.** $\dfrac{3}{5}i$
53. 10 **55.** $\dfrac{1}{5} + \dfrac{2}{5}i$ **57.** $-\dfrac{2}{13} + \dfrac{23}{13}i$
59. $8 + 15i$ **61.** $\dfrac{9}{25} + \dfrac{12}{25}i$ **63.** $-\dfrac{3}{2} - \dfrac{9}{2}i$
65. $-\dfrac{\sqrt{3}}{2} - \dfrac{3}{2}i$ **67.** $182.7 + j193.2$ ohms
69. $39500 + j49400$ ohms   **71.** In words   **73.** 0
75. $8 + 6i$ **77.** $(x - 5i)(x + 5i)$ **79.** Yes   **81.** $\dfrac{1}{x}$
83. $y = -\dfrac{7}{2}$ **85.** $2x^2 - 2x + 4 - \dfrac{10}{2x + 3}$
87. 13 grandchildren

Chapter 4 Review Exercises

1. 5 **3.** 1 **5.** 0.9 **7.** $m^{9/5}$ **9.** $x^{13/10}$
11. $\dfrac{1}{p^{5/6}}$ **13.** $16t^{1/8}$ **15.** $-18y^{9/8}$ **17.** $16xy^5$
19. $6a^{3/2}b^{1/4}$ **21.** $4x - x^{2/3}$ **23.** $3\sqrt[3]{12}$ **25.** $2\sqrt[3]{y}$
27. $\dfrac{\sqrt{y}}{3y}$ **29.** $4x\sqrt[3]{x}$ **31.** $4w^{1/2}$ **33.** $6(6s)^{1/2}$ or
$6^{3/2}s^{1/2}$ **35.** $25^{1/3}x^{2/3}y^{4/3}$ **37.** $14a^2$ **39.** $12xy^8$
41. $6a^2$ **43.** $9|m|$ **45.** $11s^2|t^3|$ **47.** $|x - 3|$
49. $10\sqrt{5}$ **51.** $2\sqrt{30}$ **53.** $5\sqrt{7}$ **55.** $10|a|\sqrt{2}$
57. $10|xy|\sqrt{3}$ **59.** $2|m|\sqrt[4]{2n^2}$ **61.** 22.3607
63. 10.9545 **65.** 13.2288 **67.** $\sqrt[4]{10w}$ **69.** $\sqrt{2a}$
71. $31\sqrt{2}$ **73.** $185\sqrt{5}$ **75.** $19\sqrt{5}$ **77.** $17\sqrt{5y}$
79. $3\sqrt[3]{5mn}$ **81.** $3 + 3\sqrt{2} + 3\sqrt{3}$ **83.** $\dfrac{3}{2}\sqrt{3}$
85. $2\sqrt[3]{3}$ **87.** $\dfrac{3}{2}\sqrt[3]{2}$ **89.** 15 **91.** 14
93. $5\sqrt[3]{2y}$ **95.** $40 - 6\sqrt{6}$
97. $wy\sqrt{w} + w^2y - wy\sqrt{wy}$ **99.** $12 - 4\sqrt{5}$
101. $35 - 12\sqrt{6}$ **103.** $y\sqrt{3} - 4\sqrt{3}$ **105.** 0
107. $\sqrt{6}$ **109.** $7\sqrt{2}$ **111.** $2\sqrt{2}$ **113.** $2\sqrt{6}$
115. $\dfrac{\sqrt{33}}{6}$ **117.** $\dfrac{4\sqrt{3}}{3}$ **119.** $\dfrac{8\sqrt{6}}{45}$ **121.** $\dfrac{3}{5}$
123. $-\dfrac{3\sqrt[4]{2}}{2}$ **125.** $\dfrac{20 + 5\sqrt{5}}{11}$ **127.** $3 + \sqrt{6}$

129. $\sqrt{5} - 2 - \sqrt{30} + 2\sqrt{6}$ **131.** $\dfrac{11 - 4\sqrt{7}}{3}$
133. $\dfrac{\sqrt{x} - \sqrt{y}}{x - y}$ **135.** $\dfrac{x^3 + 2x\sqrt{xy} + y}{x^3 - y}$ **137.** $\{144\}$
139. $\emptyset$ **141.** $\{-5, -4\}$ **143.** $\emptyset$ **145.** $\left\{\dfrac{64}{9}\right\}$
147. $0 + 22i$ **149.** i **151.** $-i$ **153.** $17 + 15i$
155. $-1 - 5i$ **157.** $11 - i\sqrt{2}$ **159.** $22 + 46i$
161. $54 + 0i$ **163.** $-64 + 120i$ **165.** $0 - 3i$
167. $\dfrac{6}{5} - \dfrac{3}{5}i$ **169.** $\dfrac{3}{4} + \dfrac{5}{4}i$ **171.** $\dfrac{7}{10} + \dfrac{19}{10}i$

Chapter 4 True–False Concept Review

1. False   **2.** False   **3.** True   **4.** True   **5.** False
6. False   **7.** False   **8.** True   **9.** True   **10.** False
11. False   **12.** True   **13.** True   **14.** True
15. True   **16.** True   **17.** True   **18.** True
19. True   **20.** False   **21.** False   **22.** True
23. False   **24.** True   **25.** True

Chapter 4 Test

1. $\dfrac{6\sqrt{2} - \sqrt{6}}{6}$ **2.** $6|a^5|b^4\sqrt{3}$ **3.** $2 - 66i$
4. $\dfrac{x^{1/2}}{z^{7/16}}$ **5.** $9x^4$ **6.** $\dfrac{81}{58} + \dfrac{15}{58}i$ **7.** $2\sqrt[3]{yz^2}$
8. 15.5 **9.** $\dfrac{a^{1/3}}{b^{3/8}}$ **10.** $-\sqrt{6}$ **11.** $\{-62\}$
12. $10b - b\sqrt{b} - 21b^2$ **13.** $16\sqrt{y}$ **14.** $11^{1/5}x^{2/5}y^{4/5}$
15. $11a^2\sqrt{2a}$ **16.** $-\dfrac{33}{65} - \dfrac{9}{65}i$ **17.** $77 + 2i$
18. $18\sqrt{2}$ **19.** $12\sqrt{2} + 6\sqrt{5}$ **20.** $\{25, 4\}$
21. $\dfrac{2c + 7\sqrt{c} - 15}{25 - c}$ **22.** $-3 - 13i$ **23.** $-y\sqrt{5}$
24. $-15 - 112i$ **25.** i **26.** $88 - 22\sqrt{15}$
27. 1950 plants   **28.** \$279.84

CHAPTER 5

Exercise 5.1

1. $\{\pm 1\}$ **3.** $\{\pm 0.4\}$ **5.** $\left\{\pm\dfrac{2}{3}\right\}$ **7.** $\{\pm 2\sqrt{7}\}$
9. $\{-2, 0\}$ **11.** $\{-1, 3\}$ **13.** $\{-2, 12\}$ **15.** $\{-2, 4\}$
17. $\{-3, 1\}$ **19.** $\{3, 4\}$ **21.** $\{-5 \pm i\}$
23. $\{-6 \pm i\sqrt{5}\}$ **25.** $\{-3 \pm \sqrt{2}\}$ **27.** $\{2 \pm 4\sqrt{3}\}$
29. $\{6 \pm 3i\sqrt{2}\}$ **31.** $\{-6 \pm 4\sqrt{2}\}$ **33.** $\left\{\dfrac{1 \pm \sqrt{5}}{2}\right\}$
35. $\left\{\dfrac{3}{2} \pm \dfrac{\sqrt{11}}{2}i\right\}$ **37.** $\left\{\dfrac{7 \pm \sqrt{57}}{2}\right\}$

39. $\left\{\dfrac{3 \pm \sqrt{89}}{4}\right\}$ **41.** $\{4, -1\}$ **43.** $\{-16 \pm 8i\}$

45. $\left\{\dfrac{2}{3} \pm 3i\right\}$ **47.** $\left\{\dfrac{7 \pm 2\sqrt{10}}{2}\right\}$ **49.** $\left\{\dfrac{5}{2} \pm i\sqrt{6}\right\}$

51. $\left\{-\dfrac{1}{2} \pm \dfrac{\sqrt{6}}{2}i\right\}$ **53.** $\left\{\dfrac{-1 \pm \sqrt{19}}{3}\right\}$

55. $\left\{\dfrac{4 \pm \sqrt{10}}{3}\right\}$ **57.** $\left\{-\dfrac{1}{4} \pm \dfrac{\sqrt{15}}{12}i\right\}$

59. $\left\{\dfrac{3 \pm \sqrt{19}}{5}\right\}$ **61.** $\left\{\dfrac{2}{3} \pm \dfrac{\sqrt{11}}{3}i\right\}$ **63.** $3\sqrt{2}$ ft

65. 5   **67.** 12   **69.** 12 ft   **71.** 20 ft
73. 12 stations   **75.** 16 sides   **77.** 12 and 9
79. $3 + i, 3 - i$ **81.** $c = \pm\sqrt{a^2 + b^2}$
83. $b = \pm\sqrt{c^2 - a^2}$ **85.** $x = \pm4\sqrt{c}$ **87.** $x = 3 \pm a$
89. $r = \pm\sqrt{\dfrac{V}{\pi h}}$ **91.** $-13, 1$ **93.** In words
95. 20 amperes   **97.** $\{-2a \pm \sqrt{4a^2 - a}\}$
99. $\left\{\dfrac{-b \pm \sqrt{b^2 - 4ac}}{2a}\right\}$ **101.** $a^{2/3}b^{1/3}$
103. $(3x^2 + 4)^{1/2}$ **105.** $2c^2d^3\sqrt{41d}$ **107.** $2x^2y^3\sqrt[3]{45}$

Exercise 5.2

1. $\{-7, -2\}$ **3.** $\{-8, 3\}$ **5.** $\{-4, 9\}$ **7.** $\{-5 \pm 2i\}$
9. $\{12, 8\}$ **11.** $\left\{\dfrac{-7 \pm \sqrt{13}}{6}\right\}$ **13.** $\left\{-\dfrac{3}{2}, -\dfrac{1}{2}\right\}$
15. $\left\{\dfrac{1}{5}, 3\right\}$ **17.** $\{-2 \pm \sqrt{2}\}$ **19.** $\{2 \pm i\}$

21. $\{2 \pm \sqrt{11}\}$ **23.** $\{-4 \pm 3\sqrt{2}\}$ **25.** $\left\{\dfrac{5}{2} \pm \dfrac{\sqrt{3}}{2}i\right\}$

27. $\left\{\dfrac{9 \pm \sqrt{65}}{2}\right\}$ **29.** $\left\{\dfrac{3 \pm \sqrt{65}}{4}\right\}$

31. $\left\{\dfrac{2}{3} \pm \dfrac{\sqrt{2}}{3}i\right\}$ **33.** $x = \dfrac{3a}{2}$ or $x = \dfrac{5a}{3}$

35. $x = \dfrac{(3 \pm \sqrt{5})b}{2}$ **37.** $\left\{-2, -\dfrac{1}{3}\right\}$ **39.** $\{-2i, 4i\}$

41. $\left\{-\dfrac{5}{2}, -\dfrac{2}{3}\right\}$ **43.** $\left\{\dfrac{9}{8} \pm \dfrac{\sqrt{7}}{8}i\right\}$ **45.** $\left\{-\dfrac{5}{2}, \dfrac{1}{3}\right\}$

47. $\left\{-\dfrac{4}{5}, \dfrac{5}{2}\right\}$ **49.** $\left\{\dfrac{-5 \pm \sqrt{73}}{4}\right\}$

51. $\left\{-\dfrac{9}{10} \pm \dfrac{\sqrt{39}}{10}i\right\}$ **53.** $x = \dfrac{3ab \pm b\sqrt{9a^2 - 8a}}{2a}$

55. $x = \dfrac{3cd \pm \sqrt{9c^2d^2 - 16d}}{4}$ **57.** $\left\{-2\sqrt{2}, \dfrac{3}{2}\sqrt{2}\right\}$

59. $\{-8i, i\}$ **61.** $\left\{-\dfrac{3}{2}i, \dfrac{1}{3}i\right\}$ **63.** 4 in.   **65.** 10%

67. 2 sec   **69.** 5 or -16 **71.** In words

73. 324 feet   **75.** $b^{5/24}$ **77.** $\dfrac{1}{p^{5/8}q^{1/2}}$

79. $b^{4/5} - 225$ **81.** $\dfrac{(a - 4c)(a - 2c)}{a(a - 3c)}$

Exercise 5.3

1. 36, two real roots   **3.** -27, two complex roots
5. 0, two equal real roots   **7.** -135, two complex roots
9. 45, two real roots   **11.** $x^2 + 8x + 7 = 0$
13. $x^2 - 2x - 35 = 0$ **15.** $x^2 + 1 = 0$
17. $\{-4 \pm 4\sqrt{2}\}$ **19.** $\{-6 \pm 6\sqrt{2}\}$ **21.** -68, two complex roots   **23.** 89, two real roots   **25.** -24, two complex roots   **27.** -311, two complex roots
29. 401, two real roots   **31.** $14x^2 + 11x - 15 = 0$
33. $x^2 - 2x - 4 = 0$ **35.** $x^2 - 4x + 5 = 0$
37. $\left\{1 \pm \dfrac{\sqrt{5}}{5}i\right\}$ **39.** $\left\{-\dfrac{5}{4}, \dfrac{7}{3}\right\}$ **41.** -279, two complex roots   **43.** 76, two real roots   **45.** 0, two equal real roots   **47.** 1368, two real roots   **49.** 1216, two real roots   **51.** $x^2 - 12x + 31 = 0$
53. $9x^2 - 12x - 1 = 0$ **55.** $9x^2 + 30x + 29 = 0$
57. $x^2 + 6x + 29 = 0$ **59.** $\{-10, 2\}$ **61.** 21 or more units, 10 employees   **63.** $\pm2\sqrt{21}$
65. $\left[\dfrac{-b + \sqrt{b^2 - 4ac}}{2a}\right] \cdot \left[\dfrac{-b - \sqrt{b^2 - 4ac}}{2a}\right] =$ $\dfrac{(-b)^2 - (\sqrt{b^2 - 4ac})^2}{4a^2} = \dfrac{b^2 - b^2 + 4ac}{4a^2} = \dfrac{4ac}{4a^2} = \dfrac{c}{a}$
67. $b = \pm4\sqrt{6}$ **69.** $b = \pm8\sqrt{3}$ **71.** $b = -7$, $r_2 = \dfrac{5}{4}$ **73.** 1   **75.** 5   **77.** $\sqrt[5]{(a^2 - 3)^3}$
79. $27a^2b^4\sqrt{b}$ **81.** $64ab^2$ **83.** $-\sqrt{5}$

Exercise 5.4

1. $\{\pm3\}$ **3.** $\{\pm18\}$ **5.** $\{-1, 9\}$ **7.** $\emptyset$ **9.** $\{-5\}$
11. $\{-33, 7\}$ **13.** $\{-11, 29\}$ **15.** $\emptyset$ **17.** $\{-7, 17\}$
19. $\{-16, -14\}$ **21.** $\{-16, 10\}$ **23.** $\{-9, 31\}$

25. $\{-5, 1\}$ **27.** $\{-8, 19\}$ **29.** $\emptyset$ **31.** $\left\{\dfrac{14}{5}\right\}$

33. $\left\{-\dfrac{17}{2}, \dfrac{3}{2}\right\}$ **35.** $\{-1, 2\}$ **37.** $\{-1, 7\}$

39. $\{3, 15\}$ **41.** $\{3\}$ **43.** $\emptyset$ **45.** $\{4\}$ **47.** $\{13\}$

49. $\{-9, 1\}$ **51.** $\emptyset$ **53.** $\left\{-\dfrac{11}{10}, \dfrac{15}{4}\right\}$ **55.** $\emptyset$

57. $\{-2, 4\}$ **59.** $\{1, 3\}$ **61.** Maximum, 7200 crowbars; minimum, 4800 crowbars   **63.** Highest $45, lowest $31   **65.** In words   **67.** $\{1\}$ **69.** $\{x \mid x \le -5\}$
71. $3(x - 3)(x + 5)$ **73.** $(x + 2)(x - 4)^2(x + 4)$
75. $(x - 2)^2(x + 2)^2$ **77.** $(x + 3)(x + 5)(2x - 3)$

Exercise 5.5

1. $\{\pm\sqrt{10}\}$ **3.** $\{\pm1\}$ **5.** $\{-1, 2\}$ **7.** $\{-3, 1\}$
9. $\{-7, 6\}$ **11.** $\{9, -7\}$ **13.** $\{\pm2, \pm2i\}$
15. $\{\pm\sqrt{3}, \pm i\sqrt{5}\}$ **17.** $\{\pm2\sqrt{3}, \pm\sqrt{10}\}$

19. $\left\{\pm2, \pm\dfrac{3}{2}\right\}$ **21.** $\left\{-\dfrac{1}{2}, \dfrac{1}{2}\right\}$ **23.** $\{3, 10\}$

25. $\left\{-1, -\dfrac{1}{3}\right\}$ **27.** $\left\{-\dfrac{1}{2}, 2\right\}$ **29.** $\{4\}$ **31.** $\{2\}$

33. $\left\{-\dfrac{2}{5}, 1\right\}$ **35.** $\left\{\dfrac{7 \pm \sqrt{37}}{4}\right\}$ **37.** $\{\pm 2, \pm\sqrt{2}\}$

39. $\{\pm\sqrt{17}i, \pm 2i\}$ **41.** $\left\{\pm\dfrac{\sqrt{22}}{2}, \pm\dfrac{1}{2}i\right\}$

43. $\{-3, 1, 5\}$ **45.** $\{3, 7, 5 \pm \sqrt{35}\}$ **47.** $\{0, 8\}$
49. $\{1 \pm \sqrt{11}\}$ **51.** $\{5\}$ **53.** $\{-6 \pm 4\sqrt{3}\}$
55. $\left\{\dfrac{-20 \pm \sqrt{445}}{5}\right\}$ **57.** $\left\{-\dfrac{1}{5}, -\dfrac{1}{6}\right\}$

59. $\left\{-\dfrac{1}{7}, \dfrac{1}{12}\right\}$ **61.** $\{81\}$ **63.** $\{121\}$

65. $\left\{\pm\dfrac{1}{2}, \pm\dfrac{\sqrt{6}}{6}i\right\}$ **67.** $\left\{\pm\dfrac{1}{3}, \pm\dfrac{1}{2}\right\}$

69. $\left\{\pm\dfrac{1}{2}i, \pm i\right\}$ **71.** $5, 4 **73.** Jane $5, Sally $6

75. 8 m, 64 m   **77.** In words   **79.** 240 mph
81. $126\sqrt{6}$ **83.** $5 - 2\sqrt{6}$ **85.** $6\sqrt{5} - 18$
87. $120s^3\sqrt{6}$

Exercise 5.6

1.

3.

5. $\{x \mid -5 \le x \le 5\}$ **7.** $\{x \mid x \le -3 \text{ or } x \ge 3\}$
9. $\{x \mid -4 < x < 4\}$ **11.** $(-\infty, 1] \cup [4, \infty)$
13. $(-9, 5)$ **15.** $\emptyset$ **17.** $(-\infty, \infty)$ **19.** $\emptyset$
21. $[-4, 1]$ **23.** $(-\infty, 0] \cup [4, \infty)$ **25.** $\left(-\dfrac{7}{3}, -1\right)$
27. $\left(-\infty, \dfrac{3}{5}\right) \cup (1, \infty)$ **29.** $(-\infty, \infty)$ **31.** $\left(-\dfrac{5}{2}, \infty\right)$

33. $\{x \mid x \in R\}$ **35.** $\left\{x \mid x > \dfrac{6}{7}\right\}$

37. $\{x \mid x \le 0\} \cup \{x \mid x \ge 8\}$ **39.** $\{x \mid -2 < x < 10\}$
41. $\{x \mid -2 \le x \le 8\}$ **43.** $\{x \mid x < 1\}$

45. $\{x \mid x < -1\} \cup \left\{x \mid x > -\dfrac{1}{3}\right\}$ **47.** $\emptyset$

49. $\{x \mid 1 < x < 2\}$ **51.** $\left\{a \mid a \ge \dfrac{9}{10}\right\} \cup \left\{a \mid a \le \dfrac{7}{10}\right\}$

53. $\{x \mid x > 6\} \cup \{x \mid x < -2\}$ **55.** $\left\{x \mid -\dfrac{3}{2} \le x \le \dfrac{15}{2}\right\}$

57. $(-\infty, +\infty)$ **59.** $\{x \mid -2 < x < -1\}$

61. $|x - 4| < 12, (-8, 16)$
63. $|x - 16| > 9, (-\infty, 7) \cup (25, \infty)$
65. $|x - 18| \ge 8, (-\infty, 10] \cup [26, \infty)$
67. $|x - 12| \le 7, [5, 19]$ **69.** $-2°F$ to $14°F$
71. 113 lb to 117 lb   **73.** $|x + 7| \le 5$
75. $|4x + 2| \ge 8$ **77.** $|2x + 3| \le 5$ **79.** In words
81. $\{x \mid -a - b < x < b - a\}$
83. $\{x \mid x > b - a\} \cup \{x \mid x < -a - b\}$ **85.** $\{x \mid x \in R\}$

87. $\{x \mid x = a\}$ **89.** $(3b + 8)(7a - 4b)$
91. $(11m + 6n)^2$ **93.** $(y^2 - 3)(y^2 + 3)(y^4 + 9)$
95. 4 hr

Exercise 5.7

1. $(-\infty, -2) \cup (5, \infty)$ **3.** $[-3, 4]$
5. $\left(-\infty, -\dfrac{3}{5}\right] \cup \left(\dfrac{1}{2}, \infty\right)$ **7.** $\left(-\dfrac{1}{2}, 2\right)$
9. $(-\infty, -2] \cup [0, 3]$ **11.** $(-\infty, -4] \cup [-1, 2]$
13. $\left[-\dfrac{3}{2}, 0\right] \cup (5, \infty)$ **15.** $\left(-\infty, -\dfrac{2}{3}\right] \cup \left[\dfrac{1}{2}, \dfrac{5}{4}\right]$
17. $(-\infty, -2) \cup (0, 3)$ **19.** $[-1, 0] \cup [2, 3]$
21. $(-\infty, -2) \cup \left(-\dfrac{3}{2}, \infty\right)$ **23.** $(-\infty, 0) \cup \left(\dfrac{5}{2}, \infty\right)$
25. $(-\infty, -1] \cup (0, \infty)$ **27.** $\left[-\dfrac{1}{2}, 3\right]$ **29.** $\emptyset$
31. $(-\infty, \infty)$ **33.** $\left\{-\dfrac{5}{4}\right\}$ **35.** $\left(-\infty, -\dfrac{1}{2}\right] \cup [0, 2]$
37. $(-6, -5)$ **39.** $(2, 3)$ **41.** $\emptyset$ **43.** $(-1, \infty)$
45. $\left(-\infty, \dfrac{1}{2}\right) \cup (1, 2)$ **47.** $[-7, -2) \cup (3, \infty)$
49. $(-\infty, -6) \cup (5, \infty)$
51. $(-\infty, -4) \cup (-3, -2) \cup (0, \infty)$
53. $(-\infty, -4) \cup \left[-3, \dfrac{1}{2}\right] \cup (1, \infty)$
55. $(-\infty, -1) \cup (1, \infty)$ **57.** $(-\infty, \infty)$
59. $(-\infty, 1) \cup (1, \infty)$ **61.** 0 units to 80 units
63. $b \le -8$ or $b \ge 8$ **65.** In words   **67.** $\{x \mid x \le 2\}$
69. $\{x \mid x < -1\} \cup \{x \mid -1 < x < 1\}$ **71.** 4   **73.** 9
75. $\dfrac{2\sqrt{5}}{5}$ **77.** $-5 + 5\sqrt{2}$

Chapter 5 Review Exercises

1. $\{\pm 16\}$ **3.** $\{-10, 2\}$ **5.** $\{\pm 9i\}$ **7.** $\{4 \pm 4\sqrt{3}\}$
9. $\{6 \pm 3i\}$ **11.** $\{-6, 2\}$ **13.** $\{-4, 9\}$
15. $\{-1 \pm \sqrt{3}\}$ **17.** $\left\{-3, \dfrac{5}{3}\right\}$ **19.** $\{-1 \pm \sqrt{7}i\}$
21. $\left\{-3, \dfrac{2}{5}\right\}$ **23.** $\left\{-\dfrac{7}{5}, \dfrac{5}{7}\right\}$ **25.** $\left\{\dfrac{3 \pm 2\sqrt{6}}{3}\right\}$
27. $\{-1 \pm \sqrt{3}i\}$ **29.** $\{-4i, -3i\}$ **31.** 2 rational
33. 2 rational   **35.** 2 equal rational   **37.** 2 irrational
39. 2 complex   **41.** $x^2 - 3x + 2 = 0$
43. $x^2 + 6x + 8 = 0$ **45.** $20x^2 + 31x + 12 = 0$
47. $x^2 - 4x - 14 = 0$ **49.** $x^2 + 2x + 10 = 0$
51. Yes   **53.** No   **55.** Yes   **57.** Yes   **59.** Yes
61. $\{\pm 8\}$ **63.** $\{-4, 2\}$ **65.** $\emptyset$ **67.** $\{7\}$
69. $\{-1, 2\}$ **71.** $\left\{-\dfrac{3}{5}, 2\right\}$ **73.** $\left\{-\dfrac{5}{4}, 2\right\}$
75. $\{3, 10\}$ **77.** $\{-1, 3\}$ **79.** $\emptyset$ **81.** $\{\pm 2, \pm 3\}$
83. $\{-4, -3\}$ **85.** $\{-5, -2, -1, 2\}$ **87.** $\left\{-\dfrac{1}{6}, 1\right\}$

89. $\left\{\dfrac{4}{9}, 9\right\}$ **91.** $(-50, 50)$ **93.** $(-\infty, -8] \cup [2, +\infty)$

95. $\emptyset$ **97.** $(-\infty, +\infty)$ **99.** $\emptyset$

101. $\{x \mid -8 < x < 2\}$ **103.** $\{x \mid x \in R\}$

105. $\{x \mid -5 \le x \le 6\}$ **107.** $\{x \mid 2 < x < 3\}$

109. $\left\{x \mid -\dfrac{3}{4} \le x \le \dfrac{4}{3}\right\}$ **111.** $(-\infty, 2) \cup [3, +\infty)$

113. $(-\infty, 3) \cup \left(\dfrac{7}{2}, +\infty\right)$ **115.** $(-\infty, -3) \cup [-2, 2]$

117. $(-2, 0)$ **119.** $(-2, +\infty)$

Chapter 5 True–False Concept Review

1. True   **2.** False   **3.** True   **4.** True   **5.** False
6. False   **7.** True   **8.** True   **9.** False   **10.** False
11. False   **12.** True   **13.** True   **14.** True
15. False   **16.** False   **17.** True

Chapter 5 Test

1. $\{-3, -13\}$ **2.** $x^2 - 4x - 165 = 0$
3. $(-\infty, 0) \cup [3, \infty)$

4. $\{-9 \pm \sqrt{11}\}$
5. $x \le -2$ or $x \ge 0$

6. $\{\pm 2\sqrt{3}\}$ **7.** $\{7 \pm \sqrt{7}\}$ **8.** $4x^2 - 8x - 5 = 0$
9. $\left\{-4, \dfrac{3}{4}\right\}$ **10.** $\emptyset$ **11.** $\left\{-\dfrac{5}{2}, \dfrac{1}{6}\right\}$ **12.** $\{4, 30\}$

13. $\{-22, 12\}$ **14.** $\{2\}$ **15.** $\left\{\dfrac{7}{2}\right\}$

16. $\left\{x \mid -3 < x < \dfrac{3}{2}\right\}$

17. 0, two equal real roots   **18.** 17 and 19 or -17 and -19 **19.** 45 minutes

CHAPTER 6

Exercise 6.1

1. $(0, 5)$ **3.** $\left(\dfrac{5}{2}, 0\right)$ **5.** $(0, 4]$ **7.** $\left(-\dfrac{7}{2}, -3\right)$
9. $(-4, -3)$ **11.** $(-2, 0)$ **13.** $(0, -5)$ **15.** $(-2, -10)$

17.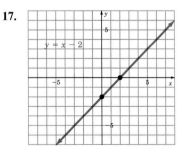

19.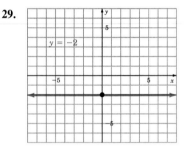

21. $(0.1, -3.5)$ **23.** $(4, 6.25)$ **25.** $\left(\dfrac{1}{2}, \dfrac{25}{18}\right)$
27. $\left(-3, \dfrac{10}{3}\right)$

29.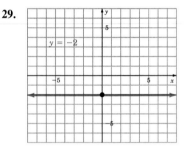

31.

33.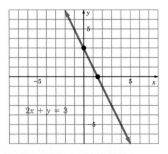

35.

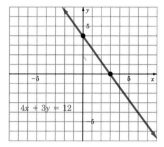

$4x + 3y = 12$

45.

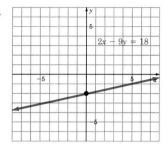

$2x - 9y = 18$

37.

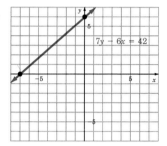

$7y - 6x = 42$

47.

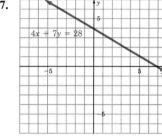

$4x + 7y = 28$

39.

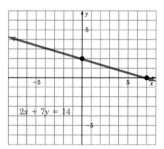

$2x + 7y = 14$

49.

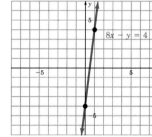

$8x - y = 4$

41.

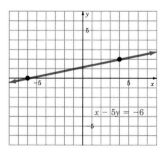

$x - 5y = -6$

51.

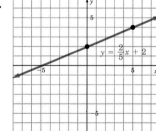

$y = \frac{2}{5}x + 2$

43.

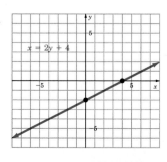

$x = 2y + 4$

53.

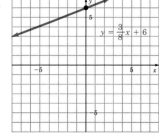

$y = \frac{3}{8}x + 6$

55.

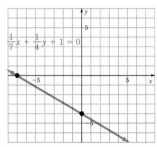

$\frac{1}{5}x - y = 2$

57.

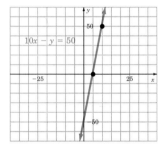

$\frac{1}{7}x + \frac{1}{4}y + 1 = 0$

59.

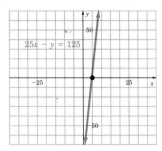

$10x - y = 50$

61.

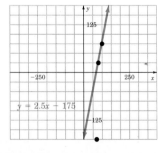

$25x - y = 125$

63.

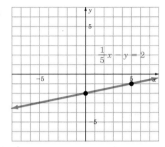

$y = 2.5x - 175$

65.

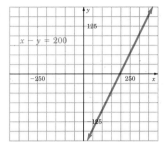

$x - y = 200$

67.

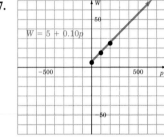

$W = 5 + 0.10p$

69.

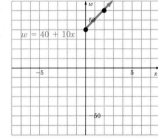

$w = 40 + 10x$

71.

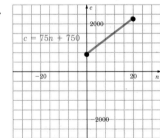

$c = 75n + 750$

73.

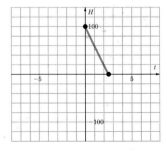

75.

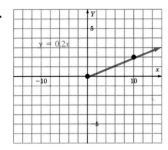

77. In words   **79.** Answers will vary

81.

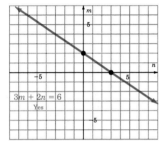

83. $\dfrac{x^2 - 23}{x^2 - 9x + 18}$ **85.** $\left\{\pm\dfrac{15}{2}\right\}$ **87.** $\{-9, 2\}$

89. $\left\{-\dfrac{2}{3}, 6\right\}$

Exercise 6.2

1. $\dfrac{1}{2}$ **3.** $\dfrac{3}{2}$ **5.** -1 **7.** $(7, 0)(0, -3)$

9. $(14, 0)(0, -2)$ **11.** No   **13.** Yes   **15.** 2

17. $\sqrt{13}$ **19.** $2\sqrt{2}$ **21.** $\left(\dfrac{11}{2}, 0\right), \left(0, -\dfrac{11}{3}\right)$

23. $(9, 0), \left(0, -\dfrac{3}{2}\right)$ **25.** $(0, 4), (3, 0)$ **27.** $d = \sqrt{221}$

29. $m = \dfrac{3}{5}, d = \sqrt{34}$ **31.** $m = 0$, distance $= 3$

33. Undefined slope, distance $= 6$ **35.** Yes, since

$(1)(-1) = -1$ **37.** No, since $\left(\dfrac{2}{7}\right)\left(-\dfrac{2}{7}\right) \neq -1$

39. Parallel   **41.** $(-9, 0), (0, 12)$ **43.** No x-intercept

45. $m = -\dfrac{2}{3}, d = \dfrac{5}{6}\sqrt{13}$ **47.** $m = -7, d = \dfrac{25}{14}\sqrt{2}$

49. $m = -0.25$, distance $= \dfrac{\sqrt{153}}{20}$ **51.** $m = -\dfrac{1}{3}$,

distance $= \sqrt{0.1} = \dfrac{\sqrt{10}}{10}$ **53.** Perpendicular

55. Parallel   **57.** Parallel   **59.** Perpendicular

61. Rate of growth is 10, seventh-year enrollment is 110

63. One pair of sides has slope 4, and the second pair of sides has slope $-\dfrac{7}{2}$.   **65.** The sides have lengths $3\sqrt{2}$, $3\sqrt{2}$, 6   **67.** The sides have lengths 3, 4, 5

69. $m_{AC} = 1, m_{BD} = -1$ since $(1)(-1) = -1 \rightarrow AC \perp BD$

71. $a^2 + b^2 \neq c^2$ $(\sqrt{74})^2 + (\sqrt{74})^2 \neq (10)^2$

73. In words.   **75.** The slopes are reciprocals.

77. 470   **79.** $2x^2y^4\sqrt{30x}$ **81.** $2a^2b^3\sqrt{20ab}$

83. $-9st\sqrt{3st}$ **85.** $\{\pm 3\sqrt{6}\}$

Exercise 6.3

1. $4x - y = 6$ **3.** $2x + y = 6$ **5.** $x + 3y = 17$

7. $0x + y = -4$ or $y = -4$ **9.** $m = \dfrac{3}{7}, (0, -5)$

11. $m = -\dfrac{1}{9}, (0, 4)$

13.

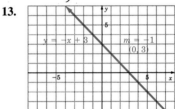

15.

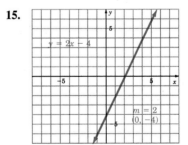

17.

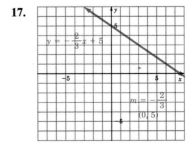

19.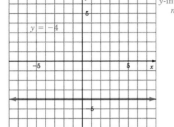

y-intercept $= -4$
$m = 0$

39.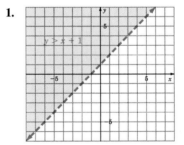

no y intercept
undefined slope

21. $3x + y = -19$ **23.** $4x + 5y = -5$
25. $x - y = 11$ **27.** $3x + 8y = 27$ **29.** $y = -5$
31. $5x - 9y = 83$

33.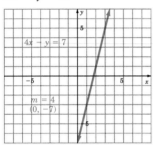

41. $y = -\dfrac{7}{12}x - \dfrac{449}{6}$ **43.** $y = \dfrac{4}{21}x + \dfrac{73}{98}$

45. $y = 4$ **47.** $x = -2.5$ **49.** $6x + 12y = 11$

51. $12x - 6y = 5$ **53.** $3x - 2y = 13$, $y = \dfrac{3}{2}x - \dfrac{13}{2}$

55. $5x - 2y = 12$, $y = \dfrac{5}{2}x - 6$ **57.** $4x - 2y = 15$,

$y = 2x - \dfrac{15}{2}$ **59.** $5x + 2y = 17$, $y = -\dfrac{5}{2}x + \dfrac{17}{2}$

61. $p = \$38{,}500$ **63.** $A = 100t$ **65.** $2x - 3y = -19$
67. $3x - 4y = 13$ **69.** $y = -5$ **71.** $5x + y = 85$
73. $V = 10R - 40$, 160 volts   **75.** In words

77. $(1, 4)$ **79.** $a = -\dfrac{15}{4}$ **81.** $\dfrac{4a + 1}{4a - 1}$

83. $5\sqrt{3} + 2\sqrt{30} - 3\sqrt{10} - 12$ **85.** $\dfrac{6xy}{2y + 7}$

87. $\{-1 \pm 2\sqrt{2}\}$

35.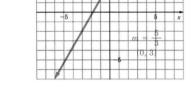

Exercise 6.4

1.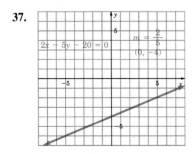

37.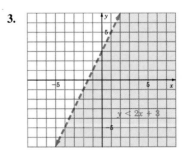

3.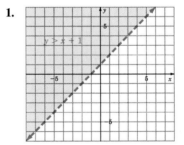

5.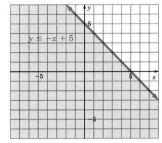

$y \leq -x + 5$

7.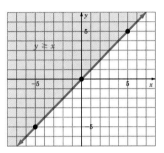

$y \geq x$

9.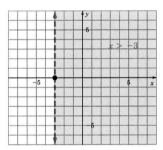

$x > -3$

11.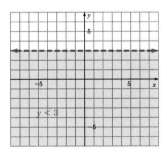

$y < 3$

13.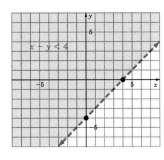

$x - y < 4$

15.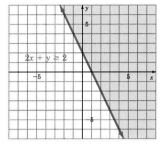

$2x + y \geq 2$

17.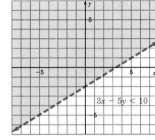

$3x - 5y < 10$

19.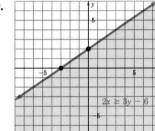

$2x \geq 3y - 6$

21.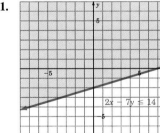

$2x - 7y \leq 14$

23.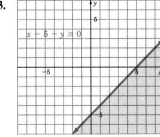

$x - 5 - y \geq 0$

25.

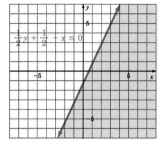

$\frac{1}{2}y + \frac{1}{2}x \le 0$

27.

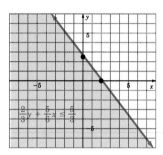

$\frac{2}{3}y + \frac{5}{6}x \le \frac{5}{3}$

29.

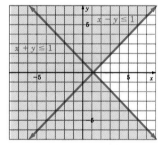

$x - y \le 1$

$x + y \le 1$

31.

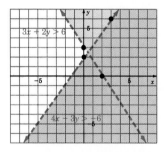

$3x + 2y > 6$

$4x - 3y > -6$

33.

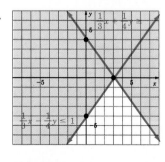

$\frac{1}{3}x + \frac{1}{4}y \ge$

$\frac{1}{3}x - \frac{1}{4}y \le 1$

35.

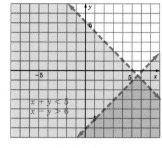

$x + y < 5$

$x - y > 6$

37.

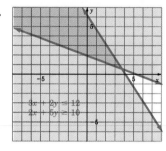

$3x + 2y \le 12$

$4x + 5y \ge 10$

39.

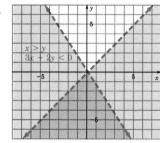

$x > y$

$3x + 2y < 0$

41.

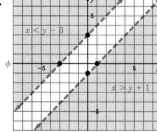

$x < y - 3$

ϕ

$x > y + 1$

43.

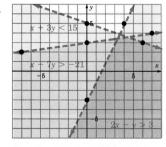

$x + 3y < 15$

$x + 7y > -21$

$2x - y > 3$

45.

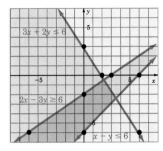

47.

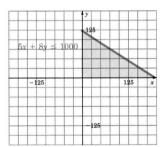

49.

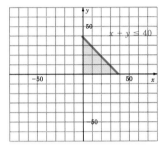

51.

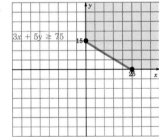

53.

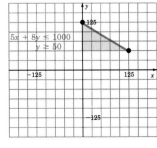

55.

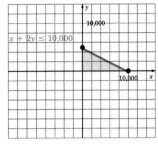

57.

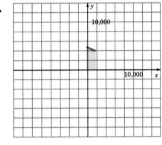

59. In words

61.

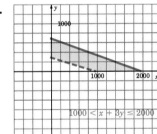

63.

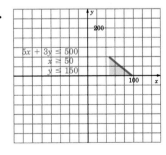

65. $\dfrac{(x-6)(x-5)}{(x+3)(x+8)}$ **67.** $-\dfrac{\sqrt{2}+\sqrt{5}}{3}$

69. $\dfrac{12}{5}-\dfrac{4}{5}i$ **71.** $\{-8, 6\}$

Exercise 6.5

1. 75 **3.** 90 **5.** 9.6 **7.** 62.5 **9.** 196
11. 576 **13.** 14.2 **15.** 11 hr **17.** 14.5 amperes
19. 14.4 **21.** \$239,460 **23.** \$3565 **25.** 630 cm^3

27. 70 pounds **29.** $\dfrac{P}{T} = \dfrac{k}{V}$ or $\dfrac{PV}{T} = k$; $k = 1.5$; 21 lbs/

in^2 **31.** $k = \dfrac{L\sqrt{P}}{MN^2}$ **33.** $y = 15.75$

35.

x	3	1.8	4.5
y	12.6	21	8.4

37. 1 **39.** $\dfrac{4}{81x^{14}y^{12}}$ **41.** 7.843×10^9 **43.** 0.7843

Chapter 6 Review Exercises

1. $(0, -6)$ **3.** $(7, -3)$ **5.** $\left(5, \dfrac{9}{2}\right)$ **7.** $\left(3, -\dfrac{9}{2}\right)$

9.

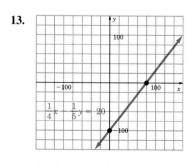

11.

13.

15. $(7, 0), (0, -5)$ **17.** $(-4, 0), (0, -8)$

19. $(63, 0), (0, 147)$ **21.** $\dfrac{1}{7}$ **23.** $\dfrac{1}{2}$ **25.** $-\dfrac{1}{4}$

27. 4 **29.** Parallel **31.** Parallel **33.** No

35. $3\sqrt{10}$ **37.** $\sqrt{137}$ **39.** $\dfrac{\sqrt{89}}{20}$

41. $\sqrt{0.65} \approx 0.81$ **43.** $x - 3y = 4$
45. $3x + 8y = 18$ **47.** $24x + 36y = -35$
49. $0.5x + y = -0.225$ or $20x + 40y = -9$

51. $m = \dfrac{2}{5}, (0, -3)$ **53.** $m = 4, \left(0, -\dfrac{9}{2}\right)$

55. $m = -\dfrac{1}{3}, (0, 7)$ **57.** $m = 6, (0, 24)$

59.

61.

63. $5x - y = 7, y = 5x - 7$ **65.** $2x + 3y = 54,$
$y = -\dfrac{2}{3}x + 18$ **67.** $3x - 7y = -14, y = \dfrac{3}{7}x + 2$

69.

71.
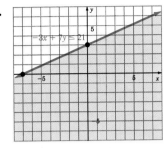

73.
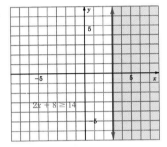

75.
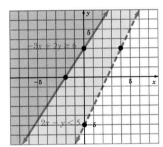

77.
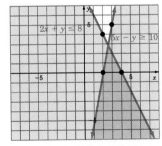

79.
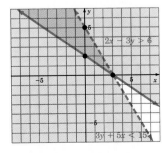

81.
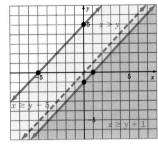

83. 24   **85.** $2734   **87.** 9   **89.** 8.7   **91.** 12.3 hr

Chapter 6 True–False Concept Review

1. True   **2.** False   **3.** True   **4.** True   **5.** False
6. True   **7.** True   **8.** True   **9.** True   **10.** False
11. True   **12.** True   **13.** False   **14.** False
15. False   **16.** True   **17.** False   **18.** False

Chapter 6 Test

1. 1   **2.** $4\sqrt{13}$ **3.** $y = -\dfrac{7}{15}x + \dfrac{1}{6}$ **4.** -1

5.
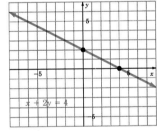

6. No   **7.** $(-4, 70)$ **8.** $14x + 30y = 5$ **9.** $(8, 0)$
and $\left(0, -\dfrac{28}{3}\right)$ **10.** Yes

11.
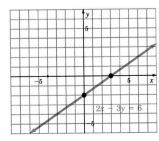

12. $-\dfrac{3}{2}$ **13.** $2x + 3y = 0$ **14.** 10

15. $m = \dfrac{4}{5}, (0, -4)$

16.

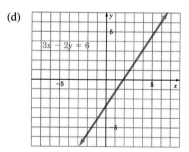

17. $\left(\dfrac{10}{3}, 0\right), (0, -4)$ **18.** $-\dfrac{1}{10}$ **19.** $2x + 3y = 16$

20. $y = -\dfrac{2}{3}x - 4$ **21.** $-\dfrac{1}{3}$ **22.** $m = \dfrac{3}{5}, (0, -3)$

23.

24.

25. No   **26.** 6 hr   **27.** 187 cm

Chapters 4–6 Cumulative Review

1. 3 **2.** $\dfrac{8}{27}$ **3.** $3\sqrt[3]{2}$ **4.** $3\sqrt{2}$ **5.** $28 - \sqrt{10}$

6. $\dfrac{5\sqrt{6}}{6}$ **7.** $2\sqrt{5} + 4$ **8.** $2t$ **9.** $27x^{3/4}y^6$

10. $2ab\sqrt[3]{2b^2}$ **11.** $3x\sqrt{2xy}$ **12.** $\dfrac{a\sqrt{a} + ab}{a - b^2}$

13. $\dfrac{x - y}{x - \sqrt{xy}}$ **14.** $0 + 8i$ **15.** $0 + 5i\sqrt{2}$

16. $0 + 5i\sqrt{2}$ **17.** i **18.** 0 **19.** $6 - 4i$

20. $5 - 10i$ **21.** $10 - 5i$ **22.** $3 - 3i$

23. $\left\{\dfrac{3}{2}, -2\right\}$ **24.** $\left\{\dfrac{\sqrt{5}}{2}, -\dfrac{\sqrt{5}}{2}\right\}$

25. $\left\{\dfrac{1 + i\sqrt{5}}{3}, \dfrac{1 - i\sqrt{5}}{3}\right\}$

26. $\left\{\dfrac{5 + \sqrt{37}}{6}, \dfrac{5 - \sqrt{37}}{6}\right\}$ **27.** $\left\{3, -\dfrac{7}{3}\right\}$

28. $\{-1, 4\}$ **29.** $\{0, 3\}$ **30.** $\{3, -1\}$

31. $\left\{3, -3, \dfrac{1}{2}, -\dfrac{1}{2}\right\}$ **32.** $\{0, 1 + i\sqrt{2}, 1 - i\sqrt{2}\}$

33. $\left\{0, \dfrac{1}{2}, \dfrac{-1 + i\sqrt{3}}{4}, \dfrac{-1 - i\sqrt{3}}{4}\right\}$ **34.** $\{-5\}$

35. $\left\{\dfrac{1 + \sqrt{97}}{4}, \dfrac{1 - \sqrt{97}}{4}\right\}$ **36.** $\left\{\dfrac{1}{3}, -\dfrac{1}{3}, \dfrac{1}{2}, -\dfrac{1}{2}\right\}$

37. $(-\infty, -2) \cup (3, +\infty)$ **38.** $[0, 3]$

39. $\left(-4\dfrac{1}{3}, -1\right)$ **40.** $\left(-\infty, -\dfrac{1}{2}\right] \cup [3, +\infty)$

41. $\left(-5, \dfrac{3}{2}\right)$ **42.** $(-\infty, -5] \cup [0, 3]$ **43.** $(-5, 3]$

44. $(-3, 1) \cup (1, 3)$ **45.** $(-\infty, -5) \cup (-2, -1)$

46. (a) $(2, 0)$; (b) $(0, -3)$; (c) any 3rd solution;

(d)

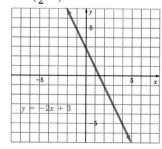

47. (a) $\left(\dfrac{3}{2}, 0\right)$; (b) $(0, 3)$; (c) any 3rd solution;

(d)

48. (a) $(0, 0)$; (b) $(0, 0)$; (c) any 3rd solution;
(d)

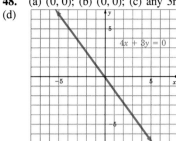

49. (a) $(-3, 0)$; (b) none; (c) Any 3rd solution;
(d)

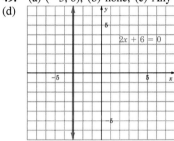

50. (a) none; (b) $(0, -3)$; (c) ;
(d)

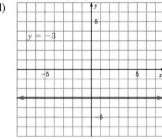

51. (a) $(3, 0)$; (b) $(0, -4)$; (c) ;
(d)

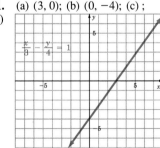

52. (a) $m = \dfrac{2}{3}$; (b) $d = \sqrt{13}$; (c) $2x - 3y = -16$

53. (a) $m = -3$; (b) $d = 3\sqrt{10}$; (c) $3x + y = 10$

54. (a) $m =$ no slope; (b) $d = 5$; (c) $x = -3$ **55.** line
#1: $m = -\dfrac{2}{3}$; line #2: $m = \dfrac{3}{2}$; the lines are perpendicular

56. line #1: $m = -\dfrac{5}{3}$; line #2: $m = \dfrac{5}{3}$; the lines are
neither

57.

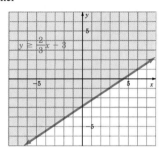

58.

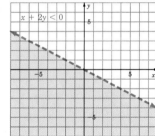

59.

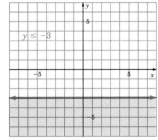

60.

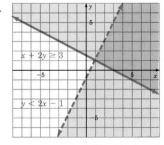

61.

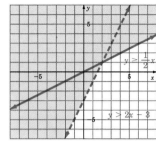

62. (a) $a = kt^2$; (b) $k = 5$; (c) $a = 125$

63. (a) $r = \dfrac{km}{n^2}$; (b) $k = 12$; (c) $r = 10.8$

64. (a) $V = kr^3$; (b) $k = \dfrac{4}{3}\pi$, (c) $V = \dfrac{500\pi}{3}$ **65.** 9

66. (a) $h = 144$ ft; (b) $h = 136$ ft   **67.** (a) $y = \dfrac{7}{2}x$;

(b) $140

68.

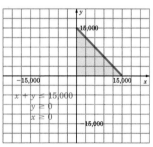

69. 14 cars

CHAPTER 7

Exercise 7.1

1.

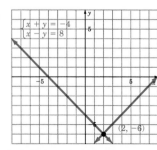

3.

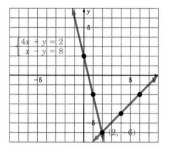

5.

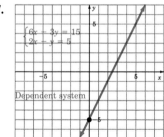

7.

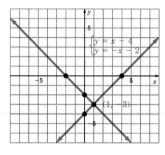

9. $\{(4, 6)\}$ **11.** $\{(4, -1)\}$ **13.** $\{(3, 1)\}$
15. Inconsistent system, $\varnothing$ **17.** $\{(5, 7)\}$ **19.** $\{(-1, 1)\}$
21. $\{(3, -4)\}$ **23.** Dependent system,
$\left\{(x, y) \mid y = \dfrac{3}{2}x + 4\right\}$ **25.** $\{(-5, -4)\}$ **27.** $\{(4, -1)\}$
29. $\{(-1, 1)\}$ **31.** $\{(-6, -2)\}$ **33.** $\{(0, -3)\}$
35. $\left\{\left(\dfrac{1}{2}, -\dfrac{1}{2}\right)\right\}$ **37.** $\left\{\left(\dfrac{61}{14}, \dfrac{3}{14}\right)\right\}$ **39.** $\{(3, -4)\}$
41. $\left\{\left(\dfrac{3}{4}, -\dfrac{5}{4}\right)\right\}$ **43.** $\left\{\left(-\dfrac{3}{5}, -\dfrac{2}{5}\right)\right\}$
45. $\left\{\left(-\dfrac{7}{3}, \dfrac{4}{3}\right)\right\}$ **47.** $\left\{\left(-\dfrac{3}{10}, -\dfrac{1}{5}\right)\right\}$
49. $\left\{\left(-2, \dfrac{1}{3}\right)\right\}$ **51.** $\left\{\left(\dfrac{5}{2}, -\dfrac{3}{2}\right)\right\}$ **53.** $\left\{\left(\dfrac{1}{2}, \dfrac{3}{4}\right)\right\}$
55. $\left\{\left(\dfrac{8}{5}, \dfrac{17}{10}\right)\right\}$ **57.** $\{(3, 4)\}$ **59.** $\left\{\left(-\dfrac{216}{13}, \dfrac{75}{13}\right)\right\}$
61. $7000 @ 18%; $33,000 @ 12%   **63.** 9 nickels, 18
quarters   **65.** 650 lb @ $0.82, 350 lb @ $0.62.
67. 350 resistors @ $2.00, 675 resistors @ $1.00

69. The ordered pair would be reversed, the solution is the same;

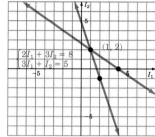

71. amp $150, speaker $50   **73.** In words   **75.** In words

77. $\left\{\left(\dfrac{1}{2}, -3\right)\right\}$;

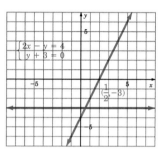

79. $\{(6, -2, -1)\}$ **81.** $\{\pm 1, \pm 5\}$
83. $\{-5, -1, -3 \pm 2\sqrt{2}\}$
85.

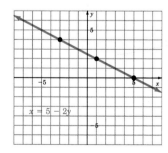

87.

Exercise 7.2

1. $\{(4, 1)\}$ **3.** $\{(1, 1)\}$ **5.** $\emptyset$ **7.** Dependent system, $\{(x, y) \mid 2x - y = 18\}$ **9.** $\{(3, 1)\}$ **11.** $\{(4, -7)\}$

13. $\{(-4, 3)\}$ **15.** $\{(8, -11)\}$ **17.** $\{(0, -3)\}$
19. $\{(-1, -1)\}$ **21.** $\{(5, 2)\}$ **23.** $\{(-3, 4)\}$
25. $\{(-2, 1)\}$ **27.** $\{(8, 2)\}$ **29.** $\{(x, y) \mid 2x - 7y = -1\}$
31. $\{(-10, 3)\}$ **33.** $\{(7, -5)\}$ **35.** $\{(-2, -5)\}$
37. $\{(2, -3)\}$ **39.** Inconsistent system, $\emptyset$

41. $\left\{\left(\dfrac{2}{5}, -\dfrac{3}{5}\right)\right\}$ **43.** $\left\{\left(-\dfrac{5}{9}, \dfrac{4}{9}\right)\right\}$ **45.** $\left\{\left(-\dfrac{5}{6}, \dfrac{1}{6}\right)\right\}$

47. $\left\{\left(\dfrac{-5}{4}, \dfrac{10}{3}\right)\right\}$ **49.** $\left\{\left(\dfrac{1}{2}, -\dfrac{1}{2}\right)\right\}$ **51.** $\left\{\left(\dfrac{3}{4}, -\dfrac{1}{4}\right)\right\}$

53. $\left\{\left(\dfrac{1}{3}, -\dfrac{1}{3}\right)\right\}$ **55.** $\left\{\left(-\dfrac{2}{3}, \dfrac{3}{4}\right)\right\}$ **57.** $\{(-1.3, 0.8)\}$

59. $\{(0.3, -0.5)\}$ **61.** 30 cc 50%, 50 cc 10%

63. 7 lb shrimp, 13 lb crab   **65.** $6\dfrac{2}{3}$ oz of 35% A and 20% B, $4\dfrac{1}{6}$ oz of 40% A and 16% B   **67.** 8 bars @ $1.75, 20 bars @ $2.25   **69.** $9\dfrac{3}{5}$ yd³ cement, $38\dfrac{2}{5}$ yd³ gravel   **71.** $R_1 = 4$ ohms, $R_2 = 2$ ohms   **73.** In words   **75.** $\{(1, b - a)\}$ **77.** $\left\{\left(\dfrac{1}{3}, 1\right)\right\}$ **79.** $\left(\dfrac{18}{5}, 0\right), (0, -6)$

81. $\left(\dfrac{1}{5}, 0\right), \left(0, -\dfrac{1}{3}\right)$ **83.** $\sqrt{73}$ **85.** No

Exercise 7.3

1. $\{(3, -1, 2)\}$ **3.** $\{(1, 0, 1)\}$ **5.** $\{(-2, -3, 1)\}$
7. $\{(3, -4, 0)\}$ **9.** $\{(0, 5, 5)\}$ **11.** $\{(-1, 1, 1)\}$
13. $\{(-1, -2, -1)\}$ **15.** $\{(-3, 1, 1)\}$ **17.** $\{(7, 4, -1)\}$
19. $\{(4, -5, -2)\}$ **21.** $\{(-1, 2, 3)\}$ **23.** $\left\{\left(\dfrac{1}{2}, \dfrac{1}{2}, \dfrac{1}{2}\right)\right\}$

25. $\{(-5, -6, 5)\}$ **27.** $\left\{\left(-\dfrac{1}{3}, \dfrac{1}{3}, -\dfrac{1}{3}\right)\right\}$

29. $\left\{\left(\dfrac{1}{5}, -\dfrac{3}{5}, \dfrac{4}{5}\right)\right\}$ **31.** $y = x^2 + 4x + 4$ **33.** 150—$8 tickets, 450—$5 tickets, 300—$6.50 tickets   **35.** 10 of Box A, 8 of Box B, 6 of Box C   **37.** $I_1 = 2$ amps, $I_2 = 4$ amps, $I_3 = 2$ amps   **39.** $1500 in stocks, $6000 in bonds, $2500 in C.D.'s   **41.** In words

43. $\left\{\left(\dfrac{1}{3}, \dfrac{1}{2}, -1\right)\right\}$ **45.** $\left\{\left(\dfrac{ce - bf}{ae - bd}, \dfrac{af - cd}{ae - bd}\right)\right\}$

47. $y = 3x + 25$ **49.** $3x - y = 25$ **51.** $3x + 7y = 9$

53.

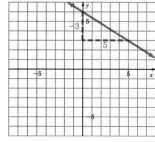

Exercise 7.4

1. 0   **3.** 0   **5.** 5   **7.** -10 **9.** -5 **11.** -12
13. 8.66   **15.** $\{-5\}$ **17.** $\{3\}$ **19.** $\{3\}$ **21.** -3
23. -17 **25.** -3 **27.** -350 **29.** $23ab$
31. -6 **33.** 0   **35.** 8   **37.** $\left\{\dfrac{43}{2}\right\}$ **39.** $\{\pm 6\}$
41. -21 **43.** $-\dfrac{3}{8}$ **45.** -22 **47.** 299   **49.** 72
51. -3 **53.** 41   **55.** $6a$ **57.** $\{1\}$ **59.** $\{-1, -2\}$
61. $\dfrac{13}{2}$ sq units   **63.** In words   **65.** 0
67. $10, -10, -10$ **69.** $b = \pm 3$ **71.** abc
73. $\left\{\dfrac{3 \pm \sqrt{33}}{2}\right\}$ **75.** $\{-2\}$ **77.** $(-\infty, -5] \cup (-4, 0]$
79. $(5, 6)$

Exercise 7.5

1. $\{(5, 1)\}$ **3.** $\{(5, 7)\}$ **5.** $\{(2, 1)\}$ **7.** $\left\{\left(\dfrac{2}{3}, \dfrac{3}{5}\right)\right\}$
9. No unique solution   **11.** $\{(0.5, -1.5)\}$
13. $\left\{\left(2, -\dfrac{1}{2}, -\dfrac{1}{2}\right)\right\}$ **15.** $\{(2, 1, -1)\}$
17. $\{(-1, 2, 3)\}$ **19.** $\{(-1, -1, -1)\}$
21. $\left\{\left(\dfrac{1}{2}, \dfrac{2}{3}, \dfrac{3}{4}\right)\right\}$ **23.** $\left\{\left(\dfrac{1}{6}, -\dfrac{5}{6}, \dfrac{1}{3}\right)\right\}$
25. $\{(-7, 8, 1)\}$ **27.** $\{(0, 2, 5)\}$ **29.** $\{(2, 3, -5)\}$
31. $\left\{\left(-\dfrac{2}{3}, \dfrac{20}{3}, -1\right)\right\}$ **33.** $\{(-1, 1, 2, -2)\}$
35. $\{(3, -4, 1, 1)\}$ **37.** $\{(1, 0)\}$ **39.** $\{(a + b, ab)\}$
41. The numbers are 772 and 687   **43.** $3\dfrac{3}{4}$ qt of 2.5 oz,
$6\dfrac{1}{4}$ qt of 3.3 oz   **45.** 200 shares at \$37.50, 450 shares at
\$14.75   **47.** 20 oz of food A, 45 oz of food B, 25 oz of
food C   **49.** $I_1 = 4.3$ amps, $I_2 = 6.2$ amps, $I_3 = 1.9$ amps
51. 20 \$5 bills, 40 \$10 bills, 25 \$20 bills   **53.** $a = -6$
55. No   **57.** $a = 0$ **59.** $\left\{\dfrac{19}{2}\right\}$ **61.** $(-\infty, 5)$
63. $\{8, 10\}$
65.

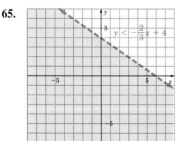

Exercise 7.6

1. $\{(3, -4)\}$ **3.** $\{(-2, -2)\}$ **5.** $\{(5, 1)\}$
7. $\{(-3, 1)\}$ **9.** $\emptyset$ **11.** $\{(1, 1)\}$ **13.** $\{(-1, -7)\}$
15. $\{(-2, 5)\}$ **17.** $\{(6, -3)\}$ **19.** $\left\{\left(\dfrac{5}{2}, -\dfrac{3}{2}\right)\right\}$
21. $\{(1, 1, 1)\}$ **23.** $\{(-2, 3, 2)\}$ **25.** $\{(-4, 5, -1)\}$
27. $\{(-2, 1, 1)\}$ **29.** $(-3, 0, -1)\}$ **31.** In words
33. $\left\{\left(\dfrac{2}{2a + 1}, \dfrac{4a + 3}{2a + 1}\right)\right\}$ **35.** $a = -\dfrac{3}{2}$
37. $(45, 0), (0, -9)$ **39.** $2x - y = -5$
41. $\left\{-\dfrac{15}{4}, 1\right\}$ **43.** $\{3, 27\}$

Chapter 7 Review Exercises

1. $\{(0, -2)\}$ **3.** $\{(1, -3)\}$ **5.** $\{(1, -2)\}$
7. $\left\{\left(\dfrac{1}{3}, -\dfrac{2}{3}\right)\right\}$ **9.** $\left\{\left(-\dfrac{1}{4}, -\dfrac{1}{4}\right)\right\}$ **11.** $\left\{\left(\dfrac{3}{5}, -1\right)\right\}$
13. $\{(4, -7)\}$ **15.** $\{(9, -2)\}$ **17.** $\{(-8, 3)\}$
19. $\left\{\left(\dfrac{1}{4}, -\dfrac{7}{4}\right)\right\}$ **21.** $\left\{\left(\dfrac{30}{11}, -\dfrac{28}{11}\right)\right\}$
23. $\{(0, -3, -9)\}$ **25.** $\{(-5, -6, -10)\}$
27. $\left\{\left(\dfrac{1}{3}, \dfrac{7}{3}, -\dfrac{8}{3}\right)\right\}$ **29.** 42   **31.** 159   **33.** -14
35. -2.915 **37.** $\dfrac{1}{8}$ **39.** 266   **41.** 44
43. -1412 **45.** 15   **47.** $\{-7\}$ **49.** $\{\pm 10\}$
51. $\{3\}$ **53.** No unique solution   **55.** $\emptyset$
57. $\{(-1, 2)\}$ **59.** $\{(2, -0.04)\}$ **61.** $\{(4, -11, 5)\}$
63. $\left\{\left(-\dfrac{5}{4}, \dfrac{1}{2}, -\dfrac{7}{4}\right)\right\}$ **65.** $\{(-5, -11, 16)\}$
67. $\{(4, -5)\}$ **69.** $\{(-7, 8)\}$ **71.** $\{(5, 10, -8)\}$

Chapter 7 True–False Concept Review

1. True   **2.** True   **3.** False   **4.** True   **5.** True
6. False   **7.** True   **8.** False   **9.** False   **10.** False
11. True   **12.** True   **13.** True   **14.** False
15. True   **16.** False   **17.** True   **18.** False
19. True   **20.** True

Chapter 7 Test

1. $\left\{\left(\dfrac{2}{3}, -\dfrac{5}{3}\right)\right\}$ **2.** $\left\{\left(-\dfrac{14}{11}, \dfrac{10}{11}\right)\right\}$ **3.** $\left\{\left(0, \dfrac{4}{5}\right)\right\}$
4. 18   **5.** $\left\{\left(\dfrac{1}{2}, -\dfrac{3}{2}, \dfrac{1}{2}\right)\right\}$ **6.** $\{(4, -2, 3)\}$
7. $\left\{\left(\dfrac{3}{4}, -\dfrac{1}{2}, 1\right)\right\}$ **8.** -40 **9.** $\left\{\left(18, \dfrac{25}{2}\right)\right\}$
10. 16   **11.** $\{(1, 0, -3)\}$ **12.** $\left\{\left(\dfrac{10}{11}, -\dfrac{14}{11}\right)\right\}$

13. 14   **14.** $\left\{\left(-\dfrac{3}{7}, 4\right)\right\}$ **15.** 30 cases at \$4.70, 25 cases at \$5.20   **16.** 12 lb of 60% alloy, 18 lb of 80% alloy

CHAPTER 8

Exercise 8.1

1. Vertex $(0, -2)$, min. pt.; x-intercepts $(-\sqrt{2}, 0)(\sqrt{2}, 0)$; y-intercept $(0, -2)$; $x = 0$;

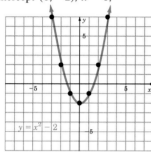

3. Vertex $(3, 0)$, min. pt.; x-intercept $(3, 0)$; y-intercept $(0, 9)$; $x = 3$;

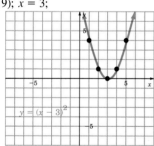

5. Vertex $(-1, -2)$, min. pt.; x-intercepts $(-1 + \sqrt{2}, 0)(-1 - \sqrt{2}, 0)$; y-intercept $(0, -1)$; $x = -1$;

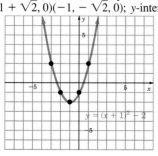

7. Vertex $(1, -2)$, min. pt.; x-intercepts $(0, 0)$, $(2, 0)$; y-intercept $(0, 0)$; $x = 1$;

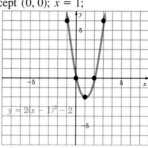

9. Vertex $(0, 3)$, max. pt.; x-intercepts $(\sqrt{3}, 0)(-\sqrt{3}, 0)$; y-intercept $(0, 3)$; $x = 0$;

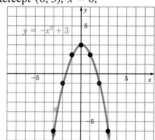

11. Vertex $(1, 0)$, max. pt.; x-intercept $(1, 0)$; y-intercept $(0, -2)$; $x = 1$;

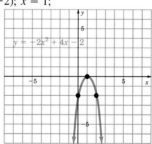

13.

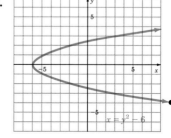

15. Vertex $(3, 3)$, max. pt.; x-intercepts $(0, 0)$, $(6, 0)$; y-intercept $(0, 0)$; $x = 3$;

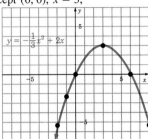

17. Vertex $(3, -2)$, max. pt.; No x-intercept; y-intercept $(0, -5)$; $x = 3$;

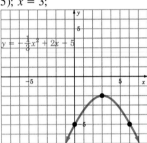

19. Vertex $(-1, -4)$, min. pt.; x-intercepts $(-3, 0)$, $(1, 0)$; y-intercept $(0, -3)$; $x = -1$;

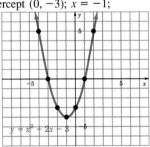

21. Vertex $(2, 3)$, min. pt.; No x-intercept; y-intercept $(0, 7)$; $x = 2$;

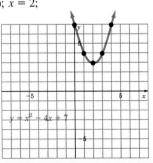

23.

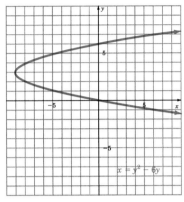

25. Vertex $(1, -6)$, min. pt.; x-intercepts $(1 + \sqrt{6}, 0)$, $(1 - \sqrt{6}, 0)$; y-intercept $(0, -5)$; $x = 1$;

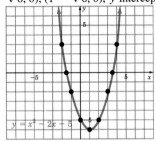

27. Vertex $(-1, -4)$, min. pt.; x-intercepts $(-1 + \sqrt{2}, 0)(-1 - \sqrt{2}, 0)$; y-intercept $(0, -2)$; $x = -1$;

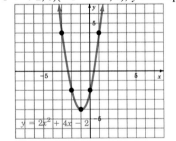

29. Vertex $(-1, 0)$, max. pt.; x-intercept $(-1, 0)$; y-intercept $(0, -3)$; $x = -1$;

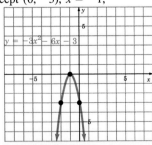

31. Vertex $(3, -1)$, max. pt.; No x-intercept; y-intercept $(0, -19)$; $x = 3$;

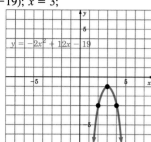

33. Vertex $(-1, -3)$, min. pt.; x-intercepts $(-1 + 2\sqrt{3}, 0)$, $(-1 - 2\sqrt{3}, 0)$; y-intercept $\left(0, -\dfrac{11}{4}\right)$; $x = -1$;

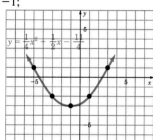

35.

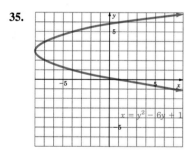

37. 4th week, 1508 T-shirts   **39.** 108 units, 18 employees   **41.** 150 ft, 3 sec
43. 364.5 ft-lb; $\sqrt{3}$ ft ≈ 1.73 ft;

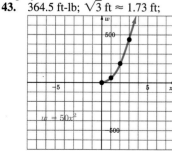

45. $C = \$225,135,000$; approximately 23,450 blocks
47. 5000 sq ft   **49.** In words   **51.** In words
53. $x = \dfrac{1}{2}y^2 - y - \dfrac{7}{2}$ **55.** approximately 25 ft

57. $y = \pm 16$ **59.** No real solution   **61.** $y = \pm\sqrt{10}$
63. No real solution

Exercise 8.2

1. $(0, 0)$, $r = 5$

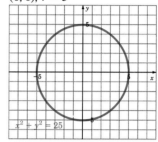

3. $(0, 0)$, $r = 6$

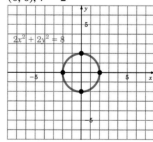

5. $(0, 0)$, $r = 2$

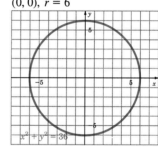

7. $x^2 + y^2 = 81$ **9.** $(x - 2)^2 + (y - 4)^2 = 49$
11. $x^2 + (y - 3)^2 = 4$ **13.** $(x - 1)^2 + (y - 1)^2 = 36$
15. $(5, -4)$, $r = 3$

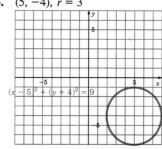

17. $(-5, 3)$, $r = 2$

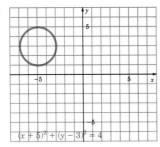

19. $(5, 4)$, $r = 8$

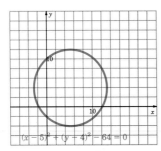

21. $(1, 2)$, $r = \dfrac{3}{2}$;

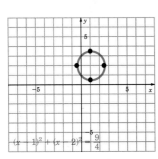

23. $(x + 5)^2 + (y - 2)^2 = 144$
25. $(x + 3)^2 + (y + 8)^2 = 256$
27. $(x - 3)^2 + (y + 1)^2 = \dfrac{4}{9}$ **29.** $(0, 0)$, $r = \dfrac{3}{2}$

31. $(3, -1)$, $r = 2$ **33.** $(x - 1)^2 + (y - 1)^2 = \dfrac{49}{4}$

35. $(x - 2)^2 + (y - 2)^2 = 2$
37. $(x - 5)^2 + (y + 2)^2 = 12$; $(5, -2)$, $r = 2\sqrt{3}$
39. $x^2 + y^2 = \dfrac{5}{2}$; $(0, 0)$, $r = \dfrac{\sqrt{10}}{2}$
41. $x^2 + y^2 + 4x + 14y + 51.56 = 0$
43.

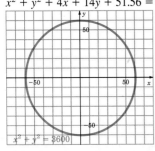

45. the point $(2, -4)$ **47.** The top half of a circle; The bottom half of a circle   **49.** $3x - 4y = 25$

51. center $(1, -2)$; radius $\dfrac{3}{2}$

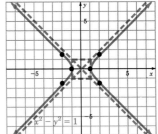

53. 16π square units   **55.** In words
57. $3x - 4y = -12$ **59.** center $(3, -2)$; radius 3; x-intercepts $(3 \pm \sqrt{5}, 0)$; y-intercepts $(0, -2)$ **61.** $\emptyset$
63. $\left\{ \dfrac{3 \pm \sqrt{29}}{5} \right\}$ **65.** $\left\{ \dfrac{3}{2}, 22 \right\}$
67. $\left(-\infty, -\dfrac{5}{2} \right) \cup (9, +\infty)$

Exercise 8.3

1.

3.

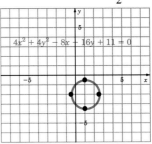

5.

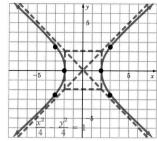

$$\frac{x^2}{4} - \frac{y^2}{4} = 1$$

15.

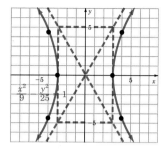

$$\frac{x^2}{9} - \frac{y^2}{25} = 1$$

7.

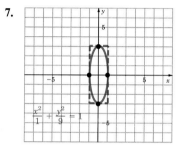

$$\frac{x^2}{1} + \frac{y^2}{9} = 1$$

17.

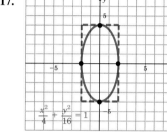

$$\frac{x^2}{4} + \frac{y^2}{16} = 1$$

9. $\dfrac{x^2}{16} + \dfrac{y^2}{9} = 1$

11.

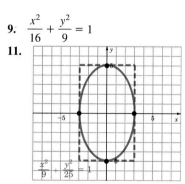

$$\frac{x^2}{9} + \frac{y^2}{25} = 1$$

19. $\dfrac{y^2}{9} - \dfrac{x^2}{1} = 1$

21.

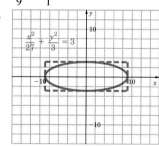

$$\frac{x^2}{27} + \frac{y^2}{3} = 3$$

13.

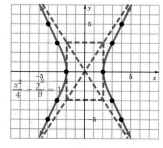

$$\frac{x^2}{4} - \frac{y^2}{9} = 1$$

23.

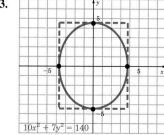

$$10x^2 + 7y^2 = 140$$

25.

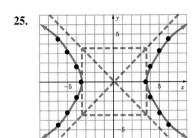

55.

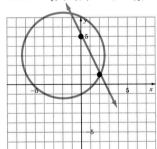

27. $\dfrac{x^2}{25} + \dfrac{y^2}{1} = 1$ **29.** $\dfrac{x^2}{4} - \dfrac{y^2}{4} = 1$

31.

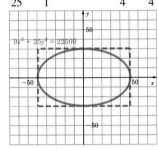

33. The top half of the ellipse; The bottom half of the ellipse   **35.** $\dfrac{x^2}{36} + \dfrac{y^2}{9} = 1$ or $x^2 + 4y^2 = 36$ **37.** Plate:

$x^2 + y^2 = 25$; hole: $\dfrac{x^2}{16} + \dfrac{y^2}{4} = 1$ or $x^2 + 4y^2 = 16$

39.

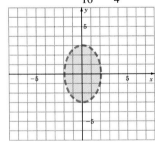

41. In words   **43.** In words   **45.** $y = \dfrac{3}{2}x,\; y = -\dfrac{3}{2}x;$

No, $\dfrac{3}{2}\left(-\dfrac{3}{2}\right) \neq -1$ **47.** $y = x$ and $y = -x$ **49.** $\left\{\dfrac{7}{2}\right\}$

51. $w = \dfrac{8a}{15}$ **53.** $w = \dfrac{1 - ab}{a}$

Exercise 8.4

1. $\{(0, 0), (4, 16)\}$ **3.** $\{(-3, 0), (2, 10)\}$
5. $\{(0, 4), (4, 0)\}$ **7.** $\{(-2, 0), (0, 2)\}$
9. $(\sqrt{5}, 2), (-\sqrt{5}, 2)\}$ **11.** $\{(3, 4), (-3, -4)\}$
13. $\left\{\left(\dfrac{5}{2}, -\dfrac{3}{2}\right)\right\}$ **15.** $\{(1, 4), (-1, -4)\}$
17. $\{(0, 2), (5, 7)\}$ **19.** No real solution
21. $\left\{\left(1, -\dfrac{3}{2}\right), \left(-1, \dfrac{3}{2}\right)\right\}$ **23.** No real solution
25. $\{(-2, -1)\}$ **27.** $\{(4, 0), (5, 1)\}$
29. $\left\{\left(-\dfrac{3\sqrt{2}}{2}, \dfrac{3\sqrt{2}}{2}\right)\right\}$ **31.** 40 cm by 80 cm
33. 100 units   **35.** The numbers are 8 and 4   **37.** Yes
$\dfrac{1 + \sqrt{5}}{2}, \dfrac{-1 + \sqrt{5}}{2}$ or $\dfrac{1 - \sqrt{5}}{2}, \dfrac{-1 - \sqrt{5}}{2}$
39. $(-3, 1)$ and $(2, 6)$ **41.** In words   **43.** $10
45. estimate $\{(2, 1), (-1.2, 7.5)\};$

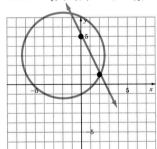

47. 2 in and 4 in   **49.** -47 **51.** -5
53. $\left\{\left(\dfrac{13}{7}, \dfrac{3}{7}\right)\right\}$ **55.** Boat 3 mph, current 1 mph

Exercise 8.5

1. $\{(2\sqrt{2}, 2), (2\sqrt{2}, -2), (-2\sqrt{2}, 2), (-2\sqrt{2}, -2)\}$
3. $\{(0, 2), (0, -2)\}$ **5.** $\{(0, 2), (0, -2)\}$
7. $\{(-2, 0), (2, 0)\}$
9. $\{(1, 2), (1, -2), (-1, 2), (-1, -2)\}$
11. $\{(4, 0), (-4, 0)\}$ **13.** $\{(0, 3)\}$ **15.** No real
solution   **17.** $\{(4, 1), (4, -1), (-4, 1), (-4, -1)\}$
19. $\{(3, \sqrt{3}, \sqrt{2}), (3\sqrt{3}, -\sqrt{2}), (-3\sqrt{3}, \sqrt{2}), (-3\sqrt{3}, -\sqrt{2})\}$

21. $\{(1, 1), (-1, 1)\}$ **23.** $\{(1, 3), (-1, 3)\}$
25. Approximate $\{(8.8, 9.6)(8.8, -9.6)(-8.8, 9.6)$
$(-8.8, -9.6)\}$ **27.** The numbers are 7 and 8 or -7 and
-8 or 7 and -8 or -7 and 8   **29.** Yes at
$(0, -4)$, $(2, 0)$ and $(-2, 0)$ **31.** The numbers are 9 and
3.   **33.** Yes, at $(5, 0)$ **35.** In words
37.

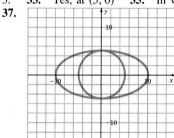

39. $(4, 1), (-4, -1); (1, 4), (-1, -4)$
41. $\dfrac{2x - 27}{x^2 - 2x - 15}$ **43.** $\dfrac{(3x + 1)(5x + 6)}{(x + 1)(5x - 6)}$
45. $\dfrac{3}{2x + 1}$ **47.** $\left\{\left(\dfrac{11}{2}, -\dfrac{1}{2}, 3\right)\right\}$

Chapter 8 Review Exercises

1. Vertex $(0, 4)$, min. pt.; No x-intercept; y-intercept $(0, 4)$;
$x = 0$;

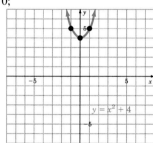

3. Vertex $(1, 3)$, min. pt.; No x-intercept; y-intercept $(0, 4)$;
$x = 1$;

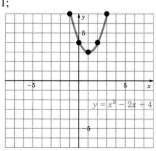

5. Vertex $(-4, -2)$, max. pt.; No x-intercept; y-intercept
$(0, -6)$; $x = -4$;

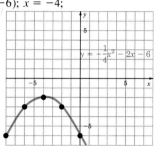

7. Vertex $\left(\dfrac{7}{2}, -\dfrac{1}{4}\right)$, min. pt.; x-intercepts $(4, 0)$, $(3, 0)$;

y-intercept $(0, 12)$; $x = \dfrac{7}{2}$;

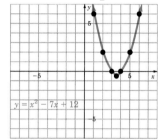

9. Vertex $\left(\dfrac{3}{4}, \dfrac{25}{8}\right)$, max. pt.; x-intercepts $\left(-\dfrac{1}{2}, 0\right)$, $(2, 0)$;

y-intercept $(0, 2)$; $x = \dfrac{3}{4}$;

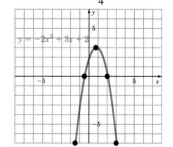

11.

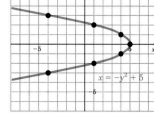

13.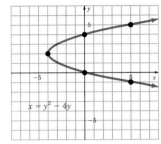

15.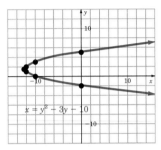

17. $(0, 0), r = 4;$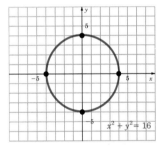

19. $(-3, 1), r = 2;$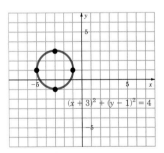

21. $(-4, -2), r = 4;$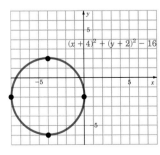

23. $(3, 7), r = 2;$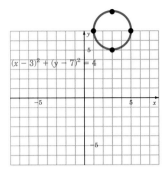

25. $(0, 4), r = 4;$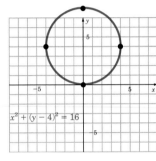

27. $x^2 + y^2 = \dfrac{1}{4}$ **29.** $(x - 3)^2 + (y + 4)^2 = 25$

31. $(x + 6)^2 + (y + 6)^2 = 16$
33. $(x + 11)^2 + (y + 12)^2 = 49$

35. $(x + 3)^2 + (y + 8)^2 = \dfrac{4}{25}$ **37.** $x^2 + y^2 = 36; (0, 0),$

$r = 6$ **39.** $x^2 + (y + 3)^2 = 9; (0, -3), r = 3$
41. $x^2 + (y - 2)^2 = 16; (0, 2), r = 4$
43. $(x + 7)^2 + (y - 3)^2 = 4; (-7, 3), r = 2$
45. $(x - 1)^2 + (y + 6)^2 = 100; (1, -6), r = 10$

47.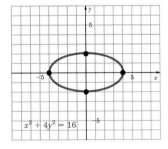

49.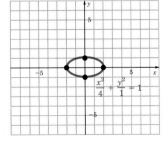

51.

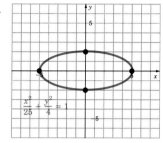

$$\frac{x^2}{25} + \frac{y^2}{4} = 1$$

53.

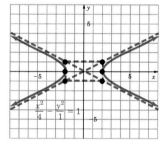

$x^2 - y^2 = 16$

55.

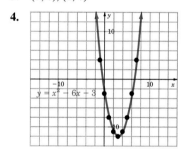

$$\frac{x^2}{4} - \frac{y^2}{1} = 1$$

57.

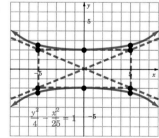

$$\frac{y^2}{4} - \frac{x^2}{25} = 1$$

59. $\dfrac{x^2}{1} + \dfrac{y^2}{16} = 1$ **61.** $\dfrac{x^2}{9} - \dfrac{y^2}{4} = 1$

63. $\left\{(0, 0), \left(-\dfrac{1}{2}, -\dfrac{1}{4}\right)\right\}$

65. $\{(1 + \sqrt{7}, 1 - \sqrt{7}), (1 - \sqrt{7}, 1 + \sqrt{7})\}$ **67.** 9 ft by 21 ft   **69.** $\{(\sqrt{5}, 0), (-\sqrt{5}, 0)\}$ **71.** $\{(-6, 0), (6, 0)\}$

Chapter 8 True–False Concept Review

1. True   **2.** False   **3.** False   **4.** False   **5.** True
6. False   **7.** True   **8.** False   **9.** False   **10.** True
11. False   **12.** False

Chapter 8 Test

1. $(6, -44)$, min. pt.

2.

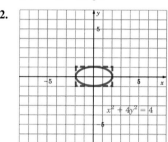

$x^2 + 4y^2 = 4$

3. $(3, 0), (1, 0)$

4.

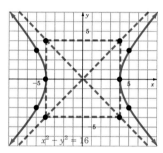

$y = x^2 - 6x - 3$

5. $\{(2, 0), (-2, 0)\}$ **6.** $x = -5$ **7.** $x^2 + y^2 = 49$
8. $(x - 2)^2 + (y + 1)^2 = 36$
9. $(1 + \sqrt{3}, 0), (1 - \sqrt{3}, 0)$ **10.** $\{(0, -1), (8, 7)\}$

11. $(0, 0), r = 9$

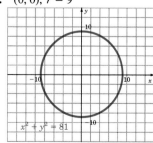

$x^2 + y^2 = 81$

12. $x = 4$

13.

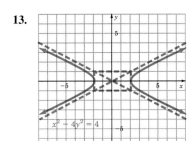

14. $(7, -4)$, max. pt.

15.

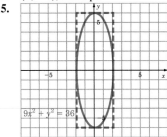

16. $(x - 2)^2 + (y - 1)^2 = 64$

17.

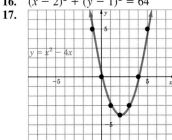

18.

19. $\{(-6, 7), (2, -1)\}$ **20.**   Center $(-3, 5)$, radius $= 3$

CHAPTER 9

Exercise 9.1

1. Domain: $\{1, 2, 3, 4\}$; Range: $\{1, 4, 9, 16\}$
3. Domain: $\{-7, -3, 0, 3\}$; Range: $\{-10, 4, 7, 8\}$

5. Domain: $\{-4, 0, 1, 4, 5\}$; Range: $\{8\}$
7. $\{(-4, 57), (-2, 21), (0, 1), (2, -3), (4, 9)\}$; Range: $\{-3, 1, 9, 21, 57\}$
9. $\{(-3, 25), (-1, 15), (1, 5), (3, 13)\}$; Range: $\{5, 13, 15, 25\}$
11. $\{(2, 13), (3, 11), (5, 7), (8, 1), (10, 5)\}$; Range: $\{1, 5, 7, 11, 13\}$
13. Domain: $\{x \mid x \in R\}$; Range: $\{y \mid y \geq 1\}$; A function
15. Domain: $\{x \mid x \in R\}$; Range: $\{y \mid y \geq -2\}$; A function
17. Domain: $\{x \mid -10 \leq x \leq 10\}$; Range: $\{y \mid -5 \leq y \leq 5\}$; Not a function
19. Domain: $\{x \mid x \in R\}$; Range: $\{y \mid y \in R\}$; Not a function
21. Domain: $\{x \mid -10 \leq x \leq 10\}$; Range: $\{y \mid 0 \leq y \leq 10\}$; A function
23. Domain: $\{x \mid x \geq 0\}$; Range: $\{y \mid y \in R\}$; Not a function
25. Domain: $\{x \mid x \in R\}$; Range: $\{y \mid y \leq -1 \text{ or } y \geq 1\}$; Not a function
27. Domain: $\{x \mid x \in R\}$; Range: $\{y \mid y \geq 0\}$; A function
29. Domain: $\{x \mid x \in R\}$; Range: $\{y \mid y \geq -7\}$; A function
31. Domain: $\{x \mid x \geq 0\}$; Range: $\{y \mid y \geq 0\}$; A function
33. Domain: $\{x \mid x \in R\}$; Range: $\{y \mid y \in R\}$; A function
35. Domain: $\{x \mid x \geq 1\}$; Range: $\{y \mid y \in R\}$; Not a function
37. Domain: $\{x \mid x \neq 0\}$; Range: $\{y \mid y \neq 0\}$; A function
39. Domain: $\{x \mid x \neq -5\}$; Range: $\{y \mid y \neq 2\}$; A function
41. Domain: $\{x \mid x \neq \pm 3\}$; Range: $\{y \mid y \in R\}$; A function
43. Domain: $\{x \mid x \in R\}$; Range: $\left\{y \mid 0 < y \leq \dfrac{3}{4}\right\}$; A function
45. Domain: $\{x \mid x \neq -2, 3\}$; Range: $\{y \mid y \in R\}$; A function
47. Domain: $\{x \mid x \in R\}$; Range: $\{y \mid y \in R\}$; Not a function

49. $\{x \mid x \in R\}$; $\{y \mid y \geq 0\}$; A function;

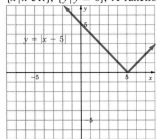

51. $\{x \mid x \geq 0\}$; $\{y \mid y \geq -4\}$; A function;

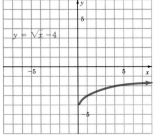

53. $\{x \mid x \in R\}$; $\{y \mid y \geq -1\}$; A function;

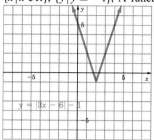

55. $\{x \mid x \in R\}$; $\{y \mid y \geq -6\}$; A function;

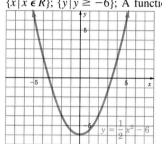

57. $\{x \mid -4 \leq x \leq 4\}$; $\{y \mid -4 \leq y \leq 4\}$; Not a function;

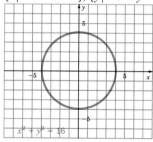

59. $\{x \mid x \leq -2 \text{ or } x \geq 2\}$; $\{y \mid y \in R\}$; Not a function;

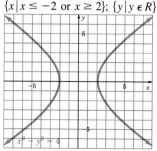

61. $\{x \mid -2 \leq x \leq 2\}$; $\{y \mid -4 \leq y \leq 4\}$; Not a function;

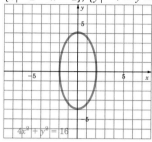

63. (5, 60), (7, 114), (8, 147), (10, 225)
65. (4, 3.5), (6, 4.7), (10, 7.3), (15, 10.8)
67. (20, 2000) (30.5, 4651.25)   **69.** (3, 76) (4, 4)
71. (4, 32) (6, 60)   **73.** In words   **75.** In words
77. Answers will vary   **79.** $D = \{x \mid x \leq 0 \text{ or } x > 1\}$
81. $x + 3y = -9$ **83.** -11 or 5   **85.** $\{(5, -2)\}$
87. $\left\{ \left(-\dfrac{33}{4}, -\dfrac{1}{8} \right) \right\}$

Exercise 9.2

1. 7   **3.** 0.5   **5.** 11   **7.** 2   **9.** 5   **11.** 4, 34
13. 22, 10   **15.** 8, -1 **17.** 2, 56   **19.** 22, 1
21. 0   **23.** 0   **25.** -18 **27.** 52   **29.** -6
31. 45   **33.** -12 **35.** $27 + 3\sqrt{2}$ **37.** 21
39. 34   **41.** 116   **43.** -21 **45.** $6 - a$
47. $|14 - 3b|$ **49.** $a^2 - 3a + 1$ **51.** -97
53. 56   **55.** $7 - a^2$ **57.** $|17 - 3a^2|$
59. $a^4 + 2a^3 - 3a^2 - 4a + 7$ **61.** In words   **63.** In
words   **65.** $f(x + h) = 2x + 2h - 3$; $\dfrac{f(x + h) - f(x)}{h} = 2$
67. $f(x + h) = x^2 + 2xh + h^2 + 1$;
$\dfrac{f(x + h) - f(x)}{h} = 2x + h$ **69.** $f(2) = 1$; $g(2) = 11$;
$(f + g)(2) = 12$; yes   **71.** $h(2) = 6$; $h(3) = 9$;
$h(2 + 3) = h(5) = 15$; yes   **73.** $g(1) = 2$; $g(3) = 10$;
$g(1 + 3) = g(4) = 17$; no   **75.** $x = 6$

77.

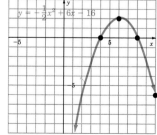

79. $x = -2$

81.

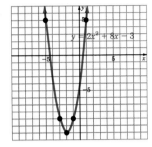

$y = 2x^2 + 8x - 3$

Exercise 9.3

1. Linear;

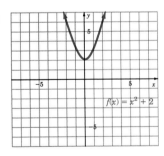

$f(x) = 3x - 5$

3. Quadratic;

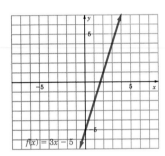

$f(x) = x^2 + 2$

5. Linear;

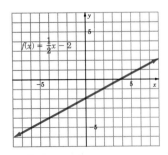

$f(x) = \frac{1}{2}x - 2$

7. Square root;

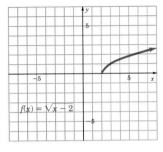

$f(x) = \sqrt{x - 2}$

9. Quadratic;

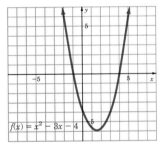

$f(x) = 2x^2 + 3$

11. Quadratic;

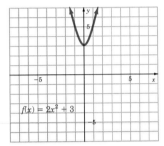

$f(x) = x^2 - 3x - 4$

13. Quadratic;

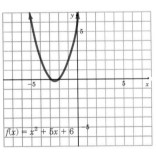

$f(x) = x^2 + 5x + 6$

15. Square root;

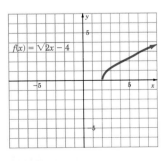

$f(x) = \sqrt{2x} - 4$

17. Absolute value;

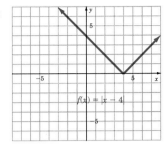

$$f(x) = |x - 4|$$

19. Absolute value;

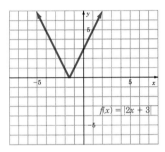

$$f(x) = |2x + 3|$$

21. Not classified;

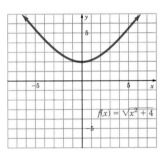

$$f(x) = \sqrt{x^2 + 4}$$

23. Absolute value;

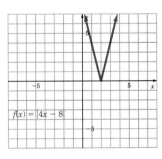

$$f(x) = |4x - 8|$$

25. Rational;

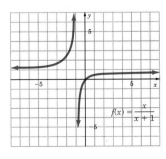

$$f(x) = \frac{x}{x + 1}$$

27. Rational;

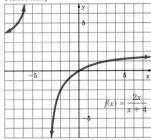

$$f(x) = \frac{2x}{x + 4}$$

29. Absolute value;

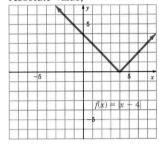

$$f(x) = |x - 4|$$

31. In words   **33.** In words

35.

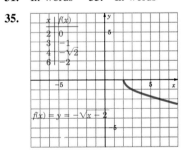

x	$f(x)$
2	0
3	-1
4	$-\sqrt{2}$
6	-2

$$f(x) = y = -\sqrt{x - 2}$$

37.

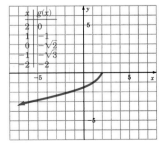

x	$g(x)$
2	0
1	-1
0	$-\sqrt{2}$
-1	$-\sqrt{3}$
-2	-2

39. $2x + 4$ is in absolute value bars

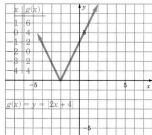

41.

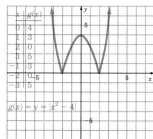

43.

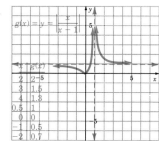

45. $x^2 + y^2 = 625$ **47.** $(-3, 2)$ **49.** Hyperbola
51. Ellipse

Exercise 9.4

1. $2x, \{x \mid x \in R\}$; $8, \{x \mid x \in R\}$; $x^2 - 16, \{x \mid x \in R\}$;
$\dfrac{x + 4}{x - 4}, \{x \mid x \in R \text{ and } x \neq 4\}$ **3.** $7x + 10, \{x \mid x \in R\}$;
$-x + 4, \{x \mid x \in R\}$; $12x^2 + 37x + 21, \{x \mid x \in R\}$;
$\dfrac{3x + 7}{4x + 3}, \left\{x \mid x \in R \text{ and } x \neq -\dfrac{3}{4}\right\}$ **5.** $x^2 + 2x - 9,$
$\{x \mid x \in R\}$; $-x^2 + 2x - 9, \{x \mid x \in R\}$; $2x^3 - 9x^2, \{x \mid x \in R\}$;
$\dfrac{2x - 9}{x^2}, \{x \mid x \in R \text{ and } x \neq 0\}$ **7.** $x^2 + 3x - 4, \{x \mid x \in R\}$;
$x^2 + x - 12, \{x \mid x \in R\}$; $x^3 + 6x^2 - 32, \{x \mid x \in R\}$;
$x - 2, \{x \mid x \in R \text{ and } x \neq -4\}$
9. $(f + g)(x) = 2x^3 - 3x^2 + 4x - 7, \{x \mid x \in R\}$;
$(f - g)(x) = -3x^2 + 4x - 7, \{x \mid x \in R\}$;
$(f \cdot g) = x^6 - 3x^5 + 4x^4 - 7x^3, \{x \mid x \in R\}$;
$\left(\dfrac{f}{x}\right)(x) = \dfrac{x^3 - 3x^2 + 4x - 7}{x^3}, \{x \mid x \in R \text{ and } x \neq 0\}$

11. 11   **13.** 0   **15.** -7 **17.** 0   **19.** 64
21. $f(x) = x^2 - 2x + 1$, $g(x) = 3x^2 - 3x + 1$
23. $f(x) = x^2$, $g(x) = 3x - 4$ **25.** $f(x) = x^2 + x + 8$,
$g(x) = -x$ **27.** $f(x) = x - 2$, $g(x) = x^2 + 2x + 4$
29. $f(x) = 4x^2 - 3x$, $g(x) = x$, $x \neq 0$
31. $f(x) = x^3 - 4x^2 + 4x - 1$, $g(x) = x - 1$, $x \neq 1$
33. 75   **35.** 300   **37.** 9   **39.** $6x - 20$, $6x + 5$
41. $2x^2 + 4$, $4x^2 + 16x + 16$ **43.** $2x^4 + 12x^2 + 13$,
$4x^4 - 20x^2 + 28$ **45.** x, x **47.** x, x **49.** -6
51. 0   **53.** 9   **55.** -4 **57.** 5   **59.** $2x + 2h - 7$;
$2h$; 2   **61.** $x^2 + 2hx + h^2$; $2hx + h^2$; $2x + h$
63. $-x - h - 9$; $-h$; -1
65. $2x^2 + 4hx + 2h^2 - 4x - 4h$; $4hx + 2h^2 - 4h$;
$4x + 2h - 4$ **67.** In words   **69.** In words
71. $-\dfrac{11}{4}$ **73.** $-\dfrac{1}{3}$ **75.** 0   **77.** $2\sqrt{6} - 3$

79. -1 **81.** $\dfrac{7}{2}$ **83.** 2   **85.** $\dfrac{5}{2}$ **87.** $x(4x - 3)$
or $4x^2 - 3x$ **89.** $\{-5\}$ **91.** $\emptyset$ **93.** $\{-4 + \sqrt{7}\}$

Exercise 9.5

1. $\{(2, 3), (3, 4), (4, 5), (5, 6)\}$
3. $\{(4.5, 9), (4, 8), (3.5, 7)\}$
5. $\{(3, 9), (2, 4), (1, 1), (0, 0), (-1, -1), (-2, -4), (-3, -9)\}$
7. $f^{-1}(x) = \dfrac{1}{3}x + \dfrac{5}{3}$ **9.** $f^{-1}(x) = -x + 3$

11. $f^{-1}(x) = \dfrac{1}{3}x + 4$ **13.** $f^{-1}(x) = \dfrac{1}{5}x - \dfrac{4}{5}$

15. $f^{-1}(x) = 2x - 4$ **17.** $f^{-1}(x) = \dfrac{3}{2}x + \dfrac{15}{2}$

19. $f^{-1}(x) = 2x - \dfrac{4}{3}$ **21.** $f^{-1}(x) = \pm\sqrt{x}, x \geq 0$
23. $f^{-1}(x) = \sqrt[3]{x - 1}$ **25.** Yes   **27.** No   **29.** No
31. No

33. $f^{-1}(x) = \dfrac{1}{4}x + \dfrac{3}{4}$, Yes;

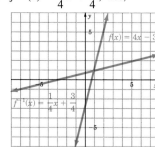

35. $x = 5$, No;

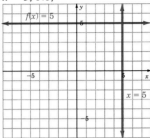

37. $y = \pm\sqrt{x - 2}$, No;

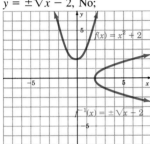

39. $y = x$, $y \geq 0$; $y = -x$, $y < 0$; No;

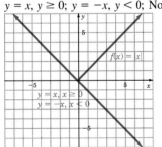

41. $f^{-1}(x) = x + 5$; $(f \circ f^{-1})(x) = x$

43. $f^{-1}(x) = -\dfrac{1}{3}x - \dfrac{5}{3}$; $(f \circ f^{-1})(x) = x$ **45.** In words

47. In words   **49.** 1; 1

51.

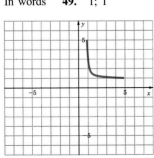

53.

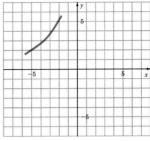

55.

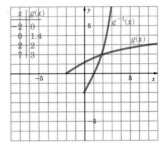

57. The inverse is a function but not a one-to-one function.   **59.** $\{-1, 10\}$ **61.** $\left\{\dfrac{-1 \pm \sqrt{3}}{4}\right\}$

63. $\{-1, \pm\sqrt{2}\}$ **65.** $x = \dfrac{f \pm \sqrt{f^2 - 4d}}{2d}$

Chapter 9 Review Exercises

1. $\{(-2, 1), (-1, 0), (0, 1), (1, 4), (2, 9)\}$;
Range $= \{0, 1, 4, 9\}$
3. $\{(0, \sqrt{2}), (1, \sqrt{3}), (2, 2), (3, \sqrt{5})\}$;
Range $= \{\sqrt{2}, \sqrt{3}, 2, \sqrt{5}\}$
5. Domain $= \{2, 3, 4, 9\}$; Range $= \{1, 2, 6\}$
7. Domain $= \{1, 2, 3, 4, 5\}$;
Range $= \{-1, -2, -3, -4, -5\}$
9. Domain $= \{x \mid x \in R\}$; Range $= \{y \mid y \geq -21\}$; Function
11. Domain $= \{x \mid x \geq \sqrt{5} \text{ or } x \leq -\sqrt{5}\}$;
Range $= \{y \mid y \geq 0\}$; Function
13. Domain $= x \mid -\sqrt{15} \leq x \leq \sqrt{15}\}$;
Range $= \{y \mid 0 \leq y \leq \sqrt{15}\}$; Function
15. Domain $= \{x \mid x \in R\}$; Range $= \{y \mid y \geq 0\}$; Function
17. Domain $= \{x \mid x \in R \text{ and } x \neq -2\}$; Range $= \{y \mid y \in R\}$;
Function
19. Domain $= \{x \mid x \in R\}$; Range $= \{y \mid y \in R\}$; Function;

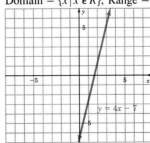

21. Domain = $\{x \mid x \geq 0\}$; Range = $\{y \mid y \geq 2\}$; Function;

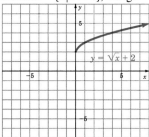

23. Domain = $x \mid x \in R\}$; Range = $\{y \mid y \geq 0\}$; Function;

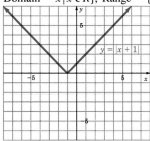

25. Domain = $\left\{x \mid -\dfrac{7}{2} \leq x \leq \dfrac{7}{2}\right\}$;

Range = $\left\{y \mid -\dfrac{7}{2} \leq y \leq \dfrac{7}{2}\right\}$;

Not a function;

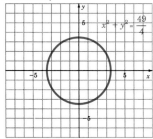

27. 63; -33 **29.** 0; 0   **31.** 7; 7   **33.** 0   **35.** 3
37. 0
39. Absolute value;

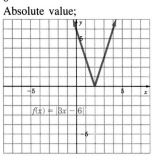

41. Quadratic;

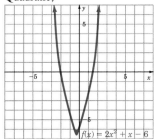

43. $2x + 3$ **45.** $x^2 + x - 4$ **47.** $-x^2 - x + 4$
49. $x^2 + 3x - 1$ **51.** $x^2 + 3x - 10$ **53.** $x^3 - 3x - 2$
55. $\dfrac{x + 5}{x - 2}$ **57.** $\dfrac{x - 2}{x^2 + 2x + 1}$ **59.** 1   **61.** 0
63. 0   **65.** Many pairs   **67.** Many pairs
69. $(f \circ g)(x) = 6x + 3$; $(g \circ f)(x) = 6x - 1$
71. $(f \circ g)(x) = x^2 + 2x + 4$; $(g \circ f)(x) = x^2 + 4$
73. $\{(2, 1), (3, 2), (4, 3), (5, 4)\}$
75. $\{(2, 8), (3, 9), (1, -8), (7, -6)\}$
77. $f^{-1}(x) = \dfrac{1}{2}x + \dfrac{3}{2}$ **79.** $f^{-1}(x) = \dfrac{1}{2}x - \dfrac{3}{2}$
81. $f^{-1}(x) = 2x - 8$ **83.** $f^{-1}(x) = \pm\dfrac{1}{2}\sqrt{2x}$
85. One-to-one   **87.** Not one-to-one   **89.** Not one-to-one   **91.** Not one-to-one
93. $f^{-1}(x) = \dfrac{1}{3}x - 1$; Function;

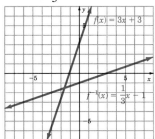

95. $f^{-1}(x) = \pm\sqrt{x}$; Not a function;

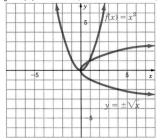

97. $f^{-1}(x) = \dfrac{3}{2}x + 2$; Function;

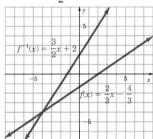

Chapter 9 True–False Concept Review

1. True   **2.** True   **3.** True   **4.** True   **5.** False
6. True   **7.** True   **8.** True   **9.** True   **10.** True
11. False   **12.** True   **13.** False   **14.** False
15. False

Chapter 9 Test

1. **a.** Domain $= \{x \mid x \in R\}$ **b.** Range $= \{y \mid y \geq -1\}$
2. **a.** $y = \pm\sqrt{x - 2}$
b.

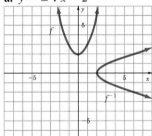

c. Not a function
3. **a.** $(f \circ g)(x) = 6x + 16$ **b.** $(g \circ f)(x) = 6x - 3$
4. Graph b
5. **a.** $(f + g)(x) = 2x^2 + x - 10$
b. $(g - f)(x) = 2x^2 - 3x - 20$
c. $(f \cdot g)(x) = 4x^3 + 8x^2 - 35x - 75$
d. $(g/f)(x) = x - 3,\ x \neq -\dfrac{5}{2}$

6. **a.** Domain $= \{-7, -5, -3, -1, 1, 3\}$
b. Range $= \{-6, -4, -2, 0, 2, 4\}$ **c.** Yes
7. **a.** 57 **b.** 16   **8.** **a.** Constant **b.** Linear **c.** Quadratic
9. $f^{-1}(x) = 4x + 8$
10. $\{(4, -7), (2, -5), (0, -3), (-2, -1), (-4, 1), (-6, 3)\}$
11. **a.** Answers will vary **b.** Answers will vary
c. Answers will vary **d.** Answers will vary   **12.** **a.** 8
b. -12 **13.** **a.** 2 **b.** -6 **14.** **a.** 11 **b.** 5

Chapters 7–9 Cumulative Review

1. $\{(3, 4)\}$ **2.** $\{(0, 3)\}$ **3.** $\left\{\left(\dfrac{11}{4}, -4\right)\right\}$
4. Dependent $\{(x, y) \mid x = -4y - 3\}$ **5.** $\{(0, 1)\}$
6. $\{(4, 3)\}$ **7.** $\emptyset$ **8.** $\left\{\left(\dfrac{19}{15}, \dfrac{37}{15}\right)\right\}$ **9.** 17
10. -28 **11.** -92 **12.** -120
13. $\left\{\left(-\dfrac{53}{9}, -\dfrac{17}{3}\right)\right\}$ **14.** $\left\{\left(-\dfrac{19}{6}, -\dfrac{17}{6}\right)\right\}$
15. $\{(1, 0 - 1)\}$ **16.** No unique solution
17. $\{(1, -1)\}$ **18.** $\{(3, -2, -1)\}$
19. $(0, -8);\ (2, 0),\ (-2, 0);\ (0, -8);\ x = 0;$

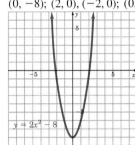

20. $(1, 8);\ (3, 0),\ (-1, 0);\ (0, 6);\ x = 1;$

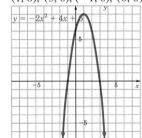

21. $\left(\dfrac{1}{2}, -\dfrac{13}{4}\right);\ (2.3, 0)\ (-1.3, 0);\ (0, -3);\ x = \dfrac{1}{2};$

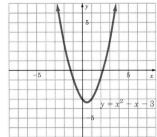

22. $(0, 1)$; $(-1, 0)$; $(0, 1)$; $y = 1$;

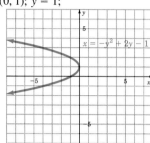

$x = -y^2 + 2y - 1$

23. $(0, 0)$; $r = 3$;

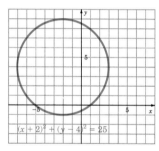

$x^2 + y^2 = 9$

24. $(-2, 4)$; $r = 5$;

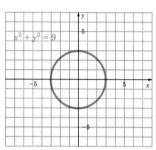

$(x + 2)^2 + (y - 4)^2 = 25$

25. $(0, 4)$; $r = 4$;

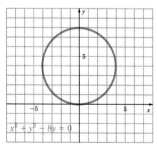

$x^2 + y^2 - 8y = 0$

26. $(3, -1)$; $r = 2$;

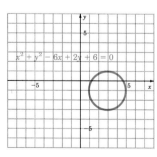

$x^2 + y^2 - 6x + 2y + 6 = 0$

27. $(2, 0)$, $(-2, 0)$; $(0, 3)$, $(0, -3)$;

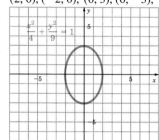

$\dfrac{x^2}{4} + \dfrac{y^2}{9} = 1$

28. $(\sqrt{5}, 0)$, $(-\sqrt{5}, 0)$; $(0, 2\sqrt{3})$, $(0, -2\sqrt{3})$;

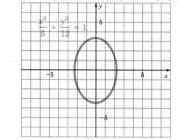

$\dfrac{x^2}{5} + \dfrac{y^2}{12} = 1$

29. $(1, 0)$, $(-1, 0)$; $(0, 5)$, $(0, -5)$;

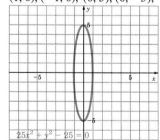

$25x^2 + y^2 - 25 = 0$

30. $(6, 0)$, $(-6, 0)$; $(0, 3)$, $(0, -3)$;

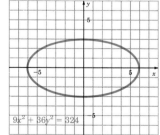

$9x^2 + 36y^2 = 324$

31. $(2, 0,), (-2, 0)$; none;

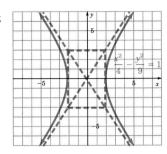

32. none; $(0, 3), (0, -3)$;

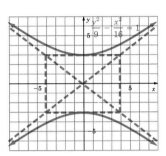

33. $(10, 0), (-10, 0)$; none;

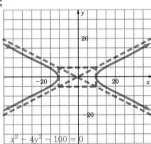

34. none; $(0, 2), (0, -2)$;

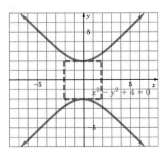

35. $\{(4, -5), (-1, 5)\}$ **36.** $\{(3, 0), (0, -3)\}$ **37.** $\emptyset$
38. $\{(2, 1), (2, -1), (-2, 1), (-2, -1)\}$
39. $\{(\sqrt{7}, \sqrt{2}), (\sqrt{7}, -\sqrt{2}), (-\sqrt{7}, \sqrt{2}), (-\sqrt{7}, -\sqrt{2})\}$
40. $D = \{2, -1, 4, 5\}$; $R = \{0, 3\}$; Yes; No

41. $D = \{-1, 5, 4\}$; $R = \{-4, 4, 3, 2\}$; No; No
42. $D = \{x \mid x \geq -3\}$; $R = \{y \mid y \in R\}$; No; Yes
43. $D = \{x \mid -2 \leq x \leq 2\}$; $R = \{y \mid -4 \leq y \leq 4\}$; No; No
44. $D = \{x \mid x \in R\}$; $R = \{y \mid y \geq -1\}$; Yes; No
45. $D = \{x \mid x \in R\}$ **46.** $D = \{x \mid x \neq -2\}$
47. $D = \{x \mid x \leq 5\}$ **48.** $D = \{x \mid x \neq \pm 3\}$
49. $D = \{x \mid x \in R\}$ **50.** $\dfrac{2}{5}$ **51.** $2\sqrt{2}$ **52.** -1
53. 1 **54.** does not exist   **55.** 6 **56.** $\sqrt{2}$
57. $\dfrac{5}{2}$ **58.** $-\dfrac{1}{2}$ **59.** $\dfrac{a + b}{a + b - 3}$
60. $a^2 - 2ab + b^2 - 3a + 3b + 1$ **61.** $4 + \sqrt{2}$
62. absolute value;

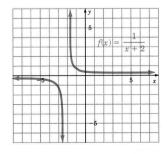

63. quadratic;

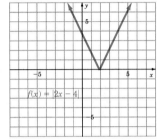

64. square root;

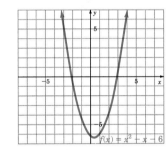

65. rational;

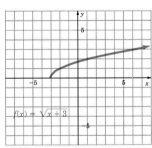

66. $y = \dfrac{1}{2}x + \dfrac{3}{2}$;

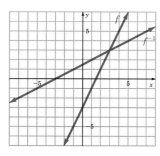

67. $y = \dfrac{3}{2}x - \dfrac{3}{2}$;

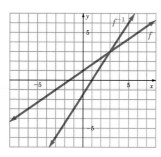

68. $y = 5x - 5$;

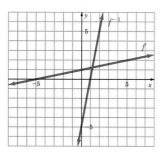

69. $y = \pm\sqrt{x + 1}$;

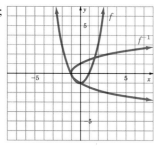

70. 100 ohms, 40 ohms   **71.** 1, 5, and -3 **72.** 38 cases; 39 cases on the twelfth day   **73.** 5 feet by 13 feet

CHAPTER 10

Exercise 10.1

1. $\dfrac{1}{27}, \dfrac{1}{9}, \dfrac{1}{3}, 1, 3, 9, 27$ **3.** $\dfrac{1}{125}, \dfrac{1}{25}, \dfrac{1}{5}, 1, 5, 25,$

125   **5.** $\dfrac{27}{8}, \dfrac{9}{4}, \dfrac{3}{2}, 1, \dfrac{2}{3}, \dfrac{4}{9}, \dfrac{8}{27}$ **7.** $\dfrac{64}{27}, \dfrac{16}{9}, \dfrac{4}{3}, 1,$

$\dfrac{3}{4}, \dfrac{9}{16}, \dfrac{27}{64}$ **9.** $\dfrac{1}{1000}, \dfrac{1}{100}, \dfrac{1}{10}, 1, 10, 100, 1000$

11. $\dfrac{8}{125}, \dfrac{4}{25}, \dfrac{2}{5}, 1, \dfrac{5}{2}, \dfrac{25}{4}, \dfrac{125}{8}$ **13.** $\{5\}$ **15.** $\{3\}$

17. Domain $= \{x \,|\, x \in R\}$; Range $= \{y \,|\, y > 0\}$;

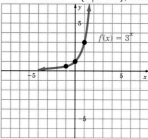

19. Domain $= \{x \,|\, x \in R\}$; Range $= \{y \,|\, y > 0\}$;

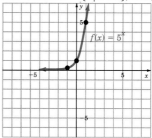

21. Domain $= \{x \,|\, x \in R\}$; Range $= \{y \,|\, y > 0\}$;

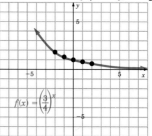

23. $\{3\}$ **25.** $\{4\}$

27. Domain = $\{x \,|\, x \in R\}$; Range = $\{y \,|\, y > 0\}$;

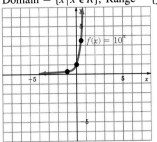

29. Domain = $\{x \,|\, x \in R\}$; Range = $\{y \,|\, y > 0\}$;

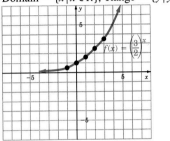

31. Domain = $\{x \,|\, x \in R\}$; Range = $\{y \,|\, y > 0\}$;

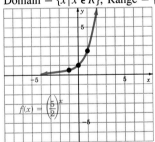

33. Domain = $\{x \,|\, x \in R\}$; Range = $\{y \,|\, y > 0\}$;

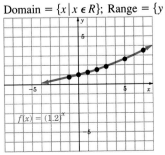

35. Domain = $\{x \,|\, x \in R\}$; Range = $\{y \,|\, y > 2\}$;

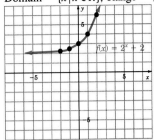

37. $\{4\}$ **39.** $\{4\}$ **41.** $\{3\}$ **43.** 0.9978 mg
45. $4 \times 10^9 = 4{,}000{,}000{,}000$; 29,600,000,000
47. 88.3 mg   **49.** \$23,770   **51.** \$2191.12
53. 4724   **55.** 64.96 lb/ft^3 **57.** 38.5%   **59.** In
words   **61.** In words   **63.** $\{-2\}$ **65.** $\{-2\}$
67. $\{7\}$ **69.** $\left\{-\dfrac{1}{3}\right\}$

71.

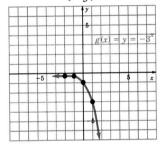

73.

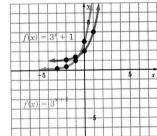

75. No, the system has no solution.   **77.** $f^{-1}(x) = x - 4$
79. $f^{-1}(x) = \dfrac{1}{3}x + \dfrac{4}{3}$ **81.** $f^{-1}(x) = 3x + 12$
83. $f^{-1}(x) = \sqrt[3]{x + 1}$

Exercise 10.2

1. $3^3 = 27$ **3.** $10^0 = 1$ **5.** $x^2 = x^2$ **7.** $5^{1/2} = \sqrt{5}$
9. $\left(\dfrac{1}{2}\right)^3 = \dfrac{1}{8}$ **11.** $\log_4 16 = 2$ **13.** $\log_2 64 = 6$
15. $\log_{1/2} \dfrac{1}{16} = 4$ **17.** $\log_8 \sqrt{8} = \dfrac{1}{2}$

19. $\log_5\left(\dfrac{1}{25}\right) = -2$ **21.** 2 **23.** $\dfrac{1}{2}$ **25.** -1

27. 2 **29.** -4 **31.** $\dfrac{1}{2}$ **33.** -2 **35.** $-2, -1,$
0, 1, 2, 3 **37.** $-2, -1, 0, 2, 3$ **39.** $-2, -1, 0, 1, 2$

41. $\dfrac{3}{2}$ **43.** $-\dfrac{1}{2}$ **45.** $\dfrac{4}{5}$ **47.** -2 **49.** -3

51. $-\dfrac{3}{2}$

53. Domain $= \{x \mid x \in R, x > 0\}$; Range $= \{y \mid y \in R\}$;

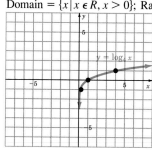

55. Domain $= \{x \mid x \in R, x > 0\}$; Range $= \{y \mid y \in R\}$;

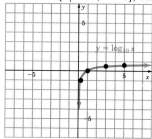

57. Domain $= \{x \mid x \in R, x > 0\}$; Range $= \{y \mid y \in R\}$;

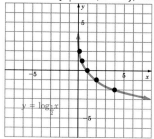

59. pH $= 2$ **61.** pH $= -5$ **63.** 66 **65.** (a) 36,000 years old (b) 54,000 years old **67.** $dB = 120$ **69.** In words **71.** In words **73.** $\{8\}$ **75.** $\left\{\dfrac{1}{64}\right\}$

77. $\left\{\dfrac{5}{2}\right\}$ **79.** $\left\{\dfrac{1}{4}\right\}$ **81.** $\{x \mid x \in R, x > 0 \text{ and } x \neq 1\}$

83.

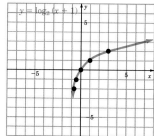

85. $-45a^8b^8c^4$ **87.** $\dfrac{16y^2}{9x^2}$ **89.** $(x + y)^6$

91. 2.71×10^{-6}

Exercise 10.3

1. $\log_b 48$ **3.** $\log_{10} 5$ **5.** $\log_5 64$ **7.** $\log_3 5$

9. $\log_4 \dfrac{1}{9}$ **11.** $\log_b 34 + \log_b 65$ **13.** $4 \log_b 132$

15. $\log_b 13 + 2 \log_b 21$ **17.** 1.39794 **19.** 1.54407
21. $\log_b 192$ **23.** $\log_b 4$ **25.** $\log_b 5$ **27.** $\log_{10} 8$
29. $\log_b z - \log_b x$ **31.** $2 \log_{10} w - \log_{10} z$
33. $2 \log_5 x - 2 \log_5 y$ **35.** $3 \log_b x - \log_b y - 2 \log_b z$
37. 1.60206 **39.** 2.07918 **41.** $\log_b 2$
43. $\log_b 1296$ **45.** $\log_b 14$ **47.** $\log_b 2$

49. $\log_c z + \dfrac{1}{2} \log_c x + \dfrac{1}{2} \log_c y$

51. $\dfrac{1}{2}(\log_7 m + \log_7 n) - 2 \log_7 p$ **53.** $\dfrac{2}{3} \log_x (a + b)$

55. $8 \log_b x + 12 \log_b y$ **57.** 1.92428 **59.** 2.49831
61. Magnitude is 2.5 **63.** 9 months **65.** pH ≈ 1.3

67. $\log_b \dfrac{P}{Q} = \log_b \dfrac{b^s}{b^t} = \log_b b^{s-t} \log_b \sqrt[n]{P} = \log_b P^{1/n} =$

$\log_b (b^s)^{1/n} = \log_b b = \dfrac{s}{n} = \dfrac{1}{n} \log_b P = (s - t)(\log_b b) =$

$s - t = \log_b P - \log_b Q$
69. $\log_{10} (\log_2(\log_3 9)) = \log_{10} (\log_2 (\log_3 3^2)) =$
$\log_{10} (\log_2 2) = \log_{10} 1 = 0$
71. $R = 6.2$ **73.** approximately 14830 years old
75. In words **77.** In words **79.** -0.6112
81. 2.7460 **83.** 0.5805 **85.** 0.5283 **87.** -0.7019
89. $(-3, 0)$ **91.** $(x + 3)^2 + (y + 4)^2 = 17$
93. $\left(\dfrac{1}{3}, \dfrac{1}{3}, -\dfrac{10}{3}\right)$ **95.** 10 oz of 70% gold, 5 oz of
85% gold

Exercise 10.4

1. -1 **3.** 2 **5.** -3 **7.** 5 **9.** $\dfrac{1}{2}$ **11.** -2

13. 10,000 **15.** $\dfrac{1}{100,000}$ **17.** 1,000,000

19. 100,000,000   **21.** 1   **23.** 0.8451   **25.** 0.3617
27. 0.8388   **29.** 0.9445   **31.** 1.4633   **33.** 2.7850
35. 4.1995   **37.** 169.9809   **39.** 0.7399   **41.** 0.0001
43. 1.8451   **45.** 2.3617   **47.** −1.1612
49. 448.9521   **51.** 0.0711   **53.** 5.99×10^4
55. 1.63×10^{-3} **57.** 7.3891   **59.** 30.5694
61. 120 decibels   **63.** 90 decibels; less   **65.** pH 5.2
67. pH 1.7   **69.** 224.1 BTU   **71.** 7489.3 years
73. $136.88   **75.** In words   **77.** In words
79. $\log 5 \approx 0.69897$ **81.** $\log 5 \approx 0.69897$
83. $\ln 4 \approx 1.38629$ **85.** $\left(\dfrac{9 \pm 5\sqrt{21}}{2}, 0 \right)$
87. $G^{-1}(x) = \dfrac{3 \pm \sqrt{9 + 8x}}{4}$ **89.** $G^{-1}(x) = \sqrt[3]{x - 2}$
91. 95

Exercise 10.5

1. $\left\{ \dfrac{5}{2} \right\}$ **3.** $\left\{ \dfrac{1}{2} \right\}$ **5.** $\left\{ \dfrac{9}{2} \right\}$ **7.** {48}   **9.** $\left\{ \dfrac{1}{2} \right\}$
11. {3}   **13.** {−23}   **15.** $\left\{ \dfrac{33}{8} \right\}$ **17.** 1.6542
19. 2.4053   **21.** 0.6986   **23.** {2.4534}
25. {1.2153}   **27.** {8}   **29.** $\left\{ \dfrac{8}{3} \right\}$ **31.** 5.6789
33. 4.1544   **35.** −2.7925   **37.** 0.6889
39. {−0.3869}   **41.** {0.2619}   **43.** {−3.1507}
45. {3}   **47.** $\left\{ \dfrac{1}{7} \right\}$ **49.** {4}   **51.** Approximately 3.5
years   **53.** 13.87 years
55. $7.39   **57.** Approximately 2000 years
59. Approximately 14 days
61. $\ln x = \dfrac{\log x}{\log e} \approx \dfrac{\log x}{0.4343} \approx \dfrac{1}{0.4343} (\log x) \approx 2.3026(\log x)$
63. 0.58 hr ≈ 35 minutes   **65.** In words   **67.** {3}
69. {8}   **71.** {0.72}   **73.** {1.51}   **75.** {8}   **77.** 16
79. $\left\{ \left(-\dfrac{5}{4}, -\dfrac{21}{2}, \dfrac{21}{4} \right) \right\}$ **81.** {(4, 3), (−4, −3)}
83. {(4, −5), (4, 5), (−4, −5), (−4, 5)}

Chapter 10 Review Exercises

1.
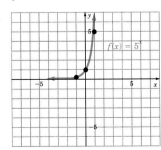

3. {4}   **5.** $\left\{ \dfrac{8}{3} \right\}$ **7.** {3}   **9.** $\log_4 1024 = 5$
11. $\log_{36} 6 = \dfrac{1}{2}$ **13.** $\log_{27} 9 = \dfrac{2}{3}$ **15.** $5^3 = 125$
17. $32^{1/5} = 2$ **19.** $16^{3/2} = 64$
21.
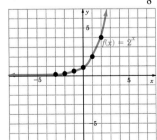

23. {2}   **25.** {−7}   **27.** $\log_b (xyz)$ **29.** $\log_b \dfrac{\sqrt{xy}}{z}$
31. $2 \log_b x + 3 \log_b y$ **33.** $\dfrac{1}{2} \log_b x + \log_b y - \log_b z$
35. 1.62324   **37.** 3.10036   **39.** 1.5563
41. −0.5447   **43.** 2.1931   **45.** −4.2687   **47.** 22.79
49. 0.81   **51.** 1642.10   **53.** 0.29   **55.** {2}
57. {5}   **59.** {2}   **61.** {1.5}   **63.** {0.5}
65. {1.1610}   **67.** 1.4307   **69.** 4.0875   **71.** 2.8614

Chapter 10 True–False Concept Review

1. False   **2.** True   **3.** True   **4.** False   **5.** True
6. False   **7.** False   **8.** False   **9.** True   **10.** False
11. True   **12.** True   **13.** True   **14.** True
15. True

Chapter 10 Test

1. $\log_a 15 = 3$ **2.** {3}   **3.** $\dfrac{1}{8}, \dfrac{1}{4}, \dfrac{1}{2}, 1, 2, 4, 8;$

4. Domain: $\{x \mid x > 0\}$; Range: R **5.** $5^3 = 125$
6. $\log_b \dfrac{\sqrt{2x}}{y}$ **7.** $\left\{ \dfrac{5}{2} \right\}$ **8.** $\log_8 2 = \dfrac{1}{3}$
9. $3 \log_a x + 2 \log_a y - \log_a z$ **10.** 4.3636
11. Domain: R; Range: $\{y \mid y > 0\}$ or R^+ **12.** 80.7049

13. -1.9101 **14.** $0, 1, 2, -1, -2, -3$;

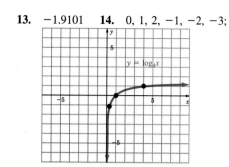

15. $b^d = c$ **16.** 2.1133 **17.** 1.25527 **18.** $\left\{-\dfrac{5}{3}\right\}$

19. 1.65321

CHAPTER 11

Exercise 11.1

1. $7, 8, 9, 10, 11$ **3.** $2, 8, 14, 20, 26$
5. $c(n) = 2n + 1$ **7.** $c(n) = 2n$ **9.** $c(n) = 3n$
11. $c(n) = n - 2$ **13.** $9 + 11 + 13 = 33$
15. $\dfrac{1}{2} + 1 + \dfrac{3}{2} + 2 + \dfrac{5}{2} + 3 = \dfrac{21}{2}$
17. $-7 - 9 - 11 - 13 - 15 = -55$
19. $c(n) = \dfrac{2n}{3} \cdot \dfrac{16}{3}$ **21.** $c(n) = \dfrac{1}{n^2} \cdot \dfrac{1}{64}$
23. $c(n) = -n^2, c(8) = -64$
25. $3 + 4 + 3 + 0 - 5 - 12 = -7$
27. $-4 + 1 + 6 + 11 + 16 + 21 + 26 = 77$
29. $0 + 4 + 18 + 48 + 100 + 180 + 294 + 448 + 648 =$
1740 **31.** $\displaystyle\sum_{n=1}^{\infty} 4n$ **33.** $\displaystyle\sum_{n=1}^{\infty} \dfrac{1}{n^3}$ **35.** $\displaystyle\sum_{n=1}^{\infty} \sqrt{n - 1}$
37. $c(n) = \dfrac{n}{n + 6}, c(9) = \dfrac{3}{5}, \displaystyle\sum_{n=1}^{10} \dfrac{n}{n + 6}$
39. $c(n) = \dfrac{x^n}{n}, c(9) = \dfrac{x^9}{9}, \displaystyle\sum_{n=1}^{10} \dfrac{x^n}{n}$
41. $c(n) = n \log n, c(9) = 9 \log 9, \displaystyle\sum_{n=1}^{10} n \log n$
43. $c(n) = \left(\dfrac{2y}{5}\right)^{n-1}, c(9) = \dfrac{256y^8}{390625}, \displaystyle\sum_{n=1}^{10} \left(\dfrac{2y}{5}\right)^{n-1}$
45. $c(n) = (6 - n)x^{3n}, c(9) = -3x^{27}, \displaystyle\sum_{n=1}^{10} (6 - n)x^{3n}$

47. $c(n) = \dfrac{1}{1000}(-10)^n, c(9) = -1,000,000,$
$\displaystyle\sum_{n=1}^{10} \dfrac{1}{1000}(-10)^n$ **49.** $-4 + 0 + 6 + 14 = 16$
51. $62 + 65 + 68 + 71 + 74 + 77 = 417$
53. $-15 - 17 - 19 - 21 - 23 = -95$
55. $-32 + 64 - 128 + 256 - 512 = -352$
57. 67 tons   **59.** 19 pennies
61. 0.0390625 ft $= \dfrac{5}{128}$ ft   **63.** \$34   **65.** \$224
67. In words   **69.** In words   **71.** $\left(\dfrac{3}{2}\right)^n; \dfrac{2187}{128}$
73. $4n - 7; 21$ **75.** $\sqrt{n^2 + 1}; 5\sqrt{2}$
77. $\dfrac{(n + 1)^2}{n}; \dfrac{64}{7}$ **79.** $\left(\dfrac{8}{3}, 0\right)$ **81.** $m = \dfrac{5}{3}$
83. $(7, 10)$ **85.** $(0, 108)$

Exercise 11.2

1. $7, 10, 13, 16, 19; a_{10} = 34$
3. $21, 28, 35, 42, 49; a_{13} = 105$
5. $7, 19, 31, 43, 55; a_{11} = 127$
7. $28, 20, 12, 4, -4; a_{10} = -44$
9. $2, 6, 10, 14, 18; a_6 = 22$ **11.** 35   **13.** 11
15. 24   **17.** $8, 6, 4, 2$ **19.** $74, 68, 62$
21. $25, 28, 31, 34$ **23.** 670   **25.** -120 **27.** 678.93
29. 0   **31.** $5, 5\dfrac{1}{2}, 6, 6\dfrac{1}{2}, 7; a_{10} = 9\dfrac{1}{2}$
33. $5, 3.5, 2, 0.5, -1; a_{15} = -16$
35. $17, 19, 21, 23, 25; a_{20} = 55$
37. $10, 8.5, 7, 5.5, 4; a_{31} = -35$
39. $n = 12, S_{12} = 115.5$ **41.** $n = 23, S_{23} = -2.3$
43. $n = 28, S_{28} = -232.54$ **45.** 675   **47.** 65
49. -380 **51.** $S_{18} = 693$ **53.** $S_{36} = -1395$
55. $S_{20} = 922.9$ **57.** 25 weeks   **59.** Joanna, \$90
61. 171 blocks   **63.** (a) 19 pipes; (b) 459 pipes
65. **a.** 12th year **b.** \$205,000 **c.** Davis \$2,130,000;
Mooney \$1,668,000   **67.** In words   **69.** (a) 7; (b) 70
71. (a) 7; (b) 55   **73.** (a) $10\sqrt{3}$; (b) $55\sqrt{3}$
75. $7^2 = 49$ **77.** $\left(\dfrac{1}{2}\right)^6 = 0.015625$
79. $\log_{4096} 4 = \dfrac{1}{6}$ **81.** $\log_{a+b} y = 9$

Exercise 11.3

1. $1, 3, 9, 27; a_6 = 243$
3. $8, 40, 200, 1000; a_8 = 625,000$ **5.** 9   **7.** 8
9. $a_2 = \pm 6, a_3 = 18, a_4 = \pm 54$
11. $\pm 28, 56, \pm 112, 224, \pm 448$ **13.** $a_3 = 3, a_4 = -9,$
$a_5 = 27, a_6 = -81$ **15.** 20,470   **17.** -364

19. 60, 6, 0.6, 0.06; $a_8 = 6 \times 10^{-6}$ **21.** 9, 6, 4, $2\frac{2}{3}$;

$a_8 = \dfrac{128}{243}$ **23.** 500, 100, 20, 4; $a_{10} = 0.000256$

25. $\dfrac{1}{2}, \dfrac{1}{6}, \dfrac{1}{18}, \dfrac{1}{54}$; $a_7 = \dfrac{1}{1458}$ **27.** 9 terms, $S_9 = \dfrac{511}{6}$

29. 6 terms, $S_6 = 106\dfrac{41}{64}$ **31.** 6 terms, $S_6 = 0.285506$

33. 5 terms, $S_5 = -1.44375$ **35.** 186   **37.** 0

39. 13.116   **41.** $1093\dfrac{40}{81}$ **43.** $-716{,}155$

45. $-\dfrac{61}{81}$ **47.** 339.99   **49.** $\dfrac{121}{162}$ **51.** \$65,610

53. \$0.89 in 4 yr, \$1.94 in 20 yr   **55.** 25,600 bacteria
57. 1024 squares   **59.** 0.006 ft, 5.994 ft
61. 3.29 ft, 204.71 ft   **63.** In words
65. 0.0625, 2.6875   **67.** $8\sqrt{3}, 7\sqrt{6} + 15\sqrt{3}$

69. $125, 156 + 31\sqrt{5}$ **71.** $\left\{\dfrac{1}{3}\right\}$ **73.** $\{4\}$

75. $\{2.6826\}$ **77.** $\{0.5681\}$

Exercise 11.4

1. 16   **3.** $18\dfrac{3}{4}$ **5.** $31\dfrac{1}{2}$ **7.** 54   **9.** 60

11. 160   **11.** 160   **13.** $S = -200$ **15.** $\dfrac{1}{2}$

17. $\dfrac{25}{2}$ **19.** 8   **21.** $1\dfrac{1}{3}$ **23.** $\dfrac{4}{21}$ **25.** $-\dfrac{110}{3}$

27. $19\dfrac{9}{29}$ **29.** Not defined   **31.** 2   **33.** 7

35. $\dfrac{5}{9}$ **37.** $\dfrac{5}{33}$ **39.** $\dfrac{16}{999}$ **41.** $\dfrac{2}{11}$ **43.** $\dfrac{4}{37}$

45. $\dfrac{1}{37}$ **47.** $\dfrac{5}{8}$ **49.** $\dfrac{4}{5}$ **51.** 600 tons   **53.** 20 ft

55. $\dfrac{1}{x-1}$ **57.** 27 ft   **59.** \$100,000,000,000

61. $2\sqrt{2}$ ft; sum of areas, 32 ft^2 **63.** In words

65. $r = \dfrac{1}{4}$ **67. a.** $1\dfrac{17}{99}$ **b.** 1 **c.** $1\dfrac{8}{10} = 1\dfrac{4}{5}$

69. $\log 2 + \log a - \log b$ **71.** $\ln 2 + \dfrac{1}{2}\ln w - 3\ln z$

73. $\ln\dfrac{y}{xz}$ **75.** $\left\{\dfrac{1}{33}\right\}$

Exercise 11.5

1. 6   **3.** 5040   **5.** 504   **7.** 1,860,480
9. $x^4 + 4x^3 + 6x^2 + 4x + 1$
11. $x^5 - 5x^4 + 10x^3 - 10x^2 + 5x - 1$

13. $x^8 + 16x^7 + 112x^6 + 448x^5 + 1120x^4 + 1792x^3 + 1792x^2 + 1024x + 256$
15. $32x^5 + 80x^4 + 80x^3 + 40x^2 + 10x + 1$ **17.** 153

19. 142,506   **21.** $81x^4 - 36x^3y + 6x^2y^2 - \dfrac{4}{9}xy^3 + \dfrac{y^4}{81}$

23. $x^{12} - 12x^{10}y^3 + 60x^8y^6 - 160x^6y^9 + 240x^4y^{12} - 192x^2y^{15} + 64y^{18}$
25. $32x^5 - 240x^4 + 720x^3 - 1080x^2 + 810x - 243$
27. $32x^{10} + 240x^8y^2 + 720x^6y^4 + 1080x^4y^6 + 810x^2y^8 + 243y^{10}$ **29.** $210x^4$ **31.** $2016x^4$
33. $(2x)^{24} + 24(2x)^{23}y + 276(2x)^{22}y^2 + 2024(2x)^{21}y^3$
35. $x^{20} - 100x^{19}y + 4750x^{18}y^2 - 142{,}500x^{17}y^3$
37. $x^{24} - 24x^{22}y^3 + 264x^{20}y^6 - 1760x^{18}y^9$
39. $x^{23}y^{23} - 69x^{22}y^{22} + 2277x^{21}y^{21} - 47{,}817x^{20}y^{20}$

41. $646{,}646x^{12}$ **43.** $\dfrac{12{,}155}{128}x^9$ **45.** $\dfrac{6435}{2}x^8$

47. $-717{,}255x^8y^{57}$ **49.** 70 ways
51. $A = A_0(1 + 5r + 10r^2 + 10r^3 + 5r^4 + r^5)$
53. $N = 1{,}126$ bacteria
55. $1 + 8(-1) + 28(-1)^2 + 56(-1)^3 + 70(-1)^4 + 56(-1)^5 + 28(-1)^6 + 8(-1)^7 + 1(-1)^8$,
$1 - 8 + 28 - 56 + 70 - 56 + 28 - 8 + 1 = 0$ **57.** 495
59. 42   **61.** In words
63. $x^5 + 5x^4h + 10x^3h^2 + 10x^2h^3 + 5xh^4 + h^5$,

$5x^4 + 10x^3h + 10x^2h^2 + 5xh^3 + h^4$ **65.** $\dfrac{63x^4}{16y^5}$

67. 3.7745   **69.** 8.6911   **71.** 0.0049   **73.** 0.0991

Chapter 11 Review Exercises

1. 3, 5, 7, 9, 11   **3.** 6, 4, 2, 0, -2 **5.** 2, 5, 10, 17, 26

7. $c(n) = \dfrac{3n}{5}, \dfrac{24}{5}$ **9.** $c(n) = \dfrac{n}{n+1}, \dfrac{8}{9}$

11. $c(n) = \dfrac{2}{n}, \dfrac{1}{4}$ **13.** $6 + 8 + 10 + 12 + 14 = 50$

15. $3 + 3\sqrt{2} + 3\sqrt{3} + 6 = 9 + 3\sqrt{2} + 3\sqrt{3}$

17. $11 + 19 + 29 + 41 = 100$ **19.** $\displaystyle\sum_{n=1}^{5} 5n$

21. $\displaystyle\sum_{n=1}^{5} n^2$ **23.** $\displaystyle\sum_{n=1}^{5} \sqrt{n+1}$

25. 10, 15, 20, 25, 30; $a_{10} = 55$
27. 2, 18, 34, 50, 66; $a_{14} = 210$
29. 4, 4.5, 5, 5.5, 6; $a_8 = 7.5$
31. $-0.8, -0.81, -0.82, -0.83, -0.84$; $a_{12} = -0.91$
33. 27, 39, 51, 63, 75, 87   **35.** $-4, 4, 12, 20, 28$
37. $-25, -31, -37, -43, -49, -55, -61$ **39.** 13
41. 14   **43.** 13   **45.** 540   **47.** -48.3 **49.** 2450
51. 5, 10, 20, 40; $a_6 = 160$ **53.** 8, 4, 2, 1; $a_8 = 0.0625$
55. $\pm 80, 40, \pm 20, 10, \pm 5$; $a_8 = 0.625$
57. 200, 40, 8, 1.6; $a_7 = 0.0128$ **59.** $\pm 20, 80, \pm 320$
61. $216, -648, 1944$ **63.** 7   **65.** 8   **67.** 3825

69. 120   **71.** 160   **73.** 8   **75.** $\dfrac{4}{33}$ **77.** $\dfrac{20}{37}$

79. 3   **81.** 479,001,600   **83.** 240   **85.** 560

87. $a^{12} + 12a^{11}b + 66a^{10}b^2 + 220a^9b^3$

89. $19,683x^9 + 118,098x^8y + 314,928x^7y^2 + 489,888x^6y^3$

91. $1024x^{10} - 15,360x^9y + 103,680x^8y^2 - 414,720x^7y^3$

93. $9375x^2$ **95.** $-1512a^5b^3$ **97.** $326,592x^4y^5$

Chapter 11 True–False Concept

1. 1. True   **2.** False   **3.** True   **4.** True
5. False   **6.** True   **7.** False   **8.** True   **9.** True
10. False   **11.** True   **12.** True

Chapter 11 Test

1. $-8, -2, 4, 10, 16$ **2.** $-2, -6, -18, -54, -162$
3. $3, 13, 23, 33, 43$ **4.** 462   **5.** $c(n) = 26 - 4n$
6. $40 + 35 + 30 + 25 + 20$ **7.** $a_{21} = 112$
8. $a_7 = \dfrac{5}{8}$ **9.** $x^7 + 21x^6 + 189x^5$ **10.** 11 terms

11. 12, 18   **12.** 308   **13.** 511   **14.** $\displaystyle\sum_{n=1}^{5}(7n - 1)$

15. $\dfrac{27}{2}$

ANSWERS TO SETS.1

1. $\{0, 2, 4, 6, 8, 10, 12, 14, 16\}$
3. $\{9, 10, 11, 12, 13, 14, 15\}$ **5.** The set of whole numbers between 10 and 16.   **7.** The set containing the number 1.   **9.** $\{x \,|\, x \in W \text{ and } x < 84\}$ **11.** $\{x \,|\, x$ is an even number and $51 < x < 133\}$ **13.** False   **15.** False
17. True   **19.** False   **21.** $\{16, 20, 24, 28, 32, 36\}$
23. $\left\{\dfrac{1}{1}, \dfrac{1}{2}, \dfrac{1}{3}, \dfrac{1}{4}, \dfrac{2}{1}, \dfrac{2}{2}, \dfrac{2}{3}, \dfrac{2}{4}, \dfrac{3}{1}, \dfrac{3}{2}, \dfrac{3}{3}, \dfrac{3}{4}\right\}$ **25.** The set of multiples of 3 from 9 to 21 inclusive.   **27.** The set of multiples of 11 from 88 to 132 inclusive.
29. $\{24, 25, 26, 27, 28\}$ **31.** $\{1, 2, 3, 6, 9, 18\}$
33. True   **35.** True   **37.** False   **39.** False
41. $\{208, 221, 234\}$ **43.** $\{210, 224, 238\}$ **45.** The set of multiples of 4 between 129 and 141.   **47.** The set of multiples of 2 or 5 between 31 and 41.
49. $\{37, 38, 39, 40\}$ **51.** $\{-4, -3, -2, -1, 0, 1, 2, 3, 4\}$
53. True   **55.** False   **57.** False   **59.** True
61. False   **63.** True   **65.** In words
67. $C = \{15, 30, 45\}$ **69.** $E = \{14, 28, 42\}$

ANSWERS TO SETS.2

1. $\{2, 10\}$ **3.** $\left\{-15.\overline{77}, -\dfrac{19}{4}, -4, -\dfrac{7}{6}, 0, 2, 10, \dfrac{42}{3}\right\}$

5. $\{2, 3, 4, 6, 8, 9, 10, 12, 15\}$
7. $\{0, 2, 3, 4, 5, 6, 8, 9, 10, 12, 15, 20\}$
9. $A = \{1, 2, 3, 4, 6, 7, 8, 9, 11, 12, 13, 14\}$ **11.** True
13. True
15.

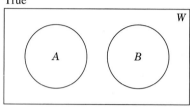

$A \cap B = \emptyset$

17. $\{0, 3, 9\}$ **19.** $\{-\sqrt{13}\}$ **21.** $\{-4, 0, 4\}$ **23.** $\{0\}$
25. $A = \{\text{all odd whole numbers}\}$ **27.** $\{\text{all whole numbers not a multiple of 5}\}$ **29.** $\{\text{all whole numbers not a multiple of 10}\}$
31. True   **33.** True
35.

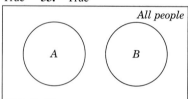

$A \cup B = \{\text{all left-handed people}\}$

37. A **39.** B **41.** $\{11, 12, 13, 14, 15, 16, 17, 18, 19\}$
43. $\{1, 2, 3, \ldots, 17\}$ **45.** J **47.** N   **49.** $A = \{\text{all females and males who are 25 years of age or older}\}$
51. $\{\text{all males 25 years of age and older whose hair is not red and } all \text{ females whose hair is not red}\}$
53. $\{\text{all males whose hair is not red, all males 25 years of age and older, and all females whose hair is not red}\}$
55.

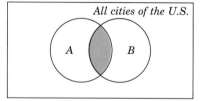

$A \cap B = \{\text{capitals that are also the largest city in the state}\}$
57. In words   **59.** C
61. $\{0, 1, 2, 3, \ldots, 40, 45, 50, 55, \ldots\}$

INDEX